Reagents for Organic Synthesis

INDEXER
Honor Ho

Fiesers'

Reagents for Organic Synthesis

VOLUME TWENTY TWO

Tse-Lok Ho

National Chiao Tung University
Republic of China

A JOHN WILEY & SONS, INC., PUBLICATION

Published by John Wiley & Sons, Inc., Hoboken, New Jersey.
Published simultaneously in Canada.

Library of Congress Cataloging in Publication Data:

ISBN 0-471-28515-3
ISSN 0271-616X

Printed in the United States of America.

10 9 8 7 6 5 4 3 2 1

PREFACE

夫上之化下，下之從上，

猶泥之在鈞，唯甄者之所為；

猶金之在鎔，唯冶者之所鑄。

In the act of transformation
Those being transformed are
Like clay in a potter's wheel
To be shaped by the artist
Or like gold in a crucible
To be molded by the metallurgist

漢書、董仲舒傳 *Han Chronicles: Biography of Tung Chung-Shu*

The work of chemists is more profound than that of a potter or a metallurgist because these other people concern mainly with physical changes of material. Systematic "reshaping" of molecules creates new material with sometimes unimaginable consequences. If we continue the comparison, chemical reagents can be equated to potter's wheel or metallurgist's crucible. Tools are fundamental to the work of the artisans, naturally, better tools make their jobs easier and products probably more elegant.

This volume of Fiesers' Reagents for Organic Synthesis surveys literature of the 2001-2002 period. It concludes my ten-year involvement in the series. Despite occasional feeling of personal deficiency and pressure of time, completion of each volume always brings a sense of elation together with earnest wish that the result of my efforts renders some service to the reader.

Since the series is well established, hardly any major change is needed in the arrangement and content. However, my unpleasant experience with results from certain research groups (irreproducible more than once with different reactions) has led me to ignore those publications from them.

TSE-LOK HO

CONTENTS

GENERAL ABBREVIATIONS

Ac	acetyl
acac	acetylacetonate
ADDP	1,1′-(azodicarbonyl)dipiperidine
AIBN	2,2′-azobisisobutyronitrile
An	*p*-anisyl
aq	aqueous
Ar	aryl
ATPH	aluminum tris(2,6-diphenylphenoxide)
9-BBN	9-borabicyclo[3.3.1]nonane
BINOL	1,1′-binaphthalene-2,2′-diol
Bn	benzyl
Boc	*t*-butoxycarbonyl
bpy	2,2′-bipyridyl
BSA	*N,O*-bis(trimethylsilyl)acetamide
Bt	benzotriazol-1-yl
Bu	*n*-butyl
Bz	benzoyl
18-c-6	18-crown-6
c-	cyclo
CAN	cerium(IV)ammonium nitrate
cat	catalytic
Cbz	benzyloxycarbonyl
Chx	cyclohexyl
cod	1,5-cyclooctadiene
cot	1,3,5-cyclooctatriene
Cp	cyclopentadienyl
Cp*	1,2,3,4,5-pentamethylcyclopentadienyl
CSA	10-camphorsulfonic acid
Cy	cyclohexyl
cyclam	1,4,8,11-tetraazacyclotetradecane
DABCO	1,4-diazobicyclo[2.2.2]octane
DAST	(diethylamino)sulfur trifluoride
dba	dibenzylideneacetone
DBN	1,5-diazobicyclo[4.3.0]non-5-ene
DBU	1,8-diazobicyclo[5.4.0]undec-7-ene
DCC	*N,N′*-dicyclohexylcarbodiimide
DDQ	2,3-dichloro-5,6-dicyano-1,4-benzoquinone

de	diastereomer excess
DEAD	diethyl azodicarboxylate
DIAD	diisopropyl azodicarboxylate
Dibal-H	diisobutylaluminum hydride
DMA	*N,N*-dimethylacetamide
DMAD	dimethyl acetylenedicarboxylate
DMAP	4-dimethylaminopyridine
DMD	dimethyldioxirane
DME	1,2-dimethoxyethane
DMF	*N,N*-dimethylformamide
DMPU	*N,N'*-dimethylpropyleneurea
DMSO	dimethyl sulfoxide
dpm	dipivaloylmethane
dppb	1,4-bis(diphenylphosphino)butane
dppe	1,2-bis(diphenylphosphino)ethane
dppf	1,2-bis(diphenylphosphino)ferrocene
dppp	1,3-bis(diphenylphosphino)propane
dr	diastereomer ratio
DTTB	4,4'-di-*t*-butylbiphenyl
E	COOMe
ee	enantiomer excess
en	ethylenediamine
er	enantiomer ratio
Et	ethyl
EVE	ethyl vinyl ether
Fc	ferrocenyl
Fmoc	9-fluorenylmethoxycarbonyl
Fu	furanyl
HMDS	hexamethyldisilazane
HMPA	hexamethylphosphoric amide
hv	light
Hx	*n*-hexyl
i	iso
Ipc	isopinocampheyl
kbar	kilobar
L	ligand
LAH	lithium aluminum hydride
LDA	lithium diisopropylamide
LHMDS	lithium hexamethyldisilazide
LTMP	lithium 2,2,6,6-tetramethylpiperidide
LN	lithium naphthalenide
lut	2,6-lutidine

M	metal
MAD	methylaluminum bis(2,6-di-*t*-butyl-4-methylphenoxide)
MCPBA	*m*-chloroperoxybenzoic acid
Me	methyl
MEM	methoxyethoxymethyl
Men	menthyl
Mes	mesityl
MOM	methoxymethyl
Ms	methanesulfonyl (mesyl)
MS	molecular sieves
MTO	methyltrioxorhodium
MVK	methyl vinyl ketone
NBS	*N*-bromosuccinimide
NCS	*N*-chlorosuccinimide
NIS	*N*-iodosuccinimide
NMO	*N*-methylmorpholine *N*-oxide
NMP	*N*-methylpyrrolidone
Np	naphthyl
Ns	*p*-nitrobenzenesulfonyl
Nu	nucleophile
Oc	octyl
PCC	pyridinium chlorochromate
PDC	pyridinium dichromate
PEG	poly(ethylene glycol)
Ph	phenyl
phen	1,10-phenenthroline
Pht	phthaloyl
Piv	pivaloyl
PMB	*p*-methoxybenzyloxymethyl
PMHS	poly(methylhydrosiloxane)
PMP	*p*-methoxyphenyl
Pr	*n*-propyl
py	pyridine
Q^+	quaternary onium ion
RAMP	(*R*)-1-amino-2-methoxymethylpyrrolidine
RaNi	Raney nickel
RCM	ring closure metathesis
R^f	perfluoroalkyl
ROMP	ring opening metathesis polymerization
s-	secondary
(s)	solid
salen	*N,N'*-ethylenebis(salicylideneiminato)

SAMP	(*S*)-1-amino-2-methoxymethylpyrrolidine
SDS	sodium dodecyl sulfate
sens.	sensitizer
SEM	2-(trimethylsilyl)ethoxymethyl
SES	2-[(trimethylsilyl)ethyl]sulfonyl
TASF	tris(dimethylamino)sulfur(trimethylsilyl)difluoride
TBAF	tetrabutylammonium fluoride
TBDPS	*t*-butyldiphenylsilyl
TBDMS	*t*-butyldimethylsilyl
TBS	*t*-butyldimethylsilyl
TEMPO	2,2,6,6-tetramethylpiperidinooxy
Tf	trifluoromethanesulfonyl
THF	tetrahydrofuran
THP	tetrahydropyranyl
Thx	*t*-hexyl
TIPS	triisopropylsilyl
TMEDA	*N,N,N′,N′*-tetramethylethylenediamine
TMS	trimethylsilyl
Tol	*p*-tolyl
Ts	tosyl (*p*-toluenesulfonyl)
TSE	2-(trimethylsilyl)ethyl
TTN	thallium trinitrate
Z	benzyloxycarbonyl
Δ	heat
))))	microwave

REFERENCE ABBREVIATIONS

ACIEE	Angew. Chem. Int. Ed. Engl.
ACR	Acc. Chem. Res.
ACS	Acta Chem. Scand.
AJC	Aust. J. Chem.
AOMC	Appl. Organomet. Chem.
ASC	Adv. Syn. Catal.
BC	Bioorg. Chem.
BCSJ	Bull. Chem. Soc. Jpn.
BMCL	Biorg. Med. Chem. Lett.
BRAS	Bull. Russ. Acad. Sci.
CB	Chem. Ber.
CC	Chem. Commun.
CEJ	Chem. Eur. J.
CI	Chem. Ind. (London)
CJC	Can. J. Chem.
CL	Chem. Lett.
CPB	Chem. Pharm. Bull.
CR	Carbohydr. Res.
EJIC	Eur. J. Inorg. Chem.
EJOC	Eur. J. Org. Chem.
H	Heterocycles
HCA	Helv. Chim. Acta
JACS	J. Am. Chem. Soc.
JCC	J. Carbohydr. Chem.
JCCS(T)	J. Chin. Chem. Soc. (Taipei)
JCR(S)	J. Chem. Res. (Synopsis)
JCS(P1)	J. Chem. Soc. Perkin Trans. 1
JFC	J. Fluorine Chem.
JHC	J. Heterocycl. Chem.
JOC	J. Org. Chem.
JOCU	J. Org. Chem. USSR (Engl. Trans.)
JOMC	J. Organomet. Chem.
MC	Mendeleev Commun.
NJC	New J. Chem.
OL	Organic Letters
OM	Organometallics
PAC	Pure Appl. Chem.

PSS	Phosphorus Sulfur Silicon
RCB	Russian Chem. Bull.
RJGC	Russ. J. Gen. Chem.
RJOC	Russian J. Org. Chem.
S	Synthesis
SC	Synth. Commun.
SCI	Science.
SL	Synlett
SOC	Synth. Org. Chem. (Jpn.)
T	Tetrahedron
TA	Tetrahedron: Asymmetry
TL	Tetrahedron Lett.
ZN	Zeitschr. Naturforsch.

Acetic anhydride. 20, 1, **21,** 1

Imino Nazarov cyclization.[1] Adducts from 1-alkoxyalkenyllithiums and acrylonitriles undergo cyclization to afford 2-acetamino-2-cyclopentenones on treatment with Ac_2O-py-DMAP. 5-Methylene derivatives are obtained from 1-alkoxyallenyllithiums.

CN + Li O OMe C Ph → THF ; Ac_2O - py - DMAP → NHAc O Ph 73%

[1]Tius, M.A., Chu, C.C., Nieves-Colberg, R. *TL* **42**, 2419 (2001).

Acetonitrile(chloro)tripyrazolylboratoruthenium.

Furans.[1] Alkynylepoxides are transformed into furans on exposure to the Ru complex.

O C_7H_{15} → TpRu(MeCN)(PPh_3)Cl, Et_3N / CH_2Cl_2, 80° → O C_7H_{15} 84%

[1]Lo, C.-Y., Guo, H., Lian, J.-J., Shen, F.-M., Liu, R.-S. *JOC* **67**, 3930 (2002).

Acetonyltriphenylphosphonium bromide. 21, 1

sym-Trioxanes.[1] The title reagent catalyzes cyclotrimerization of aldehydes.

[1]Hon, Y.-S., Lee, C.-F. *T* **57**, 6181 (2001).

Fiesers' Reagents for Organic Synthesis, Volume 22. Series editor Tse-Lok Ho
ISBN 0-471-28515-3

(*E*)-(*o*-Acetoxystyryl)diisopropylsilyl chloride.

Hydroxyl protection.[1] The title reagent is used for silylation of alcohols. Deprotection is achieved via photoisomerization of the derived phenols.

Si Cl OAc — LiNMe$_2$ / ROH → Si OR OH

→ hv / MeCN → O Si + ROH

[1]Pirrung, M.C., Fallon, L., Zhu, J., Lee, Y.R. *JACS* **5123**, 3638 (2001).

Acetylacetonato(dicarbonyl)rhodium. 21, 1

Chain elongation. Extension of the previously reported stereoselective 1,3-diol synthesis using biscrotylsilyl homoallyl ethers leads to products with a methyl branch,[1] whereas bisallylsilyl homopropargyl ethers give either keto diols or enediol derivatives, depending on work-up procedure.[2]

O Si H R R′ — (acac)Rh(CO)$_2$ / CO / PhH 60° → [O—Si—O R R′]

Bu$_4$NF → OH OH R R′

H$_2$O$_2$ - NaHCO$_3$ / THF - MeOH → OH O OH R R′

Hydroformylation. Regioselective and diastereoselective hydroformylation of allyl ethers enables synthesis of aldols of propanal.[3]

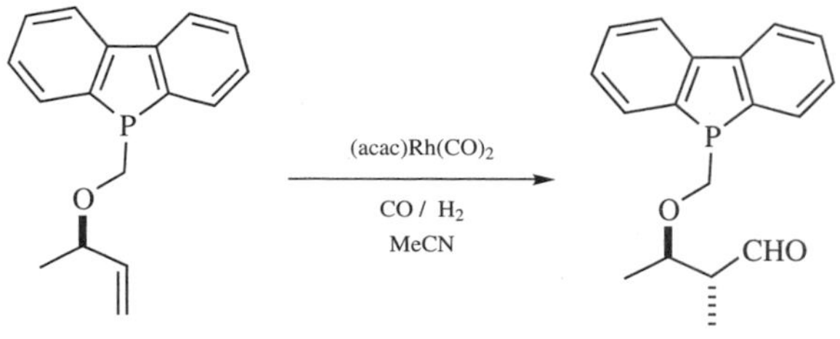
P
O
(acac)Rh(CO)$_2$
CO / H$_2$
MeCN
CHO

92% (branched:linear >98 : 2; *anti:syn* 81 : 19)

Internal alkenes afford aldehydes via an isomerization-hydroformylation process in the presence of ligands such as **1**[4] or **2**.[5]

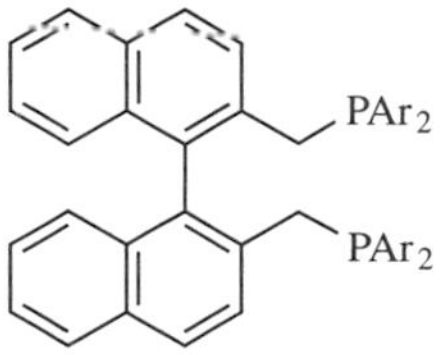
PAr$_2$
PAr$_2$

1

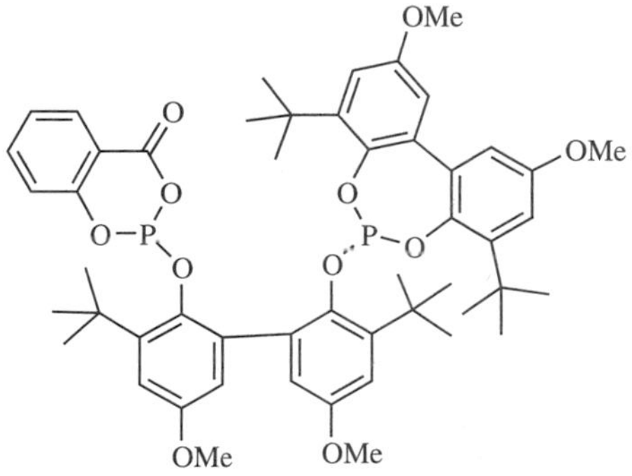
OMe
O
O
O–P
O
O
P–O
O
OMe
OMe
OMe

2

A tandem hydroformylation-allylboration-hydroformylation scheme is followed when 3-(*N*-allyl)-2-propen-1-ylboronate is subject to the Rh-catlayzed reaction.[6]

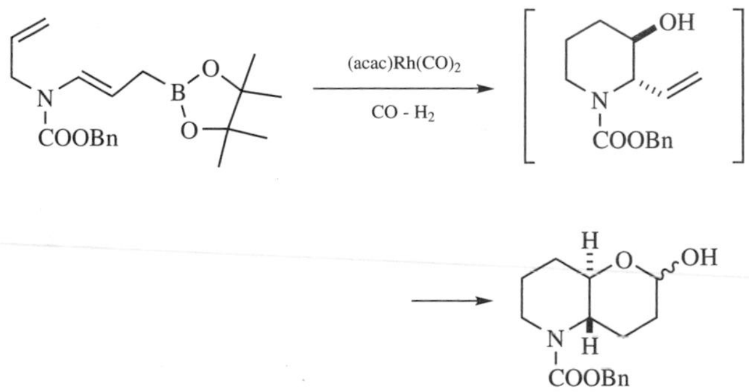
N
COOBn
B
O
O
(acac)Rh(CO)$_2$
CO - H$_2$
OH
N
COOBn
H
O
OH
N
H
COOBn

66%

[1]Zacuto, M.J., O'Malley, S.J., Leighton, J.L. *JACS* **124**, 7890 (2002).
[2]O'Malley, S.J., Leighton, J.L. *ACIEE* **40**, 2915 (2001).
[3]Krauss, I.J., Wang, C.C.-Y., Leighton, J.L. *JACS* **123**, 11514 (2001).
[4]Klein, H., Jackstell, R., Wiese, K.-D., Borgmann, C., Beller, M. *ACIEE* **40**, 3408 (2001).
[5]Selent, D., Hess, D., Wiese, K.-D., Röttger, D., Kunze, C., Börner, A. *ACIEE* **40**, 1696 (2001).
[6]Hoffmann, R.W., Brückner, D., Gerusz, V.J. *H* **52**, 121 (2000).

Acetylacetonato(diolefin)rhodium. 21, 2

Michael addition. Asymmetric arylation of conjugated esters with arylboronic acids or lithium aryltrimethylborates occurs in the presence of (acac)Rh(ethylene)$_2$ and (*S*)-BINAP.[1] The same reaction can be catalyzed by [Rh(cod)(MeCN)$_2$]BF$_4$.[2]

Hydroarylation.[3] Alkynes are converted to styrenes on reaction with arylboronic acids in the presence of (acac)Rh(ethylene)$_2$ and DPPB.

[1]Takaya, Y., Senda, T., Kurushima, H., Ogasawara, M., Hayashi, T. *TA* **10**, 4047 (1999).
[2]Sakuma, S., Sakai, M., Itooka, R., Miyaura, N. *JOC* **65**, 5951 (2000).
[3]Hayashi, T., Inoue, K., Taniguchi, N., Ogasawara, M. *JACS* **123**, 9918 (2001).

4-Acetylamino-2,2,6,6-tetramethylpiperidine-1-oxoammonium perchlorate.

Oxidation.[1] Heteroaromatic alcohols are oxidized to carbonyl compounds by this salt.

[1]Kernag, C.A., Bobbitt, J.M., McGrath, D.V. *TL* **40**, 1635 (1999).

Acylcobalt tetracarbonyls.

Acylative cyclization.[1] *N*-Tosyl-4,5-alkadienes afford 2-(*N*-tosylpyrrolidin-2-yl)-1-alken-3-ones on reaction with RCOCo(CO)$_4$.

OR; NH; Ts; C; AcCo(CO)$_4$; Et$_3$N; OR; O; N; Ts

R = Bn, Bz, MTM, TBS, TIPS 31–61%

[1]Bates, R.W., Satcharoen, V. *SL* 532 (2001).

N-Acyl-5,7-dinitroindolines.

Amides.[1] Amines are acylated by these indolines under photochemical conditions.

[1]Helgen, C., Bochet, C.G. *SL* 1968 (2001).

Acyl halides.

Amides from t-butyl carbamates.[1] An acyl halide-methanol mixture converts *t*-butyl carbamates to amides.

Pyridones and pyrimidones.[2] Acyl halides react with *N*-silyl imino ethers to furnish activated azadienes which undergo Diels-Alder reaction with alkynes and nitriles.

TBSO, N, O + (1,4-naphthoquinone) → O, HN (aza-anthraquinone)

63%

[1]Nazih, A., Heissler, D. *S* 203 (2002).
[2]Ghosez, L., Jnoff, E., Bayard, P., Sainte, F., Beaudegnies, R. *T* **55**, 3387 (1999).

4-Acyloxy-2-alkylthiopyrimidines, polymer-supported.

Acylation of amines.[1] The polymer-supported (linking via the sulfur atom) reagents deliver the acyl group to amines at room temperature.

[1]Botta, M., Corelli, F., Petricci, E, Serri, C. *H* **56**, 369 (2002).

2-Acyl-4,5-dichloropyridazin-3-ones.

Acylation of amines.[1] The reagents are formed by acylation of the heterocycles and they serve as acyl transfer agents for amines.

[1]Kang, Y.-J., Chung, H.-A., Kim, J.-J., Yoon, Y.-J. *S* 733 (2002).

Acyloxymethylzinc.

Cyclopropanation.[1] On reaction with diazomethane, zinc dicarboxylates provide $(RCOOCH_2)_2Zn$ which are active reagents for cyclopropanation. The same species are also obtained photochemically from $RCOOCH_2I$ and diethylzinc.

[1]Charette, A.B., Beauchemin, A., Francoeur, S. *JACS* **123**, 8139 (2001).

Acyl selenides, polymer-bound.

Acylation.[1] The polymer-supported species behave as acyl donors. Thus, alkynyl ketones are formed on their reaction with alkynyl coppers (9 examples, 77–92%).

[1]Qian, H., Shiao, L.-X., Huang, X. *SL* 1571 (2001).

Alkenylboronic acids and esters. 20, 3; **21,** 4

Unsaturated amino acids.[1] Pinacolyl boronic esters are useful components for the Petasis reaction. With glyoxylic acid and an amine the condensation leads to β, γ-unsaturated α-amino acids.

[1]Koolmeister, T., Södergren, M., Scobie, M. *TL* **43**, 5965 (2000).

Alkenyldimethyl(2-pyridyl)silanes.

Styrenes.[1] Desilylative arylation by a coupling reaction with ArI is promoted by $(RCN)_2PdCl_2$. Different products ensue when the catalyst is changed to $(dba)_3Pd_2$.

$(RCN)_2PdCl_2$; Bu_4NF

+ Ar-I

$(dba)_3Pd_2 \cdot CHCl_3$; Fu_3P, Et_3N

[1]Itami, K., Nokami, T., Yoshida, J. *JACS* **123**, 5600 (2001).

Alkenyldiphenylselenonium triflates.

Enones.[1] Reaction with aldehydes in the presence of NaH delivers enones.

TfO^- $\overset{+}{S}ePh_2$ + H ... NO_2

NaH, THF, 0°

49%

[1]Watanabe, S., Kusumoto, T., Yoshida, C., Kataoka, T. *CC* 839 (2001).

Alkenyliodonium tetrafluoroborates.

Fluoroalkenes.[1] Thermal decomposition of these salts gives *cis*-fluoroalkenes as kinetic products.

BF_4^- + IPh; $CHCl_3$ 60°; F; R; R = Ph, *t*-Bu; 93%

[1]Okuyama, T., Fujita, M., Gronheid, R., Lodder, G. *TL* **41**, 5125 (2000).

Alkylaluminum chlorides.

Cyclization. Intramolecular allyation of carbonyl compounds occurs in the presence of $EtAlCl_2$.[1] An elegant synthesis of (+)-onocerin is concluded at a cyclization reaction that is induced by complexation of epoxide moieties to $MeAlCl_2$.[2]

Ph; Si; Ph; O; $EtAlCl_2$ 0° PhMe; OH; 67%

Me_3Si; O; $SiMe_3$; O; $MeAlCl_2$ / CH_2Cl_2 ; Bu_4NF / THF; OH; HO; (+)-onocerin 72%

Cycloheptanones.[3] A synthesis based on [5+2]cycloaddition as mediated by $EtAlCl_2$ is shown in the following equation.

TIPSO + OTIPS; $Co(CO)_3$; $Co(CO)_3$; BzO; $EtAlCl_2$ 0° CH_2Cl_2; O; TIPSO; $Co(CO)_3$; $Co(CO)_3$; 59%

[1]Barbero, A., Castreno, P., Garcia, C., Pulido, F.J. *JOC* **66**, 7723 (2001).
[2]Mi, Y., Schreiber, J.V., Corey, E.J. *JACS* **124**, 11290 (2002).
[3]Tanino, K., Kondo, F., Shimizu, T., Miyashita, M. *OL* 2217 (2002).

Alkyldichloroboranes.

Reductive chloroalkylation.[1] Aromatic aldehydes are converted to ArCHClR by $RBCl_2$ in the presence of oxygen.

[1]Kabalka, G.W., Wu, Z., Yu, J. *TL* **42**, 6239 (2001).

***O*-Alkylisoureas.**

Esterification.[1] Carboxylic acids form esters (including the *t*-butyl version) on reaction with these reagents. The process is accelerated by microwave irradiation.

[1]Crosignani, S., White, P.D., Linclau, B. *OL* **4**, 2961 (2002).

Alkyl phenyl carbonates.

Amine protection.[1] The title reagents react with primary and secondary amines to form alkyl carbamates in DMF at room temperature. Molecules containing two primary amino groups can be derivatized at only one site.

[1]Pittelkow, M., Lewinsky, R., Christensen, J.B. *S* 2195 (2002).

Alkylsulfinyl chlorides.

Sulfonylimines.[1] Oximes are sulfinated but the products rearrange at or below room temperature (9 examples, 48–79%).

[1]Artman, III, G.D., Bartolozzi, A., Franck, R.W. *SL* 232 (2001).

Allylboronates.

Homoallylic alcohols. Various aspects of Lewis acid-catalyzed reactions have been studied.[1] The reaction with aldehydes with ROMPgel-supported boronates leads to products that are readily isolated in pure form.[2]

[1]Ishiyama, T., Ahiko, T., Miyaura, N. *JACS* **124**, 12414 (2002).
[2]Arnauld, T., Barrett, A.G.M., Seifried, R. *TL* **42**, 7899 (2001).

η^3-Allyl(chloro)-(1,3-diaryl-1,2-dihydroimidazolin-2-ylidene)palladium.

Arylation of ketones.[1] The air-stable catalyst **1** is found to promote arylation.

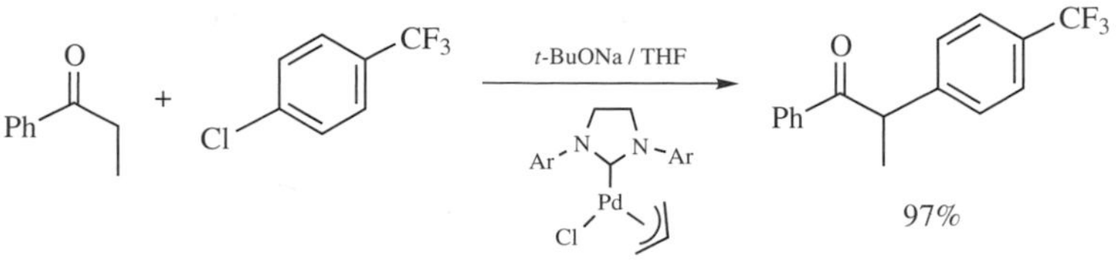

^{1}Viciu, M.S., Germaneau, R.F., Nolan, S.P. *OL* **4**, 4053 (2002).

η^3-Allyl(chloro)palladium, resin-supported diphosphine complex.

Suzuki coupling.[1] The resin-bound complex **1** is a high throughput Suzuki coupling catalyst in water.

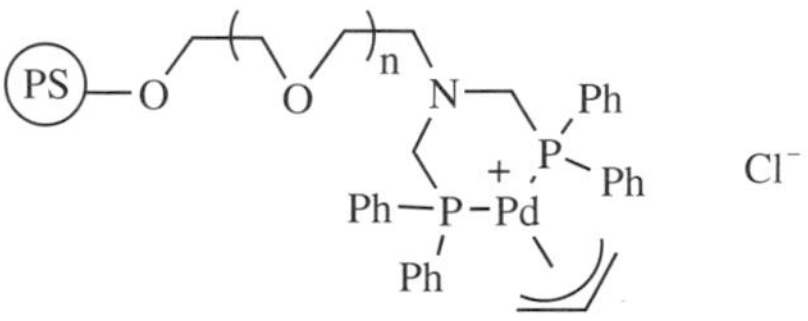

1

^{1}Uozumi, Y., Nakai, Y. *OL* **4**, 2997 (2002).

Allylphosphonates.

1,3-Dienes.[1] Condensation with aldehydes (Emmons-Wadsworth reaction) after treating these esters with BuLi is (*E*)-selective.

^{1}Wang, Y., West, F.G. *S* 99 (2002).

Allylsilanes. 21, 11

Allylation. Allylation of aldehydes in $MeNO_2$ is catalyzed by $FeCl_3$,[1] and in dichloromethane by species derived from Me_3Al and 1,1,1-tris(triflylamino)ethane.[2] Benzoylhydrazines undergo allylation with allyltrichlorosilane in DMF at room temperature.[3] Allyltrimethoxysilanes finds use in a synthesis of homoallylic alcohols in the presence of CuCl and $Bu_4NSiF_2Ph_3$.[4]

Reagent **1** is useful for preparing cyclic alkenylboronates.[5]

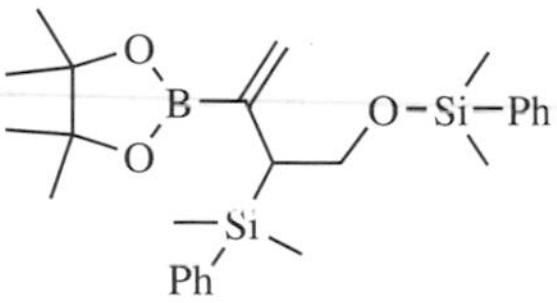

1

Anodic oxidation of dithioacetals of aromatic aldehydes in the presence of allylsilanes leads to α-allylbenzyl sulfides.[6]

MeO MeO SPh SPh + Me_3Si $LiClO_4 \cdot MeNO_2$ → MeO MeO SPh 96%

[1]Watahiki, T., Oriyama, T. *TL* **43**, 8959 (2002).
[2]Kanai, M., Kuramochi, A., Shibasaki, M. *S* 1956 (2002).
[3]Hirabayashi, R., Ogawa, C., Sugiura, M., Kobayashi, S. *JACS* **123**, 9493 (2001).
[4]Yamasaki, S., Fujii, K., Wada, R., Kanai, M., Shibasaki, M. *JACS* **124**, 6536 (2002).
[5]Suginome, M., Ohmori, Y., Ito, Y. *JACS* **123**, 4601 (2001).
[6]Chiba, K., Uchiyama, R., Kim, S., Kitano, Y., Tada, M. *OL* **3**, 1245 (2001).

Allylstannanes. 21, 13

Stannyl ethers. An allylstannane reacts with alcohols in an acid-catalyzed (TfOH) process to give R_3SnOR'. The stannyl ethers can be employed in the synthesis of glycosides from bromoglycosides.[1]

[1]Yamago, S., Yamada, T., Nishimura, R., Ito, H., Mino, Y., Yoshida, J. *CL* 152 (2002).

Alumina. 21, 14–15

Redox reactions. α-Hydroxyphosphonates are oxidized to the corresponding ketones with solvent-free alumina-supported $KMnO_4$.[1] Carbonyl compounds undergo Meerwein-Ponndorf-type reduction of with *i*-PrOH in the presence of KOH/Al_2O_3 (28 examples, 85–93%).[2]

Condensation reactions. Under microwave irradiation, the Michael reaction of enamines with conjugated enones[3] and a synthesis of α-aminophosphonates from aldehydes, diethyl phosphite, and ammonium formate[4] is promoted by alumina. Similarly, a convenient preparation of α,α'-bis(benzylidene)cycloalkanones[5] involves treatment of benzaldehyde and cycloalkanones with KF/Al_2O_3 (9 examples, 73–83%).

CHO $HPO(OEt)_2$ / $HCOONH_4$ Al_2O_3)))) → O P OEt OEt N

Tosyldiazomethane. An improved method for the preparation of this reagent via a nitrosocarbamate is concluded in a step that involves treatment with basic alumina.[6]

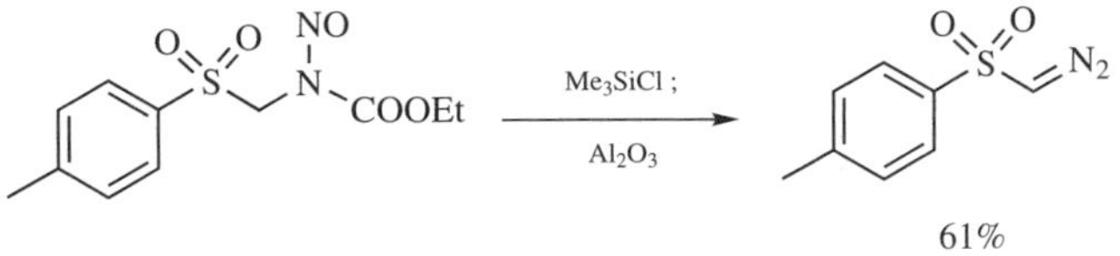

Azidoiodination. A mixture of KI and NaN_3 deposited on wet alumina is used as a source of reagent for azidoiodination of alkenes. It requires treatment with Oxone to accomplish the transformation.[7]

[1]Firouzabadi, H., Iranpoor, N., Sobhani, S. *TL* **43**, 477 (2002).
[2]Kazemi, F., Kiasat, A. R. *SC* **32**, 2255 (2002).
[3]Sharma, U., Bora, U., Boruah, R.C., Sandhu, J.S. *TL* **43**, 143 (2002).
[4]Kaboudin, B. *CL* 880 (2001).
[5]Wang, J.-X., Kang, L., Hu, Y., Wei, B. *SC* **32**, 1691 (2002).
[6]Plessis, C., Uguen, D., De Cian, A., Fischer, J. *TL* **41**, 5489 (2000).
[7]Curini, M., Epifano, F., Marcotullio, M.C., Rosati, O. *TL* **43**, 1201 (2002).

Aluminum.

Pinacol coupling. Aromatic aldehydes are coupled with Al-NaOH under ultrasound irradiation.[1]

[1]Bian, Y.-J., Liu, S.-M., Li, J.-T., Li, T.-S. *SC* **32**, 1169 (2002).

Aluminum chloride. 20, 12–13; 21, 15–17

Dealkylation. An improved method claimed for the demethylation of nitrocatechol methyl ether is by consecutive exposure to $AlCl_3$ in EtOAc and HCl.[1] The selective reaction of isopropyl esters with $AlCl_3$ is also reported.[2]

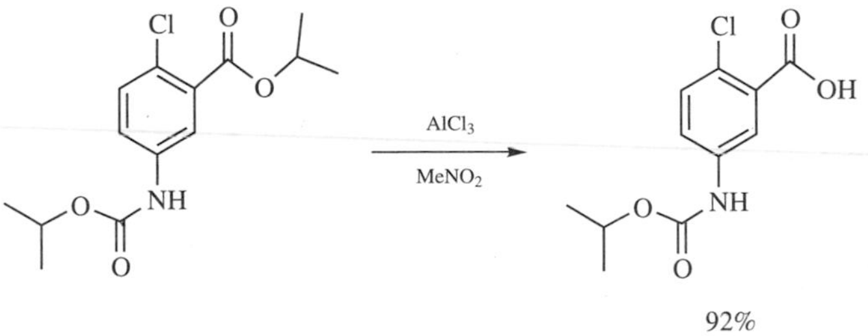

Acylations. A new formylating agent is tris(diformylamino)methane (**1**).[3] Acylation of zirconacyclopentane by aldehydes in the presence of $AlCl_3$ constitutes a useful procedure for the synthesis of γ,δ-unsaturated ketones.[4]

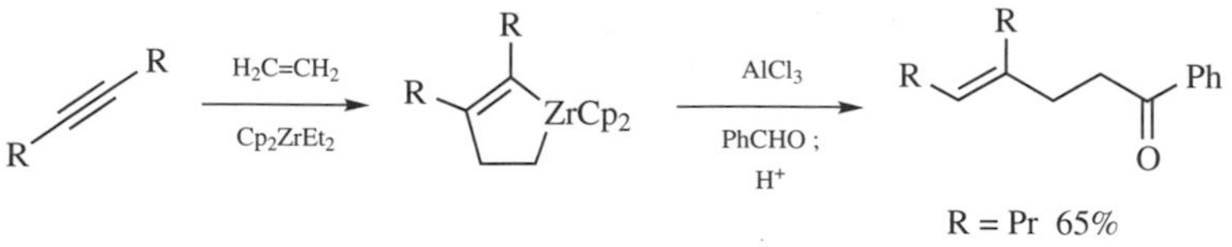

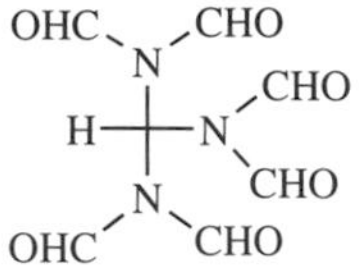

$H_2C{=}CH_2$ / Cp_2ZrEt_2; $AlCl_3$ / PhCHO ; H^+

R = Pr 65%

1

Carbazoles. When *N*-(*N,N*-diarylamino)phthalimides are treated with $AlCl_3$ diarylnitrenium ions are formed which give rise to carbazoles via an electrocyclization.[5]

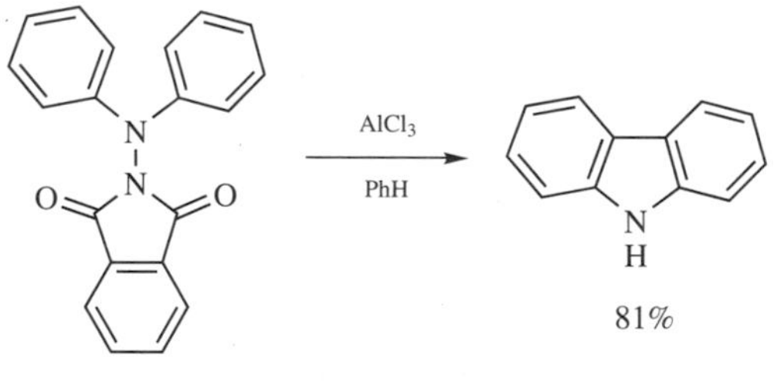

$AlCl_3$ / PhH

81%

Hydrosilylation.[6] Cyclopropanes are converted to silanes derivatives with ring opening on reaction with chlorodimethylsilanes in the presence of $AlCl_3$.

$AlCl_3$ / hexane

92%

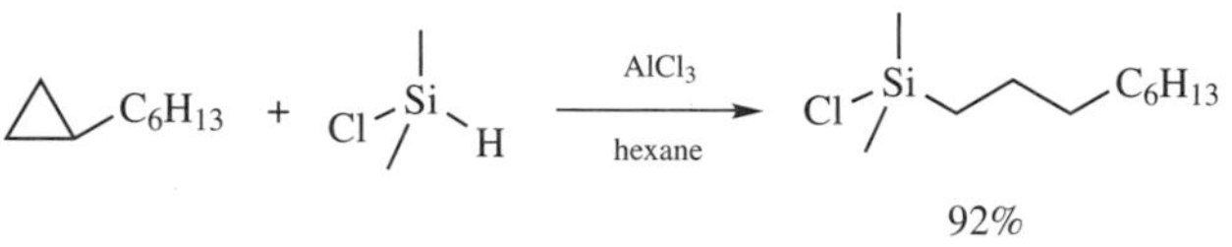

Double addition.[7] Two nucleophile units can be added to α,β-unsaturated aldimines when promoted by $AlCl_3$ and small amount of water.

Ph–CH=CH–CH=N–An + $(CH_3)_2C=C(OSiMe_3)OEt$ → ($AlCl_3$ / 4A-MS, CH_2Cl_2)

An = *p*-anisyl

81% (anti:syn >99 : 1)

Rearrangement.[8] Aryl *N,N*-dialkylsulfamates undergo a Fries-type rearrangement to afford hydroxyarenesulfonamides.

[1]Learmonth, D.A., Alves, P.C. *SC* **32**, 641 (2002).
[2]Chee, G.-L. *SL* 1593 (2001).
[3]Bagno, A., Kantlehner, W., Scher, O., Vetter, J., Ziegler, G. *EJOC* 2947 (2001).
[4]Zhao, C., Yu, T., Xi, Z. *CC* 142 (2002).
[5]Kikugawa, Y., Aoki, Y., Sakamoto, T. *JOC* **66**, 8612 (2001).
[6]Nagahara, S., Yamakawa, T., Yamamoto, H. *TL* **42**, 5057 (2001).
[7]Shimizu, M., Ogawa, T., Nishi, T. *TL* **42**, 5463 (2001).
[8]Benson, G.A., Maughan, P.J., Shelly, D.P., Spillane, W.J. *TL* **42**, 8729 (2001).

Aluminum iodide.

Reductive acylation.[1] An efficient method for transforming organic azides into amides consists of treatment with AlI_3-Ac_2O.

[1]Bez, G. *SC* **32**, 3625 (2002).

Aluminum tris(2,6-diphenylphenoxide), ATPH. 20, 14–15; **21,** 17–18

Michael reaction-alkylation tandem.[1] A synthesis of methyl jasmonate illustrates the efficient assemblage method.

cyclopentenone + LiO–C(=CH$_2$)–O–*t*-Bu → (BCl_3, dihydrofuran) → product with OH side chain and COO-*t*-Bu

[1]Saito, S., Yamazaki, S., Yamamoto, H. *ACIEE* **40**, 3613 (2001).

Aluminum biphenyl-2′-perfluorooctanesulfonamide-2-oxide isopropoxide.

Redox reactions. The new catalyst **1** corrects the inherent defects of the Meerwein-Ponndorf-Verley reduction system of (*i*-PrO)Al/*i*-PrOH that requires continuous removal of acetone to shift the equilibrium and the ability to complete the reduction at room temperature avoids side reactions.[1]

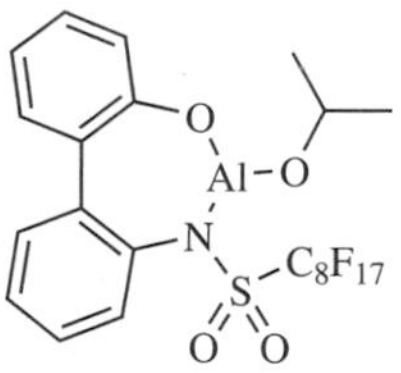

1

[1]Ooi, T., Ichikawa, H., Maruoka, K. *ACIEE* **40**, 3610 (2001).

1-Amino-3-benzylimidazolium chlorochromate, polymer-supported.

Oxidation.[1] The supported reagent **1** is recyclable oxidant for alcohols.

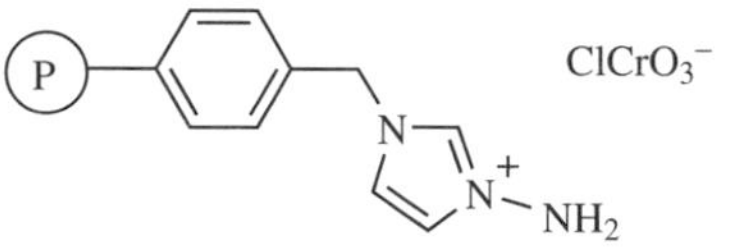

$ClCrO_3^-$

NH_2

1

[1]Linares, M.L., Sanchez, N., Alajarin, R., Vaquero, J.J., Alvarez-Builla, J. *S* 382 (2001).

***N*-Aminophthalimide.**

Aziridination.[1] Electrolytic generation of a nitrene in the presence of alkenes provides aziridines. The reaction has a broad scope (14 examples, 51–91%).

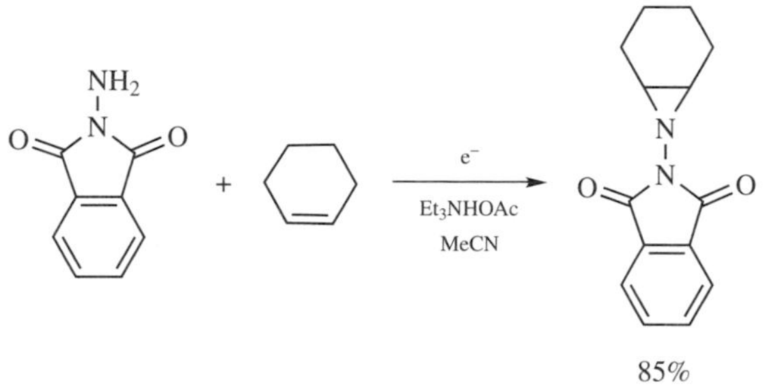

e^-, Et_3NHOAc, MeCN

85%

[1]Siu, T., Yudin, A.K. *JACS* **124**, 530 (2002).

Ammonia.

2-Heteroaryl-3-hydroxypyridines.[1] Moderate yields of the bi(heteroaryls) are obtained from 2-acylfurans on heating with aq ammonia at 150°.

γ-Hydroxy amides.[2] Ring opening of γ-lactones with ammonia is accomplished in a pressure tube at 50°. δ-Lactones react similarly.

β-Aminonaphthalene.[3] Conversion of β-naphthols to the corresponding amines is achieved by heating with aqueous ammonia and ammonium sulfite at 150–200°. The most important application is a synthesis of 2-amino-2′-hydroxy-1,1′-binaphthyl (91% yield) from BINOL.

[1]Chubb, R.W.J., Bryce, M.R., Tarbit, B. *JCS(P1)* 1853 (2001).
[2]Taylor, S.K., Ide, N.D., Silver, M.E., Stephan, M.L. *SC* **31**, 2391 (2001).
[3]Körber, K., Tang, W., Hu, X., Zhang, X. *TL* **43**, 7163 (2002).

Ammonium phenylselenosulfate.

α-Phenylseleno esters.[1] Preparation of these esters from 1-alkynes is accomplished in three simple steps. Oxidative addition by a mixture of $PhI(OAc)_2$ and PhSeSePh, followed by treatment with TsOH and then $PhSeOSO_3NH_4$ in alcohol solvents. The title reagent is obtained from PhSeSePh and ammonium persulfate.

R—≡ —(PhSeSePh, $PhI(OAc)_2$; TsOH)→ R, SePh, OTs + R′OH —($PhSeOSO_3^-$)→ R, PhSe, COOR′

[1]Tiecco, M., Testaferri, L., Temperini, A., Bagnoli, L., Marini, F., Santi, C. *SL* 706 (2001).

Antimony(V) chloride. 20, 16

Diels-Alder reaction. Improved regioselectivity is observed in certain Diels-Alder reactions with SbF_5 as compared to those catalyzed by $BF_3 \cdot OEt_2$.

PhMe, [cat]

catalyst		
$SbCl_5$, −70°	90 :	10
BF_3 , −70°	70 :	30
$TiCl_4$	38 :	62
—	30 :	70

[1]Nunes, R.L., Bieber, L.W. *TL* **42**, 219 (2001).

Arenediazonium *o*-benzenedisulfonimides. 21, 19

Substituted arenes. The diazonium salts are suitable precursors of many substituted arenes, e.g., aryl thiocyanates[1] and 1-aryl-3,3-dialkyltriazenes.[2]

[1]Barbero, M., Degani, I., Diulgheroff, N., Dughera, S., Focchi, R. *S* 585 (2001).
[2]Barbero, M., Degani, I., Diulgheroff, N., Dughera, S., Focchi, R. *S* 2180 (2001).

Arylboronic acids. 21, 20

Protection.[1] Selective masking of polyol systems using benzeneboronic acid is convenient.

HO, OH, OH, OH, OH, OH → ($PhB(OH)_2$; $C_7H_{15}COCl$; NaOH) → C_7H_{15}, O, O, OH, OH, OH, OH, O, O, C_7H_{15}

Condensation. 2-Hydroxymorpholines are accessible from 2-amino alcohols, glyoxal, and aldehydes.[2] For promoting diastereoselective Mukaiyama aldol reactions in water at room temperature diphenylboronic acid is very effective.[3] Traditional boron aldol reactions require lower temperature and anhydrous conditions.

Carboxylic acid derivatives. Conversion of carboxylic acids to acyl azides is accomplished by reaction with NaN_3 using 3,4,5-trifluorophenylboronic acid as catalyst.[4] An arylboronic acid bound to polystyrene (**1**) can serve as catalyst for amide formation from acids and amines.[5]

P, N^+, Cl^-, $B(OH)_2$

1

Reduction.[6] Selective reduction of aldehydes with Bu_3SnH is promoted by $PhB(OH)_2$.

[1]Bhaskar, V., Duggan, P.J., Humphrey, D.G., Krippner, G.Y., McCarl, V., Offermann, D.A. *JCS(P1)* 1098 (2001).
[2]Berree, F., Debache, A., Marsac, Y., Carboni, B. *TL* **42**, 3591 (2001).
[3]Mori, Y., Manabe, K., Kobayashi, S. *ACIEE* **40**, 2816 (2001).
[4]Tale, R.H., Patil, K.M. *TL* **43**, 9715 (2002).
[5]Latta, R., Springsteen, G., Wang, B. *S* 1611 (2001).
[6]Yu, H., Wang, B. *SC* **31**, 2719 (2001).

Aryllead triacetates. 20, 17

Arylation.[1] Ligand effect on the rate of arylation of β-dicarbonyl compounds has been evaluated. With 1,10-phenanthroline, almost 1000 fold increase over the uncatalyzed reaction is observed.

[1]Buston, J.E.H., Moloney, M.G., Parry, A.V.L., Wood, P. *TL* **43**, 3407 (2002).

***N*-Alkyl-*N'*-arylformamidines, polymer-supported.**

Alkyl sulfonates.[1] The polymer-bound reagents transfer their alkyl groups to sulfonic acids or sodium sulfonates.

[1]Vignola, N., Dahmen, Enders, D., Bräse, S. *TL* **42**, 7833 (2001).

1-Arenesulfinyl-2-alkanones.

4-Hydroxyalk-2-en-1-ones.[1] Condensation of these reagents with aldehydes involves formation of allylic sulfoxides that undergo rearrangement to provide the hydroxy enones.

O O Ar S + RCHO —(Et_3N / HOAc, EtCN)→ O R OH

[1]Nokami, J., Kataoka, K., Shiraishi, K., Osafune, M., Hussain, I., Sumida, S. *JOC* **66**, 1228 (2001).

***o*-Azidomethylbenzoyl chloride.**

Nucleoside protection.[1] The title reagent **1** is used as an acylating agent for substrates that require protection of the functional groups. Derivatives can be freed of the *o*-azidomethylbenzoyl group(s) by transfer hydrogenation or treatment with a phosphine.

COCl N_3

1

[1]Wada, T., Ohkubo, A., Mochizuki, A., Sekine, M. *TL* **42**, 1069 (2001).

Azidotris(diethylamino)phosphonium bromide.

Diazoalkanes and azides. The title reagent is an azo transfer agent as it reacts with deprotonated (with BuLi) hydrazones and amines to give diazoalkanes[1] and azides,[2] respectively.

[1]McGuiness, M., Shechter, H. *TL* **43**, 8425 (2002).
[2]Klump, S.P., Shechter, H. *TL* **43**, 8421 (2002).

B

Barium hydroxide. 20, 18

Arylmalonic esters.[1] *C*-Arylation of malonic esters with ArBr catalyzed by Na_2PdCl_4 and the best base is $Ba(OH)_2 \cdot 8H_2O$. Under such conditions the reaction of PhBr affords an almost quantitative yield of the product, contrary to 43% yield if Na_2CO_3 is used.

[1]Aramendia, M.A., Borau, V., Jimenez, C., marinas, J.M., Ruiz, J.R., Urbano, F.J. *TL* **43**, 2847 (2002).

Barium permanganate. 21, 21

Aromatization.[1] The reagent converts 1,4- dihydropyridines to pyridines in hot benzene.

[1]Memarian, H.R., Sadeghi, M.M., Momeni, A.R. *SC* **31**, 2241 (2001).

Benzenesulfonyl azide.

Carboazidation. Allylsilanes undergo chain elongation and functionalization with remarkable stereoselectivity on reaction with $PhSO_2N_3$ and a carbon radical source.[1] Using an iodoacetic ester to generate the radical precursors of 5-substituted pyrrolidinones are obtained.[2]

Ph–Si, OH, + EtOOC, S, OEt, S

$PhSO_2N_3$ - $Bu_3SnSnBu_3$

t-BuON=NOBu-*t*

PhH Δ

EtOOC, Ph–Si, N_3, OH

47% (*syn* : *anti* >90 : 10)

Fiesers' Reagents for Organic Synthesis, Volume 22. Series editor Tse-Lok Ho
ISBN 0-471-28515-3

$EtOOC\text{-}CH_2I + CH_2{=}CH\text{-}C_6H_{13} \xrightarrow[t\text{-BuON=NOBu-}t,\ PhH\ \Delta]{PhSO_2N_3\ -\ Bu_3SnSnBu_3} EtOOC\text{-}CH_2CH_2CH(N_3)C_6H_{13} \xrightarrow{In}$ 5-hexyl-2-pyrrolidinone, 67%

Diazo transfer.[3] A polymer-bound reagent that is readily prepared from the commercially available polymer-bound benzenesulfonyl chloride and sodium azide at room temperature can be safely handled.

[1]Chabaud, L., Landais, Y., Renaud, P. *OL* **4**, 4257 (2002).
[2]Renaud, P., Ollivier, C., Panchaud, P. *ACIEE* **41**, 3460 (2002).
[3]Green, G.M., Peet, N.P., Metz, W.A. *JOC* **66**, 2509 (2001).

Benzenethiol. 16, 327–329; **19,** 19; **20,** 20–21; **21,** 22–23

Cleavage of 2-nitrobenzenesulfonamides.[1] Primary amines protected as these sulfonamides can be regenerated on reaction with PhSH in the presence of Cs_2CO_3. For removal of the 2,4-dinitrobenzenesulfonyl group the thiol alone is effective.

[1]Nihei, K., Kato, M.J., Yamane, T., Palma, M.S., Konno, K. *SL* 1167 (2001).

Benzotriazole. 21, 24

N-Functionalization of amides.[1] A route to *t*-butyl esters of β-amino acids from carbamates uses benzotriazole to facilitate condensation and alkylation.

$PhCH_2CHO$ + benzotriazole + $H_2NCOOBn$ $\xrightarrow[PhMe\ \Delta]{TsOH}$ $PhCH_2CH(Bt)NHCOOBn$ $\xrightarrow[LDA\ /\ THF]{t\text{-BuOAc}}$ $PhCH_2CH(CH_2COO\text{-}t\text{-Bu})NHCOOBn$, 63%

[1]Katritzky, A. R., Kirichenko, K., Elsayed, A.M., Yu, J., Fang, Y., Steel, P.J. *JOC* **67**, 4957 (2002)

Benzoyl fluoride.

Benzoylation.[1] Preparation of 1,3-diketones by benzoylation of enolates with BzF is efficient.

[1]Wiles, C., Watts, P., Haswell, S.J., Pombo-Villar, E. *TL* **43**, 2945 (2002).

***O*-Benzyl benzenesulfonylformaldoxime.**

Homologation.[1] The reagent $PhSO_2CH{=}NOBn$ reacts with RI, RTePh or R_2Te to form RCH=NOBn in a tin-free radical process (initiator: AIBN).

[1]Kim, S., Song, H.-J., Choi, T.-L., Yoon, J.-Y. *ACIEE* **40**, 2524 (2001).

Benzyl 4,6-dimethoxy-1,3,5-triazin-2-yl carbonate.

Benzyloxycarbonylation.[1] Reagent **1** is useful for the protection of amino group.

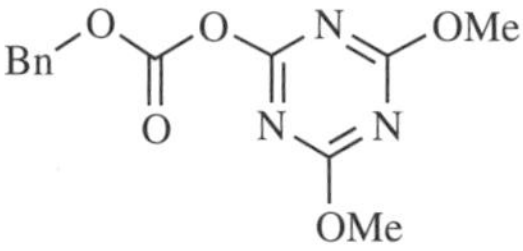

Bn O O N OMe O N N OMe

1

[1]Hioki, K., Fujiwara, M., Tani, S., Kunishima, M. *CL* 66 (2002).

***N*-Benzyl-*N′*-(1-hydroxy-2,2,2-trifluoroethyl)piperazine.**

Trifluoromethylation.[1] The stable hemiaminal of trifluoroacetaldehyde transfers the trifluoromethyl group to nonenolizable aryl carbonyl in the presence of *t*-BuOK.

1,1,1-Trifluoro-2-alken-4-ones.[2] The reagent serves as an alkylating agent for ketones. The adducts undergo β–elimination to give the fluorinated enones.

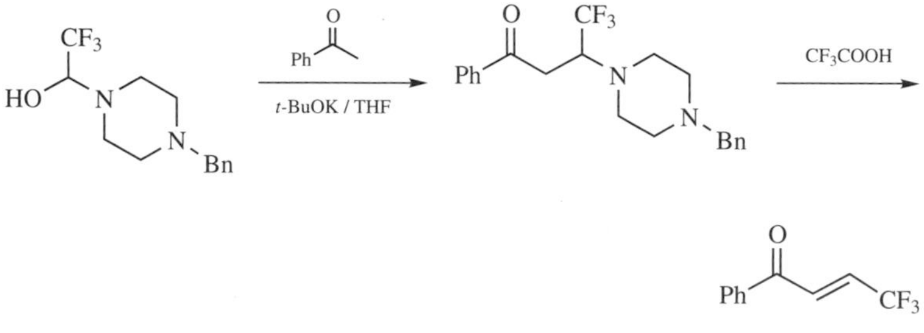

CF_3 HO N N Bn → (O, Ph; *t*-BuOK / THF) → O CF_3 Ph N N Bn → (CF_3COOH) → O Ph CF_3

70%

[1]Billard, T., Langlois, B.R., Blond, G. *EJOC* 1467 (2001).
[2]Blond, G., Billard, T., Langlois, B.R. *JOC* **66**, 4826 (2001).

N-Benzyloxycarbonylsulfamoyl triethylammonium hydroxide.

Benzyl carbamates.[1] This analogue of the Burgess reagent can be used to convert primary alcohols to benzyl carbamates ($RCH_2OH \rightarrow RCH_2NHCOOBn$) in one operation, first by mixing it with benzyl alcohol, then warming the initial adduct with the alcohols in benzene.

Cyclic sulfamidates and sulfamides. Using reagents of this type, 1,2-diols and 1,2-amino alcohols are transformed into cyclic sulfamidates[2] and sulfamides,[3] respectively. Inversion of configuration attends the replacement of the hydroxyl group.

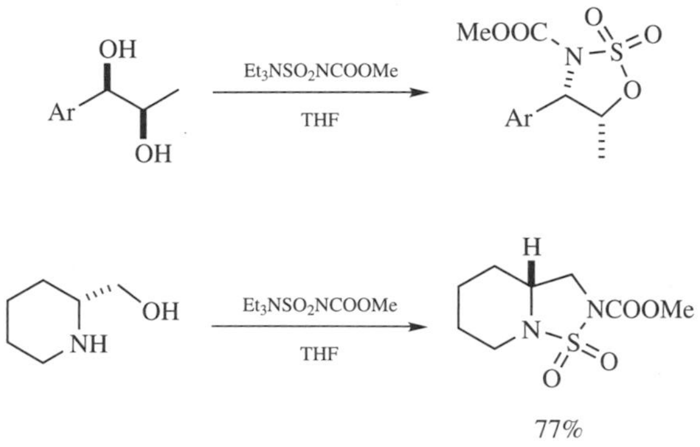

[1]Wood, M.R., Kim, J.Y., Books, K.M. *TL* **43**, 3887 (2002).
[2]Nicolaou, K.C., Huang, X., Snyder, S.A., Rao, P.B., Bella, M., Reddy, M.V. *ACIEE* **41**, 834 (2002).
[3]Nicolaou, K.C., Longbottom, D.A., Snyder, S.A., Nalbanadian, A.Z., Huang, X. *ACIEE* **41**, 3866 (2002).

Benzyltriethylammonium dichloroiodate.

Iodination. In the presence of $NaHCO_3$ anilines are iodinated with the reagent (8 examples, 58–99%).[1] The reagent is prepared by mxing NaI, NaOCl and HCl, then precipitated with $[BnNEt_3]ICl_2$.

[1]Kosynkin, D.V., Tour, J.M. *OL* **3**, 991 (2001).

Benzyltriphenylphosphonium salts. 21, 26

Oxidation. Aldehydes are formed on exposure of alcohols to the periodate salt using aluminum chloride as catalyst.[1] Benzyl trimethylsilyl and tetrahydrpyranyl ethers are directly converted to araldehydes by the chlorate salt under similar conditions.[2]

[1]Hajipour, A.R., Mallakpour, S.E., Samimi, H.A. *SL* 1735 (2001).
[2]Hajipour, A.R., Mallakpour, S.E., Mohammadpoor-Baltork, I., Malakoutikhah, M. *T* **58**, 143 (2002).

1,1′-Binaphthalene-2,2′-diamine derivatives.

C-C bond formation. Cyclic phosphoramide **1** is an excellent chiral catalyst for allylation and propargylation with allylstannane and allenylstannane reagents, respectively,[1] as well as cross-aldol reaction[2] of trichlorosilyl enol ethers with aldehydes.

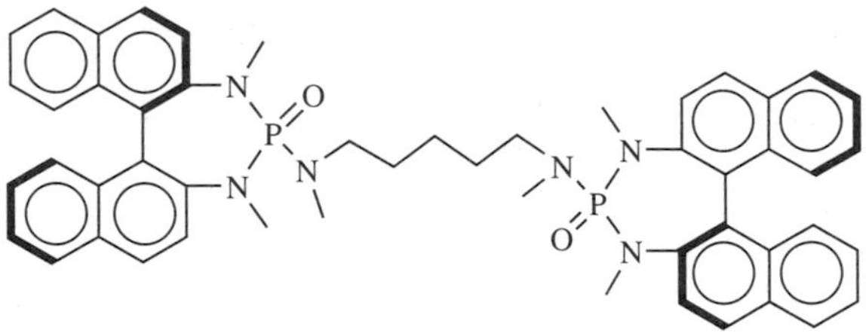

1

anti-2-Amino-3-hydroxy-3-arylpropanoic esters.[3] The imino derivative of the binaphthyldiamine and 3,5-di-*t*-butylsalicylaldehyde forms an aluminum complex (**2**) that can be used to synthesize methyl *cis*-5-aryl-2-(*p*-anisyl)oxazoline-4-carboxylates in >99% ee from 2-(*p*-anisyl)-5-methoxyoxazole and aromatic aldehydes. The products are precursors of 2-amino-3-hydroxy-3-arylpropanoic esters.

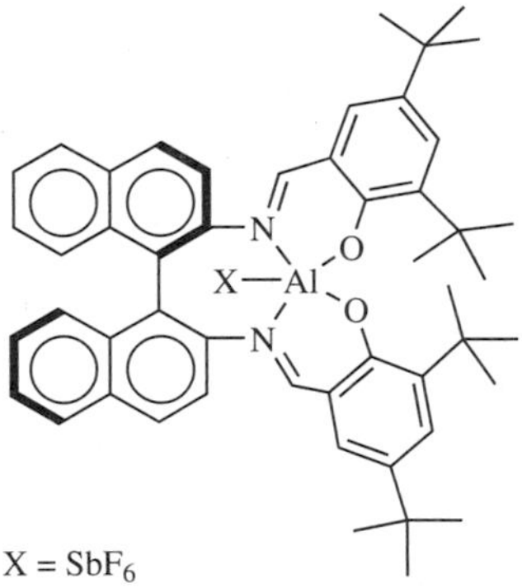

X = SbF_6

2

Neber rearrangement.[4] In the presence of the unsymmetrical quaternary ammonium salt **3** under phase-transfer conditions the rearrangement of ketoxime sulfonates proceeds via an anionic pathway.

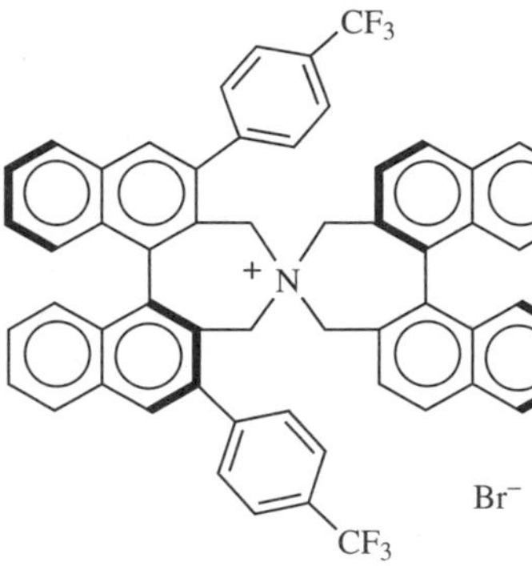

3

[1]Denmark, S.E., Wynn, T. *JACS* **123**, 6199 (2001).
[2]Denmark, S.E., Ghosh, S.K. *ACIEE* **40**, 4759 (2001).
[3]Evans,. A., Janey, J.M., Magomedov, N., Tedrow, J.S. *ACIEE* **40**, 1884 (2001).
[4]Ooi, T., Takahashi, M., Doda, K., Maruoka, K. *JACS* **124**, 7640 (2002).

1,1′-Binaphthalene-2,2′-diol, BINOL. 21, 26

Resolution.[1] Almost enantiopure BINOL is obtainable by using (*S*)-5-oxopyrrolidine-2-carboxanilide to form hydrogen-bonded complex with the (*R*)-isomer.

[1]Du, H., Ji, B., Wang, Y., Sun, J., Meng, J., Ding, K. *TL* **43**, 5273 (2002).

1,1′-Binaphthalene-2,2′-diol (modified)–aluminum complexes. 21, 27

Cyanohydrins. Aluminum complex **1** catalyzes asymmetric cyanation of aldehydes[1] with Me_3SiCN.

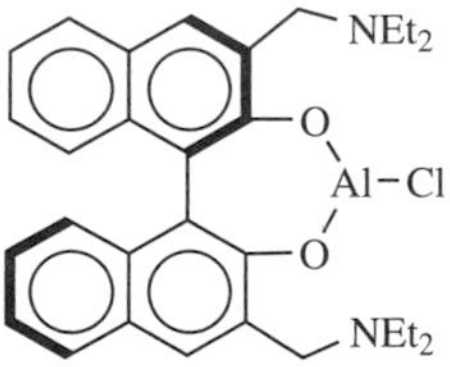

1

Reduction. BINOL[2] and its octahydro derivative[3] also mediate carbonyl reduction by borane complexes.

Nitro-Mannich reaction.[4] A route to chiral *trans*-1,2-diamines involves a catalytic addition in the presence of an Al-Li complex of BINOL followed by reduction with SmI_2.

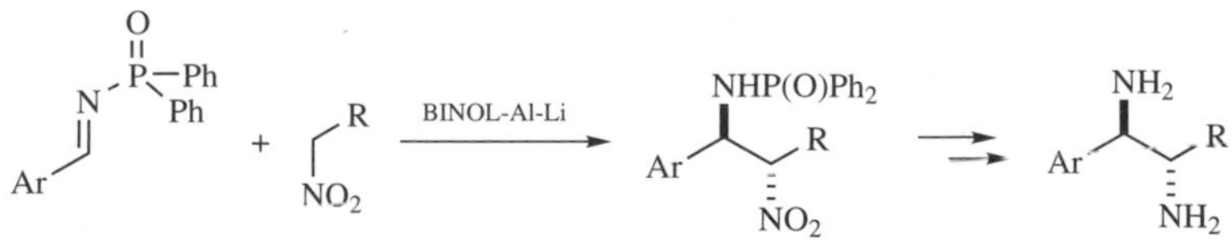

(*anti* : *syn* 3-7 : 1)

ee 60–80%

Strecker reaction.[5] The use of a polymer-bound Al-BINOLate (**2**) in promoting asymmetric Strecker reaction is demonstrated.

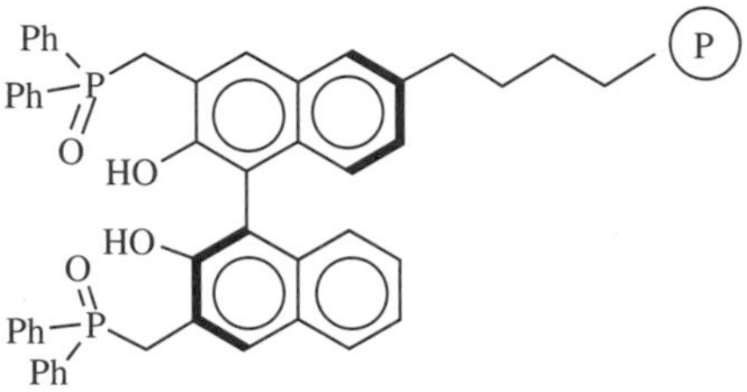

2

[1]Casas, J., Najera, C., Sansano, J.M., Saa, J.M. *OL* 2589 (2002).
[2]Fu, I.-P., Uang, B.-J. *TA* **12**, 45 (2001).
[3]Lin, Y.-M., Fu, I.-P., Uang, B.-J. *TA* **12**, 3217 (2001).
[4]Yamada, K., Moll, G., Shibasaki, M. *SL* 980 (2001).
[5]Nogtami, H., Matsunaga, S., Shibasaki, M. *TL* **42**, 279 (2001).

1,1′-Binaphthalene-2,2′-diol–calcium complex.

Michael reaction.[1] Asymmetric reactions with moderate ee are observed using this chiral complex.

[1]Kumaraswamy, G., Sastry, M.N.V., Jena, N. *TL* **42**, 8515 (2001).

1,1′-Binaphthalene-2,2′-diol–lanthanum complex.

Review.[1] A summary of asymmetric catalysis by lanthanide complexes, mainly those of BINOL and analogues, has been written.

Michael reaction.[2] A catalytic asymmetric reaction with superior ee (95→99%) uses this chiral complex.

Epoxidation. A chiral complex derived from BINOL, lanthanum isopropoxide and Ph_3AsO promotes epoxidation of conjugated ketones, aldehydes, esters, amides, and γ,δ-unsaturated β-keto esters with *t*-BuOOH.[3] A particularly useful procedure for the preparation of epoxy esters involves epoxidation of the corresponding acylimidazoles. A related study uses 6,6′-diphenyl-BINOL.[4]

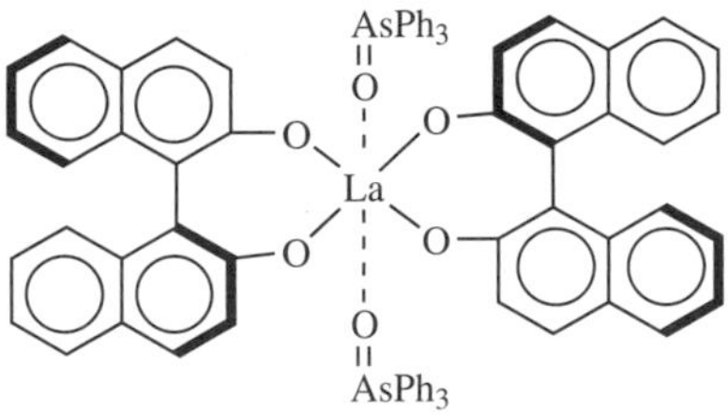

Cyanation + nitroaldol reaction.[5] With LnM_3tris(binaphthoxide) [M = Li, Na, K] as catalyst a sequential cyanation of aliphatic aldehyde and nitrol aldol reaction of aromatic aldehyde can be performed. After the first reaction, an achiral additive (e.g., $LiBF_4$) is introduced to tune the chiral environment for the second reaction.

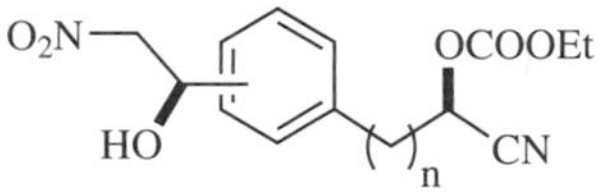

[1]Mikami, K., Terada, M., Matsuzawa, H. *ACIEE* **41**, 3554 (2002).
[2]Takita, R., Ohshima, T., Shibasaki, M. *TL* **43**, 4661 (2002).
[3]Nemoto, T., Ohshima, T., Yamaguchi, K., Shibasaki, M. *JACS* **123**, 2725, 9474 (2001).
[4]Chen, R., Qian, C., de Vries, J.G. *T* **57**, 9837 (2001).
[5]Tian, J., Yamagiwa, N., Matsunaga, S., Shibasaki, M. *ACIEE* **41**, 3636 (2002).

1,1′-Binaphthalene-2,2′-diol–lithium/scandium complex.

Strecker reaction.[1] The heterobimetallic complex $LiSc(BINOL)_2$ induces asymmetric addition to imines with Me_3SiCN in 55–95% ee.

[1]Chavarot, M., Byrne, J.J., Chavant, P.Y., Vallee, Y. *TA* **12**, 1147 (2001).

1,1′-Binaphthalene-2,2′-diol–molybdenum complexes.

Metathesis.[1,2] Modified BINOLs form complexes that find use in the synthesis of chiral cyclic ethers by olefin metathesis.

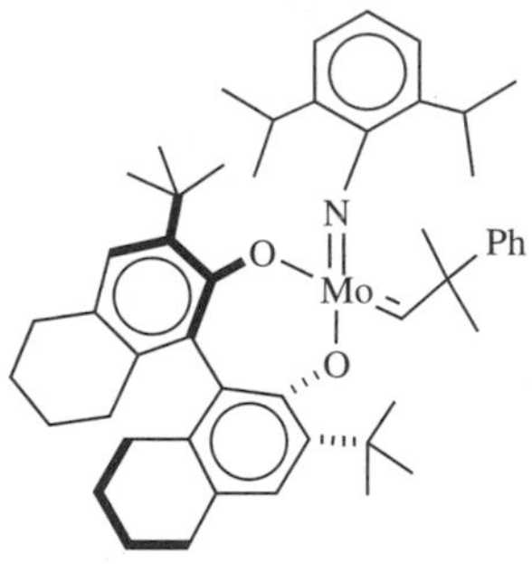

N Mo O O Ph

[1]Cefalo, D.R., Kiely, A.F., Wuchrer, M., Jamieson, J.Y., Schrock, R.R., Hoveyda, A.H. *JACS* **123**, 3139 (2001).
[2]Aeilts, S.L., Cefalo, D.R., Bonitatebus, P.J., Houser, J.H., Hoveyda, A.H., Schrock, R.R. *ACIEE* **40**, 1452 (2001).

1,1′-Binaphthalene-2,2′-diol–samarium complex.

Epoxidation.[1] A samarium-BINOL-triphenylphosphine oxide complex has been used for epoxidation of conjugated amides. Hydrogenolysis of the chiral epoxide products provides α-hydroxy amides.

[1]Nemoto, T., Kakei, H., Gnanadesikan, V., Tosaki, S., Ohshima, T., Shibasaki, M. *JACS* **124**, 14544 (2002).

1,1′-Binaphthalene-2,2′-diol–titanium complexes. 15, 26–27; **16,** 24–25; **17,** 28–30; **18,** 43–44; **19,** 25; **20,** 25–27; **21,** 28–29

Homoallylic alcohols. Various titanium compounds and BINOL combine to furnish catalysts that are effective for allyl transfer from allylstannanes to aldehydes.[1,2] For allylation of ketones a simple and reliable catalytic reaction is carried out with a catalyst system made up of $(i\text{-PrO})_4Ti$, BINOL and isopropanol (20 equiv.) under nitrogen at room temperature (isopropanol is required).[3] If the auxiliary ligand 2,2′-bis(tritylamino)-4,4′-dichlorobenzophenone is present a practical allylation with ee >97% is usually achieved.[4] Delivery of the allyl group from allyltrimethylsilane is effected using a complex derived from TiF_4.[5]

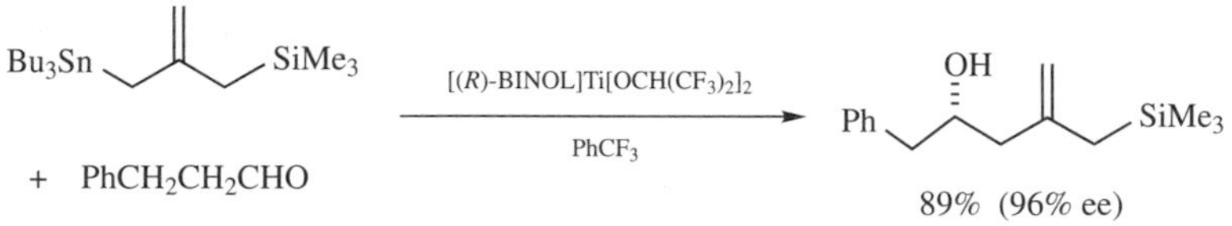

Bu_3Sn ... $SiMe_3$ + $PhCH_2CH_2CHO$ → [(*R*)-BINOL]Ti[$OCH(CF_3)_2]_2$, $PhCF_3$ → Ph ... OH ... $SiMe_3$

89% (96% ee)

By means of an ene reaction ethyl α-hydroxy-γ-methylenealkanoates are readily prepared. The catalyst is a Ti complex of octafluoro-BINOL and the reaction proceeds via a pseudoenantiomeric assembly.[6]

Propargylic alcohols. Titanium complexes of BINOL[7] and octahydro-BINOL[8] are very effective in promoting addition of 1-alkynes to aldehydes in the presence of a dialkylzinc reagent.

α-Acyloxy ketones.[9] One of the enantiomers of racemic epoxides of enol esters selectively undergoes isomerization on exposure to chiral Ti-BINOL complex. The reaction mixture gives the same enantiomer of an α-acyloxy ketone on treatment with TsOH.

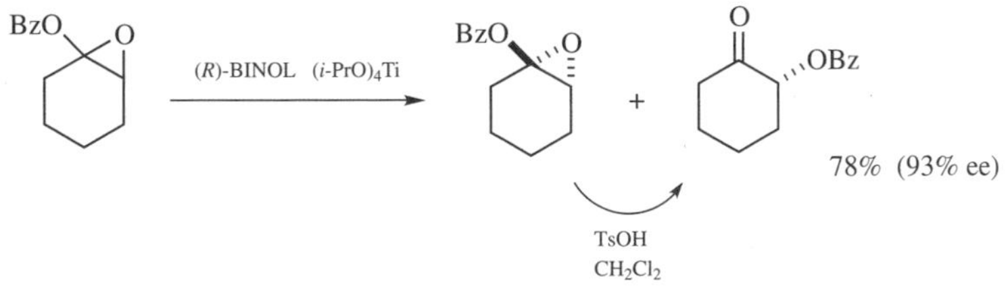

Chiral sulfoxides.[10] Asymmetric oxidation of sulfides using the renewable α,α-dimethylfurfuryl hydroperoxide, with moderate to good ee, is reported.

Addition of silyl ketene acetals to C═X. Optically active β-amino acid derivatives (80–92% ee) are obtained from reaction of silyl ketene acetals with nitrones.[11] Chiral δ-hydroxy-β-keto esters are accessible from the bisenol silyl ethers of methyl acetoacetate.[12]

Diels-Alder reactions. Various dienes and monoacetals of quinones give adducts (18 examples, 88–97%) in excellent ee (87–99%) when BINOL and $(i\text{-PrO})_2TiCl_2$ are present.[13] Catalysts for efficient hetero-Diels-Alder reactions have also been established.[14–16]

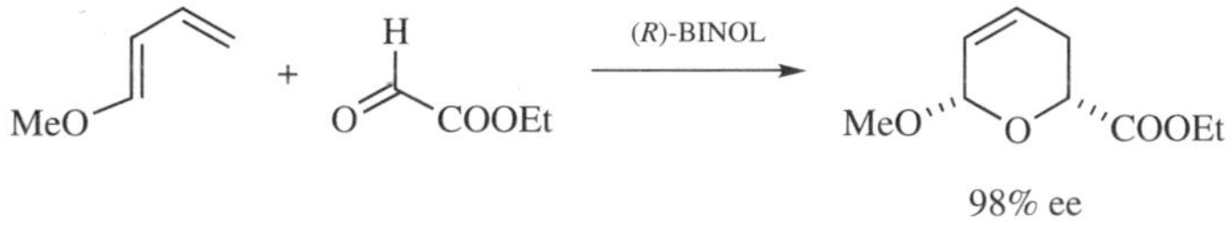

[1]Yu, C.-M., Jeon, M., Lee, J.-Y., Jeon, J. *EJOC* 1143 (2001).
[2]Yu, C.-M., Lee, J.-Y., So, B., Hong, J. *ACIEE* **41**, 161 (2002).
[3]Waltz, K.M., Gavenonis, J., Walsh, P.J. *ACIEE* **41**, 3697 (2002).
[4]Kii, S., Maruoka, K. *TL* **42**, 1935 (2001).
[5]Bode, J.W., Gauthier, D.R., Carreira, E.M. *CC* 2560 (2001).
[6]Pandiaraju, S., Chen, G., Lough, A., Yudin, A.K. *JACS* **123**, 3850 (2001).
[7]Moore, D., Pu, L. *OL* **4**, 1855 (2002).
[8]Lu, G., Li, X., Chan, W.L., Chan, A.S.C. *CC* 172 (2002).
[9]Feng, X., Shu, L., Shi, Y. *JOC* **67**, 2831 (2002).
[10]Massa, A., Lattanzi, A., Siniscalchi, F.R., Scettri, A. *TA* **12**, 2775 (2001).
[11]Murahashi, S., Imada, Y., Kawakami, T., Harada, K., Yonemushi, Y., Tomita, N. *JACS* **124**, 2888 (2002).
[12]Soriente, A., De Rosa, M., Stanzione, M., Villano, R., Scettri, A. *TA* **12**, 959 (2001).

[13]Breuning, M., Corey, E.J. *OL* **3**, 1559 (2001).
[14]Huang, Y., Feng, X., Wang, B., Zhang, G., Jiang, Y. *SL* 2122 (2002).
[15]Long, J., Hu, J., Shen, X., Ji, B., Ding, K. *JACS* **124**, 10 (2002).
[16]Quitschalle, M., Christmann, M., Bhatt, U. *TL* **42**, 1263 (2001).

1,1′-Binaphthalene-2,2′-diol (modified)–ytterbium complexes.

Epoxidation.[1] 6,6′-Diphenyl-BINOL on complexation with Yb provides a catalyst for epoxidation of conjugated ketones.

[1]Chen, R., Qian, C., De Vries, J.G. *TL* **42**, 6919 (2001).

1,1′-Binaphthalene-2,2′-diol (modified) - zinc complexes. 21, 29

Aldol reactions. A Zn/Zn-linked BINOL complex efficiently catalyzes the condensation of α-hydroxy ketones with aldehydes, affording the *syn*-2,3-dihydroxy ketones.[1] Note that heterobimetallic catalysts change the reaction pathway to afford the *anti* isomers.[2]

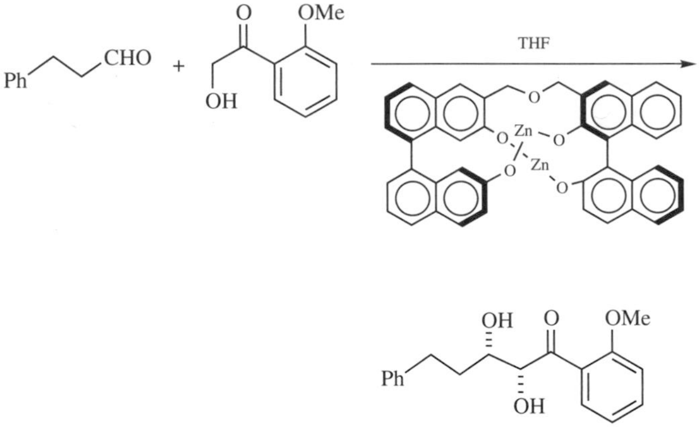

(*syn* : *anti* 89 : 11)

Propargylic alcohols.[3] Asymmetric addition of 1-alkynes to aldehydes is promoted by a complex of Et_2Zn and **1**.

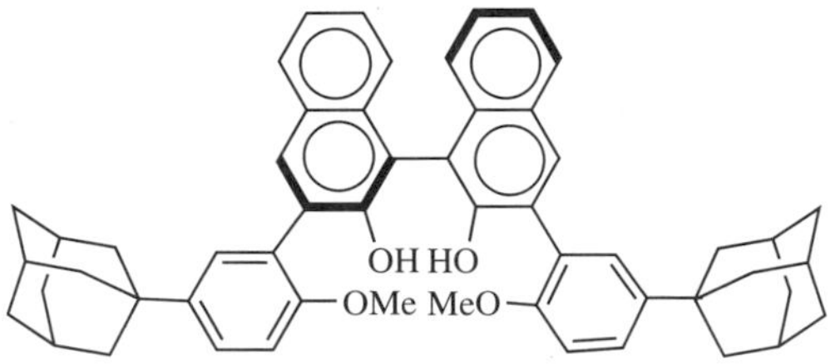

1

[1]Kumagai, N., Matsunaga, S., Yoshikawa, N., Ohshima, T., Shibasaki, M. *OL* **3**, 1539 (2001).
[2]Yoshikawa, N., Kumagai, N., Matsunaga, S., Moll, G., Ohshima, T., Suzuki, T., Shibasaki, M. *JACS* **123**, 2466 (2001).
[3]Xu, M.-H., Pu, L. *OL* **4**, 4555 (2002).

1,1′-Binaphthalene-2,2′-diol (modified) - zirconium complexes. 19, 25–26

Reduction.[1] Aldehydes are selectively reduced in the presence of a keto group by a Zr-complex of racemic BINOL in isopropanol at room temperature. Some dialdehydes are also differentiable.

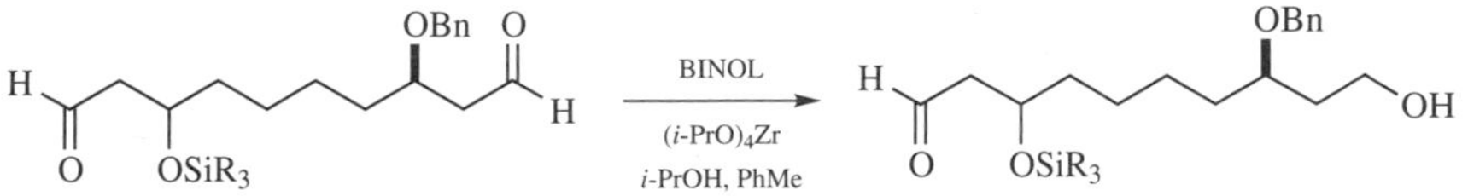

BINOL
$(i\text{-PrO})_4Zr$
i-PrOH, PhMe

Hetero-Diels-Alder reaction.[2] Zirconium complex of a chiral 3,3′-diiodo-BINOL is also found to be an effective catalyst for the smooth condensation of aldehydes with a Danishefsky diene analog.

Allylations. Methods for the formation of homoallylic alcohols[3] and amines[4] in good ee by reaction of allylstannanes to aldehydes and imines, respectively, have been established. In the latter case a 3,3′-dihalo-BINOL is used in forming an active catalyst and the stannane reagents contain an allylic hydroxy group. When such a group is absent 2 equiv of MeOH must be furnished in the catalyst preparation in order to achieve high enantioselectivity.

Addition of silyl enol ethers to C═X. Highly *anti*-selective asymmetric additions to aldehydes[5] and imines[6] are attainable. The effect of a small amount of water for catalyst preparation is noted.[5]

[1]Lorca, M., Kuhn, D., Kurosu, M. *TL* **42**, 6243 (2001).
[2]Yamashita, Y., Saito, S., Ishitani, H., Kobayashi, S. *OL* **4**, 1221 (2002).
[3]Kurosu, M., Lorca, M. *TL* **43**, 1765 (2002).
[4]Gastner, T., Ishitani, H., Akiyama, R., Kobayashi, S. *ACIEE* **40**, 1896 (2001).
[5]Yamashita, Y., Ishitani, H., Shimizu, H., Kobayashi, S. *JACS* **124**, 3292 (2002).
[6]Kobayashi, S., Kobayashi, J., Ishitani, H., Ueno, M. *CEJ* **8**, 4185 (2002).

1,1′-Binaphthalene-2,2′-dithiol derivatives.

α-Hydroxy ketones.[1] A synthetic method for this class of compounds in the optically active form involves the seven-membered cyclic dithioacetals. Reaction of the lithiated dithiazepanes with aldehydes followed by hydrolytic cleavage of the heterocycle affords the products.

Hydrogermylation.[2] Conjugated compounds such as methyl methacrylate undergo addition with germane (**1**) to give chiral adducts.

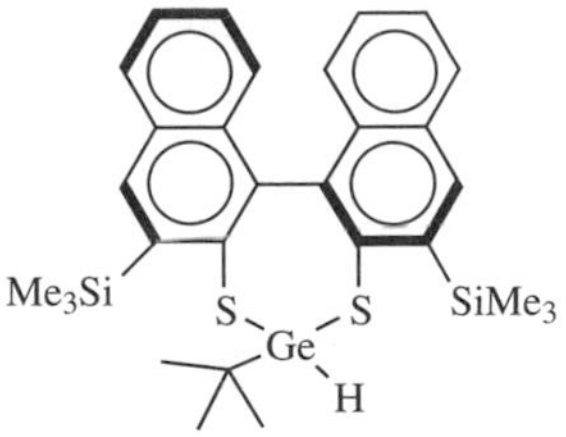

1

[1]Taka, H., Fujita, K., Oishi, A., Taguchi, Y. *H* **57**, 1487 (2002).
[2]Curran, D.P., Gualtieri, G. *SL* 1038 (2001).

1,1′-Binaphthalene-2,2′-diyl heteroesters. 21, 30–31; **21,** 30–31

Hydrogenation. Various ligands containing a phosphorus-bound BINOL residue have been investigated in asymmetric hydrogenation of enamides[1,2] and acetaminoacrylic esters that leads to α- and β-amino esters.[3–5] As demonstrated in two cases the phosphite ligand (**1**) forms a useful chiral catalyst for the hydrogenation of α-substituted acrylic esters.[6]

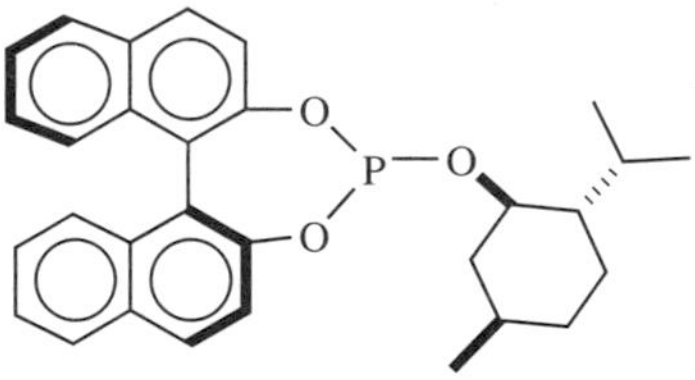

1

Kinetic resolution. By means of enantioselective allylic displacement the opening of unsaturated epoxides by dimethylzinc in the presence of $Cu(OTf)_2$ leads to resolution.[7]

Hydrosilylation.[8] Styrenes undergo highly enantioselective Pd-catalyzed hydrosilylation in the presence of a phosphoramidate in which the two ester oxygen atoms belong to BINOL.

Pauson-Khand reaction.[9] Bicyclic enones are obtained in yields and ee ranging from the very poor to excellent by this reaction in the presence of a BINOL diphosphite.

Allenecarboxylates.[10] A chain elongation process by reaction of phenyl esters with the methoxycarbonylmethylphosphonate ester of 3,3′-dimethyl-BINOL gives allenecarboxylates in variable yield and ee.

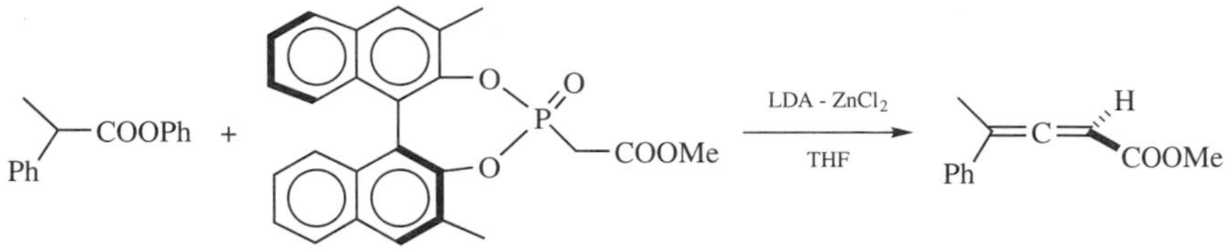

Allylic displacements.[11] Mediation of the asymmetric displacement with carbon and nitrogen nucleophiles can be performed in the presence of ligands **2** and **3,** respectively.

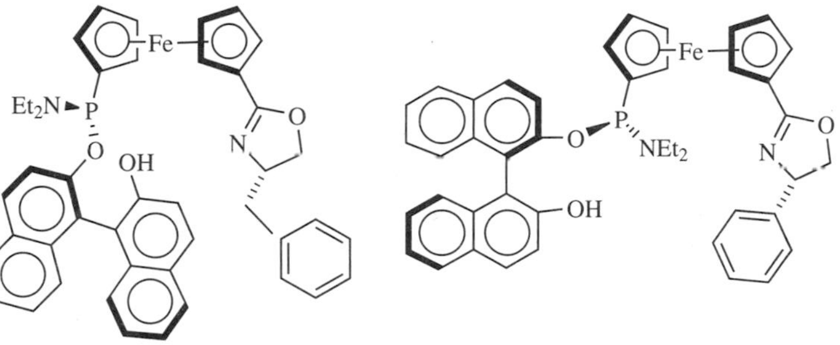

[1]Jia, X., Guo, R., Li, X., Yao, X., Chan, A.S.C. *TL* **43**, 5541 (2002).
[2]Reetz, M.T., Mehler, G., Meiswinkel, A., Sell, T. *TL* **43**, 7941 (2002).
[3]Zeng, Q., Lin, H., Cui, X., Mi, A., Jiang, Y., Li, X., Choi, M.C.K., Chan, A.S.C. *TA* **13**, 115 (2002).
[4]Zhou, Y.-G., Tang, W., Wang, W.-B., Li, W., Zhang, X. *JACS* **124**, 4592 (2002).
[5]Pena, D., Minnaard, A.J., de Vries, J.G., Feringa, B.L. *JACS* **124**, 14552 (2002).
[6]Chen, W., Xiao, J. *TL* **42**, 2897 (2001).
[7]Bertozzi, F., Crotti, P., Feringa, B.L., Macchia, F., Pineschi, M. *S* 483 (2001).
[8]Jensen, J.F., Svendsen, B.Y., La Cour, T.V., Pedersen, H.L., Johannsen, M. *JACS* **124**, 4558 (2002).
[9]Sturla, S.J., Buchwald, S.L. *JOC* **67**, 3398 (2002).
[10]Yamazaki, J., Watanabe, T., Tanaka, K. *TA* **12**, 669 (2001).
[11]You, S.-L., Zhu, X.-Z., Luo, Y.-M., Hou, X.-L., Dai, L.-X. *JACS* **123**, 7471 (2001).

1,1′-Binaphthalene-2,2′-diyl derivatives.

α-Amino acids.[1] The amino alcohol **1** is used in phase-transfer conditions to alkylate a Ni-complex of glycine.

OH
NH_2

1

C-C bond formation. In asymmetric alkenylation[2] and arylation[3] of ketone enolates, 2,2′-heterosubstituted binaphthyls containing one phosphino group and an amino or alkoxy residue show excellent activities as chiral ligands. The mixed P,N-ligand also finds application in Suzuki coupling,[4] whereas a mixed P,O-ligand which is also fluorous-modified has been developed in relation to allylic displacements.[5]

Asymmetric allyl transfer from tin reagents to ketones can be effected with the aid of binaphthalene-2-ol-2′-thiol.[6]

[1]Belokon, Y.N., Kochetkov, K.A., Churkina, T.D., Ikonnikov, N.S., Larionov, O.V., Harutyunyan, S.R., Vyskocil, S., North, M., Kagan, H.B. *ACIEE* **40**, 1948 (2001).
[2]Chieffi, A., Kamikawa, K., Ahman, J., Fox, J.M., Buchwald, S.L. *OL* **3**, 1897 (2001).
[3]Hamada, T., Chieffi, A., Ahman, J., Buchwald, S.L. *JACS* **124**, 1261 (2002).
[4]Yin, J., Buchwald, S.L. *JACS* **122**, 12051 (2000).
[5]Cavazzini, M., Pozzi, G., Quici, S., Maillard, D., Sinou, D. *CC* 1220 (2001).
[6]Cunningham, A., Woodward, S. *SL* 43 (2002).

Bis(acetonitrile)(1,5-cyclooctadiene)rhodium(I) tetrafluoroborate. 21, 33

Aryl transfer. Further use of the catalyst in aryl group transfer is from $ArSi(Me)F_2$ to aldehydes in the presence of KF.[1]

Conjugate addition. Addition aryl and alkenyl groups from stannanes[2] and siloxanes[3] to conjugated carbonyl compounds can be carried out with the cationic rhodium catalyst.

Aryltriethoxysilanes.[4] A practical preparation of the arylsiloxanes is by the Rh-catalyzed reaction between ArBr and $HSi(OEt)_3$.

[1]Oi, S., Moro, M., Inoue, Y. *OM* **20**, 1036 (2001).
[2]Oi, S., Moro, M., Ito, H., Honma, Y., Miyano, S., Inoue, Y. *T* **58**, 91 (2002).
[3]Oi, S., Honma, Y., Inoue, Y. *OL* **4**, 667 (2002).
[4]Murata, M., Ishikura, M., Nagata, M., Watanabe, S., Masuda, Y. *OL* **4**, 1843 (2002).

Bis(acetonitrile)cyclopentadienyl(triorganophosphine)ruthenium(I) tetrafluoroborate. 21, 33

Cleavage of allyl esters.[1] Carboxylic acids are recovered in excellent yields when allyl esters are treated with the cationic rhodium complex in MeOH.

[1]Kitamura, M., Tanaka, S., Yoshimura, M. *JOC* **67**, 4975 (2002).

Bis(acetonitrile)dichloropalladium(II). 13, 33, 211, 236; **14,** 35–36; **15,** 28–29; **16,** 25–26; **17,** 30–31; **18,** 44–45; **19,** 26; **21,** 33–35

Coupling reactions. With assistance of $(MeCN)_2PdCl_2$ several useful preparations are accomplished. Thus, by a Stille coupling benzylsilanes are readily procured,[1] and cross-coupling between $ArSi(OEt)_3$ and Ar′Br proceeds smoothly in aq. NaOH to afford biaryls.[2]

In the presence of molecular oxygen the Heck coupling involving a 2-hydroxymethylacrylic ester, double bond migration and replacement of the OH group by an amino residue of the base Et_3N are observed.[3] The tertiary amine is dealkylated in situ by the [I-Pd-OOH] species.

Substituted izidine compounds are assembled[4] by way of heterocyclization in conjunction with C-C bond formation. Also interesting is the union of 2,3-alkadienoic acids with 2,3-alkadienones.[5]

Cyclization reactions. β-Diketones and β-keto esters bearing a terminal double bond at proper distances furnish cyclic ketones.[6] Treatment with $(MeCN)_2PdCl_2$ converts dimethyl acetals of 2-alkynylbenzaldehydes to indenol ethers.[7]

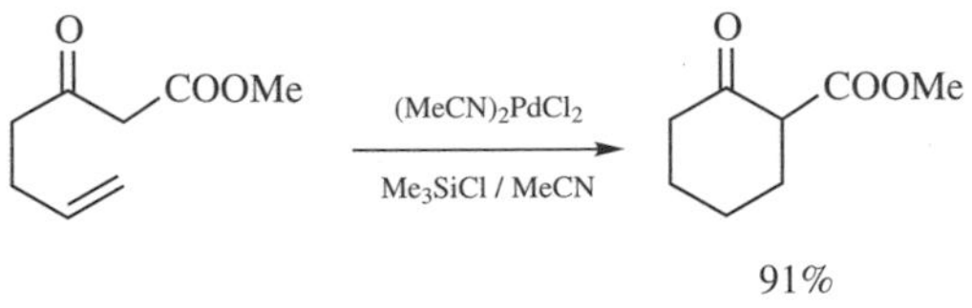

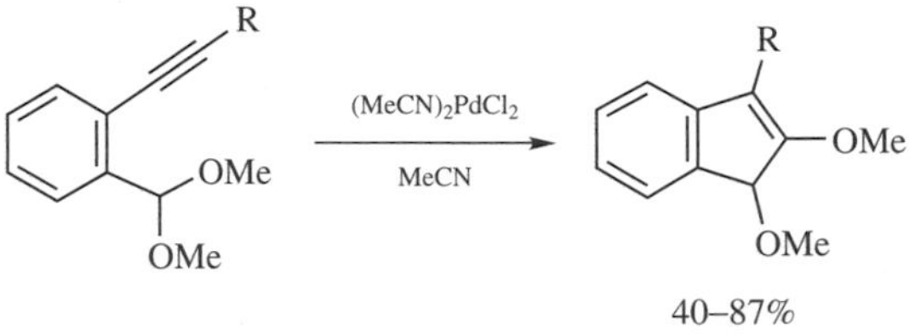

$(MeCN)_2PdCl_2$, Me_3SiCl / MeCN — 91%

$(MeCN)_2PdCl_2$, MeCN — 40–87%

Isomerization.[8] (*Z*)-Arylalkenes undergo isomerization to the (*E*)-isomers at room temperature on exposure to $(MeCN)_2PdCl_2$.

β-Amido ketones.[9] In the conjugate addition of carbamates to enones the catalyst of choice is $(MeCN)_2PdCl_2$.

[1]Itami, K., Kamei, T., Yoshida, J. *JACS* **123**, 8773 (2001).
[2]Murata, M., Shimazaki, R., Watanabe, S., Masuda, Y. *S* 2231 (2001).
[3]Hosokawa, T., Kamiike, T., Murahashi, S.-I., Shimada, M., Sugafuji, T. *TL* **43**, 9323 (2002).
[4]Karstens, W.F.J., Klomp, D., Rutjes, F.P.J.T., Hiemstra, H. *T* **57**, 5123 (2002).
[5]Ma, S., Yu, Z. *ACIEE* **41**, 1775 (2002).
[6]Pei, T., Widenhoefer, R.A. *JACS* **123**, 11290 (2001); *CC* 650 (2002).
[7]Nakamura, I., Bajracharya, G.B., Mizushima, Y., Yamamoto, Y. *ACIEE* **41**, 4328 (2002).
[8]Yu, J., Gaunt, M.J., Spencer, J.B. *JOC* **67**, 4627 (2002).
[9]Gaunt, M.J., Spencer, J.B. *OL* **3**, 25 (2001).

Bis(allyl)dichlorodipalladium. 20, 29; **21,** 35–37

Allylic displacements. A flurry of activity concerns with exploration of various chiral ligands in conjunction with the Pd complex for allylic displacements. *cis,cis,cis*-1,2,3,4-Tetrakis(diphenylphosphinomethyl)cyclopentane which is prepared in seven steps from commercially available himic anhydride, is very efficient and the catalyst system remains active for a long time.[1,2] With a substrate to catalyst ratio of 300,000 a conversion after 90 h reaches 84%. Turnover numbers as high as 680,000 can be obtained with a substrate to catalyst ratio of 1,000,000.[1]

Other ligands include a C_2-symmetric bissulfoximine (**1**)[3] and chiral *N*-(2-diphenylphosphinonaphth-1-yl)pyrrolidine[4] and piperidine derivatives.[5] 1,1-Diacetoxy-2-alkenes lose one of the acetoxy groups in an enantioselective fashion during the displacement.[6]

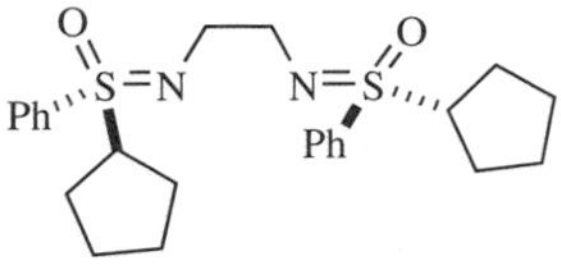

1

Coupling reactions. The air-stable Pd-complex with *cis,cis,cis*-1,2,3,4-tetrakis(diphenylphosphinomethyl)cyclopentane is an efficient catalyst for the Heck reaction[7,8] and Suzuki coupling of hindered substrates.[9] The Pd-catalyzed homocoupling of aryl halides is promoted by Bu_4NF that is generated in situ.[10]

Arylstannanes.[11] A double displacement of 2-trimethylsilylphenyl triflate occurs on the Pd-catalyzed reaction with stannane reagents.

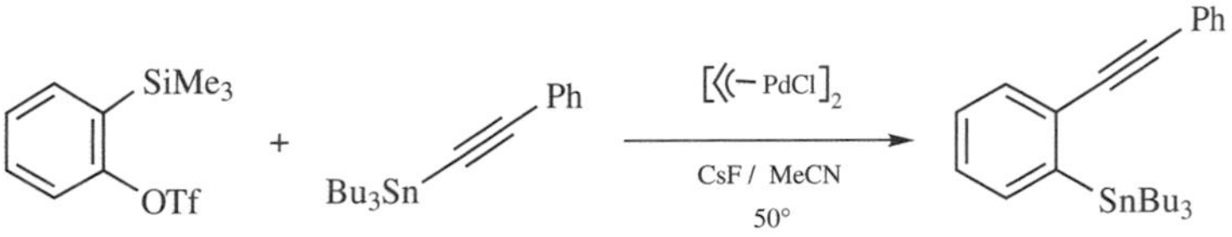

54%

Chain elongations. A synthesis of aroylamines from aryl iodides and primary amines via a *C,N*-carbonylation process uses a catalytic system consisting of bis(allyl)dichlorodipalladium, triphenylphosphine, and DABCO.[12]

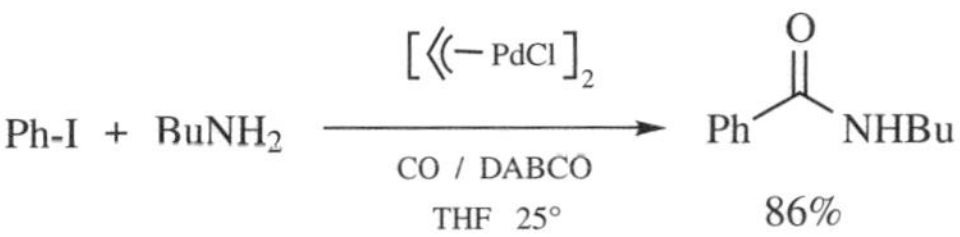

86%

Alkenoylgermanes are obtained from 1-alkynes, trifurylgermane and carbon monoxide under Pd catalysis.[13]

Cyclizations. Formation of cyclopentenes from 1,6-dienes as catalyzed by the Pd complex in the presence of a Lewis acid can be directed toward isomers with an exocyclic double bond by using a silver(I) salt.[14] Cyclopentanols and cyclohexanols are obtained in a reaction of ω-allenyl carbonyl compounds with silylstannane reagents.[15]

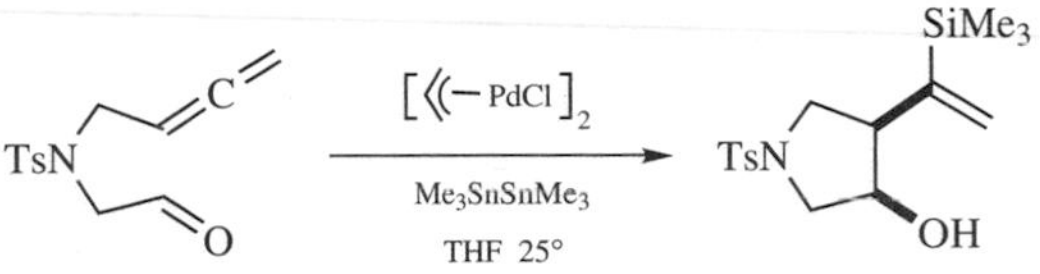

71%

[1]Laurenti, D., Feuerstein, M., Pepe, G., Doucet, H., Santelli, M. *JOC* **66**, 1633 (2001).
[2]Feuerstein, M., Doucet, H., Santelli, M. *TL* **42**, 2313 (2001).
[3]Bolm, C., Simic, O., Martin, M. *SL* 1878 (2001).
[4]Mino, T., Tanaka, Y.-I., Akita, K., Anada, K., Sakamoto, M., Fujita, T. *TA* **12**, 1677 (2001).
[5]Kondo, K., Kazuta, K., Saitoh, A., Murakami, Y. *H* **59**, 97 (2003).
[6]Trost, B.M., Lee, C.B. *JACS* **123**, 3671 (2001).
[7]Feuerstein, M., Doucet, H., Santelli, M. *JOC* **66**, 5923 (2001).
[8]Berthiol, F., Feuerstein, M., Doucet, H., Santelli, M. *TL* **43**, 5625 (2002).
[9]Feuerstein, M., Laurenti, D., Doucet, H., Santelli, M. *TL* **42**, 6667 (2001).
[10]Albanese, D. *SL* 1878 (2001)., Landini, D., Penso, M., Petricci, S. *SL* 199 (1999).
[11]Yoshida, H., Honda, Y., Shirakawa, E., Hiyama, T. *CC* 1880 (2001).
[12]Uozumi, Y., Arii, T., Watanabe, T. *JOC* **66**, 5272 (2001).
[13]Kinoshita, H., Shinokubo, H., Oshima, K. *JACS* **124**, 4220 (2002).
[14]Oh, C.H., Kim, J.D., Han, J.W. *CL* 1290 (2001).
[15]Kang, S.-K., Ha, Y.-H., Ko, B.-S., Lim, Y., Jung, J. *ACIEE* **41**, 343 (2002).

Bis(η⁶-benzenedichloro)ruthenium(II).

o-Substitution. Imines derived from ArCOR undergo arylation and alkenylation at the ortho position(s).[1] 2-Arylpyridines react in the same manner.[2]

R
N
Ar
+ Ph-Br
Cl
Ru
Cl
. 4PPh$_3$
K_2CO_3 / NMP 120°

Ph R
N
Ar
+
Ph R
N
Ar
Ph
(78:22)
91%

N
+ Ph-Br
Cl
Ru
Cl
. 4PPh$_3$
K_2CO_3 / NMP 120°

N
Ph
+
Ph
N
Ph
71%
11%

[1]Oi, S., Ogino, Y., Fukita, S., Inoue, Y. *OL* **4**, 1783 (2002).
[2]Oi, S., Fukita, S., Hirata, N., Watanabe, N., Miyano, S., Inoue, Y. *OL* **3**, 2579 (2001).

Bis(benzonitrile)(1,3-diphenylphosphinopropane)platinum tetrafluoroborate.

Cyclization.[1] Certain enynes are isomerized in contact with the cationic Pt complex.

EtOOC EtOOC → $(dppp)Pt(PhCN)_2\ [BF_4]_2$, $CHCl_3$ 25° → EtOOC EtOOC CHMe

74% (*E*:*Z* 28 : 72)

[1]Oi, S., Tsukamoto, I., Miyano, S., Inoue, Y. *OM* **20**, 3704 (2001).

Bis(benzonitrile)dichloropalladium(II). 13, 34; **15,** 29; **18,** 46–47; **19,** 27; **20,** 30–31; **21,** 38–39

Coupling reactions. The Pd-catalyzed coupling of aryl halides is promoted by tetrakis(dimethylamino)ethylene. Some easily reducible functional groups (nitro, nitrile, formyl, ester) survive the reaction conditions.[1] Alkenyldimethyl(2-pyridyl)silanes undergo cross-coupling via a transmetallation pathway.[2]

N Si → $(dba)_3Pd_2$, Fu_3P / Et_3N, ArI ; Ar′I, THF 60° → N Si Ar′ Ar → $(MeCN)_2PdCl_2$, Ar″I ; Bu_4NF → Ar″ Ar′ Ar

The Sonagashira coupling of alkenyl halides occurs at room temperature in piperidine.[3]

Arenecarboxylic acids and derivatives. Carbonylation of aryl chlorides in the presence of water, an alcohol, or an amine leads to carboxylic acid derivatives. A ferrocene-based diphosphine serves as ligand for the Pd catalyst.[4] For *n*-butoxycarbonylation of chloropyridines a 1,4-bis(dialkylphosphino)butane is used to advantage.[5]

[1]Kuroboshi, M., Waki, Y., Tanaka, H. *SL* 637 (2002).
[2]Itami, K., Nokami, T., Ishimura, Y., Mitsudo, K., Kamei, T., Yoshida, J. *JACS* **123**, 11577 (2001).
[3]Alami, M., Crousse, B., Ferri, F. *JOMC* **624**, 114 (2001).
[4]Mägerlein, W., Indolese, A.F., Beller, M. *ACIEE* **40**, 2856 (2001).
[5]Beller, M., Mägerlein, W., Indolese, A.F., Fischer, C. *S* 1098 (2001).

Bis(benzonitrile)dichloroplatinum(II).

Friedel-Crafts acylation.[1] The reaction with anhydrides is catalyzed by a mixture of $(PhCN)_2PdCl_2$ and $AgSbF_6$.

[1]Fürstner, A., Voigtländer, D., Schrader, W., Giebel, D., Reetz, M.T. *OL* **3**, 417 (2001).

Bis(bromodibutyl)tellurium oxide.

Reduction and condensation.[1] Reductive debromination of $ArCOCH_2Br$ to the methyl ketones (10 examples, 70–99%) and condensation of ArCHO with $BrCH_2COOEt$ to afford ArCH=CHCOOEt by triphenyl phosphite is catalyzed by the organotellurium oxide. Actually, the reactive species is a dibutyltellurium ylide derived from Bu_2Te that is formed in situ by reduction of the tetravalent tellurium compound by the phosphite.

[1]Huang, Z.-Z., Tang, Y. *JOC* **67**, 5320 (2000).

1,2-Bis(*N*-bromo-*p*-tosylamino)ethane.

Deoximation.[1] The title reagent is used to regenerate carbonyl compounds from oximes (11 examples, 88–97%)

[1]Khazaei, A., Vaghei, R.G., Tajbakhsh, M. *TL* **42**, 5099 (2001).

2,2-Bis(*t*-butylperoxy)butane.

Redox reaction.[1] In combination with *t*-dodecanethiol the title reagent converts cyclic benzylideneacetals to benzoates.

Ph O O RSH collidine Ph O O

[1]Roberts, B.P., Smits, T.M. *TL* **42**, 137 (2001).

Bis[(carbonyl)chloro(diphenylphosphinopropane)rhodium(I)].

Pauson-Khand reaction.[1] The annulation process can be coupled to an allylic displacement. Synthesis of bicyclic systems is achieved in one step.

MeOOCO Ar + TsN Li $[(dppp)RhCl(CO)]_2$ CO / MeCN TsN Ar H O

79%

[1]Evans, P.A., Robinson, J.E. *JACS* **123**, 4609 (2001).

Bis[chloro(1,5-cyclooctadiene)iridium(I)]. **21,** 39–40

Aromatic substitutions. Borylation of arenes with bis(pinacolato)diboron occurs in the presence of [(cod)IrCl]$_2$.[1] Note that anisole undergoes predominantly *m*-borylation. Heteroarylboronates are similarly prepared.[2]

C-C bond formation between highly activated arenes and internal alkynes is observed. Reaction occurs at the peri-position of 1-naphthol and *N*-(1-naphthyl)arenenesulfonamides. In the case of salicylaldehyde an insertion at the formyl C-H bond results in 2-hydroxyphenyl vinyl ketones.[3]

[(cod)IrCl]$_2$ *t*-Bu$_3$P
Na$_2$CO$_3$ / PhMe

R = Pr 83%

Reductive alkylation of acrylic esters. Preparation of 2-methyl-3-hydroxyalkanoic esters and β-lactams in a highly stereoselective manner is accomplished by reaction of acrylic esters with aldehydes[4] and imines,[5] respectively, with catalysis by [(cod)IrCl]$_2$.

[(cod)IrCl]$_2$
indane-pybox, Et$_2$MeSiH ;
H_2O

58% (98 : 2 er)

[(cod)IrCl]$_2$
(PhO)$_3$P, Et$_2$MeSiH
60°

58%

Exchange reactions. A highly efficient vinyl ether synthesis[6] involves exchange of the acetoxy group of vinyl acetate on reaction with alcohols under the influence of the iridium complex. Allylic phosphates are converted to β,γ-unsaturated esters on carbonylation in alcoholic solvents,[7] while displacement of allylic esters by amines proceeds with an opposite regioselectvity.[8]

Cycloisomerization and cycloadditions. The title iridium complex catalyzes cycloisomerization of enynes, for example, 1,6-enynes to 1,2-dialkylidenecyclopentanes.[9] The [2+2+2]cycloaddition involving diynes and alkynes to form fused benzene derivatives is very efficient in the presence of $[(cod)IrCl]_2$.[10]

Redox reactions. Conjugated carbonyl compounds are saturated at the double bond by hydrogen transfer from isopropanol.[11] On the other hand, a saturated carbonyl group is reduced to the hydroxy stage.

Oppenauer oxidation of secondary alcohols in aq. acetone is achieved with the iridium complex as catalyst and potassium 2,2′-biquinoline-4,4′-dicarboxylate as additive.[12]

Unsatuated amines. Addition of trimethylsilylethyne to aldimines leading to trimethylsilylated propargylamines is possible with the help of $[(cod)IrCl]_2$.[13] In a three-component coupling an *N*-allyl imine results from a mixture of an aldehyde, amine, and 1-alkyne (an internal alkyne does not react).[14] The only exception noted thus far is that trimethylsilylethyne adds to the C═N bond.

Hydrogenation. Immobilization of the title complex through ligation with a polymer-bound ferrocenyldiphosphine enables hydrogenation of hindered *N*-arylimines.[15] High turnover numbers (>100,000) are observed but the ee is below 80%.

The ferrocenyl-binaphane **1** is a ligand for the Ir-promoted hydrogenation of acyclic imines.[16]

1

[1]Ishiyama, T., Takagi, J., Ishida, K., Miyaura, N., Anastasi, N.R., Hartwig, J.F. *JACS* **124**, 390 (2002).

[2]Takagi, J.,Sato, K., Hartwig, J.F., Ishiyama, T., Miyaura, N. *TL* **43**, 5649 (2002).
[3]Nishinaka, Y., Satoh, T., Miura, M., Morisaka, H., Nomura, M., Matsui, H., Yamaguchi, C. *BCSJ* **74**, 1727 (2001).
[4]Zhao, C.-X., Duffey, M.O., Taylor, S.J., Morken, J.P. *OL* **3**, 1829 (2001).
[5]Townes, J.A., Evans, M.A., Queffelec, J., Taylor, S.J., Morken, J.P. *OL* **4**, 2537 (2002).
[6]Okimoto, Y., Sakaguchi, S., Ishii, Y. *JACS* **124**, 1590 (2002).
[7]Takeuchi, R., Akiyama, Y. *JOMC* **651**, 137 (2002).
[8]Takeuchi, R., Ue, N., Tanabe, K., Yamashita, K., Shiga, N. *JACS* **123**, 9525 (2001).
[9]Chatani, N., Inoue, H., Morimoto, T., Muto, T., Murai, S. *JOC* **66**, 4433 (2001).
[10]Takeuchi, T., Tanaka, S., Nakaya, Y. *TL* **42**, 2991 (2001).
[11]Sakaguchi, S., Yamaga, T., Ishii, Y. *JOC* **66**, 4710 (2001).
[12]Ajjou, A.N. *TL* **42**, 13 (2001).
[13]Fischer, C., Carreira, E.M. *OL* **3**, 4319 (2001).
[14]Sakaguchi, S., Kubo, T., Ishii, Y. *ACIEE* **40**, 2534 (2001).
[15]Pugin, B., Landert, H., Spindler, F., Blaser, H.-U. *ASC* **344**, 974 (2002).
[16]Xiao, D., Zhang, X. *ACIEE* **40**, 3425 (2001).

Bis[chloro(1,5-cyclooctadiene)rhodium(I)].

Primary amines.[1] A new route to primary amines is by reductive amination of carbonyl compounds with ammonia under hydrogen (pressurized).

Phenyl ketones.[2] The title Rh catalyst promotes reaction of sodium tetraphenylborate with acid anhydrides to give phenyl ketones. In the presence of norbornene the products are 2-acyl-3-phenylnorbornanes.

Macroheterocycles.[3] Under hydroformylation conditions diene-diamine pairs undergo addition to form polyazamacrocycles with ring size ranging from 13 to 17, 19, 26 to 29, and 35. Cryptands are also prepared.

Ac N + BnN(H) (H)NBn —[(cod)RhCl]$_2$, CO - H$_2$, dioxane 120°→ Ac N ... BnN NBn

47%

Beckmann rearrangement.[4] Oximes are converted to amides in the presence of [(cod)RhCl]$_2$ (0.5 mol%), tri(*p*-tolyl)phosphine and triflic acid.

Conjugate additions. Quenching the hydrosilylation products of acrylic esters with aldehydes leads to predominantly *syn*-2-methyl-3-hydroxyalkanoic esters.[5] With allyl acrylates the first step sets up substrates for the ester-Claisen rearrangement.[6]

[(cod)RhCl]$_2$ / MeDuPhos

MeSiHCl$_2$ / PhH 50°

70% (8 : 1 dr)

Transfer of aryl and alkenyl groups to conjugated carbonyl compounds from tin reagents[7,8] and from triorganobismuths[9] is catalyzed by [(cod)RhCl]$_2$ in air and water. Phenylalanine and analogs can be synthesized.[10]

+ Ph-SnMe$_3$ → [(cod)RhCl]$_2$ / H_2O sonication

82%

Coupling reactions. Stilbenes are obtained from reaction of styrenes with arylboronic acids under basic conditions in aqueous medium.[8,11] However the corresponding reaction of heteroarylethenes (pyridyl, quinolyl, pyrazyl) furnishes diarylethanes via an addition-hydrolysis pathway.[10] Hydroarylation of pyridylalkynes is observed under the same conditions.[12]

+ ArB(OH)$_2$ → [(cod)RhCl]$_2$ / Na_2CO_3 / SDS / H_2O

KO_3S ... SO_3K

β-Styrylboronates are produced from styrenes and pinacolborane as a result of dehydrogenative borylation.[13]

Allylic displacement. 2,3-Benzo-7-oxanorbornadienes undergo ring opening with arylboronic acids, giving *cis*-3-Aryl-1,2-dihydro-1-naphthols.[14]

[1]Gross, T., Seayad, A.M., Ahmad, M., Beller, M. *OL* **4**, 2055 (2002).
[2]Oguma, K., Miura, M., Satoh, T., Nomura, M. *JOMC* **648**, 297 (2002).
[3]Kranemann, C.L., Costisella, B., Eilbracht, P. *TL* **40**, 7773 (1999); Kranemann, C.L., Eilbracht, P. *EJOC* 2367 (2000).
[4]Arisawa, M., Yamaguchi, M. *OL* **3**, 311 (2001).
[5]Zhao, C.-X., Bass, J., Morken, J.P. *OL* **3**, 2839 (2001).
[6]Miller, S.P., Morken, J.P. *OL* **4**, 2743 (2002).

[7]Venkatraman, S., Meng, Y., Li, C.-J. *TL* **42**, 4459 (2001).
[8]Amengual, R., Michelet, V., Genet, J.-P. *TL* **43**, 5909 (2002).
[9]Venkatraman, S., Li, C.-J. *TL* **42**, 781 (2001).
[10]Huang, T.-S., Li, C.-J. *OL* **3**, 2037 (2001).
[11]Lautens, M., Roy, A., Fukuoka, K., Fagnou, K., Martin-Matute, B. *JACS* **123**, 5358 (2001).
[12]Lautens, M., Yoshida, M. *OL* **4**, 123 (2002).
[13]Murata, M., Kawakita, K., Asana, T., Watanabe, S., Masuda, Y. *BCSJ* **75**, 825 (2002).
[14]Murakami, M., Igawa, H. *CC* 390 (2002).

Bis[chloro(cyclooctene)rhodium(I)].

Cyclization.[1] Ring formation from alkenyl heterocycles is initiated by C-H bond activation by the Rh complex.

$[(coe)_2RhCl]_2$

Cy_3P / THF

79%

[1]Tan, K.L., Bergman, R.G., Ellman, J.A. *JACS* **123**, 2685 (2001).

Bis(chloromethyl)zinc.

Diorganosulfonium methylides.[1] The reagent, $(ClCH_2)_2Zn$, derived from reaction of Et_2Zn with $ClCH_2I$, reacts with sulfides to form ylides. Such ylides can be used as methylene-transfer agent for imines (aziridine formation).

[1]Aggarwal, V.K., Stenseon, R.A., Jones, R.V.H., Fieldhouse, R., Blacker, J. *TL* **42**, 1587 (2001).

Bis[chloro(norbornadiene)rhodium(I)].

[4+2+2]Cyclization.[1] Cyclooctatrienes are formed.

OBn

$[(nbd)RhCl]_2$

Me-DuPhos - $AgSbF_6$

CH_2OCl_2 - EtOAc

73%

[1]Gilbertson, S.R., DeBoef, B. *JACS* **124**, 8784 (2002).

Bis[(η^3-cinnamyl)trifluoroacetoxypalladium].

β,γ-Unsaturated ketones. These compounds are prepared from acylsilanes and allylic esters in the presence of the Pd complex.

Ph SiMe3 O + F3C O O Ph → (F3CCOO Pd, Ph)2 / THF Δ → Ph O Ph 60%

[1]Obora, Y., Ogawa, Y., Imai, Y., Kawamura, T., Tsuji, Y. *JACS* **123**, 10489 (2001).

Bis(*sym*-collidine)halogen(I) hexafluorophosphate. 15, 30; **17,** 155; **18,** 49; **19,** 29; **20,** 33; **21,** 42–43

3-Bromooxetanes.[1] The bromo reagent effects cyclization of cinnamyl alcohols.

Ph OH → $[(collidine)_2Br]BF_4$ / CH_2Cl_2 → O Ph Br 78%

[1]Albert, S., Robin, S., Rousseau, G. *TL* **42**, 2477 (2001).

Bis[(1,5-cyclooctadiene)hydroxyrhodium(I)].

Hydrosilylation. Conjugated carbonyl compounds are transformed into enol silyl ethers by hydrosilanes in the presence of the Rh catalyst at room temperature (13 examples, 80–99%).[1]

Arylation and hydroarylation. Aryl transfer from $ArSiX_3$ to saturated and conjugated carbonyl compounds is achieved under aqueous basic conditions.[2,3] It should be noted that a Heck-type reaction takes place with silanediols.[4]

Ph Si OH OH + COOBu → $[(cod)RhOH]_2$ / THF 70° → Ph COOBu 81%

Silanediols also react with alkynes and aldehydes by hydroarylation pathways.[5] It is implied that the hydroxy function is important since $ArSnBu_3$ are generally unreactive but in certain circumstances hydroxy-containing additives improve the results.

[1]Mori, A., Kato, T. *SL* 1167 (2002).
[2]Murata, M., Shimazaki, R., Ishikura, M., Watanabe, S., Masuda, Y. *S* 717 (2002).
[3]Koike, T., Du, X., Mori, A., Osakada, K. *SL* 301 (2002).
[4]Mori, A., Danda, Y., Fujii, T., Hirabayashi, K., Osakada, K. *JACS* **123**, 10774 (2001).
[5]Fujii, T., Koike, T., Mori, A., Osakada, K. *SL* 295, 298 (2002).

Bis[(1,5-cyclooctadiene)alkoxyrhodium(I)].

Conjugate addition-aldol reaction. The tandem process serves to assemble *syn*-2-alkyl-3-hydroxyketones in one operation.[1]

[(cod)RhOMe]$_2$

EtCHO

PhMe

72% (*syn* : *anti* 12.4 : 1)

β-Silylstyrenes. A metathetic pathway to β-silylstyrenes involves reaction of styrenes and vinylsilanes using catalytic amount of the bis[(1,5-cyclooctadiene)trimethylsiloxyrhodium(I)] complex.[2]

[1]Yoshida, K., Ogasawara, M., Hayashi, T. *JACS* **124**, 10984 (2002).
[2]Marciniec, B., Walczuk-Gusciora, E., Pietraszuk, C. *OM* **20**, 3423 (2001).

Bis[(1,5-cyclooctadiene)methoxyiridium(I)].

Borylation.[1] The iridium complex together with 4,4′-di-*t*-butyl-2,2′-bipyridine activate aromatic C-H bonds. Borylation with bis(pinacolato)diboron occurs at room temperature.

[1]Ishiyama, T., Takagi, J., Hartwig, J.F., Miyaura, N. *ACIEE* **41**, 3056 (2002).

Bis(1,5-cyclooctadiene)nickel(0). 21, 43–45

Addition reactions. Alkynes undergo hydroarylation or dimerization with arylboron reagents when promoted by Ni(cod)$_2$.[1] Both the acyl and the stannyl moieties of acylstannanes are transferred to conjugated carbonyl compounds. Further reaction is possible as shown by the formation of alkylation products in the presence of aldehydes.[2]

R–CO–$SnBu_3$ + (vinyl ketone, R′) + R″CHO → [$(dba)_3Pd_2$; Ac_2O - py] → (AcO, R″, R, R′ product)

Treatment of the carboxylation products from 1-alkynes in situ with organozinc reagents leads to β,β-disubstituted α,β-unsaturated carboxylic acids.[3] A similar reaction on 1,3-dienes gives either 1,4-dicarboxylation products (with Me_2Zn treatment) or monocarboxylic acid derivatives.[4]

$(cod)_2Ni$; CO_2 / DBU; THF 0° → [ ] → Me_2Zn; CH_2N_2 → MeOOC ... COOMe
ArZnCl; CH_2N_2 → Ar ... COOMe

With catalysis by $Ni(cod)_2$ organozinc reagents attack cyclic anhydrides to afford keto acids[5] (transfer of only one organo residue), while they add to a remote allene or alkyne moiety to initiate alkylative cyclization onto an aldehyde group.[6,7] A stereocontrolled synthesis of allylic alcohols has been developed using this reaction with α-alkynylsiloxy aldehydes and subsequent degradation of the resulting oxasilacyclopentanes.[7]

X, CHO, C, R → $(cod)_2Ni$ - Me_2Zn → X, OH, R

X = CH_2, NTs

Ring formations. Allylic substrates undergo an efficient metallo-ene reaction in aqueous dioxane (1:1) when mediated by $Ni(cod)_2$.[8] Alkenyl halides form organonickel species which add in a 1,4-fashion to conjugated carbonyl compounds when 1,2-addition is sterically prohibited.[9] A cyclic organometallic species derived from an enyne and the $Ni(cod)_2$-Zn system can be trapped by a conjugated carbonyl substrate,[10] and an intramolecular conjugate addition initiated by alkenylzirconation of a remote alkyne linkage gives rise to cyclic products.[11]

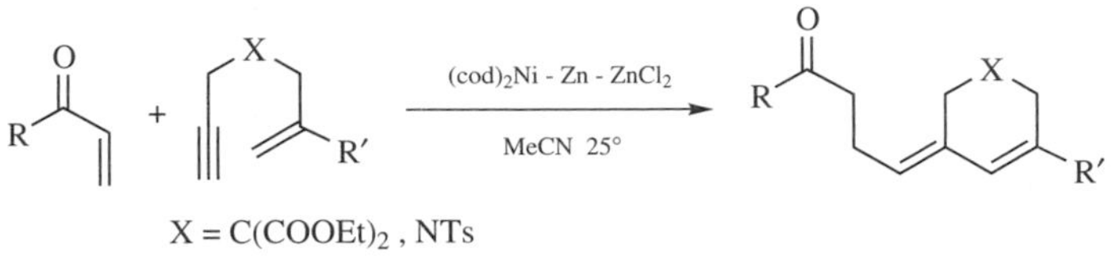

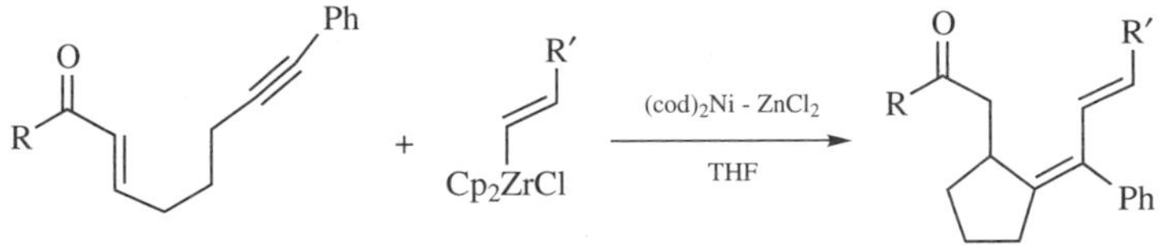

$X = C(COOEt)_2$, NTs

Carbonylative cyclization from enals that also contain a terminal double bond leads to bicyclic ketones.[12]

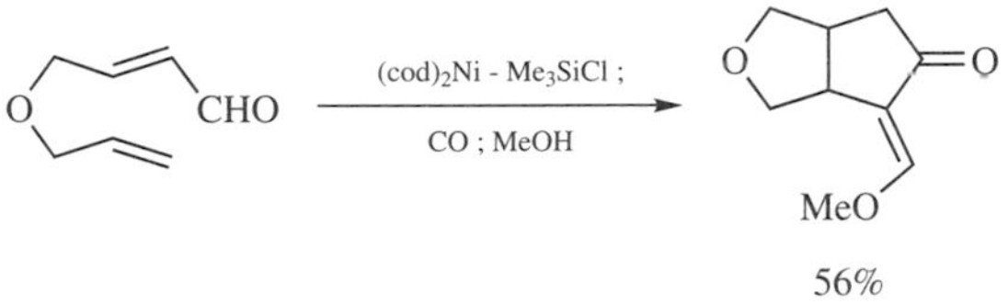

56%

Biaryls.[13] The coupling reagent for ArMgX and unactivated Ar′Cl is advantageously improved by the air-stable ligand *t*-BuP(S)H.

[1]Shirakawa, E., Takahashi, G., Tsuchimoto, T., Kawakami, Y. *CC* 2688 (2001).
[2]Shirakawa, E., Yamamoto, Y., Nakoa, Y., Tsuchimoto, T., Hiyama, T. *CC* 1926 (2001).
[3]Takimoto, M., Shimizu, K., Mori, M. *OL* **3**, 3345 (2001).
[4]Takimoto, M., Mori, M. *JACS* **123**, 2895 (2001).
[5]Bercot, E.A., Rovis, T. *JACS* **124**, 174 (2002).
[6]Montgomery, J., Song, M. *OL* **4**, 4009 (2002).
[7]Lozanov, M., Montgomery, J. *TL* **42**, 3259 (2001).
[8]Michelet, V., Galland, J.-C., Charruault, L., Savignac, M., Genet, J.-P. *OL* **3**, 2065 (2001).
[9]Nicolaou, K.C., Roecker, A.J., Follmann, M., Baati, R. *ACIEE* **41**, 2107 (2002).
[10]Ikeda, S., Miyashita, H., Taniguchi, M., Kondo, H., Okano, M., Sato, Y., Odashima, K. *JACS* **124**, 12060 (2002).
[11]Ni, Y., Amarasinghe, K.K.D., Montgomery, J. *OL* **4**, 1743 (2002).
[12]Garcia-Gomez, G., Moreto, J.M. *EJOC* 1359 (2001).
[13]Li, G.Y., Marshall, W.J. *OM* **21**, 590 (2002).

Bis(1,5-cyclooctadiene)rhodium(I) tetrafluoroborate. 21, 45–46

Conjugate additions. The reactions that transfer an aryl group from an arylsilane[1] or arylboronic acid[2] can be carried out in air and water.

Hydroamination and hydroarylation. The combination of 1-alkynes with primary amines in the Markovnikov sense is followed by rapid isomerization of the resulting enamines to imines. Addition of RLi affords secondary amines.[3]

Aryllead reagents add to conjugated carbonyl compounds[4] in the presence of $(cod)_2RhBF_4$ or $(cod)_2RhCl_2$. That this reaction tolerates water and air is advantageous.

[1]Huang, T.-S., Li, C.-J. *CC* 2348 (2001).
[2]Ramnauth, J., Poulin, O., Bratovanov, S.S., Rakhit, S., Maddaford, S.P. *OL* **3**, 2571 (2001).
[3]Hartung, C.G., Tillack, A., Trauthwein, H., Beller, M. *JOC* **66**, 6339 (2001).
[4]Ding, R., Chen, Y.-J., Wang, D., Li, C.-J. *SL* 1470 (2001).

Bis(dialkylamino)cyanoboranes.

α-Amino nitriles.[1] A convenient synthesis of the amino nitriles consists of mixing the reagents with carbonyl compounds in THF at room temperature (20 examples, 92–99%).

[1]Suginome, M., Yamamoto, A., Ito, Y. *CC* 1392 (2002).

Bis(dibenzylideneacetone)palladium(0). 21, 46–48

Arylamines. Q-phos (pentaphenylferrocenyl di-t-butylphosphine) is an excellent ligand prepared from ferrocene in 40–65% overall yield. It facilitates the substitution of both electron-rich and electron-poor aryl halides with amines (turnover number >1000) in the presence of $(dba)_2Pd$.[1] The catalytic system is also useful for those reactions involving carbon and oxygen nucleophiles.

Primary arylamines are obtained from reaction with lithium bis(trimethylsilyl)amide[2] in the presence of $(dba)_2Pd$ and t-Bu_3P. *N*-Alkenylation of azoles and phenothiazines with alkenyl halides is also accomplished by the same method.[3]

Coupling reactions. New coupling methods involving organosilanes include displacement of allylic benzoates by an aryl group of arylsiloxanes,[4] and the formation of styrenes from aryl halides, 1-alkynes, and bis(dimethyl)siloxane where hydrosilylation of the alkynes provides the coupling components.[5] Homoallylic alcohols in which the proximal sp^2-carbon atom is bonded to an aryl group are accessible via the coupling strategy.[6]

I–C6H4–COOEt + [cyclic silyl ether] → $(dba)_2Pd$, Bu_4NF / THF, 45° → HO–...–C6H4–COOEt

86%

In a styrene synthesis the tetrameric cyclic siloxane in which each silicon atom is vinylated can serve as the vinyl donor.[7] For stereoselective synthesis of (*Z*)-styrenes that bear a hydroxy group at an outlying position, cyclic siloxanes that are available through Mo-carbene-mediated ring-closing olefin metathesis represent satisfactory nonaromatic coupling partners.[8] A fluoride-free procedure for cross-coupling uses alkenylsilanols.[9]

R, OH, Si + Ar-I → (Me_3SiOK, DME) R, Ar

Polycyclic arene-1,4-epoxides undergo deoxygenative coupling to afford arene dimers on reaction with $(dba)_2Pd$, Zn, and trichlorosilane.[10]

Aryllead(IV) acetates donate their aryl residues to unite with CO in the presence of $(dba)_2Pd$ and NaOMe. Diaryl ketones are formed.[11] Alkynylarenes result on coupling of the aryllead reagents with 1-alkynes.[12]

Heck reaction of ArBr and *n*-butyl acrylate is efficiently conducted with a catalytic system consisting of $(dba)_2Pd$ and an imidazol-2-ylidene ligand.[13] In the absence of the acrylic ester a related system can be employed to reduce aryl halides.[14]

A synthesis of certain pyridine alkaloids takes advantage of two different coupling reactions at the termini of a diene.[15]

I, N, 9 + H, N, Bn, Ts → ($(dba)_2Pd$; Bu_4NCl - Na_2CO_3, DMF 100°) N, 9, N, Bn, Ts — 62%

Arylations. Ester enolates undergo α-arylation[16] that actually has been worked out with malonates and cyanoacetates[17] and in a synthesis of protected α-amino acids.[18] The method of intramolecular α-arylation of amides in the presence of *t*-BuOK leading to δ-lactams has been applied to synthesis of the alkaloids cherylline and latifine.[19]

Ph, N, Ph, O, OEt + Ph—Br → ($(dba)_2Pd$ / *t*-Bu_3P; K_3PO_4, PhMe 100°) Ph, N, Ph, Ph, O, OEt — 88%

Addition to allenes and alkynes. The terminal double bond of a 1,2-alkadiene is reactive toward compounds generally used in coupling reactions. Twofold C-C and C-B bond formation processes[20,21] convert the allenes to more complex alkenes.

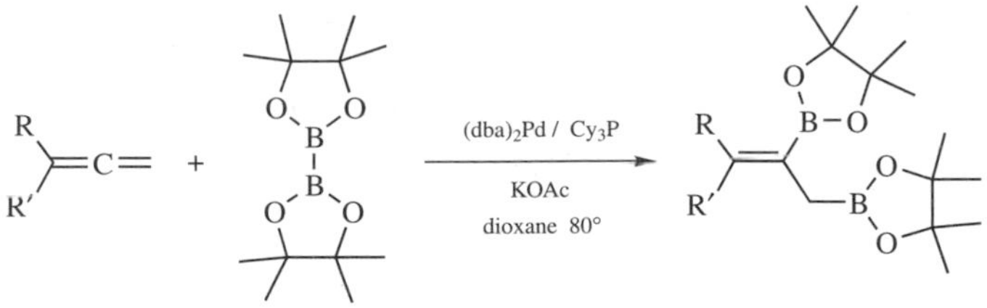

Functionalization of alkynes leading to *vic*-bimetalloalkenes is illustrated in a catalyzed reaction with stannylgermanes.[22]

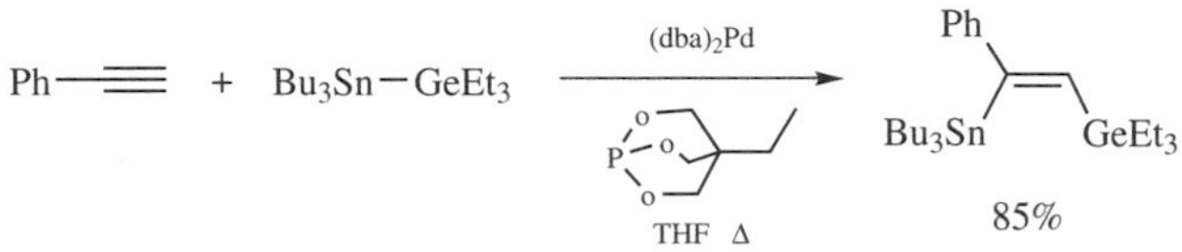

Cyclization reactions. A synthesis of dihydroquinoxaline derivatives[23] by a reductive cyclization has been demonstrated. A route to dihydrocarbazol-4-ones[24] that involves two consecutive Pd-catalyzed coupling reactions is quite remarkable. The first step in a route to 1-azaazulenes is the Pd-catalyzed process.[25]

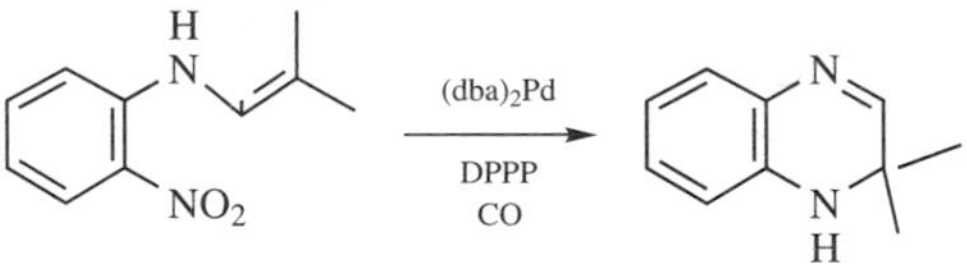

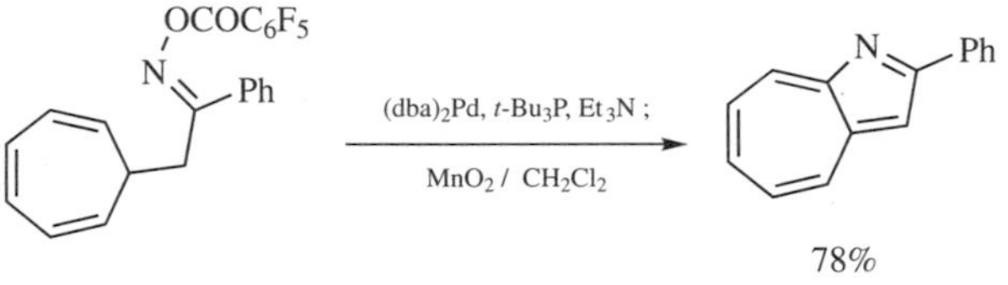

Reductive allylation. Conjugated systems that are susceptible to reduction with a tin hydride are allylated in situ with an allylic ester in the presence of $(dba)_2Pd$.[26]

[1]Kataoka, N., Shelby, Q., Stambuli, J.P., Hartwig, J.F. *JOC* **67**, 5553 (2002).
[2]Lee, S., Jorgensen, M., Hartwig, J.F. *OL* **3**, 2729 (2001).
[3]Lebedev, A.Y., Izmer, V.V., Kasyul'kin, D.N., Beletskaya, I.P., Voskobonynikov, A.Z. *OL* **4**, 623 (2002).
[4]Correia, R., DeShong, P. *JOC* **66**, 7159 (2001).

[5]Denmark, S.E., Wang, Z. *OL* **3**, 1073 (2001).
[6]Denmark, S.E., Pan, W. *OL* **3**, 61 (2001).
[7]Denmark, S.E., Wang, Z. *JOMC* **624**, 372 (2001).
[8]Denmark, S.E., Yang, S.-M. *OL* **3**, 1749 (2001).
[9]Denmark, S.E., Sweis, R.F. *JACS* **123**, 6439 (2001).
[10]Shih, H.-T., Shih, H.-H., Cheng, C.-H. *OL* **3**, 811 (2001).
[11]Kang, S.-K., Ryu, H.-C., Choi, S.-C. *SC* **31**, 1035 (2001).
[12]Kang, S.-K., Ryu, H.-C., Lee, S.-H. *SC* **31**, 1059 (2001).
[13]Yang, C., Lee, H.M., Nolan, S.P. *OL* **3**, 1511 (2001).
[14]Viciu, M.S., Grasa, G.A., Nolan, S.P. *OM* **20**, 3607 (2001).
[15]Larock, R.C., Wang, Y. *TL* **43**, 21 (2002).
[16]Jorgensen, M., Lee, S., Liu, X., Wolkowski, J.P., Hartwig, J.F. *JACS* **124**, 12557 (2002).
[17]Beare, N.A., Hartwig, J.F. *JOC* **67**, 541 (2001).
[18]Lee, S., Beare, N.A., Hartwig, J.F. *JACS* **123**, 8410 (2001).
[19]Honda, T., Namiki, H., Satoh, F. *OL* **3**, 631 (2001).
[20]Huang, T.-H., Chang, H.-M., Wu, M.-Y., Cheng, C.-H. *JOC* **67**, 99 (2002).
[21]Yang, F.-Y., Cheng, C.-H. *JACS* **123**, 761 (2001).
[22]Senda, Y., Oguchi, Y., Terayama, M., Asai, T., Nakano, T., Migita, T. *JOMC* **622**, 302 (2001).
[23]Söderberg, B.C.G., Wallace, J.M., Tamariz, J. *OL* **4**, 1339 (2002).
[24]Scott, T.L., Söderberg, B.C.G. *TL* **43**, 1621 (2002).
[25]Kitamura, M., Chiba, S., Saku, O., Narasaka, K *CL* 606 (2002).
[26]Shim, J.-G., Park, J.C., Cho, C.S., Shim, S.C., Yamamoto, Y *CC* 852 (2002).

Bis[di-*n*-butyl(chloro)tin] oxide.

Epoxide opening.[1] 1,2-Diol monoethers are readily formed. Some epoxides afford chlorohydrins.

$[ClBu_2Sn]_2O$, ROH

$[ClBu_2Sn]_2O$, ROH

[1]Salomon, C.J. *SL* 65 (2001).

Bis[di-*t*-butyl(hydroxy)tin chloride]. 21, 50

Alcoholysis.[1] The reagent promotes cleavage of *N*-acyloxazolidinones to afford esters by ROH. An advantage of this highly chemoselective alcoholysis is that no aqueous workup is required, although the tin reagent is less active than $MgBr_2$ and $Sc(OTf)_3$.

[1]Orita, A., Nagano, Y., Hirano, J., Otera, J. *SL* 637 (2001).

Bis[dicarbonylchlororhodium(I)]. 21, 50–51

Cycloadditions and cycloisomerization. Cycloheptenones are prepared from 1-alkenylcyclopropane derivatives via an intermolecular [5 + 2]cycloaddition.[1] A three-component cycloaddition of the [5+2+1] mode also has been realized.[2]

OMe + ; $[Rh(CO)_2Cl]_2$, CO / dioxane, 60°; H, OH; 97%

Silyl ethers of 2-alkynylacetophenones undergo cyclization to afford α-naphthyl silyl ethers.[3] The cycloisomerization is effected also with $PtCl_2$ or $(PhCN)_2PdCl_2$.

OTBS, R; $[Rh(CO)_2Cl]_2$, PhMe 50°; OTBS, R

Intramolecular Pauson-Khand reaction. Reversal of the π-bond selectivity of an allene permits the elaboration of a 7:5-fused ring system by this reaction.[4]

COOEt, EtOOC, C, C_6H_{11}; $[Rh(CO)_2Cl]_2$, CO / PhMe, 90°; COOEt, EtOOC, O, C_6H_{11}; 60%

2-Alkylidenecycloalkanones. Alkynoylsilanes in which the triple bond is at the δ,ε or ε,ζ position cyclize with loss of the silyl group to form ketone derivatives.[5]

O, Si, Ph, Ph; $[Rh(CO)_2Cl]_2$, HOAc -PhMe, 70°; O, Ph; 77%

[1]Wender, P.A., Barzilay, C.M., Dyckman, A.J. *JACS* **123**, 179 (2001).
[2]Wender, P.A., Gamber, G.G., Hubbard, R.D., Zhang, L. *JACS* **124**, 2876 (2002).
[3]Dankwardt, J.W. *TL* **42**, 5809 (2001).
[4]Brummond, K.M., Chen, H., Fisher, K.D., Kerekes, A.D., Rickards, B., Sill, P.C., Geib, S.J. *OL* **4**, 1931 (2002).
[5]Yamane, M., Amemiya, T., Narasaka, K. *CL* 1210 (2001).

Bis[dicarbonyl(pentamethylcyclopentadienyl)iridium].

Oxidation.[1] Primary and secondary alcohols are oxidized to carbonyl products with the iridium complex and K_2CO_3 in acetone at room temperature.

[1]Fujita, K., Furukawa, S., Yamaguchi, R.*JOMC* **649**, 289 (2002).

Bis[dicarbonyl(pentamethylcyclopentadienyl)iron].

Allylic amination.[1] An arylamino group can be introduced to the allylic position of alkenes on treating them with nitroarenes under CO in the presence of $[Cp^*Fe(CO)_2]_2$.

[1]Srivastava, R.S. Kolel-Veetil, M., Nicholas, K.M. *TL* **43**, 931 (2002).

Bis[dicarbonyl(pentamethylcyclopentadienyl)rhodium].

Primary amines.[1] Carbonyl compounds are reductively aminated with ammonium formate by means of this complex.

[1]Kitamura, M., Lee, D., Hayashi, S., Tanaka, S., Yoshimura, M. *JOC* **67**, 8685 (2002).

Bis[dicarbonyl(pentamethylcyclopentadienyl)ruthenium].

Indoles.[1] Nitroarenes and alkynes combine under CO to form indole derivatives in one step.

NO_2 + Ph ; $[(Me_5C_5)Ru(CO)_2]_2$; CO / PhH ; 170° ; Ph ; N ; H ; 39%

[1]Penoni, A., Nicholas, K.M.*CC* 484 (2002).

Bis[dichloro(*p*-cymene)ruthenium]. 21, 52

Redox reactions. With the complex, oxidation of α-hydroxy carbonyl compounds occurs under an oxygen atmosphere.[1] Cyclic enones give optically active allylic alcohols (72–99% ee) by HCHO in the presence of a chiral ligand.[2]

Nitriles.[3] Aldoximes are facilitated by the title complex to undergo dehydration in refluxing MeCN. Molecular sieves (type 4A) are also required.

Reaction of conjugated carbonyl compounds. 4-Alkynones are prepared by the conjugate addition of 1-alkynes to enones. A catalytic system consists of $[(p\text{-cymene})RuCl_2]_2$ and pyrrolidine.[4]

Suzuki coupling is achieved with this complex. Its chemoselectivity in preference to the Heck reaction is shown with a bromoarylboronic acid.[5]

[1]Chang, S., Lee, M., Ko, S., Lee, P.H. *SC* **32**, 1279 (2002).
[2]Hannedouche, J., Kenny, J.A., Walsgrove, T., Wills, M. *SL* 263 (2002).
[3]Yang, S.H., Chang, S. *OL* **3**, 4209 (2001).
[4]Chang, S., Na, Y., Choi, E., Kim, S. *OL* **3**, 2089 (2001).
[5]Farrington, E.J., Brown, J.M., Barnard, C.F.J., Rowsell, E. *ACIEE* **41**, 169 (2002).

Bis[dichloro(pentamethylcyclopentadienyl)iridium].

Benzannulated heterocycles.[1] In response to the catalyst arylamines bearing an *o*-hydroxyalkyl chain undergo oxidative cyclization. This procedure is applicable to the preparation of indoles, 1,2,3,4-tetrahydroquinolines, and 2,3,4,5-tetrahydro-1-benzazepine.

[1]Fujita, K., Yamamoto, K., Yamaguchi, R. *OL* **4**, 2691 (2002).

1,3-Bis(2,6-diisopropylphenyl)imidazolium chloride.

Suzuki coupling. The coupling using an ArCl partner with potassium 9-methoxy-9-organo-BBN is accomplished in the presence of $Pd(OAc)_2$ in THF, when the imidazolium salt is added. Apparently a carbene complex is involved.

Arylamines. *N*-Arylation with aryl halides becomes more convenient when an air-stable Pd complex derived from $PdCl_2$ and the imidazol-2-ylidene is used as promoter.[2]

[1]Fürstner, A., Leitner, A. *SL* 290 (2001).
[2]Viciu, M.S., Kissling, R.M., Stevens, E.D., Nolan, S.P. *OL* **4**, 2229 (2002).

1-Bis(dimethylaminomethyl)benzotriazole-3-oxide, polystyrene-bound.

Peptide synthesis.[1] The reagent, prepared from polymer-bonded 1-hydroxybenzotriazole, exhibits activity in wet solvents.

[1]Chinchilla, R., Dodsworth, D.J., Najera, C., Soriano, J.M. *TL* **41**, 2463 (2000).

2,2′-Bis(diphenylphosphino)-1,1′-binaphthyl, BINAP. 13, 36–37; **14,** 38–44; **15,** 34; **16,** 32–36; **17,** 34–38; **18,** 39–41; **19,** 33–35; **20,** 41–44; **21,** 53–55

Kinetic resolution.[1] A secondary alcohol is resolved by the selective reaction of one enantiomer with *N*-chlorosuccinimide to give the corresponding chloride (S_N2 pathway) in the presence of a chiral BINAP, at slightly more than 50% conversion.

NCS / (S)-BINAP
THF
(55.1% conversion)
90% (97% ee)

[1]Sekar, G., Nishiyama, H. *JACS* **123**, 3603 (2001).55

Nickel (0) complex.

Arylations.[1] An aryl group can be introduced to the α-position of α-substituted γ-butyrolactones with excellent enantioselectivity in the presence of $Ni(cod)_2$ and a chiral BINAP.

[1]Spielvogel, D.J., Buchwald, S.L. *JACS* **124**, 3500 (2002).

Palladium(II) complexes

Allylic displacement. Sodium bisformylamide is a suitable nucleophile for the preparation of protected allylic amines in an enantioselective manner, in the presence of a Pd complex and a chiral BINAP.[1]

Coupling reactions. Arylboronic acids undergo homocoupling even in the presence of an α-bromo ester.[2] The Pd-catalyzed homocoupling of organozinc halides and organoboranes in the presence of racemic BINAP forms a bond between two tetrahedral carbon atoms.[3]

9-BBN-H
$PdCl_2$ - BINAP
PhCOCH(Cl)Ph
K_3PO_4

Enamines are prepared from secondary amines and enol triflates by a Pd-catalyzed reaction.[4]

[1]Wang, Y., Ding, K. *JOC* **66**, 3238 (2001).
[2]Lei, A., Zhang, X. *TL* **43**, 2525 (2002).
[3]Lei, A., Zhang, X. *OL* **4**, 2285 (2002).
[4]Willis, M.C., Brace, G.N. *TL* **43**, 9085 (2002).

Rhodium complexes

Hydroaminomethylidenation.[1] Vinyl sulfoxides, sulfones, and phosphonates are converted to (*E*)-enamino derivatives bearing a methyl group geminal to the heteroatom functionality. The reaction is carried out under the hydroformylation conditions with amines and a Rh-BINAP complex added.

R′R″NH - BINAP, CO - H_2, THF 80°

n = 1, 2

R′R″NH - BINAP, CO - H_2, THF 80°

Cyclizations. β-Substituted α-alkylidene-γ-butyrolactones are the cyclization products from a reaction of allyl 2-alkynoates that is catalyzed by a Rh-complex and $AgSbF_6$ with BINAP as ligand.[2] Excellent enantioselectivity (>99% ee) is observed.

$[(cod)RhCl]_2$ - (*S*)-BINAP, $AgSbF_6$, $ClCH_2CH_2Cl$

93% (er > 99.5 : 0.5)

Conjugate additions. Conjugate addition with aryltitanium reagents[3] is highly enantioselective when a Rh-binap complex is added to the reaction medium. Introduction of a β-aryl group to conjugated amides creates a new sterogenic center whose absolute configuration can be controlled with the aid of BINAP.[4] Potassium organotrifluoroboronates are adequate surrogates of the commonly used organoboronic acids for conjugate addition.[5]

BINAP bearing two guanidinomethyl units (at C-6 and C-6′) forms Rh complex that is effective in catalyzing asymmetric conjugate addition of boronic acids to enones.[6]

Rearrangement. 4-Alkynals behave very differently toward Rh(binap) complex than Rh(dppe) complex in that 2-alkenyl-2-alkenals and 2-cyclopentenones are the respective major products.[7]

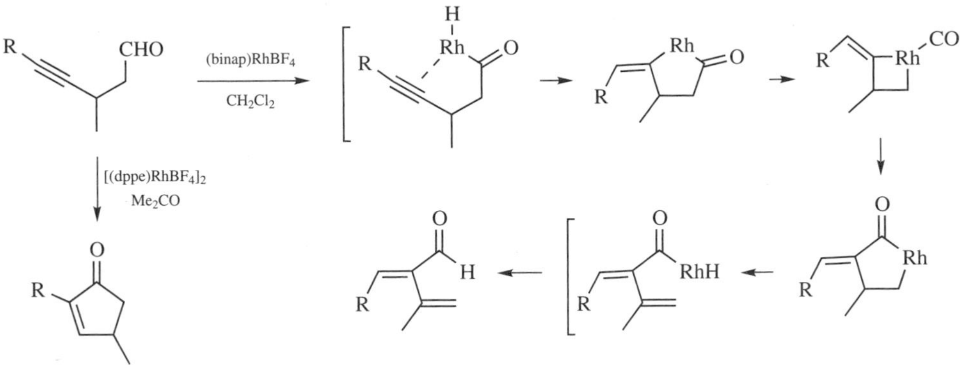

(binap)$RhBF_4$, CH_2Cl_2; $[(dppe)RhBF_4]_2$, Me_2CO

[1]Lin, Y.-S., El Ali, B., Alper, H. *JACS* **123**, 7719 (2001).
[2]Lei, A., He, M., Zhang, X. *JACS* **124**, 8198 (2002).
[3]Hayashi, T., Tokunaga, N., Yoshida, K., Han, J.W. *JACS* **124**, 12102 (2002).
[4]Sakuma, S., Miyaura, N. *JOC* **66**, 8944 (2001).
[5]Pucheault, M., Darses, S., Genet, J.-P. *TL* **43**, 6155 (2002).
[6]Amengual, R., Michelet, V., Genet, J.-P. *SL* 1791 (2002).
[7]Tanaka, K., Fu, G.C. *CC* 684 (2002).

Ruthenium(II) complexes

Asymmetric hydrogenation. Ethylphosphonic acids containing an α-aryl substituent are prepared by asymmetric hydrogenation of α-styrylphosphonic acids.[1]

Under base-free conditions ketones are reduced in almost quantitative yields and excellent ee.[2] Allylic alcohols result from enones.

Hydrogenation using the Noyori binap-Ru-diamine catalyst may involve binding of cation to enhance cleavage of the coordinated hydrogen, because alkali metal cations have profound effects on the catalytic turnover.[3]

A BINAP analogue **1** has been evaluated as ligand for asymmetric hydrogenation.[4]

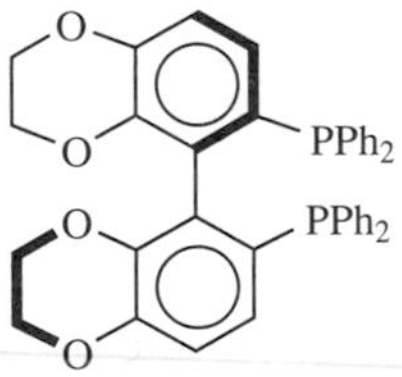

PPh_2 PPh_2

1

Diels-Alder reaction. A Ru complex of a modified BINAP (**2**) is a superior catalyst for the Diels-Alder reaction.[5]

2

[1]Goulioukina, N.S., Dolgina, T.M., Beletskaya, I.P., Henry, J.-C., Lavergne, D., Ratovelomanana-Vidal, V., Genet, J.-P. *TA* **12**, 319 (2001).
[2]Ohkuma, T., Koizumi, M., Muniz, K., Hilt, G., Kabuto, C., Noyori, R. *JACS* **124**, 6508 (2002).
[3]Hartmann, R., Chen, P. *ACIEE* **40**, 3581 (2001).
[4]Pai, C.-C., Li, Y.-M., Zhou, Z.-Y., Chan, A.S.C. *TL* **43**, 2789 (2002).
[5]Faller, J.W., Grimmond, B.J., D'Alliessi, D.G. *JACS* **123**, 2525 (2001).

Silver(I) complexes

Aldol reactions. A combination of AgOTf and (*R*)-BINAP promotes the enantioselective addition of allyltributylstannane to aldehydes (9 examples, 59–95% yield, 88–97% ee).[1]

[1]Yanagisawa, A., Nakasjhima, H., Nakatsuka, Y., Ishiba, A., Yamamoto, H. *BCSJ* **74**, 1129 (2001).

Bis(diphenylphosphinoethane)rhodium(I) chloride.

Pauson-Khand reaction.[1] A catalytic process using aldehydes as a CO source is developed. No solvent is required.

[1]Shibata, T., Toshida, N., Takagi, K. *OL* **4**, 1619 (2002).

Bis(diphenylphosphinoethane)rhodium(II) tetrafluoroborate.

Cycloisomerization. Access to cyclopentenones from 4-alkynals by an unusual Rh-catalyzed process involves intramolecular *trans* hydroacylation.[1]

Redox reaction. In the presence of phenol the above reaction is diverted to a redox pathway, leading to phenyl (*Z*)-4-alkenoates.[2]

[(dppe)RhBF$_4$]$_2$, PhOH, ClCH$_2$CH$_2$Cl

[1]Tanaka, K., Fu, G.C. *JACS* **123**, 11492 (2001).
[2]Tanaka, K., Fu, G.C. *ACIEE* **41**, 1607 (2002).

2,2′-Bis(di-*p*-tolylphosphino)-1,1′-binaphthyl. 21, 56

Copper(I) complexes

Addition to imines. With allenylstannanes as reagents, chiral homopropargylamines including 2-tosylamino-4-alkynoic acids are accessible.[1] Friedel-Crafts type reaction of activated arenes and heteroaromatic compounds occurs readily with α-imino esters.[2]

Hetero Diels-Alder reaction. This method for assembling dihydro-α-pyrones in good enantioselectivity is applicable to a synthesis of the Prelog-Djerassi lactone.[3]

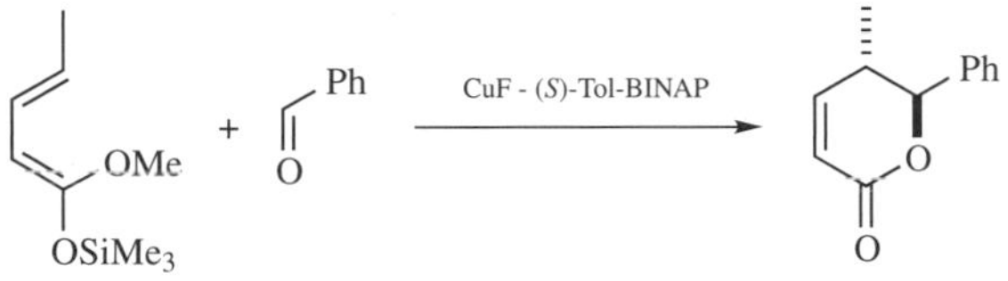

OMe, OSiMe$_3$ + Ph, O — CuF - (*S*)-Tol-BINAP → Ph, O, O

85% (87% ee)

Allylic displacement. Chiral allylic amines are obtained via a Pd-catalyzed displacement of allylic acetates with potassium phthalimide in the presence of the Tol-BINAP.[4]

Asymmetric reduction. In the asymmetric reduction of β-substituted enones with a hydrosilane in the presence of *t*-BuONa a follow-up alkylation provides α,β-disubstituted ketones in a convenient manner.[5] 2,4-Disubstituted cyclopentanones are formed in a dynamic kinetic resolution process.[6]

[1]Kagoshima, H., Uzawa, T., Akiyama, T. *CL* 298 (2002).
[2]Saaby, S., Bayon, P., Aburel, P.S., Jorgensen, K. A.. *JOC* **67**, 4352 (2002).
[3]Bluet, G., Bazan-Tejeda, B., Campagne, J.-M. *OL* **3**, 3807 (2001).
[4]Kodama, H., Taiji, T., Ohta, T., Furukawa, I. *SL* 385 (2001).

[5]Yun, J., Buchwald, S.L. *OL* **3**, 1129 (2001).
[6]Jurkauskas, V., Buchwald, S.L. *JACS* **124**, 2892 (2002).

Palladium(II) complexes

Fluorination.[1] (*R*)-Dimethyl-BINAP complexed to Pd constitutes a chiral catalyst for fluorination (6 examples, 83–92% ee) of β-keto esters with $(PhSO_2)_2NF$.

[1]Hamashima, Y., Yagi, K., Takano, H., Tamas, L., Sodeoka, M. *JACS* **124**, 14530 (2002).

Ruthenium(II) complexes

Hydrogenation.[1] A complex with $Ru(OAc)_2$ that performs asymmetric hydrogenation in an ionic liquid and the catalyst can be recovered by extraction with supercritical CO_2.

[1]Brown, R.A., Pollet, P., McKoon, E., Eckert, C.A., Liotta, C.L., Jessop, P.G. *JACS* **123**, 1254 (2001).

Silver(I) complexes

Aldol reaction. An asymmetric reaction of silyl enol ethers with aldehydes that favors formation of *syn*-aldols is readily achieved at low temperatures.[1]

[1]Yanagisawa, A., Nakatsuka, Y., Asakawa, K., Kageyama, H., Yamamoto, H. *SL* 69 (2001).

Bis[ethylene(chloro)rhodium(I)].

Deborylative acylation. Reaction of boronic acids with anhydrides produces ketones when the Rh-catalyst is present.[1]

[1]Frost, C.G., Wadsworth, K.J. *CC* 2316 (2001).

Bis[(ethylenedichloro)platinum(II)]. 21, 57

1-Aryl-1,2-ethanediols.[1] A two-staged hydrosilylation of ethynylarenes, first induced by the Pt-complex then with bis(allylchloropalladium), and followed by oxidative desilylation, affords the diols.

Ph–C≡CH $\xrightarrow[\text{HSiCl}_3\text{, [(C}_3\text{H}_5\text{)PdCl]}_2\text{ ; KF - KHCO}_3\text{ - H}_2\text{O}_2]{\text{HSiCl}_3\text{, [(C}_2\text{H}_4\text{)PtCl}_2\text{]}_2\text{ ;}}$ Ph–CH(OH)–CH$_2$OH

75% (95% ee)

[1]Shimada, T., Mukaide, K., Shinohara, A., Han, J.W., Hayashi, T. *JACS* **124**, 1584 (2002).

Bis(iodozincio)methane. 21, 57–58

Methylenation. Methylenation of the keto group in 1-hydroxy-2-alkanones and 1-alkoxy-2-alkanones by the title reagent is facile. Reactions of internal the corresponding acyloxy ketones and alkoxy ketones are sluggish.[1] Some other α-heterosubstituents show ameliorating effects.

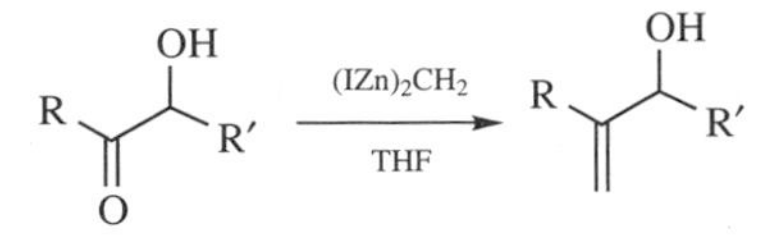

OH
R
R′
O
$(IZn)_2CH_2$
THF
OH
R
R′

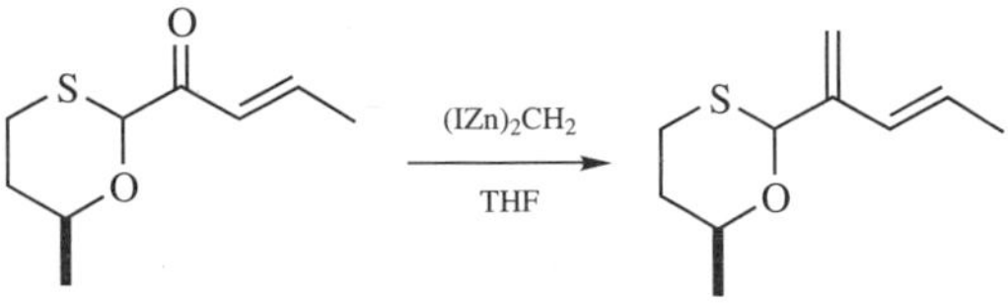

O
S
O
$(IZn)_2CH_2$
THF
S
O

82%

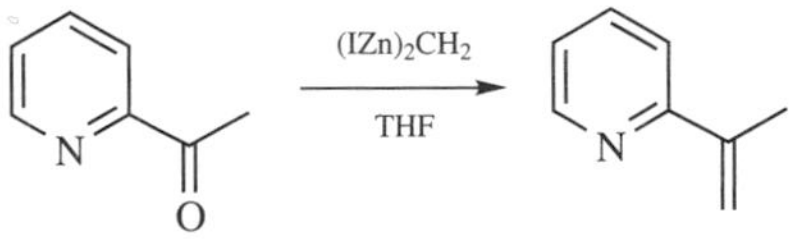

N
O
$(IZn)_2CH_2$
THF
N

86%

Rearrangement. A pinacol-type rearrangement occurs when 2,3-epoxy alcohols are treated with the reagent, and subsequent methylenation follows.[2]

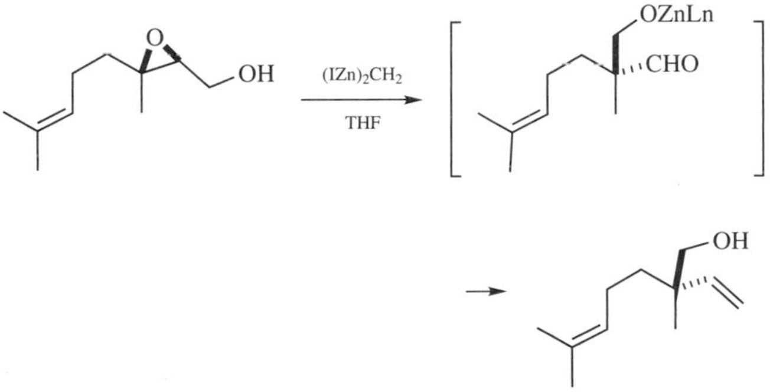

O
OH
$(IZn)_2CH_2$
THF
OZnLn
CHO

OH

[1]Ukai, K., Arioka, D., Yoshino, H., Fushimi, H., Oshima, K., Utimoto, K., Matsubara, S. *SL* 513 (2001).
[2]Matsubara, S., Yamamoto, H., Oshima, K. *ACIEE* **41**, 2837 (2002).

Bis(methanesulfonyl)methane.

Unsymmetrical gem-disulfones. A series of alkylation steps transforms the parent disulfone to its homologs.[1]

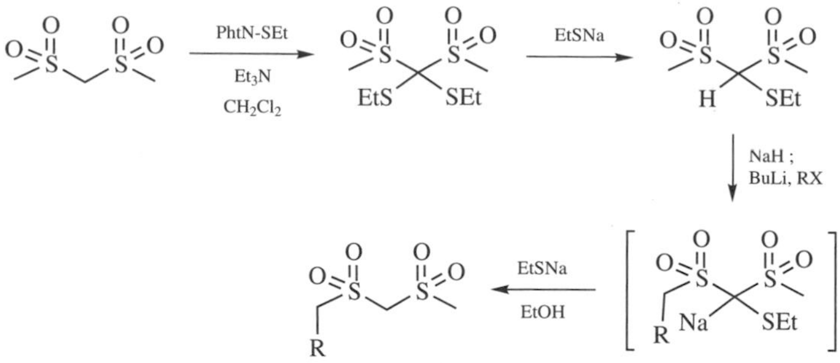

[1]Zhu, Y., Drueckhammer, D.G. *TL* **43**, 1377 (2002).

Bis(2-methoxyethyl)aminosulfur trifluoride. 21, 58

Thiazolines. Exposure of *N*-(2-hydroxyalkyl) thioamides to the reagent leads to cyclization.[1]

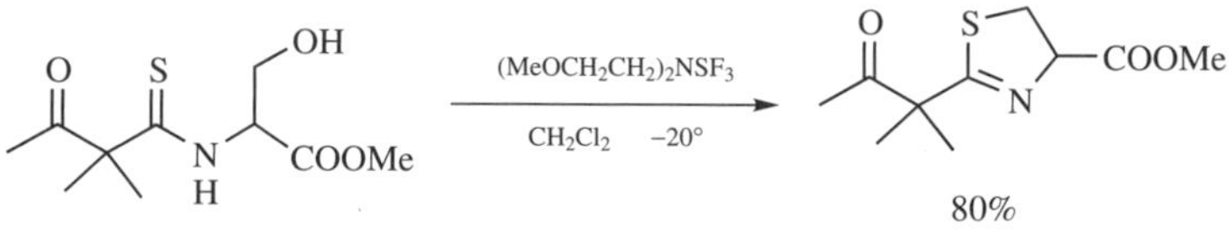

Difluorosulfuranes. Fluorodeoxygenation of sulfoxides is readily accomplished.[2]

[1]Mahler, S.G., Serra, G.L., Antonow, D., Manta, E. *TL* **42**, 8143 (2001).
[2]Laali, K.K., Borodkin, G.I. *JFC* **115**, 169 (2002).

Bis(η[3]-methylallyl)ruthenium.

Addition to alkynes. Isomerization of propargyl alcohols via catalyzed addition of benzoic acid constitutes an improved route to enals.[1]

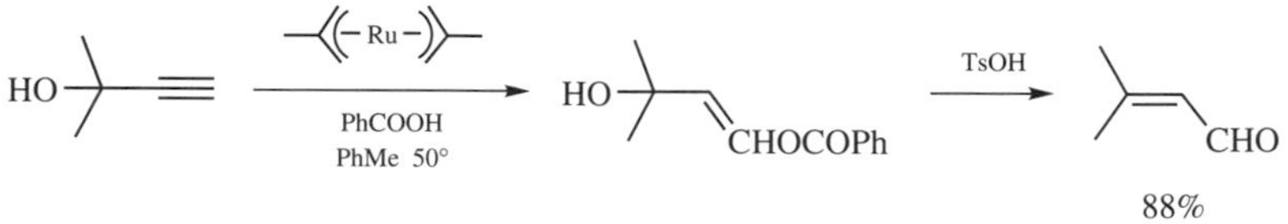

[1]Picquet, M., Fernandez, A., Bruneau, C., Dixneuf, P.H. *EJOC* 2361 (2000).

Bismuth. 20, 44; **21,** 58

α-Diketones.[1] Oxidative transformation of internal epoxides to the diketones takes place when they are treated with bismuth and $Cu(OTf)_2$ in hot DMSO.

Bi - $Cu(OTf)_2$ / DMSO 100°

74%

[1]Antoniotti, S., Dunach, E. *CC* 2566 (2001).

Bismuth(III) bromide.

Reductive etherification.[1] Carbonyl compounds are transformed into ethers on reaction with alkyl diisopropylsilyl ethers $i\text{-}Pr_2Si(H)OR$ using $BiBr_3$ as catalyst.

[1]Jiang, X., Bajwa, J.S., Slade, J., Prasad, K., Repic, O., Blacklock, T.J. *TL* **43**, 9225 (2002).

Bismuth(III) chloride. 15, 37; **18,** 52; **19,** 37; **20,** 45–46, **21,** 59

Ferrier rearrangement.[1] Alcohols react with glycals in the presence of $BiCl_3$ to furnish 2,3-unsaturated glycopyranosides.

Heterocycles.[2] A new synthesis of six-membered heterocycles (pyrans, pyridines and pyridazines) by the reaction of zirconacyclopentadienes with compounds containing C═O, C═N, and N═N moieties which are bonded to electron-withdrawing substituents is catalyzed by $BiCl_3$.

R, R, R, R, $ZrCp_2$ — $BiCl_3$ / THF ; $CO(COOEt)_2$ → R, R, R, R, O, COOEt, COOEt

R = Et 80%

Reductive etherification.[3] Symmetrical and unsymmetrical ethers can be prepared by the reaction of carbonyl compounds with alcohols and a hydrosilane in the presence of $BiCl_3$.

Hydroarylation.[4] An aryl group is introduced to the β-position of nitroalkenes in a Pd-catalyzed reaction with Ar_4Sn. The Bi(III) salt is a facilitator.

[1]Swamy, N.R., Venkateswarlu, Y. *S* 598 (2002).
[2]Takahashi, T., Li, Y., Ito, T., Xu, F., Nakajima, K., Liu, Y. *JACS* **124**, 1144 (2002).
[3]Wada, M., Nagayama, S., Mizutani, K., Hiroi, R., Miyoshi, N. *CL* 248 (2002).
[4]Ohe, T., Uemura, S. *TL* **43**, 1269 (2002).

Bismuth(III) nitrate. 20, 46–47

Deoximation. When mixed with wet silica gel[1] or together with copper(II) nitrate and montmorillonite K,[2] regeneration of ketones from oximes is promoted by $Bi(NO_3)_3$.

Oxidation. $Bi(NO_3)_3$ on montmorillonite is a highly efficient oxidant for alcohols.[3]

[1]Samajdar, S., Basu, M.K., Becker, F.F., Banik, B.K. *SC* **32**, 1917 (2002).
[2]Nattier, B.A., Eash, K.J., Mohan, R.S. *S* 1010 (2001).
[3]Samajdar, S., Becker, F.F., Banik, B.K. *SC* **31**, 2691 (2001).

Bismuth(III) triflate. 20, 47; **21,** 59

Preparation.[1] A simple preparative method of $Bi(OTf)_3$ from Bi_2O_3 and TfOH is carried out in aqueous EtOH. The reaction mixture is freeze-dried.

Homoallyl ethers.[2] $Bi(OTf)_3$ catalyzes the allylation of acetals with allylsilanes.

Acylals.[3] A chemoselective preparation of acylals from aromatic aldehydes in the presence of a ketone group using $Bi(OTf)_3$ as catalyst is reported.

Rearrangement.[4] Epoxides are prone to rearrangement on contact with $Bi(OTf)_3$, furnishing carbonyl compounds as a result.

Ph O Ph — $Bi(OTf)_3$ /.xH_2O, CH_2Cl_2 → Ph, Ph, CHO

92%

O — $Bi(OTf)_3$ /.xH_2O, CH_2Cl_2 → O + CHO

(89 : 11)
74%

[1]Repichet, S., Zwick, A., Vendier, L., Le Roux, C., Dubac, J. *TL* **43**, 993 (2002).
[2]Wieland, L.C., Zerth, H.M., Mohan, R.S. *TL* **43**, 4597 (2002).
[3]Carrigan, M.D., Eash, K.J., Oswald, M.C., Mohan, R.S. *TL* **42**, 8133 (2001).
[4]Bhatia, K.A., Eash, K.J., Leonard, N.M., Oswald, M.C., Mohan, R.S. *TL* **42**, 8129 (2001).

Bis(pentamethylcyclopentadienyl)samarium.

Cycloalkanemethanols.[1] Cyclization and hydroboration of 1,5- and 1,6-dienes take place when they encounter Cp^*_2Sm and a borane. Oxidative workup gives cyclopentanemethanols and cyclohexanemethanols, respectively.

[1]Molander, G.A., Pfeiffer, D. *OL* **3**, 361 (2001).

Bis(pentamethylcyclopentadienyl)titanium chloride.

Cycloalkanones. Reaction of allyl bromide with the reagent followed by the addition of an alkylating agent leads to a 3-substituted titanacyclobutane. Carbonylative demetallation liberates either a substituted 2-hydroxycyclopentanone[1] or cyclobutanone,[2] depending on the reaction temperature and CO pressure. Higher temperature and lower pressure favor monocarbonylation and therefore formation of the cyclobutanones.[2]

[1]Carter, C.A.G., Castry, G.L., Stryker, J.M. *SL* 1046 (2001).
[2]Carter, C.A.G., Greidanus, G., Chen, J.-X., Stryker, J.M. *JACS* **123**, 8872 (2001).

1,2-Bis(phenylsulfonyl)ethene.

Vinyl ethers. By a Michael reaction-elimination pathway one of the sulfonyl residue of the title reagent is readily replaced by an alkoxy group on reaction with an alcohol. Vinyl ethers are produced on desulfonylation with sodium amalgam.[1] Certain diols (e.g., 1,3-diols) form monovinyl ethers.[2]

[1]Cabianca, E., Chery, F., Rollin, P., Cossu, S., De Lucchi, O. *SL* 1962 (2001).
[2]Cabianca, E., Chery, F., Rollin, P., Tatibouet, A., De Lucchi, O. *TL* **43**, 585 (2002).

Bis(trialkylphosphine)palladiums.

Negishi coupling.[1] There are several advantages in using $(t\text{-}Bu_3P)_2Pd$ to promote cross coupling of aryl and alkenyl chlorides with organozinc chlorides. Sterically hindered biaryls are accessible and the presence of nitro groups in the substrates causes no problem. High turnover numbers (>3000) are generally observed.

(Z)-3-Phenylthio-2-alkenoates.[2] *S*-Phenyl thiocarbonates are split and caused to add to 1-alkynes in the presence of $(Cy_3P)_2Pd$, adducts are usually obtained in good yields.

[1]Dai, C., Fu, G.C. *JACS* **123**, 2719 (2001).
[2]Hua, R., Takeda, H., Onozawa, S., Abe, Y., Tanaka, M. *JACS* **123**, 2899 (2001).

Bis(tributyltin) oxide. 13, 41–42; **15,** 39; **18,** 54; **19,** 40; **20,** 50; **21,** 50

Alkyl hydroperoxides.[1] Primary alkyl hydroperoxides ROOH are obtained in reasonable yields (~60%) on hydrolysis of peresters R′COOOR. The protocol involves addition of the title reagent at room temperature and workup with CO_2-saturated water.

Selective protection of alditols.[2] Monobenzyl ethers can be prepared from tetritols, penitols, and some hexitols by preemptive formation of stannyl derivatives, leaving a free OH group to be benzylated. The tin residue is then removed with KF.

[1]Baj, S., Chrobok, A. *SL* 623 (2001).
[2]Halila, S., Benazza, M., Demailly, G. *JCC* **20**, 467 (2001).

Bis(tricarbonyldichlororuthenium).

Cyclic imines. Oxidative cyclization of alkenylamines is observed with the title complex.[1]

[1]Kondo, T., Okada, T., Mitsudo, T. *JACS* **124**, 186 (2002).

Bis(2,2,2-trichloroethyl) azodicarboxylate.

Sulfonamides. A method for synthesizing aliphatic and aromatic sulfonamides involves reaction of sodium sulfonates with the title reagent followed by reductive cleavage with zinc dust.[1]

Troc = $COOCH_2CCl_3$

[1]Chan, W.Y., Berthelette, C. *TL* **43**, 4537 (2002).

3,3-Bis(trifluoromethyl)dioxirane.

N-Oxidation.[1] *N*-Boc amines are converted to the corresponding hydroxylamine derivatives on treatment with the title reagent.

[1]Detomaso, A., Curci, R. *TL* **42**, 755 (2001).

***N,O*-Bis(trimethylsilyl)acetamide. 21,** 62

Silyl enol ethers.[1] Preparation of these ethers from carbonyl compounds and the title reagent in ionic liquids (Bu_4NBr, Bu_4PX; 105°) has been demonstrated.

Desulfation.[2] Selective removal of a primary sulfate group in the presence of secondary units is attained when a fully protected substrate is heated with the title reagent.

RO–CH$_2$–CH(OSO$_3$SiMe$_3$)–CH$_2$–OSO$_3$SiMe$_3$ + CH$_3$C(OSiMe$_3$)=NSiMe$_3$ $\xrightarrow[\text{NaOH}]{\text{MeCN ;}}$ RO–CH$_2$–CH(OSO$_3$Na)–CH$_2$–OH

[1]Smietana, M., Mioskowski, C. *OL* **3**, 1037 (2001).
[2]Horibe, M., Oshita, H. *BCSJ* **74**, 181 (2001).

3,6-Bis(triphenylphosphonio)cyclohexene peroxodisulfate.

Preparation.[1] The title reagent (**1**) is available from 3,6-dibromocyclohexene on reaction with triphenylphosphine followed by anion exchange.

$^+PPh_3$ / $^+PPh_3$ cyclohexene $S_2O_8^{=}$

1

Oxidation. Benzylic alcohols are rapidly oxidized to carbonyl products with this reagent.[1] At higher temperature (refluxing MeCN) alkylarenes also undergo oxidation at the benzylic position.[2]

[1]Badri, R., Shalbaf, H., Heidary, M.A. *SC* **31**, 3473 (2001).
[2]Badri, R., Soleymani, M. *SC* **32**, 2385 (2002).

9-Borabicyclo[3.3.1]nonane, 9-BBN. 14, 52–53; **15,** 43–44; **17,** 49–50; **20,** 52; **21,** 64

2H-Chromenes.[1] After treatment with 9-BBN and the usual oxidative workup, chromones afford 2H-chromenes in variable yields.

Aldol reaction.[2] Enol boronates are generated from α-iodo ketones with 9-BBN in the presence of a base such as 2,6-lutidine. Subsequent addition of aldehydes completes the aldol reaction with extremely high *syn*-selectivity.

[1]Eguchi, T., Hoshino, Y. *BCSJ* **74**, 967 (2001).
[2]Mukaiyama, T., Imachi, S., Yamane, K., Mizuta, M. *CL* 698 (2002).

Borane-amines. 13, 42; **18,** 58; **19,** 43–44; **20,** 53; **21,** 65

O-Alkylhydroxylamines.[1] The borane-pyridine complex is much more effective than the $NaBH_3CN$-HOAc system for reducing oxime ethers.

Reductive amination.[2] Secondary *N*-benzylamines are formed when aldehydes or methyl ketones and $BnNH_2 \cdot BH_3$ (with some excess $BnNH_2$) are mixed with molecular sieves.

[1]Gustafsson, M., Olsson, R., Andersson, C.-M. *TL* **42**, 133 (2001).
[2]Peterson, M.A., Bowman, A., Morgan, S. *SC* **32**, 443 (2002).

Borane-dimethyl sulfide.

N-Alkylsulfoximines.[1] These compounds are readily prepared from the unsubstituted parents by a two-step protocol, i.e., *N*-acylation followed by reduction with the borane complex.

Homoallylic alcohols.[2] In a three-component synthesis starting from an alkene via hydroboration which is followed by reaction with propargyl chloride, the way of subsequent handling appears critical to the predominant generation of one regioisomer or the other.

$BH_3.SMe_2$

R'CHO ; R_4NCl ; HOONa

(equilibrate allylboranes before R′CHO addition)

[1]Bolm, C., Hackenberger, C.P.R., Simic, O., Verrucci, M., Muller, D., Bienewald, F. *S* 879 (2002).
[2]Lombardo, M., Morganti, S., Trombini, C. *SL* 601 (2001).

Borane-tetrahydrofuran.

Reduction. Cyclic ketones afford thermodynamically more stable alcohols by reduction with the borane-tetrahydrofuran complex in refluxing THF.[1] Highly stereoselective reduction of β-keto amides to *syn*-1,4-hydroxyamines is achieved in the presence of $TiCl_4$.[2]

Reductive cleavage. Benzylideneacetals undergo regioselective opening upon treatment with borane-tetrahydrofuran and a metal triflate.[3]

Ph–benzylidene acetal (BnO, BnO, OMe) → [$BH_3 \cdot THF$ / $VO(OTf)_2$, CH_2Cl_2] → HO, BnO, BnO, BnO, OMe product, 94%

[1]Cha, J.S., Moon, S.J., Park, J.H. *JOC* **66**, 7514 (2001).
[2]Bartoli, G., Bosco, M., Dalpozzo, R., Marcantoni, E., Massaccesi, M., Rinaldi, S., Sambri, L. *TL* **42**, 8811 (2001).
[3]Wang, C.-C., Luo, S.-Y., Shie, C.-R., Hung, S.-C. *OL* **4**, 847 (2002).

Boron tribromide. 13, 43; **14,** 53–54; **18,** 59; **19,** 45; **20,** 54; **21,** 65

Baylis-Hillman adducts. The boron tribromide-dimethyl sulfide complex is useful for promoting condensation of enones and aldehydes. However, direct aqueous workup without adding a base such as trimethylamine leads to transposed bromo compounds.[1]

Ar-CHO + (methyl vinyl ketone) → [$BBr_3 \cdot SMe_2$ / CH_2Cl_2] → Ar–CH=C(CH$_2$Br)–C(O)CH$_3$

[1]Iwamura, T., Fujita, M., Kawakita, T., Kinoshita, S., Watanabe, S., Kataoka, T. *T* **57**, 8455 (2001).

Boron trichloride. 13, 43; **14,** 54; **15,** 44; **18,** 59–60; **19,** 45–46; **20,** 54–55; **21,** 66

Esterification. Carboxylic acids are esterified on successive treatment with BCl_3 and an alcohol (best with MeOH).[1]

Condensations. Aromatic aldehydes and ethynylarenes condense in the presence of a boron trihalide to give 1,5-dihalo-1,3,5-triaryl-1,4-pentadienes. The products differing in the configuration of one double bond are generated according to whether the reaction occurs in the presence of boron tribromide or boron trichloride.[2] The analogous reaction of aldehydes and styrenes furnishes 1,3-dihaho-1,3-diarylpropanes.[3]

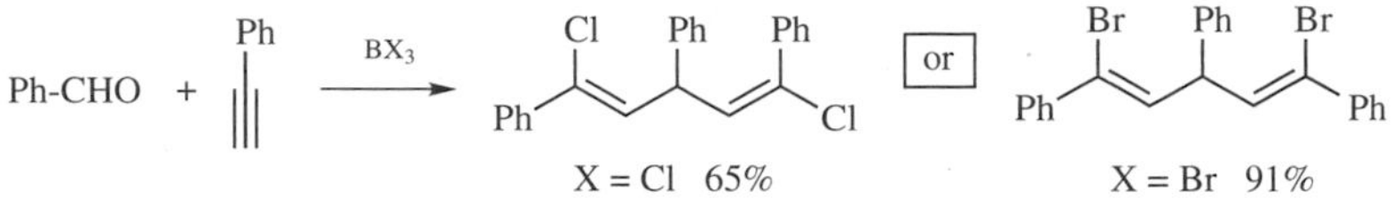

Fries rearrangement. On treatment with BX_3 (X=Cl, Br) aryl formates are transformed into salicylaldehydes.[4] Rearrangment induced by TfOH may lead to different regioisomers.

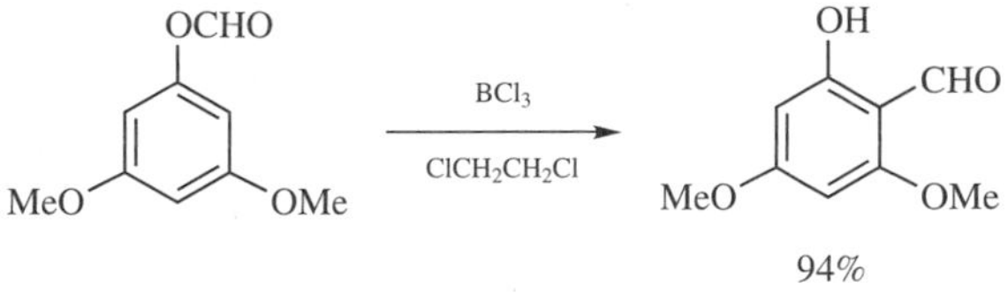

[1]Dyke, C.A., Bryson, T.A. *TL* **42**, 3959 (2001).
[2]Kabalka, G.W., Wu, Z., Ju, Y. *OL* **4**, 1491 (2002).
[3]Kabalka, G.W., Wu, Z., Ju, Y. *TL* **42**, 5793 (2001).
[4]Ziegler, G., Haug, E., Frey, W., Kantlehner, W. *ZN* **56B**, 1178 (2001).

Boron trifluoride-dimethyl sulfide.

Si-O bond scission. 1,3-Dioxa-2,2-(di-*t*-butyl)-2-silacyclohexanes are cleaved to expose the primary alcohol site by the title reagent.[1]

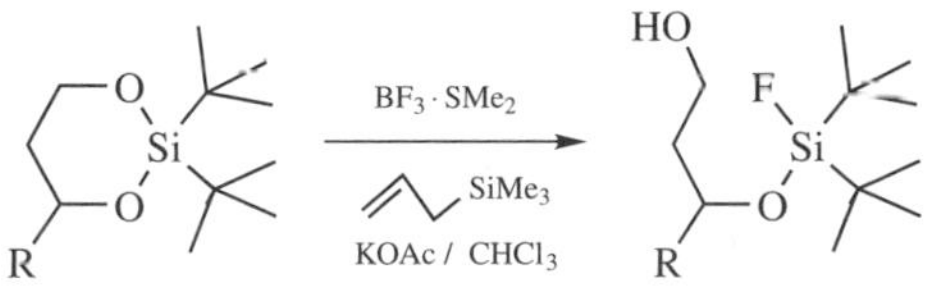

[1]Yu, Y., Pagenkopf, B.L. *JOC* **67**, 4553 (2002).

Boron trifluoride etherate. 13, 43–47; **14,** 54–56; **15,** 45–47; **16,** 44–47; **17,** 52–53; **18,** 60–63; **19,** 46–48; **20,** 55–57; **21,** 66–70

Reactions of 3-membered heterocycles. γ-Butyrolactones are obtained from terminal epoxides and *N*-trimethylsilylethynylmorpholine in a $BF_3{\cdot}OEt_2$–catalyzed reaction that is followed by fluoride treatment (desilylation).[1] The $BF_3{\cdot}THF$ complex is said to be a superior inducer for alkylation of lithium alkynides with epoxides.[2]

Conversion of epoxides to β-lactones by CO insertion between the oxygen atom and the less substituted carbon site is promoted by $BF_3{\cdot}OEt_2$ and a cobalt complex.[3]

$[(Ph_3P)_2N]Co(CO)_4$, $CO / BF_3 \cdot OEt_2$, DME 80°

$BF_3 \cdot OEt_2$ induces intramolecular cycloaddition reaction of compounds bearing an epoxide ring and an alkyne moiety that is substituted by a tungsten residue directly or at an propargylic position. It gives rise to bicyclic lactones or dihydropyrans.[4]

$(OC)_3W$ Cp; $BF_3 \cdot OEt_2$, CH_2Cl_2; $W(CO)_3$ 51%; I_2, BnOH; COOBn

OAc; Cp W $(CO)_3$; $BF_3 \cdot OEt_2$, CH_2Cl_2

β-*N*-Tosylamino hydroxylamines are obtained from opening of *N*-tosylaziridines. When the aziridine contains an unsaturation in a proper position its adduct is a useful precursor of pyrrolidine or piperidine *N*-oxide, by virtue of its susceptibility to undergo the retro-Cope rearrangement.[5] Some aziridines are converted to imines in analogy to the pinacol rearrangement.[6]

TsN + Bn–N(OH)–H; $BF_3 \cdot OEt_2$, Et_2O; OH NHTs Bn–N 68%

Cyclization reactions. An alkynyltungsten unit and ester enolate are synthetically equivalent, but they operate under different conditions. The following equation shows trapping of an acyliminium ion.[7] Alternatively, participation of an allylic silyl ether in a related process (hydride shift) is useful for elaboration of hydroisoquinolones.[8]

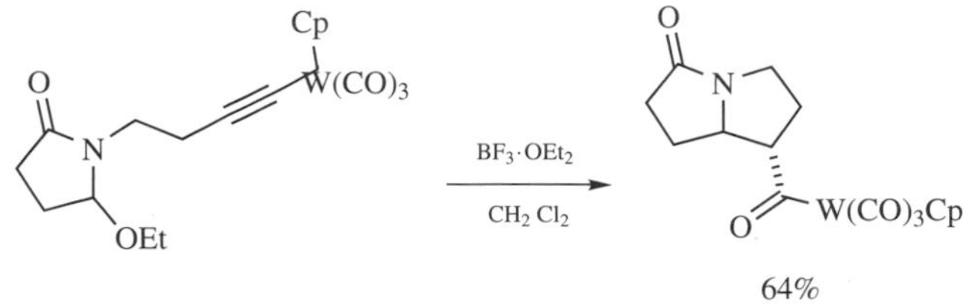

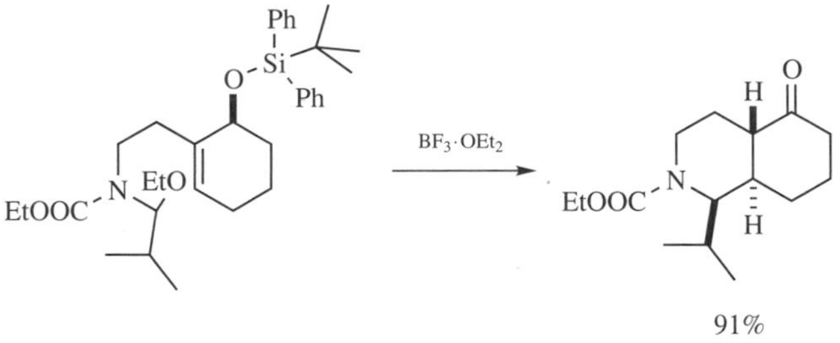

By involving an aldol reaction with an enol ether as donor and a properly located alllylsilane to terminate the charge, functionalized oxacycles are built in an expedient way. This method is applicable to a formal synthesis of leucascandrolide A.[9]

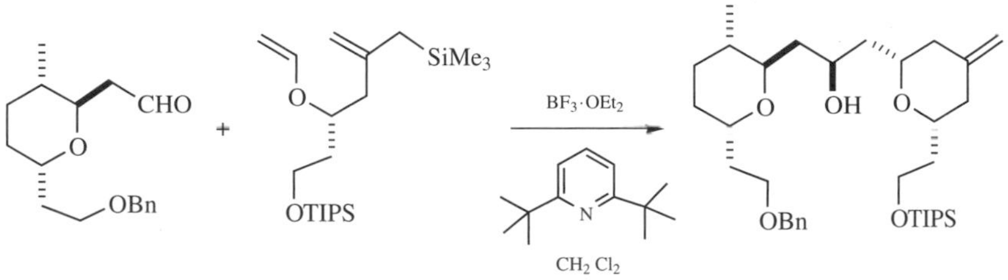

Tributylborane.[10] A simple preparation of the borane consists of slow addition of BuLi to $BF_3{\cdot}OEt_2$ in ether under nitrogen.

Activation. Activation of tertiary amines on complexation to BF_3 enables alkylation at the α-carbon atom.[11] Enaminones and pyrazoles are conveniently prepared from 1,3-diketones by sequential treatment with $BF_3{\cdot}OEt_2$ and amines and hydrazines, respectively.[12]

Building a chain segment via etherification can take advantage of the facile ionization of cobalt-coordinated propargylic alcohols on treatment with $BF_3{\cdot}OEt_2$.[13] Cycloalkanones by a rearrangement route are reported.[14]

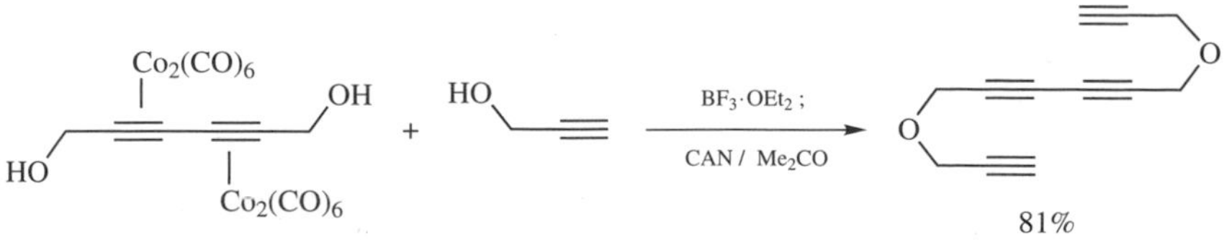

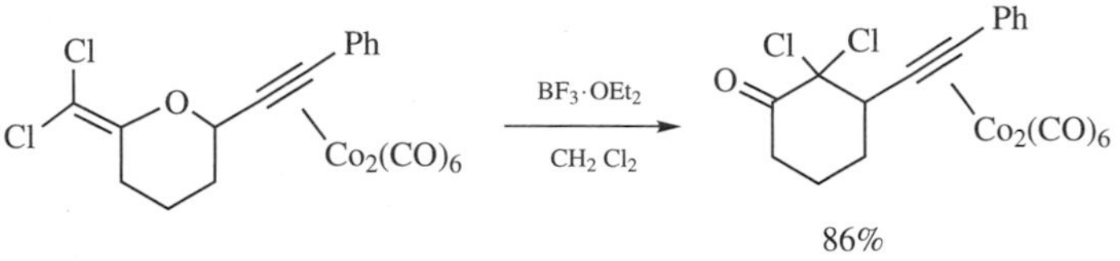

81%

86%

A synthesis of 2,5-disubstituted tetrahydrofurans by Grignard reaction of five-membered ring hemiacetal *O*-acetates is facilitated by $BF_3{\cdot}OEt_2$.[15]

The matter of activation by selective complexation of certain carbonyl groups is further clarified by a study on fluoro-substituted compounds against the normal substrates.[16]

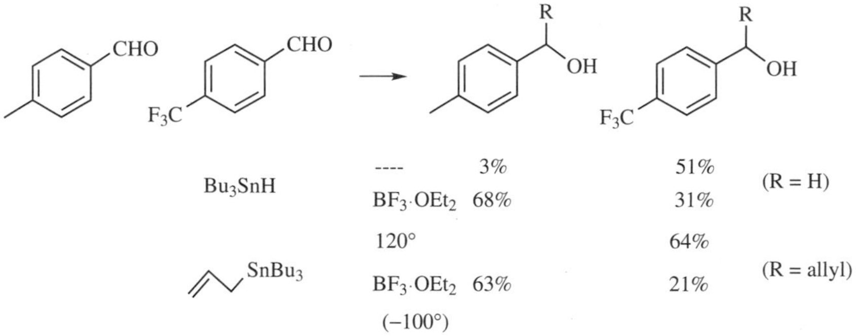

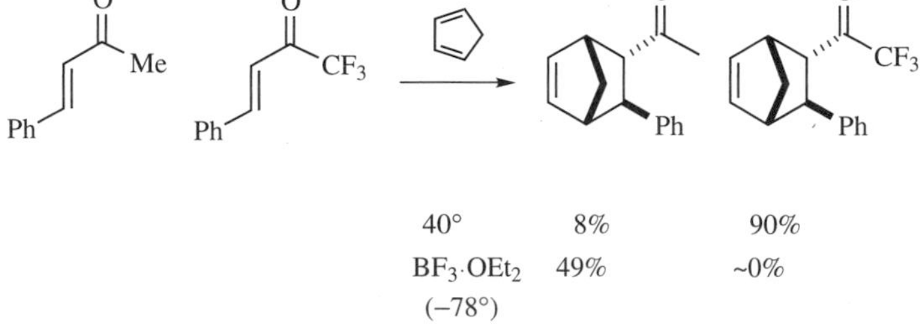

Bu_3SnH	----	3%	51%	(R = H)
	$BF_3{\cdot}OEt_2$	68%	31%	
allyl-$SnBu_3$	120°		64%	(R = allyl)
	$BF_3{\cdot}OEt_2$ (−100°)	63%	21%	

40°	8%	90%
$BF_3{\cdot}OEt_2$ (−78°)	49%	~0%

Aldol-type reaction. α-Iodostyrenes and some bromo analogues react with aromatic aldehydes to give 1,3-diarylpropenones.[17]

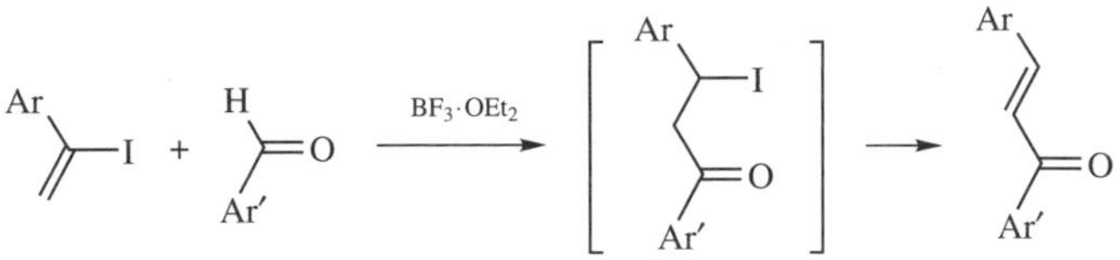

For homologation of α-methylstyrenes by reaction with paraformaldehyde, the use of $BF_3 \cdot OEt_2$ in combination with molecular sieves as promoter is recommended.[18]

Desulfurization. Sulfur atoms such as those present in cysteine peptides can be removed photochemically in the presence of $(EtO)_3P$ together with $BF_3 \cdot OEt_2$.[19]

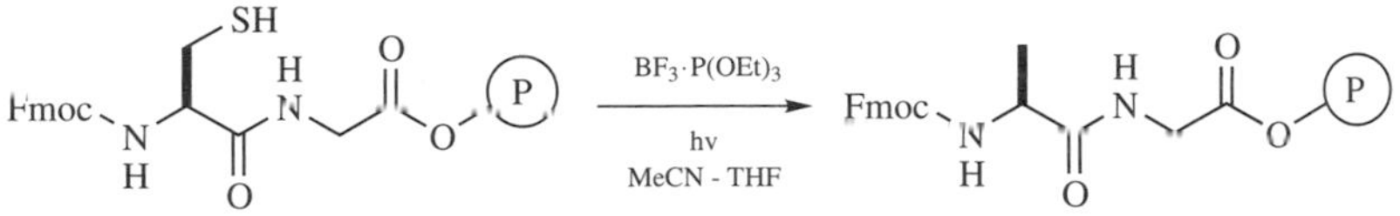

Baylis-Hillman reaction. Together with sulfolane to promote Baylis-Hillman reaction of methyl vinyl ketone it requires 1.5 equivalents of $BF_3 \cdot OEt_2$.[20]

[1]Movassaghi, M., Jacobsen, E.N. *JACS* **124**, 2456 (2002).
[2]Evans, A.B., Knight, D.W. *TL* **42**, 6947 (2001).
[3]Lee, J.T., Thomas, P.J., Alper, H. *JOC* **66**, 5424 (2001).
[4]Madhushaw, R.J., Li, C.-L., Shen, K.-H., Hu, C.-C., Liu, R.-S. *JACS* **123**, 7427 (2001).
[5]O'Neil, I.A., Woolley, J.C., Southern, J.M., Hobbs, H. *TL* **42**, 8243 (2001).
[6]Sugihara, Y., Iimura, S., Nakayama, J. *CC* 134 (2002).
[7]Huang, H.-H., Sung, W.-H., Liu, R.-S. *JOC* **66**, 6193 (2001).
[8]Kamatani, A., Overman, L.E. *OL* **3**, 1229 (2001).
[9]Kopecky, D.J., Rychnovsky, S.D. *JACS* **123**, 8420 (2001).
[10]Wu, Y., Shen, X. *SC* **31**, 2939 (2001).
[11]Kessar, S.V., Singh, P., Singh, K.N., Singh, S.K. *SL* 517 (2001).
[12]Stefane, B., Polanc, S. *NJC* **26**, 28 (2002).
[13]Diaz, D., Martin, T., Martin, V.S. *OL* **3**, 3289 (2001).
[14]Carbery, D.R., Reignier, S., Myatt, J.W., Miller, N.D., Harrity, J.P.A. *ACIEE* **41**, 2584 (2002).
[15]Franck, X., Hocquemiller, R., Figadere, B. *CC* 160 (2002).
[16]Asao, N., Sano, T., Yamamoto, Y. *ACIEE* **40**, 3206 (2001).
[17]Chan, P.W.H., Kamijo, S., Yamamoto, Y. *SL* 910 (2001).
[18]Okachi, T., Fujimoto, K., Onaka, M. *OL* **4**, 1667 (2002).
[19]Arsequell, G., Gonzalez, A., Valencia, G. *TL* **42**, 2685 (2001).
[20]Walsh, L.M., Winn, C.L., Goodman, J.M. *TL* **43**, 8219 (2002).

Bromine. 13, 47; **14,** 56–57; **15,** 47; **18,** 64; **19,** 48; **20,** 57–58; **21,** 70

Desilylation.[1] A *t*-butyldiphenylsilyl group on amines, ethers and esters can be removed with a methanolic solution of bromine. Trimethylsilyl and benzyloxymethyl ethers also suffer cleavage under the conditions.

Bromination.[2] An increase in selectivity in the bromination of arenes is noted when bromine is adsorbed on silica. Benzylic bromination is observed.

Oxidations. A combination of bromine and hydrogen peroxide converts sulfides to sulfoxides[3] and amides to imides.[4]

H_2O_2-Br_2, $ClCH_2CH_2Cl$

96%

[1]Barros, M.T., Maycock, C.D., Thomassigny, C. *SL* 1146 (2001).
[2]Ghiaci, M., Asghari, J. *BCSJ* **74**, 1151 (2001).
[3]Bravo, A., Dordi, B., Fontana, F., Minisci, F. *JOC* **66**, 3232 (2001).
[4]Bjorsvik, H.-R., Fontana, F., Liguori, L., Minisci, F. *CC* 523 (2001).

Bromine trifluoride. 19, 48; **21,** 71

Replacement of methylthio groups.[1] 1,1,1-Trismethylthioalkanes are converted to bromopolyfluoroalkanes on reaction with BrF_3.

BrF_3, $CFCl_3$ 0°

3 equiv BrF_3 70%
10 equiv BrF_3 60%

[1]Hagooly, A., Ben-David, I., Rozen, S. *JOC* **67**, 8430 (2002).

Bromoacetyl bromide.

Ether cleavage.[1] The reaction of ethers with $BrCH_2COBr$ gives bromoalkanes and bromoacetic esters, sometimes the presence of $ZnCl_2$ is beneficial.

[1]Schneider, D.F., Viljoen, M.S. *SC* **32**, 721 (2002).

***N*-Bromosuccinimide, NBS. 13,** 49; **14,** 57–58; **15,** 50–51; **16,** 49; **18,** 65–67; **19,** 50–51; **20,** 58–59; **21,** 72–73

Acylations. All sorts of alcohols including tertiary alcohols can be acetylated by acetic anhydride at room temperature in the presence of NBS.[1] Transesterification of β-keto esters is also achieved using NBS as a catalyst.[2]

Deoxygenation.[3] Sulfoxides are reduced to sulfides with NBS and 1,3-dithiane.

Bromination. *o*-Alkenylphenylboronic acids form boronolactones on treatment with NBS.[4] Reaction conditions determine the course of dibromination of methyl 2-methylfuran-3-carboxylate.[5]

B(OH)$_2$ —NBS, THF - H$_2$O→ Br, O, B, OH

85%

COOMe + NBS —BzOOBz, CCl$_4$ Δ→ COOMe, Br, O, Br

—DMF 0°→ COOMe, Br, O, Br

Quinones.[6] Polycyclic 1,4-dimethoxyarenes and 4-methoxy-1-aminoarenes are oxidized to quinones in high yields by NBS in aqueous THF with catalytic amount of H_2SO_4. This transformation is applicable to substituted quinolines. Hydroquinone dimethyl ether is largely unreactive (dibromination occurs to a slight extent).

[1]Karimi, B., Seradj, H. *SL* 519 (2001).
[2]Bandgar, B.P., Uppalla, L.S., Sadavarte, V.S. *SL* 1715 (2001).
[3]Iranpoor, N., Firouzabadi, H., Zhaterian, H.R. *JOC* **67**, 2826 (2002).
[4]Falck, J.R., Bondlela, M., Venkataraman, S.K., Srinivas, D. *JOC* **66**, 7148 (2001).
[5]Khatuya, H. *TL* **42**, 2643 (2001).
[6]Kim, D.W., Choi, H.Y., Lee, K.-J., Chi, D.Y. *OL* **3**, 445 (2001).

Bromotris(triphenylphosphine)rhodium(I).

Alkenylphosphine oxides.[1] The title complex catalyzes regioselective addition of diphenylphosphine oxide to 1-alkynes (15 examples, 76–94%).

[1]Han, L.-B., Zhao, C.-Q., Tanaka, M. *JOC* **66**, 5929 (2001).

1-*t*-Butoxy-2-*t*-butoxycarbonyl-1,2-dihydroisoquinoline.

Protection.[1] The title reagent is obtained from isoquinoline on reaction with Boc_2O. It can be used to derivatize alcohols and amines to form carbonates and carbamates.

[1]Ouchi, H., Saito, Y., Yamamoto, Y., Takahata, H. *OL* **4**, 585 (2002).

***N*-*t*-Butoxy-2-nitrobenzenesulfonamide.**

Hydroxamates.[1] The title reagent is prepared by sulfonylation of *O*-*t*-butylhydroxylamine. Its use for synthesizing hydroxamates involves *N*-alkylation, desulfonylation with a thiol, *N*-acylation, and acid treatment.

RX ; R'SH → R−NH−O-t-Bu; R'COX' ; CF_3COOH → R−N(OH)−C(=O)−R'

[1]Reddy, P.A., Schall, O.F., Wheatley, J.R., Rosik, L.O., McClurg, J.P., Marshall, G.R., Slomczynska, U. *S* 1086 (2001).

***N*-*t*-Butylbenzenesulfenamide.**

Oxidation.[1] The title compound *t*-BuNHSPh (5 mol%) mediates efficient oxidation of primary and secondary alcohols with NCS.

[1]Mukaiyama, T., Matsuo, D., Iida, D., Kitagawa, H. *CL* 864 (2001).

***t*-Butyl bromoacetate.**

t-Butyl ethers.[1] 1,1,1-Tris(hydroxymethyl)alkanes undergo *t*-butylation with *t*-butyl bromoacetate under phase transfer conditions.

t-$BuOCOCH_2Br$; 50% NaOH - CH_2Cl_2; Bu_4NBr 10–20°

[1]Dupuy, C., Viguier, R., Dupraz, A. *SC* **31**,1307 (2001).

_t_-Butyl _N,N_-dibromocarbamate.

2-Bromo-1-alkanamines.[1] Reaction of the title reagent with 1-alkenes, followed by de-*N*-bromination with sodium sulfite and finally acid hydrolysis delivers the salts of 2-bromo alkyl amines.

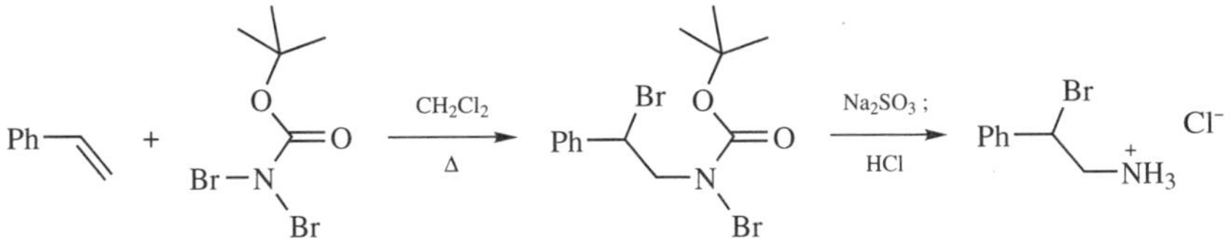

[1]Klepacz, A., Zwierzak, A. *TL* **42**, 4539 (2001).

_t_-Butyl 4-keto-3,4-dihydrobenzotriazin-3-yl carbonate.

Carboxyl activation.[1] Admixture of the title reagent (**1**) with carboxylic acids provides highly active esters which act as useful transacylating agents.

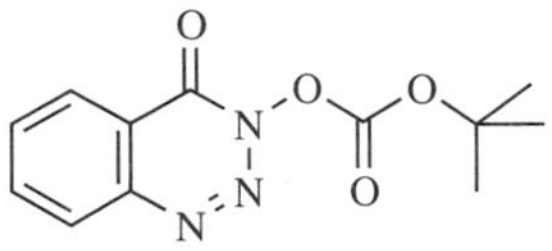

1

[1]Basel, Y., Hassner, A. *TL* **43**, 2529 (2002).

_t_-Butyl hydroperoxide. 19, 51–53

Oxidation. Cyclopropanemethanols are oxidized to the corresponding aldehydes by *t*-BuOOH in the presence of chloroperoxidase; much higher enantioselectivity is observed for the oxidation of *cis*-2-substituted cyclopropanemethanols than the *trans* isomers.[1]

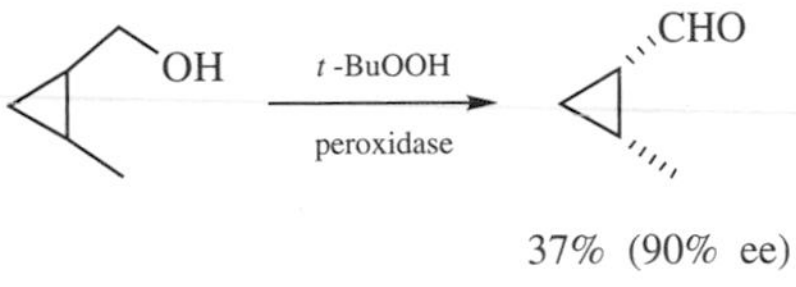

37% (90% ee)

[1]Hu, S., Dordick, J.S. *JOC* **67**, 314 (2002).

t-Butyl hydroperoxide–metal salts. 19, 51–53; **20,** 61–62; **21,** 76

Epoxidation. A very useful oxidation system constituting *t*-BuOOH and $Pd(OAc)_2$–K_2CO_3 can convert certain alkenes into epoxides but others into allylic peroxy ethers (structural dependence).[1] Enones are obtained from substrates of the latter type when Pd-C is employed instead of $Pd(OAc)_2$.[1]

t-BuOOH

$Pd(OAc)_2$ - K_2CO_3

CH_2Cl_2

80%

t-BuOOH

Pd - C

85%

$VO(acac)_2$ microencapsulated by a thin film of polystyrene is capable of effecting epoxidation with *t*-BuOOH.[2] Cyclization attends a titanocene-catalyzed epoxidation of bishomoallylic alcohols.[3]

Oxidation. Secondary allylic and benzylic alcohols are oxidized by *t*-BuOOH and catalytic $Rh_2(OAc)_4$ at room temperature, but product yields are modest.[4] 2-Substituted 1,3-dithianes undergo diastereoselective oxidation to afford sulfoxide products with a system consisting of Cp_2TiCl_2 and *t*-BuOOH.[5] Conversion of organometallic compounds RM to alcohols is accomplishable with $(i\text{-}PrO)_4Ti$ and *t*-BuOOH or just *t*-BuOOLi.[6]

t-BuOOH

Cp_2TiCl_2 - 4A MS

86% (*trans* : *cis* 98 : 2)

The hypervalent iodine reagent (**1**) in combination with *t*-BuOOH opens cyclic acetals oxidatively.[7]

1

[1]Yu, J.-Q., Corey, E.J. *OL* **4**, 2727 (2002).
[2]Lattanzi, A., Leadbeater, N.E. *OL* **4**, 1519 (2002).
[3]Lattanzi, A., Della Sala, G., Russo, M., Scettri, A. *SL* 1479 (2001).
[4]Moody, C.J., Palmer, F.N. *TL* **43**, 139 (2002).
[5]Della Sala, G., Labano, S., Lattanzi, A., Tedesco, C., Scettri, A. *S* 505 (2002).
[6]Möller, M., Husemann, M., Boche, G. *JOMC* **624**, 47 (2001).
[7]Sueda, T., Fukuda, S., Ochiai, M. *OL* **3**, 2387 (2001).

Butyllithium. 13, 56; **14,** 63–68; **15,** 59–61; **17,** 59–60; **18,** 74–77; **19,** 54–59; **20,** 62–65; **21,** 77–79

Lithiation. Halogen/lithium exchange and reaction with $(i\text{-PrO})_3B$ constitute a viable protocol for the preparation of some pyridylboronic acids from the corresponding halopyridines.[1] Aryne precursors can be synthesized from *o*-bromophenols via the silyl ethers. Lithiation is attended by an O→C silyl shift and the resulting phenoxides are promptly triflated.[2]

OSiMe$_3$ / Br —BuLi / THF→ OLi / SiMe$_3$ —Tf$_2$O→ OTf / SiMe$_3$

A route to arylbutadiynes involving the Fritsch-Buttenberg-Wicchell rearrangement[3] and that converts 1,4-diiodo-1,3-butadienes to cyclopentadienes proceed with halogen/lithium exchange (and subsequent reaction with carbonyl compounds).[4]

OMe, Br, Br, SiMe$_3$ —BuLi, −78 ~ −40°→ OMe, SiMe$_3$ 43%

Certain β-stannyl ketones undergo both Sn/Li exchange and enolization and the dilithio species thus generated are β-keto anion equivalents.[5]

Cyclization reactions. Cyclopentane formation initiated by lithiation through tin exchange and subsequent attack on a remote functionality is realized.[6,7]

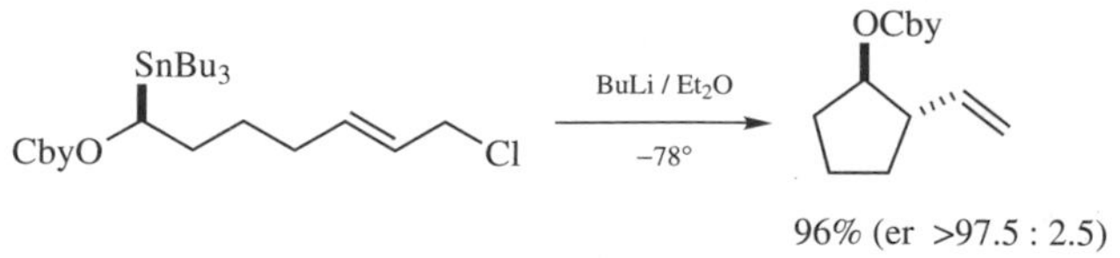

SnBu3 CbyO Cl BuLi / Et_2O −78° OCby

96% (er >97.5 : 2.5)

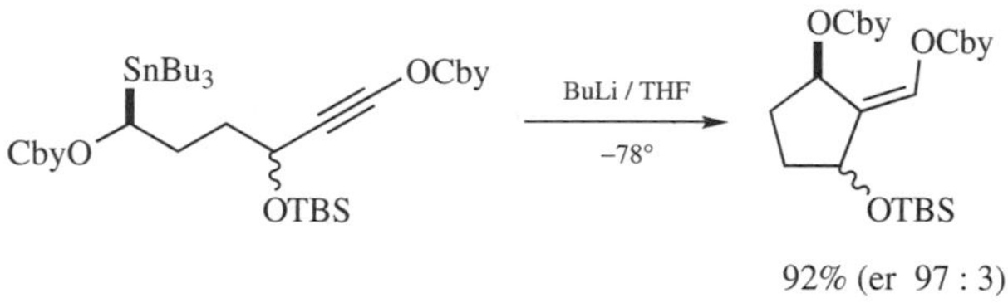

SnBu3 CbyO OCby OTBS BuLi / THF −78° OCby OCby OTBS

92% (er 97 : 3)

A diastereoselective synthesis of oxazolidinones, precursors of *anti*-1,2-amino alcohols, from (α-stannylalkyl) carbamates has been discovered.[8] Elimination likely precedes the annulation of an oxazolidine ring from an *N*-oxide[9] but an example of cyclopropanation is initiated by an addition process.[10]

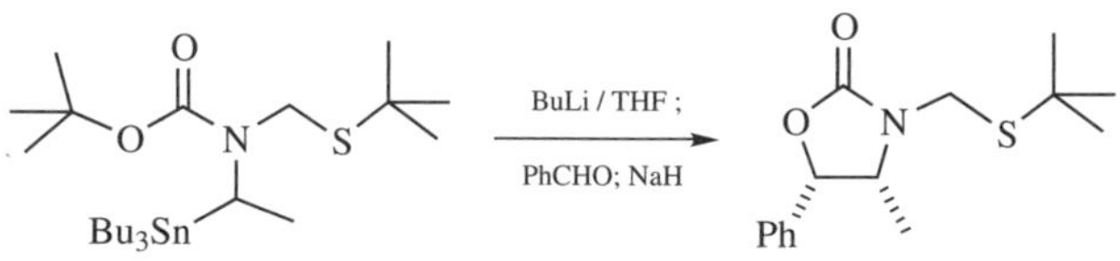

O O N S Bu3Sn BuLi / THF ; PhCHO; NaH O O N S Ph

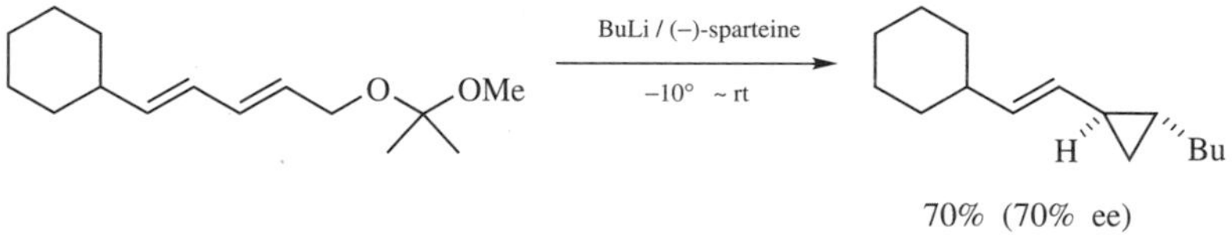

O OMe BuLi / (−)-sparteine −10° ~ rt H Bu

70% (70% ee)

3-Ethoxyvinylbenzofuran is obtained from 2-iodophenyl 4,4-diethoxy-2-butyn-1-yl ether. The lithiated species adds across the triple bond and induces LiOEt elimination. Subsequent migration of the double bond completes the transformation.[11] However, construction of the corresponding indole system cannot be carried out before isomerization to the allenic isomer.

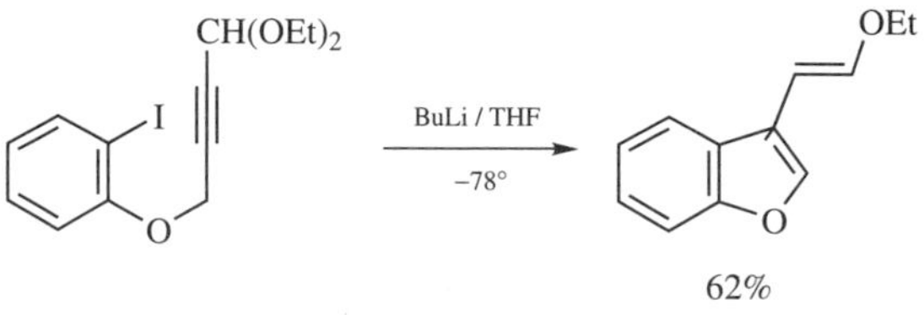

$CH(OEt)_2$ I O BuLi / THF −78° OEt O

62%

Eliminations. BuLi converts epoxides of dihydrofuran and dihydropyran to open-chain diols.[12] A method for generation of α-tributylstannyl silyl enol ethers involves an initial treatment of hindered aryl esters with BuLi to form ketenes.[13]

BuLi / THF

Bu

HO OH

90%

BuLi / THF

− 78°

R R C=O

Bu_3SnLi ; $PhMe_2SiCl$

I R′

CuCl / DMF

R R $OSiMe_2Ph$ R′

R R $OSiMe_2Ph$ $SnBu_3$

Rearrangements. 1,3-Transposed allylic amines are accessible from allylic alcohols via a [2,3]sigmatropic rearrangement of the derived *O*-allyl hydroxylamines.[14]

A convenient method for generating propyne for laboratory use (e.g., Sonogashira coupling) is by reaction of 1-bromopropene with BuLi.[15]

Alkylation. After trimethylsilyl butenoate is deprotonated with BuLi, alkylation with aldehydes takes place at the α-position.[16] During alkylation of sulfones that contain a β-trimethylsilyl group with carbonyl compounds, concomitant elimination (desulfonylation) immediately follows C→O silyl transfer.[17]

Ph Me_3Si SO_2Ph

BuLi / THF ; RR′C=O

Ph $OSiMe_3$ R R′

Selenoamides $RC(=Se)NHCH_2R'$ readily undergo *C*-alkylation via their dianions.[18]

[1]Li, W., Nelson, D.P., Jensen, M.S., Hoerrner, R.S., Cai, D., Larsen, R.D., Reider, P.J. *JOC* **67**, 5394 (2002).

[2]Pena, D., Cobas, A., Perez, D., Guitian, E. *S* 1454 (2002).

[3]Chernick, E.T., Eisler, S., Tykwinski, R.R. *TL* **42**, 8575 (2001).
[4]Xi, Z., Song, Q., Chen, J., Guan, H., Li, P. *ACIEE* **40**, 1913 (2001).
[5]Ryu, I., Ikebe, M., Sonoda, N., Yamamoto, S., Yamamura, G., Komatsu, M. *TL* **43**, 1257 (2002).
[6]Christoph, G., Hoppe, D. *OL* **4**, 2189 (2002).
[7]Gralla, G., Wibbeling, B., Hoppe, D. *OL* **4**, 2193 (2002).
[8]Ncube, A., Park, S.B., Chong, J.M. *JOC* **67**, 3625 (2002).
[9]Gamez-Montano, R., Chavez, M.I., Roussi, G., Cruz-Almanza, R. *TL* **42**, 9 (2001).
[10]Majumdar, S., de Meijere, A., Marek, I. *SL* 423 (2002).
[11]Le Strat, F., Maddaluno, J. *OL* **4**, 2791 (2002).
[12]Hodgson, D.M., Stent, M.A.H., Wilson, F.X. *OL* **3**, 3401 (2001).
[13]Barbero, A., Pulido, F.J. *SL* 827 (2001).
[14]Ishikawa, T., Kawakami, M., Fukui, M., Yamashita, A., Urano, J., Saito, S. *JACS* **123**, 7734 (2001).
[15]Abraham, E., Suffert, J. *SL* 328 (2002).
[16]Bellasoued, M., Grugier, J., Lensen, N., Catheline, A. *JOC* **67**, 5611 (2002).
[17]Menichetti, S., Stirling, C.J.M. *JCS(P1)* 28 (2002).
[18]Murai, T., Aso, H., Kato, S. *OL* **4**, 1407 (2002).

Butyllithium–potassiumt *t*-butoxide. 21, 80

Lithiation. Diphenylethyne forms an *o,o′*-dilithio derivative and it is readily functionalized.[1]

1,1-Disubstituted 1,3-butadienes. A Pd-catalyzed coupling (Suzuki-type) of species derived from an acetal of crotonaldehyde after consecutive treatment with the strong base and a triorganoborane delivers the diene products.[2]

OEt OEt *t*-BuOK - BuLi ; R_3B ; $(Ph_3P)_4Pd$, PhI [EtO EtO B R R Pd Ph] → R R

[1]Kowalik, J., Tolbert, L.M. *JOC* **66**, 3229 (2001).
[2]Tivola, P.B., Deagostino, A., Prandi, C., Venturello, P. *CC* 1536 (2001).

Butyllithium–*N,N,N′N′*-tetramethylethylenediamine (TMEDA). 19, 60; 20, 66–67; 21, 80–81

Homopropargylic alcohols. Subjecting propargyl bromide to deprotonation and Br/Li exchange with the title reagent and quenching the resulting solution with carbonyl compounds lead to homopropargylic alcohols. If paraformaldehyde is added after the alkylation the alkyne terminus is also hydroxymethylated.[1]

Br BuLi - TMEDA ; $Ph_2C{=}O$ → Ph Ph OLi Li $(HCHO)_n$ → Ph Ph OH OH 82%

Lithiation. Steric bulk of the complexed *n*-BuLi causes the transition state for lithiation of (η^6-arene)dicarbonyltriphenylphosphine chromium(0) oxazoline complexes to be different from that assumed by that involving *s*-BuLi.[2]

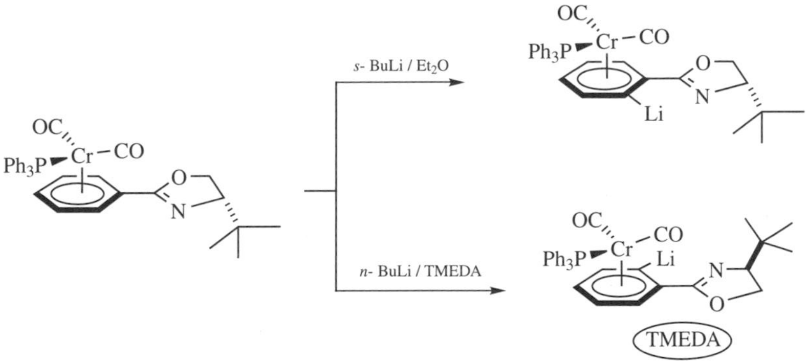

A method for preparation of monodentate cyclic phosphines related to BINAP involves formation of 2,2′-bislithiomethyl-1,1′-binaphthyl.[3]

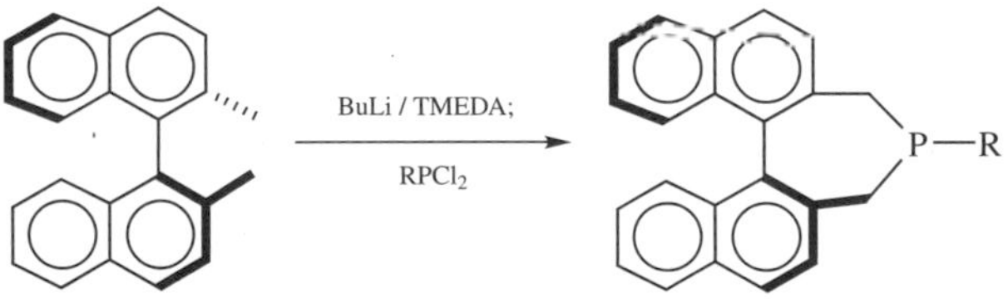

[1]Cabezas, J.A., Pereira, A.R., Amey, A. *TL* **42**, 6819 (2001).
[2]Overman, L.E., Owen, C.E., Zipp, G.G. *ACIEE* **41**, 3884 (2002).
[3]Junge, K., Oehme, G., Monsecs, A., Riermeier, T., Dingerdissen, U., Beller, M. *TL* **43**, 4977 (2002).

***s*-Butyllithium. 14,** 69; **16,** 56; **18,** 77–79; **19,** 60–61; **20,** 67; **21,** 81–82

Substituted cyclopentenones.[1] α-Keto phosphoranes are deprotonated by *s*-BuLi. Reaction of the resulting dianionic species with enediones gives cyclopentenones via a tandem conjugated addition and Wittig reaction process.

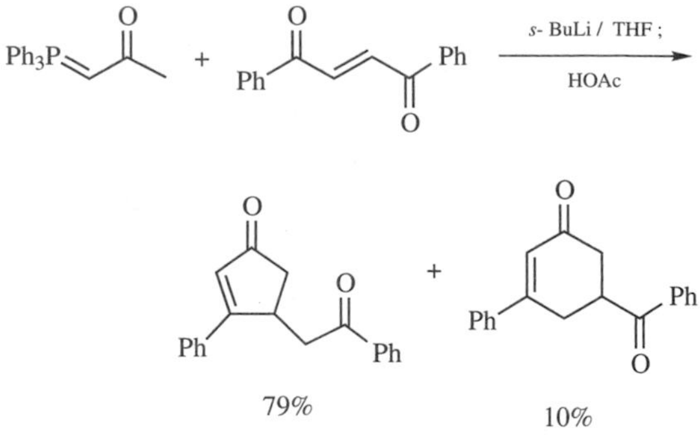

Lithiations. Lithiation at an ortho-position or at the proximal benzylic position of oxazolin-2-ylarenes by *s*-BuLi depends on whether TMEDA is present.[2]

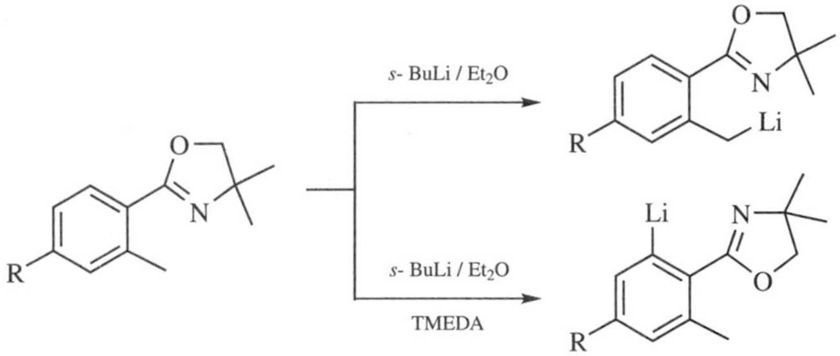

Terminal epoxides are silylated after lithiation with *s*-BuLi.[3]

[1]Kitano, H., Minami, S., Morita, T., Matsumoto, K., Hatanaka, M. *S* 739 (2002).
[2]Tahara, N., Fukuda, T., Iwao, M. *TL* **43**, 9069 (2002).
[3]Hodgson, D.M., Norsikian, S.L.M. *OL* **3**, 461 (2001).

s-Butyllithium–sparteine. 20, 67

Desymmetrization.[1] Lithiation of *meso*-epoxides by the reagent system followed by addition of electrophiles provides chiral products in moderate to good ee.

[1]Hodgson, D.M., Gras, E. *ACIEE* **41**, 2376 (2002).

***t*-Butyllithium. 13,** 58; **15,** 64–65; **16,** 56–57; **18,** 79–81; **19,** 61–63; **20,** 68–69; **21,** 82–84

Halogen-lithium exchanges. The synthesis of several types of cyclic compounds from haloalkenes starts with halogen-lithium exchange. Intramolecular Michael reaction leading to unsaturated bicyclic ketones[1] and intermolecular trapping that gives

3-cyclopentenones,[2,3] pyridines,[4] and 6-substituted phenanthridines[5] are some of the examples.

t-BuLi ; CO
H_2O

t-BuLi ; HMPA;
PhCN

75%

N-Methyl-2,3-dilithioindole has been prepared from the diiodoindole by the I/Li-exchange reaction.[6]

Alkenylsilanes.[7] Using *t*-BuLi to lithiate trimethyl silylmethyl (ethoxy)dimethylsilyl methane and reacting the ensuing species with carbonyl compounds lead to alkenyltrimethylsilanes, as a result of siloxane elimination from the adducts.

t-BuLi
pentane

– 78° ;
H_2O

71%

[1]Piers, E., Harrison, C.L., Zetina-Rocha, C. *OL* **3**, 3245 (2001).
[2]Song, Q., Chen, J., Jin, X., Xi, Z. *JACS* **123**, 10419 (2001).
[3]Song, Q., Li, Z., Chen, J., Wang, C., Xi, Z. *OL* **4**, 4627 (2002).
[4]Chen, J., Song, Q., Wang, C., Xi, Z. *JACS* **124**, 6238 (2002).
[5]Pawlas, J., Begtrup, M. *OL* **4**, 2687 (2002).
[6]Liu, Y., Gribble, G.W. *TL* **42**, 2949 (2001).
[7]Bates, T.F., Dandekar, S.A., Longlet, J.J., Thomas, R.D. *JOMC* **625**, 13 (2001).

t-Butyllithium–amine.

3,4-Disubstituted pyrroles.[1] Bis(2-bromoallyl)amines such as that shown in the following equation undergo cyclization on treatment with *t*-BuLi-TMEDA. Products obtained after reaction with electrophiles are pyrrole derivatives.

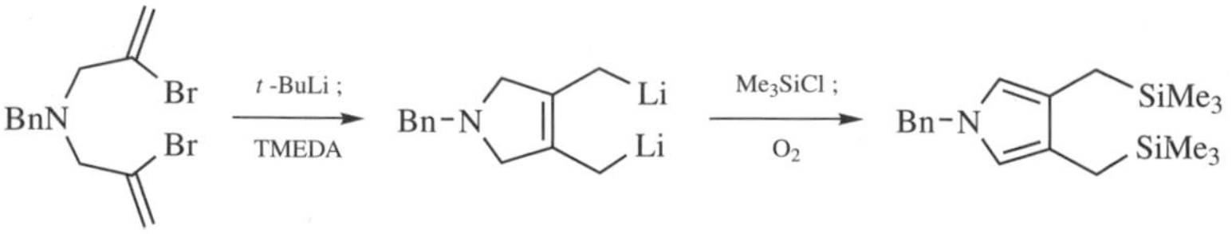

2-Cyclopropylphenols.[2] *o*-Bromoaryl allyl ethers on lithiation are transformed into arylcyclopropanes by an intramolecular addition and elimination process. (−)-Sparteine engenders the reaction course enantioselective.

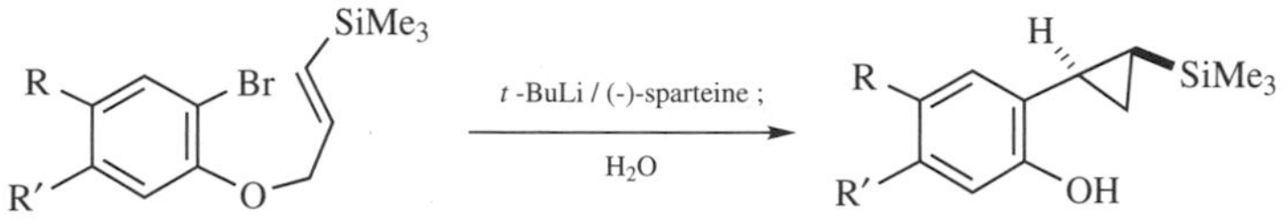

[1]Fananas, F.J., Granados, A., Sanz, R., Ignacio, J.M., Barluenga, J. *CEJ* 2896 (2001).
[2]Barluenga, J., Fananas, F.J., Sanz, R., Marcos, C. *OL* **4**, 2225 (2002).

N-(t-Butyl)methanesulfinamide.

Oxidation of alkyl triflates.[1] Aldehydes are obtained from primary alkyl triflates on reaction with *t*-BuNHSOMe, and subsequent treatment with DBU (7 examples, 71–95%).

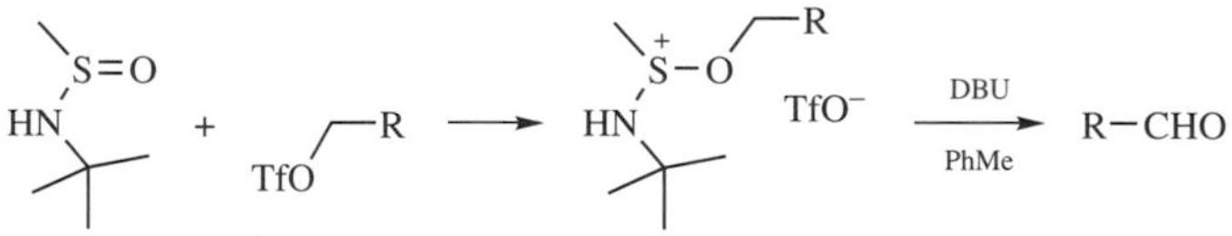

[1]Kitagawa, H., Matsuo, J.-I., Iida, D., Mukaiyama, T. *CL* 580 (2001).

1-Butyl-3-methylimidazolium salts and 1-ethyl-3-methylimidazolium salts, (bmim & emim). 20, 70; **21,** 85

Small quantity of an ionic liquid enables solvents such as hexane and toluene to reach a higher reaction temperature than their original boiling points, therefore they can be used as media for microwave-assisted reactions.[1]

Acylation. Rapid acylation of carbohydrates[2] and enhanced enantioselectivity for lipase-catalyzed transacetylation in ionic liquid[3,4] show advantages over conventional solvents. The reactions are carried out at room temperature and the lipase-catalyzed process is up to 25 times more enantioselective. Methyl carbamates are formed on treatment of amines with dimethyl carbonate in ionic liquids.[5] Unimpaired activity of lipase in catalyzing transfer acetylation of racemic allylic alcohols in ionic liquid is apparent.[6]

Acylative cleavage of ethers by RCOCl is observed in [emim]Al_2X_7.[7]

Substitutions. Exhange reactions of alkyl halides to nitriles,[8] chlorides to fluorides,[9] alcohols to chlorides[10] are readily performed in ionic liquids. For the conversion of aromatic bromides and iodides to nitriles with CuCN, halide-based [bmimX] is employed as the medium[11], copper catalysts immobilized in ionic liquids can be reused.

Ionic liquid media also favor the alkylation of malonic esters and acetoacetic esters.[12]

Condensation reactions. Among this type of reactions that employ ionic liquids are oximation of cyclic ketones,[13] Knoevenagel reaction,[14] Emmons-Wadsworth reaction,[15] asymmetric aldol reaction,[16,17] and Baylis-Hillman reaction.[18] The Ce(III)-mediated formation of 4-hydroxytetrahydropyrans from homoallylic alcohols and aldehydes,[19] condensation of 2-aminopyridine with α-tosyloxy ketones,[20] radical annulation of 1,3-cyclohexanedione and analogues,[21] and von Pechmann condensation[22] have been studied in ionic liquids. The Pauson-Khand reaction proceeds rapidly in ionic liquids at room temperature.[23]

Additions. Halogenation of alkenes and alkynes,[24] dihydroxylation of alkenes,[25] and silylstannylation of alkynes[26] are well managed in ionic liquids, such protocols offer a great advantage in the latter two reactions because the valuable catalysts are easily recovered.

Pd-catalyzed hydroarylation of alkynes in ionic liquid is further extended to a synthesis of 3-arylquinolines by virtue of a tandem cyclization reaction.[27] Iodine atom-transfer reactions in ionic liquids leading to cyclic products are feasible.[28]

Reductions. Reduction of carbonyl compounds[29] with $NaBH_4$ and [bmim] salts containing small amounts of water are used to immobilize baker's yeast to carry out reduction of ketones.[30]

Coupling reactions. Heck and Suzuki coupling reactions are realizable in ionic liquids.[31,32] After the highly regioselective coupling of ArX (X=Br, I) with butyl vinyl ether, hydrolysis of the products affords aryl methyl ketones.[33] By Suzuki coupling under these conditions, organoboronic acids are converted into selenides,[34] while linking of the aryl iodide component to a polymer backbone does not show any adverse effect.[35]

2-Arylethyl ketones are obtained from Heck reaction of Baylis-Hillman adducts. A Pd-benzothiazole carbene complex serves as catalyst.[36]

OH R COOMe + Ar-Br → [N-methylbenzothiazolylidene Pd-I]$_2$, Bu_4NBr - HCOONa → R-CO-CH$_2$CH$_2$-Ar

Naturally, Stille coupling in room temperature ionic liquids is expected to proceed and has the same advantages of catalyst and solvent recycling. Noteworthy is the report[37] of catalyst selectivity, i.e., $(Ph_3P)_4Pd$ for ArBr and $(PhCN)_2PdCl_2$–Ph_3As–CuI for ArI.

Amide formation and reactions. Many ketoximes undergo Beckmann rearrangment in [bmim]BF_4. However, cyclopentanone oxime suffers hydrolysis only.[38] Bischler-Napieralski cyclization leading to 3,4-dihydroisoquinolines is achieved with amides of arylethylamines if the aromatic ring is strongly activated.[39]

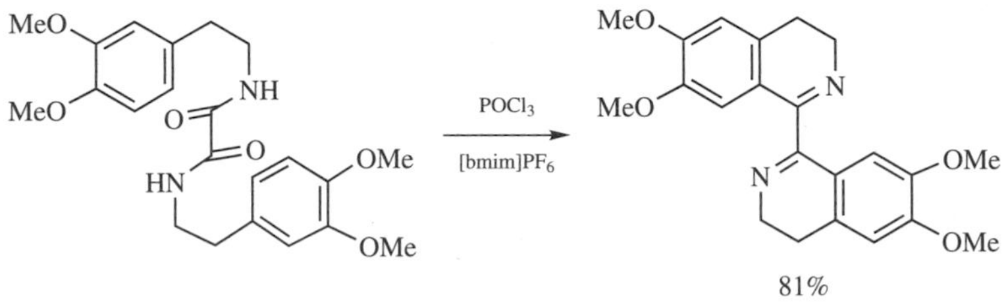

Diels-Alder reactions. Significant enhancement in rates and and selectivity in the reaction catalyzed by $Sc(OTf)_3$ and performed in ionic liquids is noted. The catalyst is also easily recycled.[40]

The Lewis acid-catalyzed Diels-Alder reaction of furan in ionic liquids is *endo*-selective,[41] contrary to that observed in conventional solvents.

Silylformylation. *syn*-Addition of formyl and silyl groups to the triple bond of 1-alkynes by Rh-catalyzed reaction with hydrosilanes and CO leads to α-alkyl-β-silylacroleins.[42] Reaction carried out in ionic liquids is also aimed at catalyst recovery.

Ring-closing metathesis. Ionic liquids are also suitable media for RCM reactions.[43]

[1]Leadbeater, N.E., Torenius, H.M. *JOC* **67**, 3145 (2002).
[2]Forsyth, S.A., MacFarlane, D.R., Thomson, R.J., von Itzstein, M. *CC* 714 (2002).
[3]Kim, K.-W., Song, B., Choi, M.-Y., Kim, M.-J. *OL* **3**, 1507 (2001).
[4]Park, S., Kazlauskas, R.J. *JOC* **66**, 8395 (2001).
[5]Sima, T., Guo, S., Shi, F., Deng, Y. *TL* **43**, 8145 (2002).
[6]Itoh, T., Akasaki, E., Kudo, K., Shirakami, S. *CL* 262 (2001).
[7]Green, L., Hemeon, I., Singer, R.D. *TL* **41**, 1343 (2000).
[8]Wheeler, C., West, K.N., Liotta, C.L., Eckert, C.A. *CC* 887 (2001).
[9]Kim, D.W., Song, C.E., Chi, D.Y. *JACS* **124**, 10278 (2002).
[10]Ren, R.X., Wu, J.X. *OL* **3**, 3727 (2001).
[11]Wu, J.X., Beck, B., Ren, R.X. *TL* **43**, 387 (2002).
[12]Kryshtal, G.V., Zhdankina, G.M., Zlotin, S.G. *MC* 57 (2002).
[13]Ren, R.X., Ou, W. *TL* **42**, 8445 (2001).
[14]Harjani, J.R., Nara, S.J., Salunkhe, M.M. *TL* **43**, 1127 (2002).
[15]Kryshtal, G.V., Zhdankina, G.M., Zlotin, S.G. *MC* 176 (2002).

[16]Kotrusz, P., Kmentova, I., Gotov, B., Toma, S., Solcaniova, E. *CC* 2510 (2002).
[17]Loh, T.-P., Feng, L.-C., Yang, H.-Y., Yang, J.-Y. *TL* **43**, 8741 (2002).
[18]Rosa, J.N., Afonso, C.A.M., Santos, A.G. *T* **57**, 4189 (2001).
[19]Keh, C.C.K., Namboodiri, V.V., Varma, R.S., Li, C.-J *TL* **43**, 4993 (2002).
[20]Xie, Y.-Y., Chen, Z.-C., Zheng, Q.-G. *S* 1505 (2002).
[21]Bar, G., Parsons, A.F., Thomas, C.B. *CC* 1350 (2001).
[22]Potdar, M.K., Mohile, S.S., Salunkhe, M.M. *TL* **42**, 9285 (2001).
[23]Becheanu, A., Laschat, S. *SL* 1865 (2002).
[24]Chiappe, C., Capraro, D., Conte, V., Pieraccini, D. *OL* **3**, 1061 (2001).
[25]Yanada, R., Takemoto, Y. *TL* **43**, 6849 (2002).
[26]Hemeon, I., Singer, R.D. *CC* 1884 (2002).
[27]Cacchi, S., Fabrizi, G., Goggiamani, A., Moreno-Manas, M., Vallribera, A. *TL* **43**, 5537 (2002).
[28]Yorimitsu, H., Oshima, K. *BCSJ* **75**, 853 (2002).
[29]Howarth, J., James, P., Ryan, R. *SC* **31**, 2935 (2001).
[30]Howarth, J., James, P., Dai, J. *TL* **42**, 7517 (2001).
[31]Hagiwara, H., Shimizu, Y., Hoshi, T., Suzuki, T., Ando, M., Ohkubo, K., Yokoyama, C. *TL* **42**, 4349 (2001).
[32]Matthews, C.J., Smith, P.J., Welton, T. *CC* 1249 (2000).
[33]Xu, L., Chen, W., Ross, J., Xiao, J. *OL* **3**, 295 (2001).
[34]Kabalka, G.W., Venkataiah, B. *TL* **43**, 3703 (2002).
[35]Revell, J.D., Ganesan, A. *OL* **4**, 3071 (2002).
[36]Calò, V., Nacci, A., Lopez, L., Napola, A. *TL* **42**, 4701 (2001).
[37]Handy, S.T., Zhang, X. *OL* **3**, 233 (2001).
[38]Peng, J., Deng, Y. *TL* **42**, 403 (2001).
[39]Judeh, Z.M.A., Ching, C.B., Bu, J., McCluskey, A. *TL* **43**, 5089 (2002).
[40]Song, C.E., Shim, W.H., Roh, E.J., Lee, S., Choi, J.H. *CC* 1122 (2001).
[41]Hemeon, I., DeAmicis, C., Jenkins, H., Scammells, P., Singer, R.D. *SL* 1815 (2002).
[42]Okazaki, H., Kawanami, Y., Yamamoto, K. *CL* 650 (2000).
[43]Buijsman, R.C., van Vuuren, E., Sterrenburg, J.G. *OL* **3**, 3785 (2001).

3-Butyl-*x*-methylthiazolium tetrafluoroborates.

Benzoin condensation. These ionic liquids (x=4, 5) promotes the condensation of ArCHO.[1]

[1]Davis, J.H., Forrester, K.J. *TL* **40**, 1621 (1999).

N-(*t*-Butyl) phenylsulfinimidoyl chloride. 21, 85

Oxidation of alcohols. A polymer-supported version of the reagent has been prepared and used in oxidation.[1] Another procedure of alcohol oxidation specifies the addition of zinc oxide.[2] Addition of NCS allows the catalytic use of the title reagent.[3]

Imines.[4] Various secondary amines are transformed into imines with *t*-BuN═S(Cl)Ph and DBU at –78°.

[1]Matsuo, J., Kawana, A., Pudhom, K., Mukaiyama, T. *CL* 250 (2002).
[2]Matsuo, J., Kitagawa, H., Iida, D., Mukaiyama, T. *CL* 150 (2001).
[3]Mukaiyama, T., Matsuo, J., Iida, D., Kitagawa, H. *CL* 846 (2001).
[4]Mukaiyama, T., Kawana, A., Fukuda, Y., Matsuo, J., *CL* 390 (2001).

(*N′*-*t*-Butylphosphinimylidyne)tris(*N,N,N′,N′,N″,N″*-hexamethyl)phosphorimidic triamide.

Phenanthrene. The title compound, known in an abbreviated name of P_4-t-Bu, promotes condensation of 2′-methylbiphenyl-2-carbaldehydes to afford phenanthrene derivatives.[1]

MeO OMe CHO OMe → (t-Bu-N=P[N=P(NMe₂)₃]₃) → MeO OMe OMe 62%

[1]Kraus, G.A., Zhang, N., Melekhov, A., Jensen, J.H. *SL* 521 (2001).

1-Butylpyridinium chloroaluminate.

Condensation reactions. The title compound is an ionic liquid that serves in the von Pechmann condensation[1] and Fischer indole synthesis.[2] However, the reported reaction conditions are quite harsh (at 130° and 180°, respectively).

[1]Khandekar, A.C., Khadikar, B.M. *SL* 152 (2002).
[2]Robeiro, G.L., Khadikar, B.M. *S* 370 (2001).

Butyltin trichloride.

Propargylation.[1] A highly diastereoselective reaction between α-alkoxy-propargylstannanes and aldehydes is catalyzed by $BuSnCl_3$.

Bu_3Sn SiMe₃ OMe + CHO → ($BuSnCl_3$, CH_2Cl_2, 0°) → OH SiMe₃ OMe 96%

[1]Savall, B.M., Powell, N.A., Roush, W.R. *OL* **3**, 3057 (2001).

C

Cadmium perchlorate.

Allylation.[1] The cadmium-catalyzed reaction of carbonyl compounds is ligand-accelerated, e.g., by methylbis(2-dimethylaminoethyl)amine.

[1]Kobayashi, S., Aoyama, N., Manabe, K. *SL* 483 (2002).

Calcium chloride.

Aldol reaction.[1] Used as a Lewis base $CaCl_2$ is useful for promoting aldol reactions of dimethylsilyl enol ethers in aqueous DMF.

3,4-Dihydro-2H-benzopyrans.[2] Phenols condense with conjugated carbon compounds to afford the benzopyrans in the presence of KOH and $CaCl_2$.

$CaCl_2 \cdot 2H_2O$

KOH / MeOH

52%

[1]Miura, K., Nakagawa, T., Hosomi, A. *JACS* **124**, 536 (2002).
[2]Saimoto, H., Ogo, Y., Komoto, M., Morimoto, M., Shigemasa, Y. *H* **55**, 2052 (2001).

Calcium fluoride.

Fluorination.[1] A porous modification of CaF_2 is obtained by treating sodalime with anhydrous HF. The reagent can be used as a fluorinating agent and as a support in heterogeneous hydrogenation.

[1]Quan, H.-D., Tamura, M., Gai, R.-X., Sekiya, A. *JFC* **116**, 65 (2002).

Carbon dioxide.

Carbonates, carbamates and ureas. Heating primary aliphatic amines and dimethyl carbonate in supercritical CO_2 gives methyl carbamates.[1] Benzyl carbamates are obtained in a reaction of primary amines with benzyl halides and DBU under carbon dioxide,[2] whereas other carbamates are similarly prepared, only by adding alkyl halides after the formation of the carbamate salts.[3]

Fiesers' Reagents for Organic Synthesis, Volume 22. Series editor Tse-Lok Ho
ISBN 0-471-28515-3

In solution or solid-phase, formation of carbamates and carbonates from CO_2 is achieved in the presence of Cs_2CO_3 and a phase-transfer catalyst.[4] Electrolytically generated tetraethyl carbonate is readily alkylated to furnish symmetrical carbonates.[5]

Five-membered cyclic carbamates and carbonates are readily obtained by adopting some of the methods listed above, e.g., electrolytically,[6,7] and in the presence of Bu_2SnO.[8] Insertion of CO_2 into epoxides is efficiently catalyzed by Cr(III)-salen complexes[9] or in molten Bu_4NBr–Bu_4NI which serves as both catalyst and reaction medium.[10] The formation of 1*H*-quinazoline-2,4-diones[11] from *o*-aminobenzonitriles and CO_2 probably involves an isomerization step from iminoanhydride intermediates.

Condnesation of secondary amines and CO_2 (1 atm) to generate tetraalkylureas starts from reaction catalyzed by 1,8-bis(methylamino)naphthalene and by CCl_4 in the second step.[12] A single-step reaction at high CO_2-pressure (low R_2NH concentrations) does not produce urea products.

Homologation. Under electrolytic conditions 1-alkynes can be converted into 2-alkynoic acids.[13] Hydroformylation of alkenes with carbonyl-ligated Rh catalysts can be employed in supercritical CO_2.[14]

Coupling reactions. Supercritical CO_2 suppresses isomerization of the exocyclic double bond of benzodihydropyrans that are initially formed in an intramolecular Heck reaction.[15] In other solvents, migration of the double bond into the ring is serious.

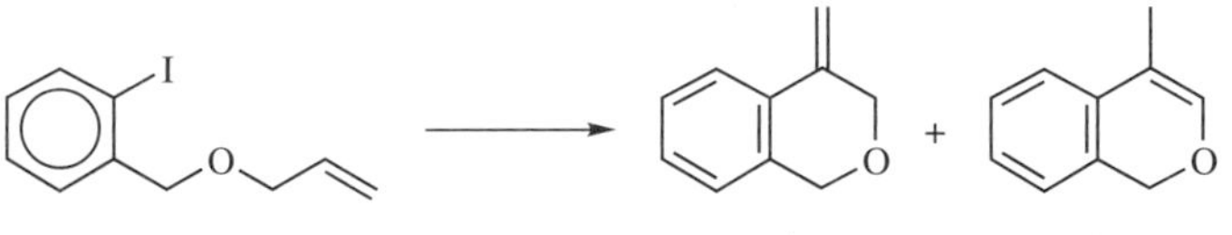

(major) (minor)

Hydroacylation of quinones and enones with aldehydes to give 2-acylhydroquinones and 1,4-diketones, respectively, is accomplished photochemically in supercritical CO_2.[16]

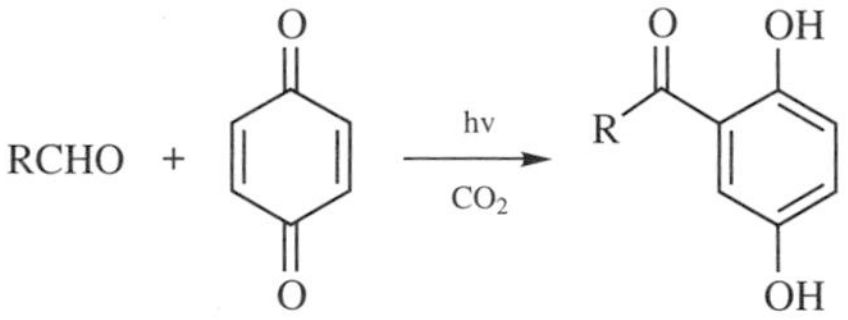

RCHO + [quinone] —hν, CO_2→ [2-acylhydroquinone]

Aldol and Mannich reactions.[17] The $Ln(OTf)_3$-catalyzed reactions in supercritical CO_2 are accelerated by poly(ethylene glycol) which form emulsions.

[1]Selva, M., Tundo, P., Perosa, A. *TL* **43**, 1217 (2002).
[2]Shi, M., Shen, Y.-M. *HCA* **84**, 3357 (2001).
[3]Perez, E.R., da Silva M.O., Costa, V.C., Rodrigues-Filho, U.P., Franco, D.W. *TL* **43**, 4091 (2002).
[4]Salvatore, R.N., Chu, F., Nagle, A.S., Kapxhiu, E.A., Cross, R.M., Jung, K.W. *T* **58**, 3329 (2002).
[5]Mucciante, V., Rossi, L., Feroci, M., Sotgiu, G. *SC* **32**, 1205 (2002).

[6]Casadei, M.A., Feroci, M., Inesi, A., Rossi, L., Sotgiu, G. *JOC* **65**, 4759 (2000).
[7]Feroci, M., Gennaro, A., Inesi, A., Orsini, M., Palombi, L. *TL* **43**, 5863 (2002).
[8]Tominaga, K., Sasaki, Y. *SL* 307 (2002).
[9]Paddock, R.L., Nguyen, S. *JACS* **123**, 11498 (2001).
[10]Calo, V., Nacci, A., Monopoli, A., Fanizzi, A. *OL* **4**, 2561 (2002).
[11]Mizuno, T., Ishino, Y. *T* **58**, 3155 (2002).
[12]Tai, C.-C., Huck, M.J., McKoon, E.P., Woo, T., Jessop, P.G. *JOC* **67**, 9070 (2002).
[13]Koster, F., Dinjus, E., Dunach, E. *EJOC* 2507 (2001).
[14]Hu, Y., Chen, W., Xu, L., Xiao, J. *OM* **20**, 3206 (2001).
[15]Shezad, N., Clifford, A.A., Rayner, C.M. *TL* **42**, 323 (2001).
[16]Pacut, R., Grimm, M.L., Kraus, G.A., Tanko, J.M. *TL* **42**, 1415 (2001).
[17]Komoto, I., Kobayashi, S. *CC* 1842 (2001).

Carbon monoxide. 20, 72, 21, 89

Friedel-Crafts reactions. Carbon monoxide under pressure serves as a Lewis acid for benzylation (to give diarylmethanes) and acylation of arenes.[1] A mixture of CO and $Fe(CO)_5$ is marginally active (yields of ca. 10% in comparison to 95% with CO alone) and as control experiments show nitrogen and carbon dioxide are not effective at all.

Coupling reaction. Silylformylation followed by a Wittig reaction accomplishes a diene synthesis from 1-alkynes.[2]

Et_3SiH + R-C≡CH + Ph_3P=CH-COOEt —[Rh] / CO, CH_2Cl_2→ (R)(SiEt₃)-diene-COOEt

[1]Ogoshi, S., Nakashima, H., Shimonaka, K., Kurosawa, H. *JACS* **123**, 8626 (2001).
[2]Eilbracht, P., Hollmann, C., Schmidt, A.M., Barfackar, L. *EJOC* 1131 (2000).

Carbon disulfide.

Thiophenes. Quenching (Z,Z)-1,4-dilithiodienes with CS_2 at room temperature leads to thiophenes.[1] However, low temperature (−78°) reaction that is followed by addition of MeI gives 5,5-bismethylthiocyclopentadienes.

R₄-1,4-dilithiobutadiene + CS_2 —−78° ; MeI→ 5,5-bis(SMe)-tetra-R-cyclopentadiene
—25°→ tetra-R-thiophene, 52–68%

[1]Chen, J., Song, Q., Xi, Z. *TL* **43**, 3533 (2002).

Carbon tetrahalide. 20, 72–73

Deprotection. Several common *O*-protection groups (methoxyethoxymethyl,[1] 2-(trimethylsilyl)ethoxymethyl,[2] primary *t*-butyldimethylsilyl[3]) are removed on treatment with CBr_4 in an alcohol.

Ramberg-Bücklund rearrangement. A new route to *exo*-glycals takes advantage of the rearrangement that is effected in CCl_4, and the ready availability of the substrates.[4]

BnO OBn BnO O BnO SO_2Me → CCl_4 - KOH / H_2O - *t*-BuOH → BnO OBn BnO O BnO

72%

[1]Lee, A.S.-Y., Hu, Y.-J., Chu, S.-F. *T* **57**, 2121 (2001).
[2]Chen, M.-Y., Lee, A.S.-Y. *JOC* **67**, 1384 (2002).
[3]Chen, M.-Y., Lu, K.-C., Lee, A.S.-Y., Lin, C.-C. *TL* **43**, 2777 (2002).
[4]Griffin, F.K., Paterson, D.E., Murphy, P.V., Taylor, R.J.K. *EJOC* 1305 (2002).

Carbonyl(chloro)bis(triphenylphosphine)iridium.

Cyclopentadienones.[1] Carbonylative coupling of two alkyne units leads to the cyclic dienones.

EtOOC EtOOC Ph Ph → CO / $(Ph_3P)_2Ir(CO)Cl$ / xylene 120° → EtOOC EtOOC Ph O Ph

86%

[1]Shibata, T., Yamashita, K., Ishida, H., Takagi, K. *OL* **3**, 1217 (2001).

Carbonyl(chloro)hydridobis(triisopropylphosphine)osmium.

Alkyne dimerization.[1] The more unusual aspect of the dimerization is the formation of 1,2,3-alkatrienes.

R—≡ → $(i\text{-}Pr_3P)_2OsH(CO)Cl$ → RCH=C=C=CHR (R = t-Bu, Cy) → (R = $SiMe_3$) → Me_3Si Me_3Si

[1]Esteruelas, M.A., Herrero, J., Lopez, A.M., Olivan, M. *OM* **20**, 3202 (2001).

Carbonyl difluoride.

Alkyl fluorides.[1] A preparation of COF_2 is from $CO(OCCl_3)_2$ and KF/18-c-6 in MeCN. Fluoroformates derived from COF_2 and alcohols decompose to afford alkyl fluorides on heating with hexabutylguanidinium fluoride.

[1]Flosser, D.A., Olofson, R.A. *TL* **43**, 4275 (2002).

Carbonylchlorohydridobis(tricyclohexylphosphine)ruthenium–fluoroboric acid etherate.

Hydrovinylation.[1] A ruthenium hydride complex generated in situ promotes hydrovinylation (9 examples, 55–98%).

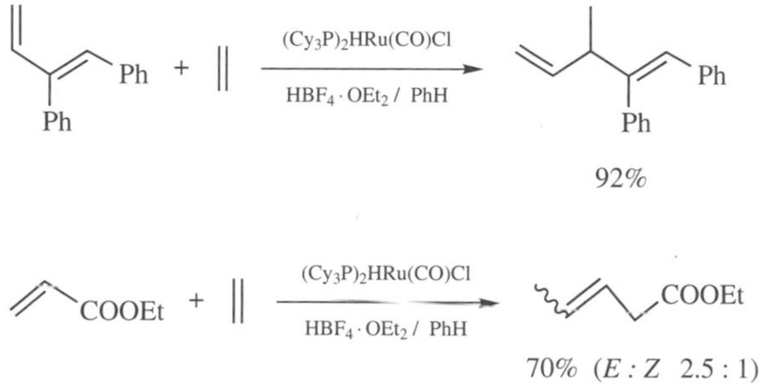

$(Cy_3P)_2HRu(CO)Cl$, $HBF_4 \cdot OEt_2$ / PhH

92%

$(Cy_3P)_2HRu(CO)Cl$, $HBF_4 \cdot OEt_2$ / PhH

COOEt + → COOEt

70% ($E : Z$ 2.5 : 1)

[1]Yi, C.S., He, Z., Lee, D.W. *OM* **20**, 802 (2001).

Carbonyldihydridotris(triphenylphosphine)ruthenium. 19, 65–66; **21,** 89–90

Vinylogous hydroformylation.[1] β-Heterosubstituted enals including 3-formylindole and 3-formylthiophene split their β-H bond and add to alkenes when catalyzed by the Ru complex.

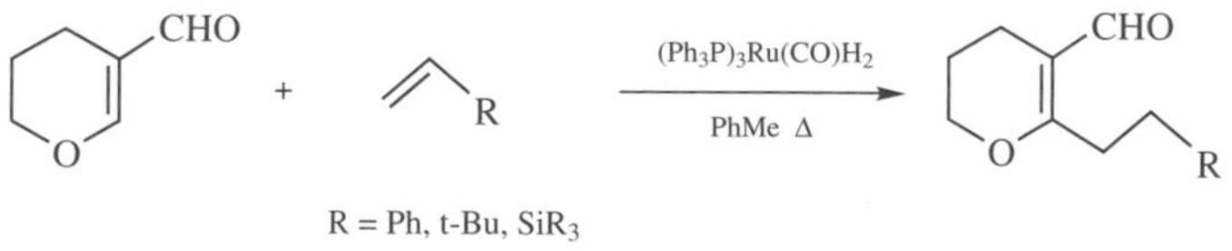

CHO + R → $(Ph_3P)_3Ru(CO)H_2$, PhMe Δ → CHO, R

R = Ph, t-Bu, SiR_3

[1]Kakiuchi, F., Sato, T., Igi, K., Chatani, N., Murai, S. *CL* 386 (2001).

Carbonylhydridotris(triphenylphosphine)rhodium.

Alkylation.[1] Hydroformylation catalyzed by the Rh complex, when followed with Knoevenagel condensation and hydrogenation, constitutes a carbon chain elongation method of considerable utility.

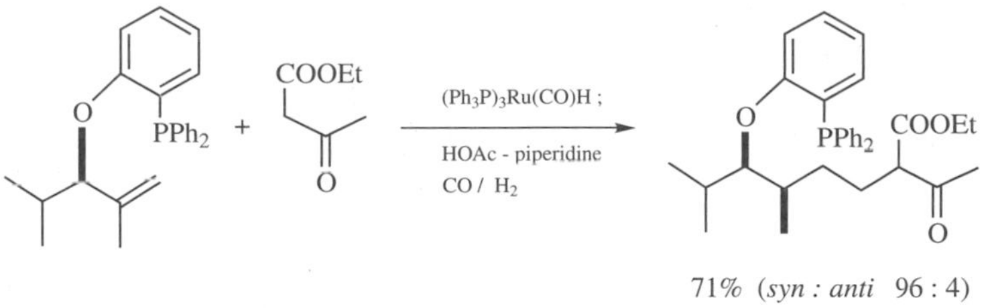
COOEt
PPh2
(Ph3P)3Ru(CO)H ;
HOAc - piperidine
CO / H2
PPh2
COOEt
71% (syn : anti 96 : 4)

[1]Breit, B., Zahn, S.K. *ACIEE* **40**, 1910 (2001).

1,1′-Carbonyldiimidazole. 13, 66; **16,** 64; **18,** 85; **20,** 73, 21, 90

Carbamates. A previous report on the direct preparation of *N*-alkyl- and *N*-arylimidazoles from the title compound and alcohols and phenols is erroneous. Instead, the products are the carbamates.[1] Mixed carbamates are readily formed by consecutive treatment of the reagent with alcohols and amines.[2]

[1]Fischer, W. *S* 29 (2002).
[2]D'Addona, D., Bochet, C.G. *TL* **42**, 5227 (2001).

2-Carboxamido-3-aryloxaziridines.

Ureas.[1] Attack of *C*-nucleophiles on the reagents afford unsymmetrical ureas in moderate yields.

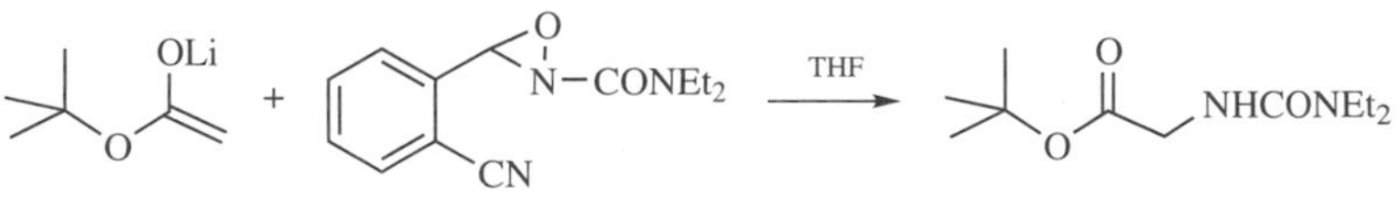
OLi
N−CONEt2
CN
THF
NHCONEt2

[1]Armstrong, A., Atkin, M.A., Swallow, S. *TL* **41**, 2247 (2000).

Cedranediolborane.

Arylboronic acids.[1] This borane is useful in a Suzuki coupling for transforming ArI into $ArB(OH)_2$.

[1]Song, Y.-L., Morin, C. *SL* 266 (2001).

Cerium(IV) ammonium nitrate. 13, 67–68; **14,** 74–75; **15,** 70–72; **16,** 66; **17,** 68; **18,** 85–87; **19,** 67–69; **20,** 73–75; **21,** 90–92

Deprotection. 2-Methoxyethoxymethyl ethers are converted to acetoxymethyl ethers by CAN in acetic anhydride, and the products are readily methanolyzed to give alcohols.[1] Cleavage of *p*-anisyloxymethyl ethers is performed with CAN and

2,6-pyridinedicarboxylic acid in aq. MeCN.[2] Selective debenzylation of tertiary amines is particularly interesting.[3]

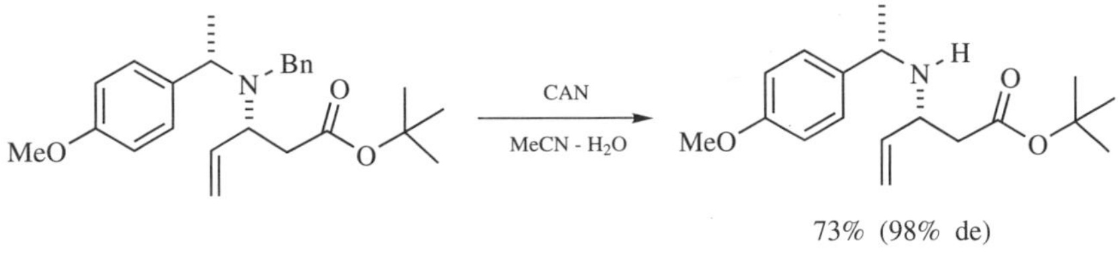

CAN
MeCN - H_2O

73% (98% de)

CAN also mediates esterification and transesterification.[4]

Alkene functionalization. Styrenes are converted to 1-aryl-1,2-bisselenocyanatoalkanes by reaction with KSeCN and CAN under argon. The presence of oxygen diverts the reaction toward formation of α-aroyl selenocyanates.[5]

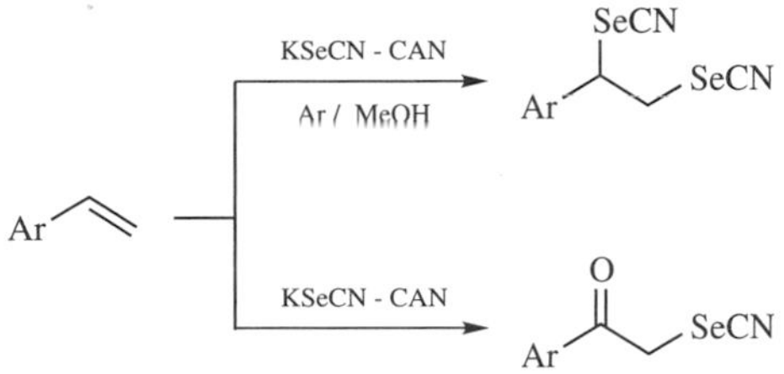

KSeCN - CAN
Ar / MeOH

KSeCN - CAN

Cinnamic esters undergo addition to afford α-bromo-β-propargyloxy adducts on admixture with CAN, LiBr, and propargyl alcohol.[6] Cinnamic acids are converted to the β-bromostyrenes.[7] In the presence of CAN oxidative sulfonylation of alkenes by sodium arenesulfonates is accomplished.[8]

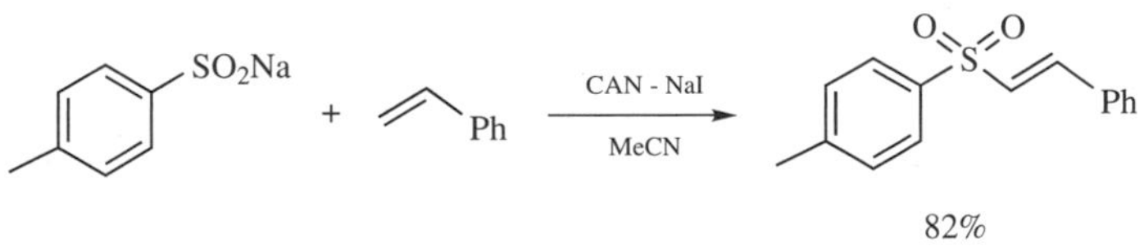

CAN - NaI
MeCN

82%

Nitration. Some coumarins are nitrated with CAN in HOAc or in conjuction with 30% H_2O_2.[9] Formation of 3-acylisoxazoles/3-acylisoxazolines from 1-alkynes/1-alkenes and methyl ketones is attributable to a reaction sequence of α-nitration, dehydration to acylnitrile oxides, and 1,3-dipolar cycloaddition.[10]

The minimal structural requirement for aromatic ipso-substitution with cleavage of C-C bond a diarylmethane unit by CAN has been determined.[11]

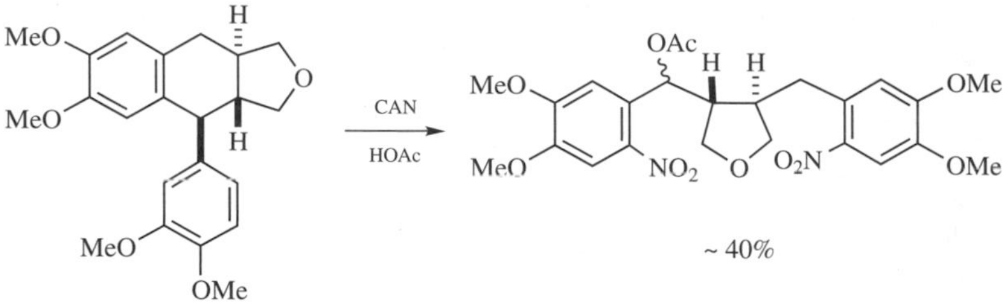

CAN, HOAc

~ 40%

Heterocycles. Condensation of arenenitriles with arenethiols on exposure to CAN to furnish 2-arylbenzothiazoles (10 examples, 78–96%).[12]

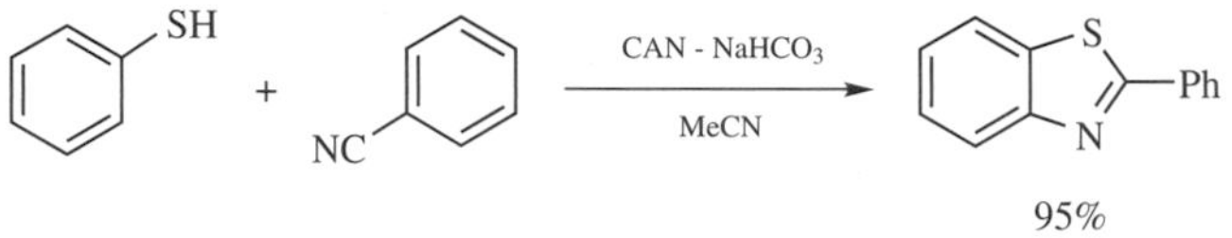

CAN - $NaHCO_3$, MeCN

95%

Oxidative dimerization. The penultimate step of a hypocarpone synthesis is oxidative dimerization of a 2-hydroxy-1,4-naphthoquinone. Apparently after much experimentation CAN was selected as the optimal single-electron transfer inducer.[13]

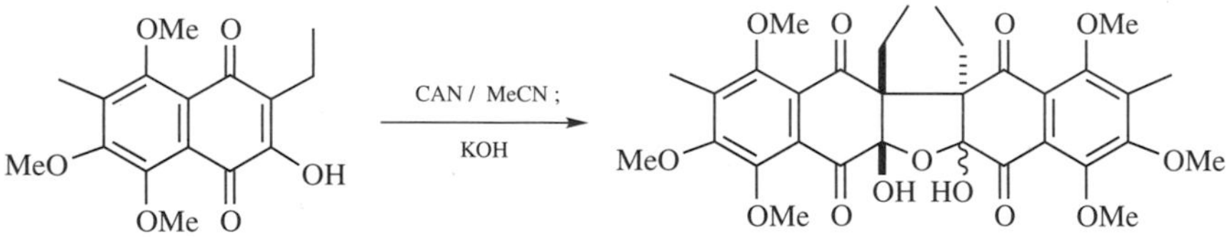

CAN / MeCN ; KOH

(*cis-anti-cis* : *cis-syn-trans* = 3 : 2)

[1]Tanemura, K., Suzuki, T., Nishida, Y., Satsumabayashi, K., Horaguchi, T. *CL* 1012 (2001).
[2]Clive, D.L.J., Sun, S. *TL* **42**, 6267 (2001).
[3]Bull, S.D., Davies, S.G., Kelly, P.M., Gianotti, M., Smith, A.D. *JCS(P1)* 3106 (2001).
[4]Stefane, B., Kocevar, M., Polanc, S. *SC* **32**, 1703 (2002).
[5]Nair, V., Augustine, A., George, T.G. *EJOC* 2363 (2002).
[6]Roy, S.C., Guin, C., Rana, K.K., Maiti, G. *SL* 226 (2001).
[7]Roy, S.C., Guin, C., Maiti, G. *TL* **42**, 9253 (2001).
[8]Nair, V., Augustine, A., George, T.G., Nair, L.G. *TL* **42**, 6763 (2001).
[9]Ganguly, N., Sukai, A.K., De, S. *SC* **31**, 301 (2001).
[10]Itoh, K., Takahashi, S., Ueki, T., Sugiyama, T., Takahashi, T.T., Horiuchi, C.A. *TL* **43**, 7035 (2002).

[11]Asghedom, H., LaLonde, R.T., Ramdayal, F. *TL* **43**, 3989 (2002).
[12]Tale, R.H. *OL* **4**, 1641 (2002).
[13]Nicolaou, K.C., Gray, D. *ACIEE* **40**, 761 (2001).

Cerium(III) chloride heptahydrate. 14, 75–77; **15,** 72–73; **16,** 67–68; **18,** 87; **20,** 75; **21,** 92–93

Protection and deprotection.[1] Chemoselective cleavage of *t*-butyldimethylsilyl ethers of nonaromatic alcohols is achieved on treatment with the cerium salt (hydrate) in refluxing MeCN. The method is complementary to that using LiOH in DMF which shows the opposite chemoselectivity.

Hydrolysis of acetonides in the presence of trityl ether, TBS ether, triflate, benzoate, acetate is accomplished with a combination of oxalic acid and $CeCl_3 \cdot 7H_2O$. Selectivity is also achieved among different types of acetonides.[2]

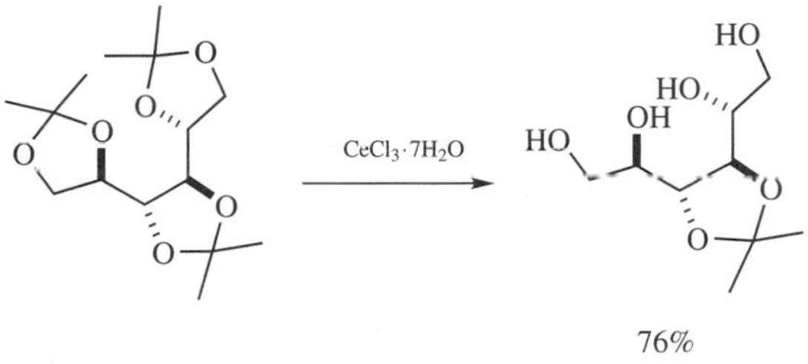

$CeCl_3 \cdot 7H_2O$

76%

Monoacetylation of *meso*- and C_2-symmetric 1,3-diols and 1,4-diols with Ac_2O is possible in the presence of $CeCl_3$.[3]

Reduction.[4] Prior complexation with $CeCl_3$ influences the stereoselectivity of α-cyano ketone reduction by $LiBH_4$, the course being opposite to that using borane-pyridine as reducing agent and involving the addition of $TiCl_4$.

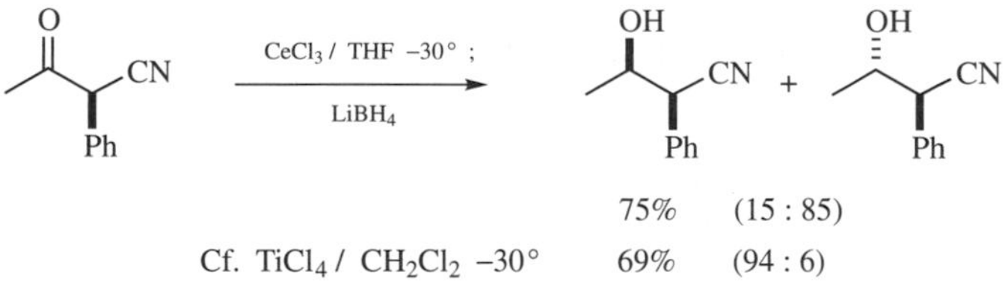

$CeCl_3$ / THF −30° ; $LiBH_4$

75% (15 : 85)

Cf. $TiCl_4$ / CH_2Cl_2 −30° 69% (94 : 6)

[1]Ankala, S.V., Fenteany, G. *TL* **43**, 4729 (2002).
[2]Xiao, X., Bai, D. *SL* 535 (2001).
[3]Clarke, P.A. *TL* **43**, 4761 (2002).
[4]Dalpozzo, R., Bartoli, G., Bosco, M., De Nino, A., Procopio, A., Sambri, L., Tagarelli, A. *EJOC* 2971 (2001).

Cerium(III) chloride heptahydrate–sodium iodide. 21, 93

Deprotection. The combined reagents direct selective cleavage of a *t*-butyl ester in the presence of an *N*-Boc group (11 examples, 75–99%).[1] Release of alcohols from allyl ethers with the same reagents in the presence of 1,3-propanedithiol is also reported.[2]

β-Amino ketones. Enones are obtained on treating β-hydroxy ketones with the reagent pair.[3] Interestingly, addition of amines to enones is promoted by $CeCl_3 \cdot 7H_2O$–NaI supported on silica gel.[4]

[1]Marcantoni, E., Massaccesi, M., Torregiani, E., Bartoli, G., Bosco, M., Sambri, L. *JOC* **66**, 4430 (2001).
[2]Bartoli, G., Cupone, G., Dalpozzo, R., De Nino, A., Marcantoni, E., Procopio, A. *SL* 1897 (2001).
[3]Bartoli, G., Bosco, M., Dalpozzo, R., Giuliani, A., Marcantoni, E., Mecozzi, T., Sambri, L., Torregiani, E. *JOC* **67**, 9111 (2002).
[4]Bartoli, G., Bosco, M., Marcantoni, E., Petrini, M., Sambri, L., Torregiani, E. *JOC* **66**, 9052 (2001).

Cerium(III) alkoxide.

Pinacol coupling.[1] Aldehydes selectively couple in a *syn*-selective fashion (*anti:syn*~ 15–20:85–80) when subjected to treatment with Mn and Me_3SiCl in THF in the presence of catalytic cerium(III) isopropoxide or *t*-butoxide. Note the previously described *anti*-selective reaction using diethylzinc-cerium(III) isopropoxide.

[1]Groth, U., Jeske, M. *SL* 129 (2001).

Cerium(IV) triflate. 20, 75; **21,** 94

Deprotection. Both trityl ethers[1] and *t*-butyldimethylsilyl ethers[2] are cleaved by $Ce(OTf)_4$.

Benzylic oxidation.[3] Aromatic ketones are obtained from alkylarenes on treatment with $Ce(OTf)_4$. The way the oxidant is prepared markedly affect the oxidation ability, as water is required to activate the anhydrous salt.

Epoxide opening.[4] Sodium carboxylates and phenolates open epoxides at room temperature when catalyzed by $Ce(OTf)_4$ in micellar media.

[1]Khalafi-Nezhad, A., Alamdari, R.F. *T* **57**, 6805 (2001).
[2]Bartoli, G., Cupone, G., Dalpozzo, R., De Nino, A., Maiuolo, L., Procopio, A., Sambri, L., Tagarelli, A. *TL* **43**, 5945 (2002).
[3]Laali, K.K., Herbert, M., Cushnyr, B., Bhatt, A., Terrano, D. *JCS(P1)* 578 (2001).
[4]Iranpoor, N., Firouzabadi, H., Safavi, A., Shekarriz, M. *SC* **32**, 2287 (2002).

Cesium carbonate–tetrabutylammonium iodide.

Derivatization of alcohols and amines. The base system is useful for converting alcohols to xanthates and amines to thiocarbamates by reaction with CS_2 and alkyl halides.[1] *N*-Alkyl carbamates are also readily prepared by sequential carboxylation with CO_2 and then double alkylation (at oxygen and nitrogen).[2]

O_2N — NH_2 → CO_2 / Cs_2CO_3 - Bu_4NI / DMF ; RBr - CO_2 ; Cs_2CO_3 - BnBr / DMF → O_2N — N(Bn)C(O)OR

The base is very beneficial to monoalkylation of primary amines, as it tends to suppress overalkylation.[3]

[1]Salvatore, R.N., Sahab, S., Jung, K.W. *TL* **42**, 2055 (2001).
[2]Salvatore, R.N., Ledger, J.A., Jung, K.W. *TL* **42**, 6023 (2001).
[3]Salvatore, R.N., Nagle, A.S., Jung, K.W. *JOC* **67**, 674 (2002).

Cesium fluoride. 13, 68; **14,** 79; **15,** 75–76; **16,** 69–70; **17,** 68; **18,** 88–89; **19,** 70–72; **20,** 77–78; **21,** 95–96

Benzannulated heterocyles.[1] Benzodiazepines and benzodiazocines are formed on treating *o*-trimethylsilylaryl triflates with CsF in the presence of a cyclic urea. The annulation step involves insertion into an amide bond of the cyclic urea by benzyne.

OTf, $SiMe_3$, OMe + O=(N,N) → CsF → OMe, N, N, O; 77%

Desilylation. While the ability of CsF to remove *O*-silyl groups is well established, its use in a tandem reaction leading to macrocyclic amino ketones is noteworthy.[2] Thus, a CsF-induced (with $PhCH_2NEt_3Cl$ also present) opening of a cyclopropyl silyl ether to unveil an enone system precedes an intramolecular Michael reaction.

NHBn, N, TBSO, MeOOC, COOMe, COOMe → CsF; Et_3NBnCl; DMF 90° → N, N, Bn, O, MeOOC, COOMe, COOMe; 49%

Ethers and esters. The solid base CsF-celite is effective to promote etherification and acylation of phenols, even the hindered 2,6-di-*t*-butylphenol.[3]

[1]Yoshida, H., Sirakawa, E., Honda, Y., Hiyama, T. *ACIEE* **41**, 3247 (2002).
[2]Patra, P.K., Reissig, H.-U. *SL* 33 (2001).
[3]Shah, S.T.A., Khan, K.M., Heinrich, A.M., Choudhary, M.I., Voelter, W. *TL* **43**, 8603 (2002).

Cesium propionate.

Epoxide inversion. Cleavage of epoxides by treatment with cesium propionate followed by mesylation and methanolysis accomplishes a configuration inversion.[1]

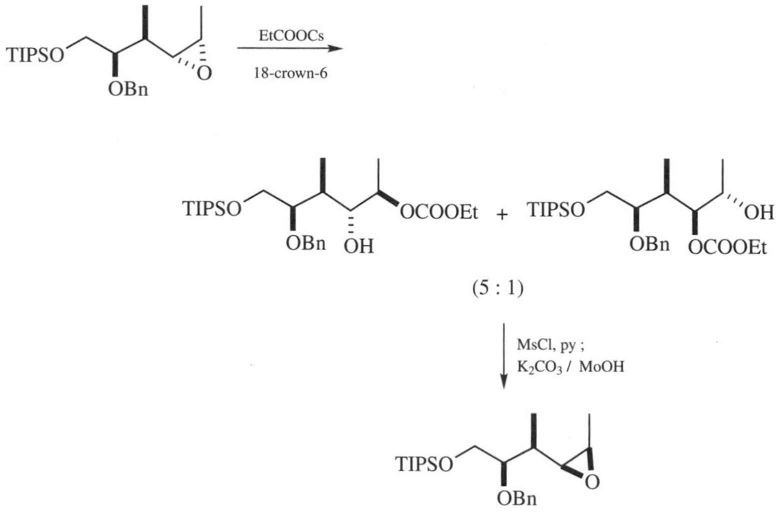

TIPSO OBn O — EtCOOCs, 18-crown-6 → TIPSO OBn OH OCOOEt + TIPSO OBn OCOOEt OH (5 : 1) — MsCl, py ; K_2CO_3 / MoOH → TIPSO OBn O

[1]Arbelo, D.O., Prieto, J.A. *TL* **43**, 4111 (2002).

Chiral auxiliaries and catalysts. 18, 89–97; **19,** 72–93; **20,** 78–103; **21,** 97–125

Reagent preparations. A practical synthesis of Koga chiral bases from (*R*)-styrene oxide and piperidine has been described.[1] An *N*-acyl group can be introduced to chiral oxazolidin-2-ones with the aid of an electrogenerated base.[2]

An improved method enables the access of Jacobsen's catalyst in both enantiomeric forms in 80–85% overall yield.[3] The C_2-symmetrical binaphthyl ketone **1** may be obtained via enzymatic resolution of the racemic alcohol followed by oxidation.[4]

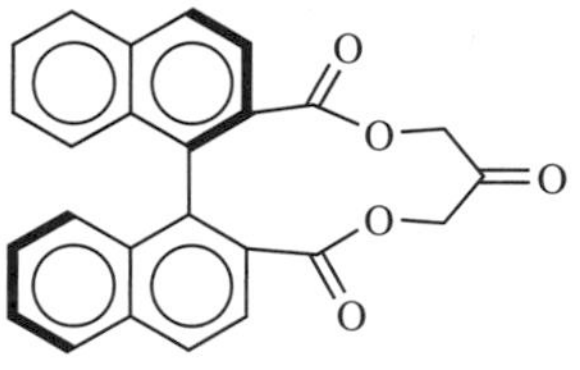

1

A chirally flexible biphenylphosphine ligand can be transformed into chiral metal complexes by using racemic Ru complexes to coordinate with enantiopure 3,3′-

dimethyldiaminobinaphthyl. The uncoordinated enantiomeric Ru-complex is epimerized to complete the conversion by the chiral additive.[5]

An excellent and comprehensive review of the tartaric acid-derived TADDOLs and analogues as chiral auxiliaries has appeared.[6] Evaluation of cinchona alkaloids and derivatives as asymmetric catalysts and ligands[7] and an article on more diverse and selected (but not too extensive) topics[8] are also available.

[1]Curthbertson, E., O'Brien, P., Towers, T.D. *S* 693 (2001).
[2]Feroci, M., Inesi, A., Palombi, L., Rossi, L., Sotgin, G. *JOC* **66**, 6185 (2001).
[3]Cepanec, I., Mikuldas, H., Vinkovi, V. *SC* **31**, 2913 (2001).
[4]Furutani, T., Hatsuda, M., Shimizu, T., Seki, M. *BBB* **65**, 180 (2001).
[5]Mikami, K., Aikawa, K., Korenaga, T. *OL* **3**, 243 (2001).
[6]Seebach, D., Beck, A.K., Heck, A. *ACIEE* **40**, 92 (2001).
[7]Kacprzak, K., Gawronski, J. *S* 961 (2001).
[8]Dalko, P.I., Moisan, L. *ACIEE* **40**, 3727 (2001).

Kinetic resolutions. Kinetic resolution of secondary alcohols by Pd(II)-catalyzed oxidation with oxygen in the presence of (−)-sparteine works very well.[1,2] Methods based on acylation employing several different ligands have also been developed, including isobutyrylation of allylic alcohols which is catalyzed by a bicyclic phosphine,[3] and benzoylation of either secondary alcohols[4] or *meso*-1,3-diols[5] involving participation of proline-based diamines.

Analogously, secondary amines are resolved by transacylation from a substituted oxazol-5-yl ester in the presence of a planar chiral ferrocene-fused DMAP,[6] and from a chiral amide[7] or polymer-supported succinimidoxy ester,[8] the last case for synthesis of Mosher amides that requires no purification.

Using (*R*)-phenylglycinol to condense with γ-substituted δ-oxo esters to provide separable chiral lactams,[9] useful precursors of enantiopure piperidine derivatives are obtained. A method for resolution of α-hydroxy carboxylic acids[10] and α-aryl α-amino acids[11] is via alcoholysis of the derived 1,3-dioxolane-2,4-diones and 1,3-oxazolidine-2,5-diones, respectively, in the presence of modified cinchona alkaloid catalysts.

Chiral amidophosphites such as **2** are effective for mediating the kinetic resolution of 2-cyclohexenones which carry a substituent at C-4 or C-5, by the $Cu(OTf)_2$-catalyzed addition of an organozinc.[12] Opening of allylic epoxides with organozinc reagents provides isomers having the secondary hydroxy group in opposite configuration.[13]

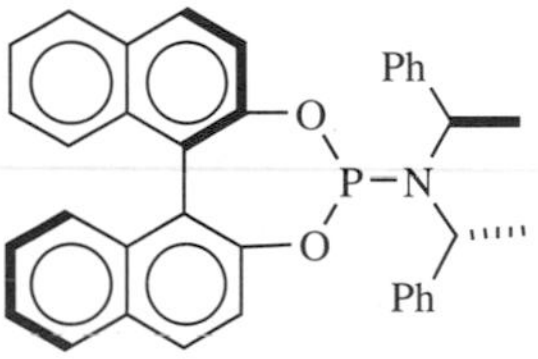

Ph
O
P–N
O
Ph

2

A method for resolution of racemic conjugated lactones involves conjugate addition of allyllithium reagents in the presence of (−)-sparteine.[14]

Rather unusual is a report on the kinetic resolution of allenylzinc species. Using α-*t*-butydimethylsiloxyphenylacetaldehyde benzylimine to react with the matched reagent, its enantiomer is left behind to be used in the desired asymmetric synthesis.[15]

A remarkably selective Oppenauer oxidation of secondary benzyl alcohols that leaves one type of enantiomers behind involves Ru and Os complexes with (*1R,2S*)-*cis*-1-amino-2-indanol.[16] Note that (*R*)-alcohols are obtained from the Ru-catalyzed reaction, whereas the (*S*)-alcohols from the Os-catalyzed reaction, the chiral ligand is the same!

Secondary alcohols are resolved by enzyme-catalyzed acetylation in the presence of a Ru complex derived from $Ru_3(CO)_{12}$ and 5-isopropylimino-1,2,3,4-tetraphenylcyclopentadiene.[17]

For dynamic kinetic resolution of *N*-acyl hemiaminals by pivaloylation an adequate resolving agent is the axially chiral twisted 3-pivaloyl-4-*t*-butylthiazolidine-2-thione.[18] Interestingly, oppositely enantiomeric pivalates predominate in the presence or absence of DMAP.

[1]Jensen, D.R., Pugsley, J.S., Sigman, M.S. *JACS* **123**, 7475 (2001).
[2]Ferreira, E.M., Stoltz, B.M. *JACS* **123**, 7725 (2001).
[3]Vedejs, E., Mackay, J.A. *OL* **3**, 535 (2001).
[4]Clapham, B., Cho, C.-W., Janda, K.D. *JOC* **66**, 868 (2001).
[5]Oriyama, T., Taguchi, H., Terakado, D., Sano, T. *CL* 26 (2002).
[6]Arai, S., Bellemin-Laponnaz, S., Fu, G.C. *ACIEE* **40**, 234 (2001).
[7]Maezaki, N., Furusawa, A., Uchida, S., Tanaka, T. *T* **57**, 9309 (2001).
[8]Arnauld, T., Barrett, A.G.M., Hopkins, B.T., Zecri, F.J. *TL* **42**, 8215 (2001).
[9]Amat, M., Canto, M., Llor, N., Ponzo, V., Perez, M., Bosch, J. *ACIEE* **41**, 335 (2002).
[10]Tang, L., Deng, L. *JACS* **124**, 2870 (2002).
[11]Hang, J., Li, H., Deng, L. *OL* **4**, 3321 (2002).
[12]Naasz, R., Arnold, L.A., Minnaard, A.J., Feringa, B.L. *ACIEE* **40**, 927 (2001).
[13]Bertozzi, F., Crotti, P., Macchia, F., Pineschi, M., Feringa, B.L. *ACIEE* **40**, 930 (2001).
[14]Lim, S.H., Beak, P. *OL* **4**, 2657 (2002).
[15]Poisson, J.-F., Normant, J.F. *JACS* **123**, 4639 (2001).
[16]Faller, J.W., Lavoie, A.R. *OL* **3**, 3703 (2001).
[17]Choi, J.H., Kim, Y.H., Nam, S.H., Shin, S.T., Kim, M.-J., Park, J. *ACIEE* **41**, 2373 (2002).
[18]Yamada, S., Noguchi, E. *TL* **42**, 3621 (2001).

Halogenation. A synthesis of chiral α-haloesters from acyl halides via ketene intermediates depends on the directing effect of benzoylquinine during the halogenation step.[1,2]

[1]Wack, H., Taggi, A.E., Hafez, A.M., Drury III, W.J., Lectka, T. *JACS* **123**, 1531 (2001).
[2]Hafez, A.M., Taggi, A.E., Wack, H., Esterbrook, J., Lectka, T. *OL* **3**, 2049 (2001).

Alkylations. Chiral α-branched carboxylic acids and ketones are synthesized from α-hydroxy ketones derived from camphor via alkylation and C-C bond cleavage that regenerates camphor.[1] A route to (*R*)-α-benzyloxy carboxylic acids consists of derivatizing D-fructose diacetonide with benzyloxyacetyl chloride, alkylation of the resulting ester, and mild hydrolysis.[2]

(*S,S*)-Cyclohexane-1,2-diol has been used as chiral auxiliary in a synthesis of α,α-disubstituted amino acids.[3]

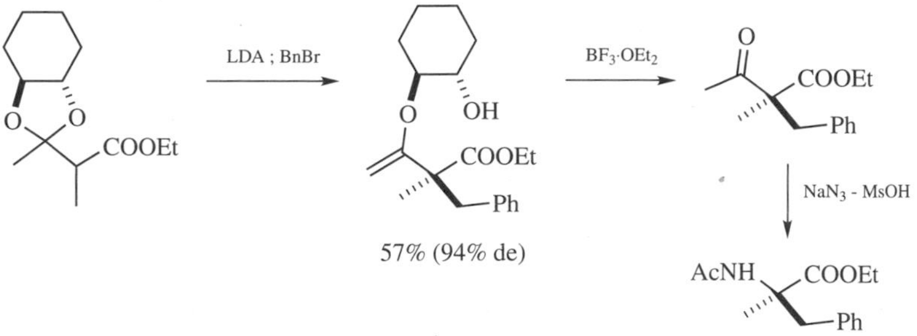

LDA ; BnBr

57% (94% de)

$BF_3 \cdot OEt_2$

NaN_3 - MsOH

An expedient way to obtain α-hydroxy acids is based on the alkylation of a cyclic derivative of glycolic acid.[4]

Chiral catalysts for alkylation of enol derivatives include those of oxazaborolidinone series[5] and C_2-symmetrical diamide **3**.[6] For Pd-catalyzed allylation of ketones diamide ligands including **3A** have been scrutinized.[7]

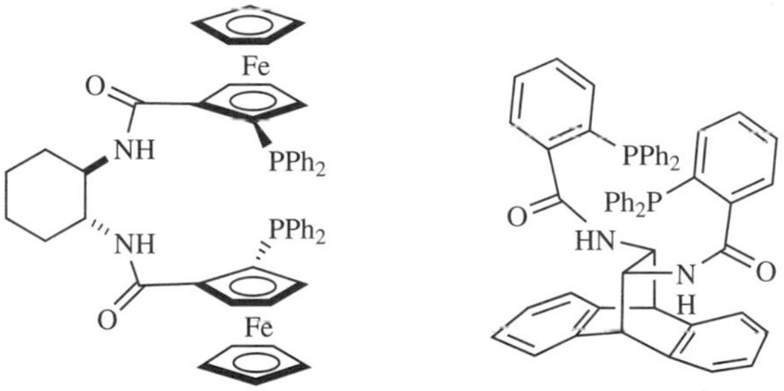

3 **3A**

For synthesis of α-amino acids, phase-transfer catalysts derived from cinchona alkaloids have been successfully employed in the alkylation of *t*-butyl *N*-diphenylmethyleneglycinate (ee >90%).[8–11] It is noteworthy that cinchona alkaloids quaternized with benzyl halides possessing an *o*-fluoro substituent drastically increase the enantioselectivity in alkylation.[10] Equally useful as catalyst is the C_2-symmetrical cyclic guanidininum chloride **4**.[12]

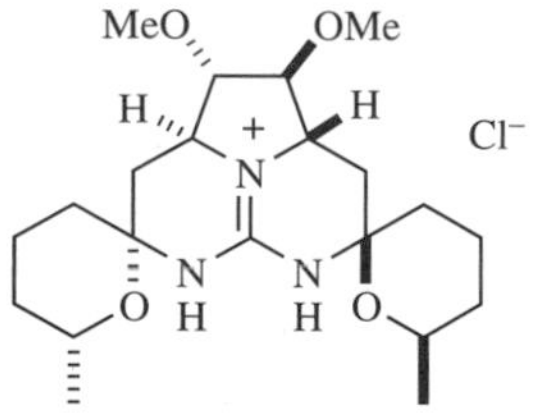

4

Alternatively, (R)-α-amino acids are obtained when imine derivatives of (+)-pseudoephedrine glycinamide are employed.[13] Alkylation and hydrolysis afford the chiral products.

Several reports concerning the asymmetric alkylation at carbon atom adjoining a sulfur functionality have appeared. Lithiated 4-isopropyl-3-(methylthiomethyl)-5,5-diphenyloxazolidin-2-one **5** is a chiral formyl anion equivalent for enantioselective synthesis of 1,2-diols, 2-amino alcohols, 2-hydroxy esters, and 4-hydroxy-2-alkenoates.[14] Alkyl 2-quinolyl sulfides undergo alkylation after deprotonation, and in the presence of a chiral bis(oxazoline) ligand, products of good ee are generated. Reduction with $NaBH_3CN$ in HOAc disengages the quinoline residue and delivers enantioenriched thiols.[15]

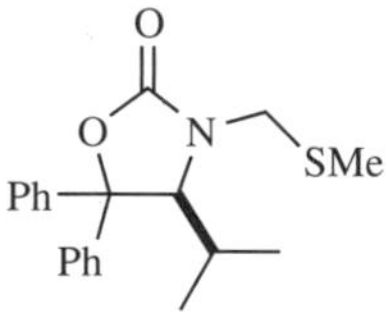

5

A first asymmetric synthesis of α-branched sulfonic esters entails alkylation of sugar derivatives such as **6**.[16]

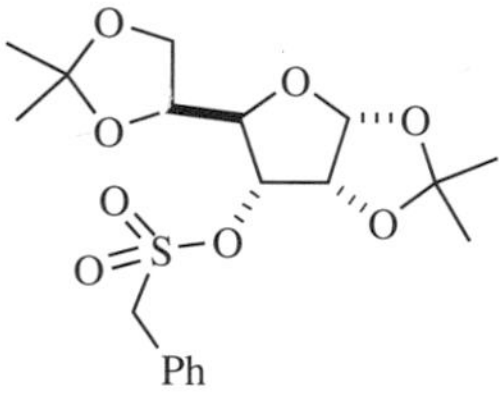

6

When benzyl amines in which the nitrogen atom is protected by Boc and TMS groups undergo carboxylation after lithiation in the presence of (−)-sparteine, *N*-Boc phenylglycine derivatives are produced.[17] However, enantioselectivity is not sufficiently high. Chiral α-(carbamoyl)alkylcuprate reagents are available from asymmetric lithiation of carbamates.[18] Coupling reaction effectively accomplishes introduction of an α-branch to the nitrogen atom.

To enantioselectively alkylate the α-position of the *N′* atom in an *N*-alkylethylenediamine, formation of the 3-Boc-imidazolidine by reaction with formaldehyde and Boc_2O precedes the critical step using *s*-BuLi/(−)-sparteine to direct the asymmetric course.[19] The same system is also capable of lithiating carbamates **7** at the carboxylated carbon atom. Transmetallation then gives organozinc and copper species. Surprisingly high enantioselectvity attends their alkylation.[20] When *meso*-epoxides are similarly treated and alkylated, products with reasonably good enantiomer ratios are observed.[21]

R O O N O

7

[1]Palomo, C., Oiarbide, M., Mielgo, A., Gonzalez, A., Garcia, J.M., Landa, C., Lecumberri, A., Linden, A. *OL* **3**, 3249 (2001).
[2]Yu, H., Ballard, C.E., Wang, B. *TL* **42**, 1835 (2001).
[3]Tanaka, M., Oba, M., Tamai, K., Suemune, H. *JOC* **66**, 2667 (2001).
[4]Diez, E., Dixon, D.J., Ley, S.V. *ACIEE* **40**, 2906 (2001).
[5]Harada, T., Yamanaka, H., Oku, A. *SL* 61 (2001).
[6]You, S.-L., Hou, X.-L., Dai, L.-X., Zhu, X.-Z. *OL* **3**, 149 (2001).
[7]Trost, B.M., Schroeder, G.M., Kristensen, J. *ACIEE* **41**, 3492 (2002).
[8]Jew, S.-S., Jeong, B.-S., Yoo, M.-S., Huh, H., Park, H.-G. *CC* 1244 (2001).
[9]Park, H.-G., Jeong, B.-S., Yoo, M.-S., Park, M.-K., Huh, H., Jew, S.-S. *TL* **42**, 4645 (2001).
[10]Park, H.-G., Jeong, B.-S., Yoo, M.-S., Lee, J.-H., Park, M.-K., Lee, Y.-J., Kim, M.-J., Jew, S.-S. *ACIEE* **41**, 3036 (2002).
[11]Jew, S.-S., Yoo, M.-S., Jeong, B.-S., Park, I.-Y., Park, H.-G. *OL* **4**, 4245 (2002).
[12]Kita, T., Georgieva, A., Hashimoto, Y., Nakata, T., Nagasawa, K. *ACIEE* **41**, 2832 (2002).
[13]Guillena, G., Najera, C. *TA* **12**, 181 (2001).
[14]Gaul, C., Schärer, K., Seebach, D. *JOC* **66**, 3059 (2001).
[15]Nakamura, S., Furutani, A., Toru, T. *EJOC* 1690 (2002).
[16]Enders, D., Vignola, N., Berner, O.M., Bats, J.W. *ACIEE* **41**, 109 (2002).
[17]Barberis, C., Voyer, N., Roby, J., Chenard, S., Tremblay, M., Labrie, P. *T* **57**, 2965 (2001).
[18]Dieter, R.K., Topping, C.M., Chandupatla, K.R., Lu, K. *JACS* **123**, 5132 (2001).
[19]Coldham, I., Copley, R.C.B., Haxell, T.F.N., Howard, S. *OL* **3**, 3799 (2001).
[20]Papillon, J.P.N., Taylor, R.J.K. *OL* **4**, 119 (2002).
[21]Hodgson, D.M., Gras, E. *ACIEE* **41**, 2376 (2002).

Displacements involving allylic systems. Chiral ligands to assist Pd-catalyzed allylic alkylation with malonate anions that have been under scrutiny include well-known as well as totally new species. Among these are a C_2-symmetric, mannitol-derived phosphite,[1] polymer-bound bis(oxazoline),[2] a bissulfoximine (**8**),[3] *o*-biphenyl-(cyclohexyl)methylphosphine,[4] *cis,cis,cis*-tetrakis(diphenylphosphinomethyl)cyclopentane,[5] *P,N*-ligands such as

9,[6] **10**,[7] **11**,[8] **12**,[9] **13**,[10] **14**,[11] **15**.[12,13] The last report shows enantiotopic displacement of one of the *gem*-diacetoxy groups. A polymer-bound ligand **15A** has also been developed[14] while the non-C_2-symmetric **15B** are found to exert maximal regioselectivity and enantioselectivity in the Mo-catalyzed displacements.[15] Other ligands that serve the purpose are **16**,[16] **17**,[17] and those for mediating displacement with organocopper reagents are **18**[18] and **19**.[19]

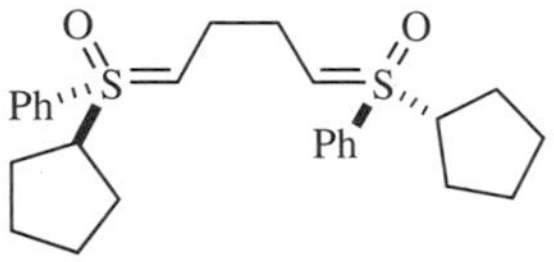

8 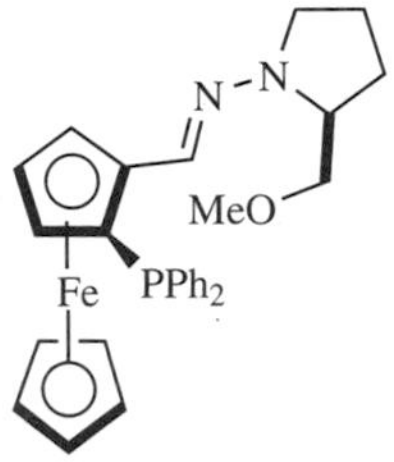 **9** 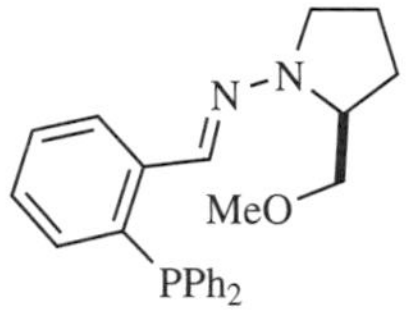 **10**

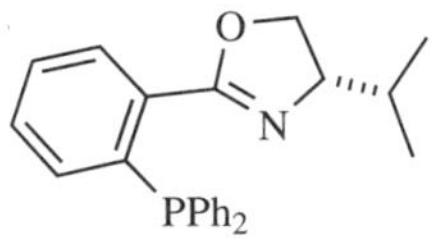

11 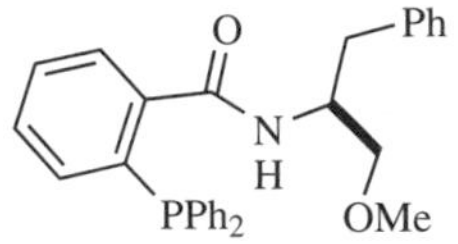 **12** 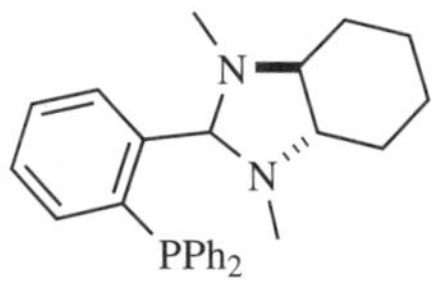 **13**

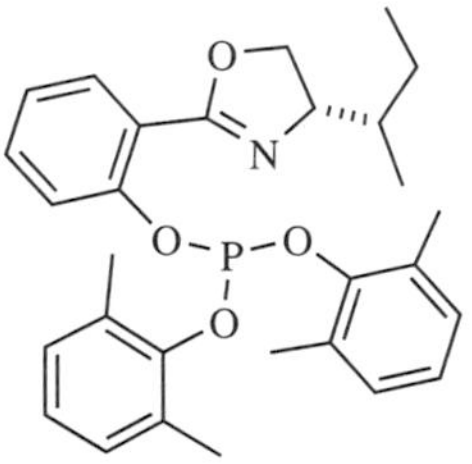

14 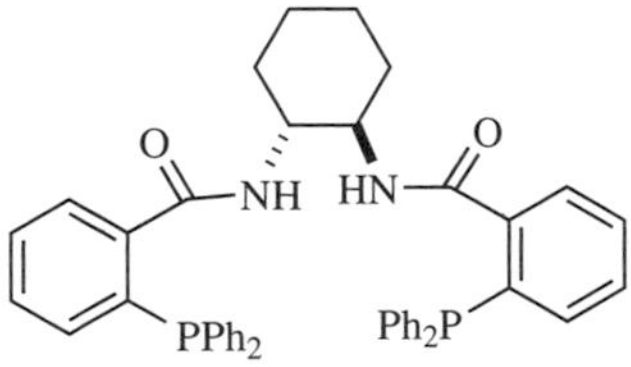 **15**

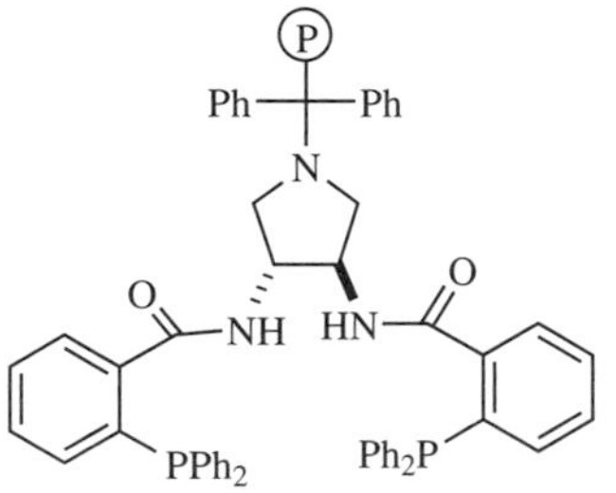

15A

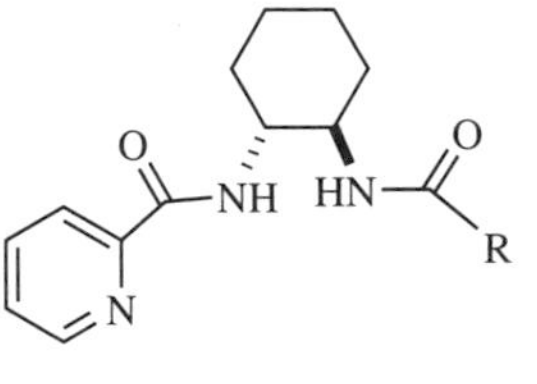

R = Ph, *t*-Bu
15B

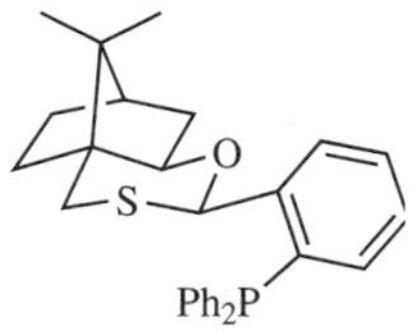

16

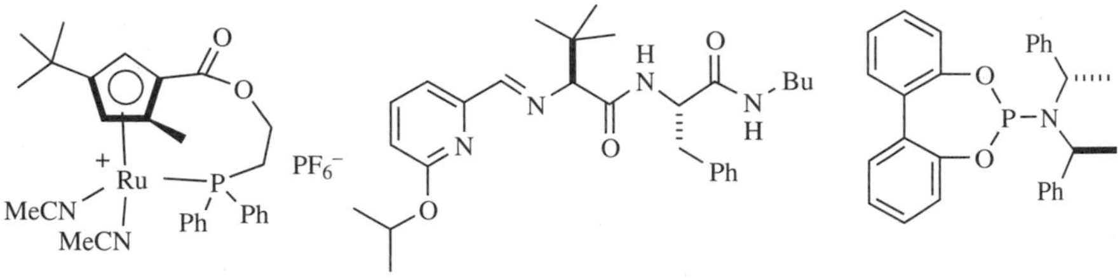

17 18 19

2-Alkynyl-1,3-dioxolanes bearing additional substituents at C-3,4 in a *cis*-relationship yield chiral propargyl ethers upon reaction with silyl enol derivatives in the presence of oxazaborolidine **20**.[20] Desymmetrization of *meso*-1,2-diols is accomplished.

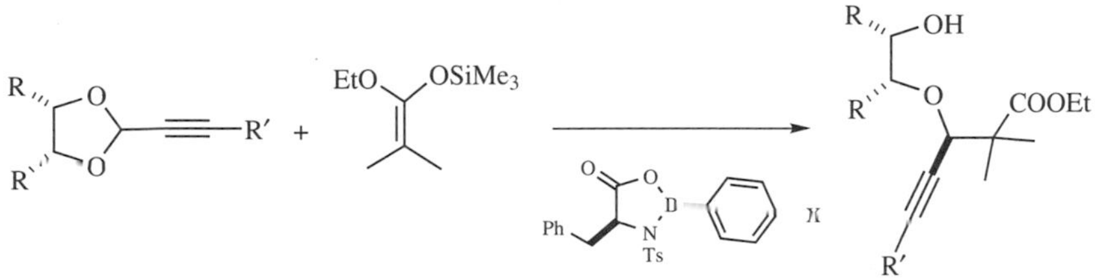

20

(−)-Sparteine induces opening of 2-aryl-1,3-dioxolanes with aryllithium reagents to varying degrees of ee.[21]

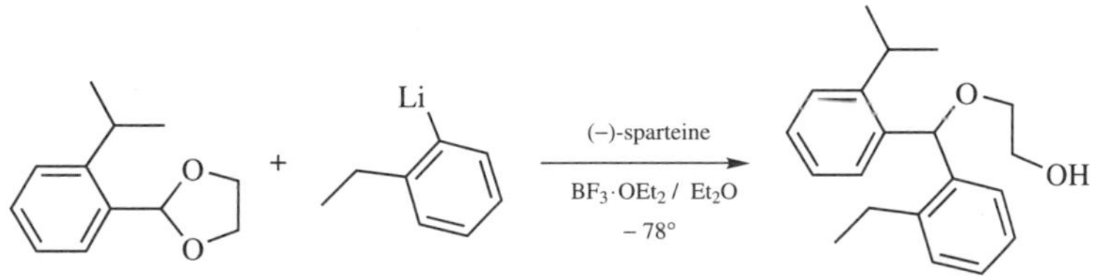

(−)-sparteine
$BF_3 \cdot OEt_2$ / Et_2O
− 78°

88% (81% ee)

[1]Zhang, R., Yu, L., Xu, L., Wang, Z., Ding, K. *TL* **42**, 7659 (2001).
[2]Hallman, K., Moberg, C. *TA* **12**, 1475 (2001).
[3]Bolm, C., Simic, O., Martin, M. *SL* 1878 (2001).
[4]Tsuruta, H., Imamoto, T. *SL* 999 (2001).
[5]Laurenti, D., Feuerstein, M., Pepe, G., Doucet, H., Santelli, M. *JOC* **66**, 1633 (2001).
[6]Mino, T., Ogawa, T., Yamashita, M. *H* **55**, 453 (2001).
[7]Mino, T., Shiotsuki, M., Yamamoto, N., Suenaga, T., Sakamoto, M., Fujita, T., Yamashita, M. *JOC* **66**, 1795 (2001).
[8]Weiss, T.D., Helmchen, G., Kazmaier, U. *CC* 1270 (2002).
[9]Mino, T., Kashihara, K., Yamashita, M. *TA* **12**, 287 (2001).
[10]Jin, M.-J., Kim, S.-H., Lee, S.-J., Kim, Y.-M. *TL* **43**, 7409 (2002).

[11]Bondarev, O.G., Lyubimov, S.E., Shiryaev, A.A., Kadilnikov, N.E., Davankov, V.A., Gavrilov, K.N. *TA* **13**, 1587 (20021).
[12]Trost, B.M., Machacek, M.R. *ACIEE* **41**, 4693 (2002).
[13]Trost, B.M., Lee, C.B. *JACS* **123**, 3671, 3687 (2001).
[14]Song, C.E., Yang, J.W., Roh, E.J., Lee, S.-g., Ahn, J.H., Han, H. *ACIEE* **41**, 3852 (2002).
[15]Trost, B.M., Dogra, K., Hachiya, I., Emura, T., Hughes, D.L., Krska, S., Reamer, R.A., Palucki, M., Yasuda, N., Reider, P.J. *ACIEE* **41**, 1929 (2002).
[16]Nakano, H., Okuyama, Y., Yanagida, M., Hongo, H. *JOC* **66**, 620 (2001).
[17]Matsushima, Y., Onitsuka, K., Kondo, T., Mitsudo, T., Takahashi, S. *JACS* **123**, 10405 (2001).
[18]Luchaco-Cullis, C.A., Mizutani, H., Murphy, K.E., Hoveyda, A.H. *ACIEE* **40**, 1456 (2001).
[19]Alexakis, A., Croset, K. *OL* **4**, 4147 (2002).
[20]Harada, T., Yamanaka, H., Oku, A. *SL* 61 (2001).
[21]Müller, P., Nury, P., Bernardinelli, G. *EJOC* 4137 (2001).

Opening of epoxides. Enantioselective opening of *meso*-epoxides to afford chlorohydrin derivatives by $SiCl_4$ in the presence of **21** has been demonstrated.[1] On the other hand, a report in year 2000 on chiral phosphoramide-mediated reaction is not reproducible, products are actually racemic.[2]

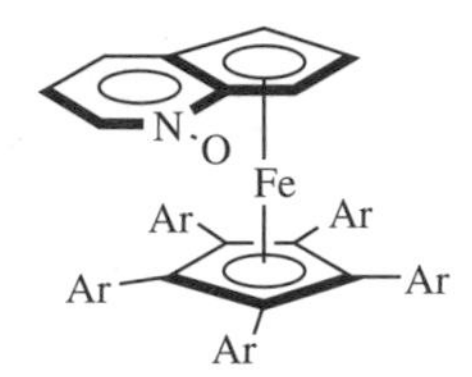

21 (Ar = 3,5-$Me_2C_6H_3$)

Fluorohydrin synthesis from epoxides and KHF_2 in the presence of 18-crown-6 and a chiral (salen)Cr complex is quite effective.[3] Highly active oligomeric (salen)Co catalysts have been prepared from diaminocyclohexane, diaryldialdehydes and cobalt acetate for epoxide ring opening (ee >94%).[4]

[1]Tao, B., Lo, M.M.-C., Fu, G.C. *JACS* **123**, 353 (2001).
[2]Denmark, S.E., Wynn, T., Jellerichs, B.G. *ACIEE* **40**, 2255 (2001).
[3]Haufe, G., Bruns, Runge, M. *JFC* **112**, 55 (2001).
[4]Ready, J.M., Jacobsen, E.N. *JACS* **123**, 2687 (2001).

Addition to C═O. Chiral cyanohydrins and derivatives continue to attract attention with respect to their formation. Thus, useful in an enantioselective process for silylcyanation of aldehydes is a Ti(IV) complex of **22**.[1] For ketones, $(i\text{-PrO})_3Al$ and a recyclable peptide ligand **23**[2] and sugar-derived phosphine oxides **24A/B**[3,4] ligated to some rare earth metal ions are highly satisfactory. Dihydroquinyl ligands which are Lewis bases play a crucial role in converting ketones to carbonates of cyanohydrins by NCCOOEt in a highly enantioselective manner.[5]

Zinc alkynides generated in situ from 1-alkynes and $Zn(OTf)_2$ in the presence of (+)-*N*-methylephedrine[6–9] or the analog **25**[10,11] add to aldehydes to afford propargylic alcohols of (*R*)-configuration. Catalyzed by a bis(oxazolinyl)pyridine (**26**) complex of $Sc(OTf)_3$ a reaction of aldehydes with allenylsilanes produces homopropargylic alcohols.[12] Increasing the steric bulk of the silane substituent diverts the reaction path to dihydrofuran formation as a result of [3+2]cycloaddition.

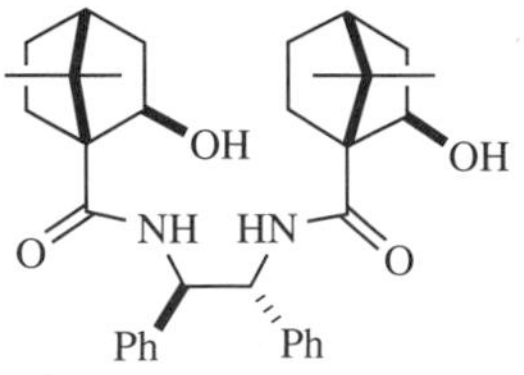

OH OH NH HN O O Ph Ph

22

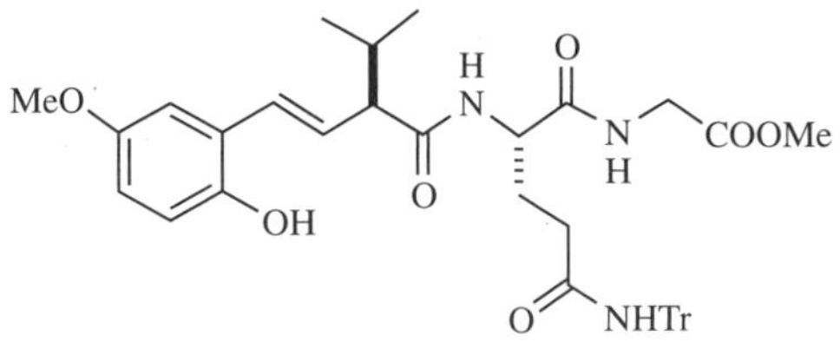

MeO OH O H N O N H COOMe O NHTr

23

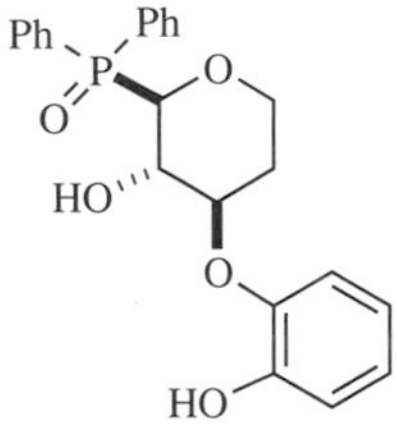

Ph Ph P O O HO O HO

24A

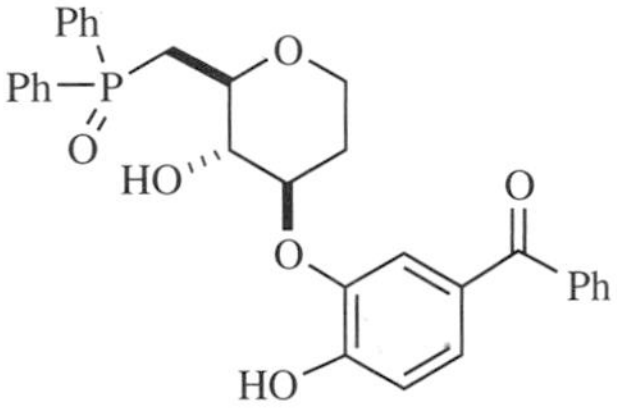

Ph Ph—P O O HO O HO O Ph

24B

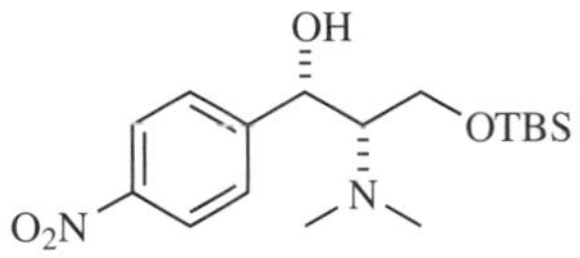

OH OTBS N O_2N

25

O N N O N Ph Ph

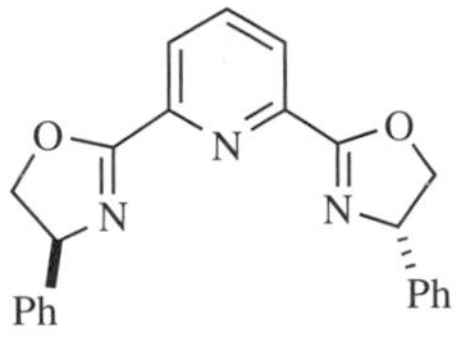

26

Alkenylzinc reagents prepared in situ from hydroborated 1-alkynes and Me_2Zn add to aldehydes to give allyl alcohols. A chiral version involves ligand **27**, the products can be manipulated to α-amino acids.[13]

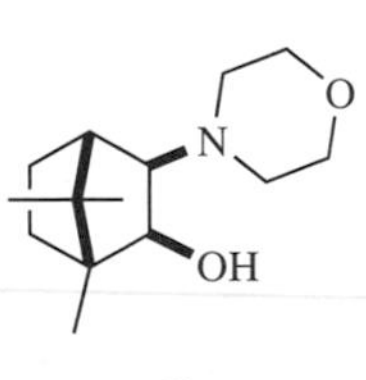

N O OH

27

Much excitement is generated from asymmetric allylation studies of aldehydes with allylsilanes. Ligands that have found applications are the axially chiral 2,2′-bipyridine

N,N′-dioxide **28**,[14] and a 2,2′-bipyrrolidine-based bisphosphoramide **29**.[15] For allylation of methyl ketones, an *O*-silylated *N*-acylnorpseudoephedrine plays a dual role of ligand and reagent.[16] Bis(oxazolinyl)phenylrhodium(III) aquo complexes catalyze allyl transfer from an allylstannane, but ee values are moderate.[17]

Si-Allyl silacycles become highly reactive toward carbonyl compounds if they are strained (≤5-membered). Reagents such as **30** and **31** are promptly developed.[18]

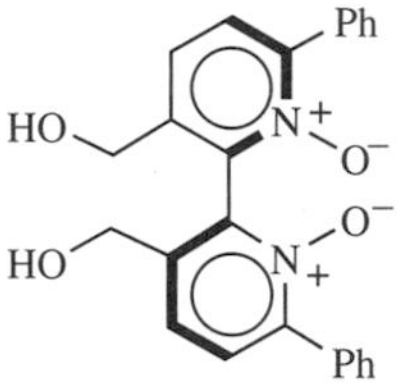

28 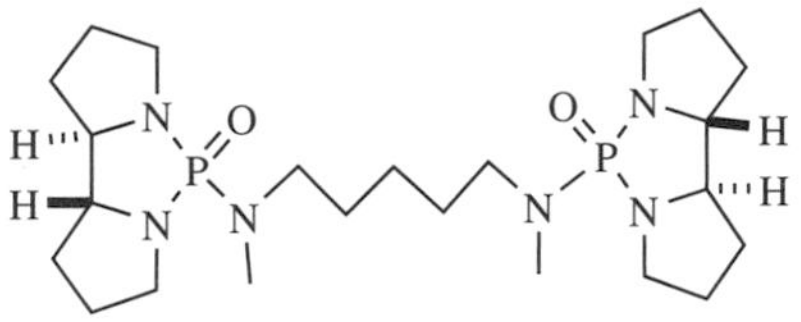 **29**

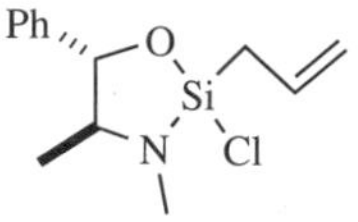

30 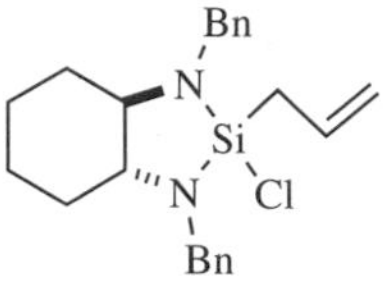 **31**

Indium-mediated allylation of *N*-(arylformyl)bornane-10,2-sultams is diastereoselective (up to >99:1).[19]

Chiral alcohols are produced from Ti(IV)-catalyzed addition of typical organozinc reagents to carbonyl compounds in the presence of various ligands such as **32**.[20,21] (*R*)-4-Carboxy[2.2]paracyclophane[22] and [n]helicene (n=5,6)[23] act as chiral initiators in the asymmetric addition of *i*-Pr_2Zn to 2-alkynylpyrimidine-5-carbaldehyde, and aryl transfer from an Ar_2Zn[24] or arylboronic acid[25] requires Et_2Zn and a chiral 2-substituted 4-*t*-butyloxazoline (**33, 34**). Chelation-assisted enantioface discrimination is noted in Reformatsky reaction of ketones in the presence of a cinchona alkaloid.[26]

Additional ligands for inducing asymmetric addition of Et_2Zn to aldehydes are the azetidine analogues of α,α-diphenylprolinol[27] and the C_2-symmetric 1,1-biindan-2,2′-diol.[28]

In a Ti(IV)-catalyzed trialkylaluminum addition to aldehydes good enantioselectivity is observed when *N*-sulfonyl amino alcohol **35** is present.[29]

1,6-Asymmetric induction is observed in a Grignard reaction on acylsilanes **36**.[30] Several 3-substituted glutaric anhydrides react with PhMgCl to give chiral keto acids, asymmetric induction being furnished by (−)-sparteine.[31]

A reagent-based hydroxymethylation of aldehydes to give chiral 1,2-diols employs **37**.[32] The process involves transmetallation with BuLi. On the other hand, a substrate-controlled asymmetric synthesis is based on organolithium addition to α-keto amides derived from (*S*)-2-methoxymethylindoline.[33]

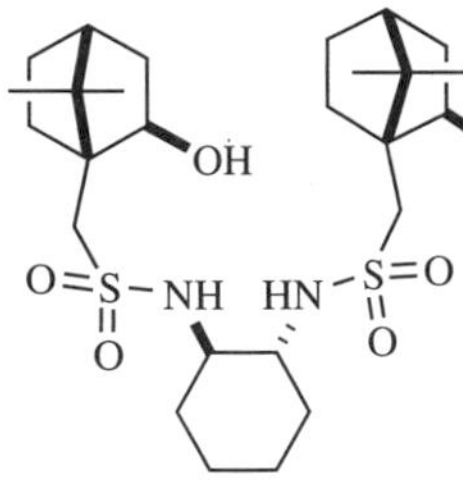

32 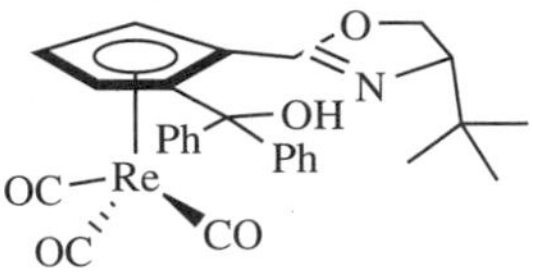 **33** 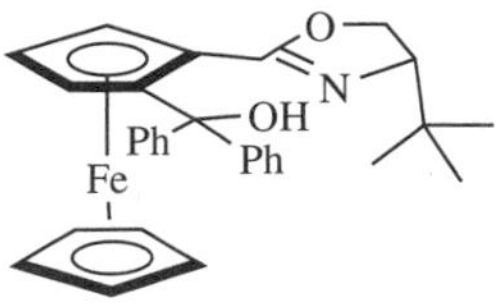 **34**

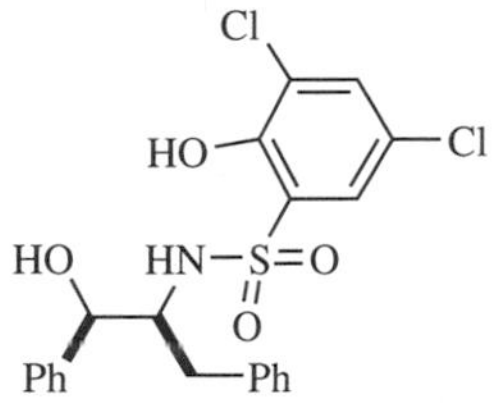

35 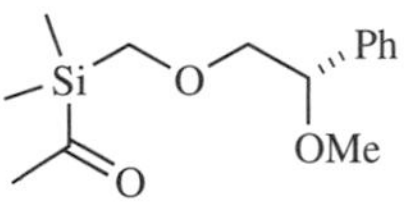 **36** 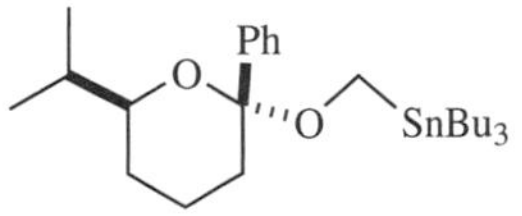 **37**

A new C_2-symmetrical bissulfide derived from (*R,R*)-tartaric acid influences the enantiotopic course of epoxide formation from aldehydes and halides, although the optical yields are moderate.[34] It is possible that sulfonium ylides are involved.

The Henry reaction is rendered in asymmetric version by using a dinuclear Zn catalyst **38**,[35,36] and those subject to Cu-catalysis combined with bis(oxazolinyl) ligand are applicable to conversion of α-keto esters.[37]

A highly selective, acid-catalyzed group transfer to aldehydes from the tertiary alcohol obtained from reaction of (−)-menthone with (*E*)-crotylmagnesium chloride delivers (*E*)-2-alken-5-ols in essentially enantiopure form.[38]

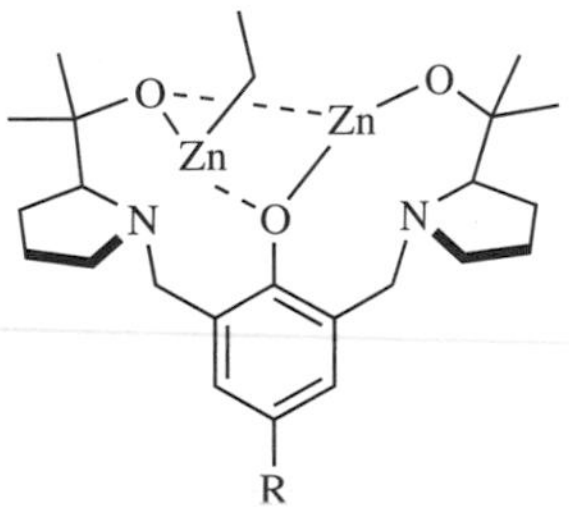

38 (R = H)
38A (R = Me)

[1]Chang, C.-W., Yang, C.-T., Hwang, C.-D., Uang, B.-J. *CC* 54 (2002).
[2]Deng, H., Isler, M.P., Snapper, M.L., Hoveyda, A.H. *ACIEE* **41**, 1009 (2002).
[3]Yabu, K., Matsumoto, S., Yamasaki, S., Hamashima, Y., Kanai, M., Du, W., Curran, D.P., Shibasaki, M. *JACS* **123**, 9908 (2001).
[4]Hamashima, Y., Kanai, M., Shibasaki, M. *TL* **42**, 691 (2001).
[5]Tian, S.-K., Deng, L. *JACS* **123**, 6195 (2001).
[6]Anand, N.K., Carreira, E.M. *JACS* **123**, 9687 (2001).
[7]Sasaki, H., Boyall, D., Carreira, E.M. *HCA* **84**, 964 (2001).
[8]Boyall, D., Frantz, D.E., Carreira, E.M. *OL* **4**, 2605 (2002).
[9]Amador, M., Ariza, X., Garcia, J., Ortiz, J. *TL* **43**, 2691 (2002).
[10]Jiang, B., Chen, Z., Xiong, W. *CC* 1524 (2002).
[11]Jiang, B., Chen, Z., Tang, X. *OL* **4**, 3451 (2002).
[12]Evans, D.A., Sweeney, Z.K., Rovis, T., Tedrow, J.S. *JACS* **123**, 12095 (2001).
[13]Chen, Y.K., Lurain, A.E., Walsh, P.J. *JACS* **124**, 12225 (2002).
[14]Shimada, T., Kina, A., Ikeda, S., Hayashi, T. *OL* **4**, 2799 (2002).
[15]Denmark, S.E., Fu, J. *JACS* **123**, 9488 (2001).
[16]Tietze, L.F., Völkel, L., Wulff, C., Weigand, B., Bittner, C., McGrath, P., Johnson, K., Schäfer, M. *CEJ* **7**, 1304 (2001).
[17]Motoyama, Y., Okano, M., Narusawa, H., Makihara, N., Aoki, K., Nishiyama, H. *OM* **20**, 1580 (2001).
[18]Kinnaird, J.W.A., Ng, P.Y., Kubota, K., Wang, X., Leighton, J.L. *JACS* **124**, 7920 (2002).
[19]Shin, J.A., Cha, J.H., Pae, A.N., Choi, K.I., Koh, H.Y., Kang, H.-Y., Cho, Y.S. *TL* **42**, 5489 (2001).
[20]Yus, M., Ramon, D.J., Prieto, O. *TA* **13**, 2291 (2002).
[21]Garcia, C., LaRochelle, L.K., Walsh, P. *JACS* **124**, 10970 (2002).
[22]Tanji, S., Ohno, A., Sato, I., Soai, K. *OL* **3**, 287 (2001).
[23]Sato, I., Yamashita, R., Kodowaki, K., Yamamoto, J., Shibata, T., Soai, K. *ACIEE* **40**, 1096 (2001).
[24]Bolm, C., Kesselgruber, M., Hermanns, N., Hildebrand, J.P., Raabe, G. *ACIEE* **40**, 1488 (2001).
[25]Bolm, C., Rudolph, J. *JACS* **124**, 14850 (2002).
[26]Ojida, A., Yamano, T., Taya, N., Tasaka, A. *OL* **4**, 3051 (2002).
[27]Hermsen, P.J., Cremers, J.G.O., Thijs, L., Zwanenburg, B. *TL* **42**, 4243 (2001).
[28]Yang, X., Shen, J., Da, G., Wang, H., Su,W., Liu, D., Wang, R., Choi, M.C.K., Chan, A.S.C. *TL* **42**, 6573 (2001).
[29]You, J.-S., Hsieh, S.-H., Gau, H.-M. *CC* 1546 (2001).
[30]Gassmann, S., Guintchin, B., Bienz, S. *OM* **20**, 1849 (2001).
[31]Shintani, R., Fu, G.C. *ACIEE* **41**, 1057 (2002).
[32]Smyj, R.P., Chong, J.M. *OL* **3**, 2903 (2001).
[33]Youn, S.W., Kim, Y.H., Hwang, J.-W., Do, Y. *CC* 996 (2001).
[34]Ishizaki, M., Hoshino, O. *H* **57**, 1399 (2002).
[35]Trost, B.M., Yeh, V.S.C. *ACIEE* **41**, 861 (2002).
[36]Trost, B.M., Yeh, V.S.C., Ito, H., Bremeyer, N. *OL* **4**, 2621 (2002).
[37]Christensen, C., Juhl, K., Hazell, R.G., Jorgensen, K.A. *JOC* **67**, 4875 (2002).
[38]Nokami, J., Ohga, M., Nakamoto, H., Matsubara, T., Hussain, I., Kataoka, K. *JACS* **123**, 9168 (2001).

Excellent results are obtained in cross-aldol reaction between aldehydes when proline is used as a chiral catalyst.[1] That involving ketones and aldehydes[2,3] and aldehydes and oxomalonic esters[4] are also reported. A polymer-supported proline promotes diastereoselective and enantioselective aldol reaction between hydroxyacetone and aldehydes, providing predominantly *anti*-3,4-dihydroxy-2-alkanones.[5] Interestingly, with *N*-protected aromatic imines as acceptors the reaction gives *syn*-3-hydroxy-4-amino-4-aryl-2-butanones as major products. Extension in the use of the dinuclear Zn catalyst **38A** to assemble *syn*-2,3-dihydroxy-1-alkanones is realized.[6]

A new class of moderately successful catalysts for asymmetric aldol reaction are calcium alkoxides derived from chiral alcohols such as (*S,S*)-hydrobenzoin.[7]

Yttryl isopropoxide in conjunction with a chiral salen ligand promotes cross-aldol reaction and Tishchenko reaction in tandem.[8] In some cases ee values of the diol monoesters are acceptable.

Polymer-bound bis(oxazoline) that complexes with Cu(II) salt to form catalyst for the Mukaiyama aldol reaction has been developed.[9] In Lewis base-catalyzed Mukaiyama aldol reaction involving trichlorosilyl enol ethers the diastereoselectivity is controlled by the chiral catalyst.[10] That between silyl allenyl ethers and aldehydes as promoted by chiral *N*-fluoroacyl oxazaborolidines furnishes products reminiscent of the Baylis-Hillman type.[11]

Asymmetric aldol reactions in aqueous media are realized with lanthanide triflate as catalyst in conjunction with a chiral azacrown ether.[12]

Reactions involving ester enolate and silyl ketene acetal donors to synthesize chiral β-hydroxy carboxylic acid derivatives are equally important. Esters and lactones that equipped with chiral auxiliaries are exemplified by **39**,[13] **40**,[14] and **41**.[15,16] Polypropionate segments with a *syn*-1,3-dimethyl pattern are readily prepared from chiral 2-methyl-3-ethylidene-3-propanolide (methylketene dimer), via ring opening and substrate-controlled stereoselective aldol reaction.[17]

Reductive condensation initiated by the generation of silyl ketene acetals in situ from acrylic esters and a hydrosilane proceeds in the presence of an iridium complex of **42**.[18] Generally, enantioselectivity and diastereoselectivity are good. The dichloroselenonium complex with a more common (pybox) ligand is a valuable catalyst for synthesis of pantolactone derivatives.[19] Catalytic, enantioselective aldol reaction of trichlorosilyl ketene acetals with ketones occurs in the presence of **43**.[20]

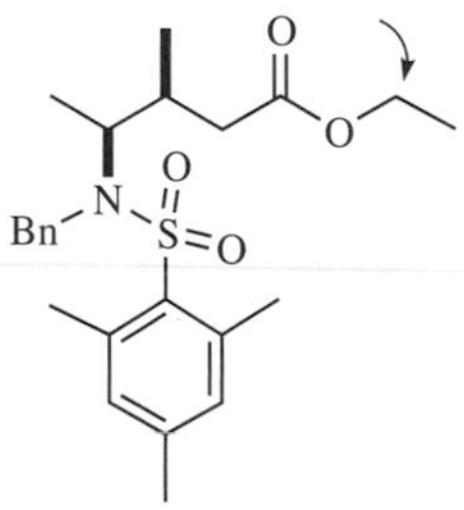

39 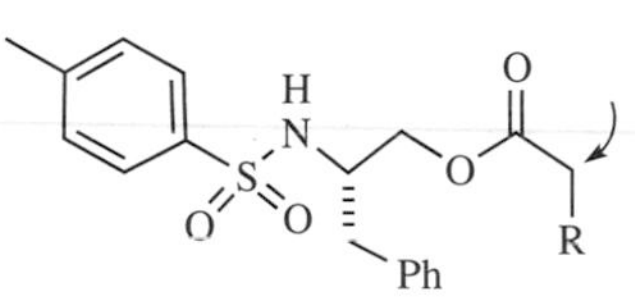 **40** 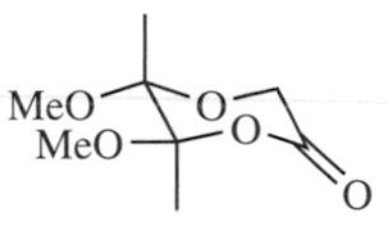 **41**

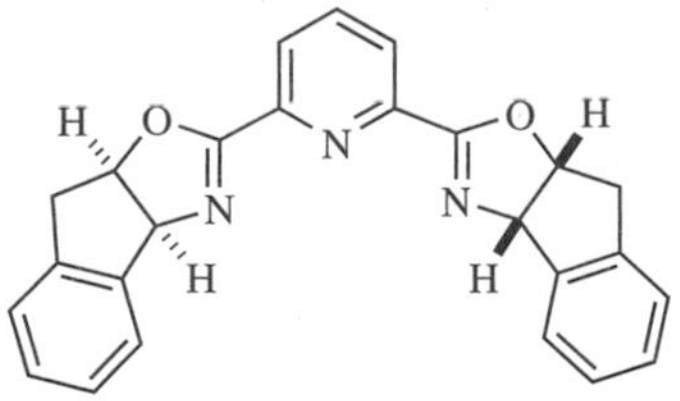

42 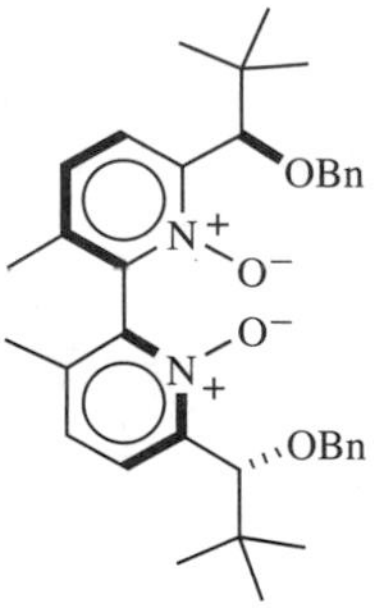 **43**

An enantioselective synthesis of β-hydroxy carboxylic acid derivatives is efficiently conducted by a one-pot aldol reaction-dynamic kinetic resolution in the presence of a Ru-catalyst, an enzyme and acyl donors.[21] The process does not require preformed silyl enol ethers.

Asymmetric benzoin condensation may be effected with the ylide of the bicyclic 1,2,4-triazolium salt **43A**.[22]

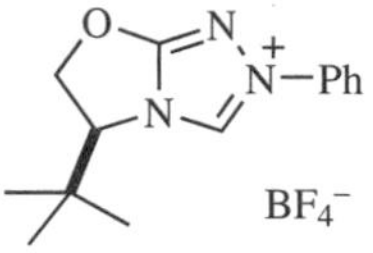

43A

[1]Northrop, A.B., MacMillan, D.W.C. *JACS* **124**, 6798 (2002).
[2]Sakthivel, K., Notz, W., Bui, T., Barbas III, C.F. *JACS* **123**, 5260 (2001).
[3]List, B., Pojarliev, P., Castello, C. *OL* **3**, 573 (2001).
[4]Bogevig, A., Kumargurubaran, N., Jorgensen, K.A. *CC* 620 (2002).
[5]Benaglia, M., Cinquini, M., Cozzi, F., Puglisi, A., Celentano, G. *ASC* **344**, 533 (2002).
[6]Trost, B.M., Ito, H., Silcoff, E.R. *JACS* **123**, 3367 (2001).
[7]Suzuki, T., Yamagiwa, N., Matsuo, Y., Sakamoto, S., Yamaguchi, K., Shibasaki, M., Noyori, R. *TL* **42**, 4669 (2001).
[8]Mascarenhas, C.M., Miller, S.P., White, P.S., Morken, J.P. *ACIEE* **40**, 601 (2001).
[9]Orlandi, S., Mandoli, A., Pini, D., Salvadori, P. *ACIEE* **40**, 2518 (2001).
[10]Denmark, S.E., Fujimori, S. *OL* **4**, 3473 (2002).
[11]Li, G., Wei, H.-X., Phelps, B.S., Purkiss, D.W., Kim, S.H. *OL* **3**, 823 (2001).
[12]Kobayashi, S., Hamada, T., Nagayama, S., Manabe, K. *OL* **3**, 165 (2001).
[13]Inoue, T., Liu, J.F., Buske, D.C., Abiko, A. *JOC* **67**, 5250 (2002).
[14]Ghosh, A.K., Kim, J.-H. *TL* **42**, 1227 (2001).
[15]Dixon, D.J., Ley, S.V., Polara, A., Sheppard, T. *OL* **3**, 3749 (2001).
[16]Dixon, D.J., Guarna, A., Ley, S.V., Polara, A., Rodriguez, F. *S* 1973 (2002).
[17]Calter, M.A., Guo, X., Liao, W. *OL* **3**, 1499 (2001).
[18]Zhao, C.-X., Duffy, M.O., Taylor, S.J., Morken, J.P. *OL* **3**, 1829 (2001).
[19]Evans, D.A., Wu, J., Masse, C.E., MacMillan, D.W.C. *OL* **4**, 3379 (2002).
[20]Denmark, S.E., Fan, Y. *JACS* **124**, 4233 (2002).
[21]Huerta, F.F., Backvall, J.-E. *OL* **3**, 1209 (2001).
[22]Enders, D., Kallfass, U. *ACIEE* **41**, 1743 (2002).

Addition to C═N. γ-Oxo-α-amino acids are readily formed in a proline-catalyzed Mannich reaction of carbonyl compounds with properly protected iminoacetic esters.[1] In one step 3-amino-2-hydroxy-1-alkanones are obtained in a highly enantioselective and diastereoselective fashion by treatment of a mixture of amines, aldehydes, and 1-hydroxy-2-alkanones with proline.[2] A chiral thiourea (**44**) has found excellent activity in promoting Mannich reactions to assemble β-aryl-β-amino acid derivatives.[3]

The asymmetric Baylis-Hillman reaction with tosylimines as acceptors is realized using the quinidine-derived cyclic ether **45**.[4] Addition of nitroalkanes to imines in the so-called aza-Henry reaction is subject to asymmetric catalysis: For example, with bis(oxazolinyl)copper complexes.[5] *O*-Silyl nitronates can be used as donors.[6]

β-Amino acids constitute a structural motif that has attracted much attention. Accordingly, development of synthetic access is of considerable interest. The Reformatsky reaction analogue is best represented, and an asymmetric version has evolved.[7]

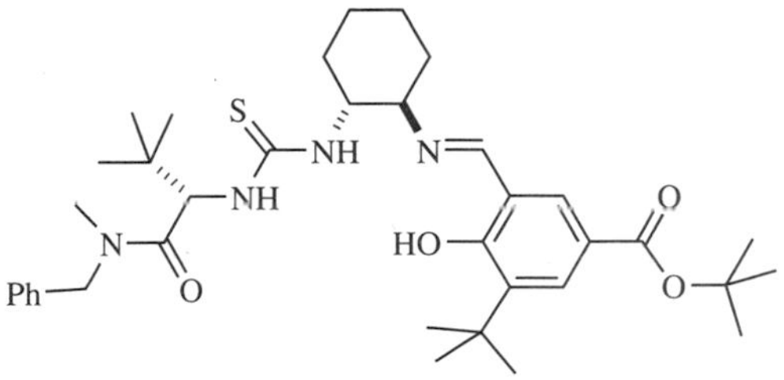

44 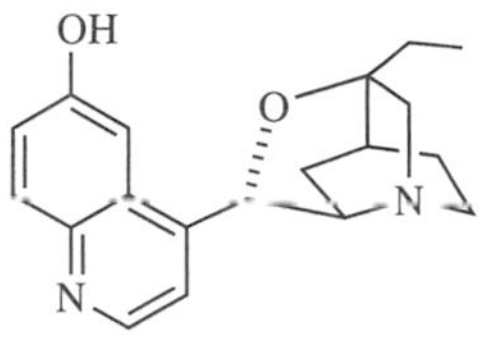 **45**

Alternatively, substrate-induced asymmetric reaction is another option. The required imines are formed in situ from aldehydes and a chiral amine and the C-C bond formation process is performed in the presence of Wilkinson's catalyst.[8] In the addition of organozinc species to chiral α-imino esters the substrates are preorganized by addition of $ZnBr_2$.[9]

A Zr-catalyzed addition of silyl ketene acetals to imines uses (*S*)-VAPOL (**46**) as chiral ligand.[10]

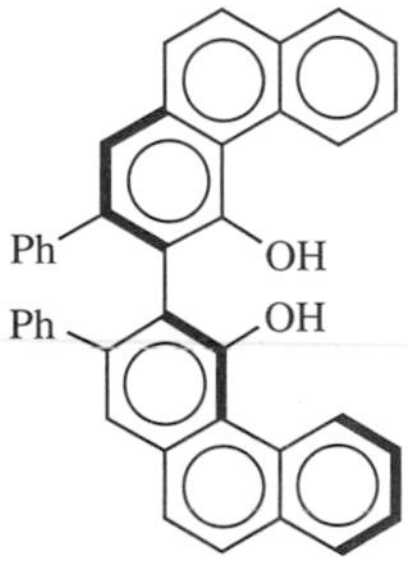

46

In a route to β-substituted aspartic acid derivatives (Note: aspartic acid is both an α-amino acid and a β-amino acid) benzoylquinine plays multiple roles.[11]

Synthesis of chiral α-branched amines from addition of organometallic reagents to imines many options are available. With organozinc species, good results are observed with a [2.2]paracyclophane-based iminc (**47**)[12] and peptide ligands.[13] A sugar-based nitrone serves as a progenitor of propargylic *N*-hydroxylamines by virtue of its ready reaction with zincioalkynes.[14] Propargylic amines are also more conveniently formed in a (pybox)-Cu(I) catalyzed reaction in water.[15] Indium-mediated reactions on *N*-(benzyloxy-iminoacetyl)bornanesultam enable the preparation of α-amino acids,[16] and allylation of chiral imines[17] or the hydrazones originated from chiral 3-amino-4-benzyloxazolidin-2-one[18] produces homoallylic amine derivatives.

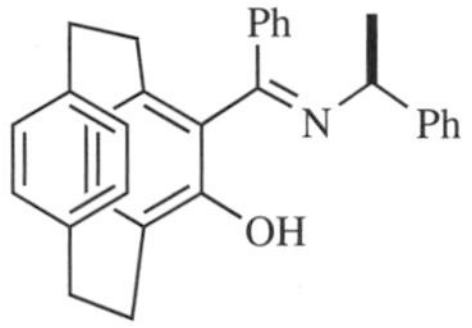

47

To synthesize C_2-symmetrical 1,*n*-diamines from dialdehydes, derivatization into the bis-SAMP-hydrazones and the subsequent reaction with organocerium reagents are the critical operations.[19] For substrate-directed asymmetric addition reactions the required imine derivatives may carry chirality in a sulfinyl group[20,21] or at the α-position (i.e., from chiral aldehyde precursors).[22]

1,3-Asymmetric induction occurs in the ring opening of 2,4-diphenyloxazolidine with silylallyllithium.[23] The products are chiral secondary amines that bear two stereocenters in proximity to the nitrogen atom.

[1]Cordova, A., Notz, W., Zhong, G., Betancort, J.M., Barbas III, C.F. *JACS* **124**, 1842, 1866 (2002).
[2]List, B., Pojarliev, P., Biller, W.T., Martin, H.J. *JACS* **124**, 827 (2002).
[3]Wenzel, A.G., Jacobsen, E.N. *JACS* **124**, 12964 (2002).
[4]Shi, M., Xu, Y.-M. *ACIEE* **41**, 4507 (2002).
[5]Nishiwaki, N., Knudsen, K.R., Gothelf, K.V., Jorgensen, K.A. *ACIEE* **40**, 2992 (2001).
[6]Knudsen, K.R., Risgaard, T., Nishiwaki, N., Gothelf, K.V., Jorgensen, K.A. *JACS* **123**, 5843 (2001).
[7]Ukaji, Y., Takenaka, S., Horita, Y., Inomata, K. *CL* 254 (2001).
[8]Honda, T., Wakabayashi, H., Kanai, K. *CPB* **50**, 307 (2002).
[9]Chiev, K.P., Roland, S., Mangeney, P. *TA* **13**, 2205 (2002).
[10]Xue, S., Yu, S., Deng, Y., Wulff, W.D. *ACIEE* **40**, 2271 (2001).
[11]Dudding, T., Hafez, A.M., Taggi, A.E., Wagerle, T.R., Lectka, T. *OL* **4**, 387 (2002).
[12]Hermanns, N., Dahmen, S., Bolm, C., Bräse, S. *ACIEE* **41**, 3692 (2002).
[13]Porter, J.R., Traverse, J.F., Hoveyda, A.H., Snapper, M.L. *JACS* **123**, 10409 (2001).
[14]Fässler, R., Frantz, D.E., Oetiker, J., Carreira, E.M. *ACIEE* **41**, 3054 (2002).
[15]Wei, C., Li, C.-J. *JACS* **124**, 5638 (2002).
[16]Miyabe, H., Nishimura, A., Ueda, M., Naito, T. *CC* 1454 (2002).

[17]Yanada, R., Kaieda, A., Takemoto, Y. *JOC* **66**, 7516 (2001).
[18]Friestad, G.K., Ding, H. *ACIEE* **40**, 4491 (2001).
[19]Enders, D., Meiers, M. *S* 2542 (2002).
[20]Barrow, J.C., Ngo, P.L., Pellicore, J.M., Selnick, H.G., Nantermet, P.G. *TL* **42**, 2051 (2001).
[21]Tang, T.P., Volkman, S.K., Ellman, J.A. *JOC* **66**, 8772 (2001).
[22]Prakash, G.K.S., Mandal, M. *JACS* **124**, 6538 (2002).
[23]Agami, C., Comesse, S., Kadouri-Puchot, C. *JOC* **67**, 1496 (2002).

Addition to multiple CC bonds. Styrenes undergo enantioselective Rh-catalyzed hydroboration in the presence of a C_2-symmetrical 1,2-diphosphane such as *(1R,2R)*-1,2-bis(dicyclohexylphosphanyl)cyclohexane.[1] Triorganoalanes originated from hydroalumination of 1-alkenes participate in Zr-catalyzed reaction with other 1-alkenes to afford methyl-branched skeletons. With (−)-(neomenthylindenyl)$_2$ZrCl$_2$ as catalyst excellent optical yields of the products are obtained.[2] Because allyl aryl ethers are susceptible to triorganoalane-catalyzed Claisen rearrangment a tandem process can be developed. For example, an *o*-allylphenol formed on treatment of the ether with Me$_3$Al is rapidly carboaluminated.[3] Oxidation gives 2-methyl-3-(*o*-hydroxyaryl)propanol.

1-Alkynes add to enamines to give propargylic amines. An asymmetric version is catalyzed by the CuBr complex of Quinap (**48**).[4]

Hydroamination of 1,3-dienes to deliver allylic amines is Pd-catalyzed in the presence of the C_2-symmetrical ligand **49**.[5] Thiocarbonylation occurs with thiols, CO, and *(R,R)*-DIOP to deliver β,γ-unsaturated thiol esters that contain an chirality center at the α-position.[6]

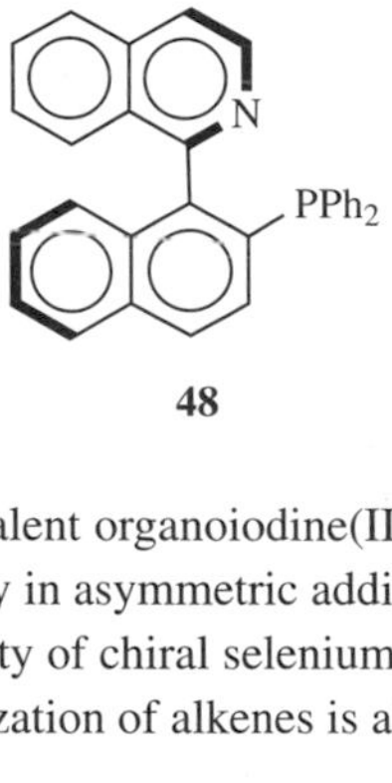

48 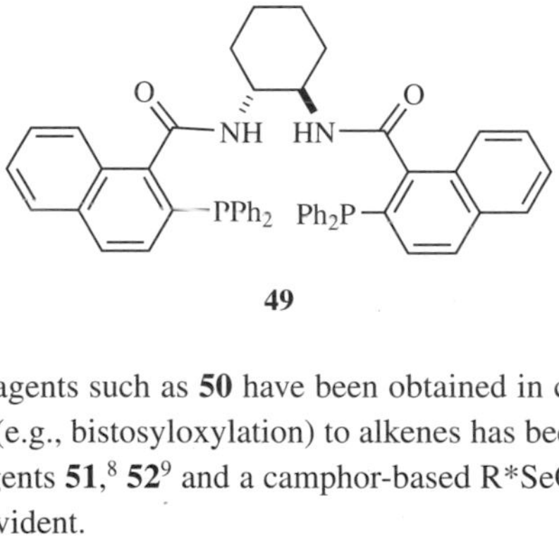 **49**

Hypervalent organoiodine(III) reagents such as **50** have been obtained in chiral form. Their utility in asymmetric addition (e.g., bistosyloxylation) to alkenes has been studied.[7] Applicability of chiral selenium reagents **51**,[8] **52**[9] and a camphor-based R*SeOSO$_3$H[10] in functionalization of alkenes is also evident.

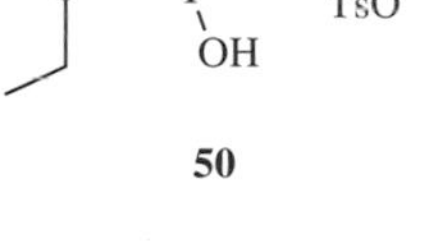

50 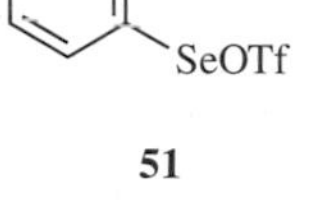 **51** 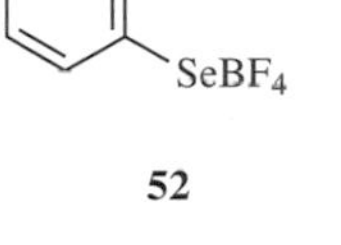 **52**

Chlorobutenolides are obtained from 2,3-alkadienoic acids on treatment of their salts with $CuCl_2$ (chlorolactonization). If the cations of such salts are derived from cinchonidine or an enantiomer of α-methylbenzylamine the products exhibit chirality (71–94% ee).[11]

Asymmetric synthesis of carboxylic acid derivatives from ketenes mediated by cinchona alkaloids[12] and by planar chiral DMAP[13] involves enantioselective protonation.

Atom-transfer radical cyclization exemplified by the formation of 3-(α-bromoalkyl)-cyclopentanones from unsaturated bromoketones is susceptible to asymmetric induction by a bis(oxazoline) ligand.[14]

A straightforward synthesis of enantiopure *P*-chiral alkenylphosphonates is by a Pd-catalyzed hydrophosphinylation of 1-alkynes with (−)-menthyl phosphonates.[15]

[1]Demay, S., Volant, F., Knochel, P. *ACIEE* **40**, 1235 (2001).
[2]Huo, S., Shi, J., Negishi, E. *ACIEE* **41**, 2141 (2002).
[3]Wipf, P., Ribe, S. *OL* **3**, 1503 (2001).
[4]Koradin, C., Polborn, K., Knochel, P. *ACIEE* **41**, 2535 (2002).
[5]Lober, O., Kawatsura, M., Hartwig, J.F. *JACS* **123**, 4366 (2001).
[6]Xiao, W.-J., Alper, H. *JOC* **66**, 6229 (2001).
[7]Hirt, U.H., Schuster, M.F.H., French, A.N., Wiest, O.G., Wirth, T. *EJOC* 1569 (2001).
[8]Tiecco, M., Testaferri, L., Silvia, S., Sternativo, S., Bagnoli, L., Santi, C., Temperini, A. *TA* **12**, 1493 (2001).
[9]Uchiyama, M., Satoh, S., Ohta, A. *TL* **42**, 1559 (2001).
[10]Tiecco, M., Testaferri, L., Santi, C., Tomassini, C., Marini, F., Bagnoli, L., Temperini, A. *TA* **13**, 429 (2002).
[11]Ma, S., Wu, S. *CC* 441 (2001).
[12]Blake, A.J., Friend, C.L., Outram, R.J., Simpkins, N.S., Whitehead, A.J. *TL* **42**, 2877 (2001).
[13]Hodous, B.L., Fu, G.C. *JACS* **124**, 10006 (2002).
[14]Yang, D., Gu, S., Yan, Y.-L., Zhu, N.-Y., Cheung, K.-K. *JACS* **123**, 8612 (2001).
[15]Han, L.-B., Zhao, C.-Q., Onozawa, S., Goto, M., Tanaka, M. *JACS* **124**, 3842 (2002).

Epoxidation and aziridination reactions. Asymmetric synthesis of epoxides still has room for improvement and many factors continue to be scrutinized. There are reports on using ruthenium complexes in conjunction with various oxidants.[1,2] The finding[3] that chiral ketone catalysts can be used with H_2O_2 is good news and a more robust catalyst in the series is **53**.[4]

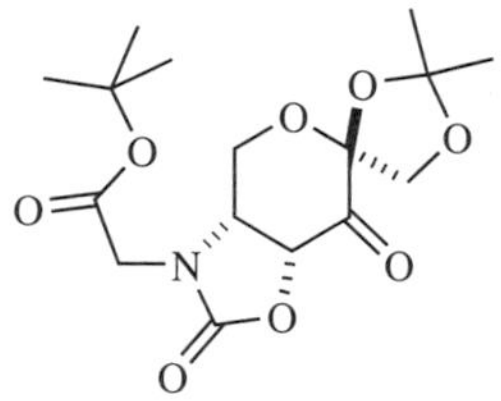

53

Salen complexes of oxochromium(V) are stable but epoxidation of an (*E*)-alkene reaches 92% ee only in the stoichiometric mode, and yield is low.[5] A polymer-bound Mn(salen) catalyst (**54**) for epoxidation has maintained C_2-symmetry.[6] To accomplish the Sharpless epoxidation a renewable oxidant is 2-furyldimethylcarbinyl hydroperoxide.[7]

Among the many catalyst-oxidant combinations for epoxidation of enones are: Chiral α-methylbenzyl hydroperoxide and a base,[8] poly-L-leucine/urea-H_2O_2,[9] diethyl tartrate, Bu_2Mg/*t*-BuOOH,[10] *N*-arylmethyl cinchona alkaloids/NaOCl[11] or LiOH-H_2O_2.[12] The quaternized cinchona alkaloid/NaOCl system converts allylic alcohols directly to the chiral epoxy ketones.[13]

Asymmetric transfer of nitrogen atom to alkenes from nitridomanganese complex of salen leads to aziridines.[14] In aziridine synthesis from imines and alkyl halides[15] or diazoalkanes[16] involving ylide intermediates, chirality is induced by camphor-based sulfides **55** and **56**, respectively.

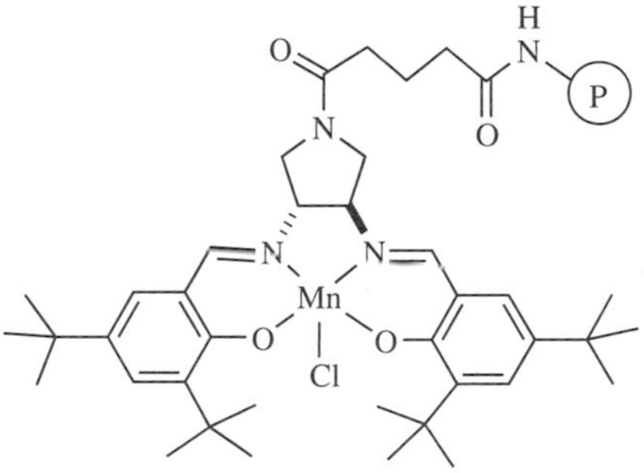

54 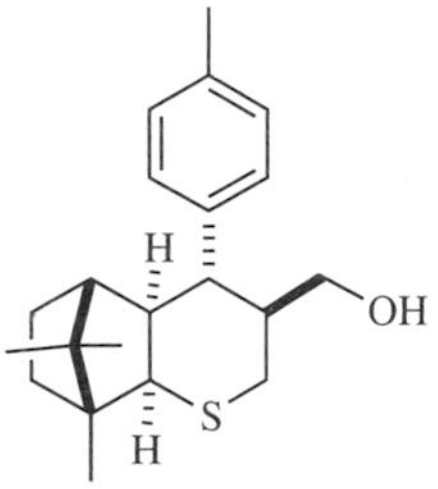 **55** 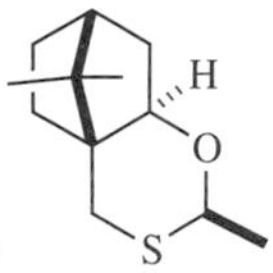 **56**

Aziridination of 2-alkoxycarbonyl-2-cycloalkenones is rendered enantioselective by choosing a chiral alkoxy group.[17]

[1]Nakata, K., Takeda, T., Mihara, J., Hamada, T., Irie, R., Katsuki, T. *CEJ* **7**, 3776 (2001).
[2]Zhang, J.-L., Liu, Y.-L., Che, C.-M. *CC* 2906 (2002).
[3]Shu, L., Shi, Y. *T* **57**, 5213 (2001).
[4]Tian, H., She, X., Shi, Y. *OL* **3**, 715 (2001).
[5]Daly, A.M., Renehan, M.F., Gilheany, D.G. *OL* **3**, 663 (2001).
[6]Song, C.E., Roh, E.J., Yu, B.M., Chi, D.Y., Kim, S.C., Lee, K.-J. *CC* 615 (2000).
[7]Lattanzi, A., Iannece, P., Scettri, A. *TL* **43**, 5629 (2002).
[8]Adam, W., Rao, P.B., Degen, H.-G., Saha-Möller, C.R. *EJOC* 630 (2002).
[9]Bentley, P.A., Bickley, J.F., Roberts, S.M., Steiner, A. *TL* **42**, 3741 (2001).
[10]Jacques, O., Richards, S.J., Jackson, R.F.W. *CC* 2712 (2001).
[11]Lygo, B., To, D.C.M. *TL* **42**, 1343 (2001).
[12]Arai, S., Tsuge, H., Oku, M., Miura, M., Shioiri, T. *T* **58**, 1623 (2002).
[13]Lygo, B., To, D.C.M. *CC* 2360 (2002).
[14]Nishimura, H., Minkata, S., Takahashi, T., Oderaotoshi, Y., Komatsu, M. *JOC* **67**, 2101 (2002).
[15]Saito, T., Sakairi, M., Akiba, D. *TL* **42**, 5451 (2001).
[16]Aggarwal, V.K., Ferrara, M., O'Brien, C.J., Thompson, A., Jones, R.V.H., Fieldhouse, R. *JCS(P1)* 1635 (2001).
[17]Fioravanti, S., Morreale, A., Pellacani, L., Tardella, P.A. *JOC* **67**, 4972 (2002).

Michael reactions. Conjugate addition with *N*-nucleophiles provides functionalized amines. Thus, in the presence of bis(oxazolinyl)copper(II) complexes *N,O*-bistrimethylsilylhydroxylamine adds enantioselectively to alkylidenemalonic esters and the adducts are converted to aziridine-1,1-dicarboxylic esters on treatment with *t*-BuOK.[1] In the addition of arylamines to *N*-alkenoyloxazolidin-2-ones affords β-amino acid precursors, **57** is a satisfactory ligand to form an active catalyst with nickel(II) perchlorate.[2]

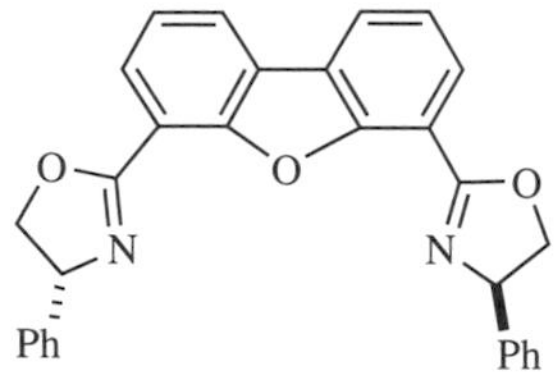

57

After deprotonation, chiral oxazolidin-2-ones add to nitroalkenes.[3,4] This route provides access to enantioenriched amines or more elaborated structures.

Tetrahydropyridine derivatives are formed when ligand **58** adds to dienals.[5] The adducts have great potentials for modification at several sites.

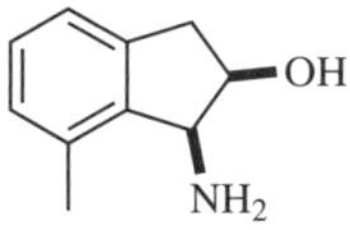

58

Thiols and thiolates are known for their high reactivity toward conjugated systems. Here the main concern is asymmetric induction during the addition processes. Applicable chiral ligands include a bis(dihydroquinidinyl)pyrimidine derivative[6] and the amino ether **59**.[7,8] As for substrate-directed asymmetric addition, 4-substituted *N*-alkenoyloxazolidin-2-thiones are good acceptors.[9] The formation of a 2-phenylthio-6-hydroxycyclohexanecarbamide from a more complex *N*-alkenoyloxazolidin-2-one[10] is due to the presence of proper functionalities that a tandem aldol cyclization follows the Michael reaction.

Copper-catalyzed addition reactions of organozinc species (mostly illustrated by use of Et_2Zn) to enones and nitroalkenes generally proceed in excellent enantioselectivity. Biphenyl-based amidophosphite **60**[11] and peptide-phosphine **61**[12,13] have been identified as good ligands.

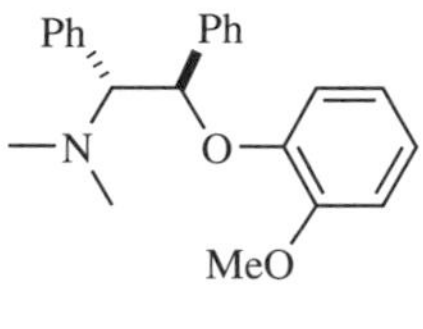

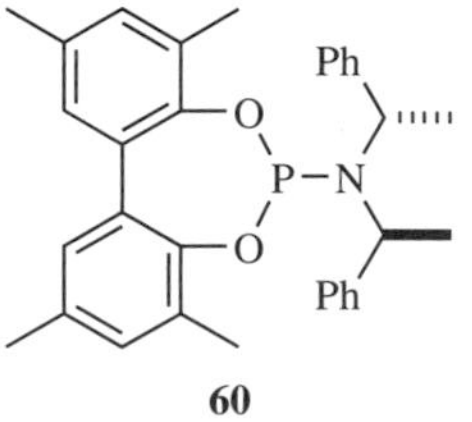

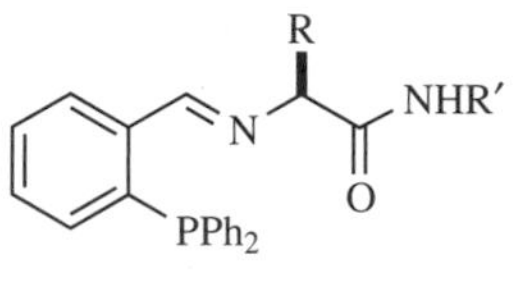

59 **60** **61** (R = *i*-Pr, *t*-Bu)

To promote Michael reaction of nitroalkanes with enones and alkenoic acid derivatives the imidazolidine **62** shows some promise.[14] Double activation by chiral Lewis acid and achiral amine in an enantioselective addition of nitromethane to Michael acceptors is demonstrated.[15]

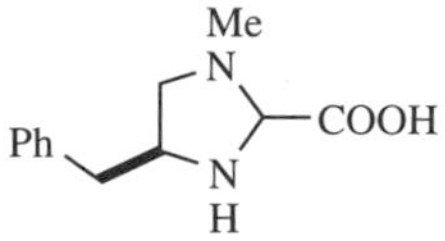

62

Asymmetric Rh-catalyzed addition of α-cyanopropionic esters to acrolein benefits from the bis(oxazoline) ligand **63**.[16] Formation of β-substituted β-arylpropanals from activated arenes and heterocycles such as *N*-methylpyrrole may be considered as a Michael reaction version, even it really is a Friedel-Crafts-type reaction. Imidazolidinone salts **64** are highly effective in promoting the reaction enantioselectively.[17]

Reports on the addition to enones and alkenoic esters with enolates or silyl enol ethers and ketene acetals include catalytic action of quaternized cinchona alkaloids under phase-transfer conditions[18] and allothreonine-derived *B*-aryloxazaborolidinones **65**.[19]

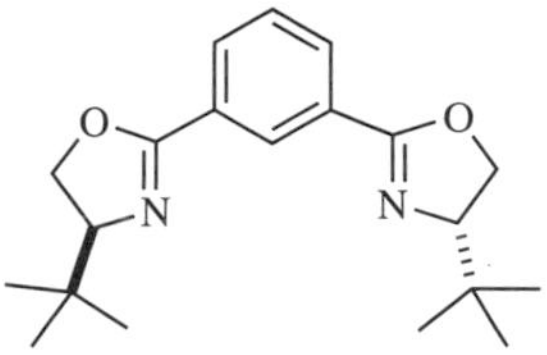

63 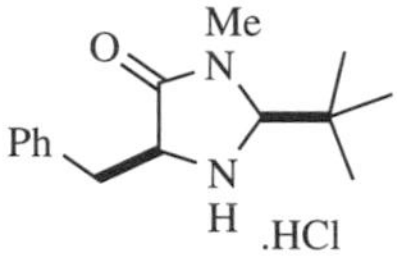 **64** 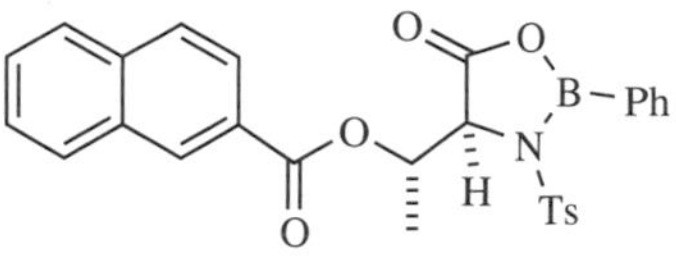 **65**

A catalyst in which Rh is ligated to **66** performs as well as the corresponding BINAP catalyst in effecting aryl transfer from $ArB(OH)_2$ to 2-cycloalkenones.[20] A bis(oxazoline)-copper(II) triflate is found to catalyze alkylation of active arenes (furan, indole, resorcinol dimethyl ether) with β,γ-unsaturated α-keto esters.[21]

Generation of a chiral quaternary center relying on the Michael reaction of β-diketones can be accomplished by introducing a chiral auxiliary via enaminone formation. Chiral *N,N*-diethyl-2-amino-3,3-dimethylbutanamide is such a derivatizing agent.[22]

Addition to acrylic esters has made use of a tripodal oxazoline (**67**).[23] Based on Michael reaction either in solution or solid-phase a route to glutamic acid derivatives has evolved.[24] Chiral benzylic anions are generated from deprotonation of tricarbonylchromium complexes using a chiral lithium amide. Michael reaction of such anions proceeds with excellent ee.[25]

As expected, alkenamides, especially *N*-alkenoyloxazolidin-2-ones, have been studied with respect to their response to asymmetric Michael reactions. The bis(oxazoline) **68**[26] and copper(II) complex of **69**[27] are included in the studies; furthermore, the service of **68** has been extended to a synthesis of homologous *N*-acyl amino acid derivatives through radical addition to *N*-acyl dehydroalanine esters.[28]

Long-range asymmetric induction by a sulfinyl group for the Michael reaction of *N*-alkenoylpyrrolcs **70**[29] is in the realm of substrate direction. Lithiation of chiral ω–toluene-sulfinyl-2,(ω-1)-alkadienoic esters (ω=7, 8) with LDA is followed by cyclization, but only the (2*Z*)-esters cyclize with high diastereoselectivity.[30]

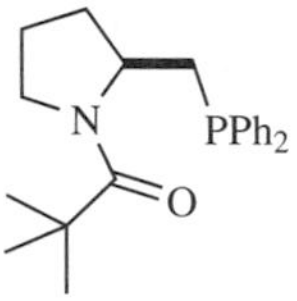

66 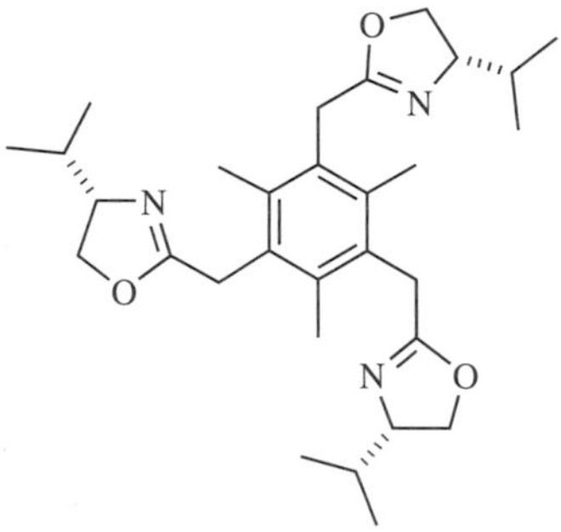 **67** 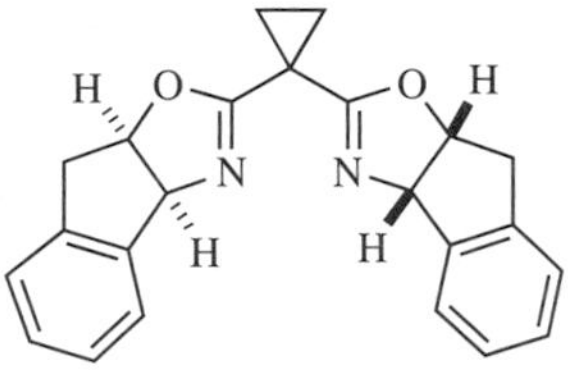 **68**

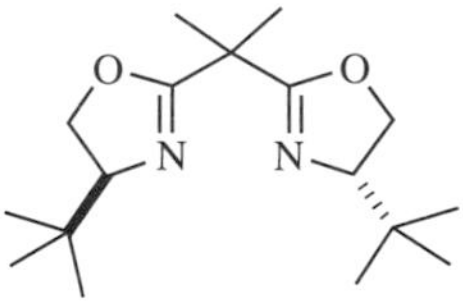

69 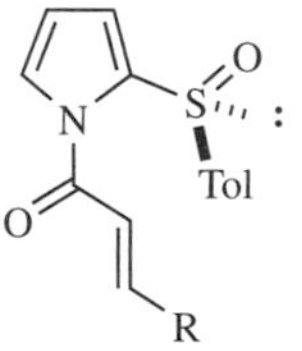 **70**

A synthesis of α-hydroxy acids containing additional functionalities is by Michael reaction of a 1,4-dioxan-2-one.[31] It reacts with many acceptors (enones, conjugated lactones, nitroalkenes).

Many chiral catalysts are used in Michael reactions of carbonyl donors on nitroalkenes. L-Proline[32] and **71**[33] are representative. Copper(II) complexes of peptide-phosphine ligands analogous to **61** promote group transfer from organozinc reagents to nitrocycloalkenes.[34]

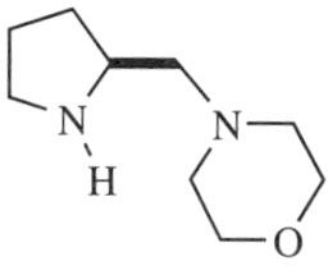

71

Related acceptors are 2-alkenyloxazolines. The steric course for addition is affected by the presence of a bulky substituent (e.g., *t*-Bu) at C-4, rendering asymmetric addition possible.[35]

α-Amination of carbonyl compounds via reaction with azodicarboxylic esters is now well developed. L-Proline appears to be a popular catalyst,[36–38] although bis(oxazoline)-copper(II) triflate type catalysts are useful also.[39]

Intramolecular Stetter reaction is rendered enantioselective by the salt **72**.[40]

BF_4^-

72

[1]Cardillo, G., Fabbroni, S., Gentilucci, L., Gianotti, M., Perciaccante, R., Tolomelli, A. *TA* **13**, 1407 (2002).
[2]Zhuang, W., Hazell, R.G., Jorgensen, K.A. *CC* 1240 (2001).
[3]Leroux, M.-L., Le Gasll, T., Mioskowski, C. *TA* **12**, 1817 (2001).
[4]Feroci, M., Inesi, A., Palombi, L., Rossi, L. *TA* **12**, 2331 (2001).
[5]Tanaka, K., Katsumura, S. *JACS* **124**, 9660 (2002).
[6]McDaid, P., Chen, Y., Deng, L. *ACIEE* **41**, 338 (2002).
[7]Nishimura, K., Ono, M., Nagaoka, Y., Tomioka, K. *ACIEE* **40**, 440 (2001).
[8]Nishimura, K., Tomioka, K. *JOC* **67**, 431 (2002).
[9]Palomo, C., Oiarbide, M., Dias, F., Ortiz, A., Linden, A. *JACS* **123**, 5602 (2001).
[10]Schneider, C., Reese, O. *CEJ* **8**, 2585 (2002).
[11]Alexakis, A., Benhaim, C., Rosset, S., Humam, M. *JACS* **124**, 5262 (2002).
[12]Mizutani, H., Degrado, S.J., Hoveyda, A.H. *JACS* **124**, 779 (2002).
[13]Degrado, S.J., Mizutani, H., Hoveyda, A.H. *JACS* **124**, 13362 (2002).
[14]Halland, N., Hazell, R.G., Jorgensen, K.A. *JOC* **67**, 8331 (2002).
[15]Itoh, K., Kanemasa, S. *JACS* **124**, 13394 (2002).
[16]Motoyama, Y., Koga, Y., Kobayashi, K., Aoki, K., Nishiyama, H. *CEJ* **8**, 2968 (2002).
[17]Paras, N.A., MacMillan, D.W.C. *JACS* **123**, 4370 (2001); *JACS* **124**, 7894 (2002).
[18]Zhang, F.-Y., Corey, E.J. *OL* **3**, 639 (2001).
[19]Harada, T., Iwai, H., Takatsuki, H., Fujita, K., Kubo, M., Oku, A. *OL* **3**, 2101 (2001).
[20]Kuriyama, M., Tomioka, K. *TL* **42**, 921 (2001).
[21]Jensen, K.B., Thorhauge, J., Hazell, R.G., Jorgensen, K.A. *ACIEE* **40**, 160 (2001).
[22]Christoffers, J., Mann, A. *CEJ* **7**, 1014 (2001).
[23]Kim, S.-G., Ahn, K.H. *TL* **42**, 4175 (2001).
[24]O'Donnell, M.J., Delgado, F., Dominguez, E., de Blas, J., Scott, W.L. *TA* **12**, 821 (2001).
[25]Beckwith, R.E.J., Heron, N., Simpkins, N.S. *JOMC* **658**, 21 (2002).
[26]Sibi, M.P., Chen, J. *JACS* **123**, 9472 (2001).
[27]Evans, D.A., Scheidt, K.A., Johnston, J.N., Willis, M.C. *JACS* **123**, 4480 (2001).
[28]Sibi, M.P., Asano, Y., Sausker, J.B. *ACIEE* **40**, 1293 (2001).
[29]Arai, Y., Ueda, K., Xie, J., Masaki, Y. *CPB* **49**, 1609 (2001).
[30]Maezaki, N., Yuyama, S., Sawamoto, H., Suzuki, T., Izumi, M., Tanaka, T. *OL* **3**, 29 (2001).
[31]Dixon, D.J., Ley, S.V., Rodriguez, F. *OL* **3**, 3753 (2001).

[32]Enders, D., Seki, A. *SL* 26 (2002).
[33]Betancort, J.M., Barbas III, C.F. *OL* **3**, 3737 (2001).
[34]Luchaco-Cullis, C.A., Hoveyda, A.H. *JACS* **124**, 8192 (2002).
[35]Basil, L.F., Meyers, A.I., Hassner, A. *T* **58**, 207 (2002).
[36]List, B. *JACS* **124**, 5656 (2002).
[37]Bogevig, A., Juhl, K., Kumaragurubaran, N., Zhuang, W., Jorgensen, K.A. *ACIEE* **41**, 1790 (2002).
[38]Kumaragurubaran, N., Juhl, K., Zhuang, W., Bogevig, A., Jorgensen, K.A. *JACS* **124**, 6254 (2002).
[39]Juhl, K., Jorgensen, K.A. *JACS* **124**, 2420 (2002).
[40]Kerr, M.S., de Alaniz, J.R., Rovis, T. *JACS* **124**, 10298 (2002).

Cycloadditions. Sulfonium ylides such as derived from **73** react with conjugated carbonyl compounds to afford chiral vinylcyclopropanes,[1] whereas asymmetric cyclopropanation of styrenes with ethyl diazoacetate is readily effected in the presence of a cobalt complex (**74A**).[2]

A variety of catalysts are effective in the promotion of enantioselective β-lactam and β-lactone formation by the [2+2]cycloaddition involving ketenes. They include cinchona alkaloid derivatives[3] and planar chiral DMAP analog **75**[4] in the cases of β-lactams, and a bis(oxazoline)-copper complex for β-lactones.[5] An alternative strategy for elaborating β-lactams involves the use of SAMP-hydrazones as the imine component.[6] *N*-Acyl derivatives of the C_2-symmetrical 2,5-dimethylpyrroldine furnish chiral cyclobutanones[7] on enol triflation and combination with alkenes.

73

74A

74B

74C

Combining nitrones with electron-deficient alkenes by a 1,3-dipolar cycloaddition is subject to asymmetric catalysis, e.g., with **76**[8] and **74B**.[9] A chiral auxiliary for generating chiral azomethine ylides from aldehydes is (*5S,6R*)-diphenylmorpholin-2-one.[10]

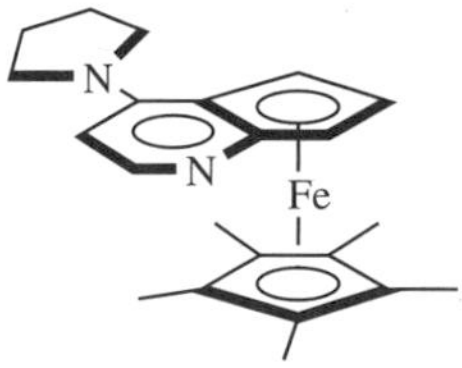

Ar = C_6F_5

75 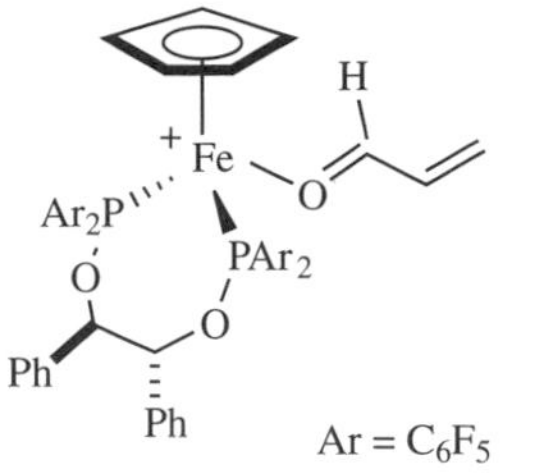 **76**

Corey's important contributions pertaining to catalytic enantioselective Diels-Alder reactions are summarized.[11] In the recent past several more catalysts are available to conduct asymmetric Diels-Alder reactions: (pybox)-Sc(III) triflate,[12] oxazaborolidinium triflates **77**,[13,14] and the C_2-symmetrical bis(sulfinylimines) **78**,[15] **79**[16] in conjuction with $(dba)_3Pd_2$ and Cu salt, respectively. The tridentate Cr(III) catalyst **80** controls hetero-Diels-Alder reactions between dienes and aldehydes[17,18] including the synthesis of dihydropyranyl ethers.[19]

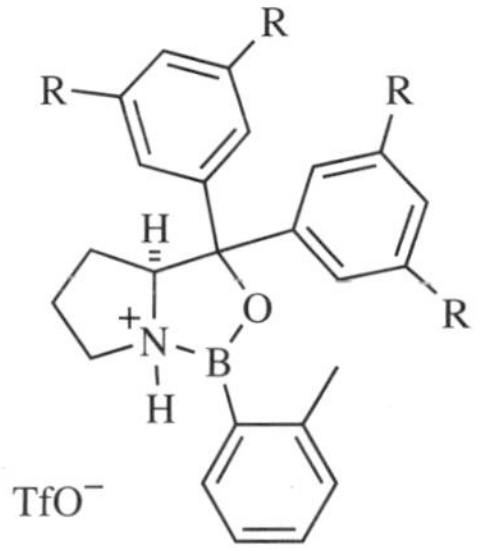

77 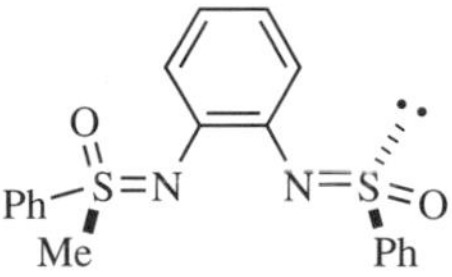 **78**

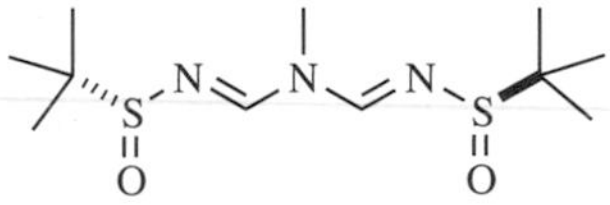

79 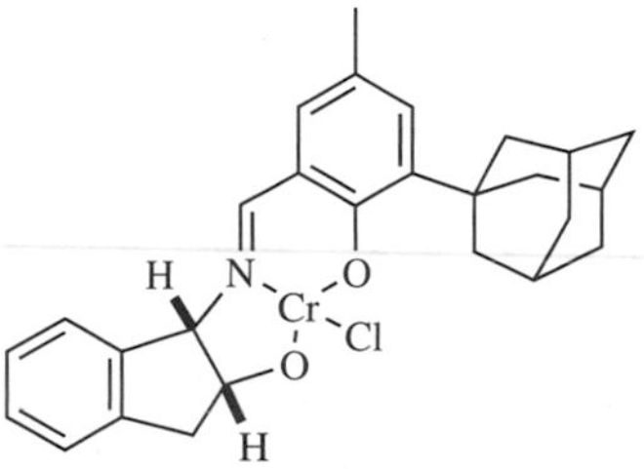 **80**

Incorporation of a dienophilic moiety in the bornane skeleton, i.e., as an *endo*-2-(2-alkenoyl)isoborneol the asymmetric Diels-Alder reaction does not require metal catalyst.[20]

Photochemical electrocyclization of the inclusion complex of **81**, 4-isopropyltropolone methyl ether and chloroform gives optically active products (high ee).[21]

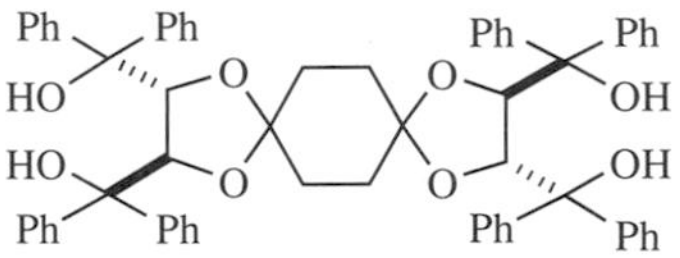

81

[1]Ye, S., Huang, Z.-Z., Xia, C.-A., Tang, Y., Dai, L.-X. *JACS* **124**, 2432 (2002).
[2]Ikeno, T., Nishizuka, A., Sato, M., Yamada, T. *SL* 406 (2001).
[3]Taggi, A.E., Hafez, A.M., Wack, H., Young, B., Ferraris, D., Lectka, T. *JACS* **124**, 6626 (2002).
[4]Hodous, B.L., Fu, G.C. *JACS* **124**, 1578 (2002).
[5]Evans, D.A., Janey, J.M. *OL* **3**, 2125 (2001).
[6]Fernandez, R., Ferrete, A., Lassaletta, J.M., Llera, J.M., Martin-Zamora, E. *ACIEE* **41**, 831 (2002).
[7]Ghosezs, L., Mahuteau-Betzer, F., Genicot, C., Vallribera, A., Cordier, J.-F. *CEJ* **8**, 3411 (2002).
[8]Viton, F., Bernardinelli, G., Kündig, E.P. *JACS* **124**, 4968 (2002).
[9]Mita, T., Ohtsuki, N., Ikeno, T., Yamada, T. *OL* **4**, 2457 (2002).
[10]Sebahar, P.R., Osada, H., Usui, T., Williams, R.M. *T* **58**, 6311 (2002).
[11]Corey, E.J. *ACIEE* **41**, 1650 (2002).
[12]Fukuzawa, S., Matsuzawa, H., Metoki, K. *SL* 709 (2001).
[13]Corey, E.J., Shibata, T., Lee, T.W. *JACS* **124**, 3808 (2002).
[14]Ryu, D.H., Lee, T.W., Corey, E.J. *JACS* **124**, 9992 (2002).
[15]Bolm, C., Simic, O. *JACS* **123**, 3830 (2001).
[16]Owens, T.D., Hollander, F.J., Oliver, A.G., Ellman, J.A. *JACS* **123**, 1539 (2001).
[17]Cox, J.M., Rainier, J.D. *OL* **3**, 2919 (2001).
[18]Jolly, G.D., Jacobsen, E.N. *OL* **4**, 1795 (2002).
[19]Gademann, K., Chavez, D.E., Jacobsen, E.N. *ACIEE* **41**, 3059 (2002).
[20]Palomo, C., Oiarbide, M., Garcia, J.M., Gonzalez, A., Lecumberri, A., Linden, A. *JACS* **124**, 10288 (2002).
[21]Tanaka, K., Nagahiro, R., Urbanczyk-Lipkowska, Z. *OL* **3**, 1567 (2001).

Ketone reductions. Corey's oxazaborolidine coupled with *N*-ethyl-*N*-isopropylaniline–borane is very effective in the reduction of α-sulfonyl and α-sulfonyloxy ketones.[1,2] Products from the latter reaction are readily converted to chiral epoxides. A new analogue of the original oxazaborolidine has 9-BBN instead of a simple alkylboron residue, and the catalyst (**82**) is particularly good for reduction of hindered and substituted alkyl aryl ketones.[3] There is also a reasonably effective structural variant, namely, (5*S*)-1,3-diaza-2-phospha-2-oxo-2-chloro-3-phenylbicyclo[3.3.0]octane (**83**).[4]

Reduction of β-keto sulfones with the $NaBH_4$-Me_3SiCl system and a polymer supported *N*-sulfonyl α,α-diphenylprolinol furnished the (*S*)-alcohols with good ee if the ketone group is moderateley hindered.[5]

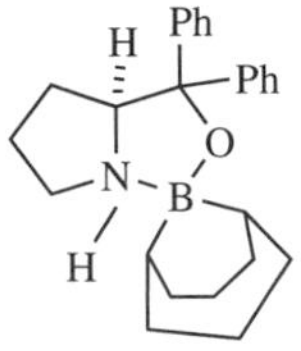

82

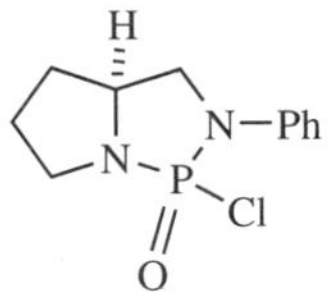

83

Boronates derived from tartaric acid (e.g., **84**)[6] assist the enantioselective delivery of hydride ion from $LiBH_4$ to aryl ketones. Hydrosilylation of aryl ketones in the presence of CuCl is accelerated by bisphosphine **85**.[7] For heteroaromatic ketones a similar ligand (with alkoxylated biphenyl moiety) is employed.[8]

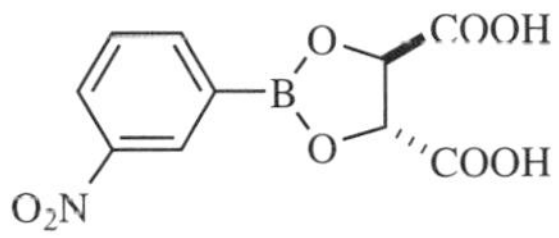

84

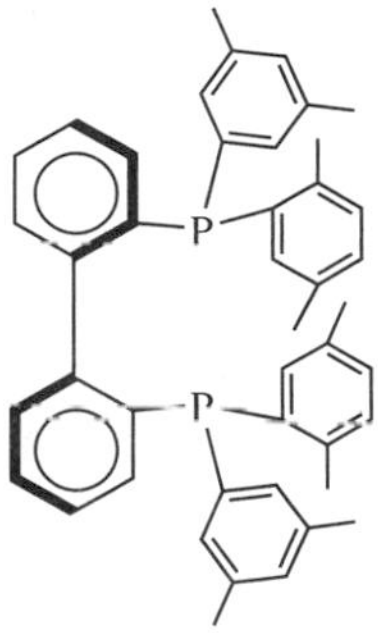

85

Transfer hydrogenation of ketones in the presence of Ru complexes is highly efficient and the attendant enantioselectivity can be very high with some well-known chiral ligands such as a proline anilide.[9] In using (*1R,2S*)-*N*-benzylnorephedrine the reduction is complementary to whole cell bioconversion therefore both enantiomeric alcohols are accessible.[10] While both Rh and Ru complexes are serviceable in promoting transfer hydrogenation of α–substituted acetophenones, Rh seems to be superior.[11] Catalyst **86** and closely related analogues serve to convert *o*-acylbenzoic esters to optically active phthalides.[12] 1,3-Diketones give mainly *anti*-1,3-diols using such a catalyst.[13]

Ru
O NR
Ph

86

On the other hand, symmetrical 2-alkyl-1,3-diketones are reduced by $NaBH_4$ in the presence of the cobalt complex **74C** to enantiomeric aldol products of the *anti* pattern.[14]

Hydrosilylation of ketones in the presence of P,N-ligands such as **87** have been developed.[15] Ketoximes furnish chiral amines in a catalytic system that contains Ru complexed to a phosphinoferrocene bearing also an oxazoline moiety.[16]

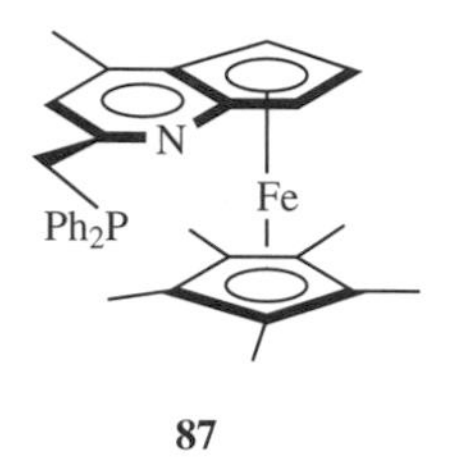

87

[1]Cho, B.T., Kim, D.J. *TA* **12**, 2043 (2001).
[2]Cho, B.T., Yang, W.K., Choi, O.K. *JCS(P1)* 1204 (2001).
[3]Kanth, J.V.B., Brown, H.C. *T* **58**, 1069 (2002).
[4]Basavaiah, D., Reddy, G.J., Chandrashekar, V. *TA* **13**, 1125 (2002).
[5]Zhao, G., Hu, J., Qian, Z., Yin, W. *TA* **13**, 2095 (2002).
[6]Suri, J.T., Vu, T., Hernandez, A., Congdon, J., Singaram, B. *TL* **43**, 3649 (2002).
[7]Lipshutz, B.H., Noson, K., Chrisman, W. *JACS* **123**, 12917 (2001).
[8]Lipshutz, B.H., Lower, A., Noson, K. *OL* **4**, 4045 (2002).
[9]Rhyoo, H.Y., Yoon, Y.-A., Park, H.-J., Chung, Y.K. *TL* **42**, 5045 (2001).
[10]Hage, A., Petra, D.G.I., Field, J.A., Schipper, D., Wijnberg, J.B.P.A., Kamer, P.C.J., Reek, J.N.H., van Leeuwen, P.W.N.M, Wever, R., Schoemaker, H.E. *TA* **12**, 1025 (2001).
[11]Cross, D.J., Kenny, J.A., Houson, I., Campbell, L., Walsgrove, T., Wills, M. *TA* **12**, 1801 (2001).
[12]Everaere, K., Scheffler, J.-L., Mortreux, A., Carpentier, J.-F. *TL* **42**, 1899 (2001).
[13]Cossy, J., Eustache, F., Dalko, P.I. *TL* **42**, 5005 (2001).
[14]Ohtsuka, Y., Koyasu, K., Ikeno, T., Yamada, T. *OL* **3**, 2543 (2001).
[15]Tao, B., Fu, G.C. *ACIEE* **41**, 3892 (2002).
[16]Takei, I., Nishibayashi, Y., Ishii, Y., Mizobe, Y., Uemura, S. *CC* 2360 (2001).

Asymmetric hydrogenation. Asymmetric hydrogenation of dehydroamino acid derivatives continues to be the focus of research. Thus, Rh complexed to phosphine ligands **88**,[1] **89**,[2] **90**,[3] and **91**[4] has been evaluated. For hydrogenation of α-aryl enamides to afford chiral α-substituted benzylamines, **92A/B** are effective ligands.[5]

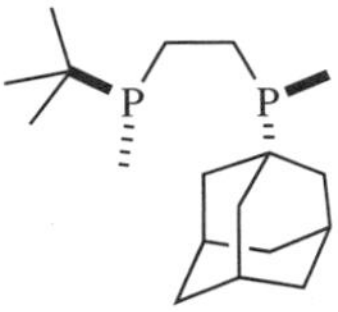

88 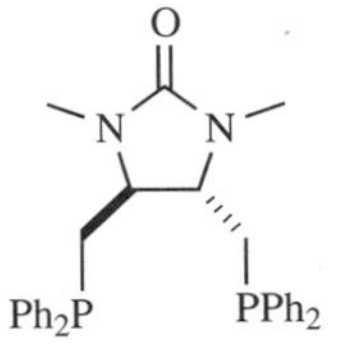 **89** 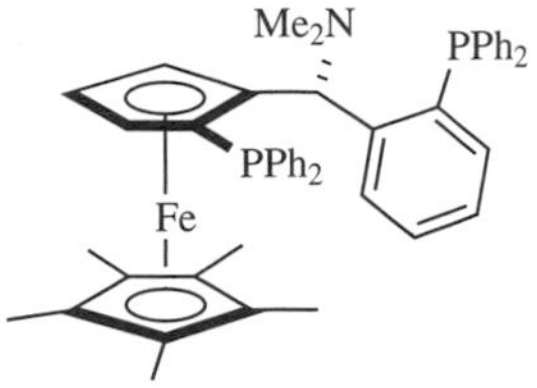 **90**

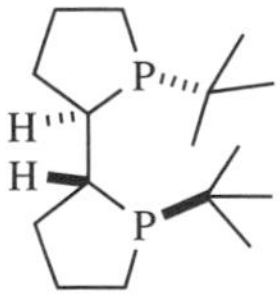

91 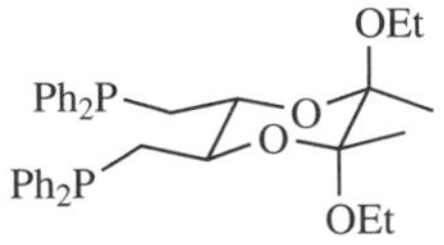 **92A** 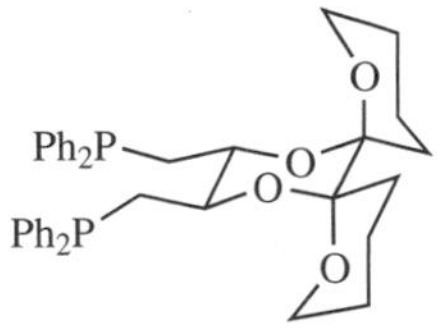 **92B**

While a system consisting of Rh-DuPHOS and ionic liquid also is shown to be effective,[6] another report indicates the superiority of a pre-catalyst containing norbornadiene instead of 1,5-cyclooctadiene.[7] A [(Me_2)DuPHOS-RhII] salt, available from the η^6-cyclooctatriene complex by hydrogenation, is a good asymmetric hydrogenation catalyst.[8]

The ability to promote asymmetric hydrogenation of arylalkenes is shown by the iridium complex **93**.[9] Also, 2-pyrones undergo enantioselective hydrogenation in the presence of cinchonine-doped Pd/TiO_2[10] and chiral ketones are obtained by exposure of α-substituted enones to baker's yeast.[11]

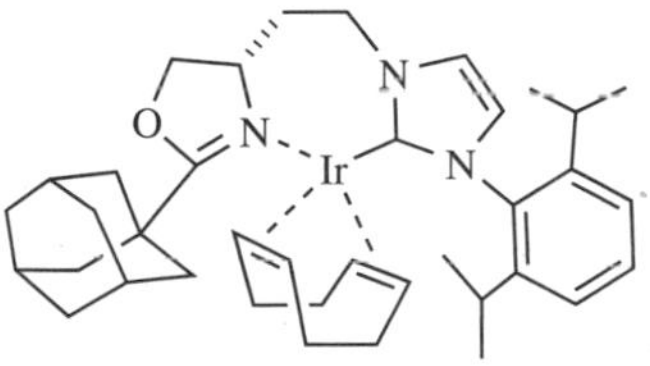

93

A spirobiindanedioxy(dimethylamino)phosphine (**94A**)[12] and the bisphosphine (**94B**)[13] based on 6,6′-locked biphenyl directs Rh-catalyzed reduction of α-acetaminostyrenes and Ru-catalyzed hydrogenation of enol acetates, respectively. In expanding the utility of established systems in asymmetric hydrogenation of new types of substrates several combinations of Ru and chiral ligands[14,15] and (*R,R*)-tartaric acid-modified Raney nickel[16] have been tried on reducing β-keto esters. Meerwein-Ponndorf-Verley-type reduction of azirenes[17] and reductive acetylation of ketoximes with an immobilized lipase[18] are also noteworthy.

94A **94B**

For a short comment on certain aspects of Ru- and Rh-catalyzed asymmetric hydrogenation, see Ref. 19.

[1]Ohashi, A., Imamoto, T. *OL* **3**, 373 (2001).
[2]Lee, S.-G., Zhang, J.Y. *OL* **4**, 2429 (2002).
[3]Ireland, T., Tappe, K., Grossheimann, G., Knochel, P. *CEJ* **8**, 843 (2002).
[4]Tang, W., Zhang, X. *ACIEE* **41**, 1612 (2002).
[5]Li, W., Waldkirch, J.P., Zhang, X. *JOC* **67**, 7618 (2002).
[6]Guernik, S., Wolfson, A., Herskowitz, M., Greenspoon, N., Gerech, S. *CC* 2314 (2001).
[7]Borner, A., Heller, D. *TL* **42**, 223 (2001).
[8]Wiles, J.A., Bergens, S.H., Vanhessche, K.P.M., Dobbs, D.A., Rautenstrauch, V. *ACIEE* **40**, 914 (2001).
[9]Powell, M.T., Hou, D.-R., Perry, M.C., Cui, X., Burgess, K. *JACS* **123**, 8878 (2001).
[10]Huck, W.-R., Mallat, T., Baiker, A. *NJC* **26**, 6 (2002).
[11]Kawai, Y., Hayashi, M., Tokitoh, N. *TA* **12**, 3007 (2001).
[12]Hu, A.G., Fu, Y., Xie, J.-H., Zhou, H., Wang, L.-X., Zhou, Q.-L. *ACIEE* **41**, 2348 (2002).
[13]Wu, S., Wang, W., Tang, W., Lin, M., Zhang, X. *OL* **4**, 4495 (2002).
[14]Madec, J., Pfister, X., Phansavath, P., Ratovelomanana-Vidal, V., Genet, J.-P. *T* **57**, 2563 (2001).
[15]Wu, J., Chen, H., Zhou, Z.-Y., Yeung, C.H., Chan, A.S.C. SL 1050 (2001).
[16]Sugimura, T., Nakagawa, S., Tai, A. *BCSJ* **75**, 355 (2002).
[17]Roth, P., Andersson, P.G., Somfai, P. *CC* 1752 (2002).
[18]Choi, Y.K., Kim, M.J., Ahn, Y., Kim, M.-J. *OL* **3**, 4099 (2001).
[19]Rossen, K. *ACIEE* **40**, 4611 (2001).

Oxidations. Oxidation of sulfides to give chiral sulfoxides by H_2O_2 or ROOH is subject to asymmetric induction by enveloping the metal catalyst in a chiral environment. The Sharpless system of $(i\text{-PrO})_4$Ti/L-diethyl tartrate is now combined with a 2-furylalkyl hydroperoxide,[1] vanadyl acetylacetonate and modified salen ligands **95**[2] and **96**[3] also form satisfactory catalysts.

95 **96**

A Ti(salen) complex in conjuction with urea–hydrogen peroxide oxidizes cyclic dithioacetals selectively.[4]

To introduce an oxygen function at the allylic position of a cycloalkene by *t*-butyl 4-nitrobenzoate a chiral Cu-bis(oxazoline) species renders the process highly enantioselective.[5]

The trityl-type salt **97** abstracts hydride from one face of tricarbonyliron complexes of *meso*-dienes, thereby generating chiral dienone complexes.[6]

+

PF_6^-

97

Asymmetric Kharasch-Sosnovsky reaction for catalytic allylic oxidation of alkenes has been reviewed.[7]

[1]Massa, A., Siniscalchi, F.R., Bugatti, V., Lattanzi, A., Scettri, A. *TA* **13**, 1277 (2002).
[2]Ohta, C., Shimizu, H., Kondo, A., Katsuki, T. *SL* 161 (2002).
[3]Pelotier, B., Anson, M.S., Campbell, I.B., Macdonald, S.J., Priem, G., Jackson, R.F.W. *SL* 1055 (2002).
[4]Tanaka, T., Saito, B., Katsuki, T. *TL* **43**, 3259 (2002).
[5]Andrus, M.B., Zhou, Z. *JACS* **124**, 8806 (2002).
[6]Magdziak, D., Pettus, L.H., Pettus, T.R.R. *OL* **3**, 557 (2001).
[7]Eames, J., Watkinson, M. *ACIEE* **40**, 3567 (2001).

Other enantioselective reactions. There is much interest in the development of enantioselective rearrangement reactions. Allylic sulfur compounds are accessible from racemic *O*-allyl thiocarbamates by a Pd-catalyzed rearrangement in the presence of **15**,[1] and in turn, allyl sulfides are transformed into the 1,3-transposed chiral *N*-tosylallylamines with TsN_3 via sulfimidation and [2,3]sigmatropic rearrangement when exposed to a Ru(CO)(salen) complex.[2] Quinine and quinidine direct steric courses in opposite directions in the Claisen rearrangement of allylic esters of *N*-trifluoroacetylglycinate.[3] A method for elaboration of 2-alkyl-4-butenoic acid derivatives involves the acid-catalyzed addition of allyl alcohol to chiral *N*-(1-alkynyl)oxazolidin-2-ones.[4]

After reaction of *N*-allylamines with acyl chlorides a Claisen rearrangement delivers 4-pentenamides in which the configurations of C-2 and C-3 are controllable by the addition of ligand **98**.[5] Asymmetric aliphatic Claisen rearrangement in the presence of chiral bis(oxazoline)-$Cu(OTf)_2$ complexes has been studied.[6]

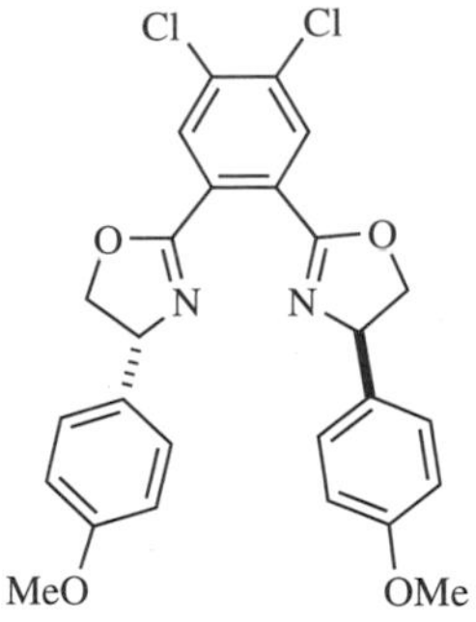

98

Phosphine **99** is new catalyst for the isomerization of allylic alcohols to chiral aldehydes.[7] Regioselective double bond migration of cyclic acetals derived from aldehydes and *cis*-2-butene-1,4-diol is readily accomplished with **100** and $LiBEt_3H$.[8]

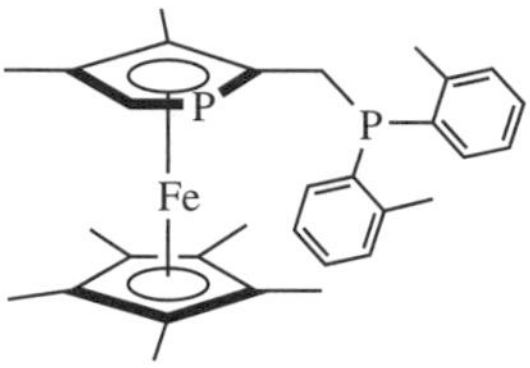

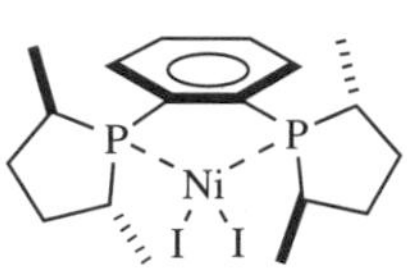

99 **100**

Chiral rhodium catalysts cause α-diazoarylacetic esters to insert into active C-H bonds (such as allylic position[9] or a methyl group adjoining an amino nitrogen atom[10]) and the products have high ee. The former process effectively provides the type of products expected of asymmetric Claisen rearragement and the latter α-chiral β-amino esters. The allylic insertion as further extended to silyl enol ethers is equivalent to an asymmetric Michael reaction.[11]

Pd-complexes of the biphenyl-based diphosphines (**101**) are promoters of ene-type cyclization of 1,6-enynes in the enantioselective manner.[12]

Asymmetric oxidative coupling of 2-naphthol to provide chiral BINOL is important because the latter is a valuable ligand. A useful catalyst is **102.**[13] A diarylation process using aryllead triacetates as reagent and brucine as asymmetry inducer gives rise to optically 2,6-diarylphenols and anilines.[14]

Introduction of a chiral sulfinyl group to C-2 of *N,N*-diisopropyl-1-naphthalamide restricts the formation of only one atropisomer. A series of such atropisomers are readily formed on exchange of the sulfinyl group to lithium followed by alkylation.[15]

Novel P,N-ligands such as **103** based on ketopinic acid prove useful in asymmetric Heck reaction.[16]

By using chiral metal complexes (e.g., **104, 105**) cyclization of dienes and unsaturated aldehydes are rendered enantioselective. Thus, chiral cyclopentanes possessing two adjacent substituents are readily prepared.[17,18]

Sulfinyl alcohols such as **106** are excellent source for asymmetric protonation.[19] As for asymmetric enolsilylation with magnesium amides in which the nitrogen atom carries both chiral and achiral substituents the steric nature of the achiral sidearm is critically important in determining the enantioselectivity of the reagent. Benzyl group appears to be the best.[20]

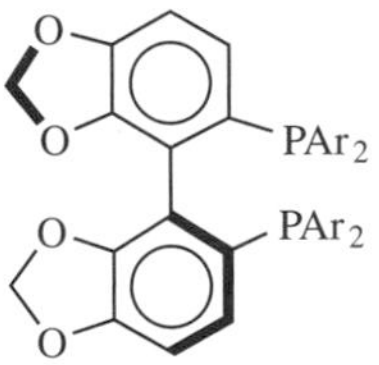

PAr_2 PAr_2

101 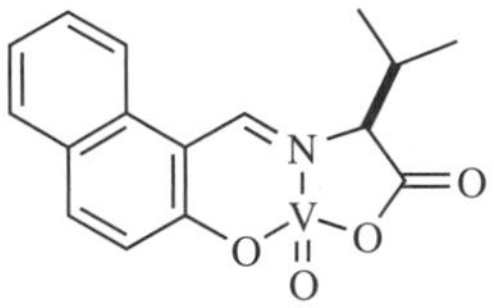 **102** 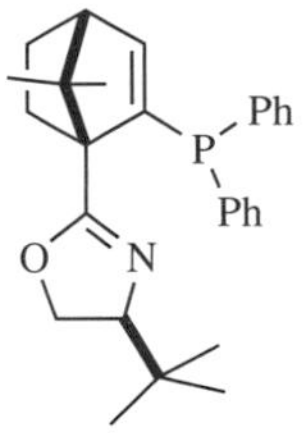 **103**

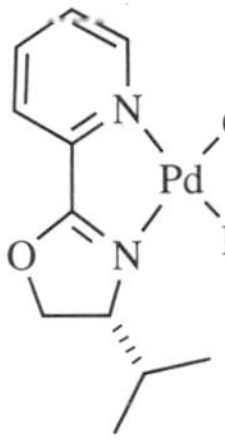

104 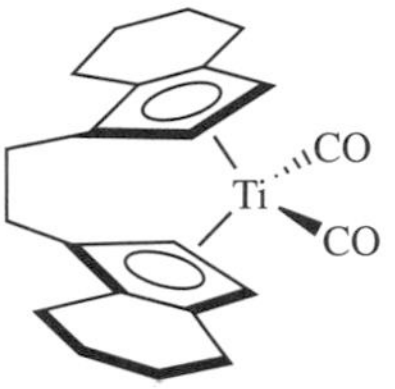 **105** 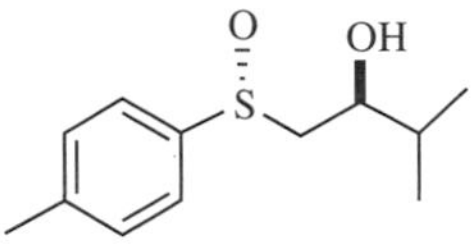 **106**

Ring opening of 1,1-diphenylsila-3-cyclopentene oxide with the lithium amide **107** leads to an allylic alcohol which is readily elaborated into (*2R,3S,4R*)-1,2,3,4-tetrahydroxyalkanes.[21]

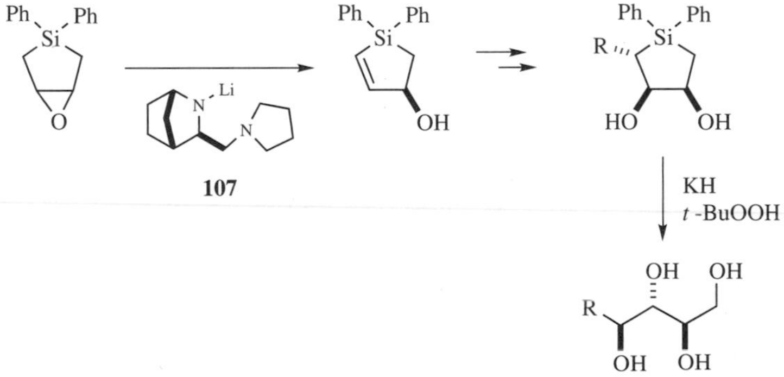

107

KH
t-BuOOH

[1]Gais, H.-J., Böhme, A. *JOC* **67**, 1153 (2002).
[2]Murakami, M., Katsuki, T. *TL* **43**, 3947 (2002).
[3]Kazmaier, U., Mues, H., Krebs, A. *CEJ* **8**, 1850 (2002).
[4]Mulder, J.A., Hsung, R.P., Frederick, M.O., Tracey, M.R., Zificsak, C.A. *OL* **4**, 1383 (2002).
[5]Yoon, T.P., MacMillan, D.W.C. *JACS* **123**, 2911 (2001).
[6]Abraham, L., Czerwonka, R., Hiersemann, H. *ACIEE* **40**, 4700 (2001).
[7]Tanaka, K., Fu, G.C. *JOC* **66**, 8177 (2001).
[8]Frauenrath, H., Brethauer, D., Reim, S., Maurer, M., Raabe, G. *ACIEE* **40**, 177 (2001).
[9]Davies, H.M.L., Ren, P., Jin, Q. *OL* **3**, 3587 (2001).
[10]Davies, H.M.L., Venkataramani, C. *ACIEE* **41**, 2197 (2002).
[11]Davies, H.M.L., Ren, P. *JACS* **123**, 2070 (2001).
[12]Hatano, M., Terada, M., Mikami, K. *ACIEE* **40**, 249 (2001).
[13]Hon, S.-W., Li, C.-H., Kuo, J.-H., Barhate, N.B., Liu, Y.-H., Wang, Y., Chen, C.-T. *OL* **3**, 869 (2001).
[14]Kano, T., Ohyashi, Y., Saito, S., Yamamoto, H. *JACS* **124**, 5365 (2002).
[15]Clayden, J., Mitjans, D., Youssef, L.H. *JACS* **124**, 5266 (2002).
[16]Gilbertson, S.R., Fu, Z. *OL* **3**, 161 (2001).
[17]Pei, T., Widenhoefer, R.A. *JOC* **66**, 7639 (2001).
[18]Mandal, S.K., Amin, S.R., Crowe, W.E. *JACS* **123**, 6457 (2001).
[19]Asensio, G., Gil, J., Aleman, P., Medio-Simon, M. *TA* **12**, 1359 (2001).
[20]Anderson, J.D., Garcia, P.G., Hayes, D., Henderson, K.W., Kerr, W.J., Moir, J.H., Fondekar, K.P. *TL* **42**, 7111 (2001).
[21]Liu, D., Kozmin, S.A. *ACIEE* **40**, 4757 (2001).

Chloral.

Formylation.[1] Alcohols are formylated with chloral in the presence of phenols.

[1]Ram, R.N., Meher, N.K. *T* **58**, 2997 (2002).

Chloramine-T. **20,** 103; **21,** 125

N-Sulfonylsulfilimines.[1] Preparation of RR′S=NTs from sulfides RSR′ and chloramine-T (15 examples, 70–99%) is carried out at room temperature.

Aziridines.[2] Under phase-transfer conditions alkenes are converted into *N*-tosylaziridines by the chloroamine-T/iodine system.

[1]Marzinzik, A.L., Sharpless, K.B. *JOC* **66**, 594 (2001).
[2]Kano, D., Minakata, S., Komatsu, M. *JCS(P1)* 3186 (2001).

Chlorine dioxide. **21,** 125

Oxidation of sulfides.[1] Oxidation of sulfides to sulfoxides or sulfones with ClO_2 depends on reaction conditions. At room temperature in a solvent sulfoxides are obtained, whereas with aqueous ClO_2 at 50° further oxidation to sulfones is observed.

[1]Kuchin, A.V., Rubtsova, S.A., Loginova, I.V., Subbotina, S.N. *RJOC* **36**, 1819 (2000).

Chlorobis(cyclopentadienyl)(dimethylaluminum)methylenetitanium. [Tebbe reagent]

1-Trimethysilyl-1-alkenes.[1] Treatment of the adducts from carbonyl compounds and lithiomethyltrimethylsilane with Tebbe reagent leads to elimination of LiOH.

[1]Kwan, M.L., Yeung, C.W., Breno, K.L., Doxsee, K.M. *TL* **42**, 1411 (2001).

Chloro(1,5-cyclooctadiene)pentamethylcyclopentadienylruthenium(I). 21, 126–127

Bicyclic pyridines and pyridones.[1,2] 1,6-Diynes and 1,ω-dinitriles undergo [2+2+2]cycloaddition in the presence of the Ru complex to give pyridine derivatives in a regioselective manner.[1] When the dinitriles are replaced with isocyanates, the products are pyridones.[2]

n = 1, 2, 3

(2-Formyl-1-cyclopentenyl)methylenemalonic esters.[3] Cycloaddition involving ketomalonic esters leads to pyrans which isomerize in situ.

X = O, NTs, $C(COOMe)_2$

[1]Yamamoto, Y., Ogawa, R., Itoh, K. *JACS* **123**, 6189 (2001).
[2]Yamamoto, Y., Takagishi, H., Itoh, K. *OL* **3**, 2117 (2001).
[3]Yamamoto, Y., Takagishi, H., Itoh, K. *JACS* **124**, 6844 (2002).

Chloro(cyclopentadienyl)(bisdiphenylphosphino)methaneruthenium(I).

Hydration of 1-alkynes.[1] Highly regioselective conversion of 1-alkynes in isopropanol to aldehydes (anti-Markovnikov hydration) is realized with the Ru complex. 1,ω-Diynes afford dialdehydes. This clean reaction does not affect ether, ester, and cyano groups.

[1]Suzuki, T., Tokunaga, M., Wakatsuki, Y. *OL* **3**, 735 (2001).

Chlorodicyclohexylborane.

Aldol reaction. The title compound is useful for mediating regioselective aldol reactions of α,α′-dioxygenated ketones with aldehydes.

TPSO–CH2–CO–CH2–OR —(Chx_2BCl - amine ; PhCHO)→ products + products

R = Bn, Bz

R = Bn 79 : 21

R = Bz >95 : 5

[1]Murga, J., Falomir, E., Carda, M., Gonzalez, F., Marco, J.A. *OL* **3**, 901 (2001).

2-Chloro-4,6-dimethoxy-1,3,5-triazine. 21, 128

Carboxyl activation. Carboxylic acids are activated on reaction with the title reagent in the presence of *N*-methylmorpholine. Weinreb amides and ketones are readily formed on subsequent reaction with MeNHOMe,[1] and Grignard reagents (with CuI),[2] respectively.

Peptide synthesis using this reagent is advantageous, for racemization is minimized. One-pot reactions generally provide products that are >97% pure.[3]

[1]De Luca, L., Giacomelli, G., Taddei, M. *JOC* **66**, 2534 (2001).
[2]De Luca, L., Giacomelli, G., Porcheddu, A. *OL* **3**, 1519 (2001).
[3]Garrett, C.E., Jiang, X., Prasad, K., Repic, O. *TL* **43**, 4161 (2002).

4-Chloro-5*H*-1,2,3-dithiazol-5-one.

N,N′-Dialkylureas.[1] The title reagent (**1**) reacts with alkylamines at room temperature to give symmetrical ureas.

1

[1]Chang, Y.-G., Lee, H.-S., Kim, K. *TL* **42**, 8197 (2001).

2-Chloroethylsulfonyl formaldoxime *O*-benzyl ether.

C-H insertion.[1] Photochemical homologation at an α-carbon of nitrogen and oxygen in certain amides and ethers is achieved with the title reagent. Adamantane gives 1-formaldoxime ether in the same way.

adamantane + Cl–CH2CH2–SO2–CH=NOBn —(hv)→ 1-adamantyl–CH=NOBn 72%

[1]Kim, S., Kim, N., Chung, W., Cho, C.H. *SL* 937 (2001).

N-Chloro-2,3,4,4,5,6-hexachloro-2,5-cyclohexadienylideneamine.

Chlorination.[1] This mild and highly regioselective chlorinating agent (for phenols) is prepared from pentachloroaniline: *N,N*-dichlorination with HOCl, and oxidation with iodine which causes one chlorine migration from nitrogen to C-4.

[1]Mamaghani, M., Zolfigol, M.A., Shojaei, M. *SC* **32**, 735 (2002).

1-Chloromethyl-4-fluoro-1,4-diazoniabicyclo[2.2.2]octane salts. 18, 100; **20,** 106; **21,** 129–130

Fluorination. Application of ionic liquid to fluorination of arenes with the title reagent has been reported.[1] Indoles are converted into 3-fluorooxindoles.[2]

[bmim]PF_6

99%

Homoallyl amines. The combination of aldehydes, primary amines, and allyltributylstannane are induced to react by the title reagent.[3]

Deprotection. A number of ethers (PMP, THP) and 1,3-dithianes are cleaved by this reagent.[4]

[1]Laali, K.K., Borodkin, G.I. *JCS(P1)* 953 (2002).
[2]Baudoux, J., Salit, A.-F., Cahard, D., Plaquevent, J.-C. *TL* **43**, 6573 (2002).
[3]Liu, J., Wong, C.-H. *TL* **43**, 3915 (2002).
[4]Liu, J., Wong, C.-H. *TL* **43**, 4037 (2002).

Chloromethyl phenyl sulfoxide. 20, 106

Homologation.[1] Transformation of 2-alkenoic acids to the homologous 3-alkenoic acids is accomplished by using them as acylating agents for the sulfoxide. The products undergo rearrangement on treatment with strong base.

LDA

72%

[1]Satoh, T., Nakamura, A., Iriuchijima, A., Hayashi, Y., Kubota, K. *T* **57**, 9689 (2001).

4-Chlorophenyl chloroformate.

Protection.[1] Mixed carbonates derived from alcohols with the title reagent are stable to conditions of deallylation, delevulinylation, and de(*p*-methoxybenzylation). However, they are decomposed on treatment with $LiOH/H_2O_2$ at 0°.

[1]Love, K.R., Seeberger, P.H. *S* 317 (2001).

***m*-Chloroperoxybenzoic acid, MCPBA. 13,** 76–79; **14,** 84–87; **15,** 86; **16,** 80–83; **17,** 76; **18,** 101; **19,** 94–95; **20,** 106–108; **21,** 130–131

Oxidation of nitrogen compounds. Secondary hydroxylamines are obtained from the corresponding amines via *N*-cyanoethylation and oxidation with MCPBA.[1] Regeneration of the nitrone functionality from isoxazolidines is easily achieved with MCPBA.[2]

COOH, CN; MCPBA / MeOH ; Me_2CO Δ; COOH, N−OH; 68%

R, OH; N, O, R; MCPBA / CH_2Cl_2; N, O

α-Aminonitriles are converted into amides (with loss of the cyano group).[3]

N-Acylaziridines.[4] Oxidation of β-amido selenides at low temperature followed by treatment with a base induces an intramolecular displacement resulting in ring closure.

β-Alkoxycarbonylethyl carbonates. 2,2-Dimethoxycyclopropanecarboxylic esters undergo oxidative ring cleavage on reaction with MCPBA.[5]

R, OMe, EtOOC, OMe; R - Me, Et; MCPBA / CH_2Cl_2; via; EtOOC, R, OMe, OMe, O, O, Ar, O, O; EtOOC, R, O, COOMe; 60–65%

Alkenenitriles.[6] Homologative functionalization of 1-alkenes by a three-step method consists of chlorosulfenylation (with PhSCl), substitution with Bu_4NCN, and oxidative elimination by treatment with MCPBA.

Baeyer-Villiger oxidation.[7] In solvent-free reactions of sterically congested ketones with McPBA rate acceleration by $NaHCO_3$ (surface phenomenon) is noted.

[1]O'Neil, I.A., Cleator, E., Tapolczay, D.J. *TL* **42**, 8247 (2001).
[2]Ishiwata, T., Hino, T., Koshino, H., Hashimoto, Y., Nakata, T., Nagasawa, K. *OL* **4**, 2921 (2002).
[3]Yokoshima, S., Kubo, T., Tokuyama, H., Fukuyama, T. *CL* 122 (2002).
[4]Ward, V.R., Copper, M.A., Ward, M.A. *JCS(P1)* 944 (2001).
[5]Piccialli, V., Graziano, M.L. *TL* **42**, 93 (2001).

[6]Temmem, O., Uguen, D., De Cian, A., Gruber, N. *TL* **43**, 3175 (2002).
[7]Yakura, T., Kitano, T., Ikeda, M., Uenishi, J. *TL* **43**, 6925 (2002).

N-Chlorosuccinimide, NCS. 13, 79–80; **15,** 86–88; **18,** 101–102; **19,** 95–96; **20,** 108; **21,** 131–132

Oxidation.[1] A combination of NCS with *n*-dodecyl methyl sulfide constitutes an odorless oxidant in the Corey-Kim oxidation.

[1]Nishide, K., Ohsugi, S., Fudesaka, M., Kodama, S., Node, M. *TL* **43**, 5177 (2002).

Chlorosulfonyl isocyanate. 13, 80–81; **18,** 102, **21,** 132–133

γ-Butyrolactones.[1] Via a formal [3+2]cycloaddition allylic silanes and chlorosulfonyl isocyanate are united and the adducts are readily transformed into lactones.

R_3Si ... Ph–Si + O=C=N–SO_2Cl → (CH_2Cl_2 ; HCl) → 79%

[1]Peng, Z.-H., Woerpel, K.A. *OL* **3**, 675 (2001).

Chloro(triphenylphosphine)gold(I). 21, 133

Carbamates.[1] Oxidative carbonylation of amines with a mixture of CO, O_2, and alcohols in the presence of $(Ph_3P)AuCl$ at high temperature results in carbamates.

[1]Shi, F., Deng, Y. *CC* 443 (2001).

Chlorotris(pyrazolylborato)bis(triphenylphosphine)ruthenium.

Alkanenitriles. Transformation of 1-alkynes into nitriles having the same number of carbon atoms is by reaction with Me_2NNH_2 using the title complex as catalyst.

[1]Fukumoto, Y., Dohi, T., Masaoka, H., Chatani, N., Murai, S. *OM* **21**, 3845 (2002).

Chlorotris(triphenylphosphine)rhodium(I). 19, 96–98; **20,** 108–109; **21,** 133–135

Hydroboration, hydrosilylation and hydrophosphonylation. Wilkinson's catalyst promotes hydroboration of thiocarbonyl compounds and alkenyl sulfides with 9-BBN.[1] A related reaction of electron-deficient alkenes (e.g., enones) with bis(pinacolato)diboron gives 1,3-difunctional products.[2]

Hydrosilylation of alkenes and alkynes with dimethyl(2-pyridyl)silanes is easily done.[3] A route from 2-alkynals to allenes involves introduction of a silyl group to an α-position through hydrosilylation.[4]

CHO → (Et_3SiH ; $(Ph_3P)_3RhCl$) → OHC, $SiEt_3$

It is interesting to note the opposite regioselectivity for hydrophosphonylation of 1-alkynes in the presence of Pd and Rh catalysts, with latter favoring formation of 1-alkenephosphonic esters.[5]

Reduction. In using this catalyst to hydrogenate alkenes in THF (with *t*-BuOH added), reduction of aromatic nitro groups can be avoided.[6] To saturate electron-deficient alkenes a polymeric ammonium formate can be used as a hydrogen donor.[7] Polychloroarenes undergo partial dehalogenation with HCOONa in isopropanol if catalyzed by the title complex.[8]

Alkenylation. A salt-free method for methylenation of carbonyl compounds is achieved by a catalyzed reaction with Me_3SiCHN_2.[9] Usually, only aldehydes are reactive but ketones containing fluorine atoms at an α-position do react.[10]

Me_3SiCH_2
$(Ph_3P)_3RhCl$
Ph_3P / THF
77%

Isomerization. Allylic alcohols give carbonyl compounds on treatment with BuLi and $ClRh(PPh_3)_3$.[11]

o-Alkylation. In the presence of $ClRh(PPh_3)_3$ alkyl groups are introduced to *o*-position(s) of aryl ketimines and dimethylhydrazones with alkenes.[12] Cyclization of 3-allylaryl ketimines to form indanes is an intramolecular version.[13]

NBn
$(Ph_3P)_3RhCl$
PhMe 125°
71%

Cycloadditions. The $ClRh(PPh_3)_3$ and AgOTf combination is able to assemble a conjugated diene and two other unsaturated components to form eight-membered ring compounds.[14]

TsN + $(Ph_3P)_3RhCl$ AgOTf PhMe Δ TsN
91%

Assembly of 3-substituted phthalides[15] from propargylic esters of propynoic acid and ethylene, and that of carbazoles[16] from *N*,2-diethynylaniline derivatives and alkynes are expedient.

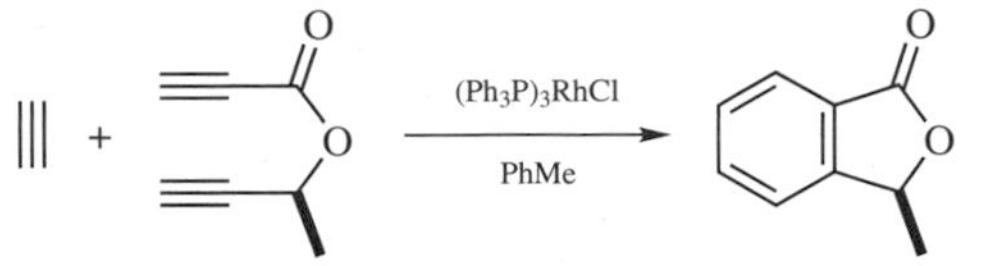

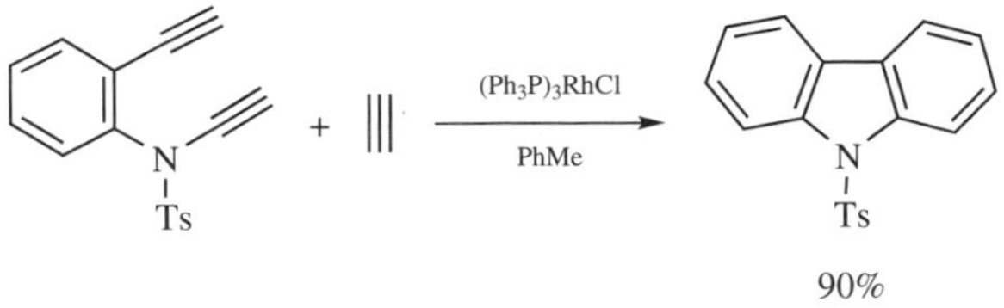

(Ph$_3$P)$_3$RhCl
PhMe

(Ph$_3$P)$_3$RhCl
PhMe
N
Ts
90%

A related cyclization (although not a cycloaddition) of 1,2,7-trienes leads to products with a 5-membered ring. The reaction pattern is strikingly different from those using Pd-catalysts.[17]

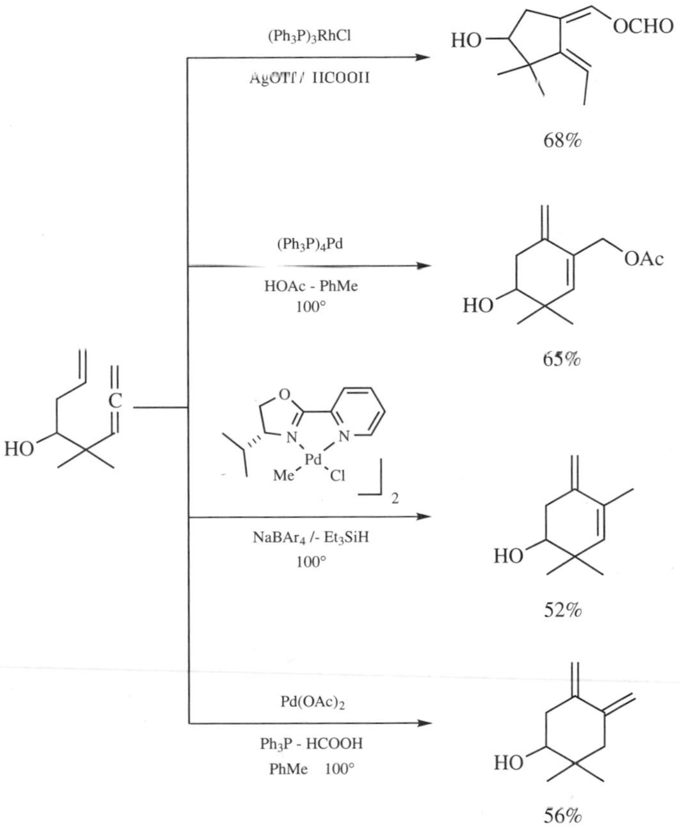

(Ph$_3$P)$_3$RhCl
AgOTf / HCOOH
HO
OCHO
68%

(Ph$_3$P)$_4$Pd
HOAc - PhMe
100°
OAc
HO
65%

HO
C

Pd, Me, Cl, N, O
NaBAr$_4$ /- Et$_3$SiH
100°
HO
52%

Pd(OAc)$_2$
Ph$_3$P - HCOOH
PhMe 100°
HO
56%

Reformatsky-type reaction.[18] Organozinc species derived from α-bromoalkanoic esters add to imines in the presence of $ClRh(PPh_3)_3$. Solvent and temperature have effects on the products: β–amino esters and β–lactams are obtained respectively from reactions in THF at 0° and in PhMe at 40°, respectively.

Enone-alkene union.[19] The Rh complex together with PhCOOH and an unhindered secondary amine induces CC bond formation between an alkene and an enone such that there is a net alkylation at the β-position of the enone molecules.

$(Ph_3P)_3RhCl$, $PhCOOH - Et_2NH$

[1]Carter, C.A.G., Vogels, C.M., Harrison, D.J., Gagnon, M.K.J., Norman, D.W., Langler, R.F., Baker, R.T., Westcott, S.A. *OM* **20**, 2130 (2001).
[2]Kabalka, G.W., Das, B.C., Das, S. *TL* **43**, 2323 (2002).
[3]Itami, K., Mitsudo, K., Nishino, A., Yoshida, J. *JOC* **67**, 2645 (2002).
[4]Tius, M.A., Pal, S.K. *TL* **42**, 2605 (2001).
[5]Zhao, C.-Q., Han, L.-B., Goto, M., Tanaka, M. *ACIEE* **40**, 1929 (2001).
[6]Jourdant, A., Gonzalez-Zamora, E., Zhu, J. *JOC* **67**, 3163 (2002).
[7]Desai, B., Danks, T.N. *TL* **42**, 5963 (2001).
[8]Atienza, M.A., Esteruelas, M.A., Fernandez, M., Hettero, J., Olivan, M. *NJC* **25**, 775 (2001).
[9]Lebel, H., Paquet, V., Proulx, C. *ACIEE* **40**, 2887 (2001).
[10]Lebel, H., Paquet, V. *OL* **4**, 1671 (2002).
[11]Uma, R., Davies, M.K., Crevisy, C., Gree, R. *EJOC* 3141 (2001).
[12]Kakiuchi, F., Tsujimoto, T., Sonoda, M., Chatani, N., Murai, S. *SL* 948 (2001).
[13]Thalji, R.K., Ahrendt, K.A., Bergman, R.G., Ellman, J.A. *JACS* **123**, 9692 (2001).
[14]Evans, P.A., Robinson, J.E., Baum, E.W., Fazal, A.N. *JACS* **124**, 8782 (2002).
[15]Witulski, B., Zimmermann, A. *SL* 1855 (2002).
[16]Witulski, B., Alayrac, C. *ACIEE* **41**, 3281 (2002).
[17]Oh, C.H., Jung, S.H., Rhim, C.Y. *TL* **42**, 8669 (2001).
[18]Kanai, K., Wakabayashi, Honda, T. *H* **58**, 47 (2002).
[19]Jun, C.-H., Moon, C.W., Kim, Y.-H., Lee, H., Lee, J.H. *TL* **43**, 4233 (2002).

Chlorotris(triphenylphosphine)rhodium(I)–2-Amino-3-picoline. 21, 135

Hydroacylation. Ketone synthesis by this method has several variations: from allylic alcohols through consecutive isomerization to aldehydes and then hydroacylation of alkenes,[1] and via cleavage of alkynes.[2] Secondary phenethyl carbinols also split to afford aldehydes for the hydroacylation.[3] Using acrylic esters as the alkene component γ-keto esters are assembled.[4]

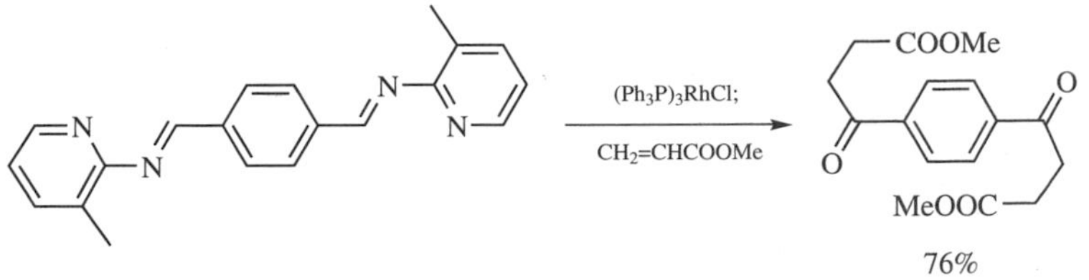

N N N N

$(Ph_3P)_3RhCl$; $CH_2{=}CHCOOMe$

COOMe O O MeOOC

76%

Primary amines are activated and the ensuing imine formation and hydroacylation of alkenes lead to ketones. The reaction of benzylamine afford products in which an ortho position is alkylated also.[5] When dienes are employed in the reaction cyclic ketones are formed, thus 2-cinnamylamino-3-picoline is a provider of CO and two hydrogen atoms to combine with the dienes.[6]

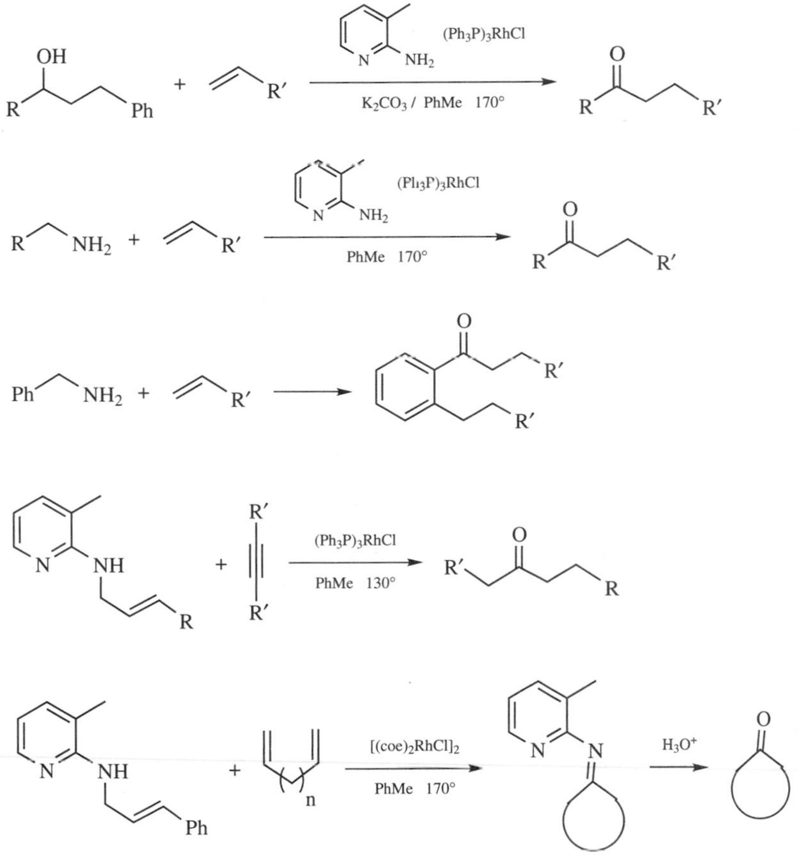

OH R Ph + R′ — N NH_2, $(Ph_3P)_3RhCl$, K_2CO_3 / PhMe 170° → O R R′

R NH_2 + R′ — N NH_2, $(Ph_3P)_3RhCl$, PhMe 170° → O R R′

Ph NH_2 + R′ → O R′ R′

N NH R + R′ R′ — $(Ph_3P)_3RhCl$, PhMe 130° → R′ O R

N NH Ph + n — $[(coe)_2RhCl]_2$, PhMe 170° → N N — H_3O^+ → O

Hydroacylation of 1-alkynes leads to 2-substituted 1-alken-3-ones.[7]

R–CHO + R'C≡CH —[2-amino-3-picoline, $(Ph_3P)_3RhCl$; PhCOOH / PhMe]→ RCOC(=CH$_2$)R' + RCOCH$_2$CH$_2$R'

R' = t-Bu

[1]Lee, D.-Y., Moon, C.W., Jun, C.-H. *JOC* **67**, 3945 (2002).
[2]Jun, C.-H., Lee, H., Moon, C.W., Hong, J.-B. *JACS* **123**, 8600 (2001).
[3]Jun, C.-H., Lee, D.-Y., Kim, Y.-H., Lee, H. *OM* **20**, 2928 (2001).
[4]Willis, M.C., Sapmaz, S. *CC* 2558 (2001).
[5]Jun, C.-H., Chung, K.-Y., Hong, J.-B. *OL* **3**, 785 (2001).
[6]Lee, H., Kim, I.-J., Jun, C.-H. *ACIEE* **41**, 3031 (2002).
[7]Jun, C.-H., Lee, H., Hong, J.-B., Kwon, B.-I. *ACIEE* **41**, 2146 (2002).

Chromium–carbene complexes. 13, 82–83; **14,** 91–93; **15,** 93–95; **16,** 88–92; **17,** 80–84; **18,** 103–104; **19,** 98–101; **20,** 110–111; **21,** 136–138

Cycloadditions. 2-Alkenyloxazolines form 2-cyclopropyloxazolines on reaction with Fischer carbene complexes.[1] Conjugated Fischer carbene complexes participate in cycloaddition with imines to produce 3-pyrrolines.[2]

Ph(MeO)C=$Cr(CO)_3$ + 2-propenyl-4,4-dimethyloxazoline —[THF, 60°]→ cyclopropyloxazoline (Ph, MeO)

Condensation of amino-carbene complexes with 2-alkenamides affords amidine derivatives which on mild thermolysis are transformed into pyrroles.[3] Fused α-pyrones are formed in a reaction of the carbene complexes with 3-alkynyl-2-heteroaraldehydes.[4]

3-(R-ethynyl)-2-furaldehyde (CHO) + $(OC)_3Cr$=C(OMe)CH$_3$ —[Ph_3P, dioxane, Δ]→ furo-α-pyrone (R, OMe)

Highly functionalized 5-membered carbocycles are assembled on heating aminoalkenylcarbene complexes with alkynes.[5]

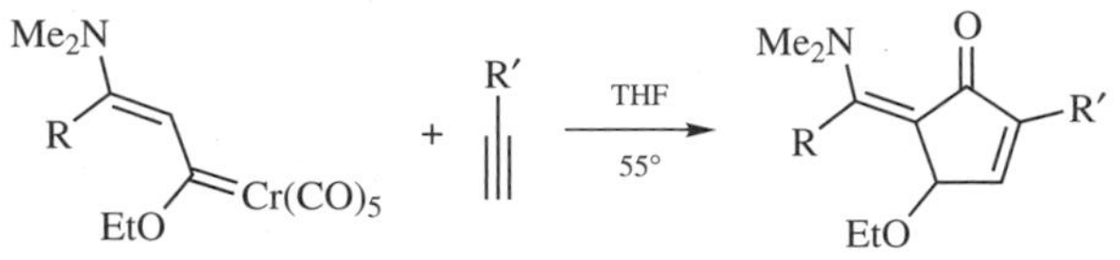

(2-Furyl)carbene complexes generated in situ undergo cyclopropanation[6], an intramolecular version exemplified by **1** react, on heating.[7]

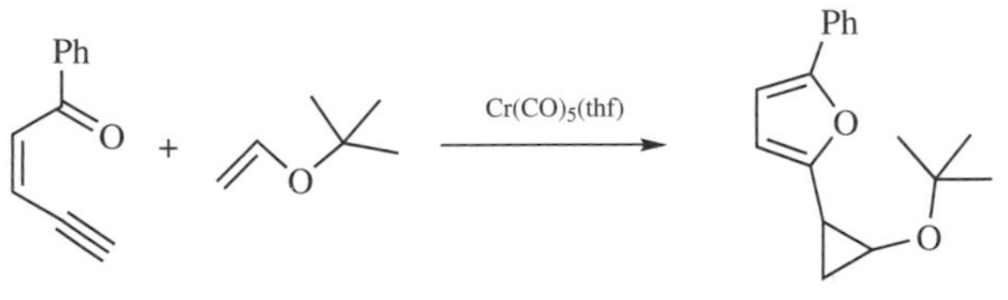

63% (*cis/trans* 76 : 24)

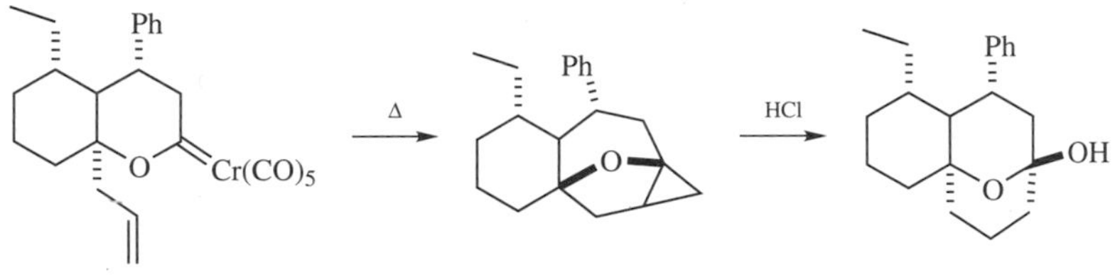

1

Reaction with imines. Heating Fischer carbene complexes with imines results in demetallative addition.[8]

Reductive dimerization/coupling. Conjugated carbene complexes undergo dimerization on treatment with C_8K. In the case of an alkynylcarbene complex cyclization occurs in tandem.[9] On the other hand, bis(carbene complexes) suffer demetallative coupling when exposed to $Pd(OAc)_2$.[10]

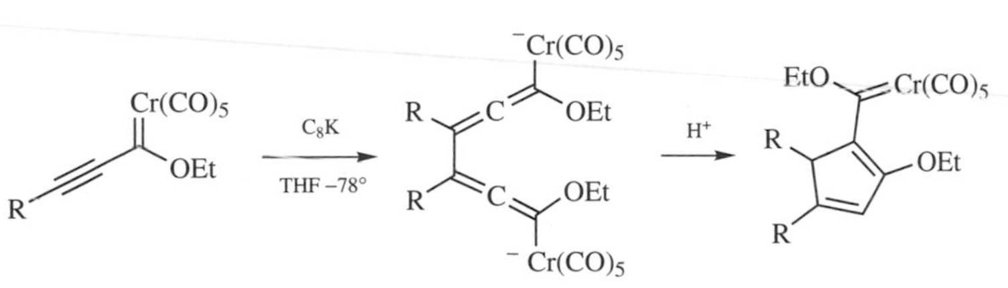

Pd(OAc)$_2$, Et$_3$N

n = 0–3

Pd(OAc)$_2$

[1]Barluenga, J., Suarez-Sobrino, A.L., Tomas, M., Garcia-Granda, S., Santiago-Garcia, R. *JACS* **123**, 10494 (2001).
[2]Kagoshima, H., Okamura, T., Akiyama, T. *JACS* **123**, 7182 (2001).
[3]Aumann, R., Vogt, D., Fröhlich, R. *OM* **21**, 1819 (2002).
[4]Zhang, Y., Herndon, J.W. *JOC* **67**, 4177 (2002).
[5]Flynn, B.L., Schirmer, H., Duetsch, M., de Meijere, A. *JOC* **66**, 1747 (2002).
[6]Miki, K., Nishino, F., Ohe, K., Uemura, S. *JACS* **124**, 5260 (2002).
[7]Barluenga, J., Dieguez, A., Rodriguez, F., Florez, J., Fananas, F.J. *JACS* **124**, 9056 (2002).
[8]Sangu, K., Kagoshima, H., Fuchibe, K., Akiyama, T. *OL* **4**, 3967 (2002).
[9]Sierra, M.A., Ramirez-Lopez, P., Gomez-Gallego, M., Lejon, T., Mancheno, M.J. *ACIEE* **41**, 3442 (2002).
[10]Sierra, M.A., del Amo, J.C., Mancheno, M.J., Gomez-Gallego, M. *JACS* **123**, 851 (2001).

Chromium(II) acetate–EDTA complex.

Alkylations. Organometallic species derived from alkyl halides and the Cr(edta)$^{2+}$ complex are used to react with carbonyl compounds in a neutral aqueous medium.[1]

[1]Micskei, K., Kiss-Szikszai, A., Gyarmati, J., Hajdu, C. *TL* **42**, 7711 (2001).

Chromium(II) chloride. 13, 84; 14, 94–97; 15, 95–96; 16, 93–94; 17, 84–85; 18, 104; 19, 101; 20, 111–113; 21, 138–140

(Z)-Chloroalkenes. 1,1,1-Trichloroalkanes undergo both elimination and dehydrochlorination with CrCl$_2$ at room temperature.[1] In the presence of an aldehyde further condensation occurs to furnish (*Z*)-2-chloroalk-2-en-1-ols.[2,3]

CrCl$_2$

R = H, OCOOMe

A tandem reaction involving one molecule of CCl_4 and two molecules of an aldehyde gives the same type of products (with two identical terminal groups).[4]

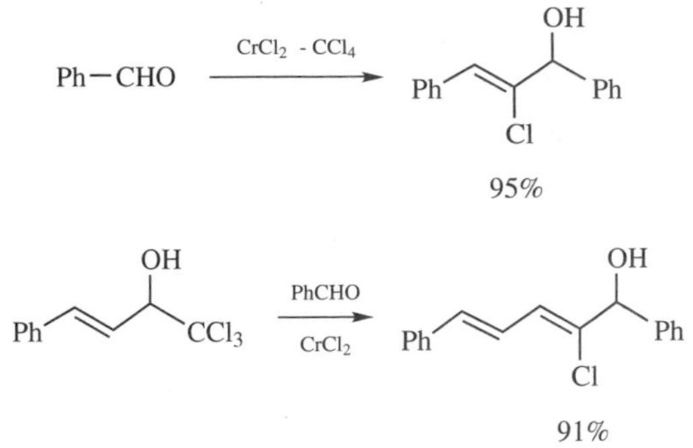

[1]Baati, R., Barma, D.K., Krishna, U.M., Mioskowski, C., Falck, J.R. *TL* **43**, 959 (2002).
[2]Barma, D.K., Baati, R., Valleix, A., Mioskowski, C., Falck, J.R. *OL* **3**, 4237 (2001).
[3]Takai, K., Kokumai, R., Nobunaka, T. *CC* 1128 (2001).
[4]Baati, R., Barma, D.K., Falck, J.R., Mioskowski, C. *TL* **43**, 2179 (2002).

Chromium(II) chloride–lithium iodide.

Debenzylation. Combining with LiI, $CrCl_2$ removes a benzyl group from protected monosaccharides.[1] The reaction is regioselective.

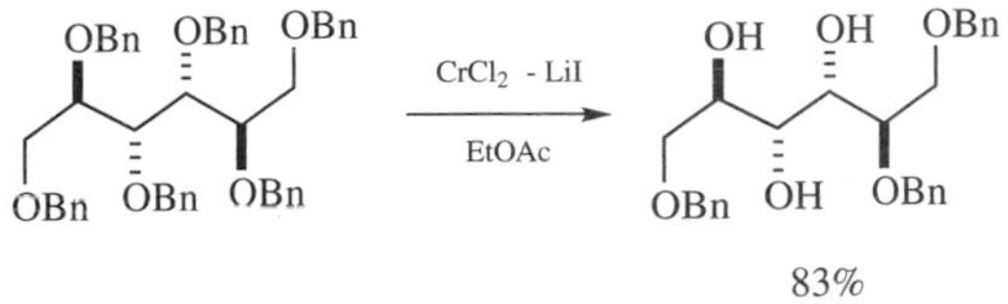

Selective cleavage of 2,6-dimethoxybenzyl ethers is achieved while retaining simple benzyl ethers.[2] However, it should be emphasized that ordinary benzyl ethers are not totally immune to the treatment. Note also that hydrogenolysis (Pd/C) reverses the chemoselectivity.

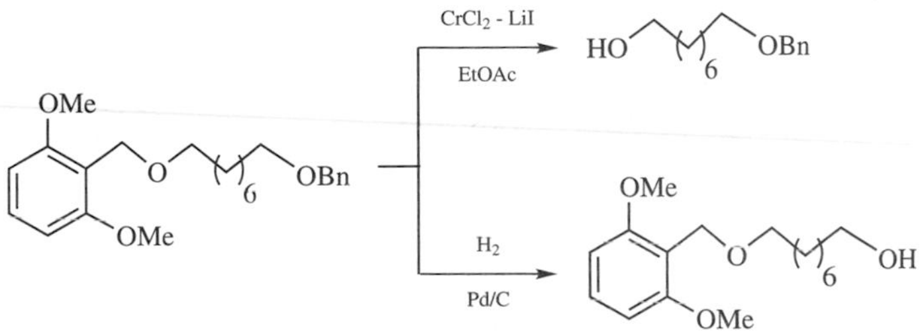

[1]Falck, J.R., Barma, D.K., Venkataraman, S.K., Baati, R., Mioskowski, C., *TL* **43**, 963 (2002).
[2]Falck, J.R., Barma, D.K., Baati, R., Mioskowski, C., *ACIEE* **40**, 1281 (2001).

Chromium(II) chloride–nickel(II) halide. 14, 97–98; **15,** 96–97; **17,** 86; **18,** 105; **19,** 102; **20,** 113–114; **21,** 140

Cyclization.[1] An intramolecular alkylation of (*Z*)-δ-bromo-γ,δ-unsaturated carbonyl compounds leads to cyclopentenols on treatment with $CrCl_2$-$NiCl_2$.

Br; O; $CrCl_2$ - $NiCl_2$; DMF 25°; OH; 70%

1,3-Amino alcohols. *O*-Acetyl oximes generate chromioenamines on reaction with $CrCl_2$-$NiCl_2$, and such species attack carbonyl compounds.[2] Direct reduction of the products affords 1,3-amino alcohols.

Ph—CHO + NOAc; $CrCl_2$ - $NiCl_2$; LAH ; Ac_2O; OH NHAc; Ph

[1]Trost, B.M., Pinkerton, A.B. *JOC* **66**, 7714 (2001).
[2]Takai, K., Katsura, N., Kunisada, Y. *CC* 1724 (2001).

Chromium(II) chloride–trialkylsilyl chloride.

Furans.[1] 1-Alkyn-3-ones react with aldehydes in the presence of $CrCl_2$-Me_3SiCl/H_2O in THF at room temperature. 2,5-Disubstituted furans are formed in moderate to good yields.

R—CHO + O R′; $CrCl_2$ - Me_3SiCl; H_2O / THF; R O R′

Pinacol coupling. Cross coupling between enones and aldehydes is mediated by $CrCl_2$-Et_3SiCl. Temperature and substrate dependence of diastereoselectivity are noted.[2]

$CrCl_2$ - Et_3SiCl / DMF ; Bu_4NF

at 0°	99%	93 : 7
at 75°	85%	10 : 90

[1]Takai, K., Morita, R., Sakamoto, S. *SL* 1614 (2001).
[2]Takai, K., Morita, R., Toratsu, C. *ACIEE* **40**, 1116 (2001).

Chromium(III) chloride–lithium aluminum hydride.

Dechlorination.[1] With isopropanol as proton source the reagent combination removes allylic chlorine atoms, usually with double bond transposition. The method is valuable for generating acid-sensitive compounds such as 2-methylenedihydrobenzofuran.

$CrCl_3$ - $LiAlH_4$ / *i*-PrOH

79% 16%

[1]Omoto, M., Kato, N., Sogon, T., Mori, A. *TL* **42**, 939 (2001).

Chromium(VI) oxide.

Oxidation.[1] Oxidation of primary benzylic alcohols to aldehydes with CrO_3 under solvent-free conditions is described.

[1]Lou, J.-D., Xu, Z.-N. *TL* **43**, 6095 (2002).

Cobalt. 21, 141

Pauson-Khand reaction. Aqueous colloidal nanoparticles of cobalt catalyzes the Pauson-Khand reaction (6 examples, 88–96%).[1] A one-pot synthesis of fenestrane derivatives[2] is based on this cycloaddition.

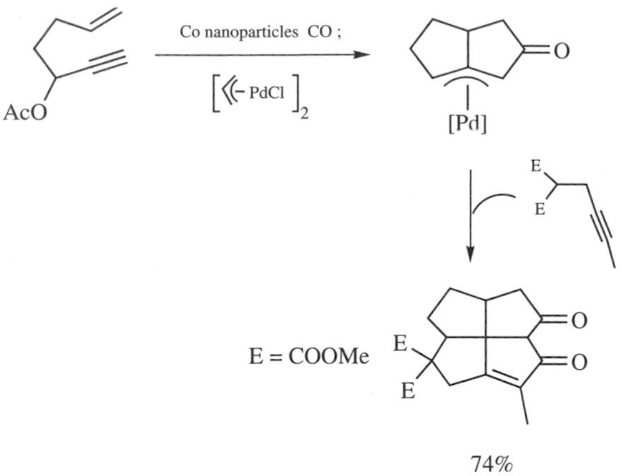

Co nanoparticles CO ; $[(\eta^3\text{-allyl})PdCl]_2$

[Pd]

E = COOMe

74%

N-Aryl alkanamides.[3] Hydrocarbonylation of 1-alkenes in the presence of $ArNH_2$ under CO delivers the amide products, with Co/C as catalyst.

Biaryls.[4] Electroreductive cross-coupling of two different aryl halides with cobalt catalyst is quite successful in certain cases.

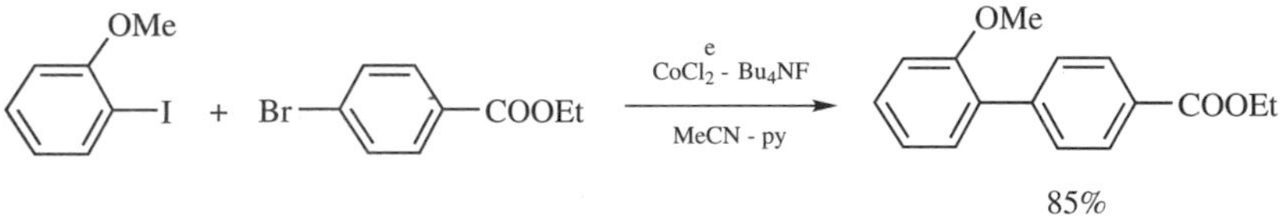

e, $CoCl_2$ - Bu_4NF, MeCN - py

85%

[1]Son, S.U., Lee, S.I., Chung, Y.K., Kim, S.-W., Hyeon, T. *OL* **4**, 277 (2002).
[2]Son, S.U., Park, K.H., Chung, Y.K. *JACS* **124**, 6838 (2002).
[3]Lee, S.I., Son, S.U., Chung, Y.K. *CC* 1310 (2002).
[4]Gomes, P., Fillon, H., Gosmini, C., Labbe, E., Perichon, J. *T* **58**, 8417 (2002).

Cobalt(II) bromide phosphine–zinc iodide.

Diels-Alder reactions.[1,2] The salt combination together with Bu_4NBr is effective to induce unactivated terminal and internal alkynes to undergo the cycloaddition as dienophiles.

Hydroalkenylation.[3,4] 1,4-Dienes are generated from reaction of conjugated dienes with terminal alkenes. Norbornadiene gives tricyclic adducts.

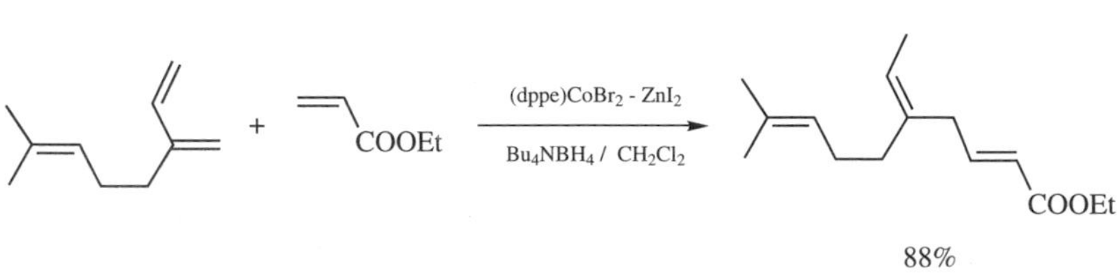

(dppe)$CoBr_2$ - ZnI_2, Bu_4NBH_4 / CH_2Cl_2

88%

(dppe)CoBr$_2$ - ZnI$_2$

Bu$_4$NBH$_4$ / CH$_2$Cl$_2$

COOEt

82%

[1]Hilt, G., Korn, T.J. *TL* **42**, 2783 (2001).
[2]Hilt, G., Smolko, K.I., Lotsch, B.V. *SL* 1081 (2002).
[3]Hilt, G., du Mesnil, F.-X., Lüer, S. *ACIEE* **40**, 387 (2001).
[4]Hilt, G., Lüer, S. *S* 609 (2002).

Copper.

Aromatic substitution. Copper catalyzes the transformation of 2-chloroaroic acids to 2-hydroxyaroic acids in water.[1] An improved preparation of aryl thiocyanates consists of reacting arenediazonium *o*-benzenedisulfonimides with NaSCN in MeCN at room temperature in the presence of Cu.[2]

Cyclopropanes.[3] Cycloaddition of copper carbenoids derived from *gem*-dichloroalkanes and 1,1,1-trichloroalkanes to activated alkenes can be achieved with electrogeneerated metal species.

Hydrogenations.[4] Cyclohexanones are reduced to cyclohexanols by hydrogenation with Cu/SiO$_2$ which is prepared from cupriammonium solution and porous silica. Exceptionally mild conditions (atmospheric pressure, 90°) are established.

[1]Comdom, R.F.P., Palacios, M.L.D. *SC* **32**, 2055 (2002).
[2]Barbero, M., Degani, I., Diulgheroff, N., Dughera, S., Fochi, R. *S* 585 (2001).
[3]Sengmany, S., Leonel, E., Paugam, J.P., Nedelec, J.-Y. *S* 533 (2002).
[4]Ravasio, N., Psaro, R., Zaccheria, F. *TL* **43**, 3943 (2002).

Copper(II) acetate. **18,** 109–110; **19,** 106; **20,** 117; **21,** 142–143

Arylamines. Arylation of amines by arylboronic acids is achieved by a vigorous agitation with copper(II) acetate, myristic acid, and 2,6-lutidine in the air.[1] Formation of *N*-arylbenzimidazolones[2] under similar conditions shows amides are also reactive.

Destannylative amination of ArSnMe$_3$ is readily achieved at room temperature,[3] while triarylbismuth diacetates react with hydrazines to give arylhydrazines.[4]

Coupling reactions. *N*-Arylation of tetrazoles[5] by reaction of *N*-tributylstannyltetrazoles with diaryliodonium salts is catalyzed by Cu(OAc)$_2$. A new method for the preparation of aryl vinyl ethers is through the reaction of phenols with tetravinylstannane, catalyzed by Cu(OAc)$_2$ under oxygen.[6] The Cu(OAc)$_2$–mediated intramolecular *O*-arylation of phenols with arylboronic acid residue represents a way to access functionalized macrocyclic diaryl ethers.[7]

A hybrid of the Heck and Suzuki coupling uses $ArB(OH)_2$ and acrylic esters to form cinnamic esters, in the presence of $Pd(OAc)_2$ and $Cu(OAc)_2$.[8] Alternatively, $[(p\text{-cymene})RuCl_2]_2$ can be used instead of $Pd(OAc)_2$.[9]

Indoles. *o*-Alkynylanilines cyclize under the influence of $Cu(OAc_2)$ to provide indoles.[10]

R′

$Cu(OAc)_2$

$ClCH_2CH_2Cl$

Δ

NHR

R′

N

R

R = Ms, COOEt

R′ = H, Bu, Ph

[1]Antilla, J.C., Buchwald, S.L. *OL* **3**, 2077 (2001).
[2]Lam, P.Y.S., Vincent, G., Clark, C.G., Deudon, S., Jadhav, P.K. *TL* **42**, 3415 (2001).
[3]Lam, P.Y.S., Vincent, G., Bonne, D., Clark, C.G. *TL* **43**, 3091 (2002).
[4]Tsubrik, O., Mäeorg, U., Ragnarsson, U. *TL* **43**, 6213 (2002).
[5]Davydov, D.V., Beletskaya, I.P., Semenov, B.B., Smushkevich, Y.I. *TL* **43**, 6217 (2002).
[6]Blouin, M., Frenette, R. *JOC* **66**, 9043 (2001).
[7]Decicco, C.P., Song, Y., Evans, D.A. *OL* **3**, 1029 (2001).
[8]Nishikata, T., Hagiwara, N., Kawata, K., Okeda, T., Wang, H.F., Fugami, K., Kosugi, M. *OL* **3**, 3313 (2001).
[9]Farrington, E.J., Brown, J.M., Barnard, C.F.J., Rowsell, E. *ACIEE* **41**, 169 (2002).
[10]Hiroya, K., Itoh, S., Ozawa, M., Kanamori, Y., Sakamoto, T. *TL* **43**, 1277 (2002).

Copper(II) acetylacetonate. 18, 110; **20,** 117–118

1,4-Oxathianes. In the presence of $Cu(acac)_2$ diazo compounds form carbenoids which insert into 1,3-oxathiolanes.[1]

3-Alken-1-ynes.[2] The preparation involves hydroboration of 1-alkynes and subsequent coupling of the alkenylboranes with trimethylsilylethynyl bromide under the influence of $Cu(acac)_2$. Optional Desilylation by NaOMe is avoided by using $LiOH \cdot H_2O$.

B

R

+

$SiMe_3$

Br

$Cu(acac)_2$; NaOMe;

H_2O_2 - NaOH

R

70%

N-Alkylation. Arylamines undergo monoalkylation with R_2Zn using $Cu(acac)_2$ as catalyst.[3]

[1]Ioannou, M., Porter, M.J., Saez, F. *CC* 346 (2002).
[2]Hoshi, M., Shirakawa, K. *SL* 1101 (2002).
[3]Brielles, C., Harnett, J.J., Doris, E. *TL* **42**, 8301 (2001).

Copper(I) bromide. 21, 143

Coupling.[1] *N*-Propargyl carbamates are obtained by a CuBr-mediated reaction of 1-alkynes with *N*-α-sulfonylalkyl carbamates in water, via desulfonylation.

1,2-Alkadien-4-ols.[2] Homologation of propargylic alcohols is accomplished in a CuBr-mediated reaction with paraformaldehyde.

R, OH + $(HCHO)_n$ —CuBr, $i\text{-}Pr_2NH$, dioxane→ R, C, OH

[1]Zhang, J., Wei, C., Li, C.-J. *TL* **43**, 5731 (2002).
[2]Ma, S., Hou, H., Zhao, S., Wang, G. *S* 1643 (2002).

Copper(II) bromide. 14, 100; **15,** 100; **18,** 111; **19,** 106; **21,** 143–144

t-Butyl esters.[1] *N*-Phenyl hydrazides are converted to *t*-butyl esters by *t*-BuOLi and $CuBr_2$ in THF.

Si-Bromination.[2] Hydrosilanes are brominated by $CuBr_2$ in the presence of CuI.

Glycosylation.[3] Oxazolines derived from 2-amino sugars give 2-acylamino sugars while introducing new glycosidic groups, by activation with $CuBr_2$.

OAc, AcO, AcO, O, N, O + HO, HO —$CuBr_2$, THF→ OAc, AcO, AcO, O, O, AcHN, HO

87%

2-Silatetrahydrofurans.[4] Insertion of an aldehyde into the silacyclopropane ring gives the 5-membered heterocycle. The more highly substituted C-Si bond is attacked.

Si + Ph–CHO —$CuBr_2$, CH_2Cl_2→ Ph, O, Si

82%

[1]Yamaguchi, J., Aoyagi, T., Fujikura, R., Suyama, T. *CL* 466 (2001).
[2]Kunai, A., Ochi, T., Iwata, A., Ohshita, J. *CL* 1228 (2001).
[3]Wittmann, V., Lennartz, D. *EJOC* 1363 (2002).
[4]Franz, A.K., Woerpel, K.A. *JACS* **121**, 949 (1999).

Copper(I) chloride. 13, 85; **15,** 101; **18,** 112–113; **19,** 107–108; **20,** 118–120; **21,** 144–146

Halomethylenation. Hydrazones condense with $CHBr_3$ and CX_4 to furnish haloalkenes and dihaloalkenes, respectively, in DMSO and in the presence of CuCl.[1]

Coupling reactions. The Ullmann diaryl ether synthesis using CuCl-Cs_2CO_3 as promoters is accelerated by dipivaloylmethane,[2] apparently due to solubility of the copper chelate. The first step in a synthesis of aryloxyamines from *N*-hydroxyphthalimide and arylboronic acid is *O*-arylation mediated by CuCl and pyridine.[3]

Coupling of alkynylsilanes to alkenyl iodides to afford conjugated enynes and dienes proceeds well.[4] A mixture of CuCl and $Cu(OAc)_2$ has been employed in the synthesis of a twisted [2.2]paracyclophane.[5]

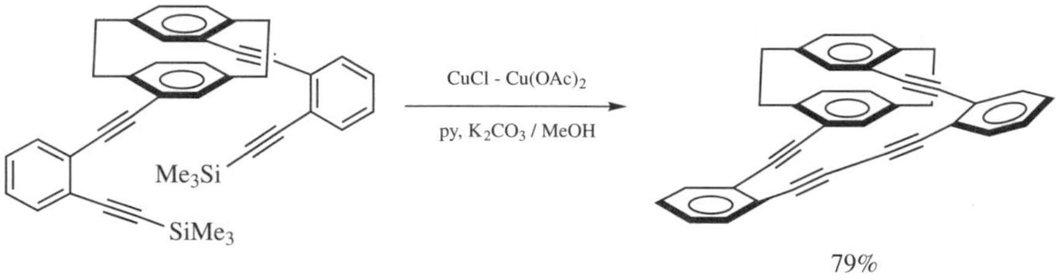

Me$_3$Si SiMe$_3$ — CuCl - Cu(OAc)$_2$ / py, K$_2$CO$_3$ / MeOH → 79%

Azacycles. Several cyclization reactions are mediated by CuCl: transformation of δ,ε-unsaturated *N*-chloramines into 3-chloropiperidines,[6] reaction of tosyl enynes with amines to give pyrroles,[7] and the assembly of quinolines from three components (arylamines, 1-alkynes, aldehydes).[8] Cycloisomerization of alkynylimines to pyrroles is extendable to 2-alkynylpyridines which deliver octadehydroindolizidines.[9] CuI is also effective for the purpose.

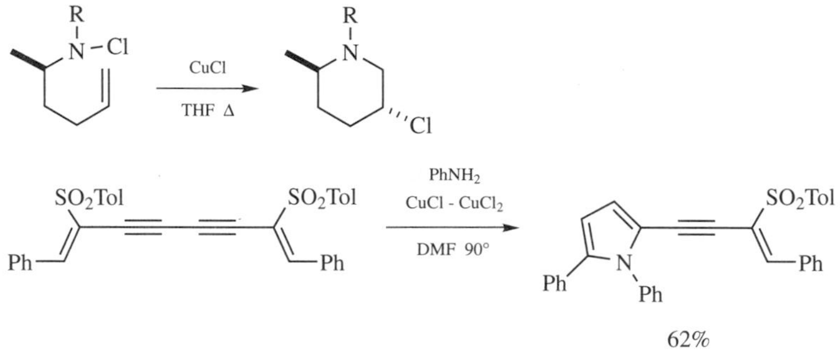

R N–Cl — CuCl / THF Δ → R N Cl

SO$_2$Tol Ph SO$_2$Tol Ph — PhNH$_2$ / CuCl - CuCl$_2$ / DMF 90° → Ph N Ph SO$_2$Tol Ph 62%

Bu

CuCl - Et_3N

DMA 130°

N

Bu

63%

monomorine

For aziridination of alkenes with PhI=NTs, a new catalyst, copper(I) tris(3,5-dimethylpyrazolyl)borate, is prepared from CuCl.[10]

Hydroboration. Borylcopper species are involved in a formal hydroboration of conjugated carbonyl compounds to afford β-boryl derivatives,[11] when they are mixed with bis(pinacolato)diboron and CuCl in DMF.

1,3-Dienyl-1,4-dicoppers.[12] Zirconacyclopentadienes undergo transmetallation with CuCl. The resulting species are derivatizable, e.g., into bisphosphines. One such product, 1,4-bis(diphenylphosphino)-1,2,3,4-tetraphenyl-1,3-butadiene, is an excellent ligand to constitute a catalyst with $PdCl_2$ for coupling of Grignard reagents with alkyl bromides.

Ph + [$ZrCp_2$] → (Ph)$_4$C$_4$$ZrCp_2$ —CuCl ; Ph_2PCl→ Ph_2P(Ph)C=C(Ph)–C(Ph)=C(Ph)PPh_2

Addition of silyl ketene acetals to enones.[13] The promotion of 1,2-addition by the bis(triphenylphosphine) complex of CuCl is remarkable.

[1]Korotchenko, V.N., Shastin, A.V., Nenajdenko, V.G., Balenkova, E.S. *JCS(P1)* 883 (2002).
[2]Buck, E., Song, Z.J., Tschaen, D., Dormer, P.G., Volante, R.P., Reider, P.J. *OL* **4**, 1623 (2002).
[3]Petrassi, H.M., Sharpless, K.B., Kelly, J.W. *OL* **3**, 139 (2001).
[4]Marshall, J.A., Chobanian, H.R., Yanik, M.M. *OL* **3**, 4107 (2001).
[5]Boydston, A.J., Bondarenko, L., Dix, I., Weakley, T.J.R., Hopf, H., Haley, M.M. *ACIEE* **40**, 2986 (2001).
[6]Heuger, G., Kalsow, S., Göttlich, R. *EJOC* 1848 (2002).
[7]Takeda, M., Matsumoto, S., Ogura, K. *H* **55**, 231 (2001).
[8]Huma, H.Z.S., Halder, R., Kalra, S.S., Das, J., Iqbal, J. *TL* **43**, 6485 (2002).
[9]Kel'in, A.V., Sromek, A.W., Gevorgyan, V. *JACS* **123**, 2074 (2001).
[10]Handy, S.T., Czopp, M. *OL* **3**, 1423 (2001).
[11]Takahashi, K., Ishiyama, T., Miyaura, N. *JOMC* **625**, 47 (2001).
[12]Doherty, S., Knight, J.G., Robins, E.G., Scanlan, T.H., Champkin, P.A., Clegg, W. *JACS* **123**, 5110 (2001).
[13]Mitani, M., Ishimoto, K., Koyama, R. *CL* 1142 (2002).

Copper(II) chloride–sodium persulfate.

γ-Aryl-γ-butyrolactones.[1] The combined reagents are useful for transforming γ-arylbutanoic acid into the lactones.

[1]Mahmoodi, N.O., Jazayri, M. *SC* **31**, 1467 (2001).

Copper(II) fluoride.

Fluorination.[1] Arenes such as benzene undergo monofluorination in the vapor phase on reacting with CuF_2 with great selectivity. The reagent is regenerated on treatment of recovered Cu with HF and O_2.

[1]Subramanian, M.A., Manzer, L.E. *SCI* **297**, 1665 (2002).

Copper(I) iodide. 16, 98; **18,** 114–115; **19,** 109–110; **20,** 120–121; **21,** 147–148

Arylations. A great variety and number of these reactions have been developed by means of CuI and a base (e.g., amine). Reports comprise *N*-arylation of amides,[1,2] amines,[1,3] hydrazides,[1,4] and indoles.[1,5] With hydrazides the nature of the acyl residue can influence regiochemistry.[4]

Ar–I + HN(R)–NH_2 —(CuI - Cs_2CO_3, DMF)→ Ar–N(R)–NH_2 + Ar–NH–HN–R

R = Boc favored

R = Bz favored

Intramolecular arylation of properly constituted compounds leads to indolines.[6]

Similarly, aryl ethers and sulfides are formed from ArI and RXH.[7–9]Arylation of diethyl malonate proceeds in refluxing THF with ArI with a catalytic system consisting of CuI, Cs_2CO_3 and 2-hydroxybiphenyl.[10] Amines and alcohols are readily discriminated under specified conditions, as shown by the selective *N*-arylation of β-amino alcohols.[11]

Exchange reactions. Halogen exchange in aryl halides (ArBr to ArI) takes place when the substrates are heated with CuI, NaI and *trans*-1,2-bis(methylamino)cyclohexane in dioxane.[12] Replacement of an alkenyl boryl group with the tributylstannyl residue is accomplished readily. The transmetallation is stereoretentive.[13]

Chx_2B(Me₃Si)C=CH–R + Bu_3Sn–Cl —(CuI - NaI, NaOH / H_2O)→ Bu_3Sn(Me_3Si)C=CH–R

Allylic displacements. The process of substituting allylic carbonates with alkoxide ions as catalyzed by Wilkinson's catalyst also depends on CuI for high stereoselectivity.[14]

Coupling reaction. With *t*-BuOCu formed in situ from CuI and *t*-BuOLi to initiate cupration the silyl group of a γ-silylated allylic alcohol can be replaced by allyl, alkenyl, or aryl substitutent on reaction with the corresponding organohalide.[15] Clean reaction is observed with the (*Z*)-isomer, indicating the importance of C→O silyl transfer.

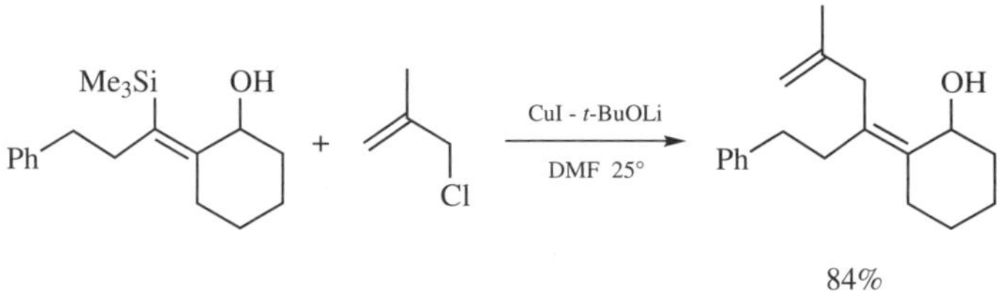

[1]Klapars, A., Antilla, J.C., Huang, X., Buchwald, S.L. *JACS* **123**, 7727 (2001).
[2]Kang, S.-K., Kim, D.-H., Park, J.-N. *SL* 427 (2002).
[3]Kwong, F.Y., Klapars, A., Buchwald, S.L. *OL* **4**, 581 (2002).
[4]Wolter, M., Klapars, A., Buchwald, S.L. *OL* **3**, 3803 (2001).
[5]Zhou, T., Chen, Z.-C. *SC* **32**, 903 (2002).
[6]Yamada, K., Kubo, T., Tokuyama, H., Fukuyama, T. *SL* 231 (2002).
[7]Wolter, M., Nordmann, G., Job, G.E., Buchwald, S.L. *OL* **4**, 973 (2002).
[8]Kwong, F.Y., Buchwald, S.L. *OL* **4**, 3512 (2002).
[9]Bates, C.G., Gujadhur, R.K., Venkataraman, D. *OL* **4**, 2803 (2002).
[10]Hennessy, E.J., Buchwald, S.L. *OL* **4**, 269 (2002).
[11]Job, G.E., Buchwald, S.L. *OL* **4**, 3703 (2002).
[12]Klapars, A., Buchwald, S.L. *JACS* **124**, 14844 (2002).
[13]Hoshi, M., Shirakawa, K., Takeda, K. *SL* 403 (2001).
[14]Evans, P.A., Leahy, D.K. *JACS* **124**, 7882 (2002).
[15]Taguchi, H., Ghoroku, K., Tadaki, M., Tsubouchi, A., Takeda, T. *JOC* **67**, 8450 (2002).

Copper(II) nitrate. 15, 101; 18, 115–116; 19, 110; 20, 121; 21, 148

Biaryls.[1] Homocoupling of $ArB(OH)_2$ is accomplished with $(Ph_3P)_4Pd$ in the presence of copper(II) nitrate at room temperature. Other metal nitrates are practically ineffective.

[1]Koza, D.J., Carita, E. *S* 2183 (2002).

Copper(I) oxide. 16, 99; 21, 148–149

Arylamines.[1] Aryl halides are converted to arylamines (in yields usually >70%) on reaction with ammonia using Cu_2O as catalyst. Selectivity is noted, for example, 2,5-dibromopyridine gives 2-amino-5-bromopyridine.

[1]Lang, F., Zewge, D., Houpis, I.N., Volante, R.P. *TL* **42**, 3251 (2001).

Copper(II) tetrafluoroborate. 21, 149

Epoxide opening.[1] Reaction of epoxides with alcohols occurs at room temperature in the presence of $Cu(BF_4)_2$.

[1]Barluenga, J., Vazquez-Villa, H., Ballesteros, A., Gonzalez, J.M. *OL* **4**, 2817 (2002).

Copper(I) hexafluorophosphate.

Radical cyclization.[1] Unsaturated *N*-benzoyloxyamines cyclize on treatment with $CuPF_6$-$BF_3 \cdot OEt_2$ when the functional groups are properly separated.

$CuPF_6$ - $BF_3 \cdot OEt_2$, PhMe 100°

79% (3 : 1)

[1]Noack, M., Göttlich, R. *CC* 536 (2002).

Copper(I) 2-thienylcarboxylate. 19, 112; **20,** 122; **21,** 149

Coupling. The Pd-catalyzed cross-coupling of alkynyl sulfides with arylboronic acids to give alkynylarenes is mediated by the copper reagent.[1]

[1]Savarin, C., Srogl, J., Liebeskind, L.S. *OL* **3**, 91 (2001).

Copper(II) triflate. 19, 112; **20,** 122–123; **21,** 149

Condensation reactions. When aldehydes, pyruvic esters and trimethyl orthoformate are heated with copper(II) triflate in dichloromethane, β,γ-unsaturated α-keto esters are obtained.[1] Silylated *N,O*-acetals in the presence of $Cu(OTf)_2$ and Me_3SiCl behave as aminoalkylation agents for indole and activated arenes.[2]

Cl_3C, NHSiMe$_3$, Me$_3$SiO; $Cu(OTf)_2$, Me_3SiCl, CH_2Cl_2; Cl_3C, NH_2

74%

Addition reactions. $Cu(OTf)_2$ catalyzes Michael reaction between *O*-benzyl carbamate and enones at room temperature[3] and allylation of ketones with tetraallylstannane.[4] In the conjugate addition of Et_2Zn to enones rates are enhanced by stabilized carbenes such as 1,3-dimesitylimidazolidene.[5]

Oxazolidines. Treatment of imines with diazo compounds in acetone with $Cu(OTf)_2$ leads to the five-membered heterocycles.[6]

MeO, COOEt, N, Ph, N_2, Cu(OTf)$_2$, 4A -mol sieves, Me$_2$CO, MeO, COOEt, Ph, N, O

68% (*cis* + *trans*)

Aryl sulfones.[7] A synthesis of $ArSO_2R$ is based on arylation of sulfinic acid salts in DMSO with $Cu(OTf)_2$ and *N,N'*-dimethylethylenediamine to form a catalytic system.

[1]Dujardin, G., Leconte, S., Benard, A., Brown, E. *SL* 147 (2001).
[2]Sakai, N., Hamajima, T., Konakahara, T. *TL* **43**, 4821 (2002).
[3]Wabnitz, T.C., Spencer, J.B. *TL* **43**, 3891 (2002).
[4]Kamble, R.M., Singh, V.K. *TL* **42**, 7525 (2001).
[5]Fraser, P.K., Woodward, S. *TL* **42**, 2747 (2001).
[6]Lee, S.-H., Yang, J., Han, T.-D. *TL* **42**, 3487 (2001).
[7]Baskin, J.M., Wang, Z. *OTL* **4**, 4423 (2002).

Cyanogen bromide.

De-S-trimethylsilylethylation.[1] Treatment of $RSCH_2CH_2SiMe_3$ with CNBr gives thiocyanates. The reaction is applicable to nucleoside chemistry.

[1]Chambert, S., Thomasson, F., Decout, J.-L. *JOC* **67**, 1898 (2002).

Cyanomethylenetriorganophosphonium halide.

Alkylations. This salt is useful for activation of alcohols so that they become alkyl donors to amines[1] and thiols.[2] For alkylation of sulfones the derived ylide ($Me_3P{=}CHCN$) generated from the salt by treatment with KHMDS must be used.[3] Homologation of alcohols by two carbon units is achieved with the reagent.[4]

R−OH + $Me_3\overset{+}{P}$ CN I^- — *i*-Pr$_2$NEt / EtCN Δ ; H$_2$O → R CN

SO$_2$Ph + $Me_3\overset{+}{P}$ CN I^- + RCH$_2$OH → R SO$_2$Ph

[1]Zaragoza, F., Stephensen, H. *JOC* **66**, 2518 (2001).
[2]Zaragoza, F. *T* **57**, 5451 (2001).
[3]Uemoto, K., Kawahito, A., Matsushita, N., Sakamoto, I., Kaku, H., Tsunoda, T. *TL* **42**, 905 (2001).
[4]Zaragoza, F. *JOC* **67**, 4963 (2002).

2-Cyano-1-phenylethyl chloroformate.

Protection. Alcohols protected as esters on reaction with the title reagent are regenerated on exposure to base (e.g., DBU).[1] Its utility in oligoribonucleotide synthesis is demonstrated.

[1]Munch, U., Pfleiderer, W. *HCA* **84**, 1504 (2001).

Cyanuric chloride. 20, 124

Hydroxy group transformation. Alcohols are either converted to chlorides or formates with cyanuric chloride-DMF.[1] Cyanuric chloride can serve as a surrogate for oxalyl chloride in the Swern oxidation.[2]

Acyl azides. The preparation of $RCON_3$ from carboxylic acids and sodium azide is conveniently mediated by cyanuric chloride using *N*-methylmorpholine as base.[3]

Beckmann rearrangement.[4] Very mild conditions are required for the conversion of oximes to amides in the presence of cyanuric chloride-DMF.

[1]De Luca, L., Giacomelli, G., Porcheddu, A. *OL* **4**, 553 (2002); *JOC* **67**, 5152 (2002).
[2]De Luca, L., Giacomelli, G., Porcheddu, A. *JOC* **66**, 7907 (2001).
[3]Bandgar, B.P., Pandit, S.S. *TL* **43**, 3413 (2002).
[4]De Luca, L., Giacomelli, G., Porcheddu, A. *JOC* **67**, 6272 (2002).

***β*-Cyclodextrin. 21,** 151

Epoxide opening. In the presence of β-cyclodextrin epoxides are opened with HX or LiX to afford halohydrins at room temperature with the halogen atom attaching to the less hindered carbon.[1] Styrene oxide yields α-hydroxyacetophenone if NBS is also added.[2]

[1]Reddy, L.R., Surendra, K., Bhanumathi, N., Rao, K.R. *T* **58**, 6003 (2002).
[2]Reddy, L.R., Bhanumathi, N., Rao, K.R. *TL* **43**, 3237 (2002).

(1,5-Cyclooctadiene)bis(triphenylphosphite)iridium(I) triflate.

Alkylation. Upon rendering propargylic esters reactive toward silyl enol ethers by the iridium complex a method for synthesis of 4-alkyn-1-ones presents itself.[1] Allylation is achieved with a very similar catalyst.[2]

[1]Matsuda, I., Komori, K., Itoh, K. *JACS* **124**, 9072 (2002).
[2]Matsuda, I., Wakamatsu, S., Komori, K., Makino, T., Itoh, K. *TL* **43**, 1043 (2002).

(1,5-Cyclooctadiene)cyclopentadienylcobalt. 21, 151–152

2-Substituted pyridines.[1] The [2+2+2]cycloaddition involving a nitrile and two alkyne units is applicable to the synthesis of monosubstituted pyridines, employing ethyne.

[1]Heller, B., Sundermann, B., Buschmann, H., Drexler, H.-J., You, J., Holzgrabe, U., Heller, E., Oehme, G. *JOC* **67**, 4414 (2002).

(1,5-Cyclooctadiene)iridium(I) hexafluorophosphate.

Hydrogenation.[1] A catalytic hydrogenation system constituting (cod)Ir[PF_6] and the chiral P,N-ligand **1** has been developed.

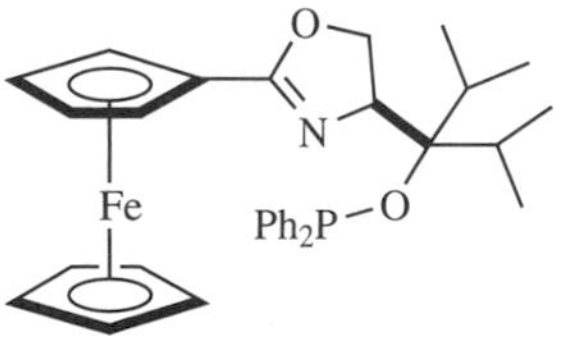

1

[1]Blankenstein, J., Pfaltz, A. *ACIEE* **40**, 4445 (2001).

(1,5-Cyclooctadiene)rhodium(I) tetraphenylborate.

Hydroaminomethylation.[1] Alkylation of amines with styrenes in the presence of the zwitterionic complex gives 2-arylpropylamines as major products.

[1]Lin, Y.-S., Ali, B.E., Alper, H. *TL* **42**, 2423 (2001).

(1,5-Cyclooctadiene)(tricyclohexylphosphine)iridium(I) hexafluorophosphate.

Deuteration.[1] This Crabtree catalyst promotes deuteration of certain substituted arenes with D_2O. Thus, *o*-deuteration is observed of esters, tertiary amides, etc. α-Tetralone is deuterated at C-8.

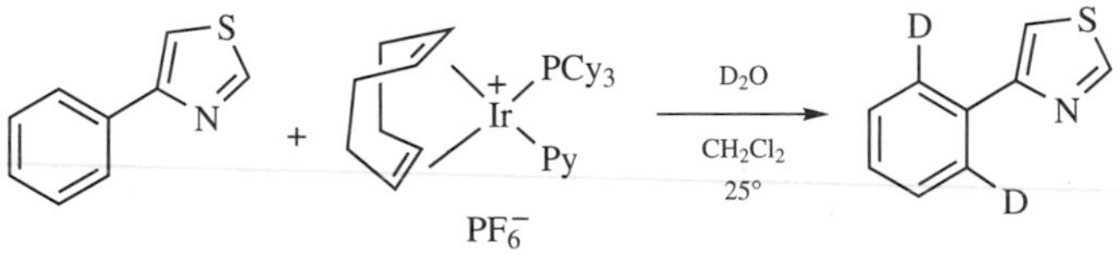

[1]Ellames, G.J., Gibson, J.S., Herbert, J.M., McNeill, A.H. *T* **57**, 9487 (2001).

1-Cyclopropylethyl trichloromethylimidate.

Alcohol protection.[1] Ether formation by reaction with the title reagent **1** is mediated by AgOTf. The acid-labile ethers are cleaved by 1% CF_3COOH at room temperature. Utility of the protecting group in polymer-supported oligosaccharide synthesis has been studied.

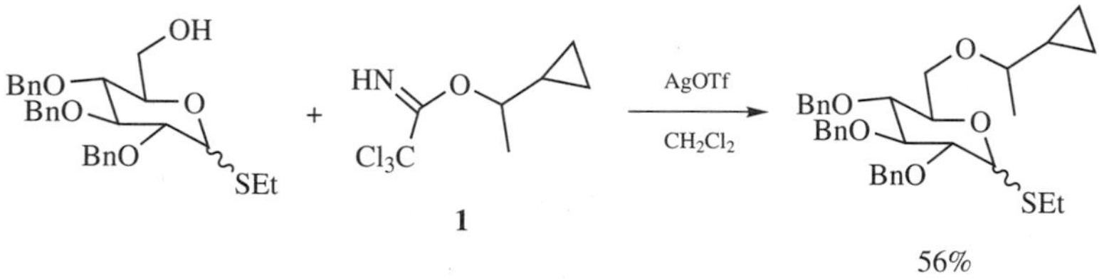

[1]Eichler, E., Yan, F., Sealy, J., Whitfield, D.M. *T* **57**, 6679 (2001).

D

Decaborane. 21, 154

Deprotection. TBS ethers are cleaved[1] by substoichiometric decaborane in THF-MeOH at room temperature. In aqueous THF it also catalyzes the hydrolysis of acetals.[2]

Dehalogenation.[3] Decaborane serves as a hydrogen transfer agent for removal of halogen atoms in α-positions of carbonyl compounds. The reducing system also contains Pd/C as catalyst.

Reductive methylation.[4] Amines undergo methylation with HCHO in methanol on treatment with decaborane.

[1]Seong, Y.J., Lee, J.H., Park, E.S., Yoon, C.M. *JCS(P1)* 12239 (2002).
[2]Lee, S.H., Lee, J.H., Yoon, C.M. *TL* **43**, 2699 (2002).
[3]Lee, S.H., Jung, Y.J., Cho, Y.J., Yoon, C.-O.M., Hwang, H.-J., Yoon, C.M. *SC* **31**, 2251 (2001).
[4]Jung, Y.J., Bae, Y.W., Yoon, C.-O.M., Yoo, B.W., Yoon, C.M. *SC* **31**, 3417 (2001).

Dess-Martin periodinane. 21, 154–155

Oxidation. The well-known capability of the title reagent to conduct mild oxidation finds further use in a synthesis of (*Z,Z*)-1,4-dienes from (*Z*)-homoallylic alcohols, i.e., oxidation and Wittig reaction.[1] Cyclic cycloalkenecarbamates are readily formed by oxidation of *N*-hydroxyalkyl carbamates.[2] 1,*n*-Diols are converted to lactol acetates or dialdehydes,[3] depending on the carbon chain length.

A three-step synthesis of 4-hydroxy-2-pyrones from 2,2-dimethyl-4-trimethylsiloxyl-6-methylene-1,3-diox-4-ene involves vinylogous aldol reaction, oxidation and pyrolytic removal of acetone.[4]

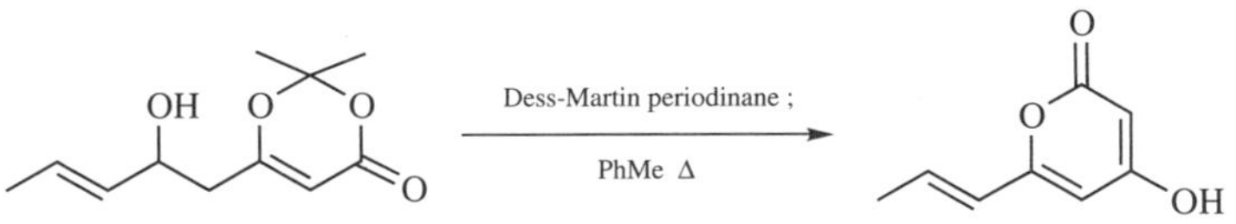

[1]Wavrin, L., Viala, J. *S* 326 (2002).
[2]Yu, C., Hu, L. *TL* **42**, 5167 (2001).
[3]Roels, J., Metz, P. *SL* 789 (2001).
[4]Bach, T., Kirsch, S. *SL* 1974 (2001).

Dialkylaluminum chloride. 20, 126–127; **21,** 155

Removal of N-benzyloxycarbonyl group. The carbamates are cleaved on short exposure to Et_2AlCl at –78°.[1]

Fiesers' Reagents for Organic Synthesis, Volume 22. Series editor Tse-Lok Ho
ISBN 0-471-28515-3

1,2-Diamines. Reductive coupling attendant by *N*-ethylation is observed when imines and Et_2AlCl are brought together.[2]

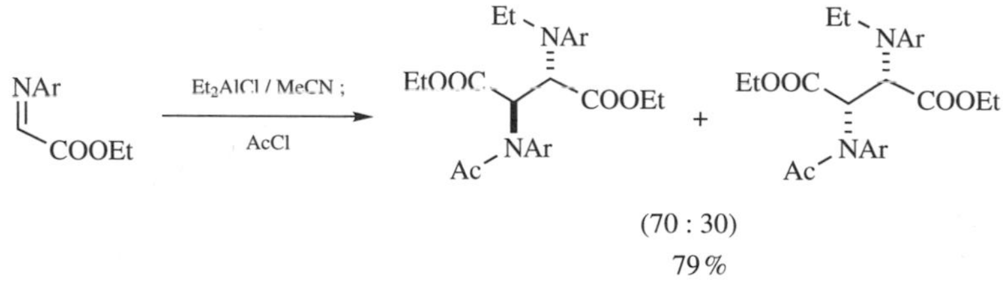

NAr COOEt

Et_2AlCl / MeCN ; AcCl

Et NAr EtOOC COOEt Ac NAr + Et NAr EtOOC COOEt Ac NAr

(70 : 30)
79 %

α-Trimethylsilylallenyl ketones. Alkynes, which bear a leaving group at one propargylic position as well as trimethylsilyl and trimethylsiloxy groups at the other, expel the leaving group while shifting the Me_3Si residue to its proximal *sp*-hybridized carbon site, when they are treated with Me_2AlCl.[3]

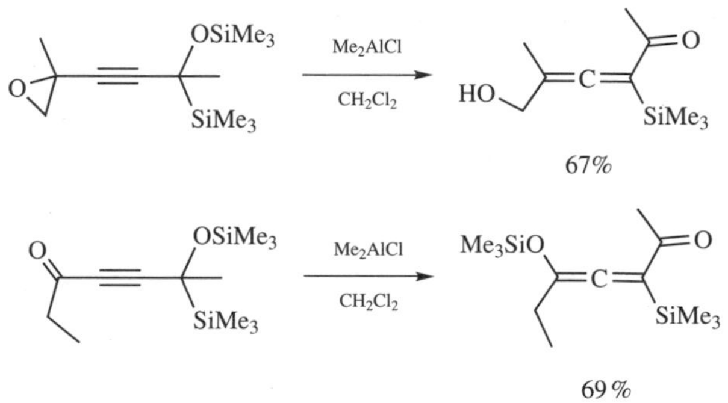

O $OSiMe_3$ $SiMe_3$

Me_2AlCl, CH_2Cl_2

HO C O $SiMe_3$

67%

O $OSiMe_3$ $SiMe_3$

Me_2AlCl, CH_2Cl_2

Me_3SiO C O $SiMe_3$

69 %

Cyclization. The influence of Et_2AlCl on the radical cyclization from predominantly a 5-*exo* mode to 6-*endo* mode is remarkable.[4]

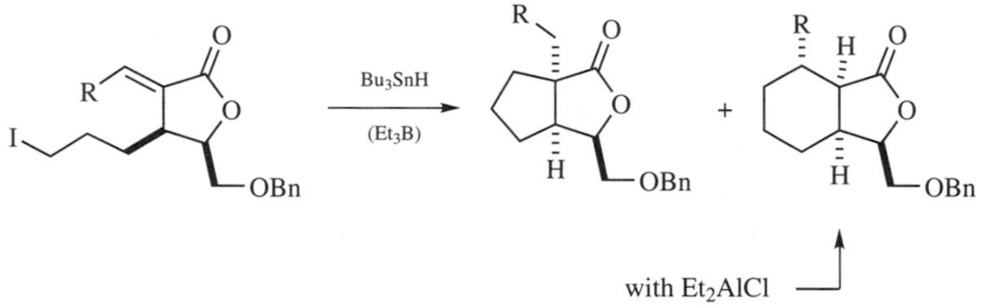

I R O O OBn

Bu_3SnH (Et_3B)

R O O H OBn + R H O O H OBn

with Et_2AlCl

Elimination.[5] Alkenylsilanes are obtained from carbonyl compounds while involving reaction with $LiCH_2SiR_3$, treatment with Et_2AlCl and aqueous workup.

[1]Tsujimoto, T., Murai, A. *SL* 1283 (2002).
[2]Shimizu, M., Niwa, Y. *TL* **42**, 2829 (2001).
[3]Cunico, R.F., Zaporowski, L.F., Rogers, M. *JOC* **64**, 9307 (1999).

[4]Kim, K., Okamoto, S., Sato, F. *OL* **3**, 67 (2001).
[5]Kwan, M.L., Battiste, M.A. *TL* **43**, 8765 (2002).

Dialkylaluminum cyanide. 21, 155–156

β-Cyano sulfoxides.[1] Addition of Et_2AlCN to alkenyl sulfoxides is diastereoselective. Accordingly, induction by chiral sulfoxides engenders establishment of asymmetric tertiary and quaternary carbon centers.

O S Tol — Et_2AlCN/THF, 65° → CN O S Tol

[1]Ruano, J.L.G., Garcia, M.C., Laso, N.M., Castro, A.M.M., Ramos, J.H.R. *ACIEE* **40**, 2507 (2001).

Dialkylaluminum iodide.

Baylis-Hillman reaction. Condensation of 2-cycloalkenones with aldehydes can be effected by Et_2AlI in CH_2Cl_2 alone.[1] The same reagent also promotes reaction of conjugated thio esters.[2]

[1]Pei, W., Wei, H.-X., Li, G. *CC* 2412 (2002).
[2]Pei, W., Wei, H.-X., Li, G. *CC* 1856 (2002).

1,3-Dialkylimidazol-2-ylidenes.

Acylation. A versatile catalyst for both acylation and transesterification (e.g., vinyl acetate as acyl source) is the stable carbene.[1,2] Formation of benzyl esters and allyl esters from the corresponding alcohols are particularly favorable.

Suzuki coupling. While the Ru complexes of this type of carbenes (e.g., 1,3-dimesityl derivatives) are well known for catalyzing alkene metathesis, the Pd-complex **1** has found use in Suzuki coupling.[3]

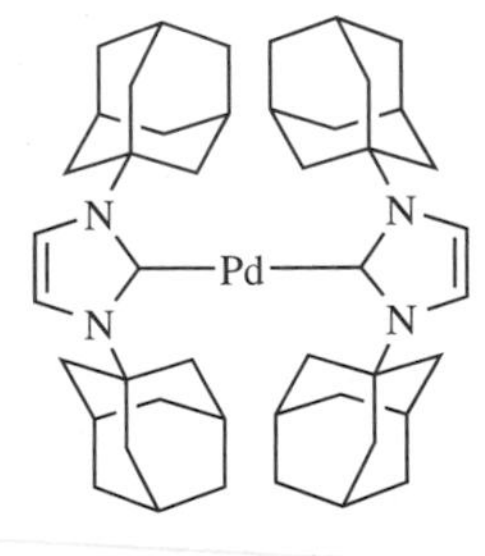

1

[1]Grasa, G.A., Kissling, R.M., Nolan, S.P. *OL* **4**, 3583 (2002).
[2]Nyce, G.W., Lamboy, Connor, E.F., Waymouth, R.M., Hedrick, J.L. *OL* **4**, 3587 (2002).
[3]Gstöttmayr, C.W.K., Böhm, V.P.W., Herdtweck, E., Grosche, M., Herrmann, W.A. *ACIEE* **41**, 1363 (2002).

Diallylaluminum 2-(*N,N*-dimethylamino)ethanoate.

Allylations. This reagent (**1**) is stable and suitable for allyl transfer to aldehydes, ketones and imines.[1]

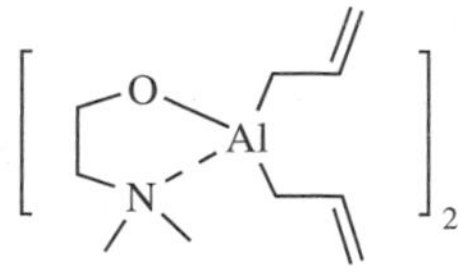

1

[1]Schumann, H., Kaufmann, J., Dechert, S., Schmalz, H.-G. *TL* **43**, 3507 (2002).

1,4-Diamino-1,4-diazoniabicyclo[2.2.2]octane nitrite.

Aziridino ketones.[1] The title reagent **1** is capable of converting enones to aziridino ketones.

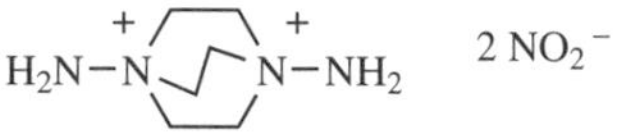

$2\ NO_2^-$

1

[1]Xu, J., Jiao, P. *JCS(P1)* 1491 (2002).

1,4-Diazabicyclo[2.2.2]octane, DABCO. 13, 92; **15,** 109; **18,** 120; **19,** 116–117; **20,** 128–129; **21,** 157–158

Baylis-Hillman reaction. Reaction of aromatic aldehydes with acrylamide is successful.[1] Stoichiometric quantities of DABCO and water (50% dioxane) are shown to be necessary to overcome problem of low yields and long reaction time.[2] The reaction rate is greatly enhanced by changing methyl acrylate to α-naphthyl acrylate.[3]

Another way of improving the reaction is to adopt an acrylic ester with a *cis*-stilbene moiety (**1**).[4] In such cases using a large excess of the acceptor aldehydes to accelerate the reaction is not severely punished by problems in product isolation, because on isomerization of the stilbene unit by iodine the adducts become very insoluble, and readily separated.

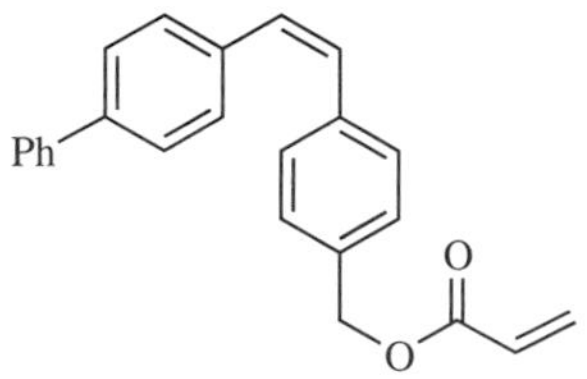

1

Alkylation of conjugated compounds by catalysis of DABCO involves a reversible Michael reaction.[5]

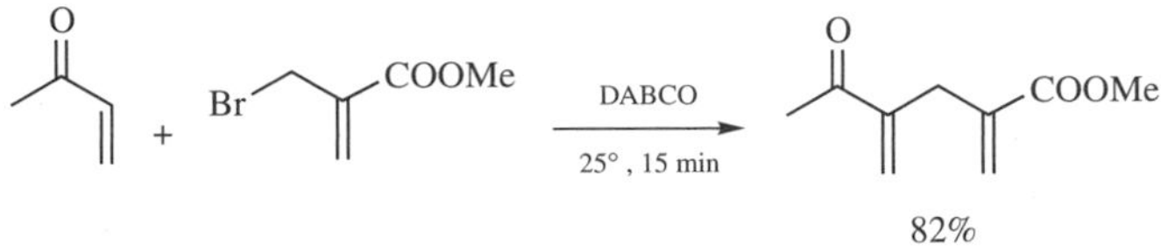

Heck reaction. Hydride transfer to the pallada intermediates from 1,2,2,6,6-pentamethylpiperidine which is the conventional base component or from dioxane if it is employed as solvent stops. Pd-catalyzed polycyclization upon formation of the first ring in the following system. To replace the piperidine base with DABCO resolves the difficulty.[6]

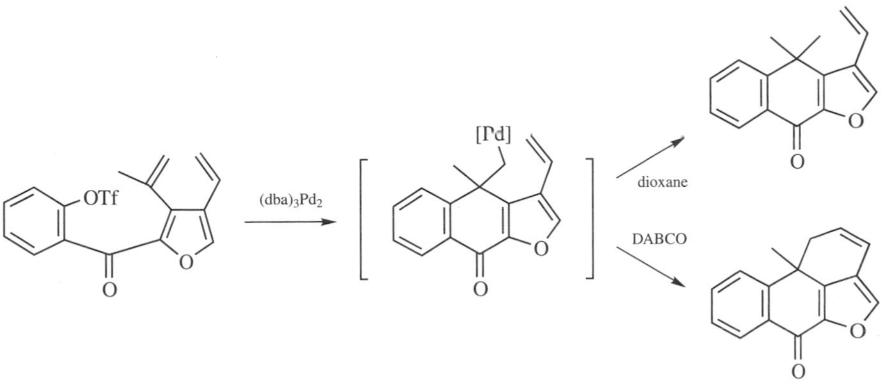

[1]Yu, C., Hu, L. *JOC* **67**, 219 (2002).
[2]Yu, C., Liu, B., Hu, L. *JOC* **66**, 5413 (2001).
[3]Lee, W.-D., Yang, K.-S., Chen, K. *CC* 1612 (2001).
[4]Bosanac, T., Wilcox, C.S. *CC* 1618 (2001).
[5]Basavaiah, D., Sharada, D.S., Kumaragurubaran, N., Reddy, R.M. *JOC* **67**, 7135 (2002).
[6]Lau, S.Y.W., Andersen, N.G., Keay, B.A. *OL* **3**, 181 (2001).

1,8-Diazabicyclo[5.4.0]undec-7-ene, DBU. 13, 92; **14,** 109; **15,** 109–110; **16,** 105–106; **17,** 99–100; **18,** 120–121; **19,** 117; **20,** 129–130; **21,** 158–160

1,4-Diketones. DBU not only promotes Michael reaction of primary nitroalkanes to enones and acrylic esters it also converts the secondary nitro group into a ketone (Nef reaction).[1] With 2-nitroethylarenes as Michael donors the reaction proceeds one step further to deliver 2-aryl-2-cyclopentenones.[2] With enediones the products are 4-alkylidene-2-cyclopentenones.[3]

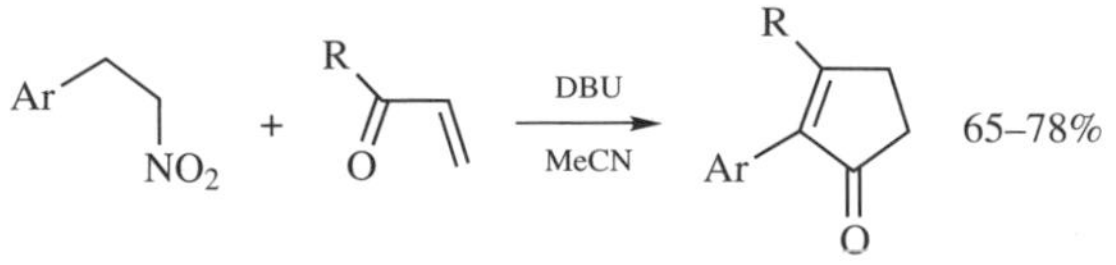

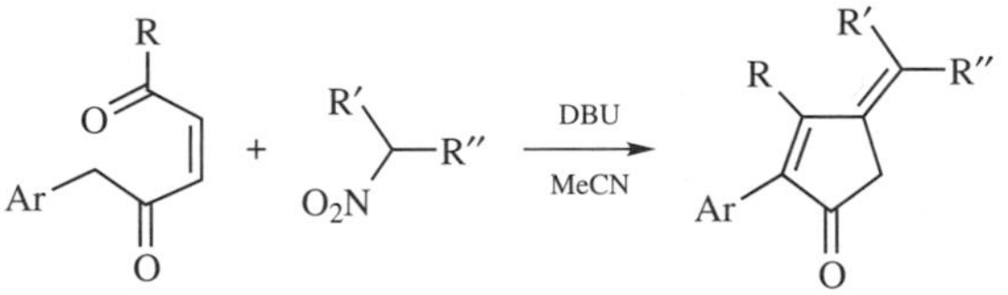

1,3-Dienes. Conjugated sulfolenes are converted to 1,3-dienes on heating with DBU in cyclohexane.[4] Deconjugation precedes cheletropic extrusion of SO_2.

Alkylations. Methylation of phenols, indoles, and benzimidazoles with $CO(OMe)_2$ using DBU as base is high-yielding. Reaction rates are enhanced by microwave irradiation.[5]

Aldol reactions. Acyldiazomethanes condense with aldehydes and imines in the presence of catalytic DBU.[6] β-Keto esters and amides undergo alkylidenation at the γ-position with aldehydes .[7]

Baylis-Hillman reaction. In a route to thiacoumarin based on Baylis-Hillman reaction of bis(*o*-formylphenyl) disulfide[8] it requires that the disulfide linkage is severed by DBU.

[1]Ballini, R., Bosica, G., Fiorini, D., Petrini, M. *TL* **43**, 5233 (2002).
[2]Ballini, R., Barboni, L., Bosica, G., Fiorini, D. *S* 2725 (2002).
[3]Ballini, R., Bosica, G., Fiorini, D., Gil, M.V., Petrini, M. *OL* **3**, 1265 (2001).
[4]Lusinchi, M., Stanbury, T.V., Zard, S.Z. *CC* 1532 (2002).
[5]Shieh, W.-C., Dell, S., Repic, O. *OL* **3**, 4279 (2001).
[6]Jiang, N., Wang, J. *TL* **43**, 1285 (2002).
[7]Charonnet, E., Filippini, M.-H., Rodriguez, J. *S* 788 (2001).
[8]Kaye, P.T., Nocanda, X.W. *S* 2389 (2001).

Dibromodifluoromethane.

Ramberg-Bäcklund rearrangement.[1] Preparation of substituted 1,3,5,7-octatetraenes is accomplished by a modified Ramberg-Bäcklund rearrangement.

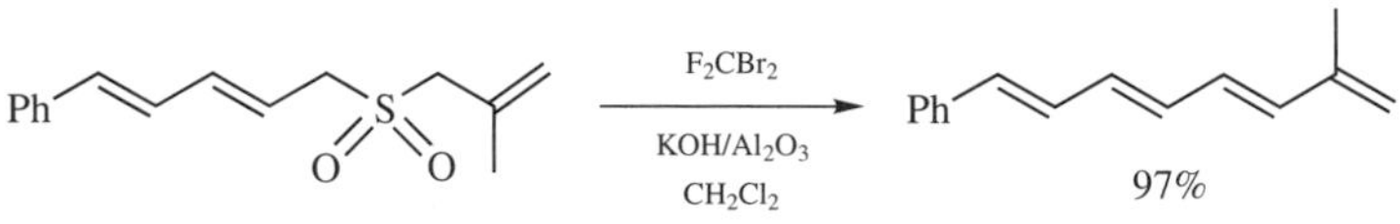

[1]Cao, X.-P. *T* **58**, 1301 (2001).

Dibutylboron triflate. 20, 132, **21,** 161

Aldol reaction.[1] Further demonstration of the usefulness of boron enolates as donors in aldol reactions is in a synthesis of *trans*-3-hydroxy-4,5-epoxy ketones.

Bu_2BOTf - *i*-Pr_2NEt

OHC

O

O

OH

[1]Righi, G., Spirito, F., Bonini, C. *TL* **43**, 4737 (2002).

2,6-Di-*t*-butylphenoxy(difluoro)borane.

4-Aryl-1,3-dioxanes.[1] The title reagent is a liquid that can be purified by vacuum distillation. It is stable at –30° for at least several weeks. Reaction of styrenes with paraformaldehyde in the presence of catalytic amount of the $ArOBF_2$ gives 1,3-dioxane derivatives. Styrenes bearing electron-withdrawing substituent(s) are not reactive.

$(HCHO)_n$

$ArOBF_2$

83%

[1]Bach, T., Löbel, J. *S* 2521 (2002).

Dibutyltin diiodide.

Aldol reactions.[1] Highly diastereoselective condensation of α-iodo carbonyl compounds with aldehydes results when the substrates are treated with Bu_2SnI_2-LiI and HMPA in THF. Enolate generation is via deiodination.

[1]Shibata, I., Suwa, T., Sakakibara, H., Baba, A. *OL* **4**, 301 (2002).

Dibutyltin diiode–magnesium bromide.

Hydrostannylation. The ate complex transforms 1-alkynes to 2-dibutyliodostannyl-1-alkenes at room temperature (5 examples, 60–86%).[1] Hydrostannylation of conjugated esters results in the reduction of the double bond.[2]

[1]Shibata, I., Suwa, T., Ryu, K., Baba, A. *JACS* **123**, 4101 (2001).
[2]Shibata, I., Suwa, T., Ryu, K., Baba, A. *JOC* **66**, 8690 (2001).

Di-*t*-butyl peroxide.

Alkylation. Radical generation initiated by C-S bond cleavage of xanthates with *t*-Bu_2O_2 is incorporated into a cyclization process leading to homophthalimides.[1] A protected glycine is subject to free-radical benzylation, affording phenylalanine derivatives.[2]

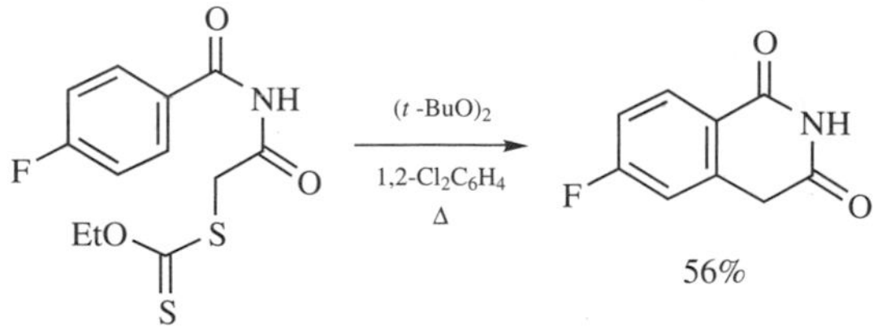

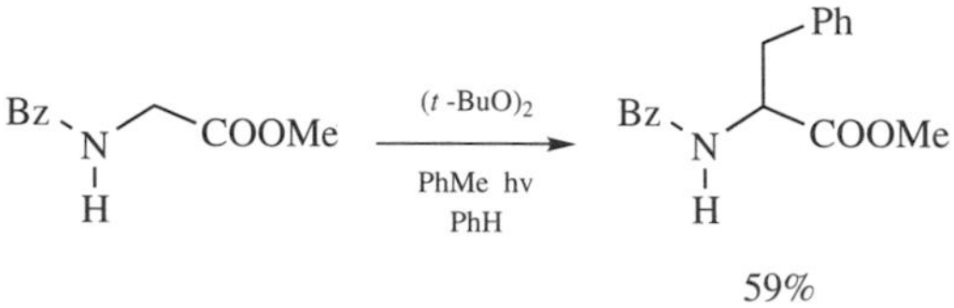

$(t\text{-BuO})_2$, 1,2-$Cl_2C_6H_4$, Δ — 56%

$(t\text{-BuO})_2$, PhMe hv, PhH — 59%

[1]Quiclet-Sire, B., Zard, S.Z. *CC* 2306 (2002).
[2]Knowles, H.S., Hunt, K., Parsons, A.F. *T* **57**, 8115 (2001).

Di-*t*-butyl peroxyoxalate.

Ketones from bromohydrins. By a free-radical elimination of HBr ketones are obtained from bromohydrins (9 examples, 38–100%).[1]

[1]Dolenc, D., Harej, M. *JOC* **67**, 312 (2002).

Dibutyltin oxide. 13, 95–96; **15,** 116–117; **16,** 112; **18,** 125; **20,** 133–134; **21,** 163

Transesterification. Ester exchange is accomplished by heating a substrate in another alcohol in the presence of Bu_2SnO.[1] A bis(tridecafluorooctyl) analogue is also effective.[2]

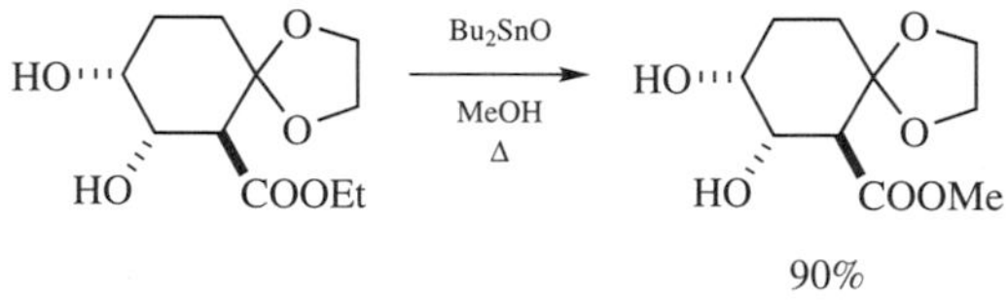

Bu_2SnO, MeOH, Δ — 90%

Deacetylation.[3] Carbohydrate acetates are readily hydrolyzed with Bu_2SnO in refluxing MeOH.

[1]Baumhof, P., Mazitschek, R., Giannis, A. *ACIEE* **40**, 3672 (2001).
[2]Xiang, J., Toyoshima, S., Orita, A., Otera, J. *ACIEE* **40**, 3670 (2001).
[3]Liu, H.-M., Yan, X., Li, W., Huang, C. *CR* **337**, 1763 (2002).

7,7-Di-*t*-butyl-7-silabicyclo[4.1.0]heptane.

Silacyclopropanation.[1] The title reagent is a sila-transfer agent for alkenes. For unactivated alkenes the reaction proceeds in the presence of AgOTf.[2]

OTIPS H + Si PhH Δ OTIPS H Si

97%

[1]Driver, T.G., Franz, A.K., Woerpel, K.A. *JACS* **124**, 6524 (2002).
[2]Cirakovic, J., Driver, T.G., Woerpel, K.A. *JACS* **124**, 9370 (2002).

Dicarbonylcyclopentadienylcobalt. 21, 163

Cycloisomerization. Treatment of 1,6-enynes with $CpCo(CO)_2$ under uv light gives five-membered ring products.[1]

MeOOC MeOOC —$SiMe_3$ $CpCo(CO)_2$ hv ; $CuCl_2$

$SiMe_3$ MeOOC MeOOC + $SiMe_3$ MeOOC MeOOC

21% 52%

[n]Phenylenes up to n=9 which are helical molecules have been synthesized by the cycloismerization process.[2]

$CpCo(CO)_2$ hv xylene Δ

12%

[6]phenylene

[1]Buisine, O., Aubert, C., Malacria, M. *CEJ* **7**, 3517 (2001).

[2]Han, S., Bond, A.D., Disch, R.L., Holmes, D., Schulman, J.M., Teat, S.J., Vollhardt, K.P.C., Whitener, G.D. *ACIEE* **41**, 3223 (2002).

Dichloroborane.

Hydroboration. Standard reaction of perfluoroalkylethenes shows reverse regioselectivity in using dichloroborane and dicyclohexylborane.[1] Dichloroborane favors attachment of its boron atom to the more highly substituted sp^2-hybrized carbon atom of the alkene.[2]

$$CH_2{=}CH{-}C_6F_{13} + R_2BH \xrightarrow[\text{[NaOOH workup]}]{} CH_3CH(OH)C_6F_{13} + HOCH_2CH_2C_6F_{13}$$

99 : 1 (R = Cl)
5 : 95 (R = Chx)

[1]Ramachandran, P.V., Jennings, M.P. *CC* 386 (2002).
[2]Ramachandran, P.V., Jennings, M.P. *OL* **3**, 3789 (2002).

2,3-Dichloro-5,6-dicyano-1,4-benzoquinone, DDQ. 13, 104–105; **14,** 126–127; **15,** 125–126; **16,** 120; **18,** 130; **19,** 121–122; **20,** 137–138; **21,** 164–165

Deprotection. Prenyl ethers[1] and *p*-phenylbenzyl ethers[2] are cleaved by DDQ. Due to the relatively high cost the use of DDQ in catalytic amounts together with a reoxidant such as $Mn(OAc)_3$ seems sensible.

Aromatization. *N*-Acyl-1,4-dihydropyridines undergo aromatization[3] with loss of the acyl residue on treatment with DDQ at room temperature.

Phenylselenenylation.[3] Diphenyl diselenide is activated by DDQ and the reaction of alkenes in hydroxylic solvents (H_2O, ROH) leads to β-seleno alcohols or ethers.

Alkyl halides from alcohols. In situ oxidation of R_4NX by DDQ to supply a positive halogen source for combination with Ph_3P so that activation of primary and secondary alcohols and conversion to halides are accomplished.[4] The reagent combination does not affect epoxides, TMS ethers, amides, thiols, and disulfides.

[1]Vatele, J.-M. *T* **58**, 5689 (2002).
[2]Sharma, G.V.M., Rakesh. *TL* **42**, 5571 (2001).
[3]Wallace, D.J., Gibb, A.D., Cottrell, I.F., Kennedy, D.J., Brands, K.M.J., Dolling, U.H. *S* 1784 (2001).
[4]Iranpoor, N., Firouzabadi, H., Aghapour, G., Vaez zadeh, A.R. *T* **58**, 8689 (2002).

1,*n*-Dichloro-1,*n*-bis(*p*-toluenesulfinyl)alkanes.

Diketones.[1] The title reagents are platforms for bidirectional chain extension while behaving as α,ω-bis(acyl anion equivalents).

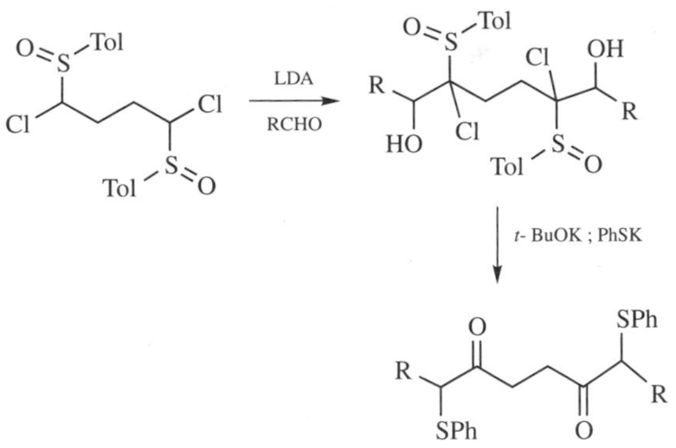

O=S-Tol, Cl, Cl, Tol-S=O

LDA / RCHO

O=S-Tol, R, HO, Cl, Cl, OH, R, Tol-S=O

t- BuOK ; PhSK

R, O, SPh, SPh, O, R

[1]Satoh, T., Taguchi, D., Kurabayashi, A., Kanoto, M. *T* **58**, 4217 (2002).

Dichlorodioxomolybdenum(VI).

Disulfides.[1] The Mo compound catalyzes dimerization of RSH by DMSO.

[1]Sanz, R., Aguado, R., Pedrosa, M.R., Arnaiz, F.J. *S* 856 (2002).

Dichloroindium hydride.

Hydrodehalogenation.[1] The indium hydride generated in situ from $InCl_3$ and Bu_3SnH is useful for the replacement of halogen atom by hydrogen in organic compounds.

(Z)-Alkenes.[2] A route to (*Z*)-alkenes from 1-alkynes is via hydroboration with $HInCl_2$ which is obtained from $InCl_3$ and Dibal-H. A follow-up Suzuki coupling concludes the synthesis.

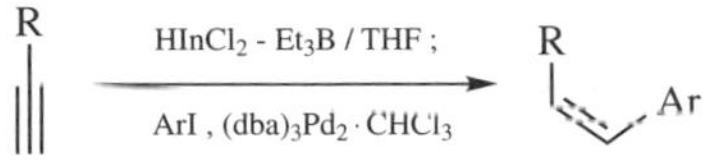

R

$HInCl_2$ - Et_3B / THF ;

ArI , $(dba)_3Pd_2 \cdot CHCl_3$

R, Ar

[1]Inoue, K., Sawada, A., Shibata, I., Baba, A. *TL* **42**, 4661 (2001).
[2]Takami, K., Yorimitsu, H., Oshima, K. *OL* **4**, 2993 (2002).

Dichlorophosphoryl isocyanate.

Cyanation.[1] The title reagent [$Cl_2P(=O)NCO$] donates a cyano group to electron-rich heterocycles and enamines very readily.

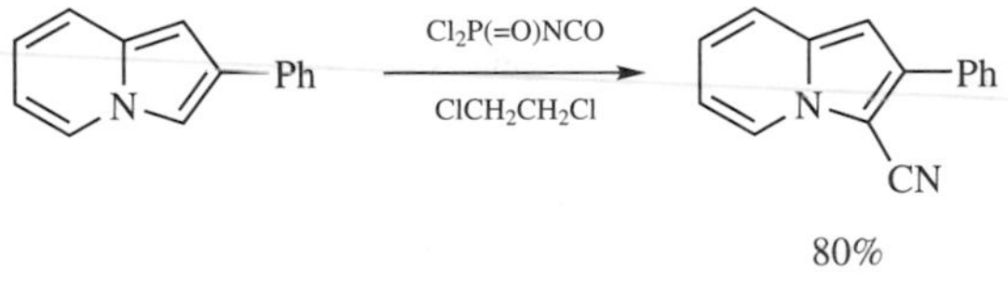

N, Ph

$Cl_2P(=O)NCO$

$ClCH_2CH_2Cl$

N, Ph, CN

80%

[1]Smaliy, R.V., Chaikovskaya, A.A., Pinchuk, A.M., Tolmachev, A.A. *S* 2416 (2002).

Dichlorotris(triphenylphosphine)ruthenium(II).

Isomerization-aldol reaction.[1] The union of allylic alcohols with aldehydes in the presence of $(Ph_3P)_3RuCl_2$ and a base involves isomerization of the alcohols to ketone enolates and aldol reaction.

Ph‑CHO + (allylic alcohol, OH) —BuLi - $(Ph_3P)_3RuCl_2$, THF Δ→ aldol product (Ph, OH, O)

62% (*syn* : *anti* 55 : 45)

Reduction-alkylation.[2] The net result of CC bond formation and carbonyl reduction on treatment of mixtures of ketones and alcohols with the Ru complex is also an intriguing process.

Ph‑CO‑CH₃ (O) + HO‑(CH₂)₃CH₃ —KOH - $(Ph_3P)_3RuCl_2$, dioxane, e 80°→ Ph‑CH(OH)‑(CH₂)₄CH₃ (OH)

Oxidation.[3] Together with a peracid (e.g., AcOOH) a polymer-anchored Ru complex effects oxidation of hydrocarbons and alcohols to ketones.

[1]Uma, R., Davies, M., Crevisy, C., Gree, R. *TL* **42**, 3069 (2001).
[2]Cho, C.S., Kim, B.T., Kim, T.-J., Shim, S.C. *JOC* **66**, 9020 (2001).
[3]Leadbeater, N.E. *JOC* **66**, 2168 (2001).

Dicobalt octacarbonyl. 13, 99–101; **14,** 117–119; **15,** 117–118; **16,** 113–115; **17,** 102–105; **18,** 132; **19,** 125–126; **20,** 139–141; **21,** 166–167

Carbonylation. *N*-(β-Hydroxyacyl)morpholines are formed from a $Co_2(CO)_8$-mediated reaction of epoxides and *N*-trimethylsilylmorpholine.[1] When benzyne is generated under CO in the presence of $Co_2(CO)_8$ it is trapped to provide anthrquinone.[2]

Carbocyclization. Two different reaction pathways are observed for enynes in the reaction with $Co_2(CO)_8$.[3]

EtOOC, EtOOC (enyne) —$Co_2(CO)_8$; Et_3SiH, H_2→ EtOOC, EtOOC (dimethylcyclopentene) 62%

—$Co_2(CO)_{12}$, *i*-PrOH, H_2→ EtOOC, EtOOC (bicyclic ketone) =O 67%

1,3-Diynes. 1-Alkynes are coupled by $Co_2(CO)_8$ using 1,10-phenanthroline as adjuvant.[4]

Pauson-Khand reaction. Efficiency of the P-K reaction is enhanced by Lewis bases,[5] and further under microwave irradiation.[6] Of particular interest is a synthesis of bicyclo[3.1.0]hex-3-en-2-ones from cyclopropene.[7]

$Co_2(CO)_8$ → $(OC)_3Co-Co(CO)_3$; NMO → 93%

An interesting stereochemical aspect concerning the P-K reaction of 3-alkoxyalk-6-yn-1-enyl sulfones is that the bicyclic ketone products contain an *endo*-oriented alkoxy substituent.[8]

The tether technique may be deployed in P-K reaction to secure monocyclic cyclopentenones.[9] Noteworthy is that the vinylsiloxy tether is a surrogate for ethylene, the regioselective reaction does not require high pressure.

Pr, Ph, O, Si; $Co_2(CO)_8$, 1% H_2O / MeCN, Δ → Pr, Ph, O; 62%

A significant contribution of the P-K reaction to synthesis is the tandem process for elaboration of fenestrane ketones. Certain triynes are shown to afford products with a novel skeleton.[10]

X, O, TIPS; X = O, CR_2; $Co_2(CO)_8$, CO / CH_2Cl_2 → [TIPS, X, O, O] → TIPS, X, O, O, O

A new stable version of P-K reaction catalyst, $(Ph_3P)Co_2(CO)_7$, is obtained by replacing one of the CO ligand by triphenylphosphine.[11] The catalytic P-K reaction [reagents: $Co_2(CO)_8$, Bu_3PS, CO] has been applied to a synthesis of cyclopenta[*c*]proline derivatives.[12]

Regarding substrates for the P-K reaction alkynes substituted with a trimethylgermyl group are better than those with trialkylstannyl group because the C-Ge bond is stronger and the group is more readily replaced on reaction with electrophiles.[13]

Heterobimetallic (Co-W) complexes of alkynes participate in the P-K reaction. Interesting stereoselectivity in combining with norbornadiene is noted when alkynes bear substituents of different electronic nature.[14]

$Co_2(CO)_8$ / $Na[CpW(CO)_3]$; norbornadiene / PhMe 90°

W(CO)$_2$Cp, Co(CO)$_3$

(R′ = H) (R′ = COX)

Cycloaddition that goes beyond P-K reaction occurs when a 1,6-diyne and 2 equivalents of another 1-alkyne (phenylethyne in tested cases) are submitted to the conventional conditions. A cobalt-mediated [2+2+2]cycloaddition involving the nascent cyclopentadienone ensues.[15]

$Co_2(CO)_8$ / CO / CH_2Cl_2 130°

74%

Rearrangement. 4-Isoxazolines undergo rearrangement to afford 2-acylaziridines with $Co_2(CO)_8$.[16]

$Co_2(CO)_8$ / MeCN Δ

(2.5 : 1)
86%

[1]Goodman, S.N., Jacobsen,, E.N. *ACIEE* **41**, 4703 (2002).
[2]Chatani, N., Kamitani, A., Oshita, M., Fukumoto, Y., Murai, S. *JACS* **123**, 12686 (2001).
[3]Krafft, M.E., Bonaga, L.V.R., Wright, J.A., Hirosawa, C. *JOC* **67**, 1233 (2002).
[4]Krafft, M.E., Hirosawa, C., Dalal, N., Ramsey, C., Stiegman, A. *TL* **42**, 7733 (2001).
[5]Krafft, M.E., Bonaga, L.V.R., Hirosawa, C. *JOC* **66**, 3004 (2001).

[6]Fischer, S., Groth, U., Jung, M., Schneider, A. *SL* 2023 (2002).
[7]Marchueta, I., Verdaguer, X., Moyano, A., Pericas, M.A., Riera, A. *OL* **3**, 3193 (2001).
[8]Adrio, J., Rivero, M.R., Carretero, J.C. *CEJ* **7**, 2435 (2001).
[9]Reichwein, J.F., Iacono, S.T., Patel, M.C., Pagenkopf, B.L. *TL* **43**, 3739 (2002).
[10]Son, S.U., Yoon, Y.A., Choi, D.S., Park, J.K., Kim, B.M., Chung, Y.K. *OL* **3**, 1065 (2001).
[11]Comely, A.C., Gibson, S.E., Stevenazzi, A., Hales, N.J. *TL* **42**, 1183 (2001).
[12]Jiang, B., Xu, M. *OL* **4**, 4077 (2002).
[13]Mukai, C., Kazaka, T., Suzuki, Y., Kim, I.J. *TL* **43**, 8575 (2002).
[14]Rios, R., Pericas, M.A., Moyano, A. *TL* **43**, 4903 (2002).
[15]Son, S.U., Choi, D.S., Chung, Y.K., Lee, S.-G. *OL* **2**, 2097 (2000).
[16]Ishikawa, T., Kudoh, T., Yoshida, J., Yasuhara, A., Manabe, S., Saito, S. *OL* **4**, 1907 (2002).

N,N′-Dicyclohexylcarbodiimide, DCC. **14,** 131–132; **16,** 128; **18,** 133–134; **21,** 169

Esterification. Esterification of by the agency of DCC alone, without base and without solvent, is advantageous.[1] *t*-Butyl esters can be prepared this way.[2]

2,5-Dihydrofurans.[3] In a CuCl-catalyzed cyclization process for (*Z*)-2-alkene-1,4-diols DCC is the dehydrant.

Acyl fluorides.[4] A preparation of RCOF from carboxylic acids is by their treatment with HF-pyridine and DCC in dichloromethane (8 examples, 60–100%).

[1]Streinz, L., Koutek, B., Saman, D. *SL* 809 (2001).
[2]Nahmany, M., Melman, A. *OL* **3**, 3733 (2001).
[3]Duffy, M.G., Grayson, D.H. *JCS(P1)* 1555 (2002).
[4]Chen, C., Chien, C.-T., Su, C.-H. *JFC* **115**, 75 (2002).

[2-(Dicyclohexylphosphino)biphenyl](4-oxa-1,6-heptadiene)palladium(0).

Biaryls.[1] Pd complex (**1**) is more efficient than conventional phosphine-ligated Pd(II) catalysts for the Suzuki coupling.

O Pd—P

1

[1]Andreu, M.G., Zapf, A., Beller, M. *CC* 2475 (2000).

Diethyl *N*-(*t*-butoxycarbonyl)oxaziridine-3,3-dicarboxylate.

Sulfimides.[1] The title reagent is obtained in 50% overall yield from condensation of diethyl oxomalonate with BocN=PPh_3 and subsequent oxidation with MCPBA. It transfers the *N*-Boc group to sulfides very readily. Reaction of allylic sulfides affords allylamine derivatives as a result of [2,3]sigmatropic rearrangement.

PhS, R, R' + EtOOC, COOEt, O–NBoc $\xrightarrow{CH_2Cl_2}$ PhS, N, Boc, R', R

73–94%

[1]Armstrong, A., Cooke, R.S. *CC* 904 (2002).

Diethyl chlorophosphite.

Dehydration and deoxygenation.[1] Aldoximes suffer dehydration (to nitriles) by this reagent. It also removes oxygen from sulfoxides and *N*-oxides.

[1]Jie, Z., Rammoorty, V., Fischer, B. *JOC* **67**, 711 (2002).

Diethyl phosphite.

Hydrodebromination. Partial debromination of 1,1-dibromoalkenes to give (*E*)-1-bromoalkenes[1] and the conversion of alkylarenes polybrominated at the benzylic position(s) to monobrominated products[2] are easily achieved by reaction with $(EtO)_2POH$.

[1]Kuang, C., Senboku, H., Tokuda, M. *T* **58**, 1491 (2002).
[2]Liu, P., Chen, Y., Deng, J., Tu, Y. *S* 2078 (2001).

Diethylphosphonoacetaldehyde.

γ-Phosphono-α,β-unsaturated ketones.[1] The title reagent is the ozonolysis product of diethyl allylphosphonate. It reacts with acylphosphoranes to give synthetic intermediates valuable for building polyunsaturated compounds.

CHO, EtO–P=O, EtO + R, O, R, PPh_3 ⟶ R, O, R, EtO–P=O, EtO

[1]Kraus, G.A., Choudhury, P.K. *S* 2230 (2001).

Diethylphosphonoacetic acid.

Butenolides.[1] Involvement of the title reagent in multicomponent assembly condensation with isocyanides and α-keto aldehydes is terminated by an intramolecular Emmons-Wadsworth reaction.

PhC(O)CHO + $(EtO)_2P(O)CH_2COOH$ + t-BuN=C=O →(LiBr - Et_3N) 87%

[1]Beck, B., Magnin-Lachaux, M., Herdtweck, E., Domling, A. *OL* **3**, 2875 (2001).

2,2-Difluoro-1,3-dimethylimidazolidine.

Fluorination.[1] This is a new reagent for exchange of oxygen functionalities to fluoro substituents. Thus alcohols give alkyl fluorides, and carbonyl compounds, *gem*-difluorides.

[1]Hayashi, H., Sonoda, H., Fukumura, K., Nagata, T. *CC* 1618 (2002).

Diiodomethane. 13, 110–115; 275–276; **16,** 184–185; **17,** 155; **18,** 139–140; **19,** 128; **20,** 143–144; **21,** 171

Cyclopropanation. The CH_2I_2-Et_2Zn combination derivatizes 2-methyleneoxetanes to afford 4-oxaspiro[2.3]hexanes which are versatile intermediates for cyclobutanones or 4-methylenetetrahydrofurans.[1] On the other hand, CH_2I_2-Et_3Al delivers a methylene and two ethyl groups to alkynes.[2]

→(CH_2I_2 - Et_2Zn; Et_2O 0°) 83%

R—≡—R′ →(CH_2I_2 - Et_3Al; CH_2Cl_2 25°)

Oxazolidin-5-ones. *N*-Boc α-amino acids form the heterocycles through *N,O*-methylenation.[3]

α-Methyl-γ-keto esters.[4] The homologation of β-keto esters by a 2-carbon units is presumably the result of cyclopropanation of the enol form, ring opening to α-methyl- β-keto esters and a second cycle involving cleavage of a different CC bond.

→($EtZnCH_2I$; Me_3SiCl) COOMe 63%

Elimination. The nucleofugal tendency of RS and RSe groups is greatly enhanced by the carbenoid species derived from CH_2I_2 and Et_2Zn. Elimination occurs when such groups are set up for β-elimination.[5]

SPh NO2 — Et_2Zn, CH_2I_2 / CF_3COOH → NO2

[1]Bekolo, H., Howell, A.R. *NJC* **25**, 673 (2001).
[2]Dzhemilev, U.M., Ramazanov, I.R., Ibragimov, A.G., Djyachenko, L.I., Lukjyanova, M.P., Nefedov, O.M. *JOMC* **636**, 91 (2001).
[3]Karmakar, S., Mohapatra, D.K. *SL* 1326 (2001).
[4]Hilgenkamp, R., Zercher, C.K. *OL* **3**, 3037 (2001).
[5]Gautier, A., Garipova, G., Deleens, R., Piettre, S.R. *TL* **43**, 4959 (2002).

Diisobutylaluminum hydride, Dibal-H. 13, 115–116; **15,** 137–138; **16,** 134–135; **17,** 123–125; **18,** 140–141; **19,** 128–129; **20,** 144–146; **21,** 171

Reduction. The double bond of an α,β-unsaturated ketone is reduced by Dibal-H in the presence of $PhMe_2SiCu$.[1] The enolate can be brominated.

The major reduction (with Dibal-H or AlH_3) products of β-hydroxy imines are *syn*-1,3-amino alcohols.[2] *N*-Acyl carbamates are reduced to *N*-protected aminals[3] that are useful for introduction of a carbon chain at the reduced site.

Selective opening of acetals with Dibal-H is valuable for regioselective manipulation of diol systems.[4]

Ph O O — Dibal-H / CH_2Cl_2 → Ph O OH

98%

The ate complex derived from Dibal-H and BuLi is effective in reducing sterically hindered α,α-disubstituted α-amino esters.[5]

Aminolysis. Esters and lactones furnish amides on exposure to amine-Dibal-H complexes at room temperature.[6]

[1]Daniewski, A.R., Liu, W. *JOC* **66**, 626 (2001).
[2]Veenstra, S.J., Kinderman, S.S. *SL* 1109 (2001).
[3]Suh, Y.-G., Shin, D.-Y., Jung, J.-K., Kim, S.-H. *CC* 1064 (2002).
[4]Cossy, J., Gille, B., Bellosta, V., Duprat, A. *NJC* **26**, 526 (2002).
[5]Glunz, P.W., Rich, D.H. *SC* **29**, 8356 (1999).
[6]Huang, P.-Q., Zheng, X., Deng, X.-M. *TL* **42**, 9039 (2001).

Diisobutylaluminum phenylselenide.

Cleavage of cyclic acetals. The C-O bond scission by *i*-Bu_2Al-SePh is critical to elaboration of a ring segment in a synthetic approach to ciguatoxin.[1]

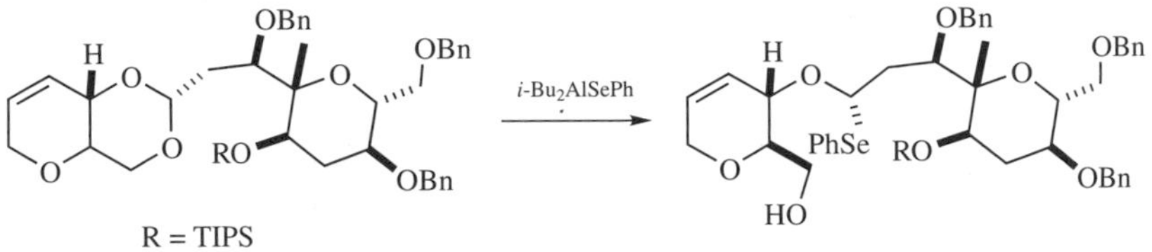

R = TIPS

[1]Sasaki, M., Noguchi, T., Tachibana, K. *JOC* **67**, 3301 (2002).

Dilauroyl peroxide. 21, 172–173

1,1,1-Trifluoro-3-alkyl xanthates.[1] *S*-Trifluoromethyl xanthates add to 1-alkenes distributively in the presence of dilauroyl peroxide.

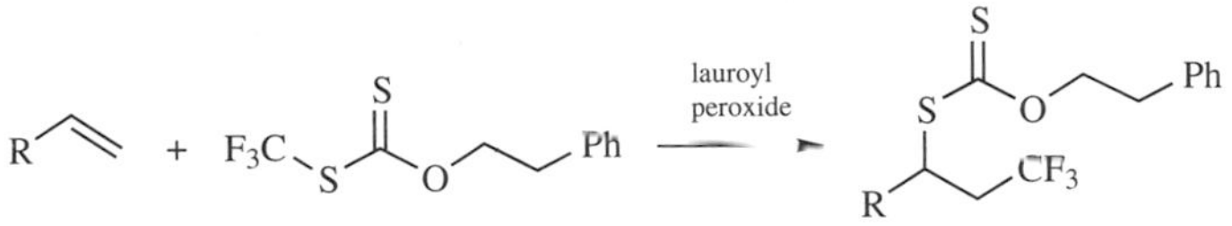

[1]Bertrand, F., Pevere, V., Quiclet-Sire, B., Zard, S.Z. *OL* **3**, 1069 (2001).

Dimanganese decacarbonyl. 21, 173

Coupling reactions. Under phase-transfer conditions and in the presence of $Mn_2(CO)_{10}$ organohalides are coupled via free radical intermediates.[1]

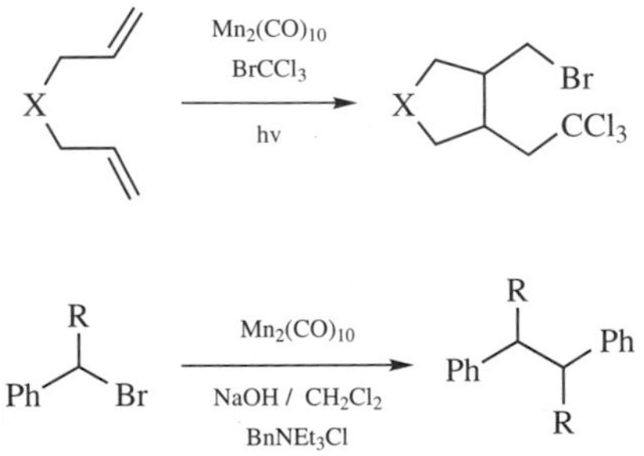

[1]Huther, N., McGrail, P.T., Parsons, A.F. *TL* **43**, 2535 (2002).

***N,N*-Dimethylalkanamide dimethyl acetals.**

Aroic esters. On refluxing with $(MeO)_2CHNMe_2$ in MeOH electron-deficient ArCOR are cleaved.[1]

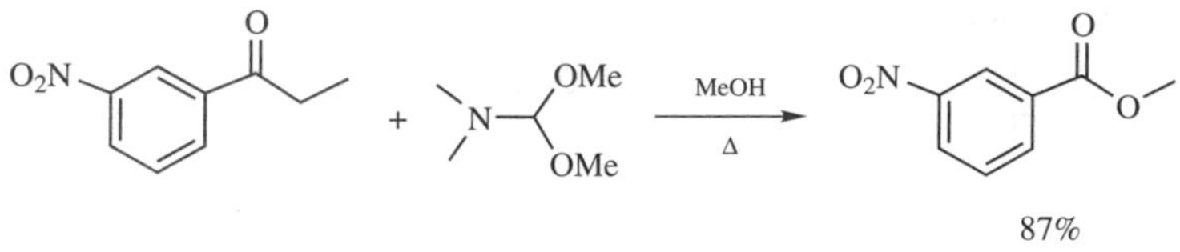

O_2N ... + Me_2N–CH(OMe)(OMe) —MeOH, Δ→ O_2N ... 87%

1,1-Dimorpholinoethene. *N,N*-Dimethylacetamide dimethyl acetal is converted to the geminal enediamine **1** on heating with morpholine at 190°. The resulting reagent can be used to prepare *N*-(4-alkenoyl)morpholines by Claisen-Eschenmoser rearrangement.[2]

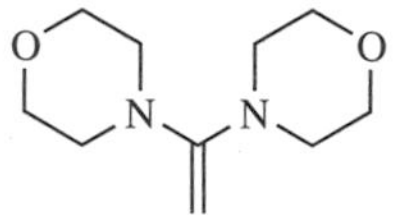

1

[1]Zhang, N., Vozzolo, J. *JOC* **67**, 1703 (2002).
[2]Gradl, S.N., Kennedy-Smith, J.J., Kim, J., Trauner, D. *SL* 411 (2002).

5-Dimethylaminomethylene-1,3-dimethylpyrimidine-2,4,6-trione.

Amino protection.[1] The dimethylamino group of reagent **1** is rapidly exchanged on contact with various amines. The derivatives are cleaved with hydrazine or other amines at room temperature.

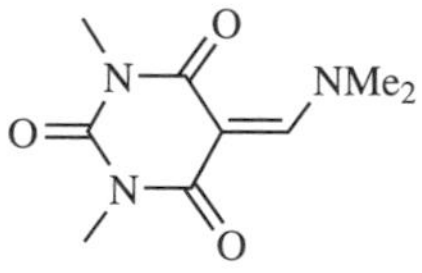

NMe_2

1

[1]Dekany, G., Bornaghi, L., Papageorgiou, J., Taylor, S. *TL* **42**, 3129 (2001).

4-Dimethylaminopyridine, DMAP. 21, 176

Amides. Catalytic DMAP promotes formation of amides in good yields from carboxylic acids and RNCO.[1] Decarboxylation is involved.

Cyclization. A combination of DMAP and its HCl salt induces cyclization of ω-formyl-α,β-unsaturated thio esters that have proper chain lengths.[2] It is an intramolecular Baylis-Hillman reaction and Me_3P can also be used as the catalyst.

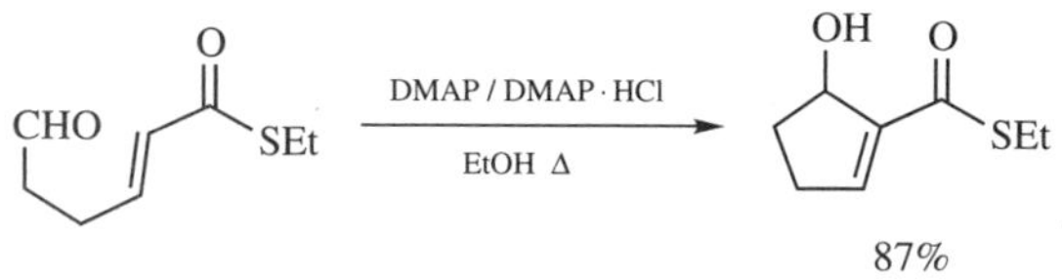

CHO ... SEt —DMAP / DMAP · HCl, EtOH Δ→ OH ... SEt

87%

Destruction of di-t-butyl pyrocarbonate.[3] Excess Boc_2O that remains from a derivatization reaction is removed by DMAP. Also effective are imidazole and 2,2,2-trifluoroethanol.

[1]Schuemacher, A.C., Hoffmann, R.W. *S* 243 (2001).
[2]Keck, G.E., Welch, D.S. *OL* **4**, 3687 (2002).
[3]Basel, Y., Hassner, A. *S* 550 (2001).

N,N-Dimethylaminosulfonyl chloride.

Condensation.[1] The title reagent mediates esterification and amidation of carboxylic acids with equimolar amounts of alcohols and amines. A tertiary amine (e.g., $BuNMe_2$) is also added during the reaction.

[1]Wakasugi, K., Nakamura, A., Tanabe, Y. *TL* **42**, 7427 (2001).

1,3-Dimethyl-1,3-diaza-2-boracyclopentane.

Cycloalkanemethanols.[1] With catalytic amounts of $Cp^*_2Sm{\cdot}thf$ the title reagent 1,3,2-diazaborolidine (but not catecholborane) reacts with 1,5-dienes and 1,6-dienes to give, after oxidative workup, cyclopentanemethanols and cyclohexanemethanols, respectively.

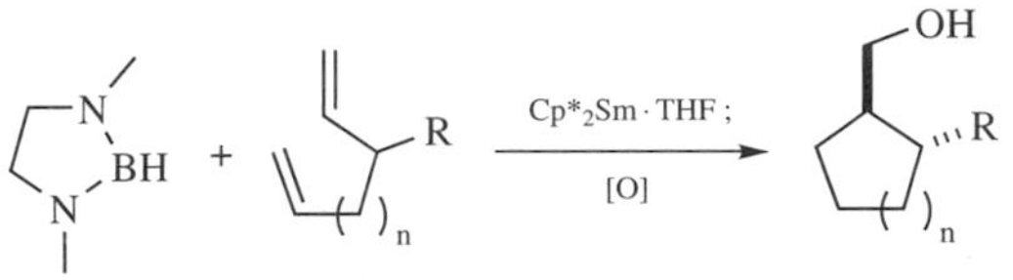

N BH N + R n → $Cp^*_2Sm \cdot THF$; [O] → OH R n

[1]Molander, G.A., Pfeiffer, D. *OL* **3**, 361 (2001).

Dimethyldioxirane, DMD. 12, 413; **13,** 120; **14,** 148; **15,** 143–144; **16,** 142–144; **18,** 144–146; **19,** 135–136; **20,** 150–152; **21,** 177–178

Epoxidation. Epoxidation of 1-amidoallenes with DMD gives zwitterionic oxyallyl species. [4+3]Cycloadducts are isolated when the reaction is carried out in the presence of furan and $ZnCl_2$.[1]

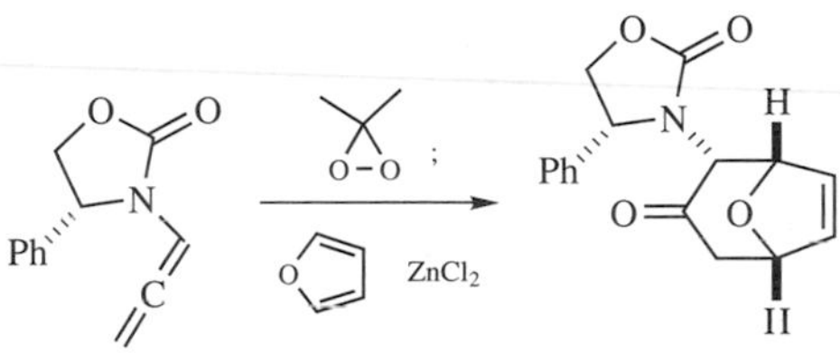

O O N Ph C → O–O ; O $ZnCl_2$ → O O N Ph H O O H

1,2-Dicarbonyl compounds. 1-Acyl-1-cyanomethylenetriphenylphosphoranes are oxidized to α-keto esters by DMD in MeOH.[2] The carbon chain is shortened.

R, CN, O, PPh_3 — O–O / $MeOH - Me_2CO$ → R, OMe, O, O

[1]Xiong, H., Hsung, R.P., Berry, C.R., Rameshkumar, C. *JACS* **123**, 7174 (2001).
[2]Wong, M.-K., Yu, C.-W., Yuen, W.-H., Yang, D. *JOC* **66**, 3606 (2001).

Dimethyl disulfide. 21, 178

Aryl methyl sulfides.[1] In the presence of MeSSMe nonaqueous diazotization of arylamines gives ArSMe in moderate yields.

[1]Allaire, F.S., Lyga, J.W. *SC* **31**, 1857 (2001).

N,N-Dimethylformamide–phosphoryl chloride. 18, 146

2-Aryl-4-chloroquinolines.[1] The Vilsmeier-Haack reagent converts 2′-azidochalcones to the quinoline derivatives. 2′-Aminochalcones give the corresponding 1-formyl-1,2-dihydroquinolines.

N_3, Cl, O — $POCl_3$ - DMF → N, Cl, Cl

72%

[1]Akila, S., Selvi, S., Balasubramanian, K. *T* **57**, 3465 (2001).

N-(1,3-Dimethylimidazolidin-2-ylidene)-*N*-alkylglycine esters.

Aziridines.[1] The title reagents are prepared from 2-chloro-1,3-dimethylimidazolinium chloride by successive treatment with amines and bromoacetic esters. The derived ylides react with aldehydes to furnish aziridine-2-carboxylic esters.

Cl, N, N^+, Cl^- — $BnNH_2$; $BrCH_2COOR$ → Bn, N^+, COOR, N, N, Br^- — NaH / DMF, ArCHO → Bn, N, Ar, COOR

[1]Hada, K., Watanabe, T., Isobe, T., Ishikawa, T. *JACS* **123**, 7705 (2001).

_N,N_-Dimethyl(methylidene)ammonium chloride.

Dimethylaminomethylation.[1] A novel reaction is the desilylative alkylation of a carbamoylsilane with the title reagent.

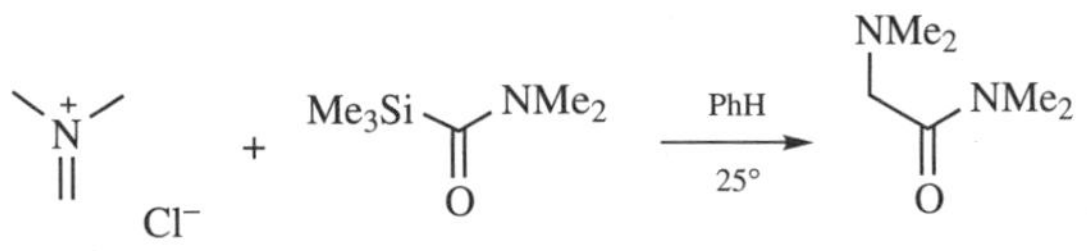

[1]Chen, J., Cunico, R.F. *TL* **43**, 8595 (2002).

1,3-Dimethyl-2-phenylbenzimidazoline–acetic acid.

β-Hydroxy ketones.[1] Photoinduced reductive cleavage of α,β-epoxy ketones is carried out with the aid of this heterocycle.

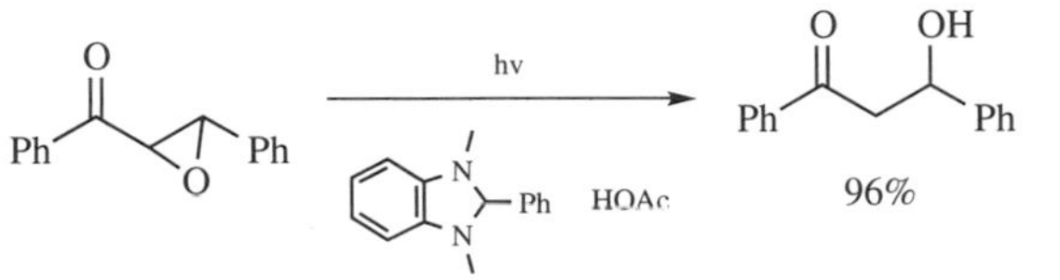

[1]Hasegawa, E., Chiba, N., Nakajima, A., Suzuki, K., Yoneoka, A., Iwaya, K. *S* 1248 (2001).

Dimethylsilyl dichloride.

Protection of amino acids. Amino acids (α and β) form cyclic derivatives with dichlorosilanes. These derivatives react with amines to give amides therefore the derivatization serves both protective and activating purposes.

[1]van Leeuwen, S.H., Quaedflieg, P.J.L.M., Broxterman, Q.B., Liskamp, R.M.J. *TL* **43**, 9203 (2002).

Dimethylsulfonium methylide. 18, 149; **20,** 154

Allylic alcohols. The reagent opens *cis*-1,2-disubstituted epoxides while donating a methylene group.

[1]Alcarez, L., Cridland, A., Kinchin, E. *OL* **3**, 4051 (2001).

Dimethylsulfoxonium methylide. 14, 152; **15,** 147; **16,** 146; **17,** 126–127; **18,** 148; **19,** 139; **20,** 155–156

Cyclopenta[b]benzofuran-3-ols. Coumarins which bear an electron-withdrawing group at C-3 are converted by the title reagent to the tricyclic system[1] via cyclopropanation, addition to the lactone carbonyl, ring expansion and heterocyclization.

=S(O)Me$_2$

X = Ac, CN, Bz, Pv, SO_2Ph, COOEt

[1]Yamashita, M., Okuyama, K., Kawajiri, T., Takada, A., Inagaki, Y., Nakano, H., Tomiyama, M., Ohnaka, A., Terayama, I., Kawasaki, I., Ohta, S. *T* **58**, 1497 (2002).

Dimethyltitanocene. 21, 181

Hydroamination. An intramolecular version of the previously reported amine-alkyne addition in the presence of Cp_2TiMe_2 with subsequent reduction is the basis of a synthesis of 2-substituted pyrrolidines and piperidines.[1] The pentamethyl analogue $Cp^*_2TiMe_2$ is thought to be better catalyst for this general reaction,[2] while the effectiveness of CpTi(=NAr)(NHAr)py, which is obtained from Cp_2TiMe_2, $ArNH_2$ and pyridine, for converting allenes to imines has been scrutinized.[3]

Methylenation. Direct methylenation of partially benzylated glycolactones is possible.[4]

[1]Bytschkov, I., Doye, S. *TL* **43**, 3715 (2002).
[2]Heutling, A., Doye, S. *JOC* **67**, 1961 (2002).
[3]Johnson, J.S., Bergman, R.G. *JACS* **123**, 2923 (2001).
[4]Li, X., Ohtake, H., Takahashi, H., Ikegami, S. *SL* 1885 (2001).

***N*-(2,4-Dinitrophenyl)hydroxylamine.**

Allylamines.[1] The title reagent is a more efficient reagent for the iron-catalyzed functionalization of alkenes.

$FeCl_2$ - $FeCl_3$, dioxane

65%

[1]Singh, S., Nicholas, K.M. *SC* **31**, 3087 (2001).

(Diphenylphosphinoethane)rhodium(I) perchlorate. 21, 182

Cycloheptenones.[1] When catalyzed by (dppe)$RhClO_4$ 4,6-alkadienals undergo cyclization. Previous reports indicate the Rh complex to be effective to cyclize 5-cyclopropyl-4-alkenals in affording eight-membered homologues.

OHC, 4, 6, R, (dppe)$RhClO_4$, O, Rh, R, O, R, 62–65%

[1]Sato, Y., Oonishi, Y., Mori, M.*ACIEE* **41**, 1218 (2002).

Diphenylphosphinomethylthio esters.

Peptide synthesis.[1] Reaction of this class of compounds with α-azidocarboxylic acid derivatives results in amides. Peptides can be assembled from the thio esters of properly protected amino acids.

Ac, H, N, O, S, PPh_2, +, Ph, N_3, NHBn, O, THF - H_2O, Ac, H, N, O, N, H, Ph, NHBn, O

[1]Soellner, M.B., Nilsson, B.L., Raines, R.T. *JOC* **67**, 4993 (2002).

2-Diphenylphosphinoyl-3-trifluoromethyloxaziridine.

Sulfoxides.[1] The title reagent oxidizes sulfides to sulfoxides with higher rates than the unfluorinated analogues.

[1]Jennings, W.B., O'Shea, J.H., Schweppe, A. *TL* **42**, 101 (2002).

Dipyridyliodonium tetrafluoroborate.

ω-Iodoalkyl ketones.[1] Cycloalkanols of various sizes are cleaved photolytically in the presence of the title reagent (Py_2IBF_4). The reaction involves generation of cycloalkoxy radicals and its preparative value is superior to protocols such as that using $PhI(OAc)_2$-I_2, usually only one product is observed (9 examples, 76–94%).

OH, hv Py_2IBF_4, Cs_2CO_3 / CH_2Cl_2, I, CHO, 94%

[1]Barluenga, J., Gonzalez-Bobes, F.G., Ananthoju, S.R., Garcia-Martin, M.A., Gonzalez, J.M. *ACIEE* **40**, 3389 (2001).

Diruthenium tetraacetate.

Carbenoid insertion.[1] For decomposition of stabilized diazo compounds in alcohols to obtain alkoxy derivatives $[Ru(OAc)_2]_2$ is the catalyst of choice.

[1]Cenini, S., Cravotto, G., Giovenzana, G.B., Palmisano, G., Penoni, A., Tollari, S. *TL* **43**, 3637 (2002).

***p*-Dodecylbenzenesulfonic acid. 21,** 183

Etherification. The sulfonic acid promotes ether formation (ROH + benzhydrol or *p*-methoxybenzyl alcohol) in aqueous media.[1] Thioetherification and dithioacetalization are also achieved.

[1]Kobayashi, S., Iimura, S., Manabe, K. *CL* 10 (2002).

Dysprosium. 21, 183–184

Diallyl carbinols.[1] This metal in the presence of $HgCl_2$ promotes reaction of allyl halides with esters.

[1]Jia, Y., Zhang, M., Tao, F., Zhou, J. *SC* **32**, 2829 (2002).

Dysprosium(III) triflate. 21, 184

Cycloaddition.[1] In the presence of $Dy(OTf)_3$ the condensation of aldehydes (including HCHO) and arylamines with cyclic or acyclic enol ethers gives tetrahydroquinolines.

+ O + HCHO — $Dy(OTf)_3$ → O N

N H

95%

[1]Chen, R., Qian, C. *SC* **32**, 2543 (2002).

E

Ethylenediamine.

Semihydrogenation.[1] For controlled hydrogenation of aminoalkynes to the corresponding (*Z*)-alkenes the Lindlar catalyst is poisoned with ethylenediamine.

[1]Campos, K.R., Cai, D., Journet, M., Kowal, J.J., Larsen, R.D., Reider, P.J. *JOC* **66**, 3634 (2001).

Ethyl(methyl)dioxirane.

Oxidation.[1] The reagent, known also as 2-butanone peroxide, is commercially available as a dimethyl phthalate solution. It is less expensive than MCPBA and *t*-BuOOH, and apparently not explosive (cf. dimethyldioxirane).

Oxidation of nucleoside phosphites into phosphates by this reagent under nonbasic and nonaqueous conditions is effective.

[1]Kataoka, M., Hattori, A., Okino, S., Hyodo, M., Asano, M., Kawai, R., Hayakawa, Y. *OL* **3**, 815 (2001).

1-Ethylpiperidinium hypophosphite - triethylborane.

Radical reactions. A method for deoxygenation of alcohols is via treatment of the derived xanthates with the reagent couple in water (and the presence of a surfactant).[1]

Conjugate additions to enones[2] and vinyl sulfones[3] with iodoalkanes as radical source is achieved under similar conditions. Cyclization of β-(3-bromopropoxy)acrylic esters affords tetrahydrofurans.[4]

$C_{10}H_{21}$ ... Br ... EtOOC + 1-ethylpiperidinium $[H_2PO_2]^-$ $\xrightarrow[EtOH]{Et_3B}$ $C_{10}H_{21}$ ... EtOOC

90% (*cis/trans* 15 : 1)

[1]Jang, D.O., Cho, D.H. *TL* **43**, 5921 (2002).
[2]Jang, D.O., Cho, D.H. *SL* 1523 (2002).
[3]Jang, D.O., Cho, D.H., Chung, C.-M. *SL* 1923 (2001).
[4]Lee, E., Han, H.O. *TL* **43**, 7295 (2002).

Fiesers' Reagents for Organic Synthesis, Volume 22. Series editor Tse-Lok Ho
ISBN 0-471-28515-3

F

Ferricenium tetrakis[3,5-bis(trifluoromethyl)phenyl]borate.

Halogenation. Arenes undergo bromination and iodination with XCl (X=Br, I) when catalyzed by the title compound.

[1]Kitagawa, H., Shibata, T., Matsuo, J., Mukaiyama, T. *BCSJ* **75**, 339 (2002).

Fluorine. 13, 135; **14,** 167; **15,** 160; **18,** 161; **19,** 146; **20,** 165; **21,** 188–189

Acylamination.[1] Fluorination of hydrocarbons with nitrogen-diluted fluorine gas in MeCN usually leads to fluoroalkanes. However, in the presence of $BF_3 \cdot OEt_2$, *N*-alkylacetamides are obtained.

F_2 / N_2, MeCN → cyclohexyl–F, 63%

F_2 / N_2, $BF_3 \cdot OEt_2$, MeCN → cyclohexyl–NHAc, 51%

[1]Chambers, R.D., Kenwright, A.M., Parsons, M., Sandford, G., Moilliet, J.S. *JCS(P1)* 2190 (2002).

***N*-Fluorobenzenesulfonimide. 20,** 165–166, **21,** 189

a-Fluoro ketones.[1] Lithium enolates of ketones abstract the fluorine atom from $(PhSO)_2NF$.

LDA, $(PhSO_2)_2NF$

[1]Enders, D., Faure, S., Potthoff, M., Runsink, J. *S* 2307 (2001).

2-Fluorobenzoyl chloride.

O-Protection.[1] Derivatization of carbohydrates with this reagent provides compounds that possess synthetic advantages over benzoyl and acetyl analogues. Highly stereoselective glycosylation with low levels of ortho ester formation are observed, yet

Fiesers' Reagents for Organic Synthesis, Volume 22. Series editor Tse-Lok Ho
ISBN 0-471-28515-3

the protecting group is easily removed with LiOH in MeOH. They can be employed in solid-phase synthesis of *O*-linked glycopeptides that are very sensitive to base-catalyzed elimination.

[1]Sjölin, P., Kihlberg, J. *JOC* **66**, 2957 (2001).

1-Fluoro-4-chloromethyl-1,4-diazoniabicyclo[2.2.2]octane bis(tetrafluoroborate).

Halogenation. Amines are fluorinated,[1] the number of fluorine atoms entering the molecules depends on the amounts of reagent used.

The time course in which sterically hindered arenes undergo iodination (catalyzed by the title reagent) can be regulated.[2]

I_2 / MeCN

F–N⁺ ... N⁺–CH$_2$Cl

$2BF_4^-$

3 hr : 90%	6.5 hr : 87%	24 hr : 73%

[1]Singh, R.P., Shreeve, J.M. *CC* 1196 (2001).
[2]Stavber, S., Kralj, P., Zupan, M. *S* 1513 (2002).

1-Fluoro-4-hydroxy-1,4-diazoniabicyclo[2.2.2]octane bis(tetrafluoroborate). 19, 146–147; **21, 190**

Fluorination.[1] α-Fluorination of ketones with the title reagent in hot MeOH usually proceeds in >70% yield.

[1]Stavber, S., Jereb, M., Zupan, M. *CC* 2609 (2002).

Fluorous reagents and ligands. 21, 191–192

Esterification and amidation. In separating enantiomeric acids by lipase-catalyzed esterification or transesterification with F_{17}-decanol,[1] the chiral fluorodecyl alkanoates stay in the fluorous phase and are recovered by simple evaporation.

Esterification has been carried out in perfluorohexanes with heavily fluorinated bis(octylchlorotin) oxide as catalyst.[2] Protecting OH groups of carbohydrates by esterification with the recyclable acid **1** is proposed.[3] Synthetic operations are simplified because various intermediates can be purified by fluorous-organic extractions.

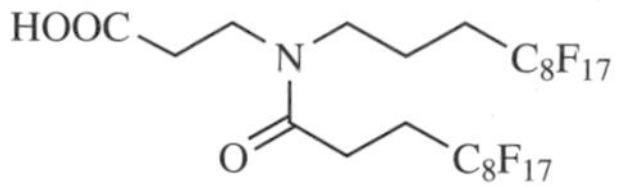

HOOC N C_8F_{17} O C_8F_{17}

1

Phenols are deoxygenated via sulfonylation and Pd-catalyzed transfer hydrogenolysis. To create a traceless linker for the first step in the solid-phase synthesis a resin-bound amide containing the $COCH_2(CF_2)_2O(CF_2)_2SO_2F$ chain is used.[4]

3,5-Bis(perfluorodecyl)phenylboronic acid is identified as a clean catalyst in amide formation.[5]

Hydrosilylation.[6] The ionic liquid **2** is a solvent for homogeneous hydrosilylation of 1-octene catalyzed by a fluorous version of Wilkinson's catalyst. Catalyst recycling is demonstrated.

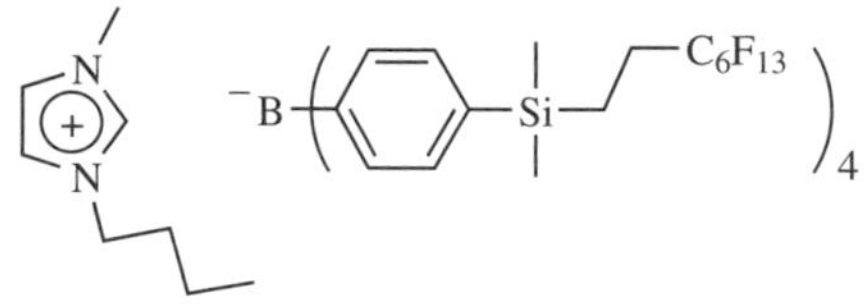

N + N $^{-}$B Si C_6F_{13} 4

2

Coupling reactions. Suzuki and Sonagashira couplings benefit from employing fluorous phosphine ligands to form the catalysts.[7,8]

Other reactions. Both Swern oxidation[9] and Wittig reaction[10] conducted with fluoroalkyl methyl sulfoxides and stabilized tris(*p*-fluoroalkoxyphenyl)phosphonium ylides, respectively, are advantageous. The spent reagents are recyclable.

A robust catalyst for epoxidation of alkenes with Oxone is tridecafluorooctyl trifluoromethyl ketone[11] which is available from tridecafluorooctyl iodide via an I/Li exchange with *t*-BuLi and then reaction with ethyl trifluoroacetate.

Azide reduction via the Staudinger reaction can be performed using an aryldiphenylphosphine in which the aryl group bears a *p*-tridecafluorooctyl chain.[12]

1-Dodecyloxy-4-perfluoroalkylbenzenes are additives that show positive effects on $Sc(OTf)_3$-catalyzed aldol and Friedel-Crafts alkylation reactions in supercritical CO_2.[13] Fluorous phase also facilitates operation of Friedel-Crafts acylation.[14,15]

[1]Beier, P., O'Hagan, D. *CC* 1680 (2002).
[2]Xiang, J., Orita, A., Otera, J. *ACIEE* **41**, 4117 (2002).
[3]Miura, T., Hirose, Y., Ohmae, M., Inazu, T. *OL* **3**, 3947 (2001).
[4]Pan, Y., Holmes, C.P. *OL* **3**, 2769 (2001).
[5]Ishihara, K., Kondo, S., Yamamoto, H. *SL* 1371 (2001).

[6]van den Broeke, J., Winter, F., Deelman, B.-J., van Koten, G. *OL* **4**, 3851 (2002).
[7]Schneider, S., Bannwarth, W. *HCA* **84**, 735 (2001).
[8]Markert, C., Bannwarth, W. *HCA* **85**, 1877 (2002).
[9]Crich, D., Neelamkavil, S. *HCA* **84**, 7449 (2001).
[10]Galante, A., Lhoste, P., Sinou, D. *TL* **42**, 5425 (2001).
[11]Legros, J., Crousse, B., Bourdon, J., Bonnet-Delpon, D., Begue, J.-P. *TL* **42**, 4463 (2001).
[12]Lindsley, C.W., Zhao, Z., Newton, R.C., Leister, W.H., Strauss, K.A. *TL* **43**, 4467 (2002).
[13]Komoto, I., Kobayashi, S. *OL* **4**, 1115 (2002).
[14]Mikami, K., Mikami, Y., Matsuzawa, H., Matsumoto, Y., Nishikido, J., Yamamoto, F., Nakajima, H. *T* **58**, 4015 (2002).
[15]Shi, M., Cui, S.-C. *JFC* **116**, 143 (2002).

1-Formamidinopyrazoles.

Guanidines. Of particular value these reagents are for rapid synthesis of guanidines when they contain a *N*-Boc group.[1]

[1]Zhang, Y., Kennan, A.J. *OL* **3**, 2341 (2001).

G

Gallium. 21, 194

Allylation. The reaction of allyl bromides with aldehydes as mediated by Ga can be performed in aqueous media.[1,2] With catalytic indium, allylgallium halides react with 1-alkynes to provide 1,4-dienes.[3] This last process is accelerated by tertiary amines.

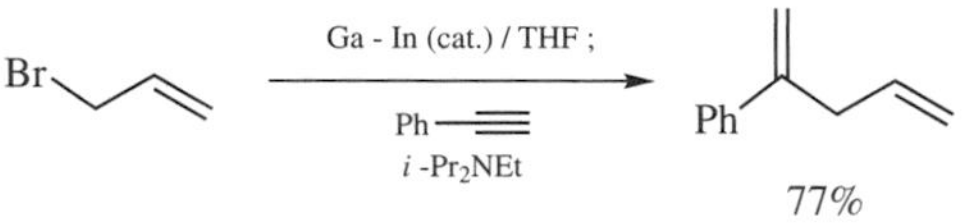

Br~allyl → Ga - In (cat.) / THF ; Ph—≡ ; i-Pr_2NEt → Ph (1,4-diene) 77%

[1]Tsuji, T., Usugi, S., Yorimitsu, H., Shinokubo, H., Matsubara, S., Oshima, K. *CL* 2 (2002).
[2]Wang, Z., Yuan, S., Li, C.-J. *TL* **43**, 5097 (2002).
[3]Takai, J., Ikawa, Y., Ishii, K., Kumanda, M. *CL* 172 (2002).

Gallium(III) halides. 20, 169–170; 21, 195–196

Vinylation. Vinylation of nucleophiles with silylalkynes in the presence of $GaCl_3$ is extended to ketene-*O,S*-acetals[1] and *O*-silylated β-dicarbonyl compounds.[2] ω-Aryl-1-alkynes provide cyclic products (applicable to 6- and 7-membered rings) in a related reaction.[3]

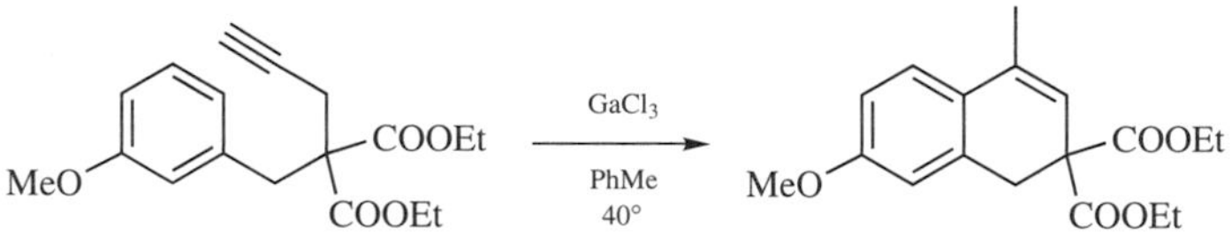

MeO, COOEt, COOEt → $GaCl_3$; PhMe ; 40° → MeO, COOEt, COOEt ; 80%

Ethynylation. Changing the electrophiles to 1-chloro-2-trialkylsilylethyne in the above reaction results in ethynylation.[4,5] Such a change indicates a substitution course instead of addition.

Alkyne-aldehyde coupling. Enones[6] or naphthalenes[7,8] are formed on union of different types of substrates.

Fiesers' Reagents for Organic Synthesis, Volume 22. Series editor Tse-Lok Ho
ISBN 0-471-28515-3

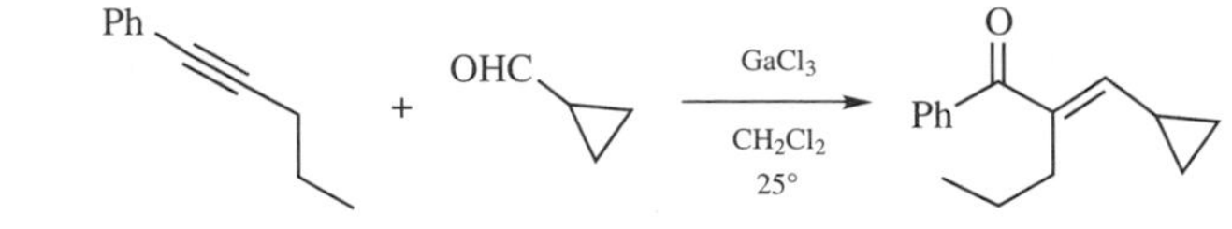

40%

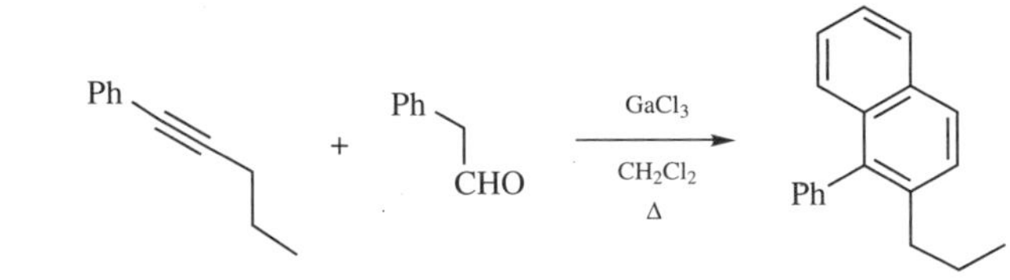

62%

Aldol reaction.[9] Either $GaCl_3$ or $Ga(OTf)_3$ in combination with **1** is effective to promote reaction of silyl enol ethers with aromatic aldehydes in aqueous media. Enantioselectivity is low for the reaction with aliphatic aldehydes.

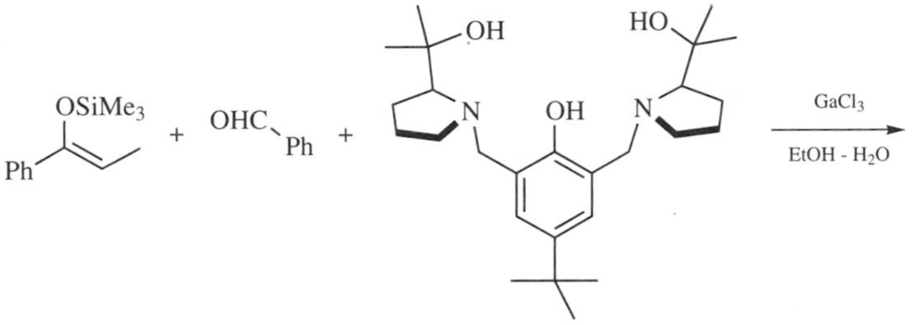

1

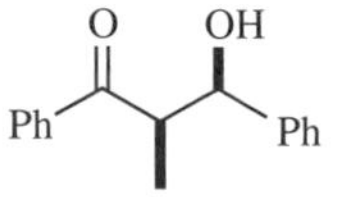

85% (*syn* : *anti* 85 : 15)

[1]Arisawa, M., Miyagawa, C., Yoshimura, S., Kido, Y., Yamaguchi, M. *CL* 1080 (2001).
[2]Arisawa, M., Akamatsu, K., Yamaguchi, M. *OL* **3,** 789 (2001).
[3]Inoue, H., Chatani, N., Murai, S. *JOC* **67**, 1414 (2002).
[4]Kobayashi, K., Arisawa, M., Yamaguchi, M. *JACS* **124**, 8528 (2002).
[5]Arisawa, M., Amemiya, R., Yamaguchi, M. *OL* **4**, 2209 (2001).
[6]Viswanathan, G.S., Li, C.-J. *TL* **43**, 1613 (2002).
[7]Viswanathan, G.S., Wang, M., Li, C.-J. *ACIEE* **41**, 2138 (2002).
[8]Viswanathan, G.S., Li, C.-J. *SL* 1553 (2002).
[9]Li, H.-J., Tian, H.-Y., Chen, Y.-J., Wang, D., Li, C.-J. *CC* 2994 (2002).

Gold(III) chloride. 21, 196

Cyclization.[1] Exposure of 2,3-alkadiyn-1-ols to gold(III) chloride affords 2,5-dihydrofurans.

COOEt OH $\xrightarrow[CH_2Cl_2, 25°]{AuCl_3}$ COOEt 74%

[1]Hoffmann-Roder, A., Krause, N. *OL* **3,** 2537 (2001).

Graphite. 20, 170; **21,** 197

Cyclic ketones.[1] Dicarboxylic acids are transformed into 5- and 6-membered cyclic ketones on heating (or microwave irradiation) with graphite. Confinement of substrates to the support decreases vaporization and that contributes to improvement of yields.

Cinnamic acids.[2] Condensation of ArCHO with malonic acid on graphite is promoted by ultrasound.

Urea pyrolysis.[3] While supported on graphite urea condenses to afford cyanuric acid on microwave heating with increased rates.

[1]Marquie, J., Laporterie, A., Dubac, J., Roques, N. *SL* 493 (2001).
[2]Li, J.-T., Zang, H.-J., Feng, J.-Y., Li, L.-J., Li, T.-S. *SC* **31**, 653 (2001).
[3]Chemat, F., Poux, M. *TL* **42**, 3693 (2001).

Grignard reagents. 13, 138–140; **14,** 171–172; **16,** 172–173; **17,** 141–142; **18,** 167–171; **19,** 151–154; **20,** 170–173; **21,** 197–202

Reagent formation by exchange reactions. Preparation of previously inaccessible Grignard reagents, i.e., that containing reactive functionality, is rendered feasible by the exchange technique (usually with *i*-PrMgBr or *i*-Pr_2Mg in THF at −25° to −40°). Alkenylmagnesium bromides with a geminal cyano group[1] or β-ester[2] are readily prepared, and arylmagnesium reagents bearing a keto group are formed by exchange reaction with *t*-$BuCH_2MgBr$.[3] Halogen/magnesium exchange can also be carried out using ate complexes such as Bu_3MgLi.[4] A general access to functionalized *o*-nitroarylmagnesium halides requires simpler operation.[5]

EtOOC, NO_2, I $\xrightarrow[PhCHO]{PhMgBr / THF -40° ;}$ EtOOC, NO_2, OH, Ph 94%

Dehydration. β-Hydroxy nitriles give conjugated nitriles on treatment with MeMgCl, instead of ketones.[6]

Alkyne synthesis. Internal alkynes can be prepared from 1-(benzotriazol-1-yl)alkynes by a Grignard reaction (14 examples, 51–83%).[7] Ruthenium-carbenoids such as **1** are susceptible to attack by Grignard reagents, the products on protonation and demetallation afford 1-alkynes.[8]

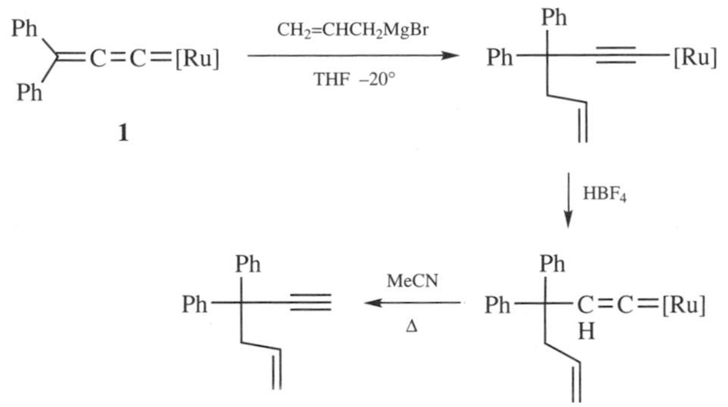

CH_2=CHCH$_2$MgBr, THF –20°; HBF_4; MeCN, Δ

1

Reaction with carbonyl compounds. High diastereoselection is observed in the Grignard reaction of 4-substituted 1,3-dioxan-2-yl ketones[9] and a 2-formyl-5,6-dimethoxy-5,6-dimethyl-1,4-dioxane.[10] Grignard reaction of allenyl ketones delivers enone products.[11]

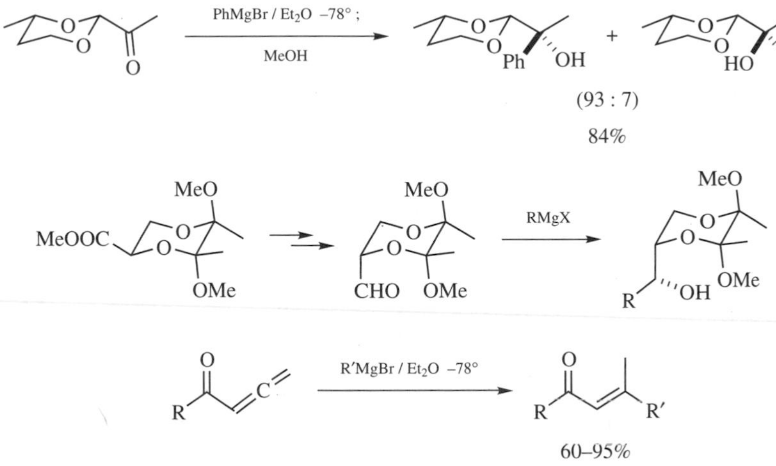

PhMgBr / Et_2O –78° ; MeOH

(93 : 7)

84%

RMgX

R′MgBr / Et_2O –78°

60–95%

An improved route to Boc-protected α-amino ketones involves successive treatment of the Weinreb amides derived from *N*-Boc amino acids with *i*-PrMgCl and then the proper RMgX.[12] The first treatment is for deprotonation.

Grignard reaction of anthranilic acids furnishes tertiary alcohols but reaction of *N*-protected (Boc and trifluoroacetyl) derivatives stops at the ketone stage.[13]

A new synthesis of 1-substituted indenes involves 2-(2-halomagnesioethyl)arylmagnesium halides with esters.[14]

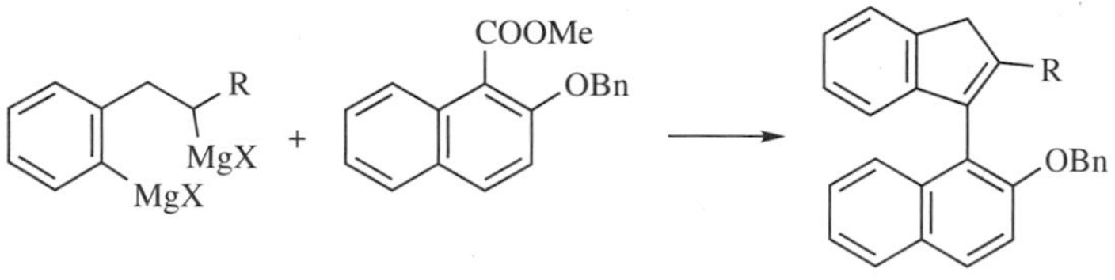

R MgX MgX + COOMe OBn → R OBn

A method for conversion of carbonyl compounds into alkenes involving two Grignard reagents operates in the following manner: Enolphosphorylation in which the enolization is effected with mesitylmagnesium bromide, and a Pd-catalyzed coupling with another Grignard reagent.[15]

Aromatic substitutions. Electron-rich arenes such as phenols become susceptible to attack by Grignard reagents on forming cationic arene-$Mn(CO)_3$ species.[16]

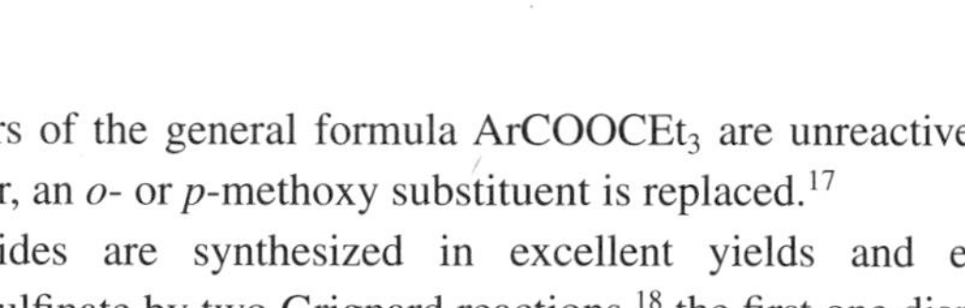

OH $^{+}Mn(CO)_3$ —(cyclohexyl-MgBr ; H_2O ; Me_3NO)→ OH

80%

Hindered esters of the general formula $ArCOOCEt_3$ are unreactive toward Grignard reagents. However, an *o*- or *p*-methoxy substituent is replaced.[17]

Chiral sulfoxides are synthesized in excellent yields and ee from menthyl *p*-bromobenzenesulfinate by two Grignard reactions,[18] the first one displaces the menthol moiety, and the second reaction, the aromatic residue. There is a double inversion of configuration.

M = MgBr 65% *syn/anti* 75 : 25
M = 9-BBN 70% *syn/anti* 18 : 82

Reaction with nitrogenous compounds. CC bond formation at the α-position of a carbamate nitrogen atom is achieved by Grignard reaction after electrochemical oxidation.[19] This method is applicable to α-silyl derivatives[20] in which the electrode process removes the silyl substituent.

Active research on Grignard reactions of imines is evident. It is noted that allylation of α-benzyloxyaldimines with allylmagnesium bromide favors production of the *syn*-isomer whereas allyl-BBN mainly pursues a different steric course.[21]

Rendering the amino nitrogen atom electrophilic by derivatizing as an iminomalonic ester enables *N*-alkylation by Grignard reaction.[22]

Latent iminium species are 1-(α-aminoalkyl)benzotriazoles[23] and aminals.[24] Such compounds readily react with Grignard reagents directly or after activation (e.g., aminals with Tf_2O).

Formation of pyrimidines by Grignard reaction of α,α-dibromo oxime ethers is intriguing,[25] and that of 7-alkylindoles from *o*-alkylnitrobenzenes constitutes a valuable synthetic method.[26]

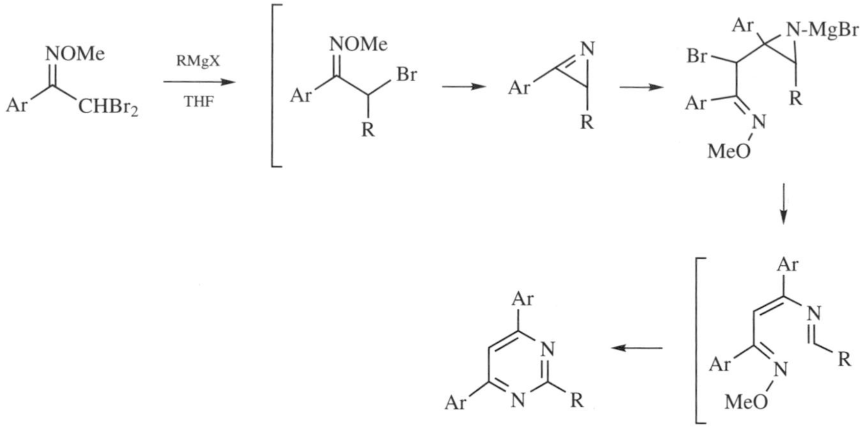

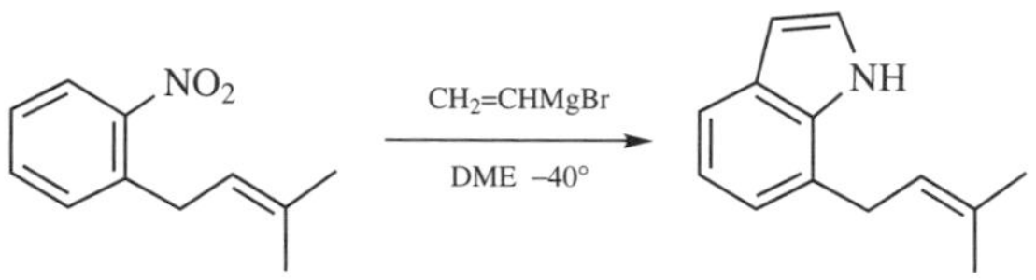

50%

Introduction of an organic substitutent into C-2 of quinoline can be performed by consecutive reactions of the *N*-oxide with isobutyl chloroformate and Grignard reagents.[27]

Substitutions and additions. A series of aryl sulfones are obtained by Grignard reaction of $ArSO_2CF_3$.[28] Chelation effects operate during conjugate addition of RMgX to 4-hydroxy-2-alkenenitriles.[29]

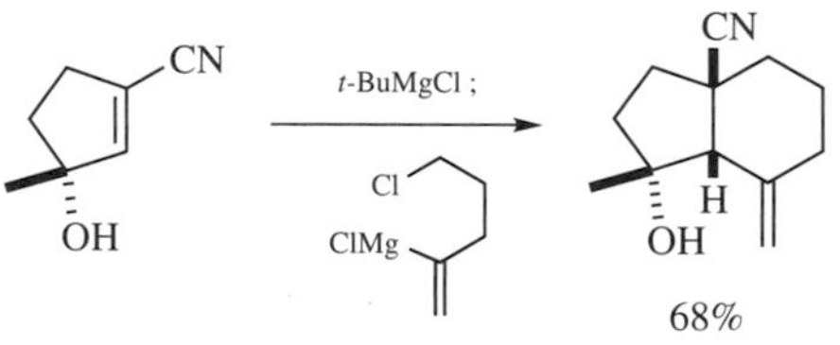

68%

Addition of RMgX to alkenyl(2-pyridyldimethyl)silanes followed by quenching with electrophiles provides silanes that can be further processed to afford alcohols.[30]

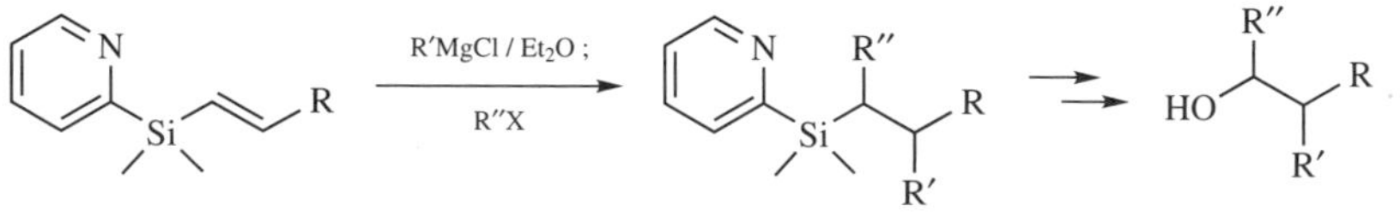

Double allylation of *N*-tributylstannylmethylimines occurs, and the products are suitable precursors of tetrahydroazepines (metathesis).[31] A coupling method for synthesis of allenes starts from carbene generation with the help of Grignard reagents.[32]

CH_2=CHCH$_2$MgBr

THF –78°

SnBu$_3$

N

H

Cl

SOTol

t-BuMgCl ;

EtMgCl

Li

Ph

SO_2Ph

Ph

63%

Hindered biaryl synthesis by the reaction of 2,6-dialkoxyphenylmagnesium bromides with 2-(*o*-methoxyaryl)oxazolines is highly dependent on the nature of the alkoxy groups in the Grignard reagents and to some extent on other nuclear substituents on the oxazoline acceptors. *p*-Methoxybenzyl ethers are most beneficial while MOM ethers are deleterious.[33]

RO

R_3SiO

Br +

OMe

MeO

OR′

$BrCH_2CH_2Br$

Mg / THF

R=R′=PMB 80%
R=R′=MOM 0%

[1]Thibonnet, J., Vu, V.A., Berillon, L., Knochel, P. *T* **58**, 4787 (2002).
[2]Sapountzis, I., Dohle, W., Knochel, P. *CC* 2068 (2002).
[3]Kneisel, F.F., Knochel, P. *SL* 1799 (2002).
[4]Inoue, A., Kitagawa, K., Shinokubo, H., Oshima, K. *JOC* **66**, 4333 (2001).
[5]Sapountzis, I., Knochel, P. *ACIEE* **41**, 1610 (2002).
[6]Fleming, F.F., Shook, B.C. *JOC* **67**, 3668 (2002).
[7]Katritzky, A.R., Abdel-Fattah, A.A.A., Wang, M. *JOC* **67**, 7526 (2002).
[8]Cadierno, V., Conejero, S., Gamasa, M.P., Gimeno, J. *OM* **21**, 3837 (2002).

[9]Bailey, W.F., Reed, D.P., Clark, D.R., Kapur, G.N. *OL* **3**, 1865 (2001).
[10]Michel, P., Ley, S.V. *ACIEE* **41**, 3898 (2002).
[11]Chinkov, N., Morlender-Vais, N., Marek, I. *TL* **43**, 6009 (2002).
[12]Liu, J., Ikemoto, N., Petrillo, D., Armstrong, J.D. *TL* **43**, 8223 (2002).
[13]Zhang, P., Terefenko, E.A., Salvin, J. *TL* **42**, 2097 (2001).
[14]Baker, R.W., Foulkes, M.A., Griggs, M., Nguyen, B.N. *TL* **43**, 9319 (2002).
[15]Miller, J.A. *TL* **43**, 7091 (2002).
[16]Seo, H., Lee, S.-G., Shin, D.M., Hong, B.K., Hwang, S., Chung, D.S., Chung, Y.K. *OM* **21**, 3417 (2002).
[17]Kojima, T., Ohishi, T., Yamamoto, I., Matsuoka, T., Kotsuki, H. *TL* **42**, 1709 (2001).
[18]Capozzi, M.A.M., Cardellicchio, C., Naso, F., Spina, G., Tortorella, P. *JOC* **66**, 5933 (2001).
[19]Suga, S., Okajima, M., Yoshida, J. *TL* **42**, 2173 (2001).
[20]Suga, S., Watanabe, M., Yoshida, J. *JACS* **124**, 14824 (2002).
[21]Badorrey, R., Cativela, C., Diaz de Villegas, M.D., Diez, R., Galvez, J.A. *EJOC* 3763 (2002).
[22]Niwa, Y., Takayama, K., Shimizu, M. *BCSJ* **75**, 1819 (2002).
[23]Katritzky, A.R., Nair, S.K., Qiu, G. *S* 199 (2002).
[24]Gommermann, N., Koradin, C., Knochel, P. *S* 2143 (2002).
[25]Kakiya, H., Yagi, K., Shinokubo, H., Oshima, K. *JACS* **124**, 9032 (2002).
[26]Pirrung, M.C., Wedel, M., Zhao, Y. *SL* 143 (2002).
[27]Fakhfakh, M.A., Franck, X., Fournet, A., Hocquemiller, R., Figadere, B. *TL* **42**, 3847 (2001).
[28]Steensma, R.W., Galabi, S., Tagat, J.R., McCombie, S.W. *TL* **42**, 2281 (2001).
[29]Fleming, F.F., Zhang, Z., Wang, Q., Steward, O. *OL* **4**, 2493 (2002).
[30]Itami, K., Mitsudo, K., Yoshida, J. *ACIEE* **40**, 2337 (2001).
[31]Pearson, W.H., Aponick, A. *OL* **3**, 1327 (2001).
[32]Satoh, T., Sakamoto, T., Watanabe, M. *TL* **43**, 2043 (2002).
[33]Fürstner, A., Stelzer, F., Rumbo, A., Krause, H. *CEJ* **8**, 1856 (2002).

Grignard reagents/chromium(III) chloride.

[2+2+2]Cycloaddition.[1] Bicyclic products are formed when allylchromiun reagents react with 1,6-diynes. Both 5:6-fused and 5:7-fused structures are accessible.

R R′ X R′ MgCl $CrCl_3$ (cat.) THF R R′ X R′ MgCl R″X R R′ X R′ R″ I_2 R R′ X R′

Cyclization. A tandem addition-cyclization pathway is pursued by 1,6-enynes under the conditions.[2]

[1]Nishikawa, T., Kakiya, H., Shinokubo, H., Oshima, K. *JACS* **123**, 4629 (2001).
[2]Nishikawa, T., Shinokubo, H., Oshima, K. *OL* **4**, 2795 (2002).

Grignard reagents/cobalt(II) chloride.

Couplings. Alkyl halides couple with allylmagnesium halides in the presence of (dppp)$CoCl_2$.[1] A Heck-type coupling between styrenes and alkyl halides also occurs.[2]

Me_3SiCH_2MgCl

(dppb)$CoCl_2$

Cyclizations. Radical addition attendant by cyclization is observed with ω-halo-alkenes on exposure to Grignard reagents in the presence of phosphine-ligated $CoCl_2$.[3] Such substrates also undergo Heck-type cyclization (induced by Me_3SiCH_2MgCl).[4]

PhMgBr - (dppe)$CoCl_2$

THF 0°

80% (55 : 45 dr)

[1]Tsuji, T., Yorimitsu, H., Oshima, K. *ACIEE* **41**, 4137 (2002).
[2]Ikeda, Y., Nakamura, T., Yorimitsu, H., Oshima, K. *JACS* **124**, 6514 (2002).
[3]Wakabayashi, K., Yorimitsu, H., Oshima, K. *JACS* **123**, 5374 (2001).
[4]Fujioka, T., Nakamura, T., Yorimitsu, H., Oshima, K. *OL* **4**, 2257 (2002).

Grignard reagents/copper salts. 18, 171–173; **19,** 154–156; **20,** 174–175; **21,** 202–203

Small ring opening. In an elaboration of the triol system of seco-pinolitoxin an epoxysilacyclopentanol was submitted to CuCN-catalyzed Grignard reaction. After regioselective opening of the epoxide ring the silyl group was replaced by OH on oxidation.[1]

PrMgCl - CuCN

(–)-pinolitoxin

Transpositional opening of 2-chloro-3,4-epoxy-1-butene with various Grignard reagents under Cu catalysis presents a method for assembling allylic alcohols with further possibility of functionalization.[2]

β-Lactones give 3-substituted carboxylic acids.[3] Thc S_N2 nature of the reaction sets up the relative/absolute configuration of the new stereocenter so established.

Cyclopropane-1,1-dicarboxylic esters are susceptible to homo-Michael-type reaction with certain Grignard reagents in the presence of CuI.[4]

MgCl + COOR COOR —CuI -HMPA / THF→ COOR COOR

Coupling reactions. A previously known *CC*-coupling of RMgX and R′OTs is extended to an intramolecular version.[5] Facile iodine/magnesium exchange techniques make this process possible.

2-Substituted 1,3-cycloalkadienes are readily obtained in two steps from 2-cycloalkenones enol triflation and Cu-catalyzed coupling with RMgX.[6] Allylic silanes bearing a long carbon chain and possibly other functional group(s) can be elaborated through silacupration of 1,3-dienes followed by trapping with electrophiles.[7]

+ Li^+ Ph–Si–$\bar{C}$uCN → Ph–Si– … Cu, Li, CN —$(PhO)_2P(O)O$–→

Ph–Si–

71%

Geminal dihaloalkylsilanes undergo substitution reactions in a pattern that shows their behavior as 1,1-zwitterions. Replacement of one halogen atom of such a silane with an organic residue from a high ordered magnesium-copper reagent is followed by the creation of a second carbon-metal bond. Subsequent addition of an electrophile completes the two-staged coupling.[8] Interestingly, if vinylmagnesium bromide is used, bond formation at the second stage can be directed to an allylic position.[9]

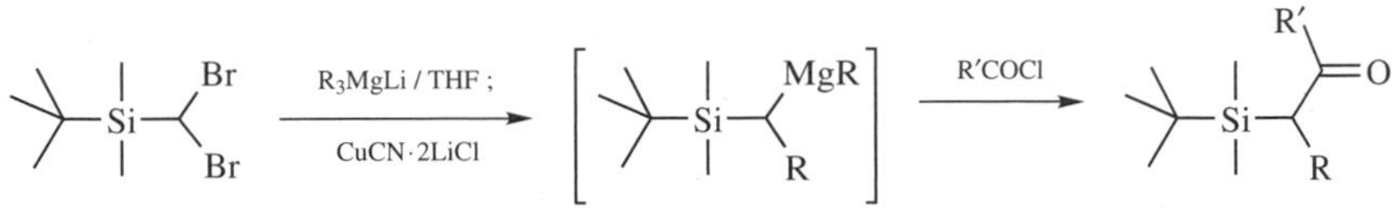

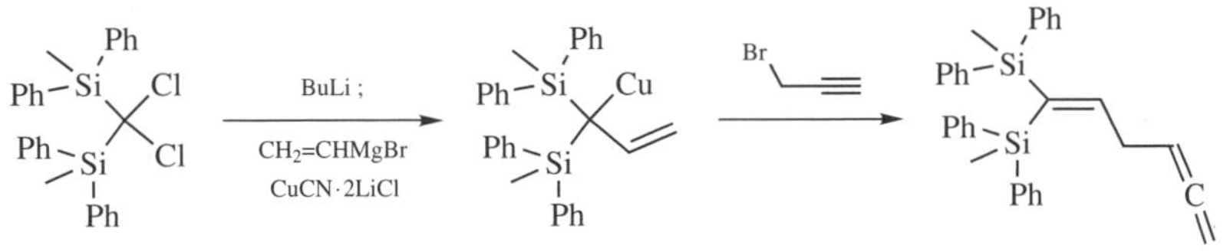

Allylic displacement. Solvent effects on regiochemistry of allylic displacement of 3-acetoxycyclopentenes are enormous.[10] The more polar THF favors direct S_N2 substitution whereas in the less polar ether products arise via a transpositional pathway.

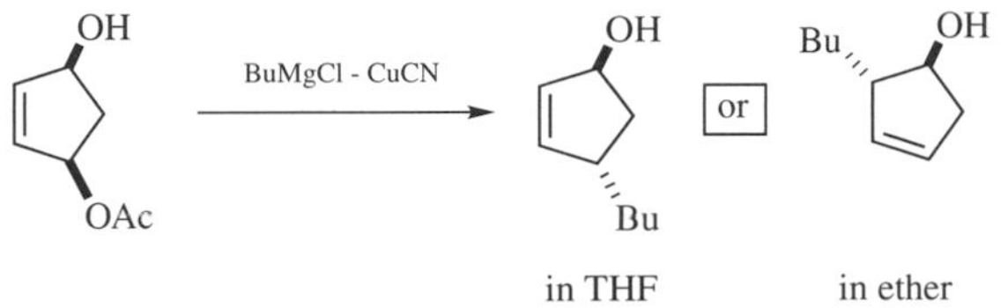

The bridged ring system of the Diels-Alder adducts from cyclopentadiene and *N*-acylnitroso compounds is broken on reaction with Grignard reagents (RMgX) in the presence of $CuCl_2$. The aminocyclopentene products contain an R group adjacent to the nitrogen substituent in a *trans* relationship as a result of an S_N2' process.[11]

[1]Liu, D., Kozmin, S.A. *OL* **4**, 3005 (2002).
[2]Taber, D.F., Mitten, J.V. *JOC* **67**, 3847 (2002).
[3]Nelson, S.G., Wan, Z., Stan, M.A. *JOC* **67**, 4680 (2002).
[4]Prowotorow, I., Wicha, J., Mikami, K. *S* 145 (2001).
[5]Kneisel, F.F., Monguchi, Y., Knapp, K.M., Zipse, H., Knochel, P. *TL* **43**, 4875 (2002).
[6]Karlström, A.S.E., Ronn, M., Thorarensen, A., Bäckvall, J.-E. *JOC* **63**, 2517 (1998).
[7]Liepins, V., Bäckvall, J.-E. *OL* **3**, 1861 (2001).
[8]Inoue, A., Kondo, J., Shinokubo, H., Oshima, K. *CEJ* **8**, 1730 (2002).
[9]Kondo, J., Inoue, A., Shinokubo, H., Oshima, K. *TL* **43**, 2399 (2002).
[10]Ito, M., Matsuumi, M., Murugesh, M.G., Kobayashi, Y. *JOC* **66**, 5881 (2001).
[11]Surman, M.D., Mulvihill, M.J., Miller, M.J. *TL* **43**, 1131 (2002).

Grignard reagents/indium(III) chloride.

Allylic alcohols. Grignard reaction of conjugated carbonyl compounds in the presence of $InCl_3$ (as low as 5%) favors 1,2-addition mode.

[1]Kelly, B.G., Gilheany, D.G. *TL* **43**, 887 (2002).

Grignard reagents/iron(III) acetylacetonate.

Hydrodebromination. β,β-Dibromostyrenes undergo hydrodebromination on exposure to *i*-PrMgCl-Fe(acac)$_3$. Cross-coupling can be carried out thereafter.[1]

Coupling reactions. Successful cross-coupling of various functionalized aryl halides, tosylates, and triflates (e.g., containing keto and ester groups) by using RMgX in the presence of Fe(acac)$_3$ is on record.[2–5] Cross-coupling can be accomplished in a follow-up operation.

Attack on aromatic nitro groups by ArMgX results in the formation of diarylamines (reaction requires addition of $FeCl_2$ and $NaBH_4$).[6]

Br–C$_6$H$_4$–NO$_2$ + ClMg–C$_6$H$_4$–COOEt —(THF –20° ; $FeCl_2$ - $NaBH_4$)→ Br–C$_6$H$_4$–NH–C$_6$H$_4$–COOEt 73%

[1]Fakhfakh, M.A., Franck, X., Hocqemiller, R., Figadere, B. *JOMC* **624**, 131 (2001).
[2]Fürstner, A., Leitner, A. *ACIEE* **41**, 609 (2002).
[3]Fürstner, A., Leitner, A., Mendez, M., Krause, H. *JACS* **124**, 13856 (2002).
[4]Quintin, J., Franck, X., Hocqemiller, R., Figadere, B. *TL* **43**, 3547 (2002).
[5]Dohle, W., Kopp, F., Cahiez, G., Knochel, P. *SL* 1901 (2001).
[6]Sapountzis, I., Knochel, P. *JACS* **124**, 9390 (2002).

Grignard reagents/nickel complexes. 18, 173; **19,** 156–157; **20,** 176–177; **21,** 204–205

Cross couplings. Unsymmetrical biphenyls are obtained by Ni-catalyzed coupling of ArMgX with heteroaryl fluorides.[1] That a highly hindered 1,3-dimesitylimidazolinium salt selectively activates ArF is noted.[2]

In the aliphatic series (RMgX+R′X), using 1,3-butadiene instead of phosphine ligand is the key to attaining high yields of the cross-coupling products.[3]

R–MgX + R′–Br —(THF; Ni)→ R–R′

Transformation of dialkenyl selenides RCH=CHSeCH=CHR to alkenes RCH=CHR' is accomplished by reaction with R'MgX in the presence of $(Ph_3P)_2NiCl_2$.[4] However, formation of RCH=CHCH=CHR as side products cannot be avoided.

Aryl cyanides show reactivity in the a nickel-catalyzed cross-coupling (with loss of CN group) to provide biaryls in good yields and selectivity when the Grignard reagents are first treated with *t*-BuOLi.[5]

The potential of allylic dithioacetals as propene-1,3-zwitterion is manifested in the reaction with an organocuprate and Ni-catalyzed Grignard reaction, intercalated with an alkylation step.[6]

SBu S Ph —MeMgI / $(dppe)NiCl_2$→ Ph 83%

[1]Mongin, F., Mojovic, L., Guillamet, B., Trecourt, F., Queguiner, G. *JOC* **67**, 8991 (2002).
[2]Böhm, V.P.W., Gstöttmayr, C.W.K., Weskamp, T., Herrmann, W.A. *ACIEE* **40**, 3387 (2001).
[3]Terao, J., Watanabe, H., Ikumi, A., Kuniyasu, H., Kambe, N. *JACS* **124**, 4222 (2002).
[4]Silveira, C.C., Santos, P.C.S., Braga, A.L. *TL* **43**, 7517 (2002).
[5]Miller, J.A. *TL* **42**, 6991 (2001).
[6]Chiang, C.-C., Luh, T.-Y. *SL* 977 (2001).

Grignard reagents/palladium complexes. 21, 205

Cross couplings. Halogenated pyridine and quinoline are susceptible to cross-coupling with ArMgX with Pd catalysts,[1] although the Ni(0)-catalyzed reaction is more suitable for preparation of pyridylpyridine and the like.[2] Other coupling pairings include benzylic Grignard reagents with allylic halides,[3] Grignard reagents with alkyl halides,[4] RMgX with enol phosphates[5] [which may be generated in situ from ketones by treatment with mesitylmagnesium bromide and $ClPO(OPh)_2$], and ArMgX with trimethylsilyl-ethyne.[6]

[1]Bonnet, V., Mongin, F., Trecourt, F., Queguiner, G., Knochel, P. *T* **58**, 4429 (2002).
[2]Bonnet, V., Mongin, F., Trecourt, F., Breton, G., Marsais, F., Knochel, P., Queguiner, G. *SL* 1008 (2002).
[3]Rosales, V., Zambrano, J.L., Demuth, M. *JOC* **67**, 1167 (2002).
[4]Frisch, A.C., Shaikh, N., Zapf, A., Beller, M. *ACIEE* **41**, 4056 (2002).
[5]Miller, J.A. *TL* **43**, 7111 (2002).
[6]Gottardo, C., Aguirre, A. *TL* **43**, 7091 (2002).

Grignard reagents/titanium(IV) compounds. 14, 121–122; **18,** 174; **19,** 158–161; **20,** 177–180; **21,** 205–210

Functionalized cyclopropanols. While the original Kulinkovich process for cyclopropanol synthesis is insensitive to the nature of the titanium complex, chlorotitanium triisopropoxide and/or methyltitanium triisopropoxide are the reagents of choice for alkene exchange modification.[1] Functionalized cyclopropylamines are obtained from α-heterosubstituted (O,N,S) nitriles.[2] Effect of chelation on the cyclopropanation step is noted.

Diastereoselection is also revealed in the Kulinkovich reaction.[3,4]

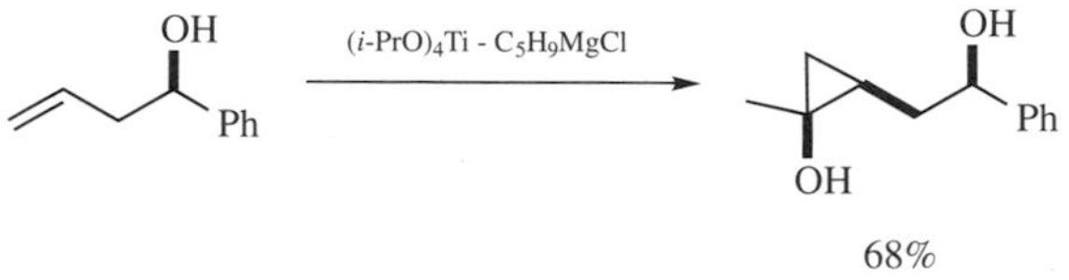

OH Ph
$(i\text{-PrO})_4Ti$ - C_5H_9MgCl
OH Ph OH
68%

Cycloadditions. Further utility of the reactive species generated from RMgX and Ti(IV) compounds is shown in the construction of cyclobutene framework[5] and benzene[6,7] or pyridine derivatives.[8]

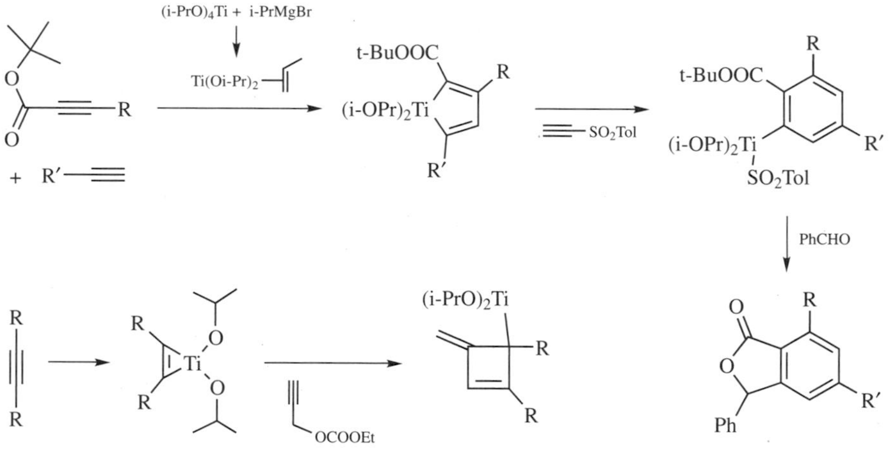

$(i\text{-PrO})_4Ti$ + i-PrMgBr
$Ti(Oi\text{-}Pr)_2$
O O R + R′
t-BuOOC R $(i\text{-OPr})_2Ti$ R′
SO_2Tol
t-BuOOC R $(i\text{-OPr})_2Ti$ R′ SO_2Tol
PhCHO
O R O Ph R′
R R
R Ti O O R
OCOOEt
$(i\text{-PrO})_2Ti$ R R

More remarkable is the reaction of terminally silylated 1,*n*-alkadiyne (n=4,5) with allyl bromide in the presence of *i*-PrMgBr/(i-PrO)$_4$Ti.[9] Another useful process is a furan synthesis from 2-alkynyl-4,4,5,5-tetramethyl-1,3-dioxolanes.[10]

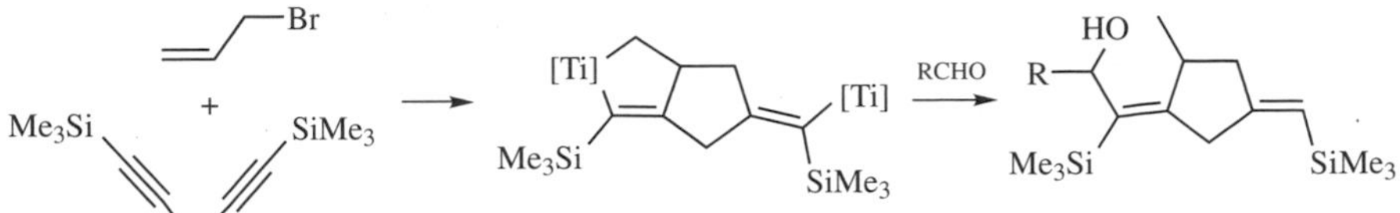

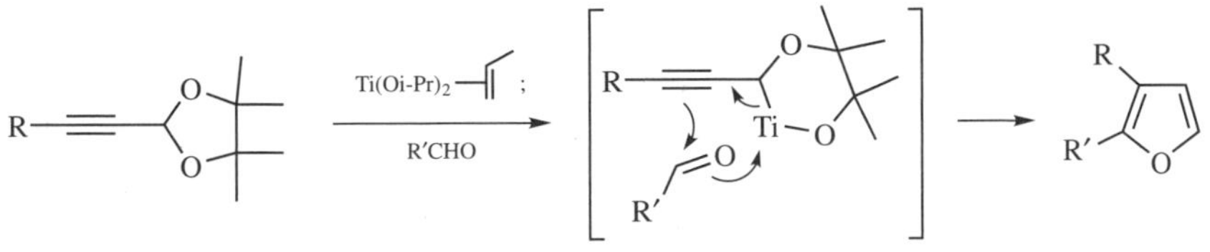

Cyclization. A facile synthesis of 1,2-bis(ethylenidene)cycloalkanes from diyne derivatives has been developed.[11] Reductive cyclization of unsaturated ketones (such as δ,ϵ-unsaturated ketones in which the double bond and carbonyl group are properly distanced) afford 1,2-disubstituted cycloalkanols.[12]

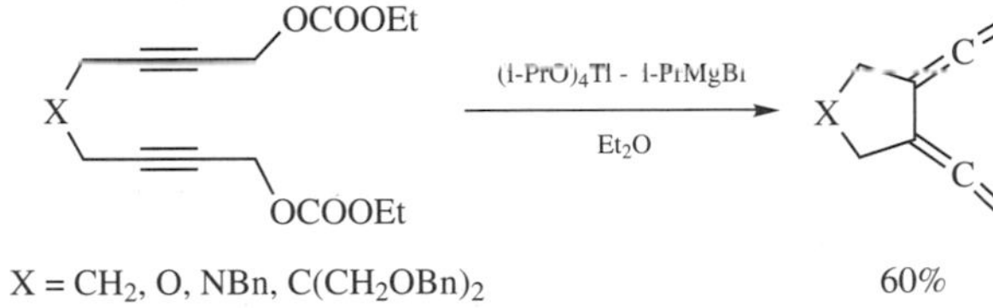

X = CH_2, O, NBn, $C(CH_2OBn)_2$ 60%

Bicyclic structures are erected from reductive cyclization of allylic acetates involving an alkynyl chain.[13]

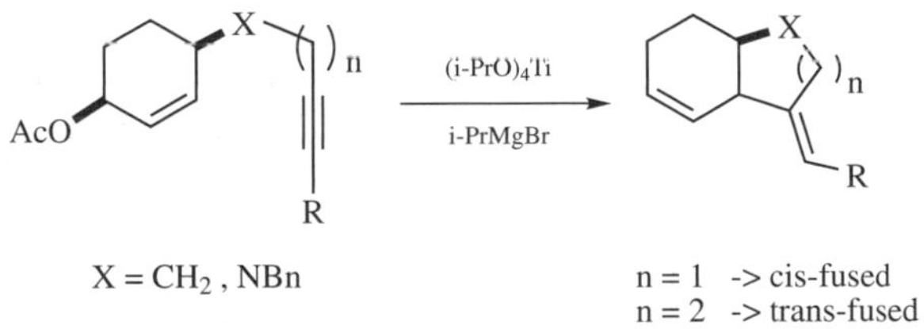

X = CH_2, NBn n = 1 -> cis-fused n = 2 -> trans-fused

Pummerer rearrangement follows cyclization that involves alkenyl sulfoxide and alkyne moieties. Accordingly, the products are 2-alkylidenecycloalkanecarbaldehydes.[14]

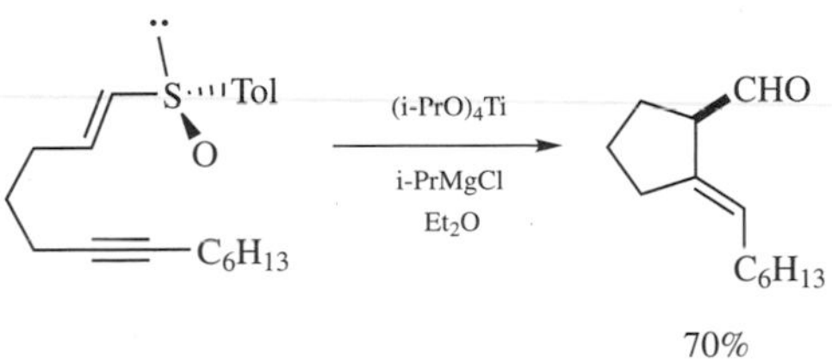

70%

Reduction and substitution. Conjugated diynes are reduced to enynes. 1-Trimethylsilyl-1,3-alkadiynes are converted to (*Z*)-1-trimethylsilyl-3-alken-1-ynes,[15] demonstrating the chemoselective and stereoselective nature of the reduction.

Me_3Si—≡—≡—C_6H_{13} —(i-PrO)$_4$Ti / i-PrMgCl ; HCl→ Me_3Si—≡—CH=CH—C_6H_{13} 85%

Optically active (configurationally inverted) propargylic stannanes are prepared from propargylic phosphates or chlorides by treatment with *i*-PrMgBr/(*i*-PrO)$_4$Ti followed by reaction with R_3SnCl. Repeated treatment but with aqueous quench leads to the corresponding (*Z*)-allylic stannanes.[16]

Bu, O=P(OEt)(EtO)O —(i-PrO)$_4$Ti / i-PrMgCl ; Bu_3SnCl→ Bu, $SnBu_3$ 78% —(i-PrO)$_4$Ti / i-PrMgCl ; H_2O→ Bu, $SnBu_3$ 91%

The chiral allenyltitanium species (formed in the initial reaction indicated above) are capable Michael reaction donors.[17]

1,4-Pentadien-3-yl anion equivalent is generated by exposure of the carbonate precursor to *i*-PrMgCl/(*i*-PrO)$_4$Ti in ether. Regioselective alkylation of such species with carbonyl compounds is observed.[18]

Allylic alcohols and ethers undergo transpositional displacement, whereby the R group of RMgBr (e.g., R=Et) is introduced to the substrates.[19]

Coupling. An intramolecular hydrovinylation occurs when vinyl ethers of unsaturated alcohols are exposed to *i*-PrMgCl/(*i*-PrO)$_4$Ti, the vinyl group is delivered to the unsaturated site.[20]

O, COOEt, Ph —(i-PrO)$_4$Ti / i-PrMgCl ; H^+→ Ph, O, O 53%

O, C_6H_{13} —(i-PrO)$_4$Ti / i-PrMgCl ; I_2→ HO, I, C_6H_{13} 61%

[1]Lee, J.C., Sung, M.J., Cha, J.K. *TL* **42**, 2059 (2001).
[2]Bertus, P., Szymoniak, J. *JOC* **67**, 3965 (2002).
[3]Racouchot, S., Sylvestre, I., Ollivier, J., Kozyrkov, Y.Yu., Pukin, A., Kulinkovich, O.G., Salaun, J. *EJOC* 2160 (2002).
[4]Quan, L.G., Kim, S.-H., Lee, J.C., Cha, J.K. *ACIEE* **41**, 2160 (2002).
[5]Delas, C., Urabe, H., Sato, F. *JACS* **123**, 7937 (2001).
[6]Suzuki, D., Urabe, H., Sato, F. *JACS* **123**, 7925 (2001).
[7]Tanaka, R., Nakano, Y., Suzuki, D., Urabe, H., Sato, F. *JACS* **124**, 9862 (2002).
[8]Suzuki, D., Tanaka, R., Urabe, H., Sato, F. *JACS* **124**, 3518 (2002).
[9]Okamoto, S., Subburaj, K., Sato, F. *JACS* **123**, 4857 (2001).
[10]Teng, X., Wada, T., Okamoto, S., Sato, F. *TL* **42**, 5501 (2001).
[11]Delas, C., Urabe, H., Sato, F. *TL* **42**, 4147 (2001).
[12]Quan, L.G., Cha, J.K. *TL* **42**, 8567 (2001).
[13]Song, Y., Okamoto, S., Sato, F. *TL* **43**, 6511 (2002).
[14]Narita, M., Urabe, H., Sato, F. *ACIEE* **41**, 3671 (2002).
[15]Delas, C., Urabe, H., Sato, F. *CC* 272 (2002).
[16]Okamoto, S., Matsuda, S.-I., An, D.K., Sato, F. *TL* **42**, 6323 (2001).
[17]Song, Y., Okamoto, S., Sato, F. *OL* **3**, 3543 (2001).
[18]Okamoto, S., Sato, F. *JOMC* **624**, 151 (2001).
[19]Kulinkovich, O.G., Epstein, O.L., Isakov, V.E., Khmel'nitskaya, E.A. *SL* 49 (2001).
[20]Nakajima, R., Urabe, H., Sato, F. *CL* 4 (2002).

Grignard reagents/zinc halide.

Substitutions and additions. A carbon-linked benzotriazolyl group is susceptible to displacement on reaction with RMgX in the presence of $ZnCl_2$ and the process has been applied to a synthesis of 4,5-disubstituted imidazolidin-2-ones.[1]

PhMgBr - $ZnCl_2$, THF Δ

68%

The tendency of alkylzinc reagents to undergo oxidation during reaction with electrophiles has been revealed. Alkoxylated instead of the expected alkylated products are obtained.[2] Accordingly, one must bear in mind this observation when such transformations are contemplated.

The stereoselective addition of allylzinc reagents to alkenyllithiums such as **1** is interesting.[3]

CH_2=$CHCH_2MgBr$, $ZnBr_2$

1

[1]Katritzky, A.R., Luo, Z., Fang, Y., Steel, P.J. *JOC* **66**, 2858 (2001).
[2]Katritzky, A.R., Luo, Z. *H* **55**, 1467 (2001).
[3]Bernard, N., Chemla, F., Ferreira, F., Mostefai, N., Normant, J.F. *CEJ* **8**, 3139 (2002).

Grignard reagents/zirconium compounds. 18, 174; **19,** 161; **20,** 180–181; **21,** 210–211

Alkynylation.[1] Alkynes undergo alkynylzirconation after successive reaction with EtMgBr-Cp_2ZrCl_2 and 1-haloalkynes. The adducts can be hydrolyzed to furnish enynes or alkynylated further with a second 1-haloalkyne.

R—≡—R; Cp_2ZrCl_2 - EtMgCl ; R′—≡—Cl; Cp_2Zr; Cl; R; R′; H_3O^+; O; R″; Ar

1,2-Ethanediyl dianion equivalent. The EtMgBr-Cp_2ZrCl_2 reagent adds to imines readily to give ethylated products (amines).[2] A second CC-bond forming process can be accomplished at the end of the addition.[3]

NPh; Ph; Cp_2ZrCl_2 - EtMgCl ; PhCHO; NHPh; Ph; OH; 60%

Alkylhydroxylation. Carbomagnesiation of styrenes introduces an alkyl group at the benzylic position. Subsequent oxidation with O_2 leads to 2-alkyl-2-phenylethanols.[4] Replacing oxygen with NBS or carbonyl compounds leads to different products.

Ph; +; MgBr; Cp_2ZrCl_2; *i*-PrOTs ; O_2; Ph; OH; 73%

Homoallylic alcohols. Alkenes are activated at an allylic position (to form allylmetals) by the title reagents and reaction with acid chlorides give homoallylic alcohols.[5] It is assumed that the allylic zirconium hydride intermediates reduce the acid chloride prior to CC bond formation.

MgBr $\xrightarrow{Cp_2ZrCl_2}$ - -ZrCp$_2$ $\xrightarrow[PhCOCl]{R\ ;}$ R Ph OH

Reductive alkylation. Certain alkenyl tosylates are transformed into cyclic products by a reductive alkylation reaction.[6] The double bond is turned into a nucleophile by RMgX-Cp_2ZrCl_2 in the process.

OTs $\xrightarrow{Cp_2ZrCl_2\text{ - BuMgCl}}$

85% (*cis/trans* 5.5 : 1)

[1]Liu, Y., Zhong, Z., Nakajima, K., Takahashi, T. *JOC* **67**, 7451 (2002).
[2]Gandon, V., Bertus, P., Szymoniak, J. *EJOC* 3677 (2001).
[3]Gandon, V., Bertus, P., Szymoniak, J. *S* 1115 (2001).
[4]de Armas, J., Hoveyda, A.H. *OL* **3**, 2097 (2001).
[5]Fujita, K., Yorimitsu, H., Shinokubo, H., Matsubara, S., Oshima, K. *JACS* **123**, 12115 (2001).
[6]Cesati III, R.R., de Armas, J., Hoveyda, A.H. *OL* **4**, 395 (2002).

H

Hafnium(IV) chloride bis(tetrahydrofuran). 20, 182; **21,** 212

Esterification.[1] Esterification of primary alcohols is selective in the presence of secondary alcohols and phenols when hafnium(IV) chloride is used as promoter. Adamantane-1-carboxylic acid can be esterified by this method.

4-Alkylidene-4,5-dihydroisoxazoles.[2] An unusual route to this class of compounds involves a double condensation via conjugated nitrones.

HfCl4 / THF; ArCHO

[1]Ishihara, K., Nakayama, M., Ohara, S., Yamamoto, Y. *SL* 1117 (2001).
[2]Dunn, P.J., Graham, A.B., Grigg, R., Higginson, P., Sridharan, V., Thornton-Pett, M. *CC* 1968 (2001).

Hafnium(IV) triflate.

Diels-Alder reaction.[1] The *endo*-selective DA reaction of furan mediated by $Hf(OTf)_4$ proceeds at 0°.

Dehydration.[2] The Hf complex is an effective catalyst for dehydration, suitable for formation of trienes and tetraenes

OH, Ph, CHO; $HfCl_4.(thf)$, MeCN Δ; Ph, CHO; 99%

Michael reaction.[3] In catalyzing Michael reaction of thiols and conjugated carbonyl compounds $Hf(OTf)_4$ is paired with a chiral ligand. However, asymmetric induction is modest only.

Fiesers' Reagents for Organic Synthesis, Volume 22. Series editor Tse-Lok Ho
ISBN 0-471-28515-3

Friedel-Crafts alkylation. Catalytic alkylation of arenes with benzyl or allyl silyl ethers is accomplished[4,5] in the presence of $Hf(OTf)_4$ or $Cl_2Si(OTf)_2$. The ethers can be generated in situ.

PhCHO + allyl-$SiMe_3$ —[$Hf(OTf)_4$, PhOMe]→ 99% (*o/p* 14 : 86)

[1]Hayashi, Y., Nakamura, M., Nakao, K., Inoue, T., Shoji, M. *ACIEE* **41**, 4079 (2002).
[2]Saito, S., Nagahara, T., Yamamoto, H. *SL* 1690 (2001).
[3]Kobayashi, S., Ogawa, C., Kawamura, M., Sugiura, M. *SL* 983 (2001).
[4]Shiina, I., Suzuki, M. *TL* **43**, 6391 (2002).
[5]Shiina, I., Suzuki, M., Yokoyama, K. *TL* **43**, 6395 (2002).

Hantzsch 1,4-dihydropyridine.

β-Hydroxy ketones.[1] Photoinduced reaction of α,β-epoxy ketones in the presence of Hantzsch dihydropyridine (**1**) leads to aldols. The well-known reagent is easily obtained from acetoacetic ester, hexamethylenetetramine and ammonium carbonate.

EtOOC, COOEt, N, H

1

Reductive amination. Aldehydes are converted into secdondary amines using $Sc(OTf)_3$ as catalyst and Hantzsch dihydropyridine as the reducing agent.[2]

[1]Zhang, J., Jin, M.-Z., Zhang, W., Yang, L., Liu, Z.-L. *TL* **43**, 9687 (2002).
[2]Itoh, T., Nagata, K., Kurihara, A., Miyazaki, M., Ohsawa, A. *TL* **43**, 3105 (2002).

Hexaalkylditin. 13, 142; 14, 173–174; 16, 174; 17, 143–144; 18, 175–176; 19, 162–163; 20, 182–184; 21, 213

Trifluoromethyl ketimines.[1] Free radicals generated from **1** via desulfonylation react with RI to afford ketimines which can serve as precursors of amines and ketones.

F_3C–C(=N–OBn)–SO_2Ph

1

cis-α,β-Bistrimethylstannyl enamines.[2] Ynamines undergo 1,2-addition with $Me_3SnSnMe_3$ under the influence of $(Ph_3P)_4Pd$. Stannylenamines offer some synthetic possibilities.

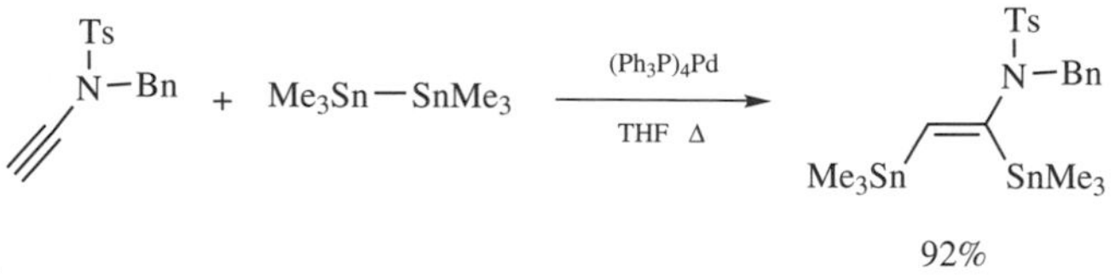

Ts
N—Bn + $Me_3Sn—SnMe_3$ $\xrightarrow[\text{THF } \Delta]{(Ph_3P)_4Pd}$ Ts N—Bn Me_3Sn $SnMe_3$

92%

Coupling reactions.[3] 2-Bromopyridines undergo selective cross-coupling with ArBr in the presence of $Me_3SnSnMe_3$ and catalytic amount of $(Ph_3P)_4Pd$. It is a variant of the Stille coupling.

[1]Kim, S., Kavali, R. *TL* **43**, 7189 (2002).
[2]Miniere, S., Cintrat, J.-C. *S* 705 (2001).
[3]Zhang, N., Thomas, L., Wu, B. *JOC* **66**, 1500 (2001).

Hexabutylguanidinium chloride.

Alkyl chlorides.[1] The guanidinium salt **1** catalyzes thermal decomposition of alkyl chloroformates to afford the corresponding chlorides.

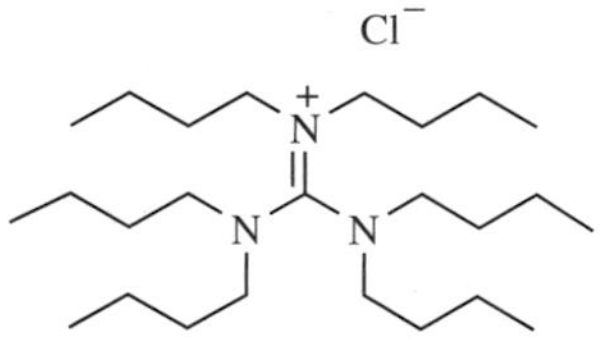

Cl^-
N^+
N N

1

[1]Violleau, F., Thiebaud, S., Borredon, E., Le Gars, P. *SC* **31**, 367 (2001).

Hexafluoroacetone.

2,2-Bis(trifluoromethyl)oxazolidin-5-ones.[1] These compounds (**1**) are prepared from α-amino acids and hexafluoroacetone in DMSO at room temperature. Peptide bond formation takes place on their mixing with amino acid esters. Accordingly, hexafluoroacetone serves as a protecting and activating agent.

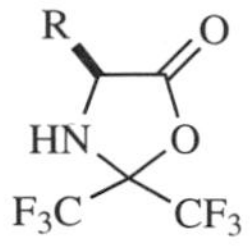

1

[1]Burger, K., Lange, T., Rudolph, M. *H* **59**, 189 (2003).

1,1,1,3,3,3-Hexafluoro-2-propanol. 21, 213–214

Epoxide opening. The reaction of epoxides with *N*-alkylarylamines in this fluorinated solvent proceeds beyond the simple substitution, as cyclization follows the initial reaction.[1]

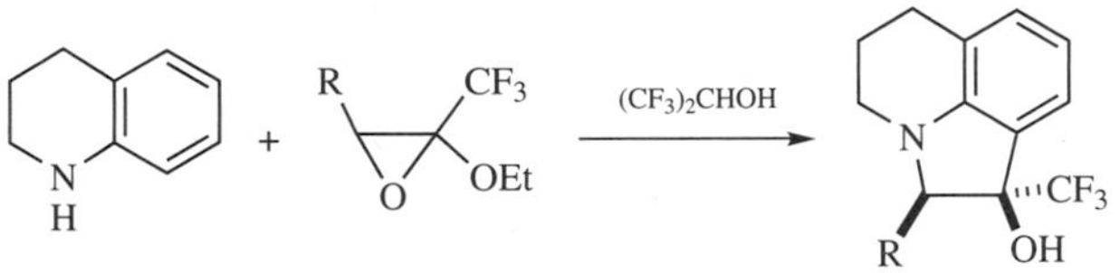

$(CF_3)_2CHOH$

[1]Rodrigues, I., Bonnet-Delpon, D., Begue, J.-P. *JOC* **66**, 2098 (2001).

Hexakis(hydridotriphenylphosphinecopper).

Reduction-aldol reaction. Formation of 5- and 6-membered 2-acylcycloalkanols from aliphatic substances containing both a conjugated and a saturated carbonyl groups is realized, when they are treated with the complexed copper hydride reagent.[1]

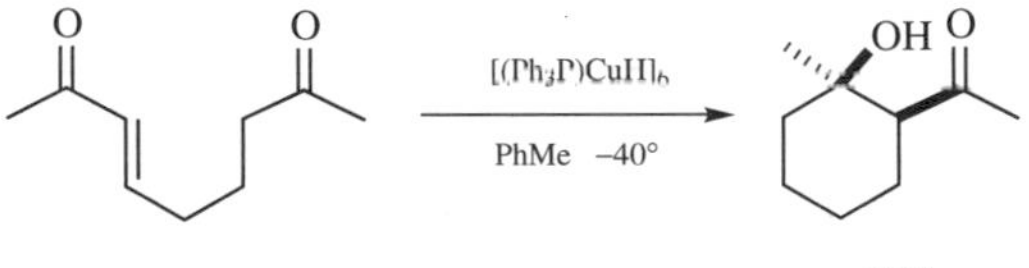

$[(Ph_3P)CuH]_6$

PhMe −40°

80%

Silyl ethers. Carbonyl compounds (selectivity: RCHO > RCOR′) are transformed to silyl ethers in a $[(Ph_3P)CuH]_6$–catalyzed hydrosilylation process. An isolated double bond does not interfere with the reaction.[2]

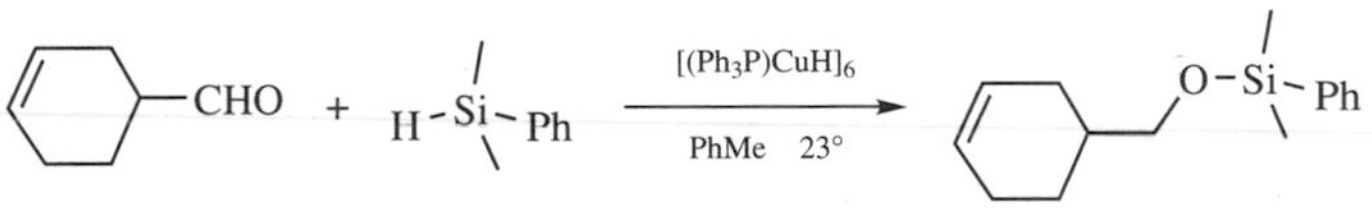

$[(Ph_3P)CuH]_6$

PhMe 23°

98%

[1]Chiu, P., Szeto, C.-P., Geng, Z., Cheng, K.-F. *OL* **3**, 1901 (2001).
[2]Lipshutz, B.H., Chrisman, W., Noson, K. *JOMC* **624**, 367 (2001).

η[4]-Hexamethylbenzene(pentamethylcyclopentadienyl)rhodium.

Borylation.[1] Introduction of a boryl group into an arene in inert solvent is promoted by the Rh complex.

F_3C F_3C + HB(pinacol) —$Cp^*{}_2Rh(\eta^6\text{-}C_6Me_6)$, cyclohexane→ 88%

[1]Tse, M.K., Cho, J.-Y., Smith III, M.R. *OL* **3**, 2831 (2001).

Hexamethyldisilazane, HMDS. 13, 141; **18,** 177–178; **19,** 163–164; **20,** 184–185; **21,** 214

Disulfides. Oxidative dimerization of thiols by DMSO is promoted by HMDS.[1]

1-Aryl-2,2,2-trifluoroethylamines.[2] The mixture of hemiaminal and its silyl ether prepared from HMDS and gaseous trifluoroacetaldehyde is reactive toward arenes in the presence of a Lewis acid.

Me_3Si NH Me_3Si + F_3C-CHO ⟶ Me_3Si N Me_3Si OH CF_3 + H N Me_3Si $OSiMe_3$ CF_3 —indole, $BF_3 \cdot OEt_2$→ F_3C NH_2 N H 61%

[1]Karimi, B., Zareyee, D. *SL* 346 (2002).
[2]Gong, Y., Kato, K. *JFC* **116**, 103 (2002).

Hexamethylphosphoric amide, HMPA.

ω-Bromo-1-alkenes. 1,ω-Dibromoalkanes undergo elimination at one terminus on heating with HMPA.[1] The procedure involves dropwise addition of HMPA to the hot vessel containing the dibromide while distilling off the product.

Hydroxyalkylation. The condensation of 1-cyanomethylnaphthalene with aromatic aldehydes in the presence of HMPA is *syn*-selective, due to equilibration.[2]

LDA / THF, HMPA ; PhCHO ; NH_4Cl aq.

(92 : 8)
76%

[1]Hoye, T.R., Van Veidhuizen, J.J., Vos, T.J., Zhao, P. *SC* **31**, 1367 (2001).
[2]Carlier, P.R., Lo, C.W.-S., Lo, M.M.-C., Wan, N.C., Williams, I.D. *OL* **2**, 2443 (2000).

Hydridobiruthenium complex.

Redox reactions. The complex **1** mediates transfer hydrogenation of imines (with *i*-PrOH as hydrogen source)[1] and dehydrogenation of *N*-alkylanilines by quinones or MnO_2.[2]

1

[1]Samec, J.S.M., Bäckvall, J.-E. *CEJ* **8**, 2955 (2002).
[2]Ell, A.H., Samec, J.S.M., Brasse, C., Bäckvall, J.-E. *CC* 1144 (2002).

Hydridotetrakis(triphenylphosphine)rhodium.

(Z)-1,2-Bisorganothioalkenes. In the presence of $(Ph_3P)_4RhH$ disulfides are split and each unit is added to the triple bond of 1-alkynes.[1]

$(Ph_3P)_4RhH$ - CF_3SO_3H ; *p*-An_3P / Me_2CO ; Δ

95%

(E)-3-Alkenyltriphenylphosphonium salts. Regioselective anti-Markovnikov addition of triphenylphosphine to 1,3-dienes under acidic conditions is accomplished when $(Ph_3P)_4RhH$ is used as catalyst. The unreactive (*Z*)-diene in the mixture is easily separated.[2]

Ph–CH₂CH₂CH=CHCH=CH₂ → (Ph₃P)₄RhH - CF₃SO₃H; Ph₃P / THF 0° ; LiPF₆ → Ph–CH₂CH₂CH=CHCH₂CH₂–$\overset{+}{P}Ph_3$ PF_6^- 89%

[1]Arisawa, M., Yamaguchi, M. *OL* **3**, 763 (2001).
[2]Arisawa, M., Momozuka, R., Yamaguchi, M. *CL* 272 (2002).

Hydriodic acid.

Amines. Aqueous HI reduces nitroarenes to arylamines without affecting halo, carbonyl, and many other substituents.[1] Azides including sulfonyl azides are also reduced.[2]

[1]Kumar, J.S.D., Ho, M.M., Toyokuni, T. *TL* **42**, 5601 (2001).
[2]Kamal, A., Reddy, P.S.M.M., Reddy, D.R. *TL* **43**, 6629 (2002).

Hydrogen fluoride–amine. 16, 286–287; 18, 181; 19, 164–165; 20, 185; 21, 215–216

Anodic fluorination. Electrolysis in MeCN containing *n*HF-Et_3N introduces fluorine atom(s) to cyclic ethers, lactones and a cyclic carbonate.[1] Similarly propargyl sulfides are also fluorinated at the methylene group.[2] Electrofluorination of phenols furnishes difluorocyclohexadienones.[3]

1-Naphthol (OH) → e / Et_3N-5HF; Me_2CO → 4,4-difluoronaphthalen-1(4H)-one (O; F F) 89%

Electrochemically generated iodonium ion initiates iodofluorination of alkenes.[4]

Selenofluorination.[5] In combination with PhSeOTf or $PhSeSbF_6$ in dichloromethane 3HF-Et_3N converts alkynes to alkenes that contain PhSe group and F atom at the two sp^2-hybridized carbon atoms.

Desilylation.[6] While practically no cleavage of secondary TBDPS ethers of carbohydrates by HF-pyridine/DMF at room temperature is observable such silyl groups are removed at high pressure (e.g., 1 GPa).

[1]Hasegawa, M., Ishii, H., Fuchigami, T. *TL* **43**, 1503 (2002).
[2]Riyadh, S.M., Ishii, H., Fuchigami, T. *T* **58**, 5877 (2002).
[3]Fukuhara, T., Akiyama, Y., Yoneda, N., Tada, T., Hara, S. *TL* **43**, 6583 (2002).
[4]Kobayashi, S., Sawaguchi, M., Ayuba, S., Fukuhara, T., Hara, S. *SL* 1938 (2001).
[5]Poleschner, H., Seppelt, K. *JCS(P1)* 2668 (2002).
[6]Matsuo, I.,Wada, M., Ito, Y. *TL* **43**, 3273 (2002).

Hydrogen peroxide.

Epoxidation. 2,2,2-Trifluoroethanol alone[1] or hexafluoroacetone in hexafluoroisopropanol,[2] provides adequate activation that allows epoxidation of alkenes with H_2O_2. There is limitation for the former system, not being effective for monosubstituted alkenes.

Oxidations. *gem*-Disilylalkanes suffer oxidation to afford carbonyl compounds.[3] This process is incorporated into a method of ketone synthesis from 1,1-disilylethenes.

BuLi / THF ; $C_8H_{17}I$ → HBF$_4$ · OEt$_2$ / CH$_2$Cl$_2$; Bu$_4$NF - H$_2$O$_2$

Bu, C_8H_{17}, O

68% overall

Baeyer-Villiger oxidation proceeds well with aqueous H_2O_2 and diselenide **1** in 2,2,2-trifluoroethanol (4 examples, 95–100%).[4]

F_3C, CF_3, Se, Se

1

[1]van Vliet, M.C.A., Arends, I.W.C.E., Sheldon, R.A. *SL* 248 (2001).
[2]van Vliet, M.C.A., Arends, I.W.C.E., Sheldon, R.A. *SL* 1305 (2001).
[3]Inoue, A., Kondo, J., Shinokubo, H., Oshima, K. *CC* 114 (2002).
[4]ten Brink, G.-J., Vis, J.-M., Arends, I.W.C.E., Sheldon, R.A. *JOC* **66**, 2429 (2001).

Hydrogen peroxide, acidic. 14, 176; **15,** 167–168; **16,** 177–178; **17,** 145; **18,** 182–183; **19,** 166; **20,** 187; **21,** 216–217

Hydroperoxides. Furfuryl hydroperoxides, which are valuable regeneratable oxygen donors in sulfoxidations, are readily prepared from the corresponding carbinols on reaction with acidic H_2O_2.[1] *gem*-Bishydroperoxides are formed when cyclic benzylic alcohols are treated with 50% H_2O_2 in the presence of sulfuric acid.[2]

OH, X, n → 50% H_2O_2 / H_2SO_4 → OH, OOH, OOH, X, n

X = O, CH_2

Epoxidation and Baeyer-Villiger oxidation.[3] In $(CF_3)_2CHOH$ epoxidation of alkenes is efficiently achieved with aqueous H_2O_2 and $PhAsO_3H_2$ as a catalyst. For Baeyer-Villiger oxidation the more common TsOH is also useful.

[1]Massa, A., Palombi, L., Scettri, A. *TL* **42**, 4577 (2001).
[2]Hamann, H.-J., Liebscher, J. *SL* 96 (2001).
[3]Berkessel, A., Andreae, M.R.M. *TL* **42**, 2293 (2001).

Hydrogen peroxide, metal catalysts. 13, 145; **14,** 177; **15,** 294; **17,** 146–148; **18,** 184–185; **19,** 166–167; **20,** 188; **21,** 217–218

Oxidation. Benzylic halides are oxidized to benzoic acids with 30% H_2O_2 in the presence of Na_2MO_4 (M=V, W, Mo) and a phase-transfer catalyst without an organic solvent.[1] Conversion of sulfides to sulfones can be achieved under similar conditions[2] or with $MnSO_4$ and a buffer solution.[3]

The capacity of the H_2O_2-V_2O_5 system to hydrolyze thioglycosides,[4] simple dithioacetals[5] and oxathiolanes[6] depends on its oxidizing power.

Stirring hydroxamic acids with H_2O_2 and a transition metal (Ru, Ir) complex generates acyl nitroso compounds in excellent yields (trapped as Diels-Alder adducts).[7,8]

The tetracyclone ring splits open on heating with H_2O_2-CaO_2 in THF.[9]

$CaO_2 \cdot 2H_2O_2$, THF 50°

95%

Epoxidation. Because of economic reasons alkene epoxidation attracts research efforts continuously. Systems involving H_2O_2 as reagent also require catalysts: manganous sulfate,[10] hydrotalcite,[11,12] and the Fe(II) complex of **1**[13] may be mentioned. Sandwich-type polyoxometalates show excellent catalytic activities for epoxidation of allylic alcohols.[14]

1

γ-Hydroxyalkyl hydroperoxides.[15] Substituted oxetanes are opened by ethereal H_2O_2 in the presence of $Yb(OTf)_3$. The products can be used to prepare 1,2,4-trioxepanes.

H_2O_2 / Et_2O, $Yb(OTf)_3$; $C_{16}H_{33}$; HO; OOH; 60%

[1]Shi, M., Feng, Y.-S. *JOC* **66**, 3235 (2001).
[2]Sato, K., Hyodo, M., Aoki, M., Zheng, X.-Q., Noyori, R. *T* **57**, 2469 (2001).
[3]Alonso, D.A., Nąjera, C., Varea, M. *TL* **43**, 3459 (2002).
[4]Barua, P.M.B., Sahu, P.R., Mondal, E., Bose, G., Khan, A.T. *SL* 81 (2002).
[5]Mondal, E., Bose, G., Sahu, P.R., Khan, A.T. *CL* 1158 (2001).
[6]Mondal, E., Sahu, P.R., Bose, G., Khan, A.T. *JCS(P1)* 1026 (2002).
[7]Iwasa, S., Tajima, K., Tsushima, S., Nishiyama, H. *TL* **42**, 5897 (2001).
[8]Iwasa, S., Fakhruddin, A., Tsukamoto, Y., Kameyama, M., Nishiyama, H. *TL* **42**, 6159 (2001).
[9]Pierlot, C., Nardello, V., Schrive, J., Mabille, C., Barbillat, J., Sombret, B., Aubry, J.-M. *JOC* **67**, 2418 (2002).
[10]Lane, B.S., Burgess, K. *JACS* **123**, 2933 (2001).
[11]Pillai, U.R., Sahle-Demessie, E., Varma, R.S. *TL* **43**, 2909 (2002).
[12]Honma, T., Nakajo, M., Mizugaki, T., Ebitani, K., Kaneda, K. *TL* **43**, 6229 (2002).
[13]White, M.C., Doyle, A.G., Jacobsen, E.N. *JACS* **123**, 7194 (2001).
[14]Adam, W., Alster, P.L., Neumann, R., Saha-Moller, C.R., Sloboda-Rozner, D., Zhang, R. *SL* 2011 (2002).
[15]Dussault, P.H., Trullinger, T.K., Noor-e-Ain, F. *OL* **4**, 4591 (2002).

Hydrosilanes. 19, 167–169; 20, 188–192; 21, 218–222

Hydrosilylation and carbosilylation. The sulfur atom of thioethers can direct Pt-catalyzed hydrosilylation of alkenes.[1] Methylenecyclopropane derivatives undergo either hydrosilylation or isomerization (to acyclic dienes), depending on the nature of the Pt catalysts.[2]

Ph; Ph; Ph_3SiH, $(Ph_3P)_2PtI_2$; Ph_3Si; Ph; Ph; 53%; Et_3SiH, $(Et_3P)_2Pt$; Ph; Ph; 70%

The ruthenium complex $[CpRu(MeCN)_3]PF_6$ directs alkyne hydrosilylation in the Markovnikov sense.[3] 1,1-Disilyl-1-alkenes are accessible from silylalkynes by hydrosilylation catalyzed with organolanthanide and group-3 metallocene complexes.[4]

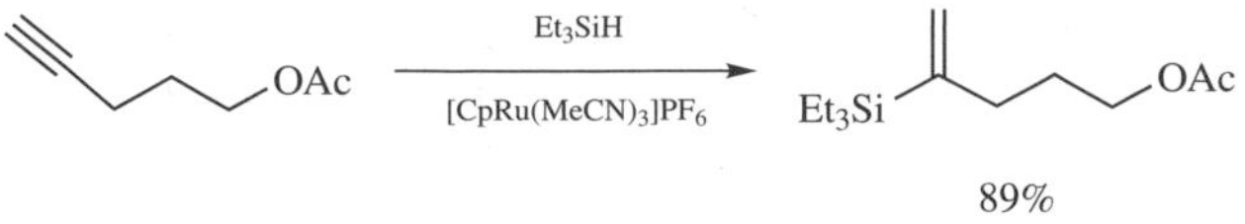

OAc —(Et_3SiH; $[CpRu(MeCN)_3]PF_6$)→ Et_3Si … OAc

89%

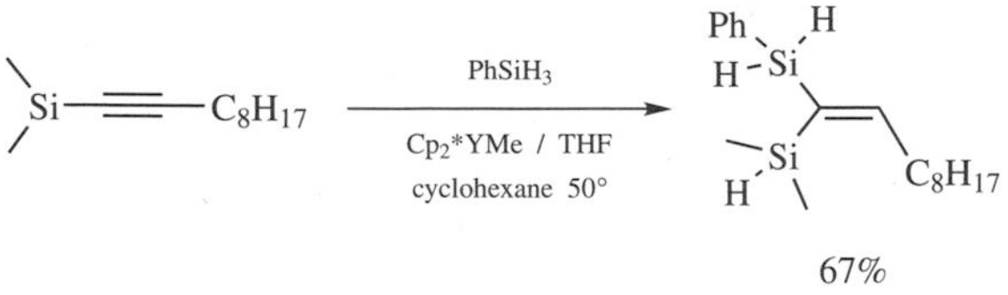

Si—≡—C_8H_{17} —($PhSiH_3$; Cp_2*YMe / THF; cyclohexane 50°)→ Ph, H, H–Si; Si, H; C_8H_{17}

67%

Reduction of macrocycloalkynones to nonconjugated (*E*)-enones is easily achieved[5] via hydrosilylation and hydrodesilylation with AgF in aqueous MeOH. Acyclic alkynones similarly furnish (*E*)-alkenones with the second step performed with CuI, Bu_4NF in THF.[6]

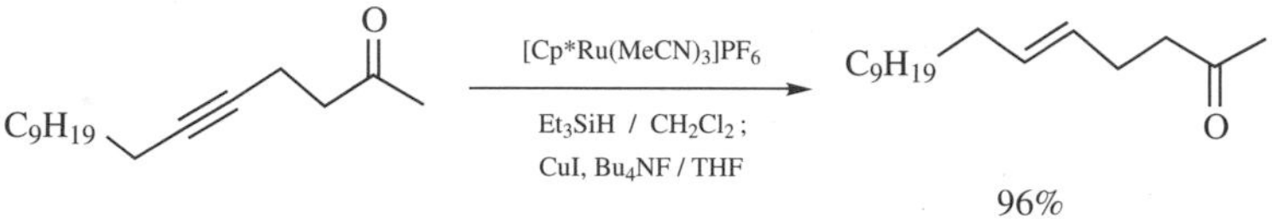

C_9H_{19} … O —($[Cp*Ru(MeCN)_3]PF_6$; Et_3SiH / CH_2Cl_2; CuI, Bu_4NF / THF)→ C_9H_{19} … O

96%

An alkyne-cobalt complex is hydrosilylated using 1,2-bis(trimethylsilyl)ethyne, the latter compound serving also as decomplexation agent.[7]

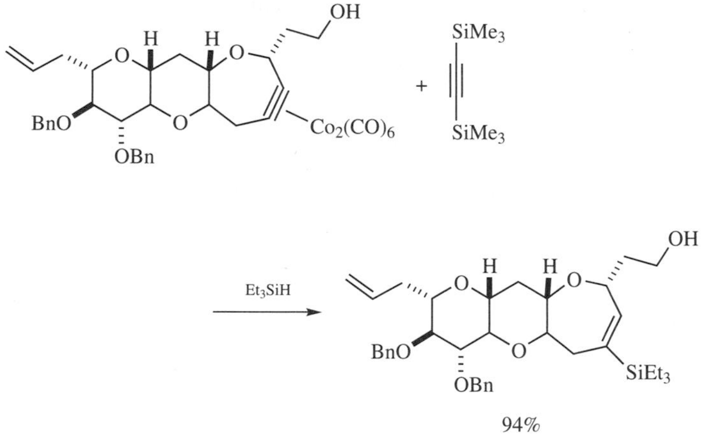

OH, H, H, O, O, BnO, O, OBn, $Co_2(CO)_6$ + $SiMe_3$ ≡ $SiMe_3$

—(Et_3SiH)→ OH, H, H, O, O, BnO, O, OBn, $SiEt_3$

94%

Hydrosilylation and carbonylsilylation with cyclization to afford silylmethylene-cyclopentanes and 2-(silylmethylene)cyclopentanylacetaldehydes from 1,6-enynes are executed with $Rh_4(CO)_{12}$ as catalyst under CO.[8] High pressure CO favors carbonylsilylation. As expected, 1,6-diynes furnish 1,2-bismethylenecyclopentanes in which one of the groups is silylated. A cationic Pt-complex containing bidentate nitrogen ligand and $(C_6F_5)_3B$ constitute the catalytic system in this case.[9]

EtOOC EtOOC → (Me_2PhSiH; $Rh_4(CO)_{12}$ - $(EtO)_3P$, CO) → EtOOC EtOOC, $SiMe_2Ph$, CHO

91%

MeOOC MeOOC + (Ph, N, Pt, N, Ph) → (Et_3SiH; $(C_6F_5)_3B$, PhMe Δ) → MeOOC MeOOC, $SiEt_3$

82% (Z/E 26 1)

Highly selective hydrosilylation of 1,ω-enynes at the triple bond is efficiently accomplished with $Pd_2(dba)_3$ as catalyst.[10]

2-(1-siloxyalkyl)imidazoles are formed in a reaction of imidazoles, aldehydes and hydrosilanes in the presence of $Ir_4(CO)_{12}$.[11]

Detritylation. To disengage an *N*-trityl group from an aziridine the Et_3SiH - CF_3COOH combination is serviceable.[12]

Reduction. Reduction of carbonyl groups with Ph_3SiH in *i*-PrOH and catalyzed by $Mn(dpm)_3$ is accelerated by oxygen [to form $HMnO_2(dpm)_2$].[13] Esters can be reduced to ethers[14] or alcohols[15] and amides to amines[15] by changing the catalyst (Me_3SiOTf–$TiCl_4$ and a triruthenium carbonyl cluster, respectively), and a general method for deoxygenation of alcohols[16] (primary, secondary, and tertiary) involves heating their trifluoroacetates with Ph_2SiH_2 and $(t\text{-}BuO)_2$ at 140°. Secondary and tertiary alcohols undergo deoxygenation on treatment with Ph_2SiClH and catalytic amount of $InCl_3$.[17]

The complete deoxygenation of carboxylic acids and derivatives in one operation is difficult, therefore the discovery that the Et_3SiH – $(C_6F_5)_3B$ combination capable of the achievement is welcome, particularly in view of mild conditions (room temperature) and high yields (5 examples, 91–94%).[18]

Intramolecular, diastereoselective reductive aldol and Michael reactions initiated by $Co(dpm)_2$-catalyzed hydrosilylation with $PhSiH_3$ lead to cycloalkyl carbonyl compounds.[19] The fact that the nature of the hydrosilane is critical for the results is shown by the formation of 1,2-diacylcyclobutane derivatives using $PhSiH_2Me$.[20]

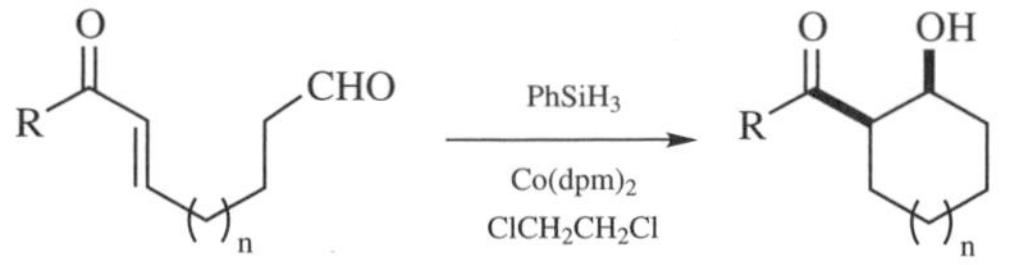

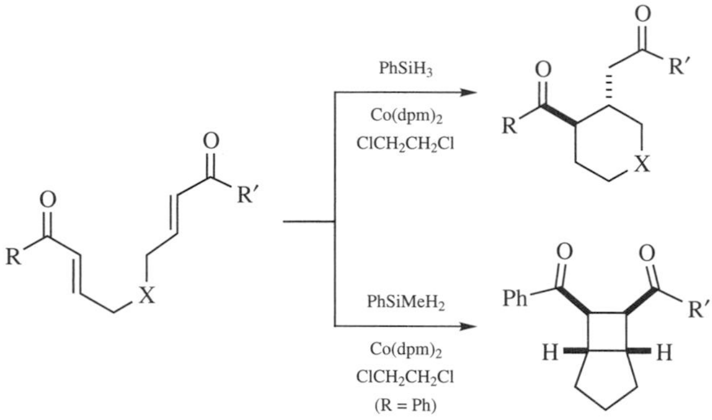

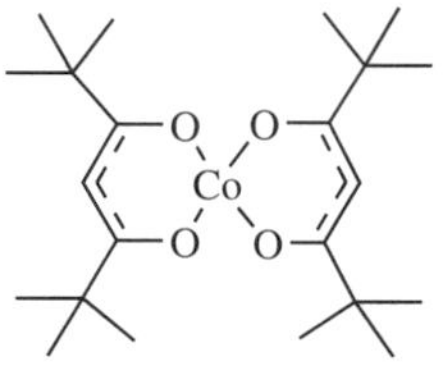

Reductive Mannich reaction[21] and carbamoylation[22] with *N*-tosylimines and aryl isocyanates, respectively, are performed in the presence of a Rh complex and Et_2MeSiH as the hydrogen source.

A method for reductive amination of carbonyl compounds employs $PhSiH_3$ and Bu_2SnCl_2,[23] whereas amides are reduced to amines under hydrosilylation conditions using Os or Ru carbonyl complexes as catalyst.[24]

Aminohydrosilanes such as $R_2NSiHMe_2$ expediently perform reductive amination of carbonyl compounds in the presence of $TiCl_4$.[25]

Ether formation from silyl ethers and aldehydes requires reduction of the latter components. The regioselectivity shown by certain silylated carbohydrates[26] is significant and synthetically valuable. The reaction is catalyzed also by $BiBr_3$.[27]

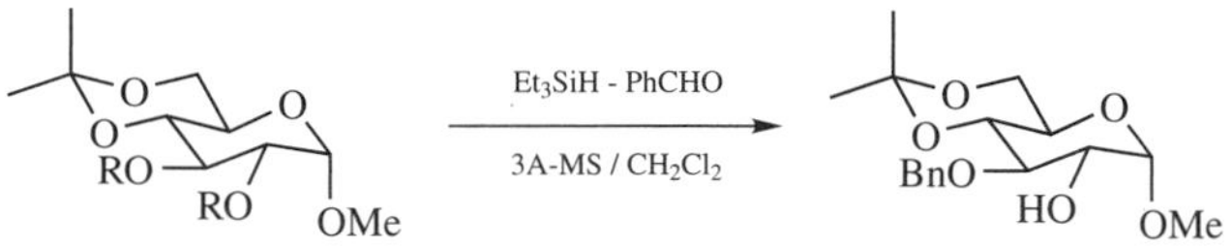

Selective protection and ordered deprotection of hydroxyl groups of carbohydrates are most important in the modifications/synthesis. Benzylidenation involving two hydroxyl groups is well established but a valuable variant is the *o*-nitrobenzylideneacetal unit.[28] Cleavage of 4,6-*O*-(*o*-nitrobenzylideneacetals) to the 6-(*o*-nitrobenzyl) ethers is accomplished by reduction with Et_3SiH and $BF_3{\cdot}OEt_2$. Such ethers are photolabile and the needed deprotection at a proper stage is performed by irradiation with 350nm light.

NO_2 AcO AcO OMe — Et_3SiH - $BF_3 \cdot OEt_2$ / CH_2Cl_2 → NO_2 HO AcO AcO OMe

A convenient access to DPPE derivatives is via reduction of the corresponding bis(phosphine oxides) which are obtained from the Pd-catalyzed twofold addition of Ph_2POH to alkynes.[29]

R + Ph_2POH — $(Ph_3P)_4Pd$ → R, $P(O)Ph_2$, $P(O)Ph_2$ — $(EtO)_3SiH$ / $(t\text{-PrO})_4Ti$ → R, PPh_2, PPh_2

Coupling reactions. Both alkyl and aryl halides can be converted to silanes.[30,31]

Ph Cl + H–$SiCl_3$ — Bu_4PCl / 140° → Ph $SiCl_3$ 95%

OMe Br — $(EtO)_3SiH$ / Mg, I_2; 5% H_2SO_4 → OMe SiH_3 28%

Ring expansion. 1-Haloalkyl-2-oxocycloalkanecarboxylic esters undergo ring expansion to give homologous 3-oxocycloalkanecarboxylic esters. This transformation involes free radical intermediates.[32]

MeOOC, Br, O + HSiPh$_2$ / Ph$_2$SiH → AIBN, PhMe Δ → O, COOMe 67%

Enolsilylation. Enolsilylation of ketones is accomplished by a Pt-catalyzed reaction with a hydrosilane.[33] The hydrosilane suffers oxidation in the process.

Silyl ω-haloalkanoates. Lactones undergo opening by a Pd-catalyzed halosilylation. Reagents include Et_3SiH and RX.[34]

[1]Perales, J.B., van Vranken, D.L. *JOC* **66**, 7270 (2001).
[2]Itazaki, M., Nishihara, Y., Osakada, K. *JOC* **67**, 6889 (2002).
[3]Trost, B.M., Ball, Z.T. *JACS* **123**, 12726 (2001).
[4]Molander, G.A., Romero, J.A.C., Corrette, C.P. *JOMC* **647**, 225 (2002).
[5]Trost, B.M., Radkowski, K. *CC* 2182 (2002).
[6]Trost, B.M., Ball, Z.T., Jöge, T. *JACS* **124**, 7922 (2002).
[7]Kira, K., Tanda, H., Hamajima, A., Baba, T., Takai, S., Isobe, M. *T* **58**, 6485 (2002).
[8]Ojima, I., Vu, A.T., Lee, S.-Y., McCullagh, J.V., Moralee, A.C., Fujiwara, M., Hoang, T.H. *JACS* **124**, 9164 (2002).
[9]Wang, X., Chakrapani, H., Madine, J.W., Keyerleber, M.A., Widenhoefer, R.A. *JOC* **67**, 2778 (2002).
[10]Motoda, D., Shinokubo, H., Oshima, K. *SL* 1529 (2002).
[11]Fukumoto, Y., Sawada, K., Hagihara, M., Chatani, N., Murai, S. *ACIEE* **41**, 2779 (2002).
[12]Vedejs, E., Klapars, A., Warner, D.L., Weiss, A.H. *JOC* **66**, 7542 (2001).
[13]Magnus, P., Fielding, M.R. *TL* **42**, 6633 (2001).
[14]Yato, M., Homma, K., Ishida, A. *T* **57**, 5353 (2001).
[15]Matsubara, K., Iura, T., Maki, T., Nagashima, H. *JOC* **67**, 4985 (2002).
[16]Jang, D.O., Kim, J., Cho, D.H., Chung, C.-M. *TL* **42**, 1073 (2001).
[17]Yasuda, M., Onishi, Y., Ueba, M., Miyai, T., Baba, A. *JOC* **66**, 7741 (2001).
[18]Gevorgyan, V., Rubin, M., Liu, J.-X., Yamamoto, Y. *JOC* **66**, 1672 (2001).
[19]Baik, T.-G., Luis, A.L., Wang, L.-C., Krische, M.J. *JACS* **123**, 5112 (2001).
[20]Baik, T.-G., Luis, A.L., Wang, L.-C., Krische, M.J. *JACS* **123**, 6716 (2001).
[21]Muraoka, T., Kamiya, S., Matsuda, I., Itoh, K. *CC* 1284 (2002).
[22]Muraoka, T., Matsuda, I., Itoh, K. *OM* **20,** 4676 (2001).
[23]Apodaca, R., Xiao, W. *OL* **3**, 1745 (2001).
[24]Igarashi, M., Fuchikami, T. *TL* **42**, 1945 (2001).
[25]Miura, K., Ootsuka, K., Suda, S., Nishikori, H., Hosomi, A. *SL* 1617 (2001).
[26]Wang, C.-C., Lee, J.-C., Luo, S.-Y., Fan, H.-F., Pai, C.-L., Yang, W.-C., Lu, L.-D., Hung, S.-C. *ACIEE* **41**, 2360 (2002).
[27]Bajwa, J.S., Jiang, X., Slade, J., Prasad, K., Repic, O., Blacklock, T.J. *TL* **43**, 6709 (2002).
[28]Watanabe, S., Sueyoshi, T., Ichihara, M., Uehara, C., Iwamura, M. *OL* **3,** 255 (2001).
[29]Allen Jr, A., Ma, L., Lin, W. *TL* **43**, 3707 (2002).

[30]Cho, Y.S., Kang, S.-H., Han, J.S., Yoo, B.R., Jung, I.N. *JACS* **123**, 5584 (2001).
[31]Minge, O., Mitzel, N.W., Schmidbauer, H. *OM* **21**, 680 (2002).
[32]Sugi, M., Togo, H. *T* **58**, 3171 (2002).
[33]Ozawa, F., Yamamoto, S., Kawagishi, S., Hiraoka, M., Ikeda, S., Minami, T., Ito, S., Yoshifuji, M. *CL* 972 (2001).
[34]Iwata, A., Ohshita, J., Tang, H., Kunai, A., Yamamoto, Y., Matsui, C. *JOC* **67**, 3927 (2002).

Hydroxylamine-*O*-sulfonic acid.

Diaziridines.[1] Adducts of the reagent with imines undergo ring closure on base treatment.

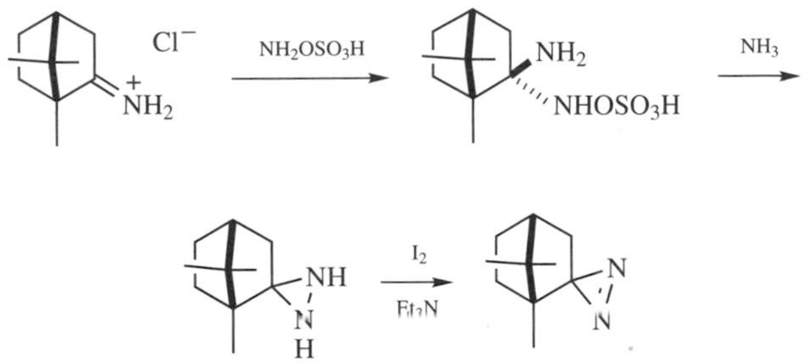

[1]Krois, D., Brinker, U.H. *S* 379 (2001).

Hydroxy-(*p*-nitrobenzenesulfonyloxy)iodobenzene.

α-Iodo ketones. Ketones are oxidized at an α-position and subsequent treatment of the products with KI (18-c-6) gives α-iodo ketones.[1] (Alkyl aryl ketones on similar treatment with hydroxy-(2,4-dinitrobenzenesulfonyloxy)iodobenzene and then Bu_4NIO_4 afford aroic acids.[2])

α-Diketones. Ketones containing an α-methylene group are subject to oxidation by microwave irradiation with the title reagent for short durations and again after addition of pyridine-*N*-oxide.[3]

[1]Lee, J.C., Jin, Y.S. *SC* **29**, 2769 (1999).
[2]Lee, J.C., Choi, J.-H., Lee, Y.C. *SL* 1563 (2001).
[3]Lee, J.C., Park, H.-J., Park, J.Y. *TL* **43**, 5661 (2002).

N-Hydroxyphthalimide, NHPI.

Ketones.[1] Radical-chain addition of aldehydes to alkenes forms ketones. NHPI together with BzOOBz form the catalytic system.

Amination and nitration of hydrocarbons. Alkylarenes and adamantane are functionalized at the benzylic position and bridgehead, respectively, by NHPI, CAN and nitriles to give *N*-benzylic amides and 1-acylaminoadamantanes.[2] A Ritter-type reaction occurs. Alkanes are nitrated with nitric acid in the presence of NHPI, and the method is well suited for preparation of nitrated cycloalkanes and adamantane.[3]

CAN - NHPI
EtCN 100°
HN
O
93%

Carbamoylation.[4] By catalysis of NHPI a [$CONH_2$] group generated in situ from formamide and CAN is readily introduced into the nucleus of a heteroaromatic base.

Oxyalkylation. Capture of carbon radicals generated from alkanes that are exposed to the NHPI-O_2 system by activated alkenes (methyl acrylate, acrylonitrile, acrylamide) leads to bifunctional compounds.[5] The work is an extension of the trapping method involving homologation of 1,3-dioxolan-2-yl radicals to yield β-hydroxy acetals[6] and conversion of secondary alcohols into α-hydroxy-γ-lactones.[7]

\+ COOMe
O_2 - NHPI
$Co(acac)_2$
PhCN 75°
OH
COOMe
78%

O O
R H
\+ CN
O_2 - NHPI
$Co(acac)_2$
25°
O O OH
R CN
R = H 60%
R = Me 77%

Adipic acid.[8] Adding small amounts of $Mn(OAc)_2$ and $Co(acac)_2$ to a modified *N*-hydroxyphthalimide enables aerobic cleavage of cyclohexane to afford adipic acid in high conversion and selectivity. Modification renders NHPI lipophilic.

[1]Tsujimoto, S., Iwahama, T., Sakaguchi, S., Ishii, Y. *CC* 2352 (2001).
[2]Sakaguchi, S., Hirabayashi, T., Ishii, Y. *CC* 516 (2002).
[3]Nishiwaki, Y., Sakaguchi, S., Ishii, Y. *JOC* **67**, 5663 (2002).
[4]Minisci, F., Recupero, F., Punta, C., Gambarotti, C., Antonietti, F., Fontana, F., Pedulli, G.F. *CC* 2496 (2002).
[5]Hara, T., Iwahama, T., Sakaguchi, S., Ishii, Y. *JOC* **66**, 6425 (2002).
[6]Hirao, K., Iwahama, T., Sakaguchi, S., Ishii, Y. *CC* 2457 (2000).
[7]Iwahama, T., Sakaguchi, S., Ishii, Y. *CC* 613 (2000).
[8]Sawatari, N., Yokota, T., Sakaguchi, S., Ishii, Y. *JOC* **66**, 7889 (2001).

Hydroxy(tosyloxy)iodobenzene. 14, 179–180; **16,** 179; **17,** 150; **18,** 187; **19,** 170; **20,** 193; **21,** 223

α-Tosyloxylation of anhydrides.[1] Heating carboxylic anhydrides with the title reagent accomplishes the transformation. The products can be esterified in conventional manner.

α-Functionalized ketones.[2] Subsequent to formation of α-tosyloxy ketones from ketones treatment with $PhSO_2Na$ gives β-keto sulfones.[2] In situ displacement of the tosyloxy group (to furnish α-hydroxy ketones) occurs when the tosyloxylation is carried out in aqueous DMSO.[3]

Analogous reagents. Polymer-bound[4] and *m*-trifluoromethylated[5] analogues of the title reagent have been prepared and their reactivities investigated.

[1]Goff, J.M., Justik, M.W., Koser, G.F. *TL* **42**, 5597 (2001).
[2]Xie, Y.-Y., Chen, Z.-C. *SC* **31**, 3145 (2001).
[3]Xie, Y.-Y., Chen, Z.-C. *SC* **32**, 1875 (2002).
[4]Abe, S., Sakurantani, K., Togo, H. *JOC* **66**, 6174 (2001).
[5]Nabana, T., Togo, H. *JOC* **67**, 4362 (2002).

Hypophosphorous acid. 21, 223

Radical reactions.[1] Combined with $NaHCO_3$ in aqueous media, H_3PO_2 mediates replacement of carbon-bound iodine atom with hydrogen.

[1]Yorimitsu, H., Shinokubo, H., Oshima, K. *BCSJ* **74**, 225 (2001).

Hypophosphorous acid–iodine. 21, 223

Reductions. Benzhydrols are deoxygenated[1] while stilbenes are reduced to 1,2-diarylethanes.[2]

[1]Gordon, P.E., Fry, A.J. *TL* **42**, 831 (2001).
[2]Fry, A.J., Allukian, M., Williams, A.D. *T* **58**, 4411 (2002).

I

Imidazole. 20, 194; **21,** 224

Protection of aldehydes.[1] Aldehydes form adducts with imidazole that are stabilized by silylation. Regeneration of aldehydes from the silyl ethers is by treating them with MeCN-49% HF (9:1) at room temperature.

2-Hydroxymethyl-2-cycloalkenones.[2] Imidazole catalyzes the Baylis-Hillman reaction of 2-cycloalkenones with various aldehydes.

[1]Quan, L.G., Cha, J.K. *SL* 1925 (2001).
[2]Gatri, R., El Gaied, M.M. *TL* **43**, 7835 (2002).

Indium. 14, 81; **16,** 181–182; **18,** 189; **19,** 171–173; **20,** 194–197; **21,** 224–227

Eliminations. Trichloroethyl carbonates[1] and carbamates[2] are cleaved by In-NH_4Cl in aqueous media. The ability of indium to initiate elimination of β-functionalized halogen compounds has been exploited in converting trihalomethyl carbinols to 1,1-dihaloalkenes via their esters (acetates, sulfonates).[3]

Reductions. In aqueous EtOH the double bond of terminal alkenes[4] and that substituted by electron-withdrawing groups[5,6] suffer reduction by indium. Various compounds containing C=N bond are reduced: imines[6,7] and oximes[5] to amines while the heterocyclic portion of the quinoline nucleus is saturated.[6] Nitro compounds are susceptible also.[7]

Glycosylation.[8] Glycosyl bromides undergo displacement with alcohols in the presence of indium.

OAc
AcO
AcO
O
Br
In / THF
ROH
OAc
AcO
AcO
O
OR

Allylations. Many new types of substrates have been studied for indium-mediated allylation: α-chloro carbonyl compounds to provide precursors of epoxides,[9] acyl cyanides to generate allyl ketones,[10] α-keto acids,[11] α-keto aldehydes,[12] quinones,[13] pyrrole-3-carbaldehyde and indole-3-carbaldehyde.[14]

Fiesers' Reagents for Organic Synthesis, Volume 22. Series editor Tse-Lok Ho
ISBN 0-471-28515-3

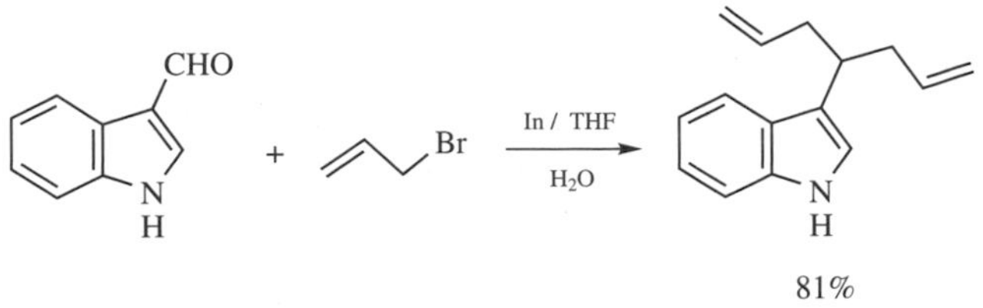

81%

The allylation mode of 2-cycloalkenones varies according to structural features.[15] Thus, in the presence of indium and Me_3SiCl the 6- and 7-membered ring substrates undergo conjugate addition unless steric factors intervene then 1,2- addition (e.g., for 4,4-dimethyl-2-cyclohexenone) predominates, which is also the course for 2-cyclopentenones and acyclic enones.

Imines behave similarly to carbonyl compounds toward allylindium reagents.[16,17] Nitroarenes form *N*-allyl bond(s).[18]

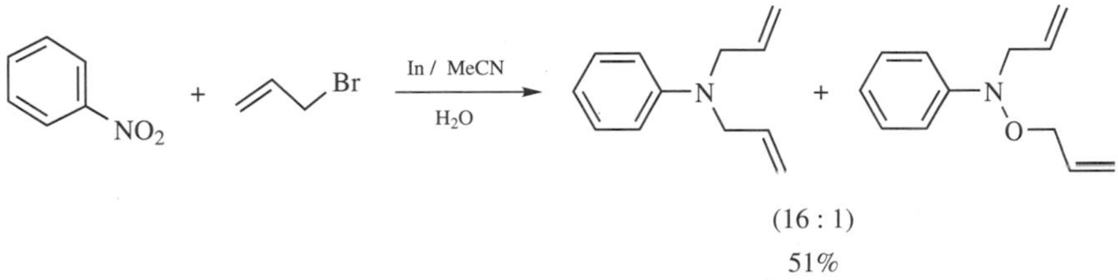

(16 : 1)
51%

A method for allylarene synthesis is via Pd-catalyzed coupling with allylindiums.[19]

Propargylations and alkynylations. The indium-mediated reaction of propargyl bromide with acyl cyanides leads to allenyl ketones, but homologous reagents (derived from bromides that possess a carbon substituent at the other *sp*-terminus) afford 3-alkynones.[20]

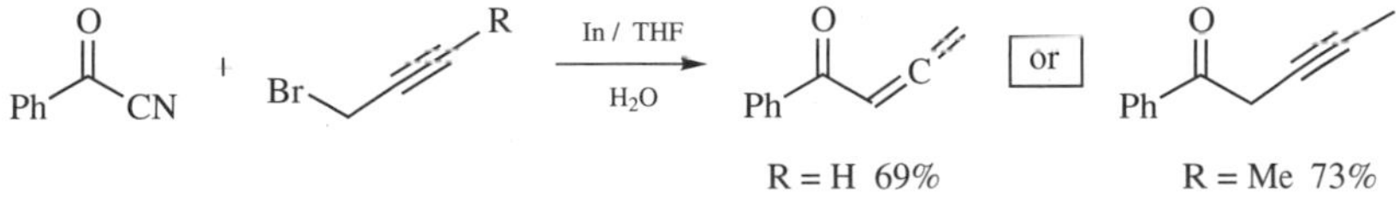

R = H 69% R = Me 73%

Alkynylindium species, generated from alkynyl iodides and indium in dichloromethane, react with carbonyl compounds to furnish propargyl alcohols.[21]

Coupling reactions. The coupling of aryl iodides with organoindium reagents derived from allyl[22] and propargyl halides (to afford allenes)[23] is Pd-catalyzed. Diaryliodonium salts afford Ar_2CO when they are treated with indium and catalytic $Pd(OAc)_2$ in DMF under CO.[24]

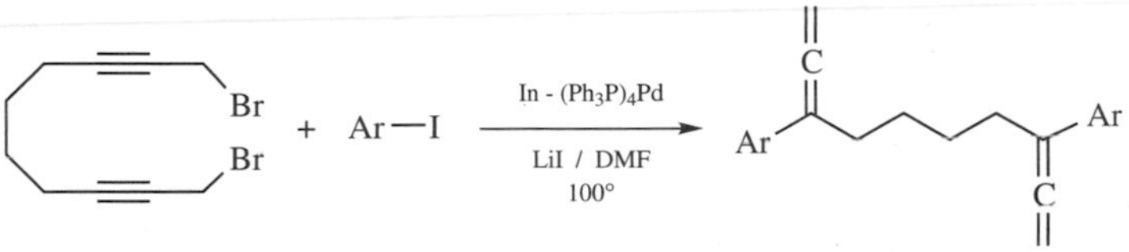

Cyclizations. Intramolecular radical addition to an alkyne moiety by an atom-transfer mechanism is realized when a proper substrate is exposed to indium and iodine.[25] The In/I ratio determines whether the exocyclic sp^2-carbon carries an iodine atom. In the absence of iodine an allylindium adds to a remote triple bond to form a cyclic structure that contains a vinyl group adjacent to the exocyclic methylene.[26]

In - I_2, MeOH

X = H
X = I (excess I_2)

In / THF, H_2O

62%

[1]Valluri, M., Mineno, T., Hindupur, R.M., Avery, M.A. *TL* **42**, 7153 (2001).
[2]Mineno, T., Choi, S.-R., Avery, M.A. *SL* 883 (2002).
[3]Ranu, B.C., Samanta, S., Das, A. *TL* **42**, 5993 (2001).
[4]Ranu, B.C., Dutta, J., Guchhait, S.K. *JOC* **66**, 5624 (2001).
[5]Ranu, B.C., Dutta, J., Guchhait, S.K. *OL* **3**, 2603 (2001).
[6]Pitts, M.R., Harrison, J.R., Moody, C.J. *JCS(P1)* 955 (2001).
[7]Banik, B.K., Hackfeld, L., Becker, F.F. *SC* **31**, 1581 (2001).
[8]Banik, B.K., Samajdar, S., Banik, I., Zegrocka, O., Becker, F.F. *H* **55**, 227 (2001).
[9]Shin, J.A., Choi, K.I., Pae, A.N., Koh, H.Y., Kang, H.-Y., Cho, Y.S. *JCS(P1)* 946 (2001).
[10]Yoo, B.W., Choi, K.-H., Lee, S.-J., Nam, G.S., Chang, K.Y., Kim, S.H., Kim, J.-H. *SC* **32**, 839 (2002).
[11]Kumar, S., Kaur, P., Chimni, S.S. *SL* 573 (2002).
[12]Kang, S.-K., Baik, T.-G., Jiao, X.-H. *SC* **32**, 75 (2002).
[13]Pan, D., Mal, S.K., Kar, G.K., Ray, J.K. *T* **58**, 2847 (2002).
[14]Kumar, S., Kumar, V., Chimni, S.S. *TL* **43**, 8029 (2002).
[15]Lee, P.H., Ahn, H., Lee, K., Sung, S.-Y., Kim, S. *TL* **42**, 37 (2001).
[16]Vilaivan, T., Winotapan, C., Shinada, T., Ohfune, Y. *TL* **42**, 9073 (2001).
[17]Lu, W., Chan, T.H. *JOC* **66**, 3467 (2001).
[18]Kang, K.H., Choi, K.I., Koh, H.Y., Kim, Y., Chung, B.Y., Cho, Y.S. *SC* **31**, 2277 (2001).
[19]Lu, W., Chan, T.H. *JOC* **67**, 8265 (2002).
[20]Yoo, B.W., Lee, S.-J., Choi, K.-H., Keum, S.-R., Ko, J.J., Choi, K.-I., Kim, J.-H. *TL* **42**, 7287 (2001).
[21]Auge, J., Lubin-Germain, N., Seghrouchni, L. *TL* **43**, 5255 (2002).
[22]Lee, P.H., Sung, S., Lee, K. *OL* **3**, 3201 (2001).
[23]Lee, K., Seomoon, D., Lee, P.H. *ACIEE* **41**, 3901 (2002).
[24]Zhou, T., Chen, Z.-C. *SC* **32**, 3431 (2002).
[25]Yanada, R., Nishimori, N., Matsumura, A., Fujii, N., Takemoto, Y. *TL* **43**, 4585 (2002).
[26]Salter, M.M., Sardo-Inffiri, S. *SL* 2068 (2002).

Indium–Indium(III) chloride. 21, 227

N-Arylhydroxylamine O,N-diacetates.[1] Nitroarenes undergo reductive acylation at room temperature by indium powder and $InCl_3$.

In - $InCl_3$; Ac_2O; $CHCl_3$ - MeOH

81%

Propargylation.[2] Regioselective and stereoselective reaction of α-chloropropargyl phenyl sulfide with carbonyl compounds is observed. The products are readily converted into epoxyalkynes. Interestingly, the indium-mediated process shows opposite stereoselectivity in the presence of $InCl_3$. (A previous comparison was made of In and $InCl_3$, essentially the same results were presented as in this report).

In; DMF - H_2O

73% (*syn/anti* 26 : 74)

+ $InCl_3$ 75% (*syn/anti* 90 : 10)

[1]Kim, B.H., Cheong, J.W., Han, R., Jun, Y.M., Baik, W., Lee, B.M. *SC* **31**, 3577 (2001).
[2]Mitzel, T.M., Palomo, C., Jendza, K. *JOC* **67**, 136 (2002).

Indium(I) bromide. 21, 228

Dichlorocyanomethylation.[1] Insertion of InBr into a C—Cl bond of trichloroacetonitrile forms an organoindium species that reacts with carbonyl compounds to afford 2,2-dichloro-3-hydroxyalkanenitriles (13 examples, 60–93%).

[1]Nobrega, J.A., Goncalves, S.M.C., Peppe, C. *TL* **42**, 4745 (2001).

Indium(III) bromide.

β-Halohydrins.[1] Indium halides induce halogenolysis of α,β-epoxyalkanoic acids in water to give α-hydroxy-β-haloalkanoic acids. Thus, the combination of NaBr and $InBr_3$ is used for preparation of the bromohydrins.

β-Alkylation of indoles. In the presence of $InBr_3$ indoles react with styrene oxides to afford 2-aryl-2-(β-indolyl)ethanols. Enantiopure products are obtained from reaction using chiral epoxides.[2] Indoles react with nitroalkenes[3] and enones[4] in the Michael reaction mode. When the latter reaction is carried out in dichloromethane instead of water, and terminated by addition of Me_3SiCl, *O*-trimethylsilylated cyanohydrins are isolated.

CN
$OSiMe_3$
$InBr_3$ / CH_2Cl_2 ;
Me_3SiCN
N
H
N
H
52%

[1]Amantini, D., Fringuelli, F., Pizzo, F., Vaccaro, L. *JOC* **66**, 4463 (2001).
[2]Bandini, M., Cozzi, P.G., Melchiorre, P., Umani-Ronchi, A. *JOC* **67**, 5386 (2002).
[3]Bandini, M., Melchiorre, P., Melloni, A., Umani-Ronchi, A. *S* 1110 (2002).
[4]Bandini, M., Cozzi, P.G., Giacomini, M., Melchiorre, P., Selva, S., Umani-Ronchi, A. *JOC* **67**, 3700 (2002).

Indium(III) chloride. 19, 173–174; **20,** 197–198; **21,** 228–230

Friedel-Crafts reactions. With $InCl_3$ impregnated in mesoporous Si-MCM-41 a moisture insensitive catalyst for Friedel-Crafts acylation is obtained.[1] Acylation of activated arenes uses catalytic $InCl_3$ and $LiClO_4$ in $MeNO_2$ at room temperature.[2]

The von Pechmann reaction that gives coumarins is related to the Friedel-Crafts reactions. A simple procedure[3] involves heating the substrate components with $InCl_3$.

HO OH
HO
EtO O
$InCl_3$
HO O O
98%

Allylations. Allyltrimethylsilane provides the allyl group source for reaction with carbonyl compounds which is carried out in the presence of $InCl_3$ and Me_3SiCl.[4] Conjugate addition to enones is observed with the same reagent system.[5] Using allyltributylstannane instead of allyltrimethylsilane the reaction with acid chlorides in the presence of Ph_3P leads to allyl ketones.[6]

An electrochemical regeneration system supplies low-valent indium species to catalyze allylation of carbonyl compounds (including esters)[7] and imines[8] has been developed. The system consists of Al anode, Pt cathode and $InCl_3$ as the active agent.

Mannich and Strecker reactions. Catalysis by $InCl_3$ a three-component reaction of amines, aldehydes, and silyl ketene acetals provides β-amino esters. The process is enantioselective if a chiral α-amino ester is used as the amine component.[9]

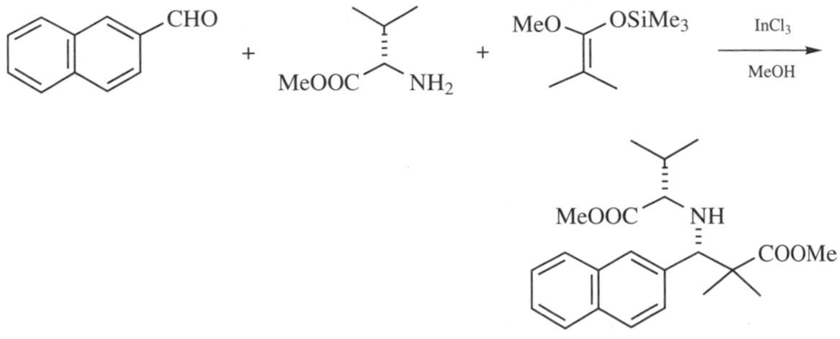

79% (*R/S* 93 : 7)

A Mannich reaction analogue is the formation of the furanobenzopyran system from salicylaldehyde, amines, and 2,3-dihydrofuran.[10] In the absence of salicylaldehyde, two equivalents of the cyclic enol ether are consumed to afford tetrahydroisoquinolines.[11]

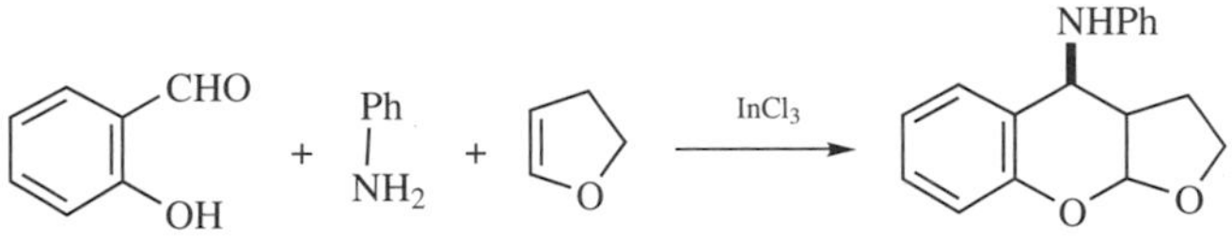

87%

α-Amino nitriles are available in one step by a catalyzed condensation of aldehydes, amines and KCN.[12]

O,S-Exchange reactions. Like so many other Lewis acids $InCl_3$ is capable of catalyzing dithioacetalization of carbonyl compounds.[13] Acetals are similarly converted,[14] but more interestingly, sulfides are formed from ethers (e.g., tetrahydrofuran gives 1,4-bisphenylthiobutane) by a reaction promoted by silica impregnated with $InCl_3$.[15]

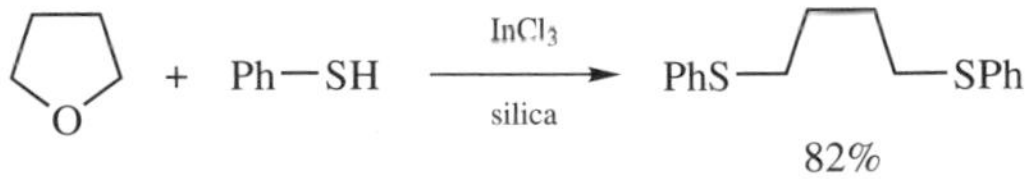

82%

Epoxide opening. Azidolysis of 2,3-epoxyalkanoic acids is easily accomplished in water with $InCl_3$ as catalyst at pH 4.[16] Epoxides are transformed, via rearrangement, into 4-chlorotetrahydropyrans in a reaction with homoallylic alcohols.[17]

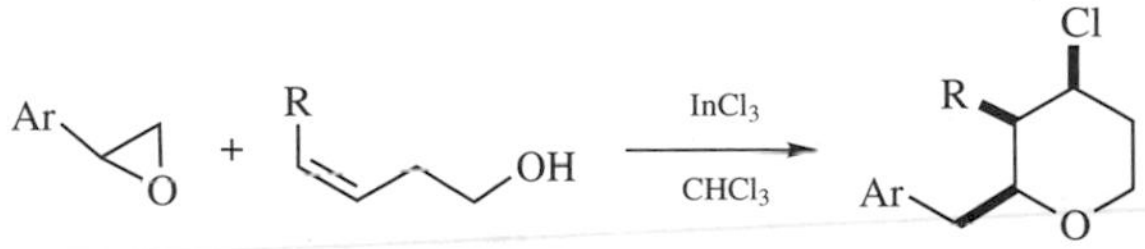

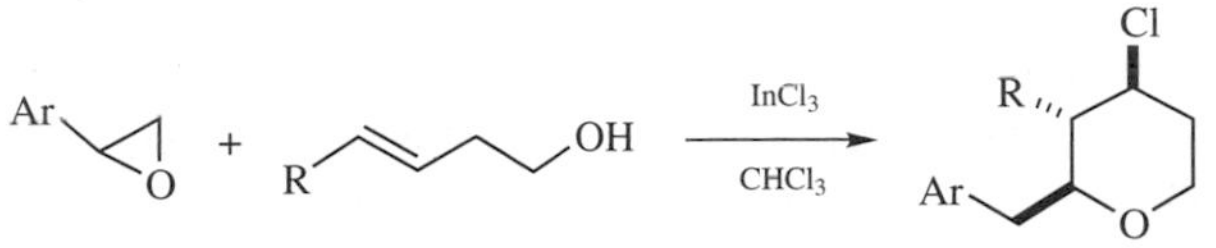

Glycosylation.[18] Alcoholysis of glycosyl bromides activated by $InCl_3$ proceeds under neutral conditions.

Dihydropyrans.[19] (*Z*)-Silylalkenes react with aldehydes in a Prins reaction-type process.

R OH SiMe3 + R′—CHO InCl3 / CH2Cl2 → R O R′

Radical addition.[20] Chiral amines are synthesized from cyclic *N*-acylhydrazones via a $Mn_2(CO)_{10}$-catalyzed addition of organoindium species.

O O N N R Ph + R′—I InCl3 - Mn2(CO)10 / hv, CH2Cl2 → O O N HN R R′ Ph

ε-Caprolactone.[21] After the 1:2 mixture of cyclohexanone and cyclohexanol reacts under oxygen in the presence of *N*-hydroxyphthalimide and AIBN in hot MeCN, treatment with $InCl_3$ at room temperature delivers the lactone in fair yield.

[1]Choudhary, V.R., Jana, S.K., Patil, N.S. *TL* **43**, 1105 (2002).
[2]Chapman, C.J., Frost, C.G., Hartley, J.P., Whittle, A.J. *TL* **42**, 773 (2001).
[3]Bose, D.S., Rudradas, A.P., Babu, M.H. *TL* **43**, 9195 (2002).
[4]Onishi, Y., Ito, T., Yasuda, M., Baba, A. *EJOC* 1578 (2002).
[5]Lee, P.H., Lee, K., Sung, S., Chang, S. *JOC* **66**, 8646 (2001).
[6]Inoue, K., Shimizu, Y., Shibata, I., Baba, A. *SL* 1659 (2001).
[7]Hilt, G., Smolko, K.I. *ACIEE* **40**, 3399 (2001).
[8]Hilt, G., Smolko, K.I, Waloch, C. *TL* **43**, 1437 (2002).
[9]Loh, T.-P., Chen, S.-L. *OL* **4**, 3647 (2002).
[10]Anniyappan, M., Muralidharan, D., Perumal, P.T. *T* **58**, 10301 (2002).
[11]Zhang, J., Li, C.-J. *JOC* **67**, 3969 (2002).
[12]Ranu, B.C., Dey, S.S., Hajra, A. *T* **58**, 2529 (2002).
[13]Muthusamy, S., Babu, S.A., Gunanathan, C. *TL* **42**, 359 (2001).
[14]Ranu, B.C., Das, A., Samanta, S. *SL* 727 (2002).
[15]Ranu, B.C., Samanta, S., Hajra, A. *SL* 987 (2002).
[16]Fringuelli, F., Pizzo, F., Vaccaro, L. *JOC* **66**, 3554 (2001).
[17]Li, J., Li, C.-J. *TL* **42**, 793 (2001).
[18]Mukherjee, D., Ray, P.K., Chowdhury, U.S. *T* **57**, 7701 (2001).
[19]Dobbs, A.P., Martinovic, S. *TL* **43**, 7055 (2002).
[20] Friestad, G.K., Qin, J. *JACS* **6123**, 9922 (2001).
[21]Fukuda, O., Iwahama, T., Sakaguchi, S., Ishii, Y. *TL* **42**, 3749 (2001).

Indium(III) hydroxide.

Deoxygenative chlorination.[1] Various carbonyl compounds (aldehydes and ketones, including β-keto esters, alkynones) are transformed into organic halides with

Me_2SiClH and catalyzed by $In(OH)_3$. To prepare the corresponding iodide one needs only adding NaI to the reaction medium.

[1]Onishi, Y., Ogawa, D., Yasuda, M., Baba, A. *JACS* **124**, 13690 (2002).

Indium(I) iodide. 21, 231

Opening of epoxides and aziridines. Together with $(Ph_3P)_4Pd$, indium(I) iodide converts vinylepoxides and vinylaziridines to allylindium species. Subsequent reaction with aldehydes furnishes 2-vinyl-1,3-diols[1] and 2-vinyl-1,3-amino alcohols,[2] respectively.

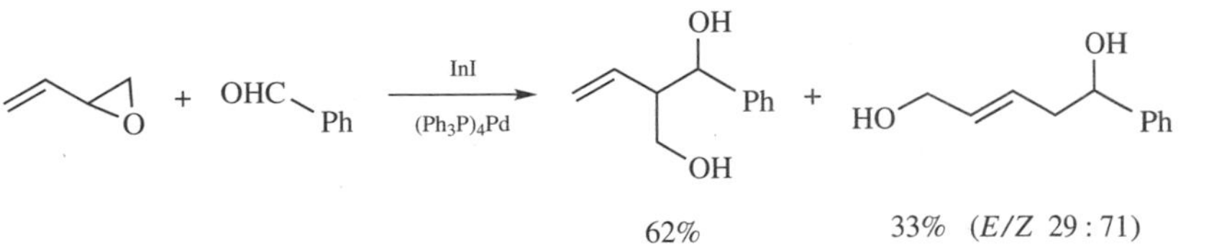

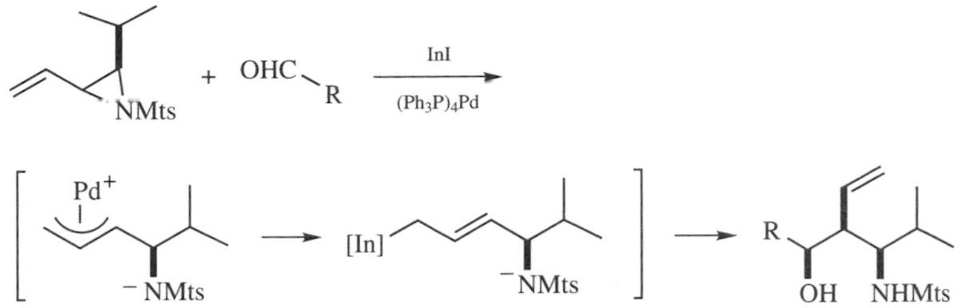

InI, $(Ph_3P)_4Pd$; 62%; 33% (*E/Z* 29 : 71)

InI, $(Ph_3P)_4Pd$; NMts; OHC–R; Pd+; [In]; OH NHMts

[1]Araki, S., Kameda, K., Tanaka, J., Hirashita, T., Yamamura, H., Kawai, M. *JOC* **66**, 7919 (2001).
[2]Takemoto, Y., Anzai, M., Yanada, R., Fujii, N., Ohno, H., Ibuka, T. *TL* **42**, 1725 (2001).

Indium(III) triflamide.

Aromatic substitutions.[1] Electrophilic nitration and acetylation can be conducted in the presence of $In(NTf_2)_3$ or $In(OTf)_3$. But in benzoylation and benzenesulfonylation $In(NTf_2)_3$ is almost inept.

[1]Frost, C.G., Hartley, J.P., Griffin, D. *TL* **43**, 4789 (2002).

Indium(III) triflate. 21, 231

Friedel-Crafts acylation and sulfonylation. 3-Acylindoles are prepared.[1] Readily achieved also are sulfonylation of arenes with RSO_2Cl and Me_2NSO_2Cl to afford sulfones[2] and sulfonamides.[3]

Isomerization. Kinetically formed, branched homoallylic alcohols are converted into the more stable linear regioisomers[4] on contact with $In(OTf)_3$. Tetrahydrofuran formation from homoallylic alcohols and carbonyl compounds also involves rearrangement.[5]

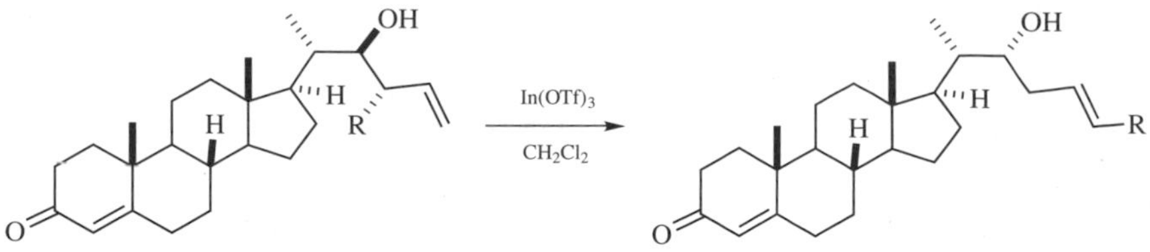

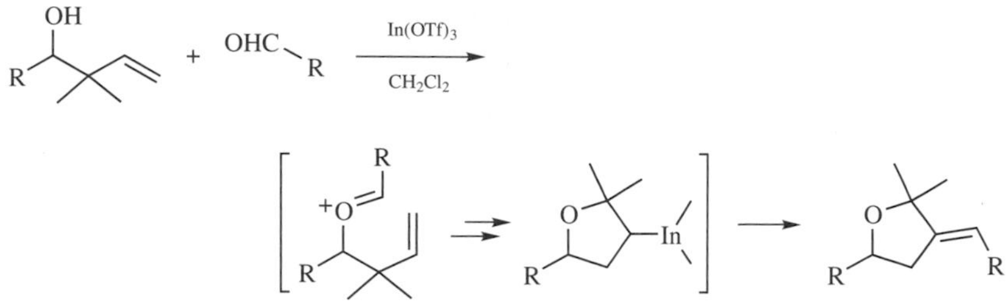

In(OTf)$_3$
CH_2Cl_2

In(OTf)$_3$
CH_2Cl_2

Functional group protection. Interconversion of alcohols and tetrahydropyranyl ethers is achieved in In(OTf)$_3$-catalyzed reactions (in anhydrous and aqueous media).[6] Derivatization of carbonyl compounds into 1,3-oxothiolanes is also conveniently carried out.[7]

[1]Nagarajan, R., Perumal, P.T. *T* **58**, 1229 (2002).
[2]Frost, C.G., Hartley, J.P., Whittle, A.J. *SL* 830 (2001).
[3]Frost, C.G., Hartley, J.P., Griffin, D. *SL* 1928 (2002).
[4]Loh, T.-P., Tan, K.-T., Hu, Q.-Y. *ACIEE* **40**, 2921 (2001).
[5]Loh, T.-P., Hu, Q.-Y., Ma, L.-T. *JACS* **123**, 2450 (2001).
[6]Mineno, T. *TL* **43**, 7975 (2002).
[7]Kazahaya, K., Hamada, N., Ito, S., Sato, T. *SL* 1535 (2002).

Iodic acid.

Carbonyl regeneration. Regeneration of carbonyl compounds from oximes with iodic acid is facile.[1] Another procedure calls for replacing a solvent with wet silica gel.[2]

For oxidative hydrolysis of dithioacetals the use of IBX in DMSO containing traces of water is effective, and the reagent system promotes cleavage of derivatives of aromatic ketones more readily.[3]

Dehydrogenation.[4] Introduction of a conjugated double bond to a carbonyl group is accomplishable with iodic acid in DMSO at 50°. The reagent is less expensive than IBX.

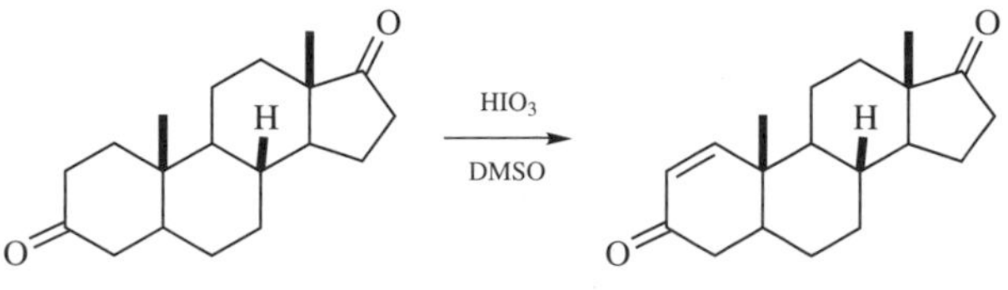

HIO_3
DMSO

90%

[1]Chandrasekhar, S., Gopalaiah, K. *TL* **43**, 4023 (2002).
[2]Shirini, F., Zolfigol, M.A., Azadbar, M.R. *SC* **32**, 315 (2002).
[3]Wu, Y., Shen, X., Huang, J.-H., Tang, C.-J., Liu, H.-H., Hu, Q. *TL* **43**, 6443 (2002).
[4]Nicolaou, K.C., Montagnon, T., Baran, P.S. *ACIEE* **41**, 1386 (2002).

Iodine. 13, 148–149; **14,** 181–182; **15,** 172–173; **16,** 182; **18,** 189–191; **19,** 174–175; **20,** 199–200; **21,** 231–232

Protection and deprotection of functional groups. Catalytic iodine is employed for acetalization,[1,2] dithioacetalization,[3–5] and esterification of carboxylic acids (also tranesterification of esters).[6]

Selective cleavage of prenyl ether with iodine is also reported.[7]

Cyclization. Five-membered and six-membered heterocycles are formed from functionalized alkynes, both *endo-dig* and *exo-dig* modes are observed. 4-Iodo-isoquinolines[8] pyrrolines,[9,10] pyrrolidones,[11] and tetrahydrofurans[12] are among those created by this method.

I_2 - $CuCl_2$, MeCN: 68%

I_2 - Na_2CO_3, CH_2Cl_2: 84%

Bicyclization from a bishomo-hydroxycarboxylic acid derived from citral leads to a lactone[13] without incorporation of iodine atom. Transformation of certain β-oxo-ω-alkenoic esters to bicyclic cyclopropyl ketones via iodination and 1,3-dehydroiodination is carried out in one step.[14]

I_2, Et_2O: 73%

I_2, $Mg(ClO_4)_2$ - Et_3N, CH_2Cl_2: 92%

To acquire a compound containing interconnecting lactone and cyclopropane units a process initiated by iodolactonization followed by intramolecular substitution is most efficient.[15]

NaH - I_2, THF Δ

80%

Epoxide opening. Iodohydrins are formed when epoxides are treated with iodine and a catalyst (e.g., phenylhydrazine,[16] 2,6-bis[2-(*o*-aminophenoxy)methyl]-4-bromo-1-methoxybenzene[17]) in an inert solvent. Iodine supported on an aminopropyl-derivatized silica gel induces alcoholysis of epoxides (and episulfides).[18]

Glycosylation. The iodine-catalyzed reaction between glycosyl bromides and serine derivatives gives glycosides that contain an amino acid residue.[19] Iodine plays a pivotal role in glycosylation via an intramolecularization tactic.[20]

I_2, AgOTf, collidine

80%

I_2, AgOTf

66%

C-Glycosylation such as 2,4,6-trimethoxyphenylation of *S*-glycosyl phosphorothiolates is promoted by iodine.[21]

Oxidation. The double bond of styrenes is cleaved by iodine supported on mesoporous silica FSM-16 under photoirradiation. Acids are obtained.[22] Benzylic and allylic alcohols undergo photochemical oxidation to aldehydes in the presence of iodine.[23]

The I_2-NH_4OH combination transforms aldehydes to nitriles at room temperature (14 examples, 57–97%).[24]

Iodination. Solvent effects are evident in iodination of aryl ketones using 1-chloromethyl-4-fluoro-1,4-diazoniabicyclo[2.2.2]octane bis(tetrafluoroborate) as activator.[25]

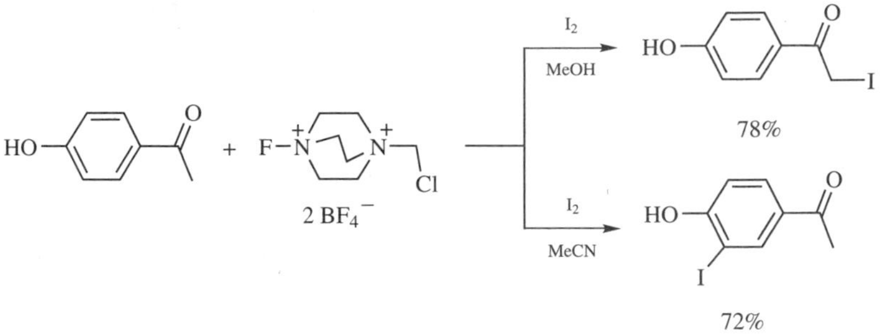

Superelectrophile such as $2AlI_3 \cdot CX_4$ [X = Cl, Br] activate iodination of alkanes and cycloalkanes.[26]

Degradation. Carbohydrates unprotected at the anomeric site undergo CC-bond cleavage on treatment with I_2-PhIO due to formation of alkoxy radicals.[27] It should be noted a related combination, I_2-$PhI(OCOCF_3)_2$, is used to convert alkenes to iodohydrins.[28]

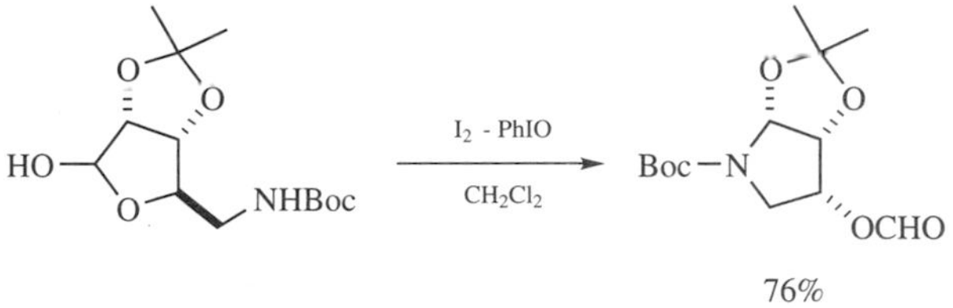

[1]Basu, M.K., Samajdar, S., Becker, F.F., Banik, B.K. *SL* 319 (2002).
[2]Karimi, B., Golshani, B. *S* 784 (2002).
[3]Firouzabadi, H., Iranpoor, N., Hazarkhani, H. *JOC* **66**, 7527 (2001).
[4]Deka, N., Sarma, J.C. *CL* 794 (2001).
[5]Samajdar, S., Basu, M.K., Becker, F.F., Banik, B.K. *TL* **42**, 4425 (2001).
[6]Ramalinga, K., Vijayalakshmi, P., Kaimal, T.N.B. *TL* **43**, 879 (2002).
[7]Vatele, J.-M. *SL* 1989 (2001).
[8]Huang, Q., Hunter, J.A., Larock, R.C. *JOC* **67**, 3437 (2002).
[9]Knight, D.W., Redfern, A.L., Gilmore, J. *JCS(P1)* 622 (2002).
[10]Stefani, H.A., de Avila, E. *SC* **32**, 2041 (2002).
[11]Terao, K., Takechi, Y., Kunishima, M., Tani, S., Ito, A., Yamasaki, C., Fukuzawa, S. *CL* 522 (2002).
[12]Ferraz, H.M.C., Sano, M.K., Nunes, M.R.S., Bianco, G.G. *JOC* **67**, 4122 (2002).
[13]Fujita, T., Hanyu, N., Mino, T., Sakamoto, M. *S* 1846 (2001).
[14]Yang, D., Gao, Q., Lee, C.-S., Cheung, K.-K. *OL* **4**, 3271 (2002).
[15]Yu, J., Lai, J.-Y., Ye, J., Balu, N., Reddy, L.M., Duan, W., Fogel, E.R., Capdevila, J.H., Falck, J.R. *TL* **43**, 3939 (2002).
[16]Sharghi, H., Eskandari, M.M. *S* 1519 (2002).
[17]Niknam, K., Nasehi, T. *T* **58**, 10259 (2002).
[18]Tamami, B., Iranpoor, N., Mahdavi, H. *SC* **32**, 1251 (2002).
[19]Kartha, K.P.R., Ballell, L., Bilke, J., McNeil, M., Field, R.A. *JCS(P1)* 770 (2001).

[20]Aloui, M., Chambers, D.J., Cumpstey, I., Fairbanks, A.J., Redgrave, A.J., Seward, C.M.P. *CEJ* **8**, 2608 (2002).
[21]Kudelska, W. *ZN* **57B**, 243 (2002).
[22]Itoh, A., Kodama, T., Masaki, Y., Inagaki, S. *SL* 522 (2002).
[23]Itoh, A., Kodama, T., Masaki, Y. *CL* 686 (2001).
[24]Talukdar, S., Hsu, J.-L., Chou, T.-C., Fang, J.-M. *TL* **42**, 1103 (2001).
[25]Stavber, S., Jereb, M., Zupan, M. *CC* 488 (2002).
[26]Akhrem, I., Orlinkov, A., Vitt, S., Chistyakov, A. *TL* **43**, 1333 (2002).
[27]Francisco, C.G., Freire, R., Gonzalez, C.C., Leon, E.I., Riesco-Fagundo, C., Suarez, E. *JOC* **66**, 1861 (2001).
[28]De Corso, A.R., Panunzi, B., Tingoli, M. *TL* **42**, 7245 (2001).

Iodine(I) azide.

Azidation.[1] Benzyl ethers are functionalized at the benzylic position by IN_3 in refluxing MeCN. The reagent promotes oxidation of benzylic alcohols but does not affect benzyl esters.

I-N_3 / MeCN, Δ

74%

[1]Viuf, C., Bols, M. *ACIEE* **40**, 623 (2001).

Iodine(I) chloride. 20, 200; 21, 233

Glycosylation.[1] The combination of ICl and AgOTf forms a useful promoter pair for glycosylation using thioglycoside donors.

5-Iodo-2-pyrones.[2] (*Z*)-2-Alken-4-ynoic esters cyclize to afford the iodinated 2-pyrones at room temperature on exposure to ICl. Benzannulated substrates react similarly.

I-Cl / CH_2Cl_2

R = Ph 94%
R = Bu 80%

[1]Ercegovic, T., Meijer, A., Magnusson, G., Ellervik, U. *OL* **3**, 913 (2001).
[2]Yai, T., Larock, R.C. *TL* **43**, 7401 (2002).

Iodine pentafluoride.

Fluorination. Alcohols, carboxylic acids, arylhydrazines, hydrazones are converted into RF, RCOF, ArF, RR′CF_2, respectively, on treatment with IF_5 and 3HF-Et_3N.[1] Active methylene groups as well as those adjacent to a divalent sulfur atom[2] are also fluorinated.

[1]Yoneda, N., Fukuhara, T. *CL* 222 (2001).
[2]Ayuba, A., Yoneda, N., Fukuhara, T., Hara, S. *BCSJ* **75**, 1597 (2002).

γ-Iodomethyl-γ-butyrolactone.

Iodination.[1] The title compound is for iodination of aryllithiums. Thus, through Br/Li exchange and the treatment aryl bromides are converted to the iodides.

[1]Harrowven, D.C., Nunn, M.I.T., Fenwick, D.R. *TL* **42**, 7501 (2001).

Iodomethylzinc trifluoroacetate.

Cyclopropanation.[1] With the title reagent [generated from CF_3COOH, Et_2Zn, and CH_2I_2] to perform cyclopropanation, ethers of *anti*-cyclopropylcarbinols are easily accessible (6 examples, 90–98%).

OR + F_3CCOO–Zn–CH_2–I → (0° / CH_2Cl_2) Ph (cyclopropane) OR

R = Bn, SiR′$_3$

[1]Charette, A.B., Lacasse, M.-C. *OL* **4**, 3351 (2002).

N-Iodosuccinimide, NIS. **16,** 185–186; **18,** 193–194; **19,** 177–178, 21, 234

Iodination.[1] Activated arenes undergo regioselective iodination with NIS and catalytic trifluoroacetic acid.

[1]Castanet, A.-S., Colobert, F., Broutin, P.-E. *TL* **43**, 5047 (2002).

Iodosylbenzene. **13,** 151; **16,** 186; **18,** 194; **19,** 178; **20,** 201; **21,** 235

Oxidations. The use of PhI=O to oxidize alcohols to carbonyl compounds is carried out in the presence of a Ru-Cu-Al-hydrotalcite.[1] Sulfinamides ($RSONH_2$) are converted into sulfonimidates [RSO(=NH)OR′] by PhIO in an alcohol in one step.[2] Note that sulfonamides undergo oxidation and in the presence of an alkene and a copper salt aziridines to form. Shown below is an intramolecular example.[3]

SO_2NH_2 + Ph—I=O → ($(MeCN)_4CuPF_6$; 3A-MS / MeCN) 69%

Benzylic oxidation of tetrahydroisoquinolines by PhIO and catalytic Bu_4NI (4 examples, 90–96%)[4] extends the possibility of CC bond formation at C-1 of such compounds. Oxidation of phosphines to phosphine oxides[5] is unexceptional.

Alkynyl phenyl iodonium salts.[6] A simple method for the preparation of the iodonium tetrafluoroborates consists of treating 1-alkynes with PhIO, HBF_4 and a catalytic amount of HgO.

[1]Friedrich, H.B., Khan, F., Singh, N., van Staden, M. *SL* 869 (2001).
[2]Leca, D., Fensterbank, L., Lacote, E., Malacria, M. *OL* **4**, 4093 (2002).
[3]Dauben, P., Saniere, L., Tarrade, A., Dodd, R.H. *JACS* **123**, 7707 (2001).
[4]Huang, W.-J., Singh, O.V., Chen, C.-H., Chiou, S.-Y., Lee, S.-S. *HCA* **85**, 1069 (2002).
[5]Mielniczak, G., Lopusinski, A. *SL* 505 (2001).
[6]Yoshida, M., Nishimura, N., Hara, S. *CC* 1014 (2002).

p-Iodotoluene difluoride. 19, 178; 21, 235–236

Iodine/fluorine exchange. Tol-IF_2, now conveniently prepared from Tol-IO with aq. HF[1] (no fluorine gas needed), is a convenient reagent for converting alkyl iodides to fluorides in Et_3N-HF.[2]

Fluorination. Introduction of a fluorine atom to α-phenylthiocarboxylic esters[3] and amides[4] with Tol-IF_2 is readily accomplished. When excess reagent is added the sulfur atom suffers oxidation. The resulting sulfoxides are versatile source of 2-fluoro-2-alkenoic esters. Under these conditions the *N*-methylanilide gives *N*-methyl-3-phenylthiooxindole, owing to rapid trapping of the sulfonium intermediate.

Tol-IF_2, Δ / CH_2Cl_2 — 53%

Tol-IF_2, Δ / CH_2Cl_2 — PhS, N(Me)Ph, F^- — 59%

[1]Sawaguchi, M., Ayuba, S., Hara, S. *S* 1802 (2002).
[2]Sawaguchi, M., Hara, S., Nakamura, Y., Ayuba, S., Fukuhara, T., Yoneda, N. *T* **57**, 3315 (2001).
[3]Greaney, M.F., Motherwell, W.B. *TL* **41**, 4463 (2000).
[4]Greaney, M.F., Motherwell, W.B. *TL* **41**, 4467 (2000).

_o_-Iodoxybenzoic acid, IBX. 19, 179; **21,** 236

Polymer-bound reagents. Several versions are available.[1–3]

Oxidations. A simple protocol for oxidation of alcohols with IBX involves heating the mixtures in EtOAc, and filtration at the end.[4] The presence of pyridone or *N*-hydroxy succinimide enables the oxidation of primary alcohols and aldehydes to carboxylic acids.[5]

Phenols give *o*-quinones,[6] whereas anilides are converted into *N*-acyl *o*-quinonimines or 2-acylamino quinones when the *p*-position of the acylamino group is substituted.[7] Methylarenes undergo oxidation (to aldehyde level) in DMSO at 120° unless the aromatic ring is deactivated or the methyl group is highly hindered.[8]

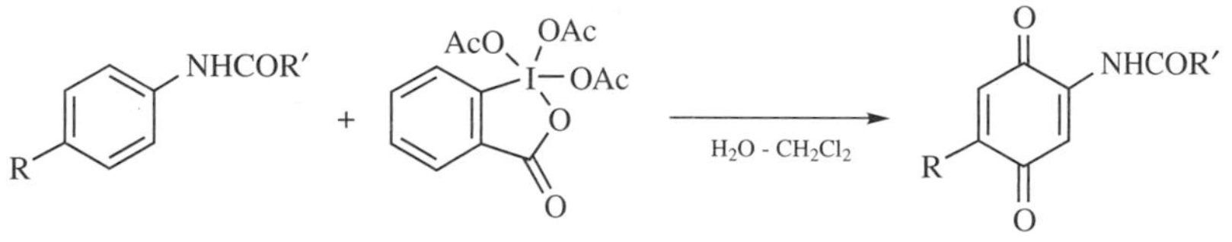

NHCOR′ + AcO, OAc, I—OAc, O, O → (H_2O - CH_2Cl_2) O, NHCOR′, R, O

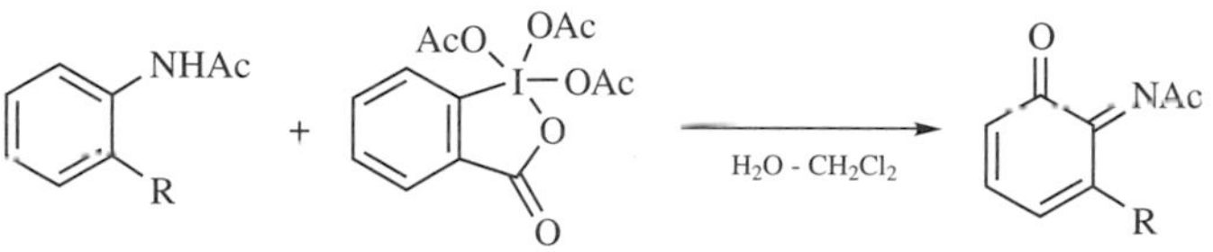

NHAc, R + AcO, OAc, I—OAc, O, O → (H_2O - CH_2Cl_2) O, NAc, R

The complex of IBX with 4-methoxypyridine *N*-oxide is an excellent reagent for converting silyl enol ethers to enones.[9]

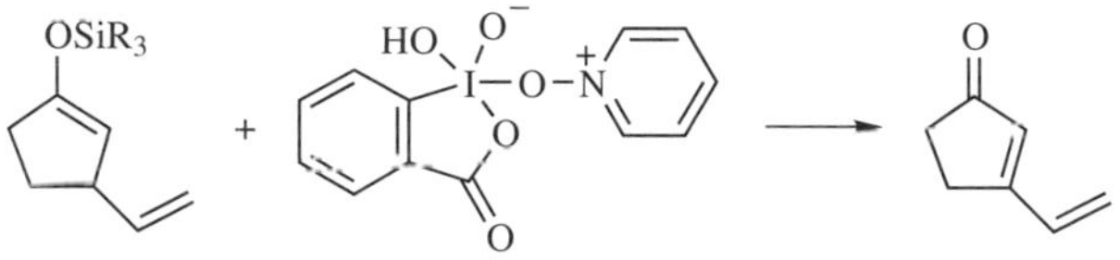

$OSiR_3$ + HO, O^-, I—O—$\overset{+}{N}$, O, O → O

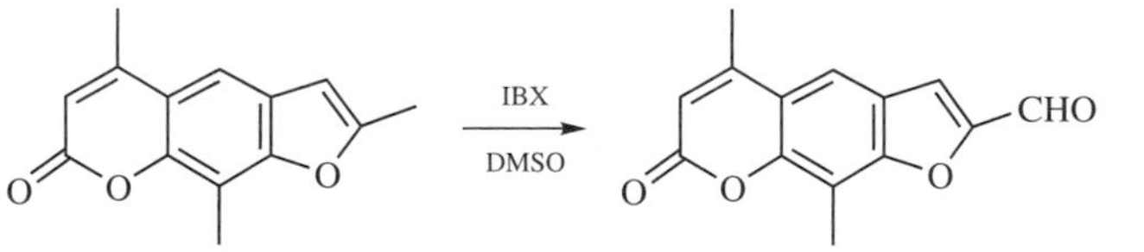

IBX / DMSO → CHO

50%

OH, O → IBX (excess) / DMSO → OHC, O, OHC

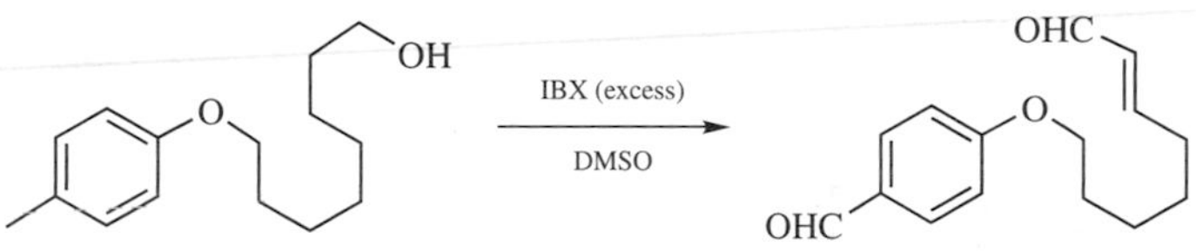

75%

A water-soluble analogue of IBX is prepared from 3-nitrophthalic acid.[10]

Deprotection. Selective cleavage of TES ethers without affecting TBS ethers is accomplished using IBX in aqueous DMSO at room temperature.[11] Preferential hydrolysis of 2-aryl/2-allyl-1,3-dithianes and dithiolanes is possible because of the much faster rates.[12]

[1]Mülbaier, M., Giannis, A. *ACIEE* **40**, 4393 (2001).
[2]Sorg, G., Mengel, A., Jung, G., Rademann, J. *ACIEE* **40**, 4395 (2001).
[3]Reed, N.N., Delgado, M., Hereford, K., Clapham, B., Janda, K.D. *BMCL* **12**, 2047 (2002).
[4]More, J.D., Finney, N.S. *OL* **4**, 3001 (2002).
[5]Mazitschek, R., Mülbaier, M., Giannis, A. *ACIEE* **41**, 4059 (2002).
[6]Magdziak, D., Rodriguez, A.A., van der Water, R.W., Pettus, T.R.R. *OL* **4**, 285 (2002).
[7]Nicolaou, K.C., Sugita, K., Baran, P.S., Zhong, Y.-L. *ACIEE* **40**, 207 (2001).
[8]Nicolaou, K.C., Baran, P.S., Zhong, Y.-L. *JACS* **123**, 3183 (2001).
[9]Nicolaou, K.C., Gray, D.L.F., Montagnon, T., Harrison, S.T. *ACIEE* **41**, 996 (2002).
[10]Thottumkara, A.P., Vinod, T.K. *TL* **43**, 569 (2002).
[11]Wu, Y., Huang, J.-H., Shen, X., Hu, Q., Tang, C.-J., Li, L. *OL* **4**, 2141 (2002).
[12]Wu, Y., Shen, X., Huang, J.-H., Tang, C.-J., Liu, H.-H., Hu, Q. *TL* **43**, 6443 (2002).

Ion exchange resins. 21, 236–237

Reactions. Amberlyst-15 resin catalyzes Mannich-type reaction,[1] formation and hydrolysis of oxathiolanes.[2,3] Different anionic forms of Dowex resins are used to hydrolyze oxazolidinones,[4] transform nitriles into thioamides,[5] oxidation,[6] and promote Michael reactions.[7] Interestingly, only 1,6-conjugate addition of nitroalkanes to methyl 2,4-pentadienoate is observed in the reaction promoted by Amberlyst A27.

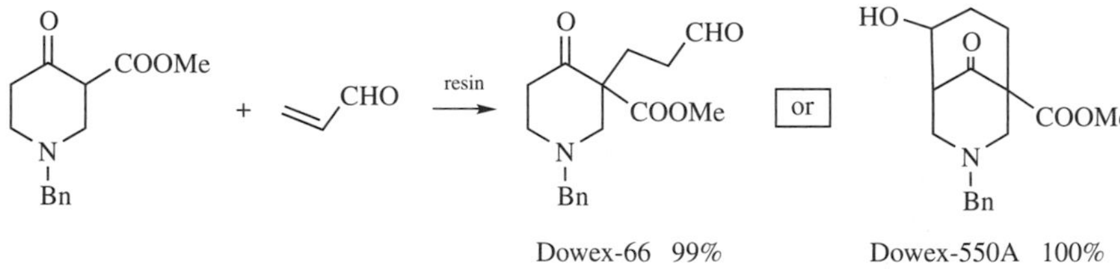

[1]Shimizu, M., Itohara, S., Hase, E. *CC* 2318 (2001).
[2]Ballini, R., Bosica, G., Maggi, R., Mazzacani, A., Righi, P., Sartori, G. *S* 1826 (2001).
[3]Chavan, S.P., Soni, P., Kamat, S.K. *SL* 1251 (2001).
[4]Katz, S.J., Bergmeir, S.C. *TL* **43**, 557 (2002).
[5]Liboska, R., Zyka, D., Bobek, M. *S* 1649 (2002).
[6]Shirini, F., Tajik, H., Jalili, F. *SC* **31**, 2885 (2001).
[7]Simon, C., Peyronel, J.-F., Clerc, F., Rodriguez, J. *EJOC* 3359 (2002).
[8]Ballini, R., Bosica, G., Fiorini, D. *TL* **42**, 8471 (2001).

Iridium complexes. 21, 237–239

Isomerization. A polymer-supported iridium promotes isomerization of double bonds.[1] Pentamethylcyclopentadienyliridium 2,2-diphenyl-2-aminoethoxide in acetone constitutes an oxidizing system that converts diols into lactones.[2]

Hydrogenation. The complex **1** is an asymmetric hydrogenation catalyst for styrenes.[3]

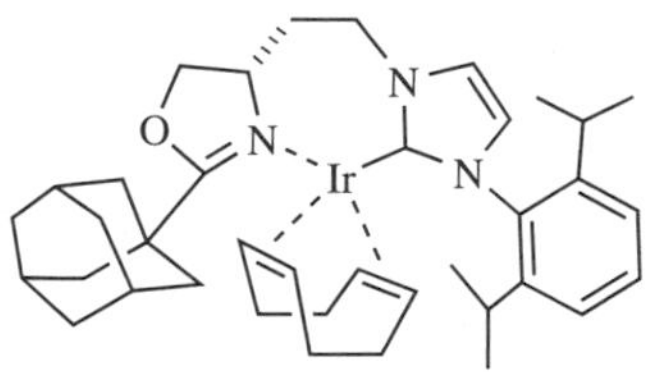

1

[1]Baxendale, I.R., Lee, A.-L., Ley, S.V. *SL* 516 (2002).
[2]Suzuki, T., Morita, K., Tsuchida, M., Hiroi, K. *OL* **4**, 2361 (2002).
[3]Powell, M.T., Hou, D.-R., Perry, M.C., Cui, X., Burgess, K. *JACS* **123**, 8878 (2001).

Iron. 19, 179–180; **20,** 203–204; **21,** 239

1H-Quinol-2-ones.[1] Reduction of the Baylis-Hillman adducts from *o*-nitrobenzaldehyde and acrylic esters with Fe leads to quinolones.

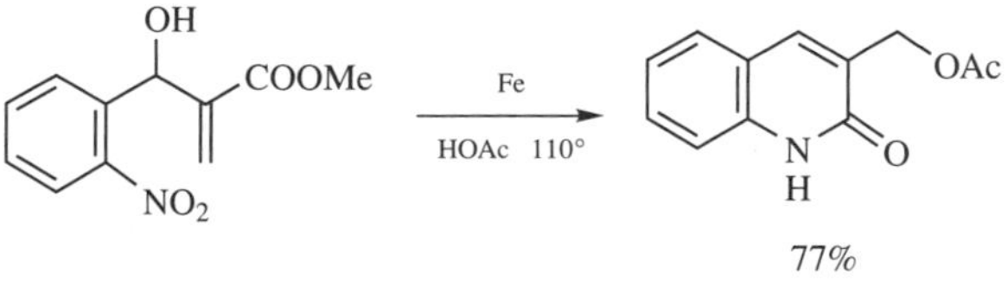

Fe, HOAc 110°

77%

[1]Basavaiah, D., Reddy, R.M., Kumaragurubaran, N., Sharada, D.S. *T* **58**, 3693 (2002).

Iron(III) acetylacetonate.

Nitration.[1] Arenes are nitrated in good yields with the Fe(acac)$_3$-N_2O_5 pair.

Hydroalkylation. Addition of RLi to alkynes is catalyzed by Fe(acac)$_3$.[2] However, the presence of an alkoxy or amino group in the substrates is required.

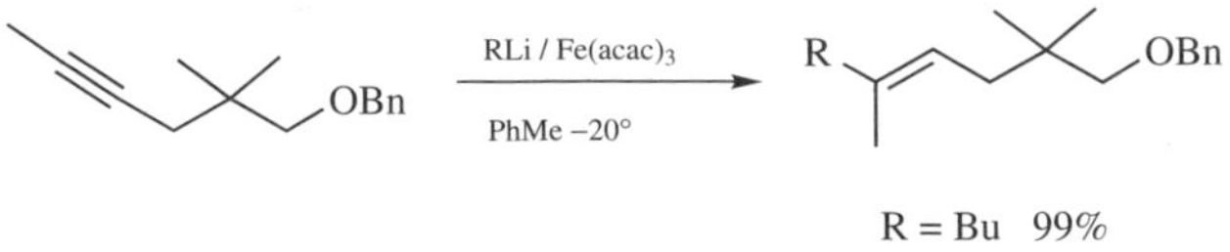

OBn

RLi / Fe(acac)$_3$

PhMe −20°

R

OBn

R = Bu 99%

[1]Bak, R.R., Smallridge, A.J. *TL* **42**, 6767 (2001).
[2]Hojo, M., Murakami, Y., Aihara, H., Sakuraji, R., Baba, Y. *ACIEE* **40**, 621 (2001).

Iron(II) chloride. 20, 204

Amidoglycosylation. Certain glycals in which the C-3 hydroxyl group is derivatized as azidoformate esters undergo aziridination and then ring opening by alcoholysis. In the absence of an alcohol protected aminoglycosyl chlorides are formed in the $FeCl_2$-mediated process.

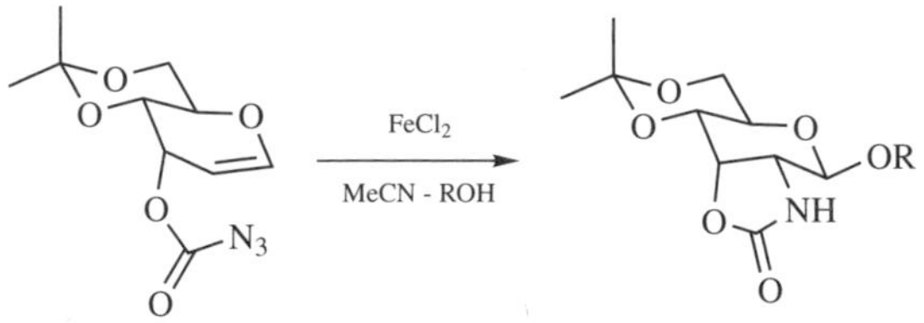

O O O O N_3 O

$FeCl_2$

MeCN - ROH

O O O OR NH O O

[1]Churchill, D.G., Rojas, C.M. *TL* **43**, 7225 (2002).

Iron(III) chloride. 13, 133–134; **14,** 164–165; **15,** 158–159; **16,** 167–169, 190–191; **17,** 138–139; **18,** 197; **19,** 180–181; **20,** 204–205; **21,** 240–241

Heterocyclization. Iodolactonization/iodoetherification[1] of unsaturated alcohols/carboxylic acids occur on treatment with $FeCl_3$-NaI. The $FeCl_3$-Ph_3P complex catalyzes diamination of enones.[2]

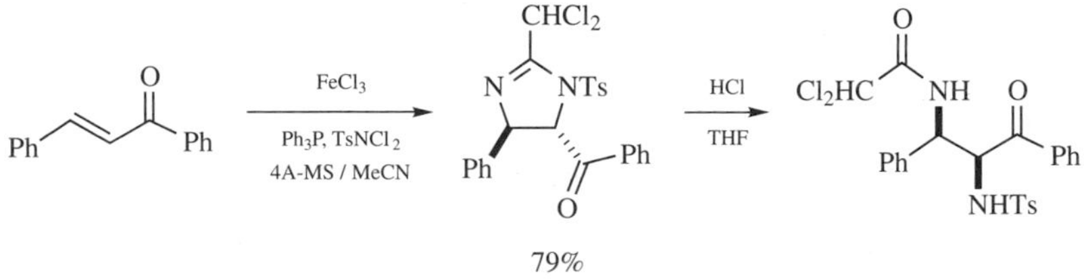

Ph O Ph

$FeCl_3$

Ph_3P, $TsNCl_2$

4A-MS / MeCN

$CHCl_2$ N NTs Ph Ph O

79%

HCl

THF

Cl_2HC O NH O Ph Ph NHTs

Coupling. Hexaoxygenated triphenylenes are formed in one step from biphenyl derivatives and catechol ethers in an oxidative coupling process.[3] Under these conditions the corresponding phenol (instead of the isopropyl ether) does not give the product.

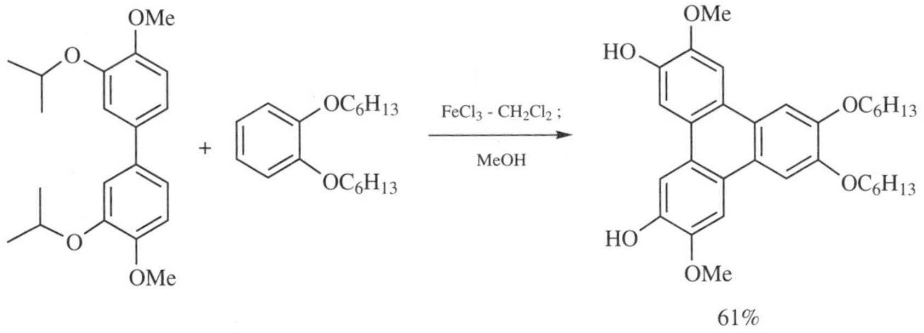

OMe OMe OC_6H_{13} OC_6H_{13} $FeCl_3$ - CH_2Cl_2 ; MeOH HO HO OMe OMe OC_6H_{13} OC_6H_{13} 61%

Rearrangements. Anhydrous $FeCl_3$ causes ketoximes to undergo Beckmann rearrangement without solvent.[4] Aryl sulfinates afford phenolic sulfoxides (thia-Fries rearrangement).[5]

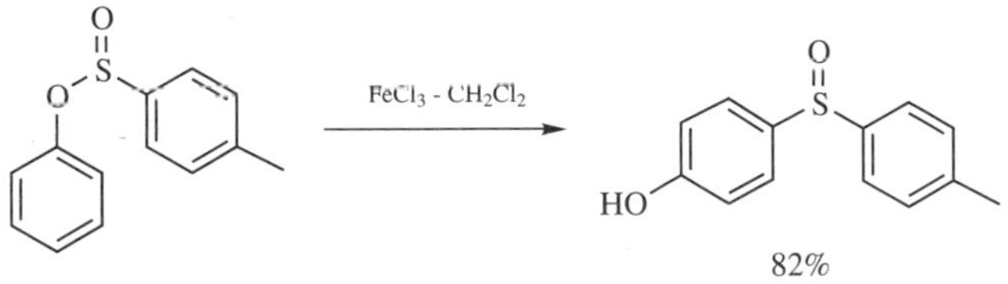

$FeCl_3$ - CH_2Cl_2 HO 82%

Oxidations. Oxidation of sulfides to sulfoxides is catalyzed by $FeCl_3$.[6] Aromatization of Hantzsch 1,4-dihydropyridine is conveniently effected by $FeCl_3 \cdot 6H_2O$.[7]

Benzylation and debenzylation. Catalyzed by $FeCl_3 \cdot 6H_2O$ nitriles undergo *N*-benzylation that is followed immediately by hydration to furnish *N*-benzylcarboxamides.[8] Benzyl esters are cleaved on heating with anhydrous $FeCl_3$.[9]

Cycloadditions. Acetals of enones are activated by $FeCl_3$ toward participation in the Diels-Alder reaction.[10] Among various Lewis acids $FeCl_3$ seems to be most suitable for both intramolecular and intermolecular [6+4]cycloaddition.[11]

$FeCl_3$ - CH_2Cl_2, 4A mol. sieves — 85%

$FeCl_3$ - CH_2Cl_2, 4A mol. sieves — 70%

Homoallylic alcohols. Allylation of aldehydes with allyltrimethylsilanes is catalyzed by $FeCl_3$.[12]

[1]Chavan, S.P., Sharma, A.K. *TL* **42**, 4923 (2001).
[2]Wei, H.-X., Kim, S.H., Li, G. *JOC* **67**, 4777 (2002).
[3]Bushby, R.J., Lu, Z. *S* 763 (2001).
[4]Khodaei, M.M., Meybodi, F.A., Rezai, N., Salehi, P. *SC* **31**, 2047 (2001).
[5]Moghaddam, F.M., Dekamin, M.G., Ghaffarzadeh, M. *TL* **42**, 8119 (2001).
[6]Kim, S.S., Nehru, K., Kim, S.S., Kim, D.W., Jung, H.C. *S* 2484 (2002).
[7]Lu, J., Bai, Y., Wang, Z., Yang, B., Li, W. *SC* **31**, 2625 (2001).
[8]Karabulut, H.R.F., Kacan, M. *SC* **32**, 2345 (2002).
[9]Davies, T.J., Jones, R.V.H., Lindsell, W.E., Miln, C., Preston, P.N. *TL* **43**, 487 (2002).
[10]Chavan, S.P., Sharma, A.K. *SL* 667 (2001).
[11]Rigby, J.H., Fleming, M. *TL* **43**, 8643 (2002).
[12]Watahiki, T., Oriyama, T. *TL* **43**, 8959 (2002).

Iron pentacarbonyl. 13, 152; 18, 196; 20, 206–207; 21, 242–243

Isomerization.[1] Heating *N*-allylcarboxamides with $Fe(CO)_5$ at 100° results in the formation of enamides.

Allylic alcohols are photochemically isomerized to ketones[2] in the presence of $Fe(CO)_5$. When aldehydes are added in the reaction milieu aldol reaction products result.[3]

OH + R‿CHO —hν, $Fe(CO)_5$, pentane→

R = H 95 : 5
R = Ph 89 : 11

Reduction.[4] With a catalytic amount of $NaBH_4$ iron pentcarbonyl react with cycloheptatriene in protic solvents to generate complex of the cyclic diene. Oxidative removal of the metal moiety completes a reduction of cycloheptatriene to 1,3-cycloheptadiene.

[1]Sergeyev, S., Hesse, M. *SL* 1313 (2002).
[2]Cherkaoui, H., Soufiaoui, M., Gree, R. *T* **57**, 2379 (2001).
[3]Crevisy, C., Wietrich, M., Le Boulaire, V., Uma, R., Gree, R. *TL* **42**, 395 (2001)
[4]Coquerel, Y., Depres, J.-P. *CC* 658 (2002).

Iron(II) sulfate. 21, 243

Transesterification.[1] Exchange of the alkoxy residues of β-keto esters is readily accomplished in the presence of $FeSO_4$ or $CuSO_4$.

Oxidative elimination.[2] The ring structure of isoxazolin-5-ones is completely destroyed on treatment with $FeSO_4$ and $NaNO_2$ in aqueous HOAc at room temperature. The transformation involves hydrolysis, diazotization and fragmentation.

$FeSO_4$ - $NaNO_2$
HOAc - H_2O

R = 4-butenyl 37%

[1]Bandgar, B.P., Sadavarte, V.S., Uppalla, L.S. *SC* **31**, 2063 (2001).
[2]Renard, D., Rezaei, H., Zard, S.Z. *SL* 1257 (2002).

Isocyanatophosphoric acid dichloride.

Cyanations.[1] A cyano group is introduced to pyrrole, indole and enamines by this reagent $Cl_2P(=O)NCO$.

[1]Smaliy, R.V., Chaikovskaya, A.A., Pinchuk, A.M., Tolmachev, A.A. *S* 2416 (2002).

Isocyanuric chloride. 21, 243

Nitrosation.[1] Secondary amines are rapidly transformed into nitrosamines on treatment with isocyanuric chloride and $NaNO_2$ on wet silica.

Oxidations. Isocyanuric chloride is a constituent of the oxidation system (with TEMPO•) for selective oxidation of primary alcohols to aldehydes,[2] and its combination with DBU and *t*-BuONa converts indolines into indoles.[3]

Various divalent organochalcogenides are oxidized: Thiols to disulfides,[4] selenols to diselenides,[5] sulfides to sulfoxides[6] including alkynyl sulfides.[7]

[1]Zolfigol, M.A., Choghamarani, A.G., Hazarkhani, H. *SL* 1002 (2002).
[2]De Luca, L., Giacomelli, G., Porcheddu, A. *OL* **3**, 3041 (2001).
[3]Tilstam, U., Harre, M., Heckrodt, T., Weinmann, H. *TL* **42**, 5385 (2001).
[4]Zhong, P., Guo, M.-P. *SC* **31**, 1825 (2001).
[5]Zhong, P., Guo, M.-P. *SC* **31**, 1507 (2001).
[6]Xiong, Z.-X., Huang, N.-P., Zhong, P. *SC* **31**, 245 (2001).
[7]Zhong, P., Guo, M.-P., Huang, N.-P. *SC* **32**, 175 (2002).

L

Lanthanum.

Couplings. Alkyl halides[1] undergo reductive coupling under the influence of lanthanum. Deoxygenative dimerization of benzylic/allylic alcohols[2] requires catalytic amount of iodine besides lanthanum, CuI and Me_3SiCl.

Reductive alkylation.[3] On treatment with lanthanum and iodine, diphenyl diselenides undergo dissociative alkylation with alkyl halides.

Benzyne.[4] Lanthanum combined with catalytic iodine induces benzyne formation from 1,2-dihaloarenes (e.g., 2-chloroiodobenzene) and 2-haloaryl triflates in quantitative yield. Cerium and ytterbium are less effective while most other rare earth metals are useless.

[1]Nishino, T., Watanabe, T., Okada, M., Nishiyama, Y., Sonoda, N. *JOC* **67**, 966 (2002).
[2]Nishino, T., Nishiyama, Y., Sonoda, N. *TL* **43**, 3689 (2002).
[3]Nishino, T., Okada, M., Kuroki, T., Watanabe, T., Nishiyama, Y., Sonoda, N. *JOC* **67**, 8696 (2002).
[4]Kawabata, H., Nishino, T., Nishiyama, Y., Sonoda, N. *TL* **43**, 4911 (2002).

Lanthanum(III) trifluoromethanesulfonate. 20, 209; 21, 245

Hetero Diels-Alder reactions.[1] Formation of 1,2,3,6-tetrahydropyridazines from conjugated dienes and azodicarboxylic esters is facile when catalyzed by $La(OTf)_3$.

EtOOC–N=N–COOEt + isoprene → ($La(OTf)_3$) → 1,2-bis(COOEt)-4-methyl-1,2,3,6-tetrahydropyridazine, 99%

Allylation. Allylation of aldehydes with allyltributylstannane[2] is mediated by $La(OTf)_3$ while PhCOOH (which is added) plays a crucial activation role. Homoallylic amines are synthesized when amines are also present.[3]

Glycidic esters.[4] Carbonyl compounds condense with diazoacetic esters to furnish glycidic esters in the presence of $La(OTf)_3$.

PhCHO + N_2CHCOOEt → ($La(OTf)_3$, CH_2Cl_2) → Ph-epoxide-COOEt, 92%

Fiesers' Reagents for Organic Synthesis, Volume 22. Series editor Tse-Lok Ho
ISBN 0-471-28515-3

[1]Curini, M., Epifano, F., Marcotullio, M.C., Rosati, O. *H* **55**, 1599 (2001).
[2]Aspinall, H.C., Bissett, J.S., Greeves, N., Levin, D. *TL* **43**, 319 (2002).
[3]Aspinall, H.C., Bissett, J.S., Greeves, N., Levin, D. *TL* **43**, 323 (2002).
[4]Curini, M., Epifano, F., Marcotullio, M.C., Rosati, O. *EJOC* 1562 (2002).

Lanthanum(III) hexamethyldisilazide.

Tishchenko reaction.[1] The title reagent catalyzes intermolecular and intramolecular Tishchenko reactions under homogeneous conditions. It also effects hydroamination of unsaturated compounds.

CHO CHO $La[N(SiMe_3)_2]_3$ O O 85%

R NH_2 $La[N(SiMe_3)_2]_3$ N R

[1]Bürgstein, M.R., Berberich, H., Roesky, P.W. *CEJ* **7**, 3078 (2001).

Lead(IV) acetate. 13, 155–156; **14,** 188; **16,** 193–194; **18,** 201–202; **19,** 184–185; **20,** 209–210; **21,** 245–246

Oxidation. The $Pb(OAc)_4$–metal halide (e.g., KBr) system oxidizes aliphatic alcohols. Primary alcohols give esters, and secondary alcohols, ketones.[1] Oxaspirannulation of 4-(3-hydroxyalkyl)phenols occurs in a reaction with $Pb(OAc)_4$.[2]

OH OMe OH $Pb(OAc)_4$ Me_2CO 0° O OMe O 69%

Oxidative rearrangement. A simple synthesis of 3,5-alkadienoic esters from 3,5-alkadien-2-ones involves treatment with $Pb(OAc)_4$–$BF_3 \cdot OEt_2$.[3]

Ph O $Pb(OAc)_4$ - $BF_3 \cdot OEt_2$ MeOH - PhH Ph COOMe 50% (*E*/*Z* 4 :1)

[1]Kapustina, N.I., Sokova, L.L., Makhaev, V.D., Borisov, A.P., Nikishin, G.I. *RCB* **49**, 1842 (2000).
[2]Plourde, G.L. *TL* **43**, 3597 (2002).
[2]Nongkhlaw, R.L., Nongrum, R., Myrboh, B. *JCS(P1)* 1300 (2001).

Lead(IV) acetate–copper(II) acetate.

Cleavage of lactols. Lactols are cleaved into ω-alkenyl formates.[1]

$Pb(OAc)_4$ - $Cu(OAc)_2$, pyridine - PhH, Δ; HCOO; 82%

[1]Rigby, J.H., Payen, A., Warshakoon, N. *TL* **42**, 2047 (2001).

Lead(II) iodide.

Allylation. A highly *syn*-selective allylation of aldehydes with allyltributylstannane in water is observed. The catalyst is PbI_2 and a phase-transfer agent is also present.[1]

R-CHO + Bu_3Sn ; PbI_2 - Bu_4NBr ; R, OH (*syn/anti* >80 : 20)

[1]Shibata, I., Yoshimura, N., Yabu, M., Baba, A. *EJOC* 3207 (2001).

Lipases. 17, 133–134; **18,** 202–204; **19,** 185–188; **20,** 211–212; **21,** 246–247

Hydrolysis. Enantioselective cleavage cyanohydrin esters by propanol occurs in the presence of a lipase, thus resolving these compounds into optical isomers.[1] In resolution of 2-phenoxypropanoic esters by the lipase-catalyzed hydrolysis the enantioselectivity is dramatically enhanced by DMSO.[2] Chiral β-hydroxyphosphonates are obtained from the corresponding esters.[3]

Acylation. 2-Aryloxy-1-propanols are resolved via acylation.[4] A polmer-supported lipase is used in transesterification of β-keto esters.[5]

Kinetic resolution of β-halo alcohols by the chemoenzymatic technique is to transfer the acyl group from *p*-chlorophenyl acetate in the presence of a lipase and a dinuclear Ru complex.[6]

[1]Veum, L., Kuster, M., Telalovic, S., Hanefeld, U., Maschmeyer, T. *EJOC* 1516 (2002).

[2] Watanabe, K., Ueji, S. *JCS(P1)* 1386 (2001).
[3]Zhang, Y., Li, Z., Yuan, C. *TL* **43**, 3247 (2002).
[4]Miyazawa, T., Yukawa, T., Koshiba, T., Sakamoto, H., Ueji, S., Yanagihara, R., Yamada, T. *TA* **12**, 1595 (2001).
[5]Cordova, A., Janda, K.D. *JOC* **66**, 1906 (2001).
[6]Pamies, O., Bäckvall, J.-E. *JOC* **67**, 9006 (2002).

Lithiopyrrolidone.

Aldol reactions.[1] The title reagent catalyzes the condensation of silyl enol ethers with aldehydes.

[1]Mukaiyama, T., Fujisawa, H. *CL* 858 (2002).

Lithium. 13, 157–158; **15,** 184; **18,** 205–206; **19,** 190–191; **20,** 212; **21,** 247

Reductive alkylation.[1] Bicyclic 2-phenyloxazolidines undergo C-O bond cleavage and the ensuing benzylic anions are alkylated stereoselectively, giving mainly the *syn*-products. However, in deuteration and reaction with less reactive halides (e.g., BuCl) stereointegrity is practically lost [with BuBr the *syn/anti* ratio is 95:5].

H O H N Ph —Li / THF −20° ; RX→ OH R N Ph + OH R N Ph

(> 95 : 5)

[1]Azzena, U., Pilo, L., Piras, E. *TL* **42**, 129 (2001).

Lithium–liquid ammonia. 13, 158; **17,** 161; **18,** 206; **20,** 213; **21,** 247

Fused cyclopropanols.[1] Some cyclohexenones undergo reductive cyclization instead of conjugate reduction.

MeOOC O —Li / NH_3 EtOH→ MeOOC OH

[1]Dumas, F., Thominiaux, C., Miet, C., d'Angelo, J., Desmaele, D., Nour, M., Cave, C., Morgant, G., Huu-Dau, M.-E.T. *TL* **43**, 9649 (2002).

Lithium alkynoxides.

Preparation.[1] A practical method for generating these ynolate ions is by treatment of ethyl 2,2-dibromoalkanoates with Li and catalytic amount of naphthalene in THF at –78°.

Aldol reactions.[2] Reaction of the title reagents with acylsilanes leads to 2-silyl-2-alkenoic esters, implying formation and subsequent cleavage of β-silyl-β-lactones.

γ-Keto esters are transformed into 2-cyclopentenones,[3] as a result of a tandem [2+2]cycloaddition and Dieckmann cyclization.

COOEt + OLi; –20°, THF; SiO_2, PhMe, Δ; 54%

[1]Shindo, M., Koretsune, R., Yokota, W., Itoh, K., Shishido, K. *TL* **42**, 8357 (2001).
[2] Shindo, M., Matsumoto, K., Mori, S., Shishido, K. *JACS* **124**, 6840 (2002).
[3]Shindo, M., Sato, Y., Shishido, K. *TL* **43**, 5039 (2002).

Lithium aluminum hydride. 14, 190–191; **18,** 207; **19,** 191; **20,** 213–214; **21,** 247–248

γ-Amino alcohols.[1] Reduction of β-hydroxy ketimines by $LiAlH_4$ is stereoselective. γ-Amino alcohols with a *syn* arrangement are major products.

Reductive desulfonylation.[2] 1-Sulfonyl glycals undergo reduction with $LiAlH_4$ to afford tetrahydropyrans.

OBn, BnO, BnO, O, SO_2Ph; $LiAlH_4$, THF Δ; OBn, BnO, BnO, O

Reduction of alkynols. (*E*)-4-Alkenols are generated in the reduction.[3] However, in the presence of Cp_2TiCl_2 the predominant products are (*Z*)-isomers.[4]

Group-selective hydroalumination of propargylic alcohols to differentiate two triple bonds is demonstrated.[5]

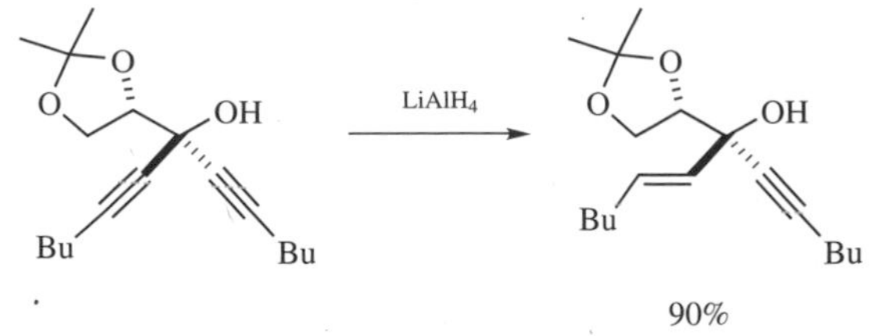

Deoxygenation. Phosphine oxides are reduced on treatment with MeOTf and then $LiAlH_4$. The reduction step is attended by inversion of configuration at phosphorus center.[6]

[1]Veenstra, S.J., Kinderman, S.S. *SL* 1109 (2001).
[2]Narama, H., Funabashi, M. *JCC* **20**, 257 (2001).
[2]Brimble, M.A., Pavia, G.S., Stevenson, R.J. *TL* **43**, 1735 (2002).
[4]Parenty, A., Campagne, J.-M. *TL* **43**, 1231 (2002).
[5]Ohmori, K., Suzuki, T., Taya, K., Tanabe, D., Ohta, T., Suzuki, K. *OL* **3**, 1057 (2001).
[6]Imamoto, T., Kikuchi, S., Miura, T., Wada, Y. *OL* **3**, 87 (2001).

Lithium aluminum hydride–selenium.

Selenylation. The reagent obtained in THF is able to convert acid chlorides into diacyl selenides (but 3-chloropropanoyl chloride into γ-selenolactone).[1] It also adds to organo-isocyanates such as PhNCO to give, after *Se*-alkylation, RSeCONHPh.

Selenoureas are formed by reacting the HCl salts of cyanamides (R_2N-CN) with the reagent.[2]

[1]Ishihara, H., Koketsu, M., Fukuta, Y., Nada, F. *JACS* **123**, 8408 (2001).
[2]Koketsu, M., Fukuta, Y., Ishihara, H. *TL* **42**, 6333 (2001).

Lithium borohydride.

Reduction.[1] Secondary amides are reduced without affecting tertiary amides.[1] Enaminones are converted into *syn*-γ-amino alcohols by reduction after $CeCl_3$-chelation.[2]

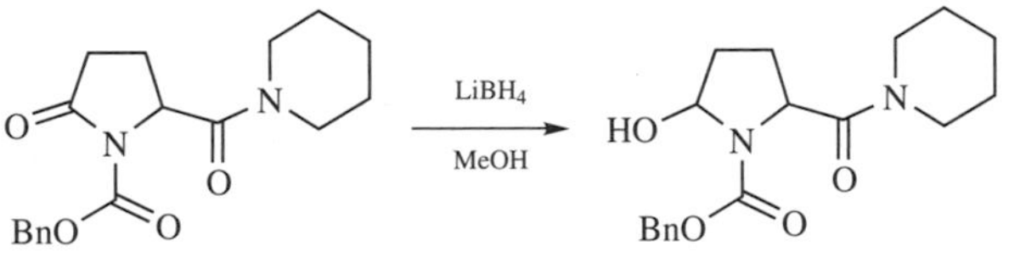

[1]Lee, B.H., Clothier, M.F. *TL* **40**, 643 (1999).
[2]Bartoli, G., Cupone, G., Dalpozzo, R., De Nino, A., Maiuolo, L., Procopio, A., Tagarelli, A. *TL* **43**, 7441 (2002).

Lithium bromide. 18, 209–210; **19,** 192; **20,** 215; **21,** 248

O,S-acetals. Replacement of an alkoxy group in an acetal by reaction with a thiol takes place with LiBr as catalyst.[1]

Rearrangement. Epoxy nitriles undergo ring opening and hydrolysis in the presence of LiBr in an aqueous solvent to afford carboxylic acids. Thus, homologation of unhindered ketones is accomplishable in two steps: Darzens condensation and the rearrangement.[2]

Condensation. LiBr catalyzes the condensation of active methylene groups with nitroso compounds, leading to functionalized imines.[3]

Me2N–C6H4–N=O + R–CH2–CN —(LiBr,))))→ Me2N–C6H4–N=C(R)–CN

1-Aryl-1-halo-1,1-dienes.[4] 1-Aryl-2,3-butadien-1-ols undergo ionization under very mild conditions, and their treatment with LiBr in HOAc is sufficient to convert them to the bromo dienes. They give the corresponding chloro dienes by changing LiBr to LiCl.

Ph–CH(OH)–CH=C=CH2 —(LiX / HOAc, 50°)→ Ph–C(X)=CH–CH=CH2

X = Cl 60%
X = Br 76%

[1]Ono, F., Negoro, R., Sato, T. *SL* 1581 (2001).
[2]Badham, N.F., Mendelson, W.L., Allen, A., Diederich, A.M., Eggleston, D.S., Filan, J.J., Freyer, A.J., Willmer, L.B., Kowalski, C.J., Liu, L., Novack, V.J., Vogt, F.G., Webb, K.S., Yang, J. *JOC* **67**, 5440 (2002).
[3]Laskar, D.D., Prajapati, D., Sandhu, J.S. *SC* **31**, 1427 (2001).
[4]Ma, S., Wang, G. *TL* **43**, 5723 (2002).

Lithium *N,N*-dialkylaminoborohydrides. 21, 249

Amination and reduction. The influence of the amino groups on the reaction course with alkyl mesylates is revealed.[1]

Ph(CH2)3OMs —($LiBH_3N(^iPr)_2$, THF)→ Ph(CH2)2CH3 80%

Ph(CH2)3OMs —($LiBH_3NMe_2$, THF)→ Ph(CH2)3NMe2 92%

Both substitution and reduction take place when 2-halobenzonitriles react with $LiBH_3NMe_2$.[2]

[1]Thomas, S., Huynh, T., Enriquez-Rios, V., Singaram, B. *OL* **3**, 3915 (2001).
[2]Thomsa, S., Collins, C.J., Cuzens, J.R., Spiciarich, D., Goralski, G.T., Singaram, B. *JOC* **66**, 1999 (2001).

Lithium 4,4′-di-*t*-butylbiphenylide. 13, 162–163; **16,** 195–196; **17,** 164; **18,** 210–211; **19,** 192–193; **20,** 216–217; **21,** 249–250

C-X bond cleavage. Further uses are found for organolithiums that are generated from C-X bond cleavage with Li-DTTB. For example, the *O,C*-dilithio species derived from isochroman has been converted to the substituted benzylzinc for Negishi coupling.[1] Imine derivatives of 3-chloropropylamine cyclizes to afford pyrrolidines[2] after the Cl/Li exchange.

Cl Ph N Li - DTBB / THF ; H_2O Ph N H

Related to the Li-DTTB system are those based on linear polyphenylene[3] and ROMPgel-supported biphenyl.[4]

Facile C-S bond cleavage generates organolithiums. Very high stereoselectivity is revealed in a bicyclic α-thialactam.[5] Cyclization occurs in unsaturated substrates.[6]

HO SPh BuLi ; Li - DTBB / THF ; PhS-SPh HO SPh

73%

O R R′ N S H Li - DTBB / THF LiO R′ N R H SLi R″I O R′ N R R″ H SR″

A C-S bond of 4-heterosubstituted dibenzothiins is severed to provide nucleophilic *o*-substituted phenyllithium species.[7]

X S Li - DTTB THF −78° SLi X Li PhCHO ; HCl SH X Ph HO

X = O 49%
X = S 52%
X = NMe 64%

Reductions. Li-DTTB induces reductive cleavage of *N*-alkoxycarboxamides (at –78°) and acyl azides.[8] In refluxing THF *N*-alkoxycarboxamides not only suffer O-N bond scission, they also lose the carbonyl group.

Nitrones behave differently toward Li-DTTB, depending whether $NiCl_2 \cdot H_2O$ is added.[9]

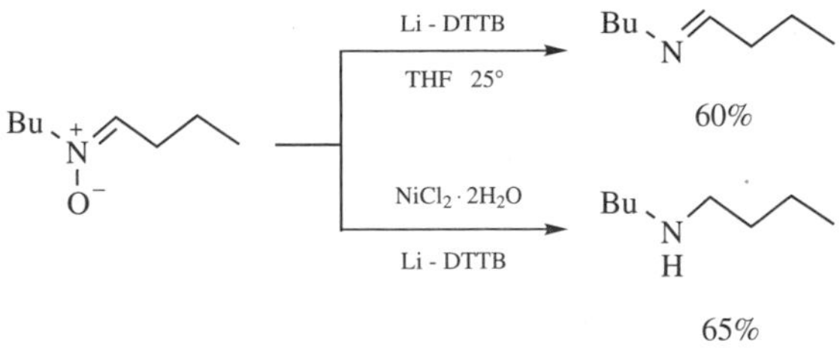

Li - DTTB, THF 25° → 60%

$NiCl_2 \cdot 2H_2O$, Li - DTTB → 65%

Dilithiation. Styrenes form dilithioarylethanes which can be doubly alkylated.[10]

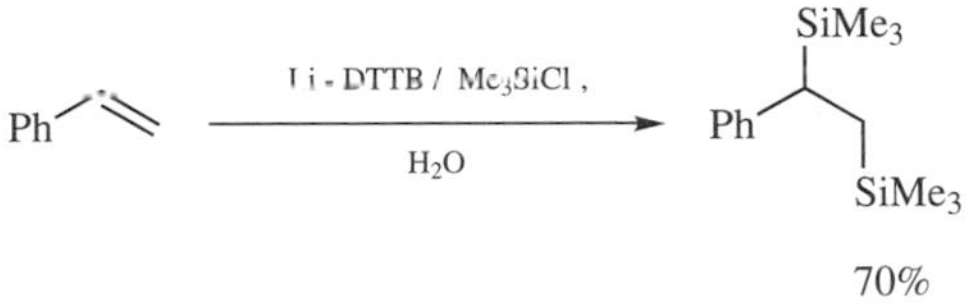

Li - DTTB / Me_3SiCl, H_2O → 70%

[1]Yus, M., Gomis, J. *TL* **42**, 5721 (2001).
[2]Yus, M., Soler, T., Foubelo, F. *JOC* **66**, 6207 (2001).
[3]Yus, M., Gomez, C., Candela, P. *T* **58**, 6207 (2002).
[4]Arnauld, T., Barrett, A.G.M., Hopkins, B.T. *TL* **43**, 1081 (2002).
[5]Manthorpe, J.M., Gleason, J.L. *ACIEE* **41**, 2338 (2002).
[6]Deng, K., Bensari, A., Cohen, T. *JACS* **124**, 12106 (2002).
[7]Yus, M., Foubelo, F., Ferrandez, J.V. *CL* 726 (2002).
[8]Yus, M., Radivoy, G., Alonso, F. *S* 914 (2001).
[9]Radivoy, G., Alonso, F., Yus, M. *S* 427 (2001).
[10]Yus, M., Martinez, P., Guijarro, A. *T* **57**, 10119 (2001).

Lithium diisopropylamide, LDA. 13, 163–164; **15,** 188–189; **16,** 196–197; **17,** 165–167; **18,** 212–214; **19,** 193–197; **20,** 218–220; **21,** 250–251

Deprotonation. Despite insufficient basicity for LDA to deprotonate most arenes, the *o*-hydrogen of a fluoroarene is abstractable.[1] More interestingly, *N*-methoxymethyl-*N*-methylformamide can be deprotonated and then silylated.[2]

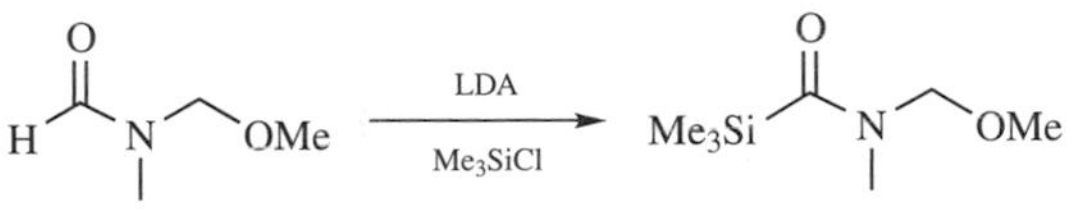

LDA, Me_3SiCl

The O->C migration of a tricarbonyliron-coordinated dienyl phosphate is initiated by deprotonation at an sp^2-hybridized carbon.[3]

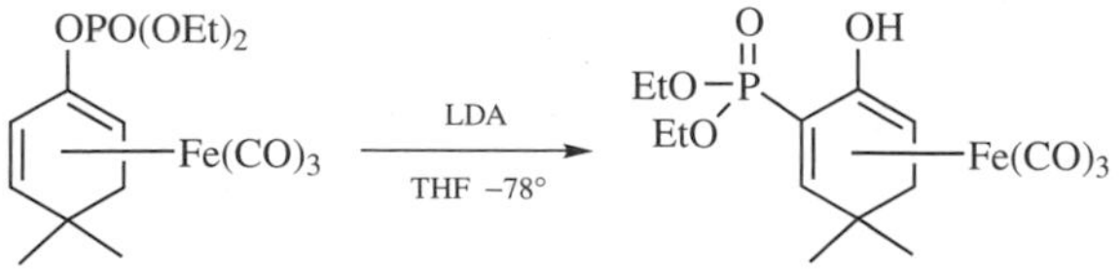

78%

Alkylations. LDA is recognized as a standard base for alkylation reactions therefore only a very few examples of application are presented here. α-Amino nitriles on deprotonation and then reaction with aldehydes afford *anti*-β-amino alcohols.[4] 2,3-Diamino alcohols are accessible if the nitrile group of the products is reduced.

α-Cyano selenides are formed on treatment of aldehyde *N,N*-dimethylhydrazones with LDA and PhSeBr. Selenylation is followed by elimination of dimethylamine.[5]

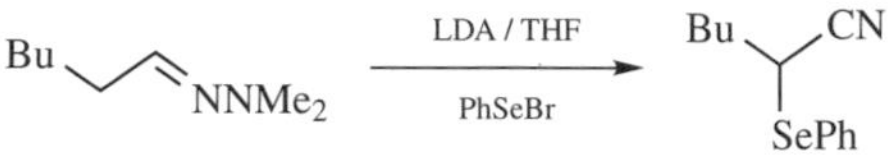

43%

A polystyrene-supported benzyl selenide undergoes alkylation in the conventional manner. Subsequent treatment with H_2O_2 dislodges the polymer chain.[6] Aldol-type reaction involving selenoamides has also been realized.[7]

Double bond migration. LDA/*i*-Pr_2NH isomerizes *N*-allylbenzamides to enamides (*E*-form major).[8]

[1]Kirsch, P., Reiffenrath, V., Bremer, M. *SL* 389 (1999).
[2]Cunico, R.F. *TL* **42**, 1423 (2001).
[3]Okauchi, T., Yeshima, T., Hayashi, K., Suetsugu, N., Minami, T. *JACS* **123**, 12117 (2001).
[4]Leclerc, E., Vranken, E., Mangeney, P. *JOC* **67**, 8928 (2002).
[5]Ternon, M., Pannecoucke, X., Outurquin, F., Paulmier, C. *T* **58**, 3275 (2002).
[6]Huang, X., Xu, W. *TL* **43**, 5495 (2002).
[7]Murai, T., Suzuki, A., Kato, S. *JCS(P1)* 2711 (2001).
[8]Ribereau, P., Delamare, M., Celanire, S., Queguiner, G. *TL* **42**, 3571 (2001).

Lithium *N,O*-dimethylhydroxylamide.

Protection. The title reagent forms adducts with ArCHO to remove interference (incompatibility) of the formyl group to reactions at other functionalities.[1]

[1]Roschangar, F., Brown, J.C., Cooley, B.E., Sharp, M.J., Matsuoka, R.T. *T* **58**, 1657 (2002).

Lithium hexamethyldisilazide, LHMDS. 13, 165; **14,** 194; **18,** 215–216; **19,** 197–198; **20,** 221–222; **21,** 251–252

[2,3]Wittig rearrangement. Treatment of γ-allyloxy-β-keto esters with LHMDS results in the production of 3-oxo-4-hydroxy-6-heptenoic esters or silyl ethers which are direct precursors of γ-allyltetronic acids.[1]

1,3-Diheteroallyl anions. These anions serve as building blocks for functionalized carbon chains.[2]

Cyclopropanation. By an addition/1,3-elimination pathway the carbanion of methyl dichloroacetate unites with 2-cycloalkenones to furnish bicyclic structures. The process is highly diastereoselective therefore synthetically useful.[3]

[1]Pevet, I., Meyer, C., Cossy, J. *TL* **42**, 5215 (2001).
[2] Lee, B.S., Gil, J.M., Oh, D.Y. *TL* **42**, 2345 (2001).
[2]Escribano, A., Pedregal, C., Gonzalez, R., Fernandez, A., Burton, K., Stephenson, G.A. *T* **57**, 9422 (2001).

Lithium hydroxide.

N-Alkylation. Amino acid esters and dipeptide esters undergo monoalkylation at the basic nitrogen with activated alkyl halides using LiOH as base.

[1]Cho, J.H., Kim, B.M. *TL* **43**, 1273 (2002).

Lithium naphthalenide, LN. 15, 190–191; **18,** 217–218; **19,** 199–200; **20,** 224–225; **21,** 252–254

Reductive alkylation. On exposure to LN γ-cyano-α,β-unsaturated ketones detach the cyano group to generate dienolate ions. Addition of electrophiles completes the reaction, but the position of new bond formation is governed by existing factors.[1]

Li - naphthalene / THF ; MeI

77%

Treatment of bis(2-methoxyethyl) acetals of aromatic aldehydes with LN results in benzylic anions. Alkylation gives α-branched benzyl ethers.[2]

Li - naphthalene / THF ; $CH_2=CHCH_2Br$

75%

Elimination. Reductive cleavage by LN of a carbon-linked benzotriazolyl group next to a leaving substituent engenders fragmentation. A synthesis of allylic alcohols is based on this premise.[3]

Li - naphthalene / THF ; H_2O

59%

F-Li exchange. Allylic and benzylic organolithium species are formed from the fluoro compounds and LN.[4]

[1]Tai, C.-L., Ly, T.W., Wu, J.-D., Shia, K.-S., Liu, H.-J. *SL* 214 (2001).
[2]von Schrader, T., Woodward, S. *EJOC* 3833 (2002).
[3]Kang, Y.H., Lee, C.J., Kim, K. *JOC* **66**, 2149 (2001).
[4]Guijarro, D., Yus, M. *JOMC* **624**, 53 (2001).

Lithium perchlorate. 18, 218–219; **19,** 200–201; **20,** 224–225

Michael reaction. Addition of alcohols to enones is completed in a pressurized system (water-freezing) and in the presence of $LiClO_4$ and DMAP.[1]

Friedel-Crafts acylation. In solvent-free conditions acylation of arenes is accomplished by the $LiClO_4$–$(RCO)_2O$ complexes.[2]

Cyclopropanes. Alkylation of stabilized organolithiums with epibromohydrin is promoted by $LiClO_4$.[3]

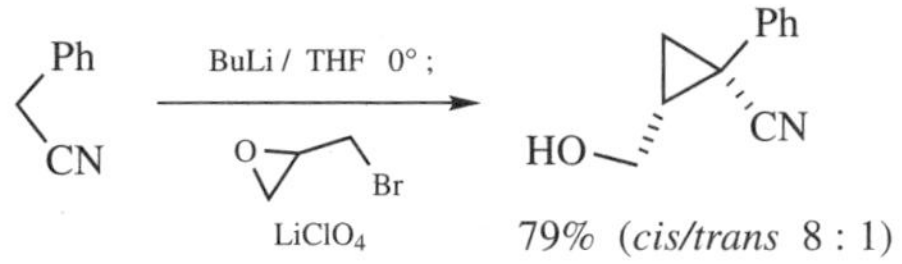

BuLi / THF 0° ; epibromohydrin, $LiClO_4$ — 79% (*cis/trans* 8 : 1)

[1]Hayashi, Y., Nishimura, K. *CL* 296 (2002).
[2]Bartoli, G., Bosco, M., Marcantoni, E., Massaccessi, M., Rinaldi, S., Sambri, L. *TL* **43**, 6331 (2002).
[3]Langer, P., Freifeld, I. *OL* **3**, 3903 (2001).

Lithium 2,2,6,6-tetramethylpiperidide, LTMP. 13, 167; **14,** 194–195; **17,** 171–172; **18,** 220–221; **19,** 202; **20,** 226–227; **21,** 256

Directed lithiation. Unstable lithio compounds such as *o*-lithiobenzoic esters which are generated by lithiation with LTMP are trapped in situ. Thus, borylation is carried out by adding the substituted arenes to a mixture of LTMP and $(i\text{-PrO})_3B$.[1]

a-Cyanoacetaldehyde anion equivalent.[2] Synthetic equivalent of the nucleophile is 2-lithio-2-cyano-1-ethoxyethene, formed by lithiation. The alkenyllithium condenses readily with carbonyl compounds.[2]

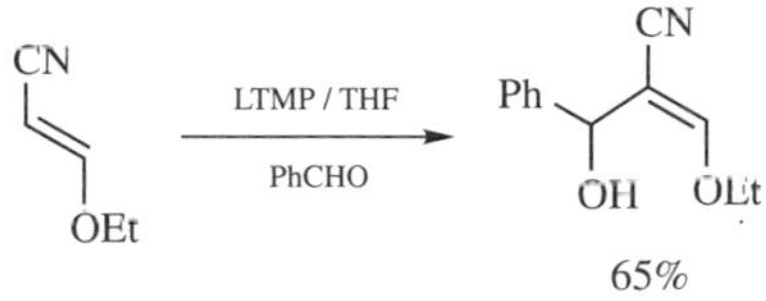

LTMP / THF; PhCHO — 65%

[1]Kristensen, J., Lysen, M., Vedso, P., Begtrup, M. *OL* **3**, 1435 (2001).
[2]Yoshimatsu, M., Yamaguchi, S., Matsubara, Y. *JCS(P1)* 2560 (2001).

Lithium 2,3,4,5-tetraphenylsilole.

Pinacol coupling.[1] This reagent and the germole analogue (**1**) promote reductive coupling of aldehydes at − 78°.

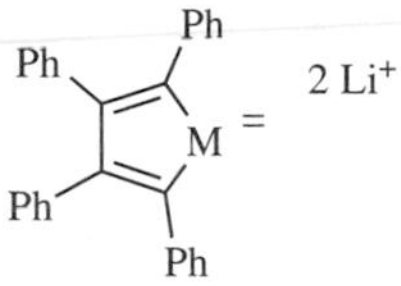

2 Li$^+$

1

[1]Liu, Y., Ballweg, D., West, R. *OM* 20, 5769 (2001).

Lithium tri-*t*-butoxyaluminum hydride. 21, 256

Reduction.[1] In the reduction of α-amino ketone derivatives the stereochemical course is determined by the amino protecting group. Cbz-substitution favors *anti*-β-amino alcohol derivatives, whereas trityl analogues, the *syn*-isomers.

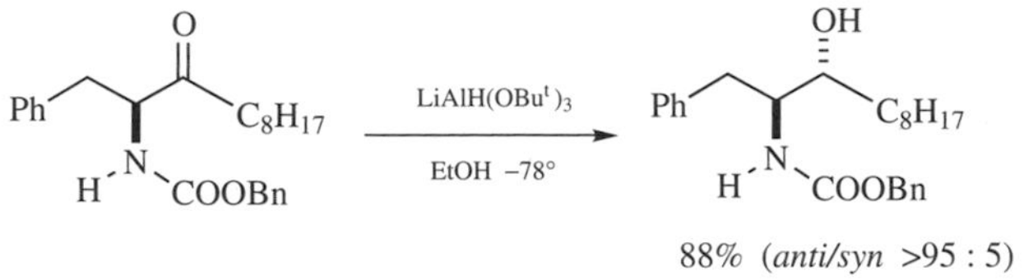

LiAlH(OBu^t)$_3$, EtOH −78°

88% (*anti/syn* >95 : 5)

[1]Hoffman, R.V., Maslouh, N., Cervantes-Lee, F. *JOC* **67**, 1045 (2002).

Lithium triethylborohydride.

Deacylation.[1] Amides and some tertiary carbamates are cleaved by $LiBHEt_3$ in THF between 0° and room temperature, the carbonyl function is reduced to an hydroxyl group.

Reduction. Formation of 2,3-dihydro-4-pyridones is observed when nitriles are treated with the title reagent and then Danishefsky's diene.[2] The imine products (perhaps still complexed to a metal) are capable of partake cycloaddition.

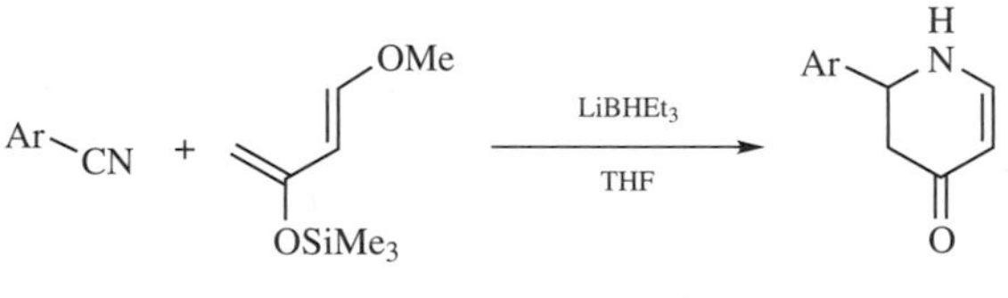

$LiBHEt_3$, THF

γ-Amino alcohols. Reduction of β-hydroxy sulfinimines with $LiBHEt_3$ gives derivatives of *anti*-γ-amino alcohols, whereas a similar reduction with catecholborane leads to predominantly *syn*-isomers.[3]

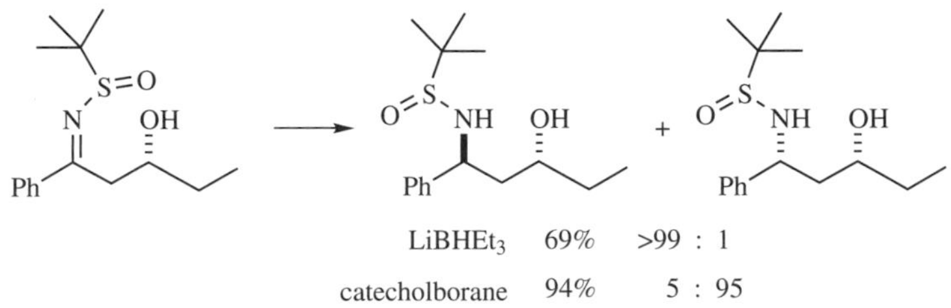

$LiBHEt_3$	69%	>99 : 1
catecholborane	94%	5 : 95

[1]Tanaka, H., Ogasawara, K. *TL* **43**, 4417 (2002).
[2]Kawecki, R. *S* 828 (2001).
[3]Kochi, T., Tang, T.P., Ellman, J.A. *JACS* **124**, 6508 (2002).

Lithium triflamide.

Small ring opening. $LiNTf_2$ catalyzes the reaction of nucleophiles with epoxides[1] and *N*-substituted aziridines[2] in dichloromethane at room temperature.

[1]Cossy, J., Bellosta, V., Hamoir, C., Desmurs, J.-R. *TL* **43**, 7083 (2002).
[2]Cossy, J., Bellosta, V., Alauze, V., Desmurs, J.-R. *S* 2211 (2002).

Lithium trifluoromethanesulfonate. 21, 257

Protection.[1] Tetrahydropyranylation of alcohols are performed with LiOTf as a catalyst.

[1]Karimi, B., Maleki, J. *TL* **43**, 5353 (2002).

M

Magnesium. **13,** 170; **15,** 194; **16,** 198–199; **18,** 224–225; **19,** 205; **20,** 229–230; **21,** 258–259

Desulfonylation. Convenient protocols for the cleavage of sulfonamides including those containing a thiophene ring[1] employ Mg in MeOH. *N*-(2-pyridyl)sulfonamides are cleaved at ice temperature.[2]

A synthesis of (*E*)-alkenylarenes involves condensation of aromatic aldehydes with *p*-tolyl alkyl sulfones and subsequent desulfonylation (Mg in DMF in the presence of Me_3SiCl).[3]

PhCH=C(Me)SO_2Tol —(Mg - Me_3SiCl, DMF)→ Ph-CH=CH-Me 72%

Other reductive defunctionalizations. 2,1,3-Benzothiadiazines are converted into 1,2-arenediamines with Mg/MeOH.[4] Together with hydrazinium monoformate, Mg achieves hydrogenolysis of Cbz-protected amino acids and peptides.[5]

(Mg / MeOH) → NH_2, NH_2 86%

A method for removing one fluorine atom from trifluoromethyl ketones and difluoromethyl ketones is by their treatment with Mg-Me_3SiCl and hydrolysis.[6]

PhC(O)CF_3 —(Mg - Me_3SiCl, THF)→ Ph(OSiMe$_3$)C=CF$_2$

Fiesers' Reagents for Organic Synthesis, Volume 22. Series editor Tse-Lok Ho
ISBN 0-471-28515-3

Coupling reactions. The Barbier protocol is a convenient alternative to the Grignard reaction of halopyridines.[7] Replacement of aromatic halogen with an amino group is accomplished on mixing ArX with acetone mesityloxyimine in the presence of Mg in THF.[8]

Ar—Br + (acetone O-mesitylsulfonyl oxime) $\xrightarrow[\text{HCl}]{\text{Mg / THF ;}}$ Ar—NH_2

Arylboronic acids are readily prepared by electroreduction of ArX in a cell equipped with Mg anode in the presence of trialkyl borates.[9] The products are isolated after hydrolysis.

Aromatic carbonyl compounds are subject to pinacol coupling by Mg in aqueous NH_4Cl[10] or Mg-$InCl_3$-Me_3SiCl.[11] 1,5-Diynes and in some cases 1,2-dien-5-ynes are formed when propargylic carbonates are treated with Mg and $(i\text{-PrO})_2TiCl_2$.[12]

Elimination. Cyclic thionocarbonates derived from dihydroxylated cinnamic esters undergo elimination to afford α-keto esters under the influence of Mg in MeOH, if a strongly electron-withdrawing group is present in the *p*-position of the aromatic nucleus.[13]

Reductive acylations. Mediated by Mg, aromatic carbonyl compounds react with acid chlorides to provide α-acyloxybenzyl ketones (20 examples, 30–98%, lowest yield is from isopropyl phenyl ketone).[14]

(3-Cl-C$_6$H$_4$)—CHO + R—COCl $\xrightarrow{\text{Mg / DMF}}$ (3-Cl-C$_6$H$_4$)—CH(COR)(OCOR)

Conjugated esters[15] and phosphonates[16] undergo reductive acylation at the β-carbon atom when they are treated with acid chlorides or anhydrides and Mg-Me_3SiCl in DMF. Stilbene and acenaphthylene derivatives are similarly acylated,[17] while silylation of α-phosphorylacrylic esters is realizable under essentially the same conditions.[18]

(4-MeO-C$_6$H$_4$)—CH=C(P(O)(OEt)$_2$)COOEt + Me_3Si—Cl $\xrightarrow{\text{Mg / DMF}}$ (4-MeO-C$_6$H$_4$)—CH(Me_3Si)—CH(P(O)(OEt)$_2$)COOEt

97%

[1]Nenajdenko, V.G., Karpov, A.S., Balenkova, E.S. *TA* **12**, 2517 (2001).
[2]Pak, C.S., Lim, D.S. *SC* **31**, 2209 (2001).
[3]Nishiguchi, I., Matsumoto, T., Kuwahara, T., Kyoda, M., Maekawa, H. *CL* 478 (2002).
[4]Prashad, M., Liu, Y., Repic, O. *TL* **42**, 2277 (2001).
[5]Gowda, D.C. *TL* **43**, 311 (2002).
[6]Prakash, G.K.S., Hu, J., Olah, G.A. *JFC* **112**, 357 (2001).
[7]Sugimoto, O., Yamada, S., Tanji, K. *TL* **43**, 3355 (2002).
[8]Erdik, E., Daskapan, T. *TL* **43**, 6237 (2002).
[9]Laza, C., Dunach, E., Serein-Spirau, F., Moreau, J.J.E., Vellutini, L. *NJC* **26**, 373 (2002).
[10]Li, J.-T., Bian, Y.-J., Zang, H.-J., Li, T.-S. *SC* **32**, 547 (2002).
[11]Mori, K., Ohtaka, S., Uemura, S. *BCSJ* **74**, 1497 (2001).
[12]Zhao, G., Ding, Y., Zhao, Z., Zheng, Y., Yang, F. *TL* **43**, 1289 (2002).
[13]Rho, H.S., Ko, B.-S. *SC* **31**, 283 (2001).
[14]Nishiguchi, I., Sakai, M., Maekawa, H., Ohno, T., Yamamoto, Y., Ishino, Y. *TL* **43**, 635 (2002).
[15]Ohno, T., Sakai, M., Ishino, Y., Shibata, T., Maekawa, H., Nishiguchi, I. *OL* **3**, 3439 (2001).
[16]Kyoda, M., Yokoyama, T., Maekawa, H., Ohno, T., Nishiguchi, I. *SL* 1535 (2001).
[17]Nishiguchi, I., Yamamoto, Y., Sakai, M., Ohno, T., Ishino, Y., Maekawa, H. *SL* 759 (2002).
[18]Kyoda, M., Yokoyama, T., Kuwahara, T., Maekawa, H., Nishiguchi, I. *CL* 228 (2002).

Magnesium bromide. 15, 194–196; **16,** 199; **17,** 174; **18,** 226–227; **19,** 206–207; **20,** 230–232; **21,** 260.

Alcoholysis and amidation. Transformation of *N*-acyloxazolidinones to esters are readily accomplished[1] in the presence of $MgBr_2 \cdot OEt_2$ although some other Lewis acids are serviceable also. Amides are formed by the catalyzed reaction of esters with amines,[2] often with selectivity.

COOMe / COOMe (indole, NH) + RNHR′ —(MgX_2, THF)→ COOMe / CONRR′ (indole, NH)

The effectiveness of $MgBr_2 \cdot OEt_2$ in the promotion of further acylation of secondary amides with anhydrides may be due to dual activation, formation of bromomagnesium derivatives of the amides and acyl bromides.[3]

Aldol reactions. Highly diastereoselective reactions are reported. α-Hetero-substituted silyl ketene acetals are included as donors.[4,5] Furthermore, asymmetric synthesis is possible.[6]

Ph–CH(OH)–CBr(CH₃)–COOEt + Bu_3Sn—H —(Et_3B; $MgBr_2 \cdot OEt_2$, CH_2Cl_2 0°)→ Ph–CH(OH)–CH(CH₃)–COOEt

87% (*syn/anti* 5 : 1)

Similarly, using imines instead of aldehydes in this reaction leads to β-amino esters.[7]

Grignard reactions. Addition of $MgBr_2$ to the reaction media is usually intended for its chelation property. Substrate invariably contain a proximal donor atom. α-(Benzyloxymethylsiloxy) carbonyl compounds have been studied with respect to this effect.[8]

$MgBr_2$; PhMgBr

86% (*dr* 72 : 28)

Addition reactions. Hydrostannylation of 1-alkynes with Bu_2SnIH is catalyzed by $MgBr_2 \cdot OEt_2$ to afford 2-stannyl-1-alkenes.[9]

In a radical cyclization performed on the (+)-isosorbide ester of *o*-(2-bromoethyl)cinnamic acid the presence of $MgBr_2 \cdot OEt_2$ or $ZnCl_2$ renders it diastereoselective to an enormous degree (>100:1 vs 2.9:1 in its absence).[10]

+ $Bu_3Sn{-}H$ — Et_3B - O_2; $MgBr_2 \cdot OEt_2$; CH_2Cl_2 −78°

85%

Allylic bromides. Tertiary cyclopropanols are converted to 2-substituted allyl bromides[11] while interesting stereochemical consequences are noted in the transformation of Baylis-Hillman adducts.[12]

+ Ac_2O — $MgBr_2$; THF

X	Yield	Ratio
X = COOMe	86%	100 : 0
X = CN	85%	7 : 93

[1]Orita, A., Nagano, Y., Hirano, J., Otera, J. *SL* 637 (2001).

[2]Guo, Z., Dowdy, E.D., Li, W.-S., Polniaszek, R., Delaney, E. *TL* **42**, 1843 (2001).

[3]Yamada, S., Yaguchi, S., Matsuda, K. *TL* **43**, 647 (2002).

[4]Fujisawa, H., Sasaki, Y., Mukaiyama, T. *CL* 190 (2001).

[5]Guindon, Y., Houde, K., Prevost, M., Cardinal-David, B., Landry, S.R., Daoust, B., Bencheqroun, M., Guerin, B. *JACS* **123**, 8496 (2001).
[6]Kiyooka, S., Shahid, K.A. *BCSJ* **74**, 1485 (2001).
[7]Ha, H.-J., Ahn, Y.-G., Woo, J.-S., Lee, G.S., Lee, W.K. *BCSJ* **74**, 1667 (2001).
[8]Trzoss, M., Shao, J., Bienz, S. *T* **58**, 5885 (2002).
[9]Shibata, I., Suwa, T., Ryu, K., Baba, A. *JACS* **123**, 4101 (2001).
[10]Enholm, E.J., Cottone, J.S., Allais, F. *OL* **3**, 145 (2001).
[11]Korzyrkov, Y.Yu., Kulinkovich, O.G. *SL* 443 (2002).
[12]Ravichandran, S. *SC* **31**, 2059 (2001).

Magnesium iodide. 20, 232; 21, 261

Aldol reactions. MgI_2 and $MgBr_2$ show similar catalytic power in aldol reactions between silyl enol ethers and aldehydes.[1] Electron-rich aromatic aldehydes react much faster.

OSiMe3 OSiMe3 + ArCHO —($MgI_2 \cdot OEt_2$, CH_2Cl_2 25°)→ O OSiMe3 Ar OR

Ring expansion. Further extension of the reported reaction concerning a pyrrolidine synthesis is the decomposition of the adducts into α-(2-aminoethyl) enones from cyclopropyl ketones, aldehydes, and amines.[2] By a slight variation of the cyclopropane structure dihydropyridone derivatives are also obtained.

O N O O —(MgI_2, THF)→ I IMg O N O O — Cl Cl TsN →

X Cl O N Ts Cl

X = I --> OAc

[1]Li, W.-D.Z., Zhang, X.-X. *OL* **4**, 3485 (2002).
[2]Bertozzi, F., Gustafsson, M., Olsson, R. *OL* **4**, 3147, 4333 (2002).
[3]Lautens, M., Han, W. *JACS* **124**, 6312 (2002).

Magnesium monoperoxyphthalate, MMPP. 14, 197; **16,** 199–200; **18,** 228; **19,** 207–208

Baeyer-Villiger reaction. Aromatic aldehydes are degraded to phenols by MMPP in MeOH.[1]

vic-Tricarbonyl compounds.[2] On contact with moist MMPP α-triphenylphosphoranyl-β-keto esters lose the phosphorus substituent.

[1]Heaney, H., Newbold, A.J. *TL* **42**, 6607 (2001).
[2]Lee, K., Im, J.-M. *TL* **42**, 1539 (2001).

Magnesium oxide.

N-Acylation. Reaction of amines with acyl chlorides can be carried out in aqueous organic solvent systems in the presence of MgO[1] (Cf. Schotten-Baumann reaction).

+ RCOCl —(MgO, THF - H_2O)→

[1]Kim, D.-H., Rho, H.-S., You, J.W., Lee, J.C. *TL* **43**, 277 (2002).

Magnesium perchlorate. 18, 228; **19,** 208

Removal of N-Boc group. One of the Boc groups in *N′,N′*-DiBoc derivatives of hydrazides can be removed with $Mg(ClO_4)_2$ in MeCN.[1]

Halogenation. Bromination of β-keto esters with NBS at the α-position is subject to Lewis acid catalysis[2] [e.g., by $Mg(ClO_4)_2$].

Aldol reactions. A chelated $Mg(ClO_4)_2$ catalyzes the formation of β-hydroxy-α-amino acid derivatives from isothiocyanatoacetic esters and aldehydes.[3]

Ph—CHO + EtOOC—N=C=S —($Mg(ClO_4)_2$ · bipy, Et_3N / THF)→

84% (*cis/trans* 65 : 35)

Ene reaction. Catalysis by $Mg(ClO_4)_2$ in conjuction with a bis(oxazoline) ligand is the key to obtain the *cis*-3,4-dialkylpyrrolidone precursor of (−)-α-kainic acid.[4] The uncatalyzed thermal ene reaction can produce the *trans*-isomer in predominance.

EtOOC O N R $\xrightarrow[\Delta]{Mg(ClO_4)_2 / CH_2Cl_2\ 23°}$ EtOOC O N R + EtOOC O N R

R = Bz : > 20 : 1
R = H 1 : 5

[1] Brosse, N., Jamart-Gregoire, B. *TL* **43**, 249 (2002).
[2] Yang, D., Yan, Y.-L., Lui, B. *JOC* **67**, 7429 (2002).
[3] Willis, M.C., Piccio, V.J.-D. *SL* 1625 (2002).
[4] Xia, Q., Ganem, B. *OL* **3**, 485 (2001).

Manganese. 20, 233–234, **21,** 261

Allylation. Active Mn is used to prepared allylmanganese and other reagents.[1] Allylation of aldehydes by reagents derived from 1,3-dichloropropene in the presence of chiral Cr(salen) complex provides optically active chlorohydrins and thence epoxides.[2]

Coupling reactions. The pinacol coupling of aldehydes mediated by Mn and catalyzed by chiral Ti complexes varies greatly with respect to asymmetric induction.[3] Cross-coupling between conjugated carbonyl compounds and aliphatic aldehydes[4] is synthetically interesting.

O + OHC Ph $\xrightarrow[Me_3SiCl\ /\ DMF\ ;\ Bu_4NF]{Mn\ -\ CrCl_2}$ HO OH Ph 80%

[1]Kakiya, H., Nishimae, S., Shinokubo, H., Oshima, K. *T* **57**, 8807 (2001).
[2]Bandini, M., Cozzi, P.G., Melchiorre, P., Morganti, S., Umani-Ronchi, A. *OL* **3**, 1153 (2001).
[3]Bensari, A., Renaud, J.-L., Riant, O. *OL* **3**, 3863 (2001).
[4]Jung, M., Groth, U. *SL* 2015 (2002).

Manganese(III) acetate. 13, 171; **14,** 197–199; **16,** 200; **17,** 175–176; **18,** 229–230; **19,** 209–210; **20,** 234–235; **21,** 261–263

Coupling reactions. Oxidation of arylhydrazines in benzene gives biaryls.[1] This reaction can be used to introduce an aryl group to C-2 of furan and thiophene.[2]

X + Ph−NHNH$_2$ HCl $\xrightarrow{Mn(OAc)_3 \cdot 2H_2O}$ X Ph

X = O 65%
X = S 70%

Alkylations. The free-radical alkylation mediated by $Mn(OAc)_3 \cdot 2H_2O$ (and sometimes $Cu(OAc)_2 \cdot H_2O$ is also added) can lead to products otherwise difficult to obtained directly.[3,4]

$Mn(OAc)_3 \cdot 2H_2O$, $Cu(OAc)_2 \cdot H_2O$, CH_2Cl_2 Δ

COOEt + OBu → OHC, COOEt 92%

Ring formation. A pyrrole ring can be created out of the enamine portion of 2-amino-1,4-naphthoquinones and carbonyl compounds by oxidation.[5] Attack of a malonyl α-radical to a sterically accessible aromatic ring completes an annulation process.[6] Macrocyclization of 3,3-diphenyl-2-propenyloxyoligomethylene 3-oxobutanoates is remarkable.[7]

NHR + O → ($Mn(OAc)_3 \cdot 2H_2O$, HOAc 45°) N R

R = Et 73%

COOEt, COOEt, COOEt → ($Mn(OAc)_3$ EtOH ; O_2 - NaI / DMF 120°) COOEt, COOEt

46%

O, (CH$_2$)n, Ph, Ph → ($Mn(OAc)_3$, HOAc) O, (CH$_2$)n, Ph, Ph

51 - 94% (n = 2, 3, 4, 6, 8)

[1]Demir, A.S., Reis, Ö., Özgül-Karaaslan, E. *JCS(P1)* 3042 (2001).
[2]Demir, A.S., Reis, Ö., Emrullahoglu, M. *T* **58**, 8055 (2002).
[3]Tanyeli, C., Özdemirhan, D. *TL* **43**, 3977 (2002).
[4]Bar, G., Parsons, A.F., Thomas, C.B. *SL* 1069 (2002).
[5]Wu, Y.-L., Chuang, C.-P., Lin, P.-Y. *T* **57**, 5543 (2001).
[6]Im, Y.J., Lee, K.Y., Kim, T.H., Kim, J.N. *TL* **43**, 4675 (2002).
[7]Jogo, S., Nishino, H., Yasutake, M., Shinmyozu, T. *TL* **43**, 9031 (2002).

Manganese(III) acetylacetonate.

Oxidation of alcohols.[1] For the oxidation of secondary alcohols to carbonyl compounds in good yields is a new system consisting of $Mn(acac)_3$, RCN, and CCl_4. However, the reaction temperature is high (200°).

[1]Khusnutdinov, R.I., Schadneva, N.A., Baiguzina, A.R., Dzhemilev, U.M. *RCB* **51**, 1065 (2002).

Manganese dioxide. 14, 200–201; **15,** 197–198; **18,** 230–231; **19,** 210; **20,** 237–238; **21,** 263

Nitrile oxides.[1] Hydroxylamines are readily oxidized to nitrones in good yields by MnO_2 at ice temperature.

Pyrans.[2] In a synthesis of epoxyquinols the last step was an oxidation by MnO_2. The product underwent electrocyclization to form a pyran ring and then dimerized.

MnO_2, CH_2Cl_2, 0°

40% epoxyquinol A + 25% epoxyquinol B

Nitrogenous compounds from alcohols. For alcohols oxidizable by activated MnO_2 the presence of alkoxylamines,[3] amines or ammonia[4], the carbonyl products would undergo further condensation. However, the synthesis of conjugated nitriles[5] is just a simple extension of the method developed for aromatic nitriles.[6]

The differential susceptibility of imines and aldehydes to reduction by cyanoborohydride reagent enables preparation of secondary amines directly from easily oxidized alcohols and primary amines. Imine formation in situ with the help of 4A-molecular sieves (after oxidation of the alcohols) is followed by reduction with polymer-supported cyanoborohydride.[7]

Oxidative cleavage. Cyclopropanone acetals suffer oxidative cleavage by MnO_2, leading to esters.[8]

MnO2 CF3SO3H / MeOH 71%

Homologous alkenes from alcohols.[9] Oxidation of alcohols carried out in the presence of a phosphonium ylide is followed by a Wittig reaction.

MnO2 THF 91%

[1]Cicchi, S., Marradi, M., Goti, A., Brandi, A. *TL* **42**, 6503 (2001).
[2]Shoji, M., Yamaguchi, J., Kakeya, H., Osada, H., Hayashi, Y. *ACIEE* **41**, 3192 (2002).
[3]Kanno, H., Taylor, R.J.K. *SL* 1287 (2002).
[4]McAllister, G.D., Wilfred, C.D., Taylor, R.J.K. *SL* 1291 (2002).
[5]Foot, J.S., Kanno, H., Giblin, G.M.P., Taylor, R.J.K. *SL* 1293 (2002).
[6]Ho, T.-L. *CI* 400 (1988).
[7]Blackburn, L., Taylor, R.J.K. *OL* **3**, 1637 (2001).
[8]Nakamura, M., Inoue, T., Nakamura, E. *JOMC* **624**, 300 (2001).
[9]Blackburn, L., Pei, C., Taylor, R.J.K. *SL* 215 (2002).

Mercury(II) acetate. 15, 198–199; **17,** 176–177; **18,** 232; **19,** 211; **20,** 238; **21,** 263–264

Detritylation. Removal of *S*-trityl groups is accomplished by $Hg(OAc)_2$ with subsequent reductive demercuration ($NaBH_4$).[1]

Transsilylation.[2] Conversion of the conventional trialkylsilyl enol ethers to trichlorosilyl analogues enables the employment of Lewis bases to promote aldol reactions. The changeover ($R_3Si \rightarrow Cl_3Si$) is performed with $SiCl_4$ and $Hg(OAc)_2$.

Hg(OAc)2 / CH2Cl2 ; PhCHO / HMPA 88%

Cleavage of aromatic ketones.[3] Both alkyl aryl ketones and diaryl ketones suffer oxidative cleavage to afford aromatic carboxylic acids on treatment with $Hg(OAc)_2$ and *N*-bromophthalimide.

[1]Maltese, M. *JOC* **66**, 7615 (2001).
[2]Denmark, S.E., Pham, S.M. *OL* **3**, 2201 (2001).
[3]Anjum, A., Srinivas, P. *CL* 900 (2001).

Mercury(II) oxide.

Hydrolysis of 1-O-acylglycosides.[1] Per-*O*-acyl sugars undergo selective hydrolysis at the glycosidic site when they are heated with HgO and $HgCl_2$ in aqueous acetone.

Oxidative cleavage of cyclopropane derivatives.[2] Cyclopropyl ketone hydrazones undergo ring cleavage with formation of enol acetates by reaction with HgO and $Hg(OAc)_2$.

NNH_2 → OAc

HgO - $Hg(OAc)_2$ / THF ; $NaBH_4$ / H_2O

84%

[1]Sambaiah, T., Fanwick, P.E., Cushman, M. *S* 1450 (2001).
[2]Di Chenna, P.H., Ferrara, A., Ghini, A.A., Burton, G. *JCS(P1)* 227 (2002).

Mercury(II) triflate-tetramethylurea.

Methyl ketones.[1] Hydration of 1-alkynes with $Hg(OTf)_2{\cdot}(TMU)_2$ in a mixture of aqueous MeOH and THF proceeds at room temperature.

[1]Nishizawa, M., Skwarczynski, M., Imagawa, H., Sugihara, T. *CL* 12 (2002).

Metaboric acid.

Beckmann rearrangement.[1] Heating ketoximes with solid metaboric acid at 140–145° gives amides.

[1]Chandrasekhar, S., Gopalaiah, K. *TL* **42**, 2455 (2001).

Methanesulfonic acid. 20, 240; 21, 264–265

Debenzylation.[1] A reagent system composed of MsOH and anisole cleaves the the exocyclic N-C bond of *N*-(1-phenylethyl)oxazolidin-2-ones, although C-4 also bears a phenyl substituent and therefore is also benzylic.

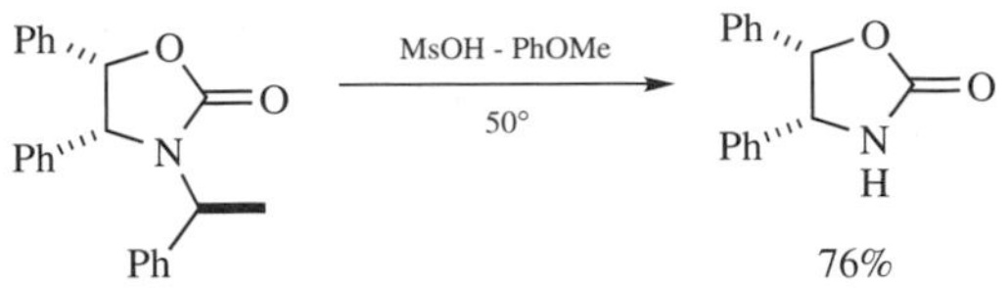

Ph,,, O ... MsOH - PhOMe, 50° ... 76%

[1]Sugiyama, S., Morishita, K., Chiba, M., Ishii, K. *H* **57**, 637 (2002).

Methanesulfonyl chloride.

Methyl esters.[1] Esterification of carboxylic acids by a mixture of MsCl and pyridine is reported.

[1]Siddiqui, B.S., Begum, F., Begum, S. *TL* **42**, 9059 (2001).

2-(4-Methoxybenzyloxy)-3-nitropyridine.

p-Methoxybenzylation.[1] The title reagent is an efficient reagent for protection of alcohols. The group transfer is completed rapidly at room temperature with Me_3SiOTf as catalyst.

[1]Nakano, M., Kikuchi, W., Matsuo, J., Mukaiyama, T. *CL* 424 (2001).

N-Methoxy-*N*-methyl-3-bromopropanamide.

Homologation.[1] The title reagent is a unit for bidirectional, functional three-carbon chain extension. For example, it selectively reacts with nucleophiles at one end or the other (with the proviso that protection at the nonreacting site is arranged).

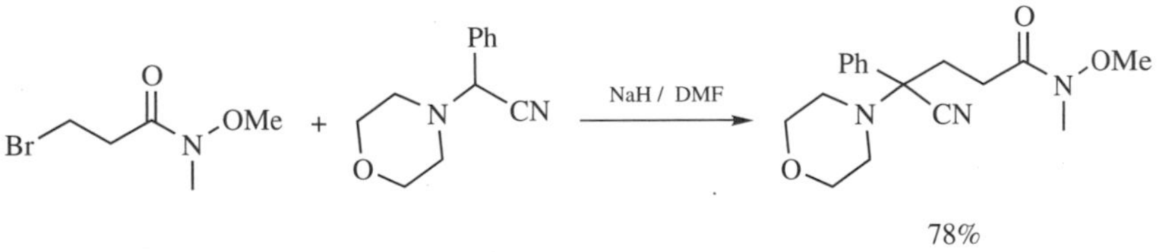

NaH / DMF

78%

[1]Selvamurugan, V., Aidhen, I.S. *T* **57**, 6065 (2001).

N-(2-Methoxycarbonylbenzoyl)succinimide.

Amino protection.[1] Phthalimides are formed on reaction of amino acids or peptides with the title reagent at room temperature. Much milder conditions are required than traditional condensation with phthalic anhydride.

[1]Casimir, J.R., Guichard, G., Briand, J.P. *JOC* **67**, 3764 (2002).

***N*-(α-Methylbenzyl)-3-alkyldiaziridines.**

Aziridination.[1] The title reagents transfer an [NH] residue to conjugated amides under mild conditions. Contrasting stereoselectivity in the reaction involving 3,3-pentamethylenediaziridine is noted.

Ph HN N

BuLi / PhMe −78°

Ph O N

Ph NH O N

(98% ee)

[1]Ishihara, H., Ito, Y.N., Katsuki, T. *CL* 984 (2001).

Methylaluminum biphenyl-2-perfluorooctanesulfonamido-2′-oxide.

Oppenauer oxidation.[1] Compound **1** catalyzes oxidation of alcohols with acetone (15 examples, 80–98%).

O AlMe N $SO_2C_8F_{17}$

1

[1]Ooi, T., Otsuka, H., Miura, T., Ichikawa, H., Maruoka, K. *OL* **4**, 2669 (2002).

***N*-Methyl-2-chloropyridinium iodide.**

Bicyclic β-lactones.[1] The pyridinium salt reacts with carboxylic acids to afford ketenes. ω-Formylalkanoic acids in which the separation of the two end groups is by 5 or 6 bonds would give bicyclic β-lactones as a result of an intramolecular [2+2]cycloaddition. This reaction is amenable to asymmetric synthesis, e.g., using a base system of *O*-acetylquinidine and *i*-Pr_2NEt.

O O COOH CHO + Cl N + I^-

Et_3N / MeCN

O O O O

66%

[1] Cortez, G.S., Tennyson, R.L., Romo, D. *JACS* **123**, 7945 (2001).

Methylidynetricobalt nonacarbonyl.

Cycloaddition reactions. Alkynes trimerize to give substituted benzenes[1] under the influence of the title complex. Under proper conditions diynes and CO combine to form cyclopentadienones.[2]

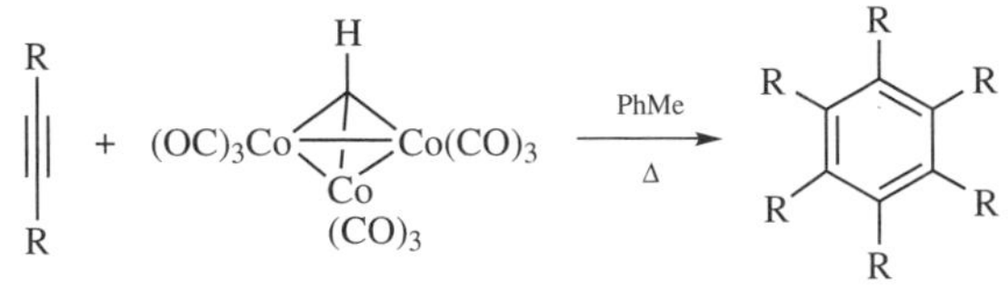

R = Pr 92%

[1]Sugihara, T., Wakabayashi, A., Nagai, Y., Takao, H., Imagawa, H., Nishizawa, M. *CC* 576 (2002).
[2]Sugihara, T., Wakabayashi, A., Takao, H., Imagawa, H., Nishizawa, M. *CC* 2456 (2001).

***O*-Methylisoureas, polymer-supported.**

O-Methylation. Carboxylic acids react with this reagent to furnish methyl esters[1].

[1]Crosignani, S., White, P.D., Linclau, B. *OL* **4**, 1035 (2002).

***N*-Methyl-3-(*p*-methoxyphenyl)-5,6-dichloroisoindolo-1-one salts.**

Glycosylation. Reaction of glycosyl fluorides with alcohols is catalyzed by these salts (**1**).[1] Remarkably, the counterion and solvent system affect the stereoselectivity of the reaction: triflate in ether–dichloromethane favors formation of the α-anomers, whereas the tetrakis(pentafluorophenyl)borate in pivalonitrile–dichloromethane, the β-anomers.

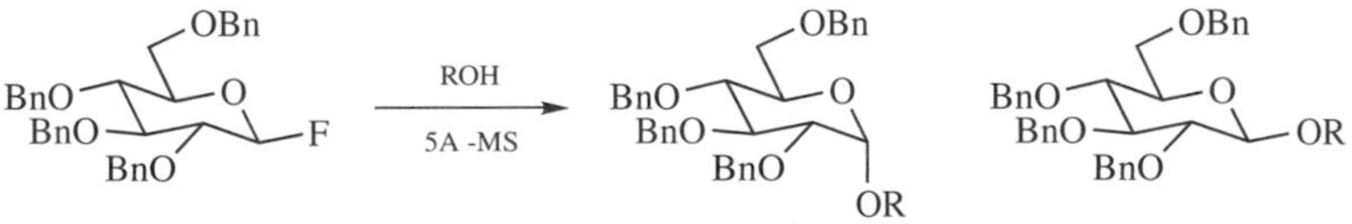

α
(X = OTf)

β
(X = $B[C_6F_5]_4$)

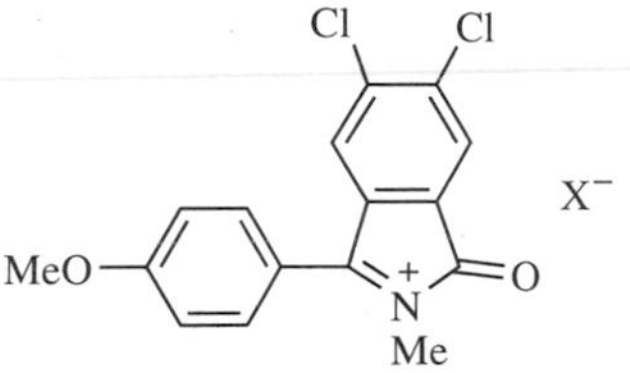

1

[1] Yanagisawa, M., Mukaiyama, T. *CL* 224 (2001).

2-Methyl-6-nitrobenzoic anhydride.

Macrolactonization. This anhydride mediates cyclization of ω-hydroxycarboxylic acids efficiently in the presence of Et_3N and DMAP at room temperature.[1]

[1]Shiina, I., Kubota, M., Ibuka, R. *TL* **43**, 7535 (2002).

***N*-Methyl-*N*-(*o*-nitrophenyl)carbamoyl chloride.**

Hydroxyl protection.[1] Carbamates derived from the title reagent are photolabile.

[1]Loudwig, S., Goeldner, M. *TL* **42**, 7957 (2001).

2-Methylsulfonylethanol.

Phenols. Aryl fluorides containing electron-withdrawing substituents at an *o*- or *p*-position are susceptible to displacement with $MeSO_2CH_2CH_2ONa$ in DMF. In situ elimination then affords ArOH.[1]

[1]Rogers, J.F., Green, D.M. *TL* **43**, 3585 (2002).

Methyltrioxorhenium–hydrogen peroxide. 19, 217; **20,** 248; **21,** 270

Oxidations. Comparing with other methods conditions for the oxidation of purines (to *N*-oxides) by this reagent system are milder.[1] Baeyer-Villiger oxidation of flavanones is achieved with this reagent pair.[2]

Epoxidations.[3] Improvements are shown when epoxidation is carried out in $(CF_3)_2CHOH$.

[1]Jiao, Y., Yu, H. *SL* 73 (2001).
[2]Bernini, R., Mincione, E., Cortese, M., Aliotta, G., Oliva, A., Saladino, R. *TL* **42**, 5401 (2001).
[3]Iskra, J., Bonnet-Delpon, D., Begue, J.-P. *TL* **43**, 1001 (2002).

Methyltriphenylphosphinegold(I).

Hydration.[1] 1-Alkynes are converted into methyl ketones in the presence of $(Ph_3P)AuMe$ and TfOH. Hydration of internal alkynes affords ketones, as regioisomeric mixtures when the alkynes are unsymmetrical.

[1]Mizushima, E., Sato, K., Hayashi, T., Tanaka, M. *ACIEE* **41,** 4563 (2002).

Methyltriphenylphosphonium borohydride.

Reduction.[1] This stable reagent achieves reductions with similar power to sodium borohydride, but it can be used in dichloromethane.

[1]Firouzabadi, H., Adibi, M., Ghadami, M. *PSS* **142**, 125 (1998).

Molybdenum carbene complexes. 17, 194–195; **18,** 242–243; **19,** 219–221; **20,** 249–251; **21,** 271–273

Alkene metathesis. Schrock's catalyst performs better than Grubb's catalyst in ring-closing metathesis (RCM) that leads to a [7.7]paracyclophane skeleton.[1] There is a self-editing process during the metathesis.

MeO OMe MeO OMe Bu Bu [Mo] Bu MeO OMe MeO OMe Bu MeO OMe [Mo] Bu MeO OMe Bu

Much recent effort has been devoted to the development of catalytic asymmetric metathesis. Chiral Mo-carbene species based on a biphenyl moiety (**1**),[2–4] including a recyclable polymer-supported version,[5] and those containing substituted BINOL (**2**)[6,7] and octahydro-BINOL (**3**)[8] enjoy great success.

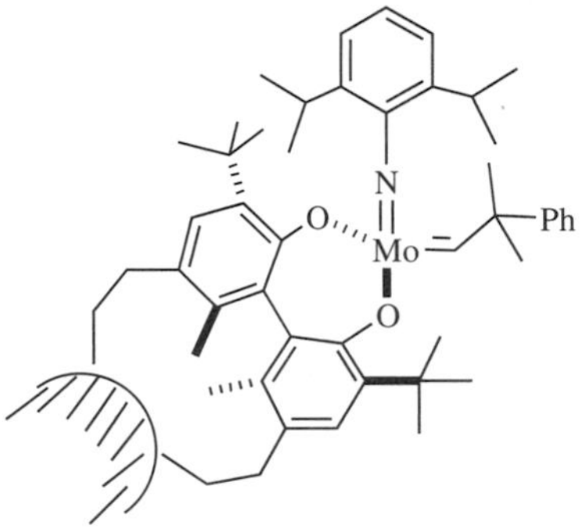

1

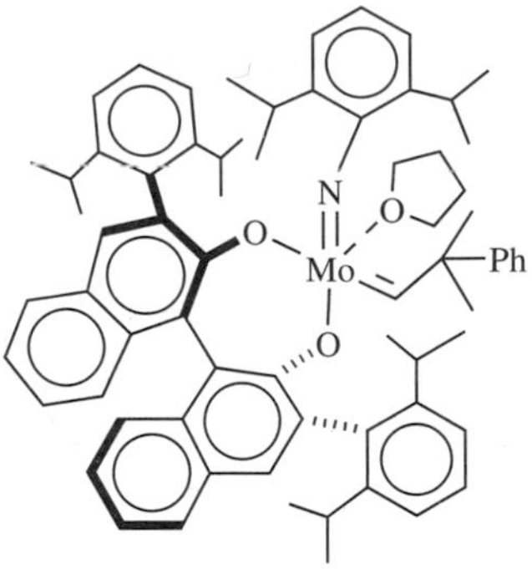

2

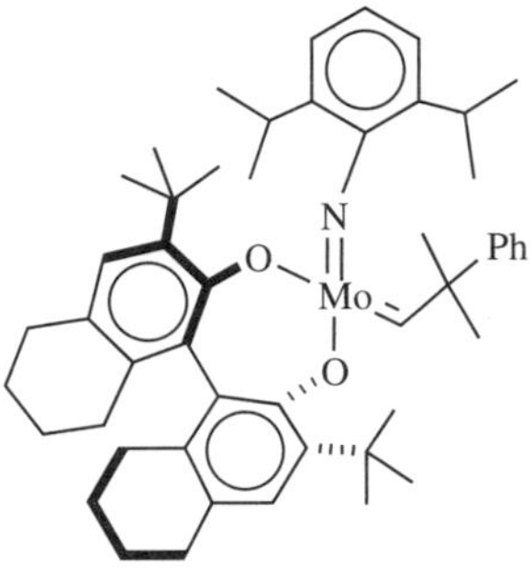

3

In situ activation of **4** by dichloromethane enables it to catalyze cross-coupling of functionalized alkynes.[9]

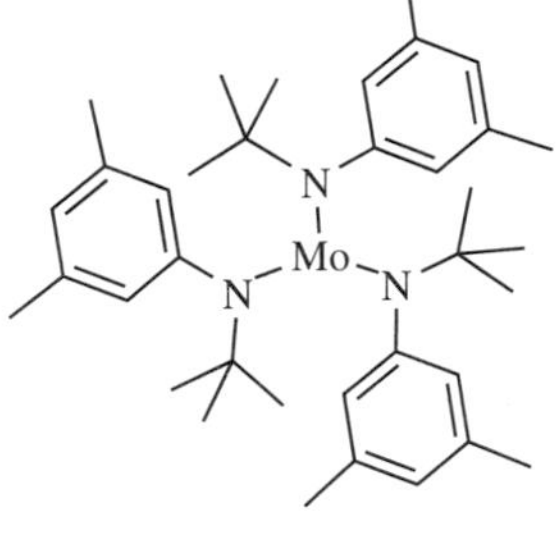

4

[1]Smith III, A.B., Adams, C.M., Kozmin, S.A., Paone, D.V. *JACS* **123**, 5925 (2001).
[2]La, D.S., Sattely, E.S., Ford, J.G., Schrock, R.R., Hoveyda, A.H. *JACS* **123**, 7767 (2001).

[3]Kiely, A.F., Jemelius, J.A., Schrock, R.R., Hoveyda, A.H. *JACS* **124**, 2868 (2002).
[4]Dolman, S.J., Sattely, E.S., Hoveyda, A.H., Schrock, R.R. *JACS* **124**, 6991 (2002).
[5]Hultzsch, K.C., Jemelius, J.A., Hoveyda, A.H., Schrock, R.R. *ACIEE* **41**, 589 (2002).
[6]Cefalo, D.S., Kiely, A.F., Wucherer, M., Jamieson, J.Y., Schrock, R.R., Hoveyda, A.H. *JACS* **123**, 3139 (2001).
[7]Teng, X., Cefalo, D.S., Schrock, R.R., Hoveyda, A.H. *JACS* **124**, 10779 (2002).
[8]Aeilts, S.L., Cefalo, D.R., Bonitatebus Jr, P.J., Houser, J.H., Hoveyda, A.H., Schrock, R.R. *ACIEE* **40**, 1452 (2001).
[9]Fürstner, A., Mathes, C. *OL* **3**, 221 (2001).

Molybdenum(V) chloride.

Ether cleavage. $MoCl_5$ catalyzes ether cleavage. Thus in the presence of an acid chloride ethers are transformed into esters ($R_2O + R'COCl \rightarrow R'COOR$). Other efficient catalysts are $NbCl_5$, $TaCl_5$, and WCl_6.[1]

2,2′Cyclolignans. Regioselective coupling to form 8-membered ring products instead of structural isomers containing a 6-membered ring is remarkable. The rationale is that Mo gathers (complexes) the oxygen substituents on the aromatic rings to render the coupling proceed in the observed manner.

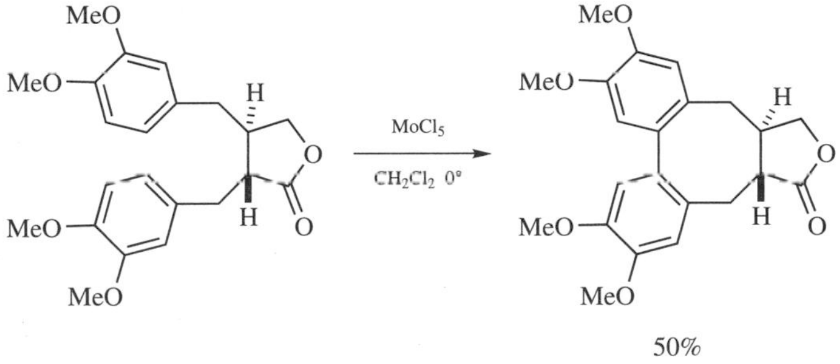

$MoCl_5$, CH_2Cl_2 0° → 50%

[1]Guo, Q., Miyaji, T., Hara, R., Shen, B., Takahashi, T. *T* **58**, 7327 (2002).
[2]Kramcr, B., Avcrhoff, A., Waldvogcl, S.R. *ACIEE* **41**, 2981 (2002).

Molybdenum hexacarbonyl.
13, 194–195; **15,** 225–226; **18,** 243–244; **19,** 221–222; **20,** 251–252; **21,** 273–274

Reductive N-O bond cleavage. *O*-Alkylhydroxylamines[1] and isoxazoles[2] are split by $Mo(CO)_6$ in aqueous MeCN, the latter into β-amino enones. Isoxazolines are cleaved in the same manner.[3]

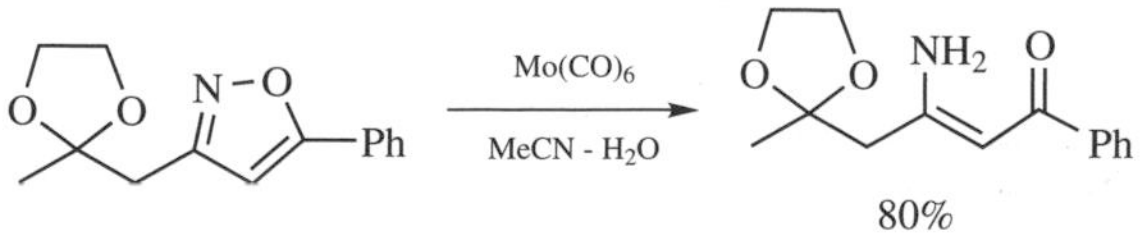

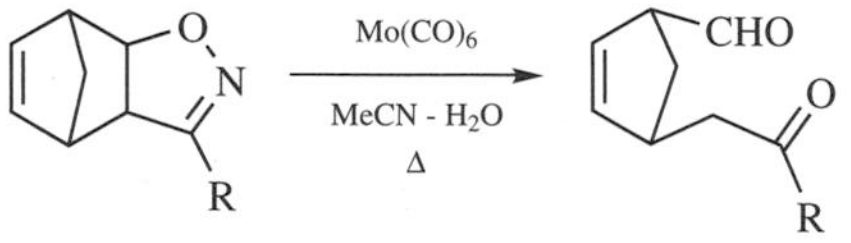

$Mo(CO)_6$, MeCN - H_2O → 80%

$Mo(CO)_6$, MeCN - H_2O, Δ

Metathesis. An active metathesis catalyst for functionalized alkynes is formed by combining $Mo(CO)_6$ with a silanol (Ph_3SiOH).[4]

[1]Cooper, T.S., Larigo, A.S., Laurent, P., Moody, C.J., Takle, A.K. *SL* 1730 (2002).
[2]Li, C.-S., Lacasse, E. *TL* **43**, 3565 (2002).
[3]Tranmer, G.K., Tam, W. *OL* **4**, 4104 (2002).
[4]Villemin, D., Heroux, M., Blot, V. *TL* **42**, 3701 (2001).

N

Nafion-H. 14, 213; **18,** 246; **20,** 253–254; **21,** 275

Sulfonylation.[1] Aryl sulfones are prepared by heating arenes and sulfonic acids with Nafion-H.

Etherifications. *O*-protected mannopyranosyl sulfoxides react with alcohols in the presence of Nafion-H to provide predominantly α-glycosides.[2] Formation of benzhydryl alkyl ethers is accomplished by a Nafion-catalyzed reaction of Ph_2CHOH with ROH in MeCN.[3]

Regeneration of hydroxamic acids.[4] Derivatization of these acids into 5,5-dimethyl-1,4,2-dioxazoles is a means of protection (both C=O and NH are masked). Heating the heterocycles with Nafion-H in isopropanol releases the hydroxamic acids.

Aldol and hetero Diels-Alder reactions.[5] Reaction of Danishefsky's diene with aldehydes and imines has different product profiles when catalyzed by Nafion-H. Aldol products and 2,3-dihydro-4-pyridones, respectively, are obtained.

Me_3SiO OMe — Nafion-H, PhCHO → Me_3SiO, Ph, O, OMe 62%
— Nafion-H, PhCH=NPh → O, NPh, Ph 72%

[1]Olah, G.A., Mathew, T., Prakash, G.K.S. *CC* 1696 (2001).
[2]Nagai, H., Sawahara, K., Matsumura, S., Toshima, K. *TL* **42**, 4159 (2001).
[3]Stanescu, M.A., Varma, R.S. *TL* **43**, 7307 (2002).
[4]Couturier, M., Tucker, J.L., Proulx, C., Boucher, G., Dube, P., Andresen, B.M., Ghosh, A. *JOC* **67**, 4833 (2002).
[5]Kumareswaran, R., Reddy, B.G., Vankar, Y.D. *TL* **42**, 7493 (2001).

Fiesers' Reagents for Organic Synthesis, Volume 22. Series editor Tse-Lok Ho
ISBN 0-471-28515-3

Nickel. 12, 355; **13,** 197; **14,** 213; **18,** 246; **19,** 224; **20,** 253–254; **21,** 276

Hydrogenation. Hydrogenation of nitriles to amines is accomplished at relatively low pressure (10 atm) with Ni-SiO_2 as catalyst (19 examples, 81–99%).[1]

[1]Takamizawa, S., Wakasa, N., Fuchikami, T. *SL* 1623 (2001).

Nickel, Raney. 13, 265–266; **14,** 270; **15,** 278; **17,** 296; **18,** 246; **20,** 254; **21,** 276–277

Hydrodehalogenations. Reduction of organic halides is easily carried out with Ra-Ni.[1,2]

F O Cl → (Ra-Ni, THF Δ) F O
30%

Cl Cl Cl Cl → (Ra-Ni, THF Δ) Cl Cl
>90%

O OMe Br → (Ra-Ni, THF Δ) O OMe
87%

3-Alkyloxindoles. Heating oxindole and alcohols with Ra-Ni in an autoclave gives 3-alkyloxindoles. Note the highly efficient *t*-butylation.[3]

O N H + OH → (Ra-Ni, 200°) O N H
96%

Reduction of nitrogenous compounds. Various hydrogen sources have been tested for the reduction. $HCOONH_4$ is a convenient candidate, e.g., for deoxygenation of *N*-oxides.[4] Hydrazinium formate also finds application in converting $ArNO_2$ into arylamines.[5] For a direct synthesis of tosylhydrazones from nitriles the substrates are treated with Ra-Ni, $TsNHNH_2$ and NaH_2PO_2 in a mixture of HOAc and pyridine in water.[6]

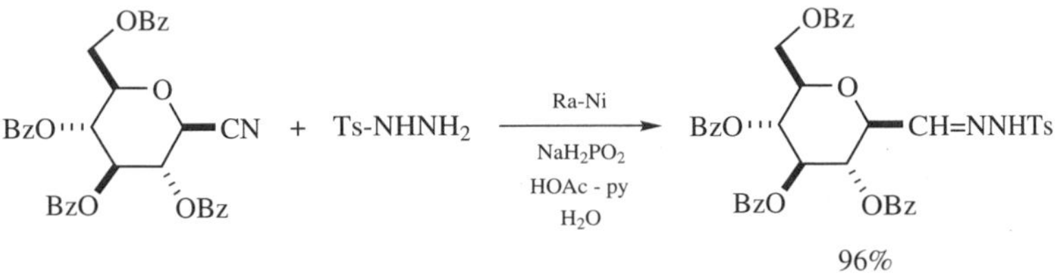

With Pd-activated Ra-Ni hydrogenation of nitriles without affecting *N*-Boc groups is feasible.[7] The breakdown of strongly complexed borane-amine systems can be achieved with Ra-Ni (or Pd-C) in methanol.[8]

[1]Barrero, A.F., Alvarez-Manzaneda, E.J., Chahboun, R., Meneses, R., Romera, J.L. *SL* 485 (2001).
[2]Mebane, R.C., Grimes, K.D., Jenkins, S.R., Deardorff, J.D., Gross, B.H. *SC* **32**, 2049 (2002).
[3]Volk, B., Mezei, T., Simig, G. *S* 595 (2002).
[4]Balicki, R., Maciejewski, G. *SC* **32**, 1681 (2002).
[5]Gowda, S., Gowda, D.C. *T* **58**, 2211 (2002).
[6]Toth, M., Somsak, L. *JCS(P1)* 942 (2001).
[7]Klenke, B., Gilbert, I.H. *JOC* **66**, 2480 (2001).
[8]Couturier, M., Tucker, J.L., Andresen, B.M., Dube, P., Negri, J.T. *OL* **3**, 465 (2001).

Nickel-carbon. 21, 277

Hydrodehalogenation. The Ni-C and Ph_3P combination catalyzes dechlorination of ArCl including polychlorinated biphenyls. Either $Me_2NH{\cdot}BH_3$ or tetramethyldisiloxane can be used as hydrogen source.[1]

Cross-coupling.[2] Ni-C is effective for coupling of alkenyldimethylaluminums and benzylic chlorides.

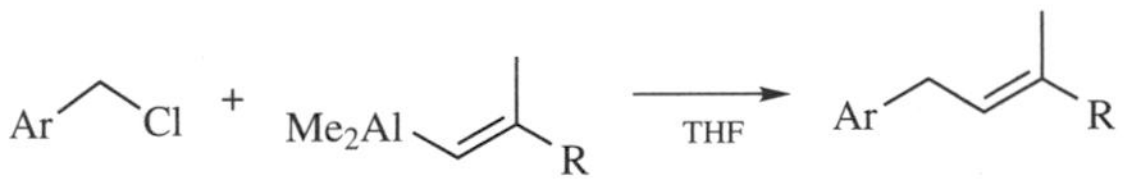

[1]Lipshutz, B.H., Tomioka, T., Sato, K. *SL* 970 (2001).
[2]Lipshutz, B.H., Frieman, B., Pfeiffer, S.S. *S* 2110 (2002).

Nickel acetate–2,2′-bipyridine. 20, 255; **21,** 277

Arylamines. The bipyridine complex of $Ni(OAc)_2$, with NaH/*t*-AmONa, catalyzes the substitution of aryl chlorides with secondary amines in THF. Stagewise substitution is realized in the reaction of polychlorinated arenes (and pyridines).[1] On the other hand, diamines can be *N*-arylated by adjusting the molar quantities of ArCl, to produce mono- or disubstitution products.[2]

Cl Cl Cl + HN O NaH - Nt(OAc)$_2$ t-AmOH 2,2′-bipy / styrene O N Cl Cl 52% piperidine NaH - Nt(OAc)$_2$ t-AmOH 2,2′-bipy / styrene O N N N 73%

HN NH NaH - Nt(OAc)$_2$ 2,2′-bipy t-AmOH - THF ; PhCl, styrene HN NPh (1 equiv PhCl) 68% or PhN NPh (2 equiv PhCl) 78%

[1]Desmarets, C., Schneider, R., Fort, Y. *T* **57**, 7657 (2001).
[2]Brenner, E., Schneider, R., Fort, Y. *T* **58**, 6913 (2002).

Nickel(II) acetylacetonate. 17, 201; **18,** 247–248; **19,** 225–226; **20,** 255–256; **21,** 277–280

Alkylations. Organosilanes are formed by the Ni-catalyzed reaction of propargyl chloride with Cl_3SiH. Subsequent reaction with aldehydes furnishes homopropargylic alcohols.[1] The corresponging CuI-catalyzed process leads to allenyl carbinols.

Cl + H SiCl$_3$ [MX] *i*-Pr$_2$NEt ; PhCHO ligand Ph OH or Ph C OH

Ni(acac)$_2$ / THF D CuCl / Et$_2$O - EtCN

Reductive alkylation accomplishes *gem*-dialkylation of ArCHO with RZnI to give $ArCHR_2$.[2]

Ni-catalyzed 1,4-addition of organozinc reagents to dienes generates nucleophiles for reaction with carbonyl compounds. Intramolecular alkylation occurs in dienals in which the two functional moieties are properly distanced (e.g., for 5-membered ring formation). Interestingly, installation of an ester group at the proximal terminus of the diene modifies the regiochemistry of the reduction, affecting the position of the remaining double bond.[3]

MeOOC, CHO — $Ni(acac)_2$ - Et_2Zn / THF → COOMe, OH — 72%

The $Ni(acac)_2$–Et_3B system is suitable for butenylation of aldehydes in aqueous solution[4] (thus reaction with water-soluble substrates such as glutaraldehyde and carbohydrates presents no problem). When the adduct of one diene unit is given opportunity to interact with another diene unit, ring formation intervenes and the active site is transmitted to another location. Termination of the reaction by an electrophile is several bonds away.[5] 1,3-Alkadien-8-ynes also follow the reaction pattern, with initial addition at an alkyne terminus.[6]

TsN — $Ni(acac)_2$ - Ph_3P - CO_2 / Ph_2Zn / THF → TsN, H, COOMe, H, Ph — 90%

Catalyzed by $Ni(acac)_2$ organozinc reagents react with thiolesters to afford ketones.[7]

(E)-Allylsilanes. The silaboration of alkenylcyclopropanes has a different stereochemical consequence from the 1,3-diene reaction in that the products are in the (*E*)-form.[8]

Ph–Si–B(O)(O) + Ph-vinylcyclopropane — $Ni(acac)_2$ - *i*-Bu_2AlH / Cy_3P / PhMe / 90° → Ph–Si, Ph, B(O)(O) — 89%

Coupling reactions. A catalyst for *N*-arylation with ArCl is made from $Ni(acac)_2$ and hindered dihydroimidazol-2-ylidene.[9] The same complex promotes reduction in the absence of amines.[10] More remarkably, the selective activation of C-F bonds (in Ar-F) to couple with Ar'MgBr is achieved.[11]

Hydrophosphinylation. The addition of Ph_2PH to 1-alkynes follows two different pathways, depending on the catalyst (Ni vs. Pd).[11]

Ni(acac)$_2$ - (EtO)$_2$POH, PhH Δ → 85%
(Ph$_3$P)$_4$Pd, MeCN 130° → 95% (Z/E 90 : 10)

Ph–C≡CH + Ph$_2$P–H

[1]Nakajima, M., Saito, M., Hashimoto, S. *TA* **13**, 2449 (2002).
[2]Hu, Y., Wang, J.-X., Li, W. *CL* 174 (2001).
[3]Shibata, K., Kimura, M., Shimizu, M., Tamaru, Y. *OL* **3**, 2181 (2001).
[4]Kimura, M., Ezoe, A., Tanaka, S., Tamaru, Y. *ACIEE* **40**, 3600 (2002).
[5]Takimoto, M., Mori, M. *JACS* **124**, 10008 (2002).
[6]Ezoe, A., Kimura, M., Inoue, T., Mori, M., Tamaru, Y. *ACIEE* **41**, 2784 (2002).
[7]Shimizu, T., Seki, M. *TL* **43**, 1039 (2002).
[8]Suginome, M., Matsuda, T., Yoshimoto, T., Ito, Y. *OM* **21**, 1537 (2002).
[9]Desmarets, C., Schneider, R., Fort, Y. *JOC* **67**, 3029 (2002).
[10]Desmarets, C., Kuhl, S., Schneider, R., Fort, Y. *OM* **21**, 1554 (2002).
[11]Böhm, V.P.W., Gstöttmayr, C.W.K., Weskamp, T., Herrmann, W.A. *ACIEE* **40**, 3387 (2001).
[12]Kazankova, M.A., Efimova, I.V., Kochetkov, A.N., Afanas'ev, V.V., Beletskaya, I.P., Dixneuf, P.H. *SL* 497 (2001).

Nickel bromide–amine complexes. 21, 280–281

Arylations. Electrochemical coupling reactions mediated by (bpy)$NiBr_2$ involve aryl halides with aldehydes[1] and with enones,[2,3] the latter process follows a 1,4-addition course.

Carboxylation. Oxacycloalk-2-ene-2-carboxylic acids are available from lactones by a two-step synthesis: enoltriflation and Ni-catalyzed electrochemical carboxylation.[4]

Transallylation.[5] Lactonization with attendant allyl transfer from allyl aroates occurs when an acceptor group is present in proximity.

CHO; e / Ni(II), Zn ; H_3O^+ → 95%

CHO; e / Ni(II), Zn ; H_3O^+ → OH, OH → 53%

[1]Budnikova, Yu.G., Keshner, T.D., Kargin, Yu.M. *RJGC* **71**, 753 (2001).
[2]Condon, S., Dupre, D., Falgayrac, G., Nedelec, J.-Y. *EJOC* 105 (2002).
[3]Condon, S., Dupre, D., Lachaise, I., Nedelec, J.-Y. *S* 1752 (2002).
[4]Senboku, H., Kanaya, H., Tokuda, M. *SL* 140 (2002).
[5]Franco, D., Dunach, E. *SL* 806 (2001).

Nickel bromide–dppe-zinc. 21, 281

Arylation. Zinc in combination with the (dppe)$NiBr_2$ complex activates aryl halides to react with various electrophiles. The presence of a suitable functional group at an *o*-position enables the 1,4-adducts with 2-alkynoic esters to enter a tandem reaction.[1]

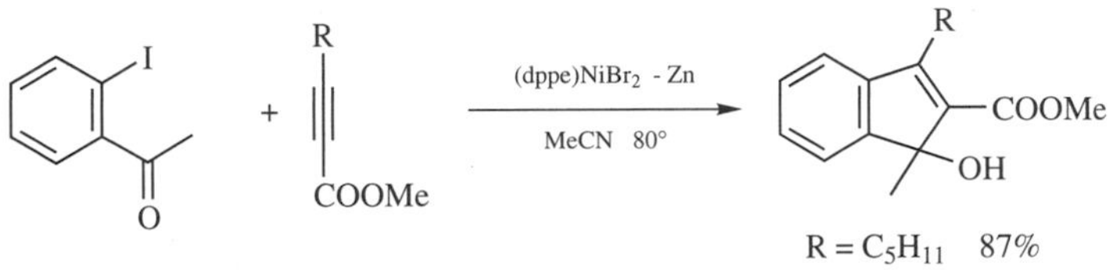

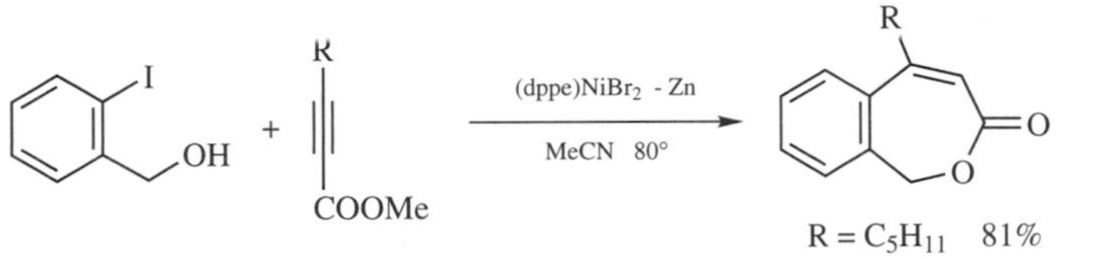

(dppe)$NiBr_2$ - Zn, MeCN 80°

$R = C_5H_{11}$ 87%

(dppe)$NiBr_2$ - Zn, MeCN 80°

$R = C_5H_{11}$ 81%

Cyclotrimerization. Allenes and alkynoic esters combine in a 1:2 ratio to afford substituted benzenes.[2] An intramolecular version affords tetralin derivatives.[3] Diynes can be used instead of allenes.

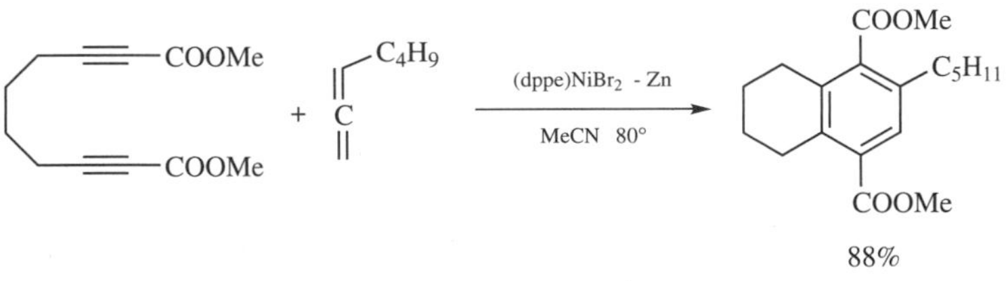

(dppe)$NiBr_2$ - Zn, MeCN 80°

88%

Isomerization/cycloaddition. 1,4-Oxa-1,4-dihydronaphthalenes undergo reaction with 2-alkynoic esters to afford benzocoumarins.

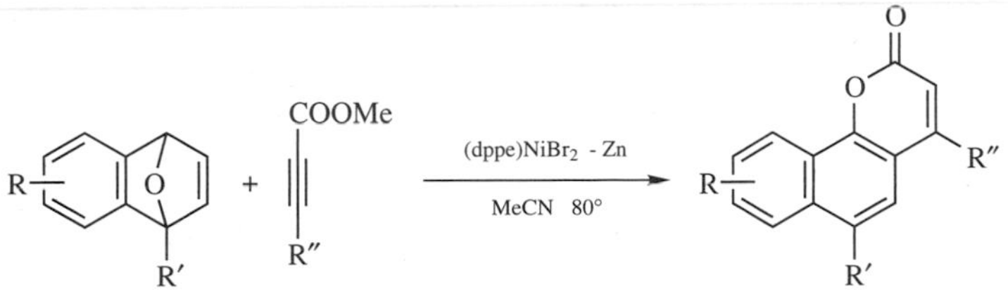

(dppe)$NiBr_2$ - Zn, MeCN 80°

[1]Rayabarapu, D.K., Cheng, C.-H. *JACS* **124**, 5630 (2002); *CC* 942 (2002).
[2]Shanmugasundaram, M., Wu, M.-S., Cheng, C.-H. *OL* **3**, 4233 (2001).
[3]Shanmugasundaram, M., Wu, M.-S., Jeganmohan, M., Huang, C.-W., Cheng, C.-H. *JOC* **67**, 7724 (2001).
[4]Jeevanandam, A., Korivi, R.P., Huang, C.-W., Cheng, C.-H. *OL* **4**, 807 (2002).
[5]Rayabarapu, D.K., Sambaiah, T., Cheng, C.-H. *ACIEE* **40**, 1286 (2001).

Nickel chloride.

Coupling reactions. Suzuki-type coupling is also catalyzed by $NiCl_2$, with or without phosphine ligand.[1,2] The cross-coupling of ArCl with diorganoaluminum 2,2-dimethylaminoethoxides is limited to those Al reagents that do not disproportionate.[3]

Biginelli reaction. Synthesis of 3,4-dihydropyrimidin-2(1*H*)-ones from β-keto esters, aldehydes, and urea is mediated by $NiCl_2$ or $FeCl_3$.[4]

Reductive metallation. With a nickel-carbene complex (prepared from 1,3-diisopropylimidazolium chloride, $NiCl_2$ and BuLi) and Et_3SiH, conjugated dienes undergo reductive metallation. The organometallic species are good nucleophiles. The ligand plays a crucial role in determining the geometry of the double bond of the products.[5]

$(Ph_3P)_2NiCl_2$ + BuLi ; *i*-$Bu_2Al(acac)$ / PhMe

OMOM CHO Ph OH Ph 62%

$NiCl_2$ + BuLi ; Pr^iN NPr^i Et_3SiH

OSiEt3 Ph OMOM 95%

[1]Zim, D., Lando, V.R., Dupont, J., Monteiro, A.L. *OL* **3**, 3049 (2001).
[2]Zim, D., Monteiro, A.L. *TL* **43**, 4009 (2002).
[3]Gelman, D., Höhne, G., Schumann, H., Blum, J. *S* 591 (2001).
[4]Lu, J., Bai, Y. *S* 466 (2002).
[5]Sato, Y., Sawaki, R., Mori, M. *OM* **20**, 5510 (2001).

Nickel chloride–phosphine complexes. **14,** 125; **15,** 122; **16,** 124; **18,** 250; **19,** 227–228; **20,** 258–259; **21,** 281–282

Reductive metallation. The phosphine-ligated Ni(0) catalyst has different effect on conjugate dienes as the latter participate in metallation and subsequent reaction with electrophiles.[1]

Defunctionalization. Benzyl carbamates are cleaved by $Me_2NH \cdot BH_3$ in the presence of Ni(0) catalyst.[2] Under these conditions aryl halides (including ArCl) suffer reduction.[3]

Carbonylation. Zirconacyclopentadienes furnish cyclopentadienones[4] by dematallative insertion of CO, under the influence of $(Ph_3P)_2NiCl_2$. Similar conditions bring about the formation of thiocarbamates from thiols and amines.[5]

Alkene formation.[6] Ni-catalyzed condensation of organozinc reagents with aldehydes in the presence of Me_3SiCl leads to alkenes directly.

CHO + ClZn⁄⁀Ph —[$(Ph_3P)_2NiCl_2$ / Me_3SiCl / THF]→ Ph-vinyl product; OMe, OMe; 79%

[1]Sato, Y., Sawaki, R., Saito, N., Mori, M. *JOC* **67**, 656 (2002).
[2]Lipshutz, B.H., Pfeiffer, S.S., Reed, A.B. *OL* **3**, 4145 (2001).
[3]Lipshutz, B.H., Tomioka, T., Pfeiffer, S.S. *TL* **42**, 7737 (2001).
[4]Takahashi, T., Tsai, F.-Y., Li, Y., Nakajima, K. *OM* **20**, 4122 (2001).
[5]Jacob, J., Reynolds, K.A., Jones, W.D. *OM* **20**, 1028 (2001).
[6]Wang, J.-X., Fu, Y., Hu, Y. *ACIEE* **41**, 2757 (2002).

Nickel chloride–zinc.

Coupling reactions. A new protocol for Ullmann coupling uses $(Ph_3P)_2NiCl_2$, Ph_3P, NaH, and Zn as reagent.[1] Biaryls that are 2,2′,6,6′-tetrasubstituted can be prepared with the system.[2]

Perfluoroalkylation. Introduction of an R_f group to alkenes, alkynes and arenes is accomplished with perfluoroalkyl chlorides using $NiCl_2$, Ph_3P, and Zn.[3]

Functional group transformations. Some alkyl aryl ethers (with chelation possibility) suffer O-C_{Alkyl} bond cleavage[4] on prolonged heating with Zn and $NiCl_2$ in xylene. However, aromatic aldoxime ethers are converted into nitriles[5] when an *o*-methoxy group is present in those molecules.

NOR′; OR —[$NiCl_2$; Zn / p-xylene, Δ]→ CN; OR

[1]Lin, G., Hong, R. *JOC* **66**, 2877 (2001).
[2]Hong, R., Hoen, R., Zhang, J., Lin, G.-Q. *SL* 1527 (2001).
[3]Huang, X.-T., Chen, Q.-Y. *JOC* **66**, 4651 (2001).
[4]Maeyama, K., Kobayashi, M., Yonezawa, N. *SC* **31**, 869 (2001).
[5]Maeyama, K., Kobayashi, M., Kato, H., Yonezawa, N. *SC* **32**, 2519 (2002).

Niobium(V) chloride.

Allylation. $NbCl_5$ mediates allyl transfer from allyltributylstannane to aldehydes and imines.[1] The reaction with crotylstannanes gives methyl-branched homoallylic alcohols.[2]

[1]Andrade, C.K.Z., Azevedo, N.R., Oliveira, G.R. *S* 928 (2002).
[2]Andrade, C.K.Z., Azevedo, N.R. *TL* **42**, 6473 (2001).

Nitric acid. 18, 251–252; 19, 228; 20, 259; 21, 281

Nitrolysis. Anhydrous HNO_3 is liberated from KNO_3 and 96% H_2SO_4 in dichloromethane.[1] This form is useful for nitration of arenes and cleavage of *t*-butyl esters. For example, benzyl ester of an *N*-Boc-amino acid is deprotected selectively, leaving the benzyl ester intact.[2]

Nitroalkenes. Nitrodecarboxylation of 2-alkenoic acids occurs when they are treated with HNO_3 and AIBN.[3]

COOH, Ph, Ph — HNO_3 - AIBN, MeCN 50° → NO_2, Ph, Ph
58%

Dimeric naphthoquinones. Oxidative dimerization of 4-alkoxy-1-naphthols occurs when they are exposed to 40% HNO_3 and Ac_2O with AgO as catalyst.[4]

OH, MeO — 40% HNO_3 -AgO, Ac_2O 25° → O, O, O, O
94%

[1]Strazzolini, P., Giumanini, A.G., Runcio, A. *TL* **42**, 1387 (2001).
[2]Strazzolini, P., Melloni, T., Giumanini, A.G. *T* **57**, 9033 (2001).
[3]Das, J.P., Sinha, P., Roy, S. *OL* **4**, 3055 (2002).
[4]Tanoue, Y., Sakata, K., Hashimoto, M., Morishita, S., Hamada, M., Kai, N., Nagai, T. *T* **58**, 99 (2002).

2-Nitrobenzenesulfonamide.

β-Amino acids.[1] Sodium salt of the title reagent opens β-lactones by an S_N2 pathway. Reaction of chiral substrates gives products of high ee.

DMSO, 4A-MS, 50°

72% (er $>$ 97.5 : 2.5)

[1]Nelson, S.G., Spencer, K.L., Cheung, W.S., Mamie, J. *T* **58**, 7081 (2002).

2-Nitrobenzenesulfonyl chloride.

Amine activation.[1] Sulfonylation of primary amines enables *N*-alkylation. The sulfonyl group is removed with 2-mercaptoethanol-DBU in DMF.

[1]Nihei, K., Kato, M.J., Yamane, T., Palma, M.S., Konno, K. *SL* 1167 (2001).

Nitrogen dioxide. 15, 219; **18,** 252–253, **20,** 260–261; **21,** 285

Oxidation of sulfides. Bound on acetylated silica NO_2 oxidizes sulfides to sulfoxides.[1] When used together with ozone, aryl sulfides are converted to nitroaryl sulfones.[2]

Arenediazonium nitrates. Arylureas undergo two-stage diazotization to give diazonium salts[3] on treatment with the dioxane complex of NO_2.

[1]Firouzabadi, H., Iranpoor, N., Heydari, R. *SC* **31**, 2037 (2001).
[2]Nose, M., Suzuki, H. *S* 1065 (2002).
[3]Zhang, Z., Zhang, Q., Zhang, S., Liu, X., Zhao, G. *SC* **31**, 329 (2001).

4-Nitrophenyl *N*-benzylcarbamate.

Unsymmetrical ureas.[1] The title reagent converts amines to ureas which can be selectively debenzylated, therefore $RNHCONH_2$ are readily accessible.

[1]Liu, Q., Luedtke, N.W., Tor, Y. *TL* **42**, 1445 (2001).

4-Nitrophenyl chloroformate.

Activation of carboxylic acids.[1] Peptide formation via mixed anhydrides derived from the title reagent is convenient because the condensation does not require protected amino acids.

[1]Gagnon, P., Huang, X., Therrien, E., Keillor, J.W. *TL* **43**, 7717 (2002).

Nitrosyl chloride.

Nitrosation.[1] Perfluoroalkyltrimethylsilanes react with NOCl/CsF to give R_f-NO.

α-Isoamyloxy oximes.[2] With NOCl generated in situ from *i*-AmONO and Me_3SiCl, alkenes are functionalized.

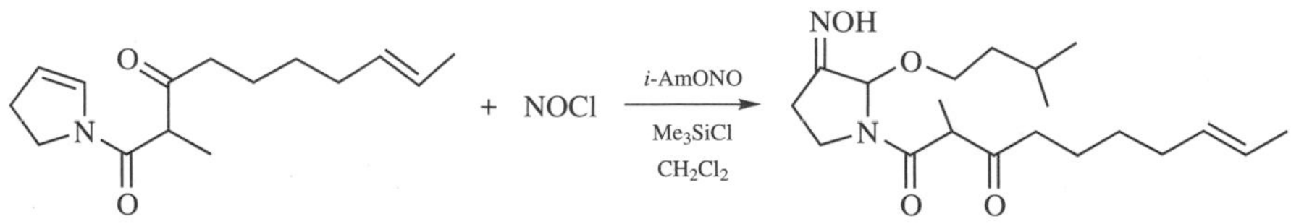

[1]Singh, R.P., Shreeve, J.M. *CC* 1818 (2002).
[2]Parsons, P.J., Karadogan, B., Macritchie, J.A. *SL* 257 (2001).
[3]Weib, K., Wagner, K. *CB* **117**, 1973 (1984).

Nitrous oxide.

Oxidation.[1] 9,10-Dihydroanthracene is oxidized in two different ways by a Ru-catalyzed reaction with N_2O, depending on whether sulfuric acid is present.

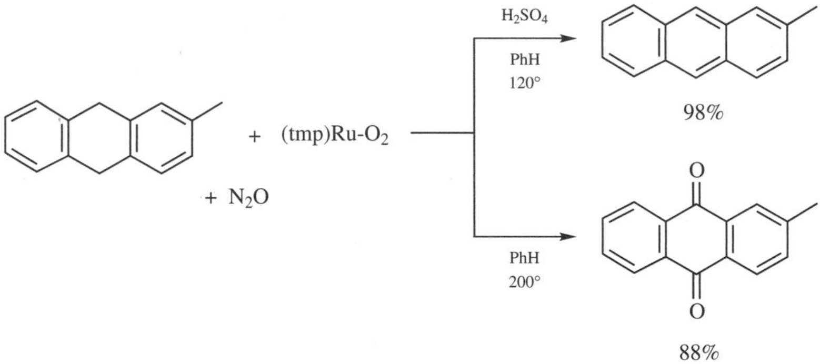

[1]Hashimoto, K., Tanaka, H., Ikeno, T., Yamada, T. *CL* 582 (2002).

Nonafluorobutanesulfonyl fluoride.

Enol nonaflates.[1] In the presence of Bu_4NF the title reagent transforms trimethylsilyl enol ethers to alkenyl nonaflates directly.

[1]Lyapkalo, I.M., Webel, M., Reissig, H.-U. *EJOC* 1015 (2002).

***exo*-Norbornyldimethylsilyl chloride.**

Hydoxyl group protection.[1] The title reagent is prepared by hydrosilylation of norbornene with $Me_2Si(H)Cl$ in the presence of a Pt catalyst. Derivatization of alcohols furnishes silyl ethers with susceptibility to cleavage by Bu_4NF in between TMS ethers and TBS ethers.

[1]Heldmann, D.K., Stohrer, J., Zauner, R. *SL* 1919 (2002).

O

Organoaluminum reagents. 21, 287

anti-1,2-Methoxyiodoalkanes. Selective coupling of 1-chloro-1-methoxy-2-iodoalkanes, which are adducts of ICl and 1-methoxyalkenes, with organoaluminum reagents (R_3Al and R_2AlCl) leads to the iodohydrin ethers.[1]

$$\text{R-CH=CH-OMe} \xrightarrow[\text{PhMe, }-78^\circ]{\text{I-Cl}} \text{MeO-CH(Cl)-CH(I)-R} \xrightarrow{\text{R'M}} \text{MeO-CH(R')-CH(I)-R}$$

$R'M = R'_2Zn, R'_3Al, R'Al_2Cl$

[1]Inoue, A., Maeda, K., Shinokubo, H., Oshima, K. *T* **55**, 665 (1999).

Organocerium reagents. 13, 206; **14,** 217–218; **15,** 221; **16,** 232; **17,** 205–207; **18,** 256; **19,** 231; **20,** 263–264

Addition to homoallylic alcohols.[1] After forming the lithium alkoxides (with LiH) homoallylic alcohols undergo addition with organocerium reagents (derived from RLi and $CeCl_3$). It seems that only when the carbinolic center of the alcohols is also substituted with an aryl group the reaction works reasonably well and CC bond formation occurs at the far end of the double bond (6 examples, 85–94%).

1,1-Dialkylperfluoroalkylamines.[2] Perfluoroalkanamides R_fCONH_2 are susceptible to attack by organocerium reagents to afford $R_2R_fCNH_2$.

[1]Bartoli, G., Dalpozzo, R., De Nino, A., Procopio, A., Sambri, L., Tagarelli, A. *TL* **42**, 8833 (2001).
[2]Zhang, N., Ayral-Kaloustian, S. *JFC* **117**, 9 (2002).

Organocopper reagents. 13, 207–209; **14,** 218–219; **15,** 221–227; **16,** 232–238; **17,** 207–218; **18,** 257–262; **19,** 232–235; **20,** 264–267; **21,** 287–290

Reagents. Formation of organocuprates from iodides is achieved through I-Cu exchange with the highly hindered lithium dineopentylcuprate. Such reagents are essential for the chemoselectivity. The protocol makes functionalized mixed cuprate

Fiesers' Reagents for Organic Synthesis, Volume 22. Series editor Tse-Lok Ho
ISBN 0-471-28515-3

reagents accessible.[1] (2-Methoxyphenyl)dimethylsilyllithium and cuprate reagents offer unique advantages in multistep synthesis.

t-Am_2CuLi → CH_2=$CHCH_2Br$ →

Various factors affecting the α-(*N*-carbamoyl)alkylcuprate reagents in their reactivities have been examined.[3]

Coupling reactions. Coupling reactions employing cuprate reagents serve to extend a carbon framework by an ethynyl group (as silylated products).[4] Its involvement in group attachment to the α-position of alkenyl sulfides effectively completes an addition to alkynyl sulfides,[5] only that a prior step of adding TsOH is required.

+ $Ph_2Cu(CN)Li_2$ —$(Ph_3P)_2PdCl_2$, THF→ 83%

A method for synthesizing pentasubstituted 1,3-dienes involves regioselective coupling of cuprate reagents with 1,4-diiodo-1,3-alkadienes.[6]

+ Me_2CuLi —Et_2O, −50°→ 90%

α-Amino ketones are readily prepared from protected amino acids via 2-pyridylthio esters,[7] while taking advantage of the differentiable reactivity of various functionalities toward cuprate reagents.

Conjugate additions. While conjugate addition by organocuprates is well known, much current work is concerned with subsequent trapping by electrophiles to increase the synthetic potential of the reaction. A route to α-exomethylene-γ-lactones starts from cuprate addition to 2-alkynoic esters and trapping with *B*-iodomethylpinacolatoborane. The resulting allylboronates condense with aldehydes to conclude the synthesis.[8]

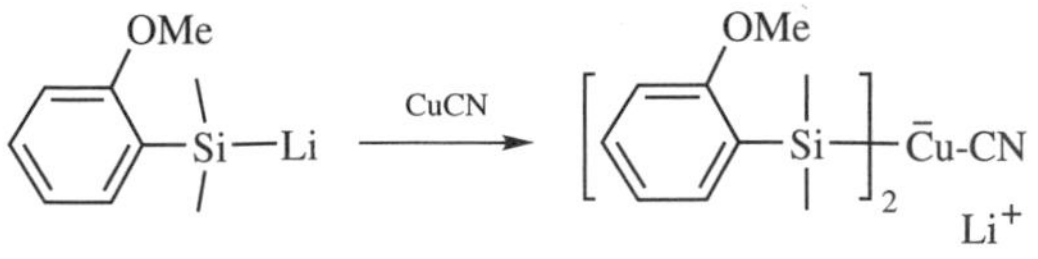

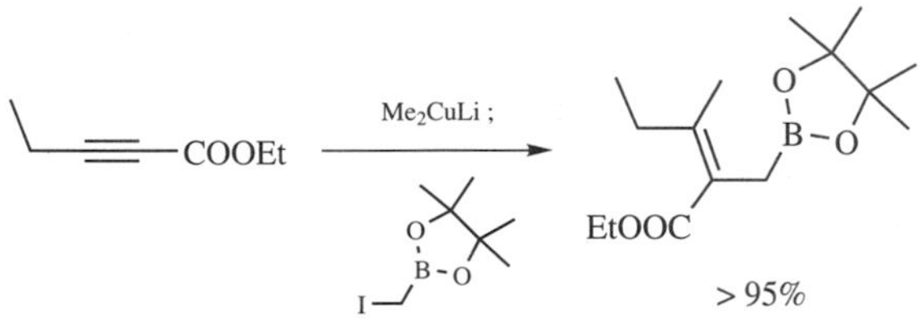

CuCN

Li+

Me_2CuLi ;

COOEt

EtOOC

> 95%

Trapping the initial adducts with PhSeBr furnishes 2-phenylseleno-2-alkenoic esters.[9]

Asymmetric conjugate addition of Li[RCuI] to chiral *N*-alkenoyloxazolidin-2-ones is promoted by Me_3SiI.[10] When the substrate addend contains an internal electrophile (e.g., formyl group) the addition is followed by cyclization.[11] Furthermore, organocopper-zinc species not only are capable of conjugate addition, cyclization onto a vinyl group is observed.[12]

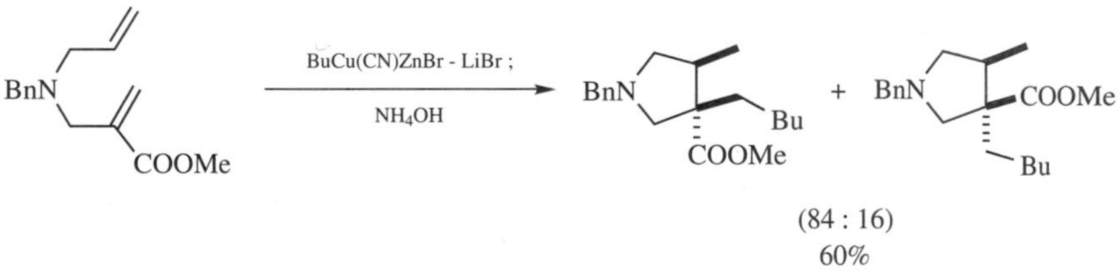

BuCu(CN)ZnBr - LiBr ;

NH_4OH

(84 : 16)

60%

A remarkable 1,2-addition vs. 1,4-addition is manifest in the addition to cyclobutenones on varying the reagent.[13] The softer high-order cuprates tend to perform conjugate addition.

Organostannylcuprates add to 2-alkynones stereoselectively, giving products with the (*Z*)-configuration.[14]

$(PhMe_2Si)_2Cu(CN)Li_2$; Ac_2O → 83%

$PhMe_2SiLi$; Ac_2O → 81%

Addition to alkynes, allenes and dienes. Silylcuprates are active for the addition. Again, trapping of the initial adducts to furnish desirable products are most significant to synthesis. Thus, using allylic electrophiles to terminate the reaction leads to dienes.[15,16] Silylalkenylcopper species resulted from the addition to allene form a 2:1 adduct with acrylonitrile.[17]

$PhMe_2SiLi$ - LiCN, THF –40° ; $BF_3 \cdot OEt_2$; CH_2=CHCN

Trisubstituted (*E*)-1,2-disilylalkenes are formed on the addition of silylcuprates to alkynylsilanes followed by alkylation.[18]

β-Alkoxy ketone dithioacetals. Following condensation of ketene dithioacetals with 1,1-dialkoxyalkanes (Mukaiyama-type condensation) the addition of organocuprate reagents completes a synthesis of the doubly protected aldol products.[19]

[1]Piazza, C., Knochel, P. *ACIEE* **41**, 3263 (2002).
[2]Lee, T.W., Corey, E.J. *OL* **3**, 3337 (2001).
[3]Dieter, R.K., Topping, C.M., Nice, L.E. *JOC* **66**, 2302 (2001).
[4]Hupe, E., Knochel, P. *ACIEE* **40**, 3022 (2001).
[5]Braga, A.L., Emmerich, D.J., Silveira, C.C., Martins, T.L.C., Rodrigues, O.E.D. *SL* 371 (2001).
[6]Nakajima, R., Delas, C., Takayama, Y., Sato, F. *ACIEE* **41**, 3023 (2002).
[7]Vazquez, J., Alberico, F. *TL* **43**, 7499 (2002).
[8]Kennedy, J.W.J., Hall, D.G. *JACS* **124**, 898 (2002).

[9]Silveira, C.C., Braga, A.L., Guerra, R.B. *TL* **43**, 3395 (2002).
[10]Pollock, P., Dambacher, J., Annes, R., Bergdahl, M. *TL* **43**, 3693 (2002).
[11]Schneider, C., Reese, O. *CEJ* **8**, 2585 (2002).
[12]Denes, F., Chemla, F., Normant, J.F. *EJOC* 3536 (2002).
[13]Murakami, M., Miyamoto, Y., Ito, Y. *JACS* **123**, 6441 (2001).
[14]Nielsen, T.E., Cubillo de Dios, M.A., Tanner, D. *JOC* **67**, 7309 (2002).
[15]Liepins, V., Karlström, A.S.E., Bäckvall, J.-E. *JOC* **67**, 2136 (2002).
[16]Liepins, V., Bäckvall, J.-E. *EJOC* 3527 (2002).
[17]Barbero, A., Blanco, Y., Pulido, F.J. *CC* 1606 (2001).
[18]Cuadrado, P., Gonzalez-Nogal, A.M., Sanchez, A. *JOC* **66**, 1961 (2001).
[19]Ichikawa, J., Saitoh, T., Tada, T., Mukaiyama, T. *CL* 996 (2002).

Organogallium reagents. 21, 290

Coupling reactions. Haloarenes undergo Pd-catalyzed coupling with alkenylgallium chlorides. Functional groups such as hydroxyl do not interfere with the reaction.

OH; I; + $GaCl_2$; $(dba)_3Pd_2 \cdot CHCl_3$; $(o\text{-}Tol)_3P$; THF - DMSO; OH; 96%

[1]Mikami, S., Yorimitsu, H., Oshima, K. *SL* 1137 (2002).

Organolithium reagents. 20, 268–269; **21,** 290–293

Special reagents. Organolithium reagents containing sensitive groups such as *t*-butyl ester can be prepared at low temperature by exchange with mesityllithium (itself derived from MesBr and *t* -BuLi).[1]

Addition reactions.[2] Vinyllithium and vinylmagnesium bromide show different diastereoselectivity in the addition to a 6-chloro-2-cyclohexenone. The product analysis suggests that chlorine atom can exert a directing effect to Mg but not Li.

O; Cl; + M; THF; HO; Cl; +; OH; Cl

M = MgBr 65% M = Li 70%

The products from reaction of lithiosilanes with acid chlorides undergo Brook rearrangement to generate *O,Si*-disubstituted carbanions, therefore further reaction with carbon electrophiles and TBAF leads to ketones.[3]

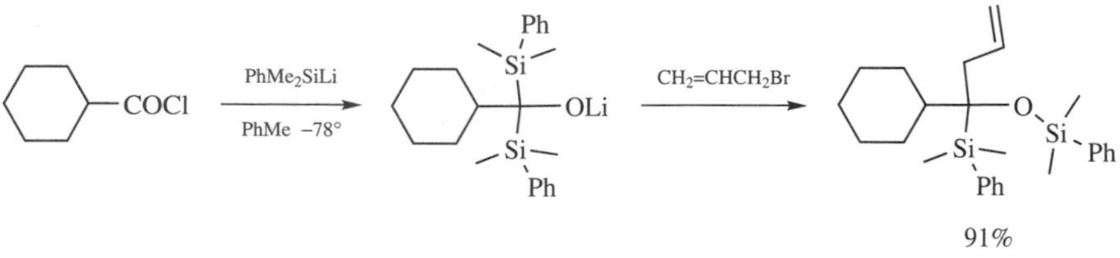

Nonpolar solvents (e.g., PhMe) increase the efficiency of PhLi addition to hindered ketones.[4] (*Z,Z*)-1,4-dilithio-1,3-butadienes (derived from the diiodides) attack carbonyl compounds to give cyclopentadienes.[5]

The addition of aryllithiums to 1-nitrocycloalkenes results in *cis*-2-aryl-1-nitrocycloalkanes.[6] Phenyllithium reacts with cinnamaldehyde by a single-electron transfer mechanism and subsequent addition of electrophiles leads to 1,3-diphenylpropanones.[7]

α-Oxyalkyllithiums. By virtue of steric factors that disfavor ring opening (β-elimination) oxazolidin-5-yllithiums such as **1** can be generated for use in synthesis.[8]

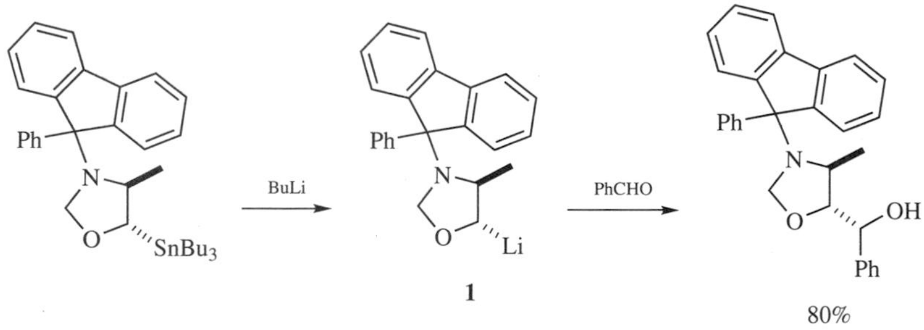

Protons of epoxide rings are abstractable, the ensuing lithio derivatives are known to decompose into carbenoid species. By this reaction cycloalkene epoxides with a proximal double bond are transformed into spiro[2.n]alkanes[9] via an intramolecular cyclopropanation.

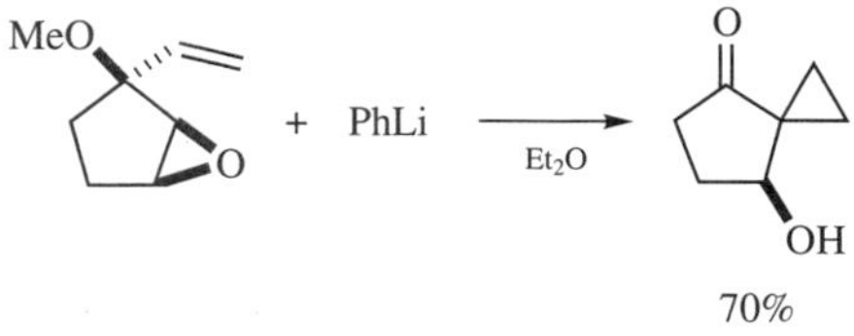

Treatment of 1,1-dichloro-2-trifluoromethyl-2-alkanols with an organolithium reagent affords lithiated epoxides which can be alkylated. Thus this reaction sequence constitutes a synthesis of epoxides homologated at the carbon atom vicinal to that bearing the trifluoromethyl group.[10] Furthermore, stereoisomers are obtained by an interchange of R′Li/R″I with R″Li/R′I in the reaction order.

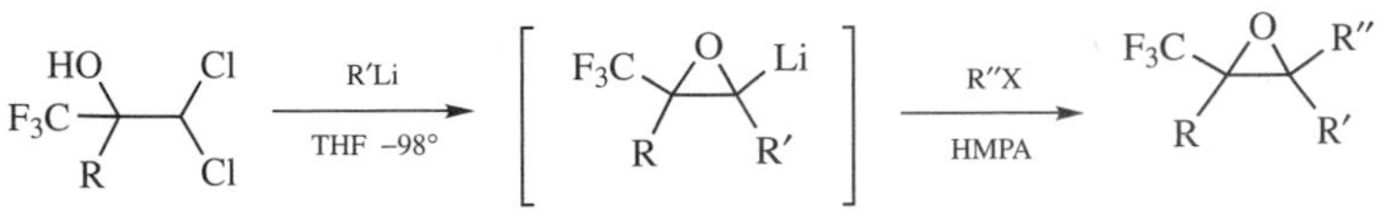

Addition-rearrangement. Generation of a cycloheptenone unit from a 1-alken-6-yn-3-ol is realized on heating with catalytic MeLi at high temperature.[11]

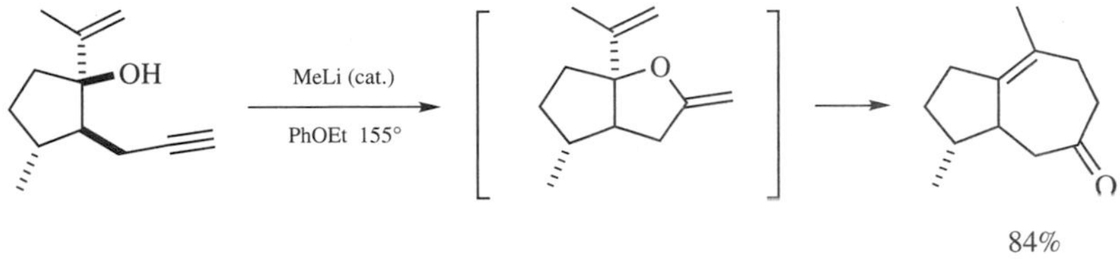

84%

[1]Kondo, Y., Asai, M., Miura, T., Uchiyama, M., Sakamoto, T. *OL* **3**, 13 (2001).
[2]Lindsay, H.A., Salisbury, C.L., Cordes, W., McIntosh, M.C. *OL* **3**, 4007 (2001).
[3]Fleming, I., Lawrence, A.J., Richardson, R.D., Surry, D.S., West, M.C. *HCA* **85**, 3349 (2002).
[4]Lecomte, V., Stephan, E., Jaouen, G. *TL* **43**, 3463 (2002).
[5]Xi, Z., Song, Q., Chen, J., Guan, H., Li, P. *ACIEE* **40**, 1913 (2001).
[6]Santos, R.P., Lopes, R.S.C., Lopes, C.C. *S* 845 (2001).
[7]Nudelman, N.S., Garcia, G.V. *JOC* **66**, 1387 (2001).
[8]Calaza, M.I., Paleo, M.R., Sardina, F.J. *JACS* **123**, 2095 (2001).
[9]Dechoux, L., Agami, C., Doris, E., Mioskowski, C. *EJOC* 4107 (2001).
[10]Shimizu, M., Fujimoto, T., Minezaki, H., Hata, T., Hiyama, T. *JACS* **123**, 6947 (2001).
[11]Kumar, J.S.R., O'Sullivan, M.F., Reisman, S.E., Hulford, C.A., Ovaska, T.V *TL* **43**, 1939 (2002).

Organomanganese reagents.

Alkenes.[1] A synthesis of trisubstituted and tetrasubstituted alkenes from trihalomethylcarbinols involves reaction of the corresponding silyl ethers with lithium triorganomanganates. It proceeds from displacement, rearrangement and elimination steps.

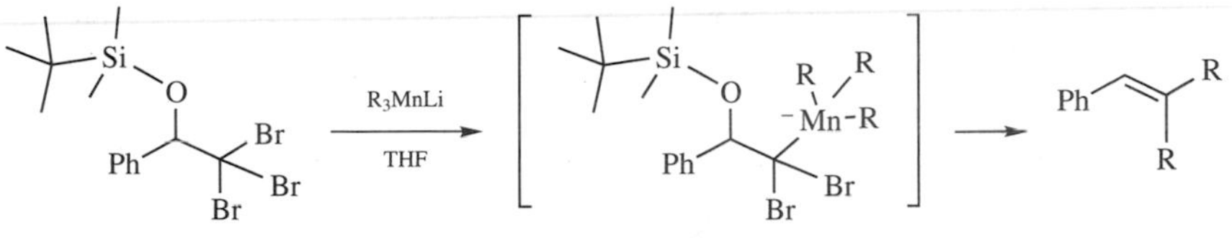

[1]Kakiya, H., Shinokubo, H., Oshima, K. *T* **57**, 10063 (2001).

Organotitanium reagents. 21, 295

Reaction with aldehydes. Pentadienyllithium species are converted into the corresponding titanium reagents on addition of (i-PrO)$_3$TiCl. Regioselective reaction with aldehydes gives bishomoallylic alcohols (reaction at the central carbon atom).[1] Functionalized organotitanium reagents show chemoselectivity such that aldehydes are attacked cleanly in the presence of ketones.[2]

OLi
Li
+ PhCHO
PhCOMe
unchanged
(i-PrO)$_3$TiCl
THF
OH
Ph OH

Coupling reaction.[3] Reagents such as (i -PrO)$_3$TiR are useful for cross-coupling to convert ArX (X=OTf, halide) to ArR which is catalyzed by Pd complexes.

[1]Zellner, A., Schlosser, M. *SL* 1016 (2001).
[2]Pastor, I.M., Yus, M. *T* **57**, 2365 (2001).
[3]Han, J.W., Tokunaga, N., Hauashi, T. *SL* 871 (2002).

Organozinc reagents. 13, 220–222; **14,** 233–235; **15,** 238–240; **16,** 246–248; **17,** 228–234; **18,** 264–265; **19,** 240–241; **20,** 270–275; **21,** 295–299

Preparation. Secondary i-PrZnR are obtained from trisubstituted alkenes via hydroboration and exchange with i-Pr$_2$Zn. These reagents are configurationally stable, convertible to various compounds with retention of configuration.[1]

Addition to C=X bond. R$_2$Zn adds to carbonyl compounds with CeCl$_3$ as catalyst.[2] Moderate asymmetric induction arised in the Et$_2$Zn addition to α-keto esters in the presence of a chiral Ti(salen) catalyst.[3]

Propargylic epoxides are prepared starting from propargyl chloride that is trimethylsilylated at C–1. The derived 3-chloro-1-trimethylsilyl-1,2-propadien-1-ylzinc bromide reacts with carbonyl compounds to deliver chlorohydrins from which the epoxides are readily formed.[4] An analogous approach to aziridines[5] is realized by merely changing the electrophiles to imines. If change is made to the propargylic component, i.e., replacing the chlorine with a MOM group, the derived alkoxyallenylzinc reagent is configurationally stable and extremely highly diastereoselective addition to α-oxy imines results, yielding the *anti-anti* amino diols.[6]

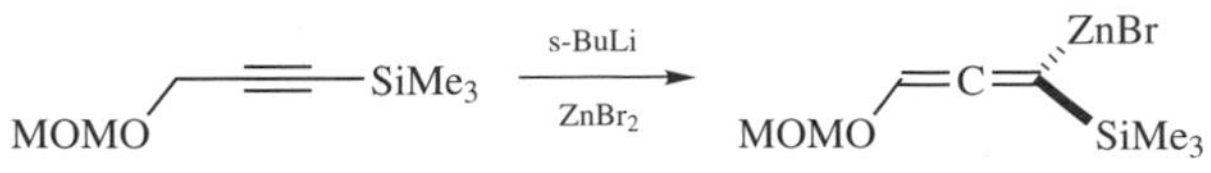

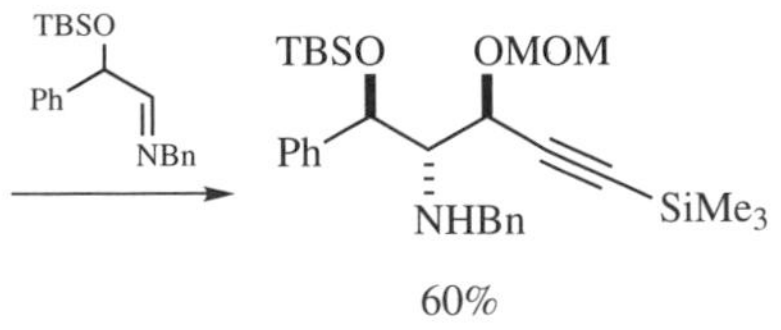

Allylic and propargylic hydroxylamines are similarly synthesized by addition of organozinc reagents to nitrones.[7,8] The allylic hydroxylamines are capable of rearrangement into (*E*)-*O*-allylhydroxylamines.[7]

Conjugate additions. Electrochemically generated arylzinc species add to activated alkenes in the presence of $CoBr_2(bipy)_2$.[9] A diastereoselective addition assisted by ultrasound is performed in water.[10]

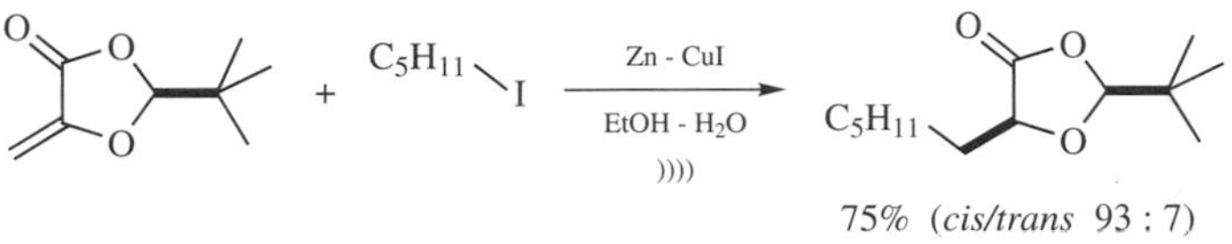

For addition to nitroalkenes both diorganozincs and organozinc cuprates are useful. Reagents of the general formula $RZnCH_2SiMe_3$ transfer the R group.[11] Vinylboronates also serve as acceptors to organozincs.[12]

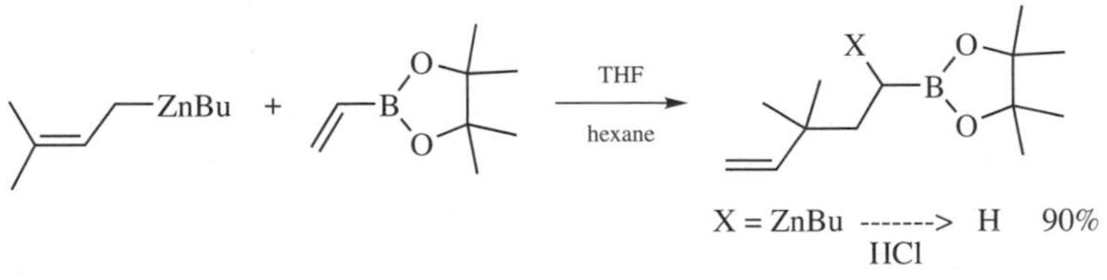

Coupling reactions. Different types of Negishi coupling have been investigated and their synthetic utility expanded. Alkenyl-alkenyl coupling is represented by the assemblage of functionalized 1,3-dienes.[13] For alkenyl-aryl coupling the organozinc reagents can be made from electrogenerated zinc.[14] Alkyl-alkenyl coupling constitutes a crucial step in a total synthesis of pumiliotoxin-A,[15] and cyclopropyl-alkenyl coupling sets up a divinylcyclopropane for thermal transformation into a 1,4-cycloheptadiene.[16]

t-BuLi / Et_2O ; $ZnCl_2$; Pd(0) OBn
60%
pumiliotoxin-A

BuLi ; $ZnCl_2$; Pd(0)
xylene Δ

A general catalyst for Negishi cross-coupling is $(t\text{-}Bu_3P)_2Pd$.[17] Secondary alkylzinc iodides used for coupling reactions are better prepared from 1,1-diiodoalkanes via reaction with R_2Zn.[18]

Cyclopropanation. Synthesis of *gem*-dimethylcyclopropanes[19] using Et_2Zn and Me_2CI_2 is an important extension of the well-known process. Acyloxymethylzinc reagents include those made from perfluorocarboxylic acids.[20]

Internal homologation of β-keto esters by $EtZnCH_2I$ (via cyclopropanation of the enol followed by fragmentation) is extendable to β-keto amides.[21] It is feasible to continue such a reaction with an aldol reaction.[22]

Et_2Zn - CH_2I_2 ; PhCHO
(12 : 1)
97%

Allylic displacements. Glycal acetates form *C*-glycosides by reaction with organozinc reagents.[23] 4,5-Epoxy-2-alkanoic esters are converted into 5-hydroxy-2-methyl-3-alkenoic esters from a CuCN-catalyzed reaction with Me_2Zn. The configuration of the epoxides correlates with the relative stereochemical relationship of the methyl and hydroxy groups in the products: *trans*-epoxides to *anti*-isomers, and *cis*-epoxides to *syn*-isomers.[24]

Me_2Zn - CuCN ; DMF
81% (*anti/syn* >95 : 5)

Nitroaldol reactions. This reaction is mediated by Et_2Zn in the presence of a promoter such as ethylenediamine or 2-aminoethanol.[25]

Elimination. 1-Mesyloxy-2-bromo-2-alkenes are converted to allenes by the $(Ph_3P)_4Pd$-Et_2Zn system.[26]

[1]Hupe, E., Knochel, P. *OL* **3**, 127 (2001).
[2]Fischer, S., Groth, U., Jeske, M., Schütz, T. *SL* 1922 (2002).
[3]DiMauro, E.F., Kozlowski, M.C.*OL* **4**, 3781 (2002).
[4]Chemla, F., Bernard, N., Ferreira, F., Normant, J.F. *EJOC* 3295 (2001).
[5]Chemla, F., Ferreira, F., Hebbe, V., Stercklen, E. *EJOC* 1385 (2002).
[6]Poisson, J.-F., Normant, J.F. *OL* **3**, 1889 (2001).
[7]Pandya, S.U., Garcon, C., Chavant, P.Y., Py, S., Vallee, Y. *CC* 1806 (2001).
[8]Pinet, S., Pandya, S.U., Chavant, P.Y., Ayling, A., Vallee, Y. *OL* **4**, 1463 (2002).
[9]Gomes, P., Gosmini, C., Perichon, J. *SL* 1673 (2002).
[10]Suarez, R.M., Sestelo, J.P., Sarandeses, L.A. *SL* 1435 (2002).
[11]Rimkus, A., Sewald, N. *OL* **4**, 3289 (2002).
[12]Nakamura, M., Hara, K., Hatakeyama, T., Nakamura, E. *SL* 3137 (2001).
[13]Su, M., Kang, Y., Yu, Y., Hua, Z., Jin, Z. *OL* **4**, 691 (2002).
[14]Jalil, A.A., Kurono, N., Tokuda, M. *SL* 1944 (2001).
[15]Aoyagi, S., Hirashima, S., Saito, K., Kibayashi, C. *JOC* **67**, 5517 (2002).
[16]Piers, E., Coish, P.D.G. *S* 251 (2001).
[17]Dai, C., Fu, G.C. *JACS* **123**, 2719 (2001).
[18]Shibli, A., Varghese, J.P., Knochel, P., Marek, I. *SL* 818 (2001).
[19]Charette, A.B., Wilb, N. *SL* 176 (2002).
[20]Charette, A.B., Beauchemin, A., Francoeur, S. *JACS* **123**, 8139 (2001).
[21]Hilgenkamp, R., Zercher, C.K. *T* **57**, 8793 (2001).
[22]Lai, S., Zercher, C.K, Jasinski, J.P., Reid, S.N., Staples, R.J. *OL* **3**, 4169 (2001).
[23]Steinhuebel, D.P., Fleming, J.J., Du Bois, J. *OL* **4**, 293 (2002).
[24]Hirai, A., Matsui, A., Komatsu, K., Tanino, K., Miyashita, M. *CC* 1970 (2002).
[25]Klein, G., Pandiaraju, S., Reiser, O. *TL* **43**, 7503 (2002).
[26]Ohno, H., Miyamura, K., Tanaka, M. *JOC* **67**, 1359 (2002).

Osmium tetroxide. 13, 222–225; **14,** 233–239; **15,** 240–241; **16,** 249–253; **17,** 236–240; **18,** 265–267; **19,** 241–242; **20,** 275–276; **21,** 301–303

Modifications. Chiral catalysts encapsulated in polymer,[1] bound to a soluble polymer,[2] or with a modified cinchona alkaloid grafted on to mesoporous molecular sieves or silica gel,[3] in which OsO_4 is layered on $Al_xMg_{1-x}(OH)_2Cl_x \cdot zH_2O$ by the ion-exchange technique,[4] have been evaluated for dihydroxylation and aminohydroxylation of alkenes. Another recyclable and reusable catalyst system is composed of OsO_4 and an ionic liquid.[5]

Dihydroxylations and aminohydroxylations. As co-oxidant selenoxides such as $PhSe({=}O)CH_2Ph$ are useful,[6] and a flavin-based biomimetic system containing OsO_4 and H_2O_2 shows proper activity.[7] Immobilization of OsO_4 as a stable tetrasubstituted diolate complex is proposed for dihydroxylation.[8]

Dihydroxylation of the ketimine derivatives of 3-amino-1-alkenes is *anti*-selective (3-7 folds).[9] Directed dihydroxylation of 2-cycloalken-1-ols and 3-trichloroacetamido-1-

cycloalkenes is observed with OsO_4-TMEDA.[10] A tandem Heck reaction and dihydroxylation are achieved to provide 2,3-dihydroxy-3-arylpropanoic esters,[11] using both Pd and Os species anchored on silica gel through a mercaptopropyl spacer and a cinchona alkaloid, respectively.

A practical aminohydroxylation protocol for 2-alkenoic acids (11 examples, 88–98%) has been delineated.[12]

Mechanistic understanding of the catalytic cycle has helped design an approach to incorporate ligands that can enforce second-cycle turnover by never leaving the catalytic center.[13]

Nitriles from primary amines. Moderate to low yields of nitriles are obtained on treatment of primary amines with OsO_4-Me_3NO/pyridine in aqueous THF at room temperature.[14]

[1]Kobayashi, S., Ishida, T., Akiyama, R. *OL* **3**, 2649 (2001).
[2]Kuang, Y.-Q., Zhang, S.-Y., Wei, L.-L. *TL* **42**, 5925 (2001).
[3]Motorina, I., Crudden, C.M. *OL* **3**, 2325 (2001).
[4]Choudary, B.M., Chowdari, N.S., Kantam, M.L., Raghavan, K.V. *JACS* **123**, 9220 (2001).
[5]Yao, Q. *OL* **4**, 2197 (2002).
[6]Krief, A., Castillo-Colaux, C. *SL* 501 (2001).
[7]Jonsson, S.Y., Farnegardh, K., Bäckvall, J.-E. *JACS* **123**, 1365 (2001).
[8]Severeyns, A., De Vos, D.E., Fiermans, L., Verpoort, F., Grobet, P.J., Jacobs, P.A. *ACIEE* **40**, 586 (2001).
[9]Oh, J.S., Park, D.Y., Song, B.S., Bae, J.G., Yoon, S.W., Kim, Y.G. *TL* **43**, 7209 (2002).
[10]Donohoe, T.J., Blades, K., Moore, P.R., Waring, M.J., Winter, J.J.G., Helliwell, M., Newcombe, N.J., Stemp, G. *JOC* **67**, 7946 (2002).
[11]Choudary, B.M., Chowdari, N.S., Jyothi, K., Kumar, N.S., Kantam, M.L. *CC* 586 (2002).
[12]Fokin, V.V., Sharpless, K.B. *ACIEE* **40**, 3455 (2001).
[13]Andersson, M.A., Epple, R., Fokin, V.V., Sharpless, K.B. *ACIEE* **41**, 472 (2002).
[14]Gao, S., Herzig, D., Wang, B. *S* 544 (2001).

Oxalyl chloride. **17,** 241–242; **18,** 267–268; **19,** 243; **20,** 277; **21,** 304

Imidoyl chlorides.[1] Imidoyl chlorides are rapidly formed from amides, they react with pyridine *N*-oxide to give 2-alkanamidopyridines.

NHMe
$(COCl)_2$
2,6-lutidine ;
pyridine oxide
O
N
N
Me
O
73%

N-Alkylisatins. The oxalyl chloride-DABCO combination transforms *N,N*-dimethylanilines to *N*-methylisatins.[2] *N*-Arylated cyclic amines give *N*-ω-chloroalkylisatins.[3]

MeO N (COCl)$_2$ DABCO $CHCl_3$ Δ MeO N $(CH_2)_5Cl$ O O

59%

[1]Manley, P.J., Bilodeau, M.T. *OL* **4**, 3127 (2002).
[2]Cheng, Y., Ye, H.-L., Zhan, Y.-H., Meth-Cohn, O. *S* 904 (2001).
[3]Cheng, Y., Zhan, Y.-H., Meth-Cohn, O. *S* 34 (2002).

Oxazolidin-2-one.

2-Aminoethyl sulfides.[1] Base-promoted reaction of thiols with 2-oxazolidin-2-one furnishes $RSCH_2CH_2NH_2$.

Esters from trityl ethers.[2] A one-step conversion of trityl ethers to esters is effected with acid chlorides (ArCOCl takes longer time) in the presence of oxazolidin-2-one, DMAP, *i* -Pr_2NEt in dichloromethane.

[1]Ishibashi, H., Uegaki, M., Sakai, M., Takeda, Y. *T* **57**, 2115 (2001).
[2]Bergmeier, S.C., Arason, K.M. *TL* **41**, 5799 (2000).

Oxygen. 18, 268–269; **19,** 243–244; **20,** 277–279; **21,** 305–308

Epoxidations. A metal catalyst and an aldehyde together with molecular oxygen form heterogeneous epoxidation system. Regarding the metal component, Pd-SiO_2[1] and amidate-bridged platinum blue complexes[2] can be used.

Catalysts prepared from (*p*-cymene)RuCl and triazepinethione derivatives are useful in epoxidation of terpenes with an aldehyde-oxygen system, with which limonene gives mainly *trans*-1,2-epoxide.[3]

Oxidations. A great number of effective methods for aerobic oxidation of alcohols and other functional groups to carbonyl compounds are available. The octahedral molecular sieve OMS-2 shows catalytic property in aerobic oxidation of alcohols, with active sites probably derived from Mn(III) ions in the tunnels.[4] Copper salts and complexes are very effective for various transformations, e.g., for oxidation of prolinol without racemization,[5] phenols to quinones (*p*-position open),[6] and benzylic alcohols to carbonyl compounds in ionic liquid.[7] α-Silylalkylcopper compounds derived from Mg-Cu exchange undergo desilylative oxidation.[8] Cobalt(II) acetate and *N*-hydroxyimides mediate oxidation of benzylic amines to carbonyl compounds[9] and hydrosilanes to silanols.[10] A mixture of Co(II) and Mn(II) nitrates with TEMPO oxidizes many alcohols (11 examples, 96–100%).[11] Ring opening oxidation of *t*-cyclopropanols in the presence of Mn(II) abietate gives epoxy ketones.[12]

HO–(1-methylcyclopropyl)–C_6H_{13} + O_2 $\xrightarrow[\text{KOH / } H_2O]{\text{Mn(II) abietate / PhH ;}}$ epoxy ketone–C_6H_{13}

85%

Aerobic oxidation of alcohols is also catalyzed by Ni-Al hydrotalcite[13] or $FeBr_3$–$Fe(NO_3)_3$,[14] whereas α-hydroxy ketones are converted to the diketones with the Fe(III)-EDTA system[15] and quinones are obtained using iron phthalocyanine complexes.[16]

Catalysts containing precious metal salts include Pd,[17,18] Os.[19,20] By far the most frequently used are various forms of Ru,[21–26] and more diverse structural types can be oxidized, e.g., primary amines to nitriles,[27] tertiary amines to *N*-oxides,[28] and diols to lactols.[29]

Of course, vanadyl complexes are of historical significance in promoting aerial oxidation. Propargyl alcohols[30] and benzylsilanes[31] are found to be converted to carbonyl compounds by such systems. Phosphonium salts undergo oxidative coupling to furnish alkenes on treatment with $VO(acac)_2$–K_2CO_3/18-crown-6 under oxygen.[32]

With the enzyme laccase, alcohols are oxidized by oxygen as mediated by TEMPO.[33]

N-Oxide formation from tertiary amines can be effected by the aldehyde-O_2 system without metal catalyst.[34] Oxidation of organoboronates and alkylboronic acids to alcohols by oxygen-Et_3N is quite satisfactory.[35]

For allylic oxidation of isophorone to 4-ketoisophorone a system consisting of *t*-BuOK and phosphomolybdic acid in DMSO is advantageous in view of increased conversion and selectivity.[36]

α-Hydroxylation of β-diketo compounds with molecular oxygen is catalyzed by $CeCl_3 \cdot 7H_2O$.[37] Reductive α-hydroxylation of α,β-unsaturated nitriles to afford cyanohydrins is accomplished with $PhSiH_3$ and oxygen in the presence of $Mn(dpm)_3$.[38] A method for preparing α′-hydroxy-α,β-unsaturated tosylhydrazones involves treatment with BuLi-TMEDA at low temperature and oxygenation.[39]

Oxidative cleavage. Baeyer-Villiger oxidation with oxygen-PhCHO can be carried out in compressed carbon dioxide.[40] Attempts at achieving an asymmetric conversion of cyclobutanones[41] using oxygen-Et_2Zn and chiral amino alcohols met with little success.

Aerobic degradation of alkyl aryl ketones to aromatic carboxylic acids occurs on exposure to a mixture of Co(II) and Mn(II) nitrates in hot HOAc[42] or under basic conditions (*t*-BuOK, *m*-dinitrobenzene).[43] A new oxidation system for effecting CC bond cleavage in α-ketols, α-hydroxy acids and epoxides is Bi(0)-O_2.[44]

A thiyl radical that is generated in situ from ArSH in the presence of a M(salen) complex [M=V, Mn] and oxygen promotes double bond cleavage,[45] whereas 1,2-diamines and 1,2-amino alcohols suffer cleavage[46] to give two molecules of imines and a mixture of imines and aldehydes, respectively, when oxidation is carried out in the presence of $BF_3 \cdot OEt_2$.

Two oxidative cleavage processes concern deprotection are the following. Splitting of C-O bond in benzylidene acetals[47] and C-N bond in aminomalonic esters[48] are observed.

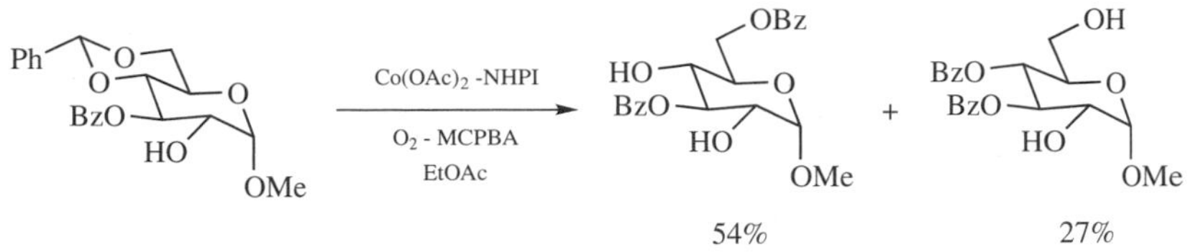

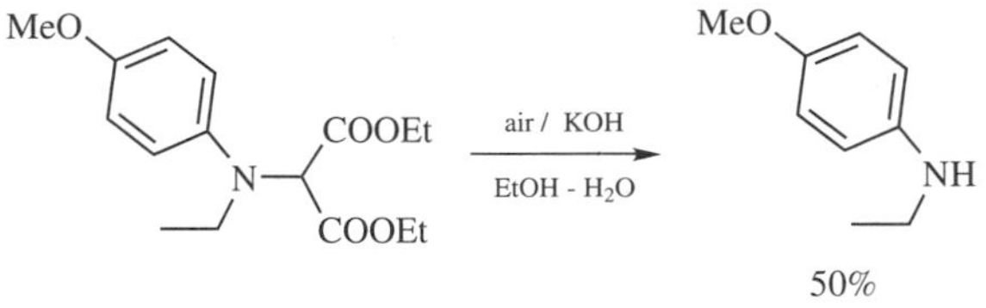

Coupling reactions. Suzuki coupling for *N*-arylation of imidazoles is performed under oxygen in the presence of Cu(II) complexes,[49] oxidative dimerization of arylboronic acids affords biaryls (also alkenylboronic acids to 1,3-dienes) when the catalyst is changed to $Pd(OAc)_2$.[50] Oxidative ligand coupling of tetraarylborate salts (e.g., $NaBAr_4$) occurs in the presence of a chlorosilane (Ph_2SiCl_2 or Ph_3SiCl) and oxygen also giving biaryls.[51] BINOLs are produced from a CuCl-mediated coupling of 3-substituted 2-naphthols in oxygen, *cis*-1,5-diazadecalin induces asymmetry but ee is not consistently high.[52]

Perhaps the more remarkable process is the conversion of benzene to biphenyl[53] using a mixture of $Pd(OAc)_2$ and $MoO_2(acac)_2$. Phenol formation is minimized.

Quinoxaline derivatives are assembled from epoxides and 1,2-diaminoarenes in one step by treatment with Bi, $Cu(OTf)_2$ and oxygen in DMSO.[54]

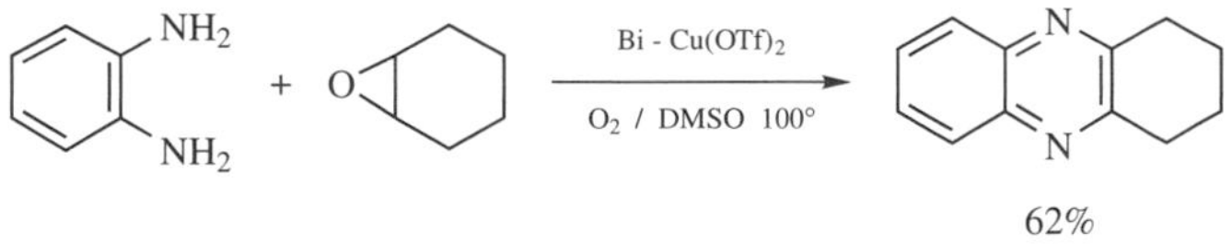

Alkylations. Radical alkylation at the active methylene group of malonic esters[55] and an α-carbon atom of ethers[56] is mediated by metal salt systems containing $Co(OAc)_2$.

Alkylation of ArCHO by R_2BCl takes place under oxygen, and the benzylic alcohols so produced are further transformed into the chlorides if $RBCl_2$ is used.[57]

Oxidative perfluoroalkylation of alkenes gives alcohols that bear a perfluoroalkyl residue.[58] The reaction involves free radical generation from pefluoroalkyl iodides by action of sodium dithionite and addition to the alkenes with trapping by oxygen.

Heterocyclization of unsaturated amine derivatives[59] catalyzed by $Pd(OAc)_2$ and oxygen is a useful synthetic operation because the unsaturation is retained for further modification.

Pd(OAc)$_2$ - py
O_2 / PhMe
80°

87%

[1]Gao, H., Angelici, R.J. *SC* **30**, 1239 (2000).
[2]Chen, W., Yamada, J., Matsumoto, K. *SC* **32**, 17 (2002).
[3]Fdil, N., Itto, Y.A., Ali, M.A., Karim, A., Daran, J.-C. *TL* **43**, 8769 (2002).
[4]Son, Y.-C., Makwana, V.D., Howell, A.R., Suib, S.L. *ACIEE* **40**, 4280 (2001).
[5]Marko, I.E., Gautier, A., Mutonkole, J.-L., Dumeunier, R., Ates, A., Urch, C.J., Brown, S.M. *JOMC* **624**, 344 (2001).
[6]Takaki, K., Shimasaki, Y., Shishido, T., Takehira, K. *BCSJ* **75**, 311 (2002).
[7]Ansari, I.A., Gree, R. *OL* **4**, 1507 (2002).
[8]Inoue, A., Kindo, J., Shinokubo, H., Oshima, K. *JACS* **123**, 11109 (2002).
[9]Cecchetto, A., Minisci, F., Recupero, F., Fontana, F., Pedulli, G.F. *TL* **43**, 3605 (2002).
[10]Minisci, F., Recupero, F., Punta, C., Guidarini, C., Fontana, F., Pedulli, G.F. *SL* 1173 (2002).
[11]Cecchetto, A., Fontana, F., Minisci, F., Recupero, F. *TL* **42**, 6651 (2001).
[12]Kulinkovich, O.G., Astahko, D.A., Tyvorskii, V.I., Ilyina, N.A. *S* 1453 (2001).
[13]Choudary, B.M., Kantam, M.L, Raman, A., Reddy, C.V., Rao, K.K. *ACIEE* **40**, 763 (2001).
[14]Martin, S.E., Suarez, D.F. *TL* **43**, 4475 (2002).
[15]Rao, T.V., Dongre, R.S., Jain, S.L., Sain, B. *SC* **32**, 2637 (2002).
[16]Villemin, D., Hammadi, M., Hachemi, M. *SC* **32**, 1501 (2002).
[17]Kakiuchi, N., Maeda, Y., Nishimura, T., Uemura, S. *JOC* **66**, 6620 (2001).
[18]Schultz, M.J., Park, C.C., Sigman, M.S. *CC* 3034 (2002).
[19]Döbler, C., Mehltretter, G.M., Sundermeier, U., Eckert, M., Militzer, H.-C., Beller, M. *TL* **42**, 8447 (2001).
[20]Muldoon, J., Brown, S.N. *OL* **4**, 1043 (2002).
[21]Miyata, A., Murakami, M., Irie, R., Katsuki, T., Uchida, T. *TL* **42**, 7067 (2001).
[22]Yamaguchi, K., Mizuno, N. *ACIEE* **41**, 4538 (2002); *NJC* **26**, 972 (2002).
[23]Csjernyik, G., Ell, A.H., Fadini, L., Pugin, B., Bäckvall, J.-E. *JOC* **67**, 1657 (2002).
[24]Choi, E., Lee, C., Na, Y., Chang, S. *OL* **4**, 2369 (2002).
[25]Wolfson, A., Wuyts, S., De Vos, D.E., Vankelecom, I.F.J., Jacobs, P.A. *TL* **43**, 8107 (2002).
[26]Ji, H., Mizugaki, T., Ebitani, K., Kaneda, K. *TL* **43**, 7179 (2002).
[27]Mori, K., Yamaguchi, K., Mizugaki, T., Ebitani, K., Kaneda, K. *CC* 461 (2002).
[28]Jain, S.L., Sain, B. *CC* 1040 (2002).
[29]Miyata, A., Furukawa, M., Irie, R., Kstsuki, T. *TL* **43**, 3481 (2002).
[30]Maeda, Y., Kakiuchi, N., Matsumura, S., Nishimura, T., Kawamura, T., Uemura, S. *JOC* **67**, 6718 (2002).
[31]Hirao, T., Morimoto, C., Takada, T., Sakurai, H. *T* **57**, 5073 (2001).
[32]Shi, M., Xu, B. *JOC* **67**, 294 (2002).
[33]Fabbrini, M., Galli, C., Gentili, P., Macchitella, D. *TL* **42**, 7551 (2001).
[34]Dongre, R.S., Rao, V.V., Sharma, B.K., Sain, B., Bhatia, V.K. *SC* **31**, 167 (2001).
[35]Cadot, C., Dalko, P.I., Cossy, J. *TL* **42**, 1661 (2001).
[36]Murphy, E.F., Schneider, M., Mallat, T., Baiker, A. *S* 547 (2001).
[37]Christoffers, J., Werner, T. *SL* 119 (2002).
[38]Magnus, P., Scott, D.A., Fielding, M.R. *TL* **42**, 4127 (2001).
[39]Baptistella, L.H.B., Aleixo, A.M. *SC* **32**, 2937 (2002).
[40]Bolm, C., Palazzi, C., Francio, G., Leitner, W. *CC* 1588 (2002).
[41]Shinohara, T., Fujioka, S., Kotsuki, H. *H* **55**, 237 (2001).

[42]Minisci, F., Recupero, F., Fontana, F., Bjorsvik, H.R., Liguori, L. *SL* 610 (2002).
[43]Bjorsvik, H.R., Liguori, L., Gonzalez, R.R., Merinero, J.A.V. *TL* **43**, 4985 (2002).
[44]Coin, C., Le Boisselier, V., Favier, I., Postel, M., Dunach, E. *EJOC* 735 (2001).
[45]Baucherel, X., Uziel, J., Juge, S. *JOC* **66**, 4504 (2001).
[46]Shimizu, M., Makino, H. *TL* **42**, 8865 (2001).
[47]Chen, W., Wang, P.G. *TL* **42**, 4955 (2001).
[48]Niwa, Y., Takayama, K., Shimizu, M. *TL* **42**, 5473 (2001).
[49]Collman, J.P., Zhong, M., Zhang, C., Costanzo, S. *JOC* **66**, 7892 (2001).
[50]Parrish, J.P., Jung, Y.C., Floyd, R.J., Jung, K.W. *TL* **43**, 7899 (2002).
[51]Sakurai, H., Morimoto, C., Hirao, T. *CL* 1084 (2001).
[52]Li, X., Yang, J., Kozlowski, M.C. *OL* **3**, 1137 (2001).
[53]Okamoto, M., Yamagi, T. *CL* 212 (2001).[54] Antoniotti, S., Dunach, E. *TL* **43**, 3971 (2002).
[55]Hirase, K., Iwahama, T., Sakaguchi, S., Ishii, Y. *JOC* **67**, 970 (2002).
[56]Hirano, K., Sakaguchi, S., Ishii, Y. *TL* **43**, 3617 (2002).
[57]Kabalka, G.W., Wu, Z., Ju, Y. *T* **58**, 3243 (2002).
[58]Yoshida, M., Ohkoshi, M., Muraoka, T., Matsuyama, H., Iyoda, M. *BCSJ* **75**, 1833 (2002).
[59]Fix, S.R., Brice, J.L., Stahl, S.S. *ACIEE* **41**, 164 (2002).

Oxygen, singlet. 13, 228–229; **14,** 247; **15,** 243; **16,** 257–258; **17,** 251–253; **18,** 269–270; **19,** 244; **21**, 309.

Amides from amidoximes. In the presence of NaOMe amidoximes undergo deoximation on reaction with photochemically generated singlet oxygen.[1]

[1]Öcal, N., Erden, I. *TL* **42**, 4765 (2001).

Ozone. 13, 229; **15,** 243–244; **17,** 253–254; **18,** 270–272; **19,** 244–246; **20,** 279

1,2-Diols. Ozonolysis of alkenylstannanes afford stable ozonides, subsequent treatment with borane leads to 1,2-diols.[1]

Ph–CH=CH–$SnBu_3$ + O_3 —(MeOH ; $H_3B \cdot SMe_2$)→ PhCH(OH)CH$_2$OH 61%

Oxidation.[2] 5-Cyanomorpholin-2-ones undergo both dehydrogenation and ring-opening degradation.

+ O_3 —(EtOAc)→ 55% + 43%

[1]Gomez, A.M., Company, M.D., Valverde, S., Lopez, J.C. *OL* **4**, 383 (2002).
[2]Namba, K., Kawasaki, M., Takada, I., Iwama, S., Izumida, M., Shinada, T., Ohfune, Y. *TL* **42**, 3733 (2001).

P

Palladacycles. 21, 310–312

Coupling reactions. Continuing exploration of Heck, Stille, Suzuki, Sonogashira, and related coupling reactions using palladacycles is evident. The readily prepared species from oximes of acetophenone, benzophenone and ring substituted analogues have received much attention, for example, Suzuki coupling in water, [1] Sonagashira coupling in copper-free and amine-free conditions.[2] A ferrocenyl analogue **1**, [3] and a new catalyst containing carbenoid and benzoquinone components **2**[4] are good for the Heck reaction. Other effective catalysts are **3**[5] and **4**[6] for Suzuki coupling, **5**[7] for Sonagashira coupling, and **6**[8] for Heck and Suzuki couplings and *N*-arylation.

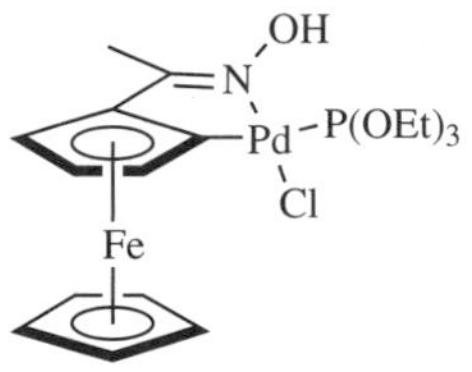

1 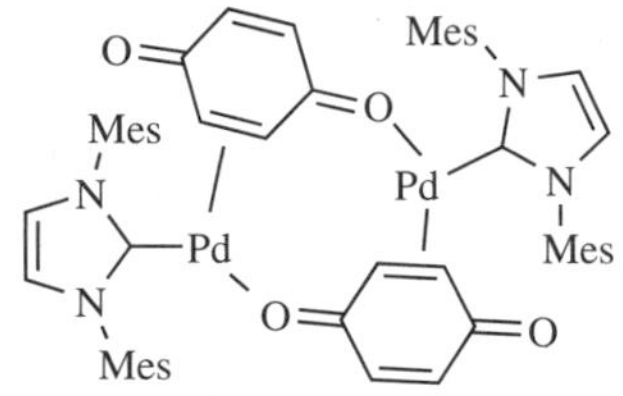 **2** 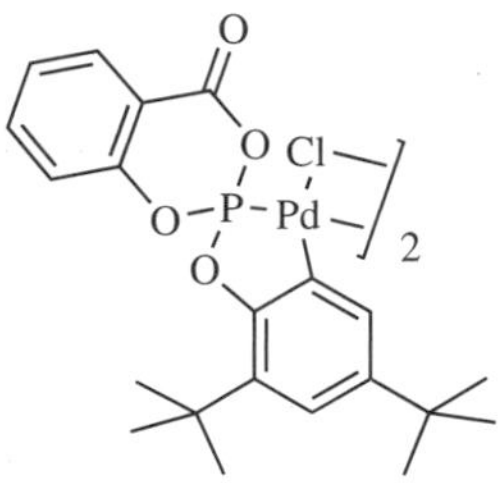 **3**

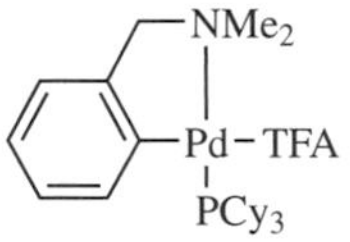

4 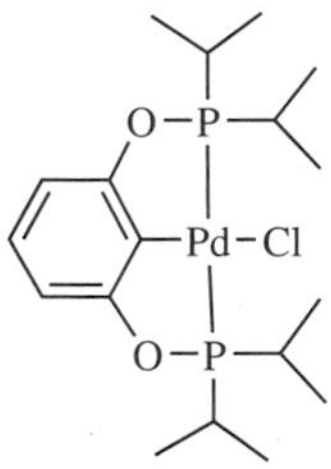 **5** 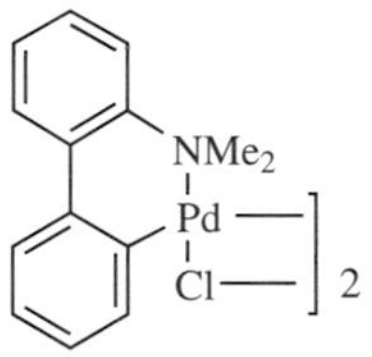 **6**

A palladacycle containing a sulfide unit promotes the Ullmann coupling.[9]

Carbonylation. The catalyst derived from benzophenone oxime and Li_2PdCl_4 is very efficient in converting ArI into carboxylic acid derivatives.[10]

Redox reaction.[11] Ketones bearing an *o*-bromobenzyl group at the α-position are hydrodebrominated as well as dehydrogenated to provide bromine-free enones. The reaction is related to the oxidation of silyl enol ethers by $Pd(OAc)_2$.

Fiesers' Reagents for Organic Synthesis, Volume 22. Series editor Tse-Lok Ho
ISBN 0-471-28515-3

Hydroamination.[12] Hydroamination of conjugated dienes in the presence of **7** proceeds at room temperature. For styrenes, more vigorous conditions are required.

7

[1]Botella, L., Najera, C. *JOMC* **663**, 46 (2002).
[2]Alonso, D.A., Najera, C., Pacheco, M.C. *TL* **43**, 9365 (2002).
[3]Iyer, S., Jayanthi, A. *TL* **42**, 7877 (2001).
[4]Selvakumar, K., Zapf, A., Beller, M. *OL* **4**, 3031 (2002).
[5]Bedford, R.B., Hazelwood, S.L., Limmert, M.E. *CC* 2608 (2002).
[6]Bedford, R.B., Cazin, C.S.J. *CC* 1540 (2001).
[7]Eberhard, M.R., Wang, Z., Jensen, C.M. *CC* 818 (2002).
[8]Schnyder, A., Indolese, A.F., Studer, M., Blaser, H.-U. *ACIEE* **41**, 3668 (2002).
[9]Silveira, P.B., Lando, V.R., Dupont, J., Monteiro, A.L. *TL* **43**, 2327 (2002).
[10]Ramesh, C., Kubota, Y., Miwa, M., Sugi, Y. *S* 2171 (2002).
[11]Högenauer, K., Mulzer, J. *OL* **3**, 1495 (2001).
[12]Minami, T., Okamoto, H., Ikeda, S., Tanaka, R., Ozawa, F., Yoshifuji, M. *ACIEE* **40**, 4501 (2001).

Palladium/carbon. 13, 230–232; **15,** 245; **18,** 273; **19,** 247; **20,** 280–281; **21,** 312–314

Hydrogenolysis. An *N*-benzoyl group can be removed by reduction with borane followed by hydrogenolysis of the resulting benzyl group.[1] Aryl halides undergo hydrogenolysis with Pd/C in the presence of solid hydrazine hydrochloride,[2] or Et_3N,[3,4] including polychlorinated biphenyls.[4] Cleavage of THP ethers is observed under Pd/C-catalyzed hydrogenolysis conditions.[5]

Hydrogenolysis of benzyl 3-oxoalkanoates in the presence of cinchonine leads to chiral ketones with moderate ee (66–75%)[6] as a result of asymmetric protonation of the enolic intermediates. Rearrangement occurs during hydrogenolysis that releases 2-cyanocyclopropanones.[7]

CN $\xrightarrow[\text{EtOAc}]{H_2 / \text{Pd-C}}$ HOOC NH$_2$

95%

Deuterium exchange (with D_2O) at a benzylic position[8] occurs in the presence of Pd/C.

Hydrogenation. Hydrogenation of Baylis-Hillman adducts is highly diastereoselective in the presence of $MgBr_2$,[9] while no such additive is required for excellent diastereoselectivity in the case of allylically substituted alkenylboronates.[10]

OH Ph COOMe $\xrightarrow[MgBr_2 / CH_2Cl_2]{H_2 / \text{Pd-C}}$ OH Ph COOMe + OH Ph COOMe

(62:1)

93%

To avoid the loss of TBS groups under hydrogenation conditions, Pd/C(en) should be used as catalyst.[11] This same system mediates the hydrogenation of aryl carbonyl compounds to benzylic alcohols.[12]

Reductive amination of ketones is accomplished by transfer hydrogenation (Pd/C, $HCOONH_4$, aq. MeOH).[13] A preparation of 2,5-diarylpyrroles from 1,4-diaryl-2-butene-1,4-diones under phase-transfer conditions (PEG-200) and with microwave irradiation involves hydrogenation of the double bond and aminocyclization.[14]

Hydrogenolysis of benzyl ethers can be avoided by using palladium black in hydrogenation.[15] Both hydrogenation and hydrogenolysis complete the conversion of α-dichloromethylenearylacetic esters to 2-arylpropionic esters[16] which are valuable intermediates of several thereapeutic agents. The substrates are products of Wittig reaction of aroylformates with Ph_3P-CCl_4.

Aromatization. 1-Aminocyclohexenes give arylamines[17] on heating with catalytic amounts of Pd/C and molecular sieves in $PhNO_2$ and PhMe. The more facile aromatization of pyrazolines and Hantzsch 1,4-dihydropyridine is complete at 80° in HOAc.[18]

Coupling reactions. A synthesis of aryldiphenylphosphines from ArX and Ph_3P[19] or Ph_2PH[20] is by simply heating them with Pd/C in DMF. As for the prevalent Pd-catalyzed coupling reactions the employment of Pd/C, whenever possible, is definitely advantageous. Indeed, the general utility has been demonstrated.[21] Contributions on Suzuki coupling, [22–24] and Sonagashira coupling [25,26] are also pointed out, as well as coupling of ArX with arylhalosilanes to give biaryls.[27]

Borylation.[28] Pd/C promotes borylation at benzylic position with bis(pinacolato)diboron or pinacolborane.

Ring opening.[29] Imidazolidines and oxazolidines open in MeOH at room temperature in the presence of Pd/C.

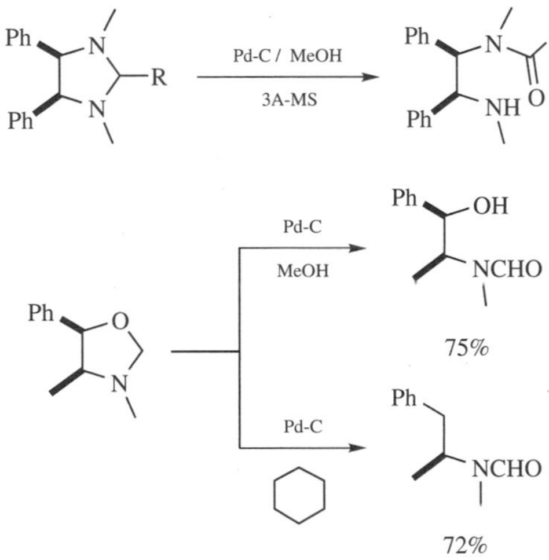
Ph
N
R
Ph
N
Pd-C / MeOH
3A-MS
Ph
N
R
Ph
NH
O
Ph
O
N
Pd-C
MeOH
Ph
OH
NCHO
75%
Pd-C
Ph
NCHO
72%

[1]Couturier, M., Andresen, B.M., Tucker, J.L., Dube, P., Brenek, S.J., Negri, J.T. *TL* **42**, 2763 (2001).
[2]Rodriguez, J.G., Lafuente, A. *TL* **43**, 9645 (2002).
[3]Faucher, N., Ambroise, Y., Cintrat, J.-C., Doris, E., Pillon, F., Rousseau, B. *JOC* **67**, 932 (2002).
[4]Sajiki, H., Kume, A., Hattori, K., Hirota, K. *TL* **43**, 7247 (2002); Sajiki, H., Kume, A., Hattori, K., Nagase, H., Hirota, K. *TL* **43**, 7251 (2002).
[5]Kaisalo, L.H., Hase, T.A. *TL* **42**, 7699 (2001).
[6]Roy, O., Riahi, A., Henin, F., Muzart, J. *EJOC* 3986 (2002).
[7]Royer, F., Felpin, F.-X., Doris, E. *JOC* **66**, 6487 (2001).
[8]Sajiki, H., Hattori, K., Aoki, F., Yasunaga, K., Hirota, K. *SL* 1149 (2002).
[9]Bouzide, A. *OL* **4**, 1347 (2002).
[10]Hupe, E., Marek, I., Knochel, P. *OL* **4**, 2861 (2002).
[11]Hattori, K., Sajiki, H., Hirota, K. *T* **57**, 2109 (2001).
[12]Hattori, K., Sajiki, H., Hirota, K. *T* **57**, 4817 (2001).
[13]Allegretti, M., Berdini, V., Cesta, M.C., Curti, R., Nicolini, L., Topai, A. *TL* **42**, 4257 (2001).
[14]Rao, H.S.P., Jothilingam, S. *TL* **42**, 6595 (2001).
[15]Maki, S., Okawa, M., Matsui, R., Hirano, T., Niwa, H. *SL* 1590 (2001).
[16]Patil, D.V., Wadia, M.S. *SC* **32**, 2821 (2002).
[17]Cossy, J., Belotti, D. *OL* **4**, 2557 (2002).
[18]Nakamichi, N., Kawashita, Y., Hayashi, M. *OL* **4**, 3955 (2002).

[19]Lai, C.W., Kwong, F.Y., Wang, Y., Chan, K.S. *TL* **42**, 4883 (2001).
[20]Stadler, A., Kappe, C.O. *OL* **4**, 3541 (2002).
[21]Heidenreich, R.G., Köhler, K., Krauter, J.G.E., Pietsch, J. *SL* 1118 (2002).
[22]Le Blond, C.R., Andrews, A.T., Sun, Y., Sowa Jr, J.R. *OL* **3**, 1555 (2001).
[23]Kabalka, G.W., Namboodiri, V., Wang, L. *CC* 775 (2001).
[24]Sakurai, H., Hirao, T., Tsukuda, T. *JOC* **67**, 2721 (2002).
[25]Lopez-Deber, M.P., Castedo, L., Granja, J.R. *OL* **3**, 2823 (2001).
[26]Kabalka, G.W., Wang, L., Pagni, R.M. *T* **57**, 8017 (2001).
[27]Huang, T., Li, C.-J. *TL* **43**, 403 (2002).
[28]Ishiyama, T., Ishida, K., Takagi, J., Miyaura, N. *CL* 1082 (2001).
[29]Alexakis, A., Aujard, I., Pytkowicz, J., Roland, S., Mangeney, P. *JCS(P1)* 949 (2001).

Palladium(II) acetate. 13, 232–233; **14,** 248; **15,** 245–247; **16,** 259–263; **17,** 255–259; **18,** 274–277; **19,** 248–251; **20,** 281–283; **21,** 314–320

Coupling reactions. Some work on using $Pd(OAc)_2$ to catalyze sp^2-sp^2 couplings emphasizes conditions without phosphine ligands. Adding a hindered 1,3-diarylimidazolium chloride[1,2] or its dihydro derivative[3] to the reaction medium as precursor of carbene ligand is very popular. *N,N′*-Dicyclohexyl-1,4-aza-1,3-butadiene[4] and 2-aryl-2-oxazolines[5] are further useful ligands.

Microwave accelerates Suzuki coupling in water.[6] Ultrasound irradiation of reaction components in an ionic liquid at ambient temperature is another protocol.[7] $(ArBF_3)K$ also finds use in the ligandless coupling with Ar′X, [8] whereas replacement of a halogen substituent in an aromatic ring by an alkyl group can be achieved by coupling with alkylboranes (from alkenes +9-H-9-BBN).[9]

Applying Suzuki coupling to peracetylated glycals leads to *C*-glycosides.[10]

By a Heck reaction 2-arylindenes are readily prepared from indene, [11] 2-benzyl-2-buten-4-olides from α-methylene-γ-butyrolactone,[12] and 1-arylalkenes in which the aliphatic carbon chain is completely fluorinated from $R_fCH{=}CH_2$.[13] As expected, carbene ligands have been tested for the Heck reaction.[14,15] Polyethyleneglycol as a solvent is also proposed.[16]

Alkenyl(aryl)iodonium fluorides couple by selective transferring the alkenyl group. Accordingly, the coupling can be used to prepared dienes.[17]

F
I—Tol
F
$CH_2{=}CHCOMe$
CH_2Cl_2
O
F
51%

The hybrid Heck-Stille coupling ($ArSnBu_3$ + alkenes) requires an oxidant ($CuCl_2$ or O_2).[18]

Ring-forming coupling reactions include those giving rise to oxazolidin-2-ones,[19] alkylidenecyclopentanes,[20] fulvenes,[21] and bicyclic acetates.[22]

MeOOC COOMe + Cl → $Pd(OAc)_2$ - LiCl, BSA / THF → MeOOC COOMe 87%

R—≡—R + I COOMe → $Pd(OAc)_2$ - Ag_2CO_3, MeCN 20° → R R R R COOMe 85%

The $Pd(OAc)_2$–catalyzed substitution of ArX leading to aryl ethers[23] and *N*-aryl-2-oxazolidinones[24] are simple operations.

Activation of an ortho C-H bond in anilides by $Pd(OAc)_2$ to participate in coupling reactions has been demonstrated.[25] Benzoquinone is used as reoxidant.

NHAc + COOBu → $Pd(OAc)_2$ - TsOH, HOAc , benzoquinone PhMe → NHAc COOBu 72%

Under CO the carbonylative coupling of iodonium salts with alkynes by $Pd(OAc)_2$ provides alkynyl aryl ketones.[26] Insertive carbonylation into the Suzuki coupling components leads to unsymmetrical benzophenones.[27]

Me_2N $\overset{+}{I}$ NMe_2 I^- → $Pd(OAc)_2$ - CO, Bu_3N / DMF → Ph O Me_2N I NMe_2 80%

$B(OH)_2$ + $Ph\text{-}N_2^+$ BF_4^- → $Pd(OAc)_2$ - CO, dioxane, MesN NMes + Cl^- → Ph O 52%

Like the reaction of Pd-catalyzed aryl exchange between Ar_3P and Ar'X to give mixed-aryl phosphines, Ph_2AsAr' can also be prepared in the same manner.[28]

Redox reactions. Alcohols are oxidized by allyl diethyl phosphate in the presence of $Pd(OAc)_2$.[29] Secondary alcohols afford ketones but primary alcohols furnish esters. Selective oxidation of bidentate phosphines to monophosphine oxides is a valuable process and it can be carried out with $Pd(OAc)_2$.[30]

Intramolecular demetallative coupling of Fischer carbene complexes leads to 1,4-dioxa-2-cycloalkenes.[31] Note the different behavior of aminocarbene complexes.

$Pd(OAc)_2$, Et_3N

M = $Cr(CO)_5$

n = 0 70%
n = 1 64%
n = 2 21%
n = 3 14%

$Pd(OAc)_2$, Et_3N

52 – 54%

Reduction of polymer-bound nitroarenes with $Pd(OAc)_2$-N_2H_4 to arylamines[32] presents no problem. As a hydrogenation catalyst $Pd(OAc)_2$ microencapsulated in polyurea has the advantage of easy recovery and reuse.[33]

Vinyl glycosides.[34] Monosaccharides in which all except the glycosidic hydroxy group are protected undergo *O*-vinylation in a Pd-catalyzed reaction with an alkyl vinyl ether.

Cyclizations. In the ring closure of alkynals to provide cyclic alkenyl ethers $Pd(OAc)_2$ plays the role of a Lewis acid besides transition metal catalyst.[35] Under pressurized carbon dioxide $Pd(OAc)_2$ catalyzes the formation of cyclic alkenyl carbamate.[36]

$Pd(OAc)_2$, Ph_2CO / MeOH, dioxane

66%

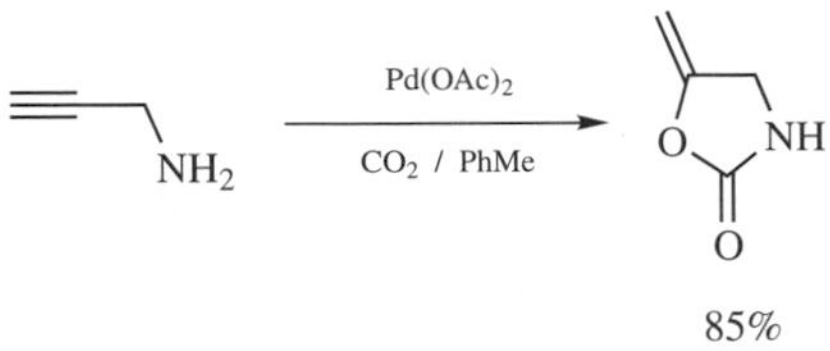

Pd(OAc)$_2$

CO_2 / PhMe

85%

Additions. Dimerization of 1-alkynes to (*E*)-enynes is catalyzed by an imidazolyl carbene-coordinated Pd system.[37] For carbonyl compounds as acceptors the addition an acetoxy group to alkynes is accompanied by CC bond formation.[38,39]

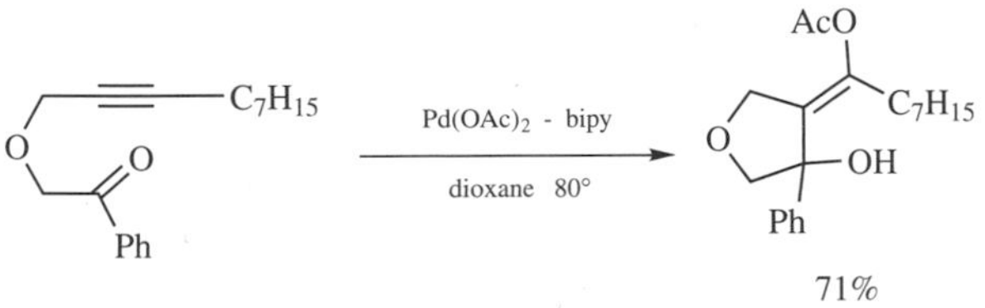

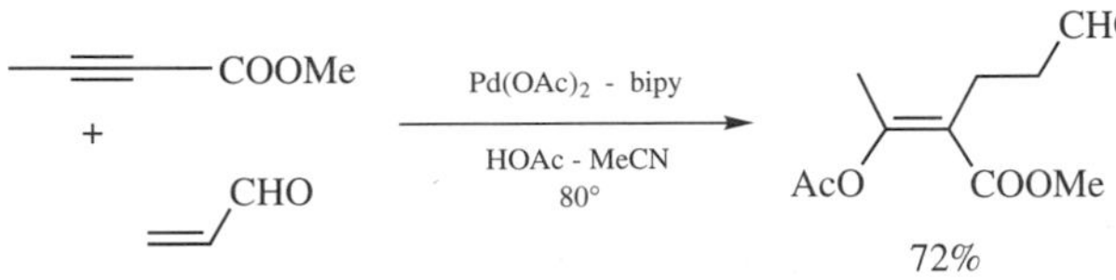

Pd(OAc)$_2$ - bipy

dioxane 80°

71%

Pd(OAc)$_2$ - bipy

HOAc - MeCN 80°

72%

Nonoxidative addition of a disilane to α,β-unsaturated carbonyl compounds in a 1,4-mode is observed with Pd(OAc)$_2$ and Me_3SiOTf.[40]

Allylic displacements. 2,3-Alkadienyl acetates are transformed into 2-bromo-1,3-alkadienes with LiBr under the influence of Pd(OAc)$_2$.[41]

Cyclopropanation. Dienones and dienylboronates are cyclopropanated regioselectively at the more electron-deficient double bond by a Pd(OAc)$_2$-catalyzed reaction with diazomethane, therefore cyclopropanes possessing highly different functional groups are accessible.[42]

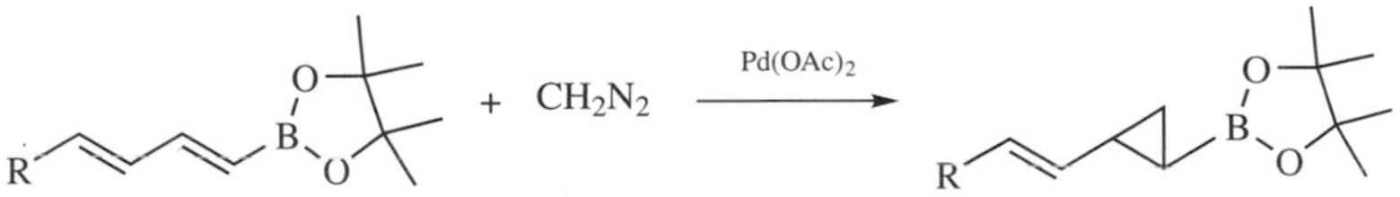

\+ CH_2N_2 —Pd(OAc)$_2$→

[1]Grasa, G.A., Nolan, S.P. *OL* **3**, 119 (2001); Grasa, G.A., Viciu, M.S., Huang, J., Zhang, C., Trudell, M.L., Nolan, S.P. *OM* **21**, 2866 (2002).
[2]Fürstner, A., Seidel, G. *OL* **4**, 541 (2002).
[3]Andrus, M.B., Song, C. *OL* **3**, 3761 (2001).
[4]Grasa, G.A., Hillier, A.C., Nolan, S.P. *OL* **3**, 1077 (2001).
[5]Tao, B., Boykin, D.W. *TL* **43**, 4955 (2002).
[6]Leadbetter, N.E., Marco, M. *OL* **4**, 2973 (2002).
[7]Dilip, R.R., Jarikote, V., Srinivasan, K.V. *CC* 6165 (2002).

[8]Molander, G.A., Biolatto, B. *OL* **4**, 1867 (2002).
[9]Fürstner, A., Leitner, A. *SL* 290 (2001).
[10]Ramnauth, J., Poulin, O., Rakhit, S., Maddaford, S..P. *OL* **3**, 2013 (2001).
[11]Nifant'ev, I.E., Sitnikov, A.A., Andriukhova, N.V., Laishevtsev, I.P., Luzikov, Y.N. *TL* **43**, 3213 (2002).
[12]Arcadi, A., Chiarini, M., Marinelli, F., Berente, Z., Kollar, I. *EJOC* 3165 (2001).
[13]Darses, S., Pucheault, M., Genet, J.-P. *EJOC* 1121 (2001).
[14]Yang, C., Nolan, S.P. *SL* 1539 (2001).
[15]Andrus, M.B., Song, C., Zhang, J. *OL* **4**, 2079 (2002).
[16]Chandrasekhar, S., Narsihmulu, C., Sultana, S.S., Reddy, N.R. *OL* **4**, 4399 (2002).
[17]Yoshida, M., Nagahara, D., Fukuhara, T., Yoneda, N., Hara, S. *JCS(P1)* 2283 (2001).
[18]Parrish, J.P., Jung, Y.C., Shin, S.I., Jung, K.W. *JOC* **67**, 7127 (2002).
[19]Liu, G., Lu, X. *OL* **3**, 3879 (2001).
[20]Liu, G., Lu, X. *TL* **43**, 6791 (2002).
[21]Kotera, M., Matsumura, H., Gao, G., Takahashi, T. *OL* **3**, 3467 (2001).
[22]Löfstedt, J., Franzen, J., Bäckvall, J.-E. *JOC* **66,** 8015 (2001).
[23]Torraca, K.E., Huang, X., Parrish, C.A., Buchwald, S.L. *JACS* **123**, 10770 (2001).
[24]Cacchi, S., Fabrizi, G., Goggiamani, A., Zappia, G. *OL* **3**, 2539 (2001).
[25]Boele, M.D.K., van Strijdonck, G.P.F., de Vries, A.H.M., Kamer, P.C.J., de Vries, J.G., van Leeuwen, P.W.N.M. *JACS* **124**,1586 (2002).
[26]Luo, S., Liang, Y., Liu, C., Ma, Y. *SC* **31**, 343 (2001).
[27]Andrus, M.B., Ma, Y., Zang, Y., Song, C. *TL* **43**, 9137 (2002).
[28]Kwong, F.Y., Lai, C.W., Chan, K.S. *JACS* **123**, 8864 (2001).
[29]Shvo, Y., Goldman-Lev, V. *JOMC* **650**, 151 (2002).
[30]Grushin, V.V. *OM* **20**, 3950 (2001).
[31]Sierra, M.A., del Amo, J.C., Mancheno, M.J., Gomez-Gallego, M. *JACS* **123**, 851 (2001).
[32]Rödel, M., Thieme, F., Buchholz, H., König, B.B. *SC* **32**, 1181 (2002).
[33]Bremeyer, N., Ley, S.V., Ramarao, C., Shirley, I.M., Smith, S.C. *SL* 1843 (2002).
[34]Handerson, S., Schlaf, M. *OL* **4**, 407 (2002).
[35]Asao, N., Nogami, T., Takahashi, K., Yamamoto, Y. *JACS* **124**, 764 (2002).
[36]Shi, M., Shen, Y.-M. *JOC* **67**, 16 (2002).
[37]Yang, C., Nolan, S.P. *JOC* **67**, 591 (2002).
[38]Zhao, L., Lu, X. *ACIEE* **41**, 4343 (2002).
[39]Zhao, L., Lu, X. *OL* **4**, 3903 (2002).
[40]Ogoshi, S., Tomiyasu, S., Morita, M., Kurosawa, H. *JACS* **124**, 11598 (2002).
[41]Horvath, A., Bäckvall, J.-E. *JOC* **66**, 8120 (2001).
[42]Marko, I.E., Giard, T., Sumida, S., Gies, A.-E. *TL* **43**, 2317 (2002).

Palladium(II) acetate–phase-transfer catalyst. 20, 284–286; 21, 320

Coupling reactions. Conditions for Heck and Suzuki couplings catalyzed by $Pd(OAc)_2$ that include addition of a phase-transfer agent are well established. Variations and refinements are those applying microwave assistance in water [1] and in the molten salt state (100°).[2] A catalyst obtained from miccroencapsulation of $Pd(OAc)_2$ in polyurea is easily recovered and reused (up to 4 times).[3]

(*E*)-Cinnamonitriles[4] and 3,3-disubstituted acrylonitriles[5] are now readily available from Heck reaction under phase-transfer conditions. A one-pot transformation of silyl enol ethers into 1,3-dienes involves in situ generation of alkenyl nonaflates and subsequent Heck reaction.[6]

$OSiMe_3$
Bu_4NF - NfF ;
$Pd(OAc)_2$ - Bu_4NCl
CH_2=CHCOOMe
DMF
COOMe
75%

A reductive Heck reaction is the key step that forms a six-membered ring during a synthesis of (+)-phorbol.[7]

I
Ph
Me_3SiO
O
OTBS
$Pd(OAc)_2$ - Bu_4NBr
HCOOK
79%
(+)-phorbol

Acting as electrolyte Bu_4NF is used in an electrochemical carbonylation of 2-amino-1-alkanols to afford oxazolidin-2-ones.[8]

[1]Leadbeater, N.E., Marco, M. *OL* **4**, 2973 (2002).
[2]Battistuzzi, G., Cacchi, S., Fabrizi, G. *SL* 439 (2002).
[3]Ley, S.V., Ramarao, C., Gordon, R.S., Holmes, A.B., Morrison, A.J., McConvey, I.F., Shirley, I.M., Smith, S.C., Smith, M.D. *CC* 1134 (2002).
[4]Zhao, H., Cai, M.-Z., Peng, C.-Y. *SC* **32**, 3419 (2002).
[5]Masllorens, J., Moreno-Manas, M., Pla-Quintana, A., Plexats, R., Roglans, A. *S* 1903 (2002).
[6]Lyapkalo. I.M., Webel, M., Reissig, H.-U. *EJOC* 3646 (2002).
[7]Lee, K., Cha, J.K. *JACS* **123**, 5590 (2001).
[8]Chiarotto, I., Feroci, M. *TL* **42**, 3451 (2001).

Palladium(II) acetate–tertiary phosphine. 13, 91, 233–234; **14,** 249, 250–253; **15,** 247–248; **16,** 264–268; **17,** 259–269; **18,** 277–281; **19,** 252–256; **20**, 286–289; **21,** 321–324

Coupling reactions. Suzuki coupling is used in preparing aryl trifluoromethyl ketones from $ArB(OH)_2$ and CF_3COOPh.[1] The Heck reaction initiates the annulation of *o*-iodoaraldehyde imines with internal alkynes to form isoquinoline and carboline derivatives.[2,3] Intramolecular coupling of an ester gives a biaryl *o,o′*-lactone that is subject to atroposelective cleavage.[4]

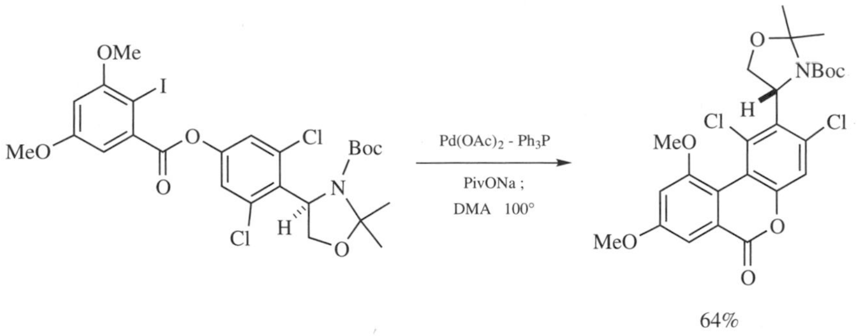

64%

With the presence of Bu_4NOAc the catalytic system effects cyclization of *N*-(*o*-iodobenzyl)-3-alkenamides. The dryness of DMF has a dramatic effect on the product structures (ring size).[5]

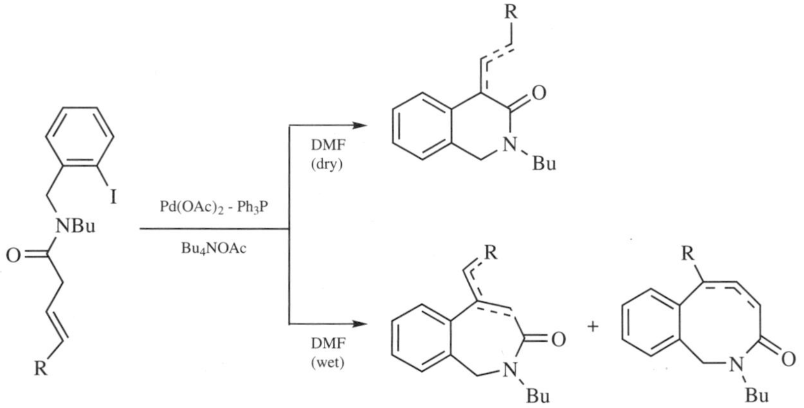

Reduction. Reduction of carboxylic acids to aldehydes with sodium hypophosphite is Pd-catalyzed.[6]

Addition reactions. A previous assignment for the addition of 1-alkynes to 2-alkynoic esters to be in the Michael sense is incorrect.[7]

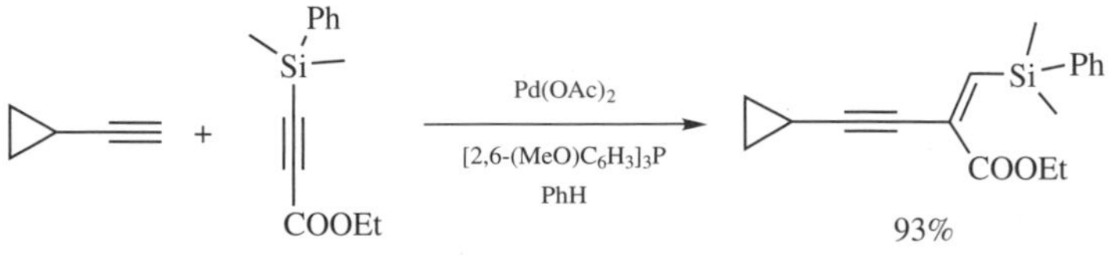

93%

Racemization. Occasionally, racemization of chiral compounds are necessary (for example, recycling the undesirable enantiomer after resolution). *N*-Acetyl amino acids can be racemized by $Pd(OAc)_2$–Ph_3P, $(dba)_3Pd_2$, or $(Ph_2P)_4Pd$.[8]

[1]Kakino, R., Shimizu, I., Yamamoto, A. *BCSJ* **74**, 371 (2001).
[2]Riesch, K.R., Zhang, H., Larock, R.C. *JOC* **66**, 8042 (2001).
[3]Zhang, H., Larock, R.C. *OL* **3**, 3083 (2001).
[4]Bringmann, G., Menche, D., Muhlbacher, J., Riechert, M., Saito, N., Pfeiffer, S.S., Lipshutz, B.H. *OL* **4**, 2833 (2002).
[5]Ferraccioli, R., Carenzi, D., Catellani, M. *SL* 1860 (2002).
[6]Goossen, L.J., Ghosh, K. *CC* 836 (2002).
[7]Trost, B.M., Gunzer, J.L., Yasukata, T. *TL* **42**, 3775 (2001).
[8]Hateley, M.J., Schichl, D.A., Fischer, C., Beller, M. *SL* 25 (2001).

Palladium(II) acetate–tertiary phosphine–base. 20, 289–292; **21,** 324–327

Aryl coupling reactions. Suzuki coupling in supercritical CO_2[1] as well as homocoupling of $ArB(OH)_2$[2] are reported. Using 9-alkyl-9-BBN to couple with alkyl tosylates and bromides represents a new way to synthesize aliphatic compounds (RX that possess β-H can be used).[3,4] By changing the boron-containing partners to $ArB(OH)_2$ the process is applicable to the synthesis of alkylarenes.[5] Here, methyl-di-*t*-butylphosphine or tricyclohexylphosphine is the ligand to employ, and ROTs and RBr possessing β-hydrogen atoms are allowed. Such complexes also catalyzes Stille coupling efficiently.[6] Other developments concern the discovery of sterically demanding, water-soluble alkylphosphines (containing a quaternary ammonium head) as ligands for high activity Suzuki coupling of ArBr in aqueous environments,[7] and replacing the potassium cation in $[ArBF_3]K$ by tetrabutylammonium ion to render them soluble in organic solvents while maintaining their stability to air and moisture.[8]

Allylamines undergo Heck reaction at the sp^2-carbon closer to the nitrogen atom.[9] A highly stereoselective Heck reaction of alkenyl sulfoxides is noted (Ag_2CO_3 present).[10] An intramolecular Heck reaction is the key step in a synthesis of (−)-galanthamine.[11]

NMe_2, C_5H_{11} + Ph-I — Pd(OAc)2 - DPPF / Ag_2CO_3 / DMF, 100° → C_5H_{11}, Ph, NMe_2 75%

MeO, Br, CN, CHO — $Pd(OAc)_2$ - DPPF / Ag_2CO_3 / PhMe, 107° → MeO, CN, CHO

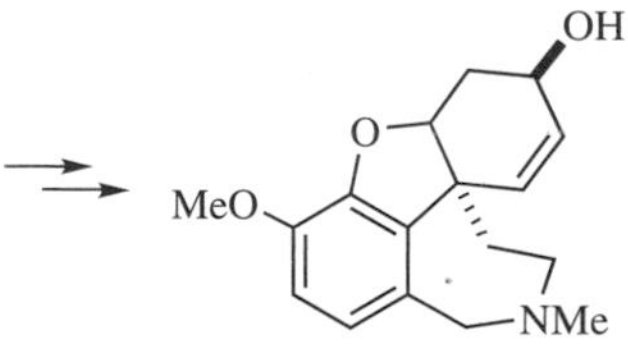

OH
MeO
O
NMe
(–)-galanthamine

Heck reaction of ArX with in situ alkylation completes an annulation process.[12,13] Arylative cyclization of allenyl carbonyl compounds is promoted by a combination of $Pd(OAc)_2$ and indium.[14] Indanone acetals are formed from *o*-tosyloxybenzaldehydes by a Heck reaction with hydroxyethyl vinyl ether, due to aldol reaction that is in tandem with the coupling.[15] Products from reaction with simple vinyl ethers are acetophenones[16] when it is conducted in aqueous DMF. A distinct advantage is that K_2CO_3 instead of thallium and silver additives is present.

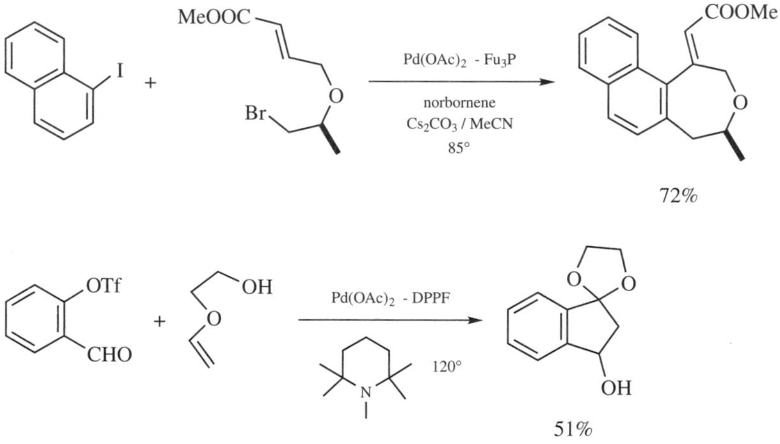

$Pd(OAc)_2$ - Fu_3P
norbornene
Cs_2CO_3 / MeCN
85°
72%

$Pd(OAc)_2$ - DPPF
120°
51%

On the other hand, coupling after opening of cyclobutanols shows moderate to high degree of asymmetric induction in the presence of a chiral P,N-ligand.[17]

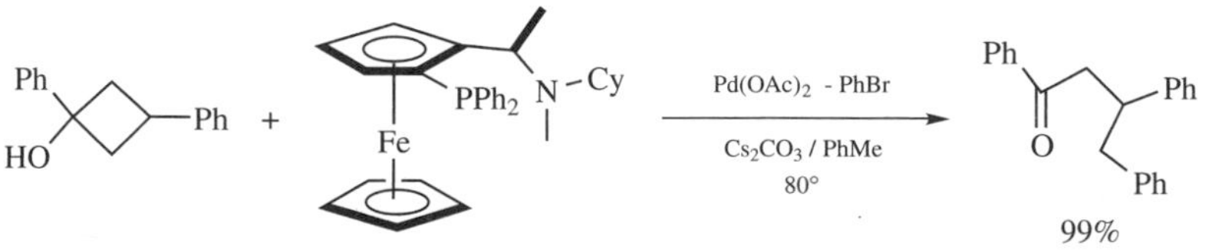

$Pd(OAc)_2$ - PhBr
Cs_2CO_3 / PhMe
80°
99%

Enyne formation by coupling of alkenyl tosylates with 1-alkynes is carried out under copper-free conditions.[18] A new variation in coupling involves in situ cleavage of tertiary benzylic alcohols and subsequent reaction with ArX.[19] CC bond formation takes place at the site vacated by the carbinol side chain.

OH + Ph-Br → $Pd(OAc)_2$ - Ph_3P / Cs_2CO_3 / o-xylene → Ph

94%

Triarylation of thiophenes with formal decarbamoylation of secondary amide group[20] is observed. Carbonyl compounds undergo arylation,[21] and extensively in the cases of enones and alkyl aryl ketones.[22]

$Pd(OAc)_2$ - Ph_3P / Cs_2CO_3 / o-xylene

R = H 29%
+ R = Ph 21%

8%

Sequential Stille and Heck couplings with bromoalkenyl tosylates are chemoselective and the overall process furnishes a method for synthesis of conjugated trienes.[23]

OTBS, Br, OTf → $Pd(OAc)_2$ - Ph_3P / DMF; LiCl, CH_2=$CHSnBu_3$; CH_2=$CHCOOBu^t$, Et_3N → OTBS, $COOBu^t$

73%

Coupling-induced rearrangement of 1-alkynylcyclobutanols results in the formation of 2-benzylidenecyclopentanones.[24]

Ph, HO + Ph-I → $Pd(OAc)_2$ - Ph_3P / Bu_4NCl - $EtNPr^i_2$ / DMF 80° → O, Ph, Ph

70%

Aryl *t*-butyl ethers are obtained from a coupling reaction with unactivated ArBr,[25] and intramolecularly the benzannulated oxacycles.[26] As for *N*-arylation a method for carbazole synthesis[27] from ArBr with *o*-chloroanilines is accomplished in one step, the cyclization after the coupling benefits from C-H bond activated by the catalyst. *N*-2-pyridylhydrazones are obtained from unsubstituted hydrazones and 2-halopyridines.[28]

Cl, NHMe + Br, OMe — $Pd(OAc)_2$ - *t*-Bu_3P / *t*-BuONa / PhMe, Δ → OMe, N, Me

47%

A system for the conversion of ArCl to ArCN consists of $Pd(OAc)_2$, KCN, a diphosphine, and TMEDA.[29]

Homoallylic amines are prepared from a three-component coupling of imines, allene and ArI.[30]

O, S, N, COOMe + C + Ph-I — $Pd(OAc)_2$ - Fu_3P / In / DMA 100° → O, S, N, H, Ph, COOMe

56%

Cyclization. Palladium catalysis solves the problem that faces the classical intramolecular Grignard reaction involving halonitriles, i.e., formation of cyclic ketones in reasonable yields. Thus, reaction of *o*-iodoarylalkanenitriles leads to ketones.[31]

Cyclization of *N*-aryl-2-trifloxybenzamides to tricyclic lactams is mediated by a catalyst prepared from $Pd(OAc)_2$ and Bu_3P while *i*-Pr_2NEt is also present.[32] A more intriguing cyclization is that which leads to fulvene derivatives.[33]

R, Br, X, E, E; E = COOEt — $Pd(OAc)_2$ - Ph_3P / K_2CO_3 / MeCN → (R = Ph, t-Bu.) X = OH: E, E, R, CHO; (R = TBS) X = H: E, E, Si

Arylation. Both α-arylation of ester enolates[34] and the coupling of $ArB(OH)_2$ and bromoacetic esters[35] are catalyzed by $Pd(OAc)_2$. The transformation of carboxylic anhydrides to aryl ketones[36] are similarly carried out, with the aryl groups also provided by $ArB(OH)_2$. If the carboxylic acids are precious, a protocol consisting of in situ formation of reactive mixed anhydrides with pivalic anhydride may be practiced.

[1]Early, T.R., Gordon, R.S., Carroll, M.A., Holmes, A.B., Shute, R.E., McConvey, I.F. *CC* 1966 (2001).
[2]Wong, M.S., Zhang, X.L. *TL* **42**, 4087 (2001).
[3]Netherton, M.R., Fu, G.C. *ACIEE* **41**, 3910 (2002).
[4]Netherton, M.R., Dai, C., Neuschutz, K., Fu, G.C. *JACS* **123**, 10099 (2001).
[5]Kirchhoff, J.H., Netherton, M.R., Hills, I.D., Neuschutz, K., Fu, G.C. *JACS* **124**, 13662 (2001).
[6]Bedford, R.B., Cazin, C.S.J., Hazelwood, S.L. *CC* 2608 (2002).
[7]Shaughnessy, K.H., Booth, R.S. *OL* **3**, 2757 (2001).
[8]Batey, R.A., Quach, T.D. *TL* **42**, 9099 (2001).
[9]Wu, J., Marcoux, J.-F., Davies, I.W., Reider, P.J. *TL* **42**, 159 (2001).
[10]Alonso, I., Carretero, J.C. *JOC* **66**, 4453 (2001).
[11]Trost, B.M., Tang, W. *ACIEE* **41**, 2795 (2002).
[12]Lautens, M., Paquin, J.F., Piguel, S., Dahlmann, M. *JOC* **66**, 8127 (2001).
[13]Lautens, M., Paquin, J.F., Piguel, S. *JOC* **67**, 3972 (2002).
[14]Kang, S.-K., Lee, S.-W., Jung, J., Lim, Y. *JOC* **67**, 4376 (2002).
[15]Bengtson, A., Larhed, M., Hallberg, A. *JOC* **67**, 5854 (2002).
[16]Vallin, K.S.A., Larhed, M., Hallberg, A. *JOC* **66**, 4340 (2001).
[17]Nishimura, T., Matsumura, S., Maeda, Y., Uemura, S. *CC* 50 (2002).
[18]Fu, X., Zhang, S., Yin, J., Schumacher, D.P. *TL* **43**, 6673 (2002).
[19]Terao, Y., Wakui, H., Satoh, T., Miura, M., Nomura, M. *JACS* **123**, 10407 (2001).
[20]Okazawa, T., Satoh, T., Miura, M., Nomura, M. *JACS* **124**, 5286 (2002).
[21]Terao, Y., Fukuoka, Y., Satoh, T., Miura, M., Nomura, M. *TL* **43**, 101 (2002).
[22]Terao, Y., Kametani, Y., Wakui, H., Satoh, T., Miura, M., Nomura, M. *T* **57**, 5967 (2002).
[23] von Zezschwitz, P., Petry, F., de Meijere, A. *CEJ* **7**, 4035 (2001).
[24]Larock, R.C., Reddy, C.K. *JOC* **67**, 2027 (2002).
[25]Parrish, C.A., Buchwald, S.L. *JOC* **66**, 2498 (2001).
[26]Kuwabe, S., Torraca, K.E., Buchwald, S.L. *JACS* **123**, 12202 (2001).
[27]Bedford, R.B., Cazin, C.S.J., Hazelwood, S.L. *CC* 2310 (2002).
[28]Arterburn, J.B., Rao, K.V., Ramdas, R., Dible, B.R. *OL* **3**, 1351 (2001).
[29]Sundermeier, M., Zapf, A., Beller, M., Sans, J. *TL* **42**, 6707 (2001).
[30]Cooper, I.R., Grigg, R., MacLachlan, W.S., Thornton-Pett, M., Sridharan, V. *CC* 1372 (2002).
[31]Pletnev, A.A., Larock, R.C. *TL* **43**, 2133 (2002).
[32]Harayama, T., Hori, A., Nakano, Y., Akiyama, T., Abe, H., Takeuchi, Y. *H* **58**, 159 (2002).
[33]Schweizer, S., Schelper, M., Thies, C., Parsons, P.J., Noltemeyer, M., de Meijere, A. *SL* 920 (2001).
[34]Moradi, W.A., Buchwald, S.L. *JACS* **123**, 7996 (2001).
[35]Goossen, L.J. *CC* 669 (2001).
[36]Goossen, L.J., Ghosh, K. *ACIEE* **40**, 3458 (2001); *EJOC* 3254 (2002).

Palladium(II) acetate–tertiary phosphine–carbon monoxide. 20, 292; **21,** 327–328

Ketones. Alkyl aryl ketones are prepared from coupling of ArOTf with R_4Sn under CO.[1] An interesting 1,2,2-triarylethanone synthesis involves two Heck reactions.[2]

Carboxylic acid derivatives. Pd-catalyzed carbonylation of alkynols that gives α-methylenelactones can be performed in ionic liquids.[2] *N*-Arylpyrrolidinones are also available from cyclocarbonylative coupling of 2-aminophenol/2-aminothiophenol, allyl bromide, and CO.[3] Mixture of internal alkenes undergo Pd-catalyzed isomerization and homologative functionalization (carbonylation) to give esters can be carried out in one step. Diphosphine **1** is the ligand to activate $Pd(OAc)_2$.[4]

1

Amide of benzotriazole is formed on reaction with diaryliodonium salts under CO.[5] Iodoferrocene gives amides or ketoamides, depending on the nature of the amines.[6] *N*,2-Diarylacrylamides are the products from coupling of arylethyne, arylamines under carbonylation conditions.[7]

[1]Nilsson, P., Larhed, M., Hallberg, A. *JACS* **123**, 8217 (2001).
[2]Consorti, C.S., Ebeling, G., Dupont, J. *TL* **43**, 753 (2002).
[3]Longo, L., Mele, G., Ciccarella, G., Sgobba, V., El Ali, B., Vasapollo, G. *AOMC* **16**, 537 (2002).
[4]Pugh, R.I., Drent, E., Pringle, P.G. *CC* 1476 (2001).
[5]Wang, L., Chen, Z.-C. *SC* **31**, 1633 (2001).
[6]Szarka, Z., Skoda-Földes, R. *TL* **42**, 739 (2001).
[7]El Ali, B., Tijani, J., El-Ghanam, A.M. *AOMC* **16**, 369 (2002).

Palladium(II) bromide.

Carbonylation. 1-Alkynes are converted into 2-alkynoic esters with CO in a primary or secondary alcohol in the presence of $PdBr_2$-$CuBr_2$ and $NaHCO_3$. Further conversion of the products leads to maleic esters.[1] Substituted maleic anhydrides are obtained if tertiary alcohols are used.

Coupling reactions. Alkenylarenes are prepared from ArOTf and alkenylsilanols using $PdBr_2$ and phosphine as catalyst.[2] Fluoride ion is present as desilylating agent and its activity is moderated by some water.

A synthesis of 3,4-disubstituted isoquinolines[3] consists of cyclization and Heck reaction in tandem.

N-Arylation of amines in good yields is rapidly achieved at room temperature with $(t\text{-}Bu_3P)_2PdBr_2$.[4]

[1]Li, J., Jiang, H., Chen, M. *SC* **31**, 199, 3131 (2001).
[2]Denmark, S.E., Sweis, R.F. *OL* **4**, 3771 (2002).
[3]Huang, Q., Larock, R.C. *TL* **43**, 3557 (2002).
[4]Stambuli, J.P., Kuwano, R., Hartwig, J.F. *ACIEE* **41**, 4740 (2002).

Palladium(II) chloride. 13, 234–235; **15,** 248–249; **16,** 268–269; **18,** 282; **19,** 257–258; **20,** 293–394; **21,** 329

Coupling reactions. Ligandless homocoupling of $ArB(OH)_2$ with catalytic $PdCl_2$ proceeds at room temperature.[1]

Heck reaction involving allylic alcohols has been carried out in molten salts (e.g., Bu_4NBr).[2] Coupling of allenyl carbinols or homoallenyl carbinols with allylic halides generates 2,5-dihydrofurans /5,6-dihydro-2*H*-pyrans containing a 3-allyl group.[3]

In situ activation of ArCOOH (e.g., by Boc_2O) makes them coupling partners analogous to ArX.[4] *p*-Nitrophenyl aroates serve the same purpose.[5] Organotellurium[6,7] and organobismuth compounds[8] participate in couplings normally.

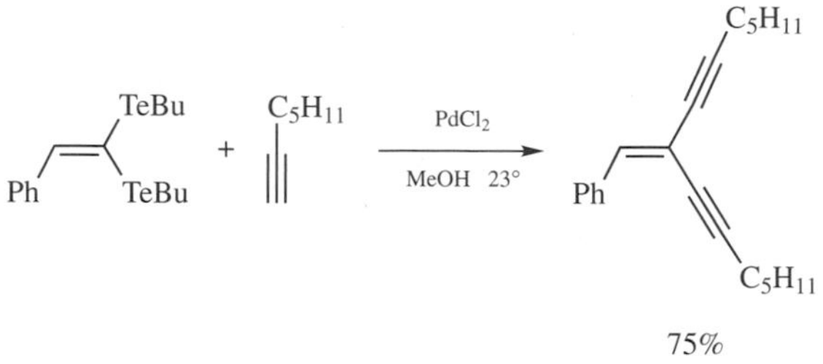

Treatment of $PdCl_2$ with NH_4OH and deposit the entity in Na-Y zeolite provide a reusable catalyst for coupling of ArBr and activated ArCl.[9]

A very unusual redox coupling of ketone acetals is effected by a mixture of $PdCl_2$, Et_3SiH and RX (MeI, allyl bromide).[10]

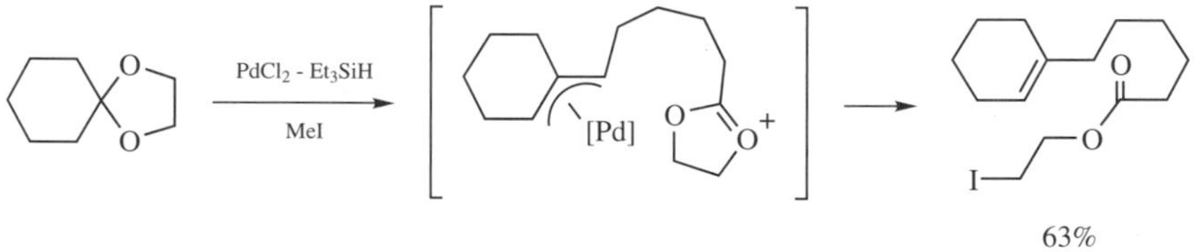

Oxidation. In molten Bu_4NBr benzylic alcohols undergo dehydrogenation[11] with catalytic $PdCl_2$. Internal alkynes afford α-diketones on heating with $PdCl_2$ in DMSO.[12]

Aziridines. $PdCl_2$ promotes aziridination of alkenes (e.g., *N,N*-dimethylacrylamide) with bromamine T as nitrogen transfer agent.[13]

[1]Kabalka, G.W., Wang, L. *TL* **43**, 3067 (2002).
[2]Bouquillon, S., Ganchegui, B., Estrine, B., Henin, F., Muzart, J. *JOMC* **632**, 153 (2001).
[3]Ma, S., Gao, W. *JOC* **67**, 6104 (2002).
[4]Goossen, L.J., Paetzold, J., Winkel, L. *SL* 1721 (2002).
[5]Goossen, L.J., Paetzold, J. *ACIEE* **541**, 1237 (2002).
[6]Zeni, G., Perin, G., Cella, R., Jacob, R.G., Braga, A.L., Silveira, C.C., Stefani, H.A. *SL* 975 (2002).
[7]Kang, S.-K., Lee, S.-W., Kim, M.-S., Kwon, H.-S. *SC* **31**, 1721 (2001).
[8]Kang, S.-K., Ryu, H.-C., Kim, M.-S., Lee, S.-W. *SC* **31**, 1021, 1027 (2001).
[9]Djakovitch, L., Koehler, K. *JACS* **123**, 5990 (2001).
[10]Iwata, A., Tang, H., Kunai, A., Oshita, J., Yamamoto, Y., Matui, C. *JOC* **67**, 5170 (2002).
[11]Ganchegui, B., Bouquillon, S., Henin, F., Muzart, J. *TL* **43**, 6641 (2002).
[12]Yusubov, M.S., Filimonov, V.D., Chi, K.-W. *RCB* **50**, 649 (2001).
[13]Antunes, A.M.M., Marto, S.J.L., Branco, P.S., Prabhakar, S., Lobo, A.M. *CC* 405 (2001).

Palladium(II) chloride–copper salts.

Coupling. A modified Sonagashira coupling enables the synthesis of symmetrical and unsymmetrical diarylethynes.[1]

(*Z*)-enynes are prepared by coupling of 1-alkynes with (*Z*)-dialkenyltelluriums with $PdCl_2$-CuI.[2]

Cyclization. An allylic silane acts as leaving group in oxidative cyclization.[3]

[1]Mio, M.J., Kopel, L.C., Braun, J.B., Gadzikwa, T.L., Grieco, P.A., Brisbois, R.G. *OL* **4**, 3199 (2002).
[2]Zeni, G., Menezes, P.H., Moro, A.V., Braga, A.L., Silveira, C.C., Stefani, H.A. *SL* 1473 (2001).
[3]Macsari, I., Szabo, K.J. *CEJ* **7**, 4097 (2001).

Palladium(II) chloride–copper(II) chloride–carbon monoxide. 20, 294; 21, 330

Carbonylation. Cyclization and carbonylation occur in the same step when 2-alkynyl-3-acyloxypyridines are treated with CO in the presence of $PdCl_2$, $CuCl_2$, and K_2CO_3 in MeOH.[1] Succinic esters are generated via vicinal dicarbonylation of 1-alkenes (catalyst: $PdCl_2$-CuOTf).[2] Moderate asymmetric induction is observed in the presence of a chiral 2,2′-bisoxazoline.

2,3-Disubstituted indoles are synthesized from *o*-alkynylanilides. Coupling accompaning cyclization can be induced.[3]

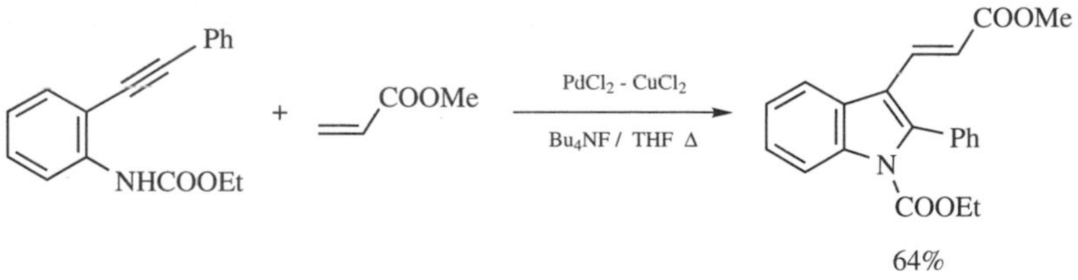

[1]Arcadi, A., Cacchi, S., Di Giuseppi, S., Fabrizi, G., Marinelli, F. *OL* **4**, 2409 (2002).
[2]Takeuchi, S., Ukaji, Y., Inomata, K. *BCSJ* **74**, 955 (2001).
[3]Yasuhara, A., Takeda, Y., Suzuki, N., Sakamoto, T. *CPB* **50**, 235 (2002).

Palladium(II) chloride–copper(I) iodide–triphenylphosphine. 21, 331–332

Sonogashira coupling. The coupling can be performed in the presence of aqueous ammonia[1] or Bu_4NX (X=F, OH)[2], advantage being that high boiling amine solvents are avoided. The alkynyl-aryl coupling (alkynes+diaryliodonium salts) has received much attention,[3] and under phase-transfer conditions (aq. NaOH also present) better results are obtained from reactions using the alkyne-acetone adducts.[4] 3,5-Diene-1,7-diynes are prepared from 1,4-diiodo-1,3-butadienes and alkynes.[5]

3,5-Dibromo-2-pyrone undergoes regioselective coupling at C-3.[6] Enones that are produced from coupling using propargyl alcohols (due to isomerization after CC bond formation) are susceptible to Stetter addition to generate 1,4-diketones. Further cyclization to give pyrrole derivatives is accomplished by adding amines.[7] In other words, two more steps can be performed without interference by various substances in the reaction mixture.

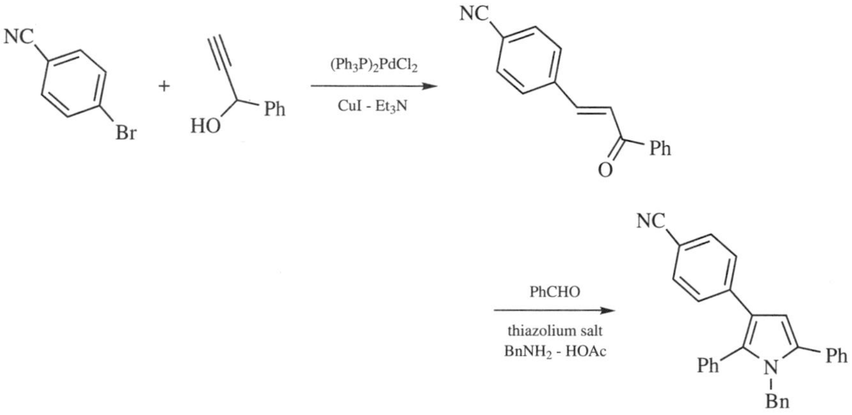

Enol tosylates of thiolesters prepared from addition of TsOH to alkynyl sulfides couple with alkynes to furnish alkylthioenynes (9 examples, 80–90%).[8]

When alkynes couple with (*Z*)-3-iodoacrylic acid the products are 4-alkylidene-2-buten-4-olides.[9] Ring formation also occurs during Sonogashira coupling of *o*-bromoacetophenone in the presence of a secondary amine. 3-Substituted 1-aminonaphthalenes are obtained.[10]

While the coupling of 9-bromoanthracene with phenylethyne proceeds normally, several other alkynes give mainly tetracyclic products (reactions in the presence of $CuSO_4$/alumina).[11]

R = CMe_2OH	91-96 : 9-4	
R = $SiMe_3$	93 : 7	
R = Ph	0 : 100	

[1]Mori, A., Ahmed, M.S.M., Sekiguchi, A., Masui, K., Koike, T. *CL* 756 (2002).
[2]Mori, A., Shimada, T., Kondo, T., Sekiguchi, A. *SL* 649 (2001).
[3]Radhakrishnan, U., Stang, P.J. *OL* **3**, 859 (2001).
[4]Chow, H.-F., Wan, C.-W., Low, K.-H., Yeung, Y.-Y. *JOC* **66**, 1910 (2001).
[5]Trostyanskaya, I.G., Titskiy, D.Y., Anufrieva, E.A., Borissenko, A.A., Kazankova, M.A., Beletskaya, I.P. *RCB* **50,** 2095 (2001).
[6]Lee, J.-H., Park, J.-S., Cho, C.-G. *OL* **4**, 1171 (2002).
[7]Braun, R.U., Zeitler, K., Müller, T.J.J. *OL* **3**, 3297 (2001).
[8]Braga, A.L., Emmerich, D.J., Silveira, C.C., Martins, T.L.C., Rodrigues, O.E.D. *SL* 371 (2001).
[9]Fiandanese, V., Bottalico, D., Marchese, G. *TL* **57**, 10213 (2001).
[10]Herndon, J.W., Zhang, Y., Wang, K. *JOMC* **634**, 1 (2001).
[11]Dang, H., Garcia-Garibay, M.A. *JACS* **123**, 355 (2001).

Palladium(II) chloride–tertiary phosphine. 19, 261; **20,** 295–298; **21,** 332–334

Coupling reactions. A new catalyst is used for Suzuki coupling involving allylic acetates.[1] A highly active and air-stable coupling catalyst is made up of $PdCl_2$ and *t*-$Bu_2P(OH)$.[2] Linking the boron-containing substrates to a polymer support does not seem to affect the coupling.[3] Primary alkylboronic acids are shown to behave perfectly in coupling with aryl and alkenyl halides.[4,5] In the latter reaction Ag_2O is added as a promoter.[5] Lithium alkylborates, which are generated from boronic esters by treatment with BuLi in situ, are useful partners for coupling.[6]

Unprotected phenylalanine-*p*-boronic acid can be used for the direct synthesis of *p*-aryl derivatives.[7] The basis of a "traceless" synthesis of biaryls is Suzuki coupling of aryl perfluoroalkanesulfonates linked to a polymer, the coupling effects cleavage of the linker.[8]

Alkenyl boronates and benzyl boronates are accessible by coupling of alkenyl triflates with bis(pinacolato)diboron[9] and benzyl halides with pinacolborane, [10] respectively. Actually, the alkenyl boronates are transformed into unsymmetrical 1,3-dienes by a second coupling with different alkenyl triflates; furthermore, the dienes can be obtained from sequential reactions in one-pot.[11]

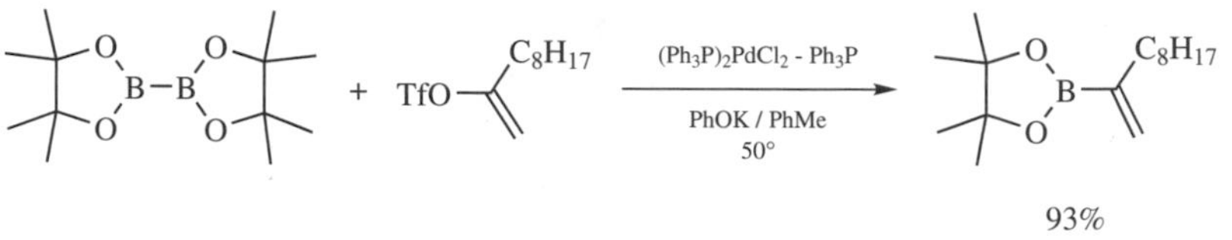

The synthetic application of Suzuki coupling also includes preparation of arylalkylaziridines from ω-alkenylaziridines.[12]

With phase-transfer agent present the Heck reaction is done in water under microwave irradiation,[13] and Sonogashira and Heck reactions also proceed in ionic liquids.[14,15] Following a Heck-type reaction of 3-alkyn-1-ol with ArX hydration gives aryl ketones.[16]

Arylation of furaldehyde occurs at C-5.[17] Pd-catalyzed cross-coupling between organoindium[18] and organomanganese compounds[19] have been established. A variation of the Sonogashira coupling employs sodium tetralkynylaluminates that are prepared from 1-alkynes and $NaAlH_4$.[20]

Diethylamine has a dramatic effect on the Stille coupling with R_4Sn (increase in yield).[21] A tandem process involving hydrostannylation of alkynes and Stille coupling affords 1,3-dienes.[22]

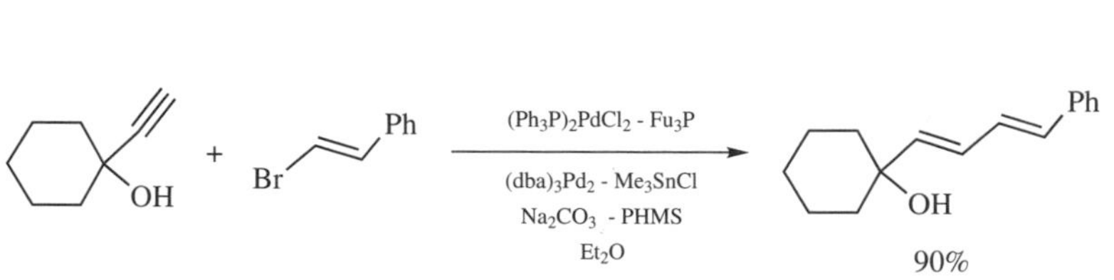

For the transformation of ArCl to heteroatom-substituted products (amines and sulfides) there is yet another catalyst system: $PdCl_2$, *t*-$Bu_2P(OH)$ and *t*-BuONa.[23]

Diaryl ketones. Pd-doped KF/alumina is an excellent catalyst for cross-coupling of ArCOCl and $[Ar'_4B]Na$ with microwave assistance.[24]

Hydrostannylation. Hydrostannylation of unsymmetrical alkynes is under the influence of structural factors. Regiochemical differences in free radical reaction and the Pd-catalyzed reaction are discerned.[25] (However, in more complex compounds a seemingly different direction is followed.[26,27])

$(Ph_3P)_2PdCl_2$ - Bu_3SnH, THF → 95%

OH, Ph, Ph, $SnBu_3$

[1]Bouyssi, D., Gerusz, V., Balme, G. *EJOC* 2445 (2002).
[2]Li, G.Y. *JOC* **67**, 3643 (2002).
[3]Pourbaix, C., Carreaux, F., Carboni, B. *OL* **3**, 803 (2001).
[4]Molander, G.A., Yun, C.-S. *T* **58**, 1465 (2002).
[5]Zou, G., Reddy, Y.K., Falck, J.R. *TL* **42**, 7213 (2001).
[6]Zou, G., Falck, J.R. *TL* **42**, 5817 (2001).
[7]Gong, Y., He, W. *OL* **4**, 3803 (2002).
[8]Pan, Y., Ruhland, B., Holmes, C.P. *ACIEE* **40**, 4488 (2001).
[9]Takagi, J., Kamon, A., Ishiyama, T., Miyaura, N. *SL* 1880 (2002).
[10]Murata, M., Oyama, T., Watanabe, S., Masuda, Y. *SC* **32**, 2513 (2002).
[11]Takagi, J., Takahashi, K., Ishiyama, T., Miyaura, N. *JACS* **124**, 8001 (2002).
[12]Lapinsky, D.J., Bergmeier, S.C. *TL* **42**, 8583 (2001).
[13]Wang, J.-X., Liu, Z., Hu, Y., Wei, B., Bai, L. *SC* **32**, 1607 (2002).
[14]Fukuyama, T., Shinmen, M., Nishitani, S., Sato, M., Ryu, I. *OL* **4**, 1691 (2002).
[15]Vallin, K.S.A., Emilsson, P., Larhed, M., Hallberg, A. *JOC* **67**, 6243 (2002).
[16]Pal, M., Parasuraman, K., Gupta, S., Yeleswarapu, K.R. *SL* 1976 (2002).
[17]McClure, M.S., Glover, B., McSorley, E., Millar, A., Osterhout, M.H., Roschangar, F. *OL* **3**, 1677 (2001).
[18]Perez, I., Sestelo, J.P., Sarandeses, L.A. *JACS* **123**, 4155 (2001).
[19]Riguet, E., Alami, M., Cahiez, G. *JOMC* **624**, 376 (2001).
[20]Gelman, D., Tsvelikhovsky, D., Molander, G.A., Blum, J. *JOC* **67**, 6287 (2002).
[21]Barros, M.T., Maycock, C.D., Madueira, M.I., Ventura, M.R. *CC* 1662 (2001).
[22]Gallagher, W.P., Terstiege, I., Maleczka Jr, R.E. *JACS* **123**, 3194 (2001).
[23]Li, G.Y., Zheng, G., Noonan, A.F. *JOC* **66**, 8677 (2001).
[24]Wang, J.-X., Yang, Y., Wei, B., Hu, Y., Fu, Y. *BCSJ* **75**, 1381 (2002).
[25]Dodero, V.I., Koll, L.C., Mandolesi, S.D., Podesta, J.C. *JOMC* **650**, 173 (2002).
[26]Nielsen, T.E., Tanner, D. *JOC* **67**, 6366 (2002).
[27]Alami, M., Liron, F., Gervais, M., Peyrat, J.-F., Brion, J.-D. *ACIEE* **41**, 1578 (2002).

Palladium(II) chloride–triphenylphosphine–carbon monoxide. 20, 298–299; 21, 334–335

Carbonylation. Preparation of β-keto esters from halomethyl ketones[1] and 3-methyleneisoindolin-1-ones from *o*-haloacetophenones[2] are among those reports concerning utility of the carbonylation process.

A useful scheme for the synthesis of 2-fluoro-2-alkenoic esters from 1-bromo-1-fluoroalkenes is delineated in the following.[3]

PhCH=C(F)Br → hv ; $(Ph_3P)_2PdCl_2$, Bu_3N - BuOH, CO → Ph–CH=C(F)–COOBu (Z/E 99 : 1)

PhCH=C(F)Br → $(Ph_3P)_2PdCl_2$, Bu_3N - HCOOH ; $(Ph_3P)_2PdCl_2$, Bu_3N - BuOH, CO → Ph–CH=C(COOBu)–F (Z/E 0 : 100)

In the presence of formamide conversion of ArBr into $ArCONH_2$ is realized.[4] Mixed imides ArCONHCOR and ArCONHTs are obtained using $RCONH_2$ in which R = Ac, Bz and $TsNH_2$, respectively.[5]

Imidates, thioimidates. In a similar reaction to carbonylation in which CO is replaced by *t*-butyl isocyanide the intermediates are trapped by nucleophiles. Alkoxides and thiolates transform them into imidates and thioimidates. In a one-pot procedure the imidates are converted into amidines.[6]

[1]Lapidus, A.L., Eliseev, O.L., Bondarenko, T.N., Sizan, O.E., Ostapenko, A.G., Beletskaya, I.P. *S* 317 (2002).
[2]Cho, C.S., Shim, H.S., Choi, H.-J., Kim, T.-J., Shim, S.C. *SC* **32**, 1821 (2002).
[3]Xu, J., Burton, D.J. *OL* **4**, 831 (2002).
[4]Schnyder, A., Beller, M., Mehltretter, G., Nsender, T., Studer, M., Indolese, A.F. *JOC* **66**, 4311 (2001).
[5]Schnyder, A., Indolese, A.F. *JOC* **67**, 594 (2002).
[6]Saluste, C.G., Whitby, R.J., Furber, M. *TL* **42**, 6191 (2001).

Palladium(II) hydroxide/carbon. 19, 262; **20,** 299–300; **21,** 335–336

Hydrogenation.[1] β-Amino-α,β-unsaturated esters undergo hydrogenation with $Pd(OH)_2/C$ in the presence of $BF_3 \cdot OEt_2$. It is subject to 1,3-asymmetric induction.

Bismetallative cyclization.[2] Enynes give cyclized products containing vinylsilane and homoallyltin moieties.

$(EtOOC)_2C$(propargyl)(allyl) + Me_3Si–$SnBu_3$ → $Pd(OH)_2$/ C, THF, r.t. → cyclopentane product bearing EtOOC, EtOOC, =CH–$SiMe_3$ and CH_2SnBu_3 (90%)

[1]Cohen, J.H., Abdel-Magid, A.F., Almond Jr, H.R., Maryanoff, C.A. *TL* **43**, 1977 (2002).
[2]Mori, M., Hirose, T., Wakamatsu, H., Imakuni, N., Sato, Y. *OM* **20**, 1907 (2001).

Palladium(II) trifluoroacetate.

1-Alken-4-ones.[1] Allylic esters react with acylstannanes to give allyl ketones in the presence of $Pd(OCOCF_3)_2$. This method complements to that involving acylsilanes.

Oxidative cyclization.[2] The following reaction is also subject to asymmetric synthesis.

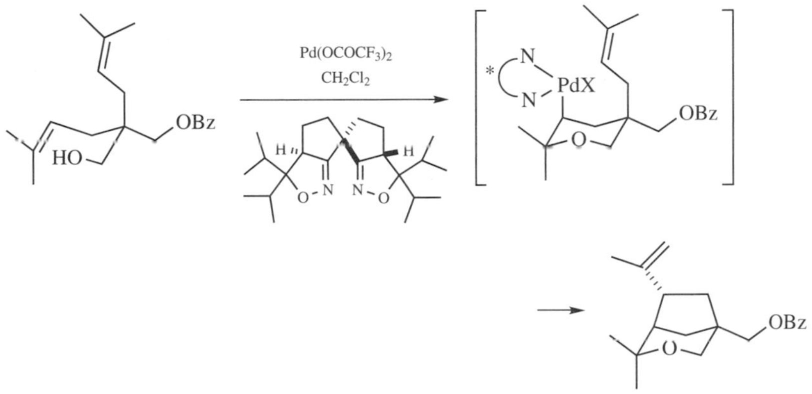

$Pd(OCOCF_3)_2$
CH_2Cl_2

Coupling.[3] In Heck reaction ArCOOH undergoes decarboxylative palladation to generate the active species.

[1]Obora, Y., Nakanishi, M., Yokunaga, M., Tsuji, Y. *JOC* **67**, 5835 (2002).
[2]Arai, M.A., Kuraishi, M., Arai, T., Sasai, H. *JACS* **123**, 2907 (2001).
[3]Myers, A.G., Tanaka, D., Mannion, M.R. *JACS* **124**, 11250 (2002).

Paraformaldehyde. 18, 284; **19,** 262–263; **20,** 301

N-Methylation.[1] Amines undergo methylation by heating with $(HCHO)_n$, oxalic acid dihydrate without solvent at 100–120°.

1-Aryl-2-dimethylaminoethanols.[2] Condensation of aromatic aldehydes with *N*-methylglycine and $(HCHO)_n$ produces *N*-methyl-5-aryloxazolidines. The heterocycles are cleaved by $NaBH_4$.

Condensation with arylamines.[3] The reaction of arylamines with $(HCHO)_n$ is dependent on acid concentration and the products can be the Tröger's base-type or acridine-type.

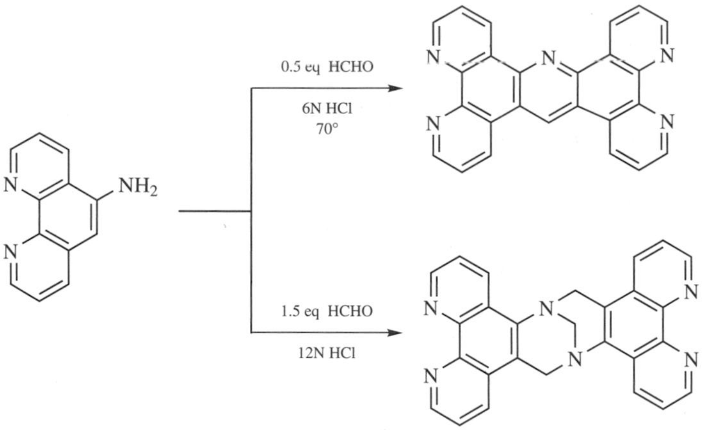

[1]Rosenau, T., Potthast, A., Röhrling, J., Hofinger, A., Sixta, H., Kosma, P. *SC* **32**, 457 (2002).
[2]Nyerges, M., Fejes, I., Viranyi, A., Groundwater, P.W., Töke, L. *S* 1479 (2001).
[3]Demeunynck, M., Moucheron, C., Kirsch-De Mesmaeker, A. *TL* **43**, 261 (2002).

Pentafluorophenyl 4-nitrobenzenesulfonate.

Peptide synthesis.[1] The title reagent, together with 1-hydroxybenzotriazole, constitutes a system for peptide bond formation.

[1]Pudhom, K., Vilaivan, T. *SC* **31**, 61 (2001).

Pentamethylcyclopentadienyl(bistrimethylsilylethene)rhodium.

o-Alkylation.[1] Regioselective introduction of an alkyl group to functionalized arenes such as aryl ketones by alkenes (trimethylsilylethene, norbornene, etc.) is catalyzed by Rh complexes.

[1]Lenges, C.P., Brookhart, M. *JACS* **121**, 6616 (1999).

Perchloric acid.

Glycosylation.[1] The stereoselectivity of glycoside formation from glycosyl fluorides is dependent on the acid catalyst. Perchloric acid favors α-anomer formation.

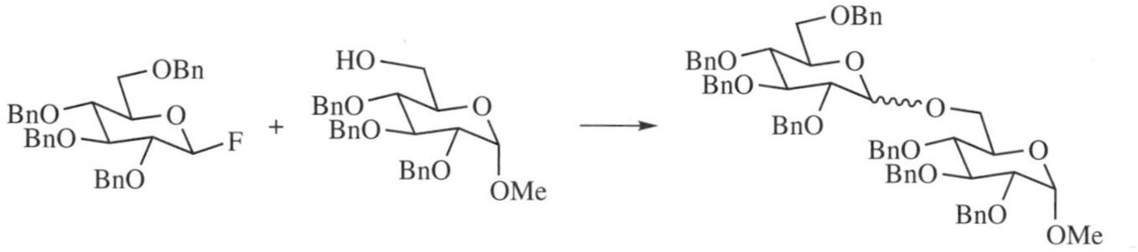

	α : β
$HClO_4$ / Et_2O	92 : 8
$HB(C_6F_5)_3$	7 : 93

[1]Jona, H., Mandai, H., Chavasiri, W., Takeuchi, K., Mukaiyama, T. *BCSJ* **75**, 291 (2002).

Periodic acid. 13, 238–239; **16,** 292; **18,** 285–286; **20,** 302

Oxidation.[1] Alcohols are oxidized to carbonyl compounds in excellent yields by periodic acid using 2,2,6,6-tetramethylpiperidinyl-1-oxyl (TEMPO) as catalyst.

Polycyclic aromatic hydrocarbons including naphthalene, anthracene, and phenanthrene are oxidized to quinones[2] by periodic acid with catalytic amounts of CrO_3 in MeCN at 5°.

Nitrosamines.[3] Secondary amines are nitrosated by a mixture of $NaNO_2$ and periodic acid on wet silica gel.

[1]Kim, S.S., Nehru, K. *SL* 616 (2002).
[2]Yamazaki, S. *TL* **42**, 3355 (2001).
[3]Zolfigol, M.A., Choghamarani, A.G., Shirini, F., Keypour, H., Salezadeh, S. *SC* **31**, 359 (2001).

Perrhenic acid.

Nitriles.[1] Aldoximes and primary amides are dehydrated by perrhenic acid in refluxing toluene or mesitylene with azeotropic distillation of water. Dibutyltin oxide has about the same activity as perrhenic acid.

[1]Ishihara, K., Furuya, Y., Yamamoto, H. *ACIEE* **41**, 2983 (2002).

Phase-transfer catalysts. 13, 239–240; **15,** 252–253; **18,** 286–289; **19,** 264–267; **20,** 302–303; **21,** 338–341

Alkylations. Phase-transfer agent-assisted *O*-methylation of phenols either with $CO(OMe)_2$[1] or MeI,[2] and silylation of alcohols with Ph_3SiH[3] are high-yielding. A poly(ethylene glycol)-supported tetrakisammonium salt that is recyclable has been demonstrated in an *O*-benzylation.[4]

Alkylation of carbonyl compounds in water is accomplished by adding a surfactant,[5] and a solid-liquid system provides the special environment for the selective monoalkylation of α-sulfonyl thioesters.[6] Chlorovinylation of α–substituted phenylacetonitriles usually proceeds well under phase-transfer conditions.[7]

Of continuing interest to synthetic chemists is the *C*-alkylation of glycine derivatives, particularly asymmetrically, as an entry to higher amino acids. Quaternized *O*-protected cinchona alkaloids are good candidates as chiral catalysts that are also phase-transfer agents. Under micellar conditions such a salt indeed functions as expected, although the ee is in the range of 64–85%.[8]

Aldol and Darzens reactions. In situ generation of quaternary ammonium fluorides (e.g., Bu_4NHSO_4 + KF) under phase-transfer conditions has definite advantages. The preparation of C_2-symmetrical ammonium fluorides enables the asymmetric version of Mukaiyama-type aldol reactions to proceed.[9]

Regioselective aldol reaction employing β-keto esters results in the loss of the ester group.[10]

Darzens reactions of α-chloroesters, nitriles, and amides are diastereoselective.[11] An asymmetric reaction involving chloromethyl phenyl sulfone proceeds reasonably well (33–81% ee).[12]

Michael reaction. Asymmetric Michael reactions use the chiral salt **1** derived from L-tartrate.[13] Other catalysts include those based on quaternized cinchona alkaloids.[14] Water-soluble calixarenes which are inverse phase-transfer catalysts are useful in Michael and aldol reactions.[15]

$ArCH_2O$ OCH_2Ar N^+ Br^- $ArCH_2O$ OCH_2Ar

1

N-Arylation. The Pd-catalyzed amination of ArCl and ArBr under PTC uses aqueous hydroxide as a base.[16]

Oxidation. Alkylarenes are oxidized to carbonyl compounds with $NaBrO_3$ in the presence of Bu_4NHSO_4.[17]

Cyclizations. Annulation of acetanilides by the Vilsmeier-Haack reagent in the presence of a phase-transfer agent in refluxing MeCN usually gives better yields of 2-chloro-3-formylquinolines.[18]

To effect radical cyclization of hydrophobic substrates in water the use of water-soluble initiator 2,2′-azobis[2-(2-imidazolin-2-yl)propane] and chain carrier 1-ethylpiperidine hypophosphite with surfactant cetryltrimethylammonium bromide are most beneficial.[19]

[1]Ouk, S., Thiebaud, S., Borredon, E., Legars, P., Lecomte, L. *TL* **43**, 2661 (2002).
[2]Vanden Eynde, J.J., Mailleux, I. *SC* **31**, 1 (2001).
[3]Le Bideau, F., Coradin, T., Henique, J., Samuel, E. *CC* 1408 (2001).
[4]Benaglia, M., Cinquini, M., Cozzi, F., Tocco, G. *TL* **43**, 3391 (2002).
[5]Cerichelli, G., Cerritelli, S., Chiarini, M., De Maria, P., Fontana, A. *CEJ* **8**, 5204 (2002).
[6]Wladislaw, B., Marzorati, L., Neves, R.M.A., Di Vitta, C. *SC* **32**, 1483 (2002).
[7]Jonczyk, A., Gierczak, A.H. *S* 93 (2001).
[8]Okino, T., Takemoto, Y. *OL* **3**, 1515 (2001).
[9]Ooi, T., Doda, K., Maruoka, K. *OL* **3**, 1273 (2001).
[10]Kourouli, T., Kefalas, P., Ragoussis, N., Ragoussis, V. *JOC* **67**, 4615 (2002).
[11]Arai, S., Suzuki, Y., Tokumaru, K., Shioiri, T. *TL* **43**, 833 (2002).
[12]Arai, S., Shioiri, T. *T* **58**, 1407 (2002).
[13]Arai, S., Tsuji, R., Nishida, A. *TL* **43**, 9535 (2002).
[14]Kim, D.Y., Huh, S.C., Kim, S.M. *TL* **42**, 6299 (2001).
[15]Shimizu, S., Shirakawa, S., Suzuki, T., Sasaki, Y. *T* **57**, 6169 (2001).
[16]Kuwano, R., Utsunomiya, M., Hartwig, J.F. *JOC* **67**, 6479 (2002).
[17]Shaabani, A., Bazgir, A., Abdoli, M. *SC* **32**, 675 (2002).
[18]Ali, M.M., Tasneem, Rajanna, K.C., Prakash, G.K.S. *SL* 251 (2001).
[19]Kita, Y., Nambu, H., Ramesh, N.G., Anilkumar, G., Matsugi, M. *OL* **3**, 1157 (2001).

Phenylimidophosphoric tris(dimethylamide).

Michael reaction.[1] The title reagent **1** catalyzes Michael reaction of β-keto esters in water.

1

[1]Bensa, D., Brunel, J.-M., Buono, G., Rodriguez, J. *SL* 715 (2001).

Phenyliodine(III) bis(trifluoroacetate). 13, 241–242; **14,** 257; **15,** 257–258; **16,** 274–275; **18,** 289–290; **19,** 267–268; **20,** 305; **21,** 342–343

Hydroxylation. With the title reagent anilides undergo *p*-hydroxylation, but if the *N*-acyl group is strongly electron-withdrawing (e.g., CF_3CO) or an electron-deficient group is present in the aromatic ring *N*-phenylation occurs.[1]

Oxidative couplings. Electron-rich α,ω-diarylalkanes cyclize on exposure to a mixture of $PhI(OCOCF_3)_2$ and $BF_3 \cdot OEt_2$ [2] or a heteropoly acid.[3]

$PhI(OCOCF_3)_2$; $H_3[PW_{12}O_{40}]$, MeCN

97%

Formation of 1,2,3,4-tetrahydroisoquinoline-1-carboxylic esters[4] from *N*-tosyl-2-arylethylamines and methanesulfenylacetic esters is mediated by $PhI(OCOCF_3)_2$.

$PhI(OCOCF_3)_2$; $ClCH_2CH_2Cl$, Δ

87%

Simultaneous oxidation of phenol and dithioacetal units to form oxaspirocyclic systems is the key step of a synthesis of aculeatin-A and –B.[5]

$PhI(OCOCF_3)_2$, MeCN - H_2O

aculeatin-A + aculeatin-B

Hofmann rearrangement. Degradation of β-hydroxyalkanamides by this reagent in MeCN at room temperature gives oxazolidin-2-ones.[6] Critical solvent effects are noted (a case that gives quantitative yield is dropped to 27% in ether and 5% in DMSO). An asymmetric synthesis of amines is terminated by this reaction.[7]

$PhI(OCOCF_3)_2$, MeCN - H_2O — .HCl, 91%

Deoximation.[8] A polymer-bound reagent is used to regenerate carbonyl compounds from their oxime derivatives.

[1]Itoh, N., Sakamoto, T., Miyazawa, E., Kikugawa, Y. *JOC* **67**, 7424 (2002).
[2]Moreno, I., Tellitu, I., San Martin, R., Dominguez, E. *SL* 1161 (2001).
[3]Hamamoto, H., Anikumar, G., Tohma, H., Kita, Y. *CC* 450 (2002).
[4]Kang, I.-J., Wang, H.-M., Su, C.-H., Chen, L.-C. *H* **57**, 1 (2002).
[5]Wong, Y.-S. *CC* 686 (2002).
[6]Yu, C., Jiang, Y., Liu, B., Hu, L. *TL* **42**, 1449 (2001).
[7]Davies, S.G., Dixon, D.J. *JCS(P1)* 1869 (2002).
[8]Chen, D.-J., Cheng, D.-P., Chen, Z.-C. *SC* **31**, 3847 (2001).

Phenyliodine(III) diacetate. 13, 242–243; **14,** 258–259; **15,** 258; **16,** 275–276; **17,** 280–281; **18,** 290–291; **19,** 268–270; **20,** 305–307; **21,** 343–344

Oxidations. The combination of $PhI(OAc)_2$ and TEMPO is a convenient oxidant for polyfluoroalkanols to the corresponding aldehydes.[1] Hydroquinones undergo oxidation with a polymer-bound iodine(III) diacetate.[2]

Serine derivatives lose a [CH_2OH] residue and the α-carbon is acetoxylated.[3] The products are electrophilic and useful in Friedel-Crafts reactions.

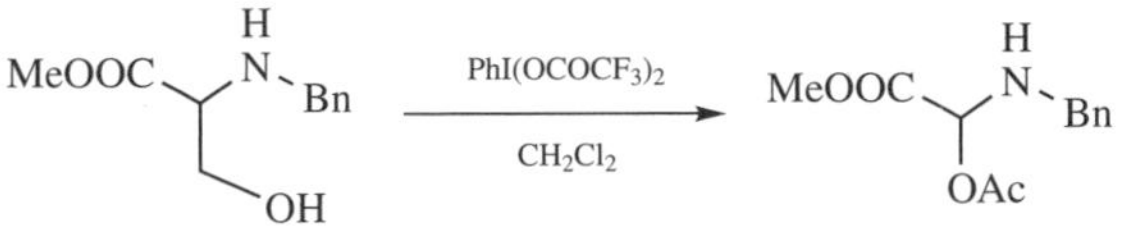

MeOOC, N(H)Bn, OH → PhI(OCOCF$_3$)$_2$ / CH_2Cl_2 → MeOOC, N(H)Bn, OAc

Heterocycles. Oxidative cyclization of indan-2-ylmethanesulfonamide to the tricyclic sultam,[4] *N*-(benzylidene)-*o*-hydroxyanilines to 2-arylbenzoxazoles,[5] and 3-aminoacrylic esters to pyrroles (dimerization)[6] is readily effected by $PhI(OAc)_2$.

Aziridination of alkenes by $Cl_3CCH_2OSO_2NH_2$ is accomplished in the presence of $PhI(OAc)_2$ and $(CF_3CONH)_4Rh_2$.[7] Aromatization of 1,4-dihydropyridines[8] by $PhI(OAc)_2$ is perhaps well anticipated.

Oxidative cyclization of 2-(2-aminobenzyl)indole affords quindoline in one step.[9] The bridged ring segment of the alkaloid TAN1251A is accessible by oxidative cyclization.[10]

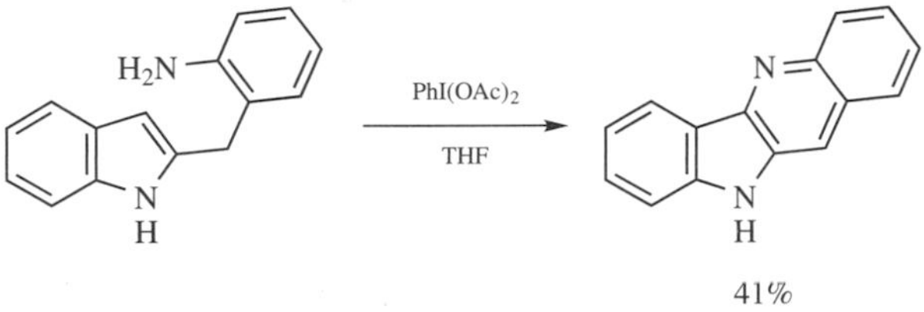

H_2N, N H → PhI(OAc)$_2$ / THF → N, N H

41%

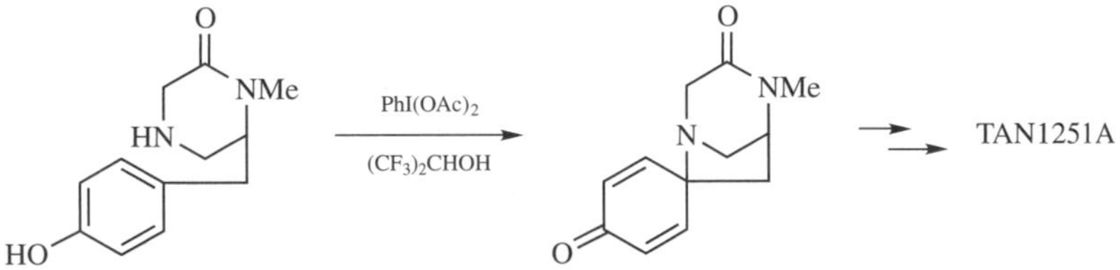

O, NMe, HN, HO → PhI(OAc)$_2$ / $(CF_3)_2CHOH$ → O, NMe, N, O → → TAN1251A

Acetoxy transfer. In the presence of $BF_3 \cdot OEt_2$ glycals are refunctionalized by $PhI(OAc)_2$ in a β-selective manner (β-glycoside formation).[11]

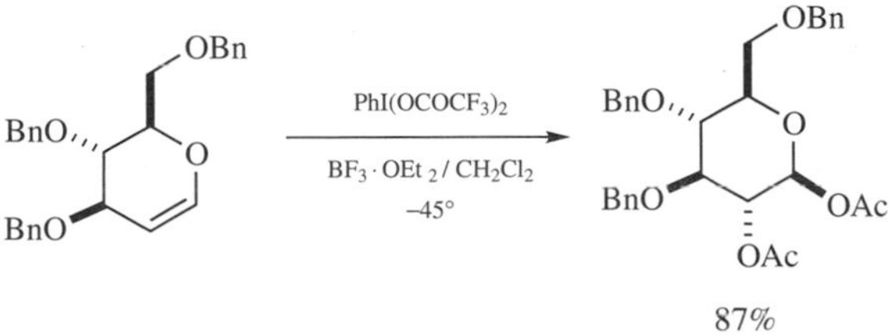

OBn, BnO, O, BnO → PhI(OCOCF$_3$)$_2$ / $BF_3 \cdot OEt_2$ / CH_2Cl_2, –45° → OBn, BnO, O, BnO, OAc, OAc

87%

Ar_3Bi receives two acetoxy groups from $PhI(OAc)_2$ to form $Ar_3Bi(OAc)_2$.[12]

[1]Pozzi, G., Quici, S., Shepperson, I. *TL* **43**, 6141 (2002).
[2]Ficht, S., Mulbaier, M., Giannis, A. *T* **57**, 4863 (2001).
[3]Boto, A., Hernandez, R., Montoya, A., Suarez, E. *TL* **43**, 8269 (2002).
[4]Liang, J.L., Yuan, S.X., Huang, J.S., Yu, W.Y., Che, C.M. *ACIEE* **41**, 3465 (2002).
[5]Zhang, J.-Z., Zhu, Q., Huang, X. *SC* **32**, 2175 (2002).
[6]Zhang, P.-F., Chen, Z.-C. *SC* **31**, 1619 (2001).
[7]Guthikonda, K., Du Bois, J. *JACS* **124**, 13672 (2002).
[8]Cheng, D.-P., Chen, Z.-C. *SC* **32**, 793 (2002).
[9]Ho, T.-L., Jou, D.-G. *HCA* **85**, 3823 (2002).
[10]Mizutani, H., Takayama, J., Soeda, Y., Honda, T. *TL* **43**, 2411 (2002).
[11]Shi, L., Kim, Y.-J., Gin, D.Y. *JACS* **123**, 6939 (2001).
[12]Bolshakov, A.V., Ganina, O.G., Shavirin, A.S., Kurskii, Y.A., Finet, J.-P., Fedorov, A.Y. *TL* **43**, 8245 (2002).

Phenyliodine(III) diacetate-iodine. 20, 307; **21,** 345–347

Iodination. Alkanes are iodinated with a combination of $PhI(OAc)_2$, I_2, and *t*-BuOH to afford different types of products thermally and photochemically.[1] The substrates are initially attacked by *t*-BuOI and further conversion of the iodoalkane products to $RI(OAc)_2$ which decompose into alkenes, and IOAc and acetic acid. The two former components combine to afford 2-iodoalkyl acetates.

$PhI(OAc)_2$ - I_2, hν *t*-BuOH → iodocyclopentane, 98%

$PhI(OAc)_2$ - I_2, 40° *t*-BuOH → 2-iodocyclopentyl acetate (OAc, I), 71%

Dealkylation. The reagent combination is useful for removal of an alkyl group from *N*-alkylsulfonamides.[2] Selectivity is shown by the clean deethylation of RSO_2NHEt in the presence of RSO_2NEt_2.

Heterocycles. 3-Quinolinecarboxylic esters are readily formed via conversion of certain Baylis-Hillman adducts to α-tosylaminomethylcinnamic esters and oxidative cyclization with the title reagent couple. A separate elimination step ensures complete aromatization of the initial products.[3]

COOEt, NHTs —[$PhI(OAc)_2$ - I_2, $ClCH_2CH_2Cl$, Δ]→ COOEt, N, Ts + COOEt, N (K_2CO_3 / DMF)

Some *N*-phosphorylated aminoalkyl glycosides are converted to spirocyclic structures by the reagent pair.[4]

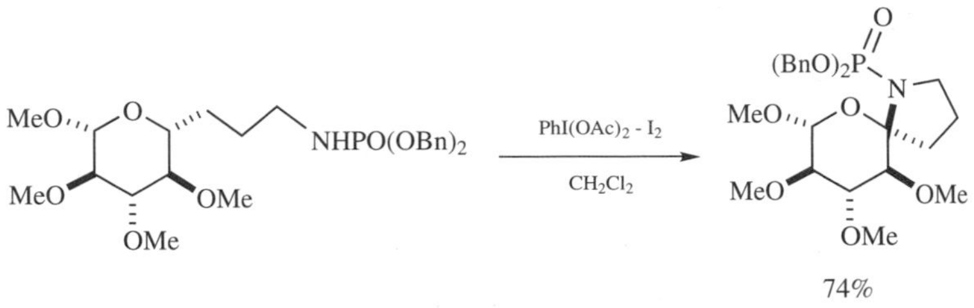

[1]Barluenga, J., Gonzalez-Bobes, F., Gonzalez, J.M. *ACIEE* **41**, 2556 (2002).
[2]Katohgi, M., Togo, H. *T* **57**, 7481 (2001).
[3]Kim, J.N., Chung, Y.M., Im, Y.J. *TL* **43**, 6209 (2002).
[4]Freire, R., Martin, A., Perez-Martin, I., Suarez, I. *TL* **43**, 5113 (2002).

Phenyliodine(III) dichloride.

Thiocyanation. Silyl enol ethers and alkynes react with $Pb(SCN)_2$ in the presence of $PhICl_2$ to give, respectively, α-thiocyanato ketones[1] and (*E*)-1,2-dithiocyanatoalkenes.[2]

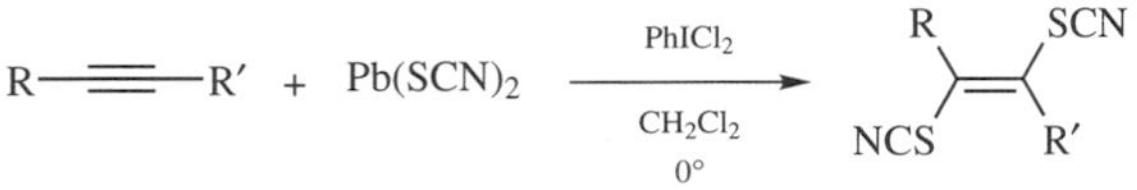

[1]Prakash, O., Kaur, H., Batra, H., Rani, N., Singh, S.P., Moriarty, R.M. *JOC* **66**, 2019 (2001).
[2]Prakash, O., Sharma, V., Batra, H., Moriarty, R.M. *TL* **42**, 553 (2001).

Phenylthio(chloro)methylene dimethylammonium chloride.

Carboxyl activation. With this salt, $[Me_2N{=}C(SPh)Cl]Cl$, and *i*-Pr_2NEt carboxylic acids are activated for derivatization into esters and amides.

[1]Gomez, L., Ngouela, S., Gellibert, F., Wagner, A., Mioskowski, C. *TL* **43**, 7597 (2002).

***N*-Phenyl trifluoromethanesulfonamide.**

Aryl triflates.[1] This reagent, $PhNTf_2$, offers a rapid way to derivatize phenols with controlled microwave irradiation.

[1]Bengtson, A., Hallberg, A., Larhed, M. *OL* **4**, 1231 (2002).

Phosphines. 21, 350–351

Reduction. The classical Staudinger reaction is a reduction of organic azides by a phosphine. Various reagents can be added to convert the iminophosphorane into other products. For example, Boc derivatives of amines are acquired by carrying out the reduction in the presence of Boc_2O under basic conditions.[1] *N*-Alkylation is also accomplishable at the end of the Staudinger reaction, on treatment of the iminophosphoranes with RX prior to the hydrolysis.[2]

$$R{-}N_3 + Me_3P \xrightarrow[\text{PhMe}]{CH_2Cl_2} R{-}N{=}PMe_3 \xrightarrow{\text{MeI}} R\text{-}NHMe$$

Regioselective reduction of azides[3] is important and its influence by neighboring group is observed.

N_3, N_3, HO, OAc, OAc + Me_3P $\xrightarrow[\text{ClCOOBn}]{\text{THF ;}}$ N_3, NHCOOBn, HO, OAc, OAc

51%

Weinreb amides.[4] A mild protocol for the preparation of these amides involves mixing the acids and *N*-methoxy-*N*-methylamine (generated in situ from the HCl salt with Et_3N) with Bu_3P and bis(2-pyridyl)disulfurane-*N,N′*-dioxide.

Deacylation.[5] Tris(2,4,6-trimethoxyphenyl)phosphine in MeOH is effective for deacylation of common esters. Acetoxy groups suffer cleavage while pivaloxy groups survive.

Alkyl halides. As a side product Ph_3PO often causes inconvenience in product purification. The situation is alleviated by changing Ph_3P to BPPE (Diphos) because the correponding phosphine oxides can be removed by filtration. Thus, the advantage is demonstrated in the conversion of alcohols to halides by a phosphine-pseudohalogen reagent.[6]

[1]Ariza, X., Pineda, O., Urpi, F., Vilarrasa, J. *TL* **42**, 4995 (2001).
[2]Kato, H., Ohmori, K., Suzuki, K. *SL* 1003 (2001).
[3]Nyffeler, P.T., Liang, C.-H., Koefler, K.M., Wong, C.-H. *JACS* **124**, 10773 (2002).
[4]Banwell, M., Smith, J. *SC* **31**, 2011 (2001).
[5]Yoshimoto, K., Kawabata, H., Nakamichi, N., Hayashi, M. *CL* 934 (2001).
[6]Pollastri, M.P., Sagal, J.F., Chang, G. *TL* **42**, 2459 (2001).

3-Phosphono-2-(*N*-cyanoimino)thiazolidines.

Phosphorylation.[1] Phosphates can be prepared from reaction of alcohols with the title reagents (**1**) at room temperature using $Zr(acac)_4$ as catalyst.

S, =NCN, N, P, OPh, OPh, O

1

[1]Maezaki, N., Furusawa, A., Hirose, Y., Uchida, S., Tanaka, T. *T* **58**, 3493 (2002).

Phosphorus(V) chloride.

Glycosyl chlorides.[1] Glycosyl acetates are transformed into the reactive chlorides. Other acetoxy groups in the monosaccharides are not affected.

[1]Ibatullin, F.M., Selivanov, S.I. *TL* **43**, 9577 (2002).

Phosphorus(V) oxide. 20, 309–310

Indene-2-carboxylic esters.[1] Some Baylis-Hillman adducts undergo cyclodehydration at room temperature.

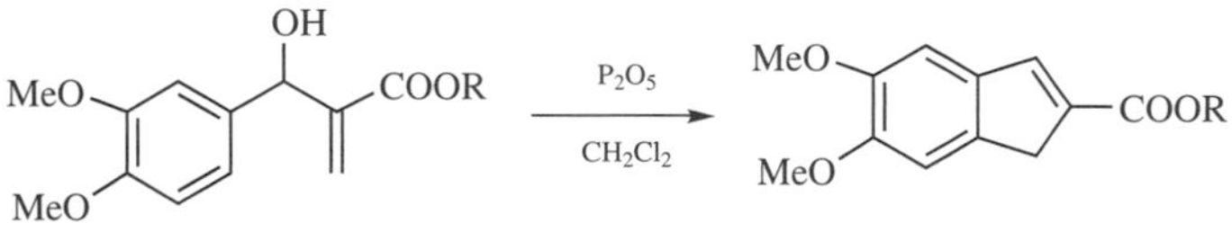

Bromoheteroarenes.[2] A hydroxy substituent on a heteroarene (e.g., 2-hydroxypyridine) can be replaced by a bromine atom by reaction with P_2O_5-Bu_4NBr.

[1]Basavaiah, D., Bakthadoss, M., Reddy, G.J. *S* 919 (2001).
[2]Kato, Y., Okada, S., Tomimoto, K., Mase, T. *TL* **42**, 4849 (2001).

Phosphorus(V) sulfide.

Thionation.[1] Esters and lactones undergo replacement of the C=O group with the C=S group on reaction with P_2S_5-$(Me_3Si)_2O$. While product yields are comparable to those obtained with Lawesson's reagent, the byproducts from this protocol are either filterable or removable by a simple hydrolytic workup.

[1]Curphey, T.J. *TL* **43**, 371 (2002).

Pinacolatoboratamethylenetriphenylphosphonium iodide.

Methylenation.[1] The title compound is a Wittig reagent precursor, ylide generation from which is carried out with LiHMDS in HMPA.

[1]Quntar, A.A., Srebnik, M. *SC* **32**, 2575 (2002).

Pinacolatoborylsilanes.

1-Boryl-1-silyl-2-alkenes.[1] Reaction of the title reagents with lithiated allylic chlorides delivers alkenes bearing two different electrofugal substituents at the same allylic position. CC bond formation with retention of one such functional group or the other can be accomplished at will.

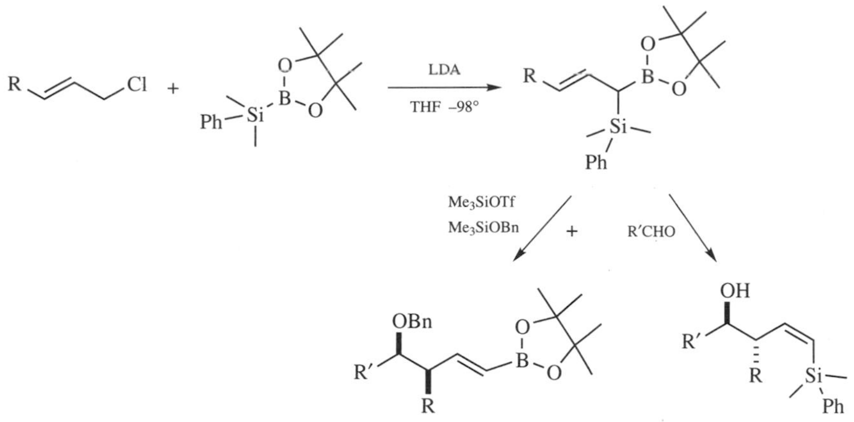

[1]Shimizu, M., Kitagawa, H., Kurahashi, T., Hiyama, T. *ACIEE* **40**, 4282 (2001).

Piperidine.

Condensations. A solvent-free coumarin synthesis from salicylaldehyde and β-oxo esters is complete at room temperature with piperidine as catalyst.[1] Clean condensation of cyanoacetamide with aromatic aldehydes is achieved by grinding the reaction components with piperidine.[2] Almost without exception, this latter neat process proceeds more rapidly and gives highly yields than conventional methods.

[1]Sugino, T., Tanaka, K. *CL* 110 (2001).
[2]McCluskey, A., Robinson, P.J., Hill, T., Scott, J.L., Edwards, J.K. *TL* **43**, 3117 (2002).

Platinum and complexes.

Hydrosilylation. Either Pt/C alone[1] or Pt(0) complex with 1,3-divinyl-1,1,3,3-tetramethyldisiloxane and tris(*t*-butyl)phosphine[2] catalyzes hydrosilylation efficiently. One-pot synthesis of (*E*)-alkenes from 1-alkynes is developed based on Suzuki coupling following the hydrosilylation.[2]

Suzuki coupling. Catalytic activity of a platinacycle derived from 2,4-di-*t*-butylphenol in Suzuki coupling is disclosed.

[1]Chauhan, M., Hauck, B.J., Keller, L.P., Boudjouk, P. *JOMC* **645**, 1 (2002).
[2]Denmark, S.E., Wang, Z. *OL* **3**, 1073 (2001).
[3]Bedford, R.B., Hazelwood, S.L. *OM* **21**, 2599 (2002).

Platinum(II) chloride. 19, 272; 21, 352–353

Allylation. Carbonyl allylation is induced by tetragonal tin(II) oxide and catalyzed by a phosphine-complexed $PtCl_2$.[1]

Cycloisomerization. Ring closure of 2-alkynylbiaryls to afford phenanthrenes is observed on tretament with $PtCl_2$.[2] 1,6-Enynes undergo cycloisomerization, cyclopentane derivatives are obtained.[3–5] Slight variation of product structures arises according to reaction conditions and additives.

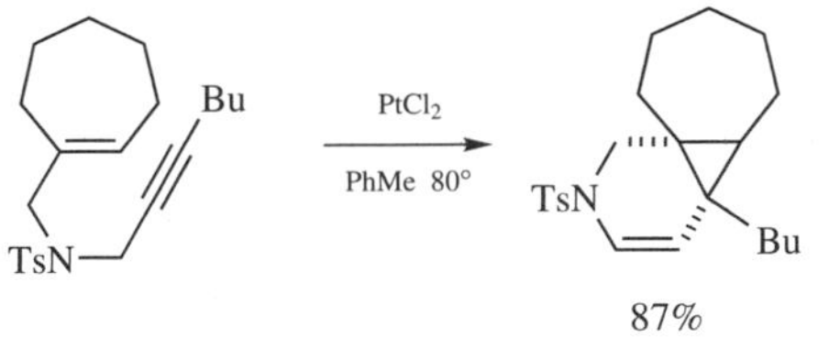

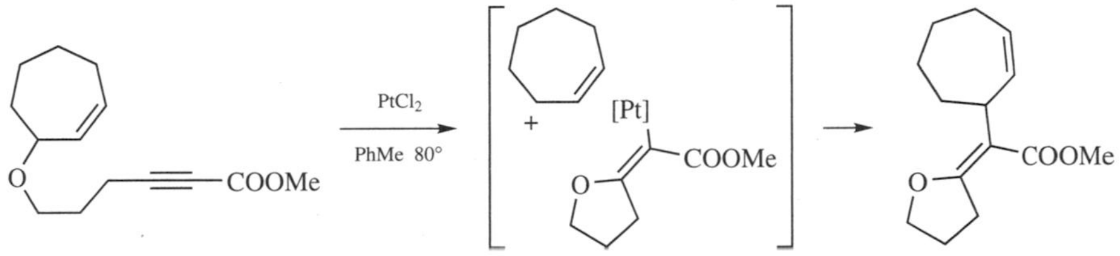

The reaction profiles of dienynes are more intriguing, dependence of products on minor variation of an oxygen substituent at a propargylic position is noted.[6]

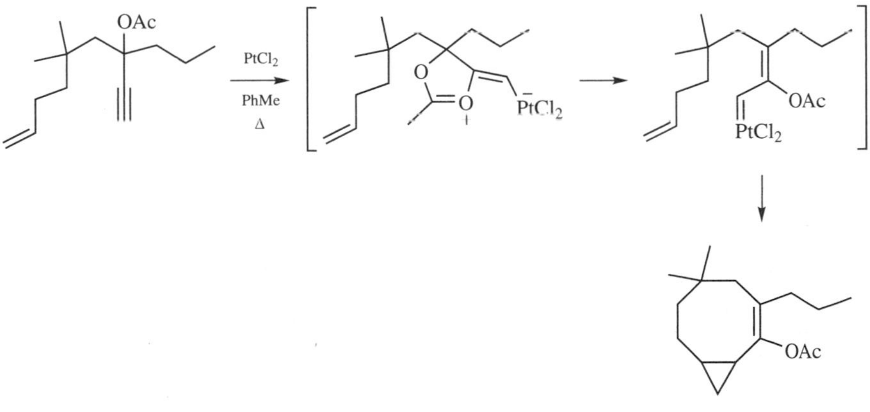

[1]Sinha, P., Roy, S. *CC* 1798 (2001).
[2]Fürstner, A., Mamane, V. *JOC* **67**, 6264 (2002).
[3]Mendez, M., Munoz, M.P., Nevado, C., Cardenas, D.J., Echavarren, A.M. *JACS* **123**, 10511 (2001).
[4]Fernandez-Rivas, C., Mendez, M., Nieto-Oberhuber, C., Echavarren, A.M. *JOC* **67**, 5197 (2002).
[5]Fürstner, A., Stelzer, F., Szillat, H. *JACS* **123**, 11863 (2001).
[6]Mainetti, E., Mouries, V., Fensterbank, L., Malacria, M., Marco-Contelles, J. *ACIEE* **41**, 2132 (2002).

Platinum(IV) oxide. 21, 353–354

Reductive hydrolysis. A convenient conversion of aromatic nitriles into aldehydes consists of reaction with HCOOH and some water in the presence of PtO_2.[1]

Hydrosilylation.[2] The excellent activity of PtO_2 to promote hydrosilylation cannot be overemphasized.

[1]Xi, F., Kamal, F., Schenerman, M.A. *TL* **43**, 1395 (2002).
[2]Saboureault, N., Mignani, G., Wagner, A., Mioskowski, C. *OL* **4**, 2117 (2002).

Poly(methylhydrosiloxane), PMHS. 20, 311; **21,** 354–355

Deoxygenation. Removal of the oxygen atom from an *N*-oxide can be done by PMHS in combination with various catalysts, e.g., Pd/C, (*i*-PrO)Ti.[1] More significantly, a carbonyl group is completely reduced with PMHS and $(C_6F_5)_3B$.[2]

Reductive cleavage. Allylic ethers, esters, and amines suffer cleavage on treatment with PMHS, $ZnCl_2$, and $(Ph_3P)_4Pd$.[3] In the presence of Pd/C, arylhydrazines and azoarenes are converted directly to *N*-Boc amines.[4]

Chloroarenes are dechlorinated at room temperature in a system containing PMHS, $Pd(OAc)_2$, and aqueous KF.[5]

[1]Chandrasekhar, S., Reddy, C.R., Rao, R.J., Rao, J.M. *SL* 349 (2002).
[2]Chandrasekhar, S., Reddy, C.R., Babu, B.N. *JOC* **67**, 9080 (2002).
[3]Chandrasekhar, S., Reddy, C.R., Rao, R.J. *T* **57,** 3435 (2001).
[4]Chandrasekhar, S., Reddy, C.R., Rao, R.J. *SL* 1561 (2001).
[5]Rahaim Jr, R.J., Maleczka Jr., R.E. *TL* **43**, 8823 (2002).

Polyphosphoric acid, PPA.

Heterocycles. Both syntheses of 1,4-dihydroisoquinolin-3-ones from arylacetonitriles and aldehydes[1] and of 2-substituted benzimidazoles from *o*-phenylenediamines and carboxylic acids[2] are mediated by PPA. The latter protocol involves microwave irradiation in the solvent-free state.

Hydrolysis. Chromonecarboxylic esters suffer hydrolysis on heating with PPA.[3]

[1]Barbry, D., Sokolowski, G., Champagne, P. *SC* **32**, 1787 (2002).
[2]Lu, J., Yang, B., Bai, Y. *SC* **32**, 3703 (2002).
[3]He, X., Li, Z., You, Q. *SC* **32**, 709 (2002).

Potassium. 21, 355

Debenzylation.[1] Benzyl ethers including benzylidene derivatives of diols are cleaved with K. This electron-transfer process does not affect THP ethers and TBS ethers.

[1]Shi, L., Xia, W.J., Zhang, F.M., Tu, Y.Q. *SL* 1505 (2002).

Potassium *t*-butoxide. **13,** 252–254; **15,** 271–272; **17,** 289–290; **18,** 296–297; **19,** 273–275; **20,** 311–313; **21,** 355–358

Eliminations. A synthesis of (*Z*)-3,3,3-trifluoro-1-propenylamines involves dehydrobromination of 2-bromo-3,3,3-trifluoropropene by *t*-BuOK in the presence of amine derivatives.[1] In situ addition to the fluorinated propyne occurs. For an access to 2-phenylcyclopropenone acetals a method involving 1,3-elimination is most convenient.[2]

t-BuOK / THF

70%

Aromatic substitutions. Aryl fluorides containing additional electron-withdrawing substituents are transformed into phenols (e.g., 1,4-difluoroanthraquinone -> 1,4-dihydroxyanthraquinone) on reaction with propargyl alcohol.[3] The base used (*t*-BuOK/DMSO) also isomerizes the initial products to allenyl aryl ethers which are readily hydrolyzed.

N-Arylation of amines by aryl chlorides can be carried out with *t*-BuOK instead of transition metal catalysts. *N*-Arylindolines are formed in a reaction of $ArNH_2$ with *o*-chlorostyrene.[4]

\+ $Ph\text{-}NH_2$ — t-BuOK / PhMe, 135° →

53%

Dealkylation. A heteroaromatic *N*-benzyl group is removed with *t*-BuOK/DMSO in THF at room temperature under oxygen.[5] The usefulness of the 2-benzenesulfonylethyl group for indole protection owes to its ease at removal, and one method involves treatment of the derivatives with *t*-BuOK in DMF.[6]

Transesterification.[7] A mild method for the preparation of *t*-butyl esters is to treat other esters with *t*-BuOK in ether at 0–20°.

\+ *t*-BuOK — Et_2O →

90%

Solvent-free reactions. Claisen and Cannizzaro reactions,[8] as well as addition of alkynes to ketones,[9] can be performed without solvent using powdered *t*-BuOK.

Addition+redox reactions. A mixture of *t*-BuOK and $(t\text{-BuO})_3Al$ mediates the Michael reaction to enones and Meerwein-Ponndorf-Verley reduction of the nascent adducts.[10] With an enone and the corresponding allylic alcohol copresent, supply of the Michael acceptor is supplanted by hydride transfer from the allylic alcohol to the adduct.

OH + O + CN NC Ph —[*t*-BuOK - $(t\text{-BuO})_3Al$ / CH_2Cl_2]→ OH CN NC Ph 81%

Isomerization. *N*-Propargyl carboxamides are isomerized by *t*-BuOK to furnish ynamides[11] which possess great synthetic potential.

[1]Jiang, B., Zhang, F., Xiong, W. *T* **58**, 265 (2002).
[2]Ando, R., Sakaki, T., Jikihara, T. *JOC* **66**, 3617 (2001).
[3]Levin, J.I., Du, M.T. *SC* **32**, 1401 (2002).
[4]Beller, M., Breindl, C., Riermeier, T.H., Tillack, A. *JOC* **66**, 1403 (2001).
[5]Haddach, A.A., Kelleman, A., Deaton-Rewolinski, M.V. *TL* **43**, 399 (2002).
[6]Bashford, K.E., Cooper, A.L., Kane, P.D., Moody, C.J. *TL* **43**, 135 (2002).
[7]Vasin, V.A., Razin, V.V. *SL* 658 (2001).
[8]Yoshizawa, K., Toyota, S., Toda, F. *TL* **42**, 7983 (2001).
[9]Miyamoto, H., Yasaka, S., Tanaka, K. *BCSJ* **74**, 185 (2001).
[10]Black, P.J., Harris, W., Williams, J.M.J. *ACIEE* **40**, 4475 (2001).
[11]Huang, J., Xiong, H., Hsung, R.P., Rameshkumar, C., Mulder, J.A., Grebe, T.P. *OL* **4**, 2417 (2002).

Potassium carbonate.

Aldoximes.[1] The *syn*-oxime from an ArCHO is obtained when reaction involves K_2CO_3 to neutralize $NH_2OH{\cdot}HCl$. On the other hand, condensation in the presence of $CuSO_4$ leads to the *anti*-isomer.

Ar-CHO + $NH_2OH \cdot HCl$ —[90°]→ (OH, N, Ar, H) + (HO, N, Ar, H)

$CuSO_4$	100	0
K_2CO_3	0	100

[1]Sharghi, H., Sarvari, M.H. *SL* 99 (2001).

Potassium dichloroiodate.

Iodination.[1] The title reagent iodinate arenes such as imidazole (to 4,5-diiodoimidazole) and isatin (to 5-iodoisatin) in water.

[1]Garden, S.J., Torres, J.C., de Souza Melo, S.C., Lima, A.S., Pinto, A.C., Lima, E.L.S. *TL* **42**, 2089 (2001).

Potassium fluoride. 13, 256–257; **15,** 272; **18,** 297–298; **19,** 275–276; **20,** 313; **21,** 359

Fluorination. Alkyl halides and sulfonates undergo substitution by KF under phase-transfer conditions.[1] A cocatalyst is Ph_3SnF. *gem*-Difluoroalkanes are obtained from reaction of bistriflates.[2]

[1]Makosza, M., Bujok, R. *TL* **43**, 2761 (2002).
[2]Bujok, R., Makosza, M. *SL* 1285 (2002).

Potassium fluoride/alumina. 20, 313; **21,** 359

Desilylation. KF/Al_2O_3 is useful for cleaving various silyl ethers in DME as well as promotion of SEM-ether formation. In other words, silyl ethers can be converted into SEM ethers in one step.[1]

Desulfonylation.[2] Elimination of TsH from *N*-tosyl-1-acyl-1,2,3,4-tetrahydroisoquinolines is accomplished on brief microwave irradiation. Longer reaction time results in the formation of the isoquinolines (dehydrogenation).

[1]Blass, B.E., Harris, C.L., Portlock, D.E. *TL* **42**, 1611 (2001).
[2]Silveira, C.C., Bernardi, C.R., Braga, A.L., Kaufman, T.S. *SL* 907 (2002).

Potassium hexamethyldisilazide, KHMDS. 21, 359–360

Unsaturated 5-membered rings. 1-Bromo-3-aminopropenes are converted into 3-pyrrolines via alkenylcarbene insertion into a C-H bond.[1] This reaction has been applied to a formal synthesis of (+)-lactacystin.[2] The ring skeleton of cephalotoxin has also been assembled using the same method to create a spirocyclic segment.[3]

$KN(SiMe_3)_2$, Et_2O - PhMe

49%

(+)-lactacystin

Isomerization and elimination.[4] Sulfenylated propargylic alcohols and derivatives undergo isomerization to allenyl alcohols and elimination to afford enynes, respectively. Interestingly, different (*E/Z*)-isomer ratios are observed from elimination with KHMDS and MeLi.

Anionic rearrangement.[5] Bicyclo[3.2.1]oct-6-en-2-ols are susceptible to rearrangement to afford 5:5-fused ring products. This process has good potential in the synthesis of polyquinane terpenes.

[1]Green, M.P., Prodger, J.C., Sherlock, A.E., Hayes, C.J. *OL* **3**, 3377 (2001).
[2]Green, M.P., Prodger, J.C., Hayes, C.J. *TL* **43**, 6609 (2002).
[3]Worden, S.M., Mapitse, R., Hayes, C.J. *TL* **43**, 6011 (2002).
[4]Ogawa, A., Sakafami, K., Shima, A., Suzuki, H., Komiya, S., Katano, Y., Mitsunobu, O. *TL* **43**, 6387 (2002).
[5]Hashimoto, H., Abe, Y., Mayuzumi, Y., Karikomi, M., Seki, K., Haga, K., Uyehara, T. *TL* **43**, 265 (2002).

Potassium hydride. 21, 360–361

Ring expansion. 1-Vinyl-1,2-cyclobutanediols are transformed into cyclopentenones through an anionic oxy-retro-ene decomposition followed by an intramolecular aldol reaction.[1]

[1]Jung, M.E., Davidov, P. *OL* **3**, 3025 (2001).

Potassium hydroxide. 21, 360–361

Condensation. In DMF 1,1-bistrimethylsiloxy-2-trimethylsilylethene condenses with aldehydes using solid KOH as reagent. The products are (*E*)-2-alkenoic acids.[1]

Degradation. 1,2-Amino alcohols are degraded by KOH in ether at room temperature to afford amides.[2]

KOH / Et_2O ; HCl / H_2O — OH, NMe_2 → O, NMe_2 73%

Hydroarylation. The conjugate hydroarylation and 1,2-addition to aldehydes with $ArSnCl_3$ or $ArBiCl_2$ in water which involve $(cod)_2RhBF_4$ as catalyst also require a base.[3] It may pertain to hydrolysis of $ArM(OH)_n$ prior to generation of active species such as $ArRh(L)_2M(OH)_n$. The effectiveness trend of KOH > NaOH > LiOH has been observed.

Amination. Condensation of cyclohexanone with nitroarenes that results in diarylamines[4] is an interesting result. It involves α-nitrosation and dehydration.

Me_2N, NO_2 + O → KOH, DMSO, 60° → Me_2N, N H, OH

Alcohols from amines.[5] Degassing diethylene glycol improves the transformation such that even hindered nitriles are converted into primary alcohols in good yields.

NC, OH → KOH / DEG (degassed) 210° → HO, OH 66%

[1]Lensen, N., Mouelhi, S., Bellassoued, M. *SC* **31**, 1007 (2001).
[2]Garcia-Valverde, M., Pedrosa, R., Vicente, M. *SL* 2092 (2002).
[3]Huang, T., Meng, Y., Venkatraman, S., Wang, D., Li, C.-J. *JACS* **123**, 7451 (2001).
[4]Moskalev, N., Makosza, M. *CC* 1248 (2001).
[5]Rahman, S.M.A., Ohno, H., Tanaka, T. *TL* **42**, 8007 (2001).

Potassium monoperoxysulfate, Oxone®. 13, 259; **14,** 267; **15,** 274–275; **16,** 285; **18,** 300; **19,** 277; **20,** 313–315; **21,** 361–362

Epoxidations. A technique of epoxidation is by mediation of oxaziridinium salts (e.g., **1**), in which an oxygen atom moves from Oxone® to iminium ions and is later transferred to alkenes.[1]

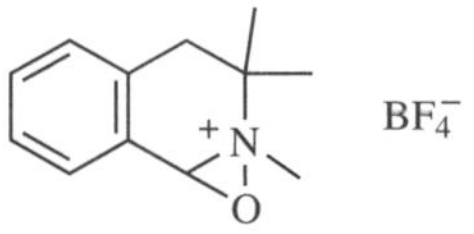

Fluorinated precursors for dioxirane intermediates are 2,2,2-trifluoroacetophenone,[2] trifluoromethyl tridecafluorooctyl ketone,[3] and *N*-carbethoxy-2-fluoro-tropinone.[4] The C_2-symmetrical dibenzosuberone **2** in conjunction with Oxone achieves asymmetric epoxidation, in some cases with high ee.[5] Successful asymmetric epoxidation of conjugated esters using Oxone and ketone **3** to form a chiral dioxirane has been delineated.[6] There is also a study of using bicyclic ketone **4** derived from prolinol,[7] and a much less efficient catalyst (in terms of asymmetric induction) is **5** generated in situ from an acetoxyprolinamide and adamantanamine.[8]

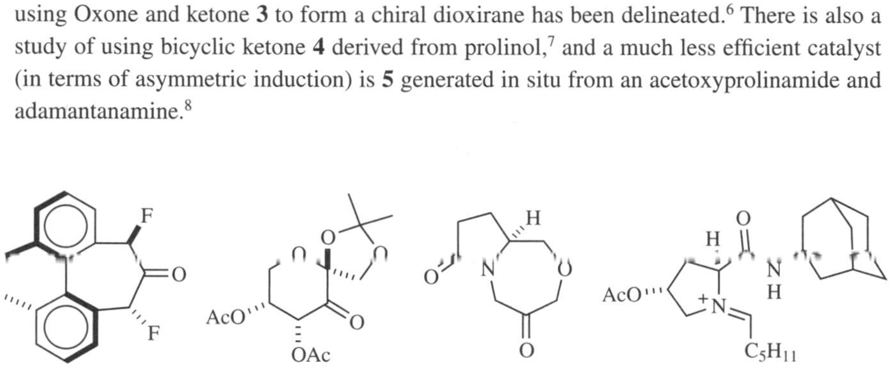

2 **3** **4** **5**

Oxidative cleavage. Cycloalkanones undergo ring cleavage to give dicarboxylic esters on consecutive reaction with hypervalent iodine reagent **6** and Oxone in MeOH.[9]

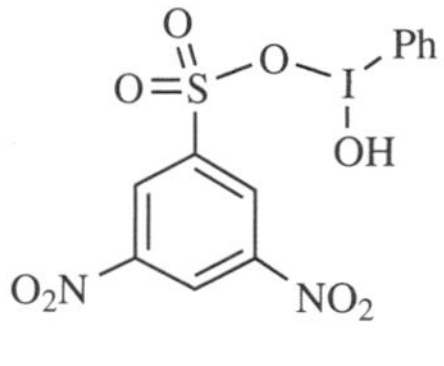

6

Oxone also performs a retro-Claisen condensation of 1,3-dicarbonyl compounds to afford carboxylic acids[10] and a Hunsdiecker reaction of 2-alkenoic acids in the presence of NaX (X=Cl, Br).[11] Cleavage of alkenes to aldehydes by Oxone is catalyzed by $RhCl_3$.[12]

Halogenation. Acetophenones afford halomethyl ketones by treatment with Oxone-NaX in refluxing NaX.[13]

[1]Bohe, L., Kammoun, M. *TL* **43**, 803 (2002).
[2]Grocock, E.L., Marples, B.A., Toon, R.C. *T* **56**, 989 (2002).
[3]Legros, J., Crousse, B., Bourdon, J., Bonnet-Delpon, D., Begue, J.-P. *TL* **42**, 4463 (2001).
[4]Armstrong, A., Ahmed, G., Dominguez-Fernandez, B., Hayter, B.R., Wailes, J.S. *JOC* **67**, 8610 (2002).
[5]Denmark, S.E., Matsuhashi, H. *JOC* **67**, 3479 (2002).
[6]Wu, X.-Y., She, X., Shi, Y. *JACS* **124**, 8792 (2002).
[7]Matsumoto, K., Tomioka, K. *TL* **43**, 631 (2002).
[8]Wong, M.-K., Ho, L.-M., Zheng, Y.-S., Ho, C.-Y., Yang, D. *OL* **3**, 2587 (2001).
[9]Lee, J.C., Ku, C.H. *SL* 1679 (2002).
[10]Ashford, S.W., Grega, K.C. *JOC* **66**, 1523 (2001).
[11]You, H.-W., Lee, K.-J. *SL* 105 (2001).
[12]Yang, D., Zhang, C. *JOC* **66**, 4814 (2001).
[13]Kim, E.-H., Koo, B.-S., Song, C.-E., Lee, K.-J. *SC* **31**, 3627 (2001).

Potassium permanganate. 13, 258–259; **14,** 267; **15,** 273–274; **18,** 301; **19,** 277–278; **20,** 315–316; **21,** 362–363

Oxidation. One of the more significant oxidation protocols involving $KMnO_4$ is the preparation of α-keto acids[1] from 1-alkynes via 1-bromoalkynes. *o*-Carboxyphenylboronic acid that is useful for Suzuki coupling is readily obtained by oxidation of *o*-tolylboronic acid under alkaline conditions.[2]

Solvent-free benzylic oxidation with $KMnO_4$ on montmorillonite K10 readily provides indanone, acetophenone, phthalide, and various aromatic ketones.[3]

The $KMnO_4$–$CuSO_4 \cdot 5H_2O$ combination rapidly oxidizes various functional groups under solvent-free conditions[4]: arylamines to azoarenes, sulfides to sulfones, alcohols to carbonyl compounds, 1,4-diols and cyclic ethers to lactones.

Nucleophilic aromatic substitutions. Nucleophiles (amines, ketones) react with some strongly deactivated arenes (e.g., *m*-dinitrobenzene) in the presence of Bu_4NF and $KMnO_4$. Under such conditions the adducts undergo rearomatization.[5]

[1]Li, L.-S., Wu, Y.-L. *TL* **43**, 2427 (2002).
[2]Tao, B., Goel, S.C., Singh, J., Boykin, D.W. *S* 1043 (2002).
[3]Shaabani, A., Bazgir, A., Teimouri, F., Lee, D.G. *TL* **43**, 5165 (2002).
[4]Shaabani, A., Lee, D.G. *TL* **42**, 5833 (2001).
[5]Huertas, I., Gallardo, I., Marquet, J. *TL* **42**, 3439 (2001).

Potassium permanganate/alumina.

Oxidation. Arylethanediols are converted into α-hydroxyacetophenones[1], and α-silylated benzyl alcohols, aroylsilanes,[2] both oxidations occurring at the benzylic position.

A simple procedure for the oxidative cleavage of styrenes to afford carbonyl compounds[3] uses the deposited oxidant in dichloromethane. Deoximation is similarly accomplished in wet ether.[4]

[1]Firouzabadi, H., Fakoorpour, M., Hazarkhani, H. *SC* **31**, 3859 (2001).
[2]Patrocinio, A.F., Moran, P.J.S. *SC* **31**, 2457 (2001).
[3]Lai, S., Lee, D.G. *S* 1645 (2001).
[4]Chrisman, W., Blankinship, M.J., Taylor, B., Harris, C.E. *TL* **42**, 4775 (2001).

Proline.

Aldol+Michael reactions. Chemoselectivity permits a well-ordered condensation of aldehydes, ketones and Meldrum's acid in the presence of proline. Thus, aldol reaction (and dehydration) precedes the Michael reaction.

CHO + L-proline / $CHCl_3$ → 83%

[1]List, B., Castello,C. *SL* 1687 (2001).

Pyridinium chlorochromate. **14,** 269; **15,** 276; **18,** 302–303

Oxidation with rearrangement. Alcohols in which a three-membered ring is present (e.g., epoxy carbinols[1] and cyclopropyl carbinols[2]) undergo oxidation with rearrangement.

Ph, OH, O → PCC / CH_2Cl_2 → O, Ph, O — 94%

HO, R → PCC / THF → R, HO — R = H 80%

[1]Ren, S.-K., Wang, F., Dou, H.-N., Fan, C.-A., He, L., Song, Z.-L., Xia, W.-J., Li, D.-R., Jia, Y.-X., Li, X., Tu, Y.-Q. *S* 2384 (2001).
[2]Cossy, J., BouzBouz, S., Laghgar, M., Tabyaoui, B. *TL* **43**, 823 (2002).

Pyridinium dichromate. 14, 269; **20,** 316

Oxidation. Certain tertiary α-silyl alcohols are oxidized with loss of the silyl group.[1]

PDC / DMF

In the context of a Strychnos alkaloid synthesis a benzindolizidine was oxidized to the lactam derivative with PDC on Celite.[2]

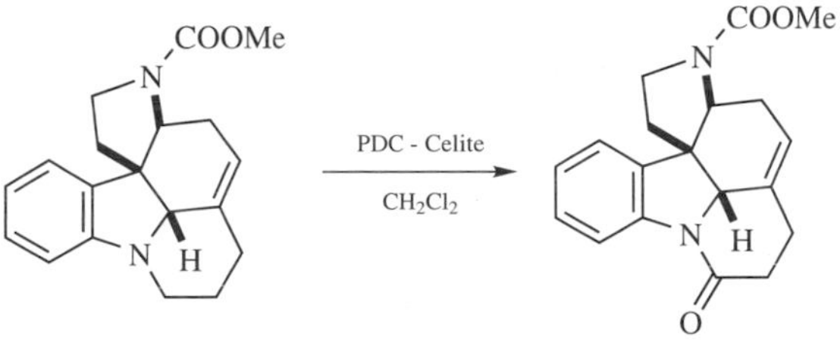

PDC - Celite

CH_2Cl_2

[1]Bonini, B.F., Comes-Franchini, M., Fochi, M., Lunazzi, L., Mazzanti, A., Mazzanti, G., Ricci, A., Varchi, G. *SL* 995 (2001).
[2]Bodwell, G.J., Li, J. *ACIEE* **41**, 3261 (2002).

Pyridinium triflate.

Silylation. The combination of an *O*-silylated benzamide (e.g., from PhCONHPh, NaH and Me_3SiCl) and PyHOTf is a powerful silylating agent that readily converts tertiary alcohols to silyl ethers.

[1]Misaki, T., Kurihara, M., Tanabe, Y. *CC* 2478 (2001).

Pyridinium trihalides.

Bromination. Bromodeformylation of ArCHO containing an *o*-OR (R=H, Me) occurs when treated with $PyHBr_3$.[1] A newly introduced stable brominating for arenes is pyridinium dichlorobromate.[2]

[1]Cordoba, R., Plumet, J. *TL* **43**, 9303 (2002).
[2]Muathen, H.A. *S* 169 (2002).

3-[*N*-(4-pyridyl)-*N*-decyl]aminopropionic acid.

Acetylation. Selective acetylation (Ac_2O, $CHCl_3$, room temperature) of *n*-octyl glucopyranosides and galactoside (both α- and β-anomers) at the primary hydroxyl group is observed using the title compound as catalyst. while DMAP promotes acetylation of secondary OH groups at C-3 and C-4 to greater extent.[1] However, the new catalyst mediates acetylation of mannopyranosides at both 4-OH and 6-OH equally well.

[1]Kurahashi, T., Mizutani, T., Yoshida, J. *T* **58**, 8669 (2002).

1-[(2-Pyridyl)dimethylsilyl]alkenes.

Pauson-Khand reaction.[1] The intermolecular P-K reaction benefits from the pyridyl group which is able to gather both the metallacyclopropenes and the proximal alkenyl substituent of the silicon atom.

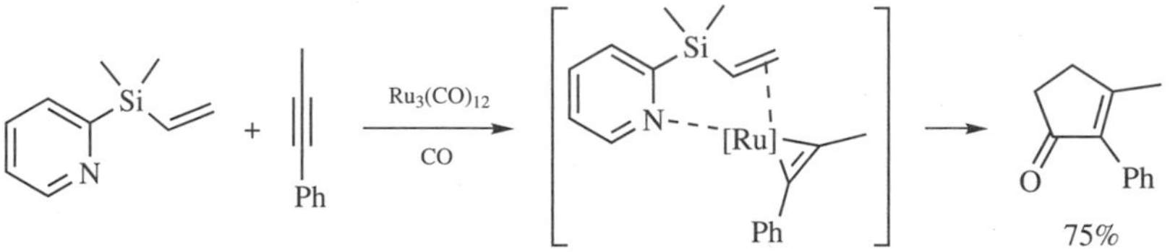

Si N + Ph $\xrightarrow[CO]{Ru_3(CO)_{12}}$ [Si N [Ru] Ph] → O Ph

75%

Diels-Alder reactions.[2] 1-[(2-Pyridyl)dimethylsilyl]-1,3-alkadienes behave well as hydrophilic dienes which undergo Diels-Alder reaction in aqueous media, under conditions that the corresponding phenyldimethylsilyl analogues do not react. The silyl moiety becomes allylic in the adducts and is easily detachable.

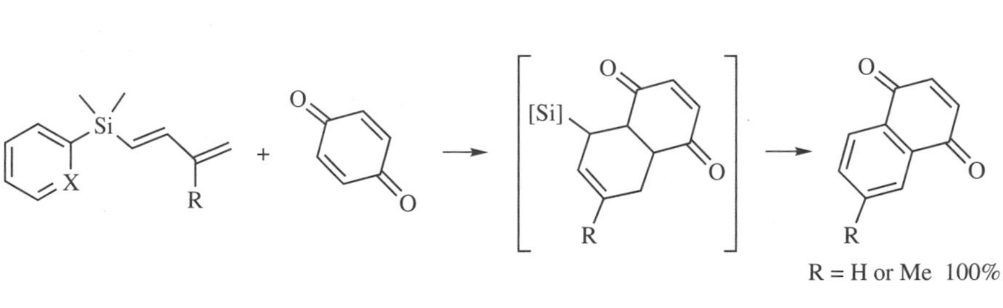

Si X R + O O → [[Si] O O R] → O O R

R = H or Me 100%

[1]Itami, K., Mitsudo, K., Yoshida, JHH. *ACIEE* **41**, 3481 (2002).
[2]Itami, K., Nokami, T., Yoshida, JHH. *ACIEE* **40**, 1074 (2001).

R

Rhenium(VI) oxide.

Epoxidation.[1] Rhenium(VI) oxide catalyzes alkene epoxidation by $(Me_3Si)_2O_2$ in the presence of a small quantity of water.

[1]Yudin, A.K., Chiang, J.P., Adolfsson, H., Coperet, C. *JOC* **66**, 4713 (2001).

Rhodium. 21, 366

Deoxygenation.[1] Partial deoxygenation of polyhydroxybenzenes is accomplished on hydrogenation with Rh/Al_2O_3 as catalyst under alkaline conditions. Thus, resorcinol is obtained from 1,3,5-trihydroxybenzene, hydroquinone from 1,2,4-trihydroxybenzene, and pyrogallol from 1,2,3,4-tetrahydroxybenzene.

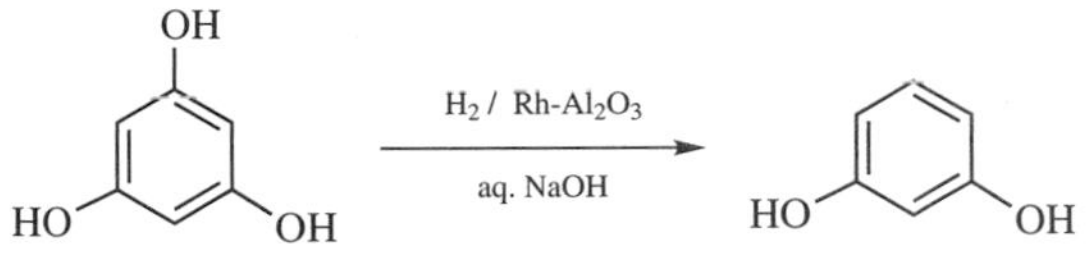

OH, HO, OH — H_2 / Rh-Al_2O_3, aq. NaOH → HO, OH

[1]Hansen, C.A., Frost, J.W. *JACS* **124**, 5926 (2002).

Rhodium carbonyl clusters. 13, 288; **15,** 334; **18,** 305–306; **19,** 280–281; **20,** 317–318; **21,** 366

α-Dehydroamino esters and pyrroles.[1] Catalytic $Rh_4(CO)_{12}$ effects condensation of ethyl isocyanoacetate with ketones and β-dicarbonyl compounds (β-diketones and β-keto esters) to provide α-dehydroamino esters and pyrrole-2-carboxylic esters, respectively.

Fiesers' Reagents for Organic Synthesis, Volume 22. Series editor Tse-Lok Ho
ISBN 0-471-28515-3

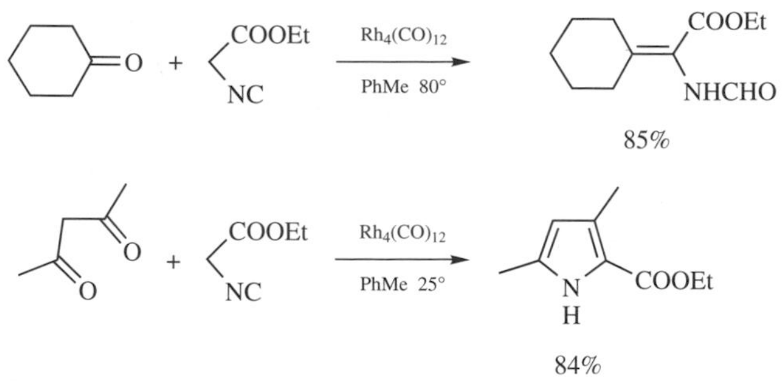

Rh$_4$(CO)$_{12}$, PhMe 80° — 85%

Rh$_4$(CO)$_{12}$, PhMe 25° — 84%

[1]Takaya, H., Kojima, S., Murahashi, S.-I. *OL* 421 (2001).

Rhodium(II) carboxamidates.

Hetero Diels-Alder reaction.[1] Chiral Rh catalysts that induce formation of dihydro-4-pyrones can have high turnover numbers.

[1]Doyle, M.P., Phillips, I.M., Hu, W. *JACS* **123**, 5366 (2001).

Rhodium(II) carboxylates. 13, 266; **15,** 278–286; **16,** 289–292; **17,** 298–302; **18,** 306–307; **19,** 281–285; **20,** 318–320; **21,** 367–369

Three-membered rings. In situ generation of diazo compounds from sodium salts of tosylhydrazones and decomposition-cycloaddition to form cyclopropanes has been demonstrated,[1] although $Rh_2(OAc)_4$ is less effective than an iron-porphyrin as catalyst. There is a report on carrying out the cyclopropanation with ethyl diazoacetate in water.[2]

Cyclopropanation during decomposition of an ylide derived from dimethyl 2-phenyliodoniamalonate[3] is successful. The use of α-heteroaryldiazoacetic esters in catalytic asymmetric cyclopropanation with $Rh_2(S\text{-DOSP})_4$ as catalyst[4] is on record.

Similar cycloadditions of carbenoids with aldehydes and imines lead to epoxides and aziridines, respectively.[5,6] Again, ylides generated in situ can be used.[7,8]

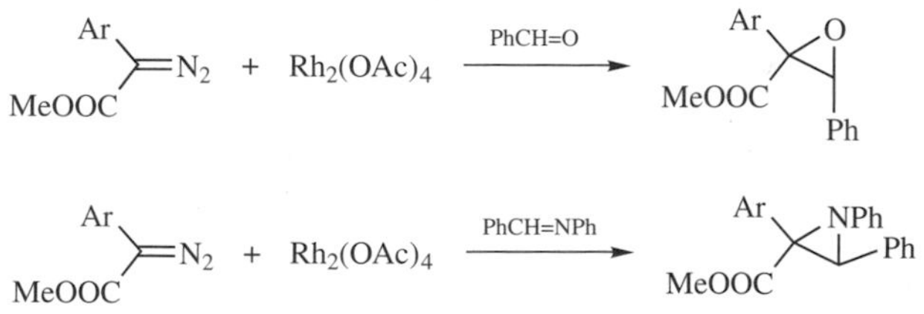

PhCH=O

PhCH=NPh

Five-membered rings. Further transformation of the 3-membered ring adducts is observed when they possess some conducive structural features. Formation of 4-acyl-2,3-dihydrofurans[9] from 2-diazo-1,3-dicarbonyl compounds and alkenes may proceed via cyclopropanes which then undergo isomerization.

Imidazolines formation from electron-deficient alkenes and $TsNCl_2$ in acetonitrile represents a new method for synthesis of differentially protected *syn*-1,2-diamines.[10]

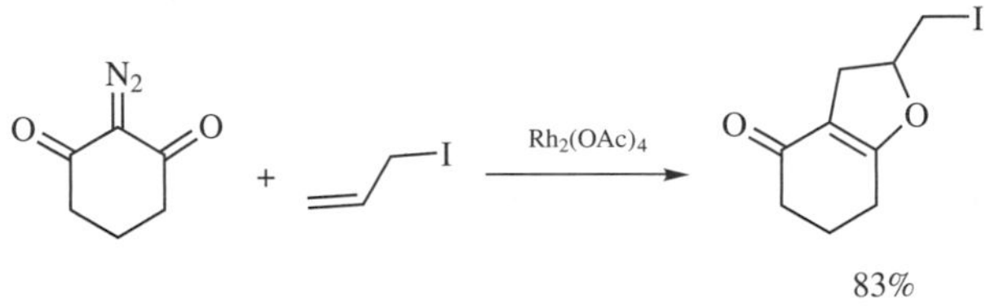

N_2 O O + I $Rh_2(OAc)_4$ O O I

83%

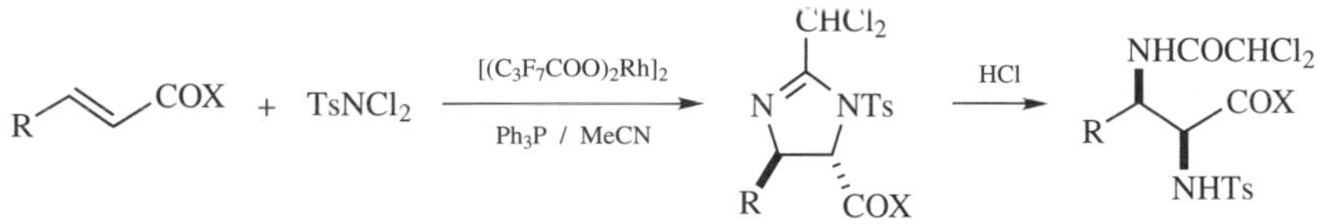

R COX + $TsNCl_2$ $[(C_3F_7COO)_2Rh]_2$ Ph_3P / MeCN $CHCl_2$ N NTs R COX HCl $NHCOCHCl_2$ R COX NHTs

A 2:1 cycloadducts are obtained from reaction of ArCHO and methyl 2-diazo-3,3,3-trifluoropropanoate.[11] Apparently the initially formed epoxides isomerize very readily to 1,3-dipoles for a second-round cycloaddition.

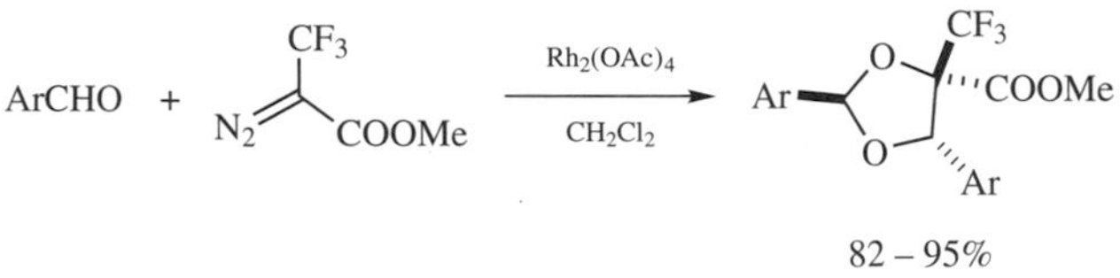

ArCHO + CF_3 N_2 COOMe $Rh_2(OAc)_4$ CH_2Cl_2 Ar O O CF_3 COOMe Ar

82 – 95%

Indanes[12] are also accessible by the formal [3 + 2]cycloaddition.

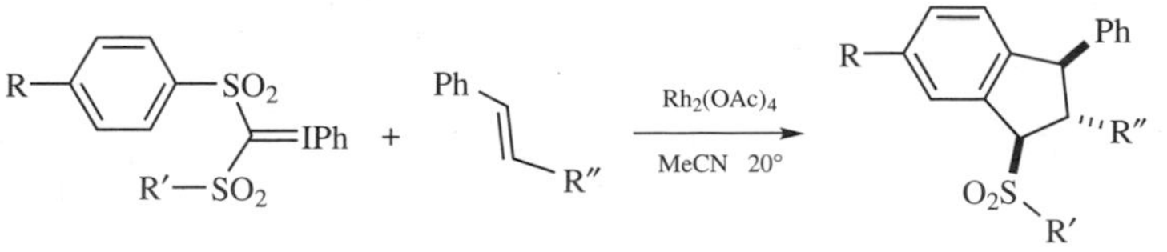

R SO_2 =IPh R′—SO_2 + Ph R″ $Rh_2(OAc)_4$ MeCN 20° R Ph R″ O_2S R′

1,3-Dipolar species derived from insertion reactions are readily trapped by cycloaddition. The tricyclic skeleton of colchicine has been assembled by such a method.[13]

Insertion prior to cycloaddition is actually prevalent in many of the observed reactions. Examples include formation of thiophenes from conjugated thiones,[14] 1,3-dioxolan-4-ones from α-silyldiazoacetic esters and carbonyl compounds.[15]

Indoles and isocoumarins are formed when *o*-heterosubstituted arylcarbonyl compounds and α-diazophosphonates are treated sequentially with $Rh_2(OAc)_4$ and a base.[16] Carbenoid insertion into the X-H bond is followed by an intramolecular Horner-Emmons-Wadsworth reaction. Actually, insertion into N-H bond provides many useful synthetic intermediates, and it is shown by another indole synthesis.[17]

By an oxidative activation followed by intramolecular C-H inversion, carbamates are converted into oxazolidin-5-ones.[18] The oxidant is $PhI(OAc)_2$ and the catalyst, $Rh_2(OAc)_4$.

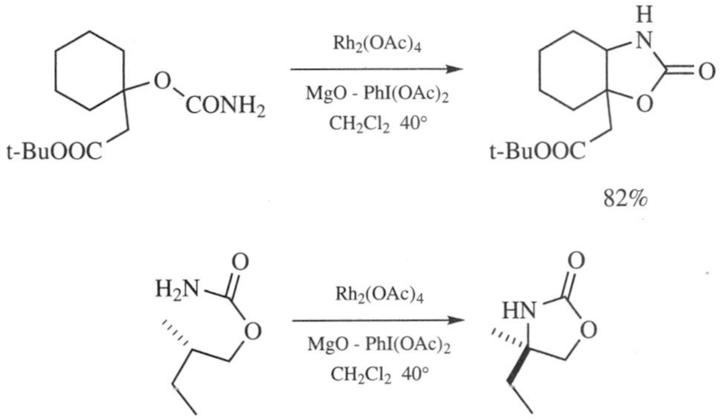
Rh2(OAc)4
MgO - PhI(OAc)2
CH2Cl2 40°
O
CONH2
t-BuOOC
H
N
O
O
t-BuOOC
82%
H2N
O
O
Rh2(OAc)4
MgO - PhI(OAc)2
CH2Cl2 40°
HN
O
O

C-H insertion. Although several examples shown above indicate formation of five-membered rings, it is not necessarily the exclusive reaction mode. Decomposition of α-phenyldiazoacetic esters leads to products that suggest a trend of tert. C-H>sec. C-H>>prim. C-H for carbenoid insertion, provided that conformational restrictions are absent.[19]

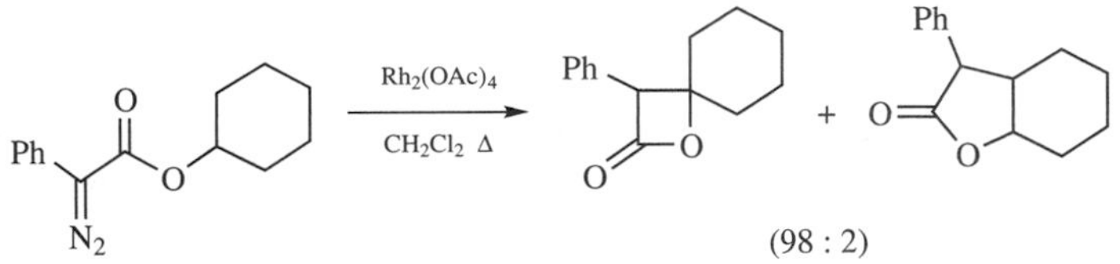
Ph
O
O
N2
Rh2(OAc)4
CH2Cl2 Δ
Ph
O
O
+
Ph
O
O
(98 : 2)

Interestingly, sulfamic esters seem to prefer formation of six-membered ring products. This insertion reaction can be exploited for stereocontolled synthesis of 1,3-amino alcohols.[20]

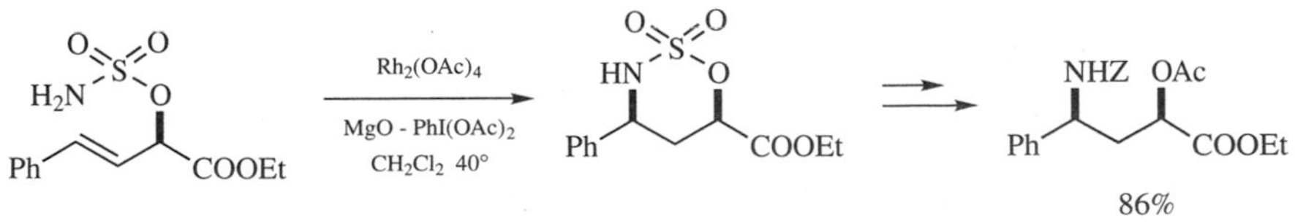
O
O
S
H2N
O
Ph
COOEt
Rh2(OAc)4
MgO - PhI(OAc)2
CH2Cl2 40°
O
O
S
HN
O
Ph
COOEt
NHZ
OAc
Ph
COOEt
86%

Benzylic and allylic C-H insertion reactions have been studied. The one with silyl enol ethers leads to compounds identical to Michael reaction products.[21] Also significant is the preferred reaction occurs at the C-H bond of a methyl group rather than benzylic position in a carbamate of benzyl(methyl)amine.[22]

OTIPS + N_2=C(Ar)COOMe —Rh_2(*S*-DOSP)$_4$→ OTIPS, H, COOMe, Ar

Rearrangement. A Rh-catalyzed tandem Bamford-Stevens reaction and Claisen rearrangement are realized.[23] The sequence is useful for synthesis of *anti*-2,3-diaryl-4-pentenals.

MeO, Ph, N, N, Ph, O —Rh_2(OAc)$_4$→ MeO, Ph, CHO
82%

α-Diazoketones react with propargyl alcohol to give allenyl ketols in the presence of a rhodium carboxylate. Remarkably, variation of the acyloxy group in the catalyst causes a change of the sigmatropic rearrangement pathway (milder catalyst favoring the [2.3] mode, and the harsher catalyst, the [3.3] mode) of the insertion products and thence different regioisomers are generated.[24]

N_2, O, COOMe + OH —Rh_2(L)$_4$, PhH Δ→ HO, O, C, COOMe; OH, O, C, COOMe
L = oct 60% L = cap 64%

In the diazo decomposition 1,2-aryl and 1,2-hydride migration can occur.[25]

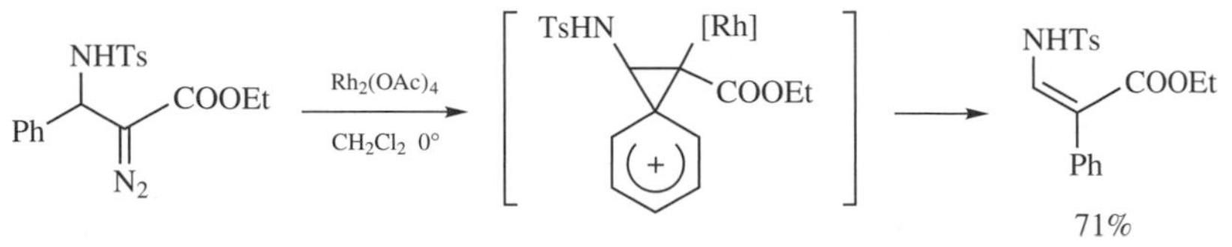

1-Acetoxyazulenes.[26] Derivatization of 3-aryl-3-bromopropanoic acids (prepared by NBS bromination or HBr addition from the proper precursors) into diazoketones, followed by consecutive treatment with $Rh_2(OCOCMe_3)_4$ and Ac_2O-DMAP leads to 1-acetoxyazulenes. The corresponding triflates are able to participate in Suzuki coupling therefore various 1-substituted azulenes are accessible.

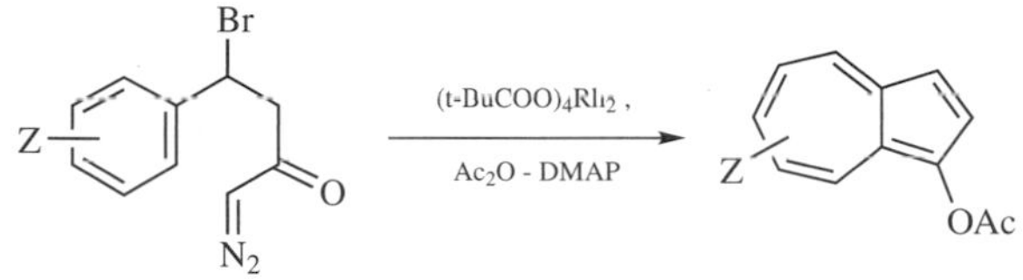

Coupling. On treatment with $Rh_2(OAc)_4$ diazoalkenes undergo a highly stereoselective coupling to furnish (*E,E,E*)-1,3,5-alkatrienes.[27]

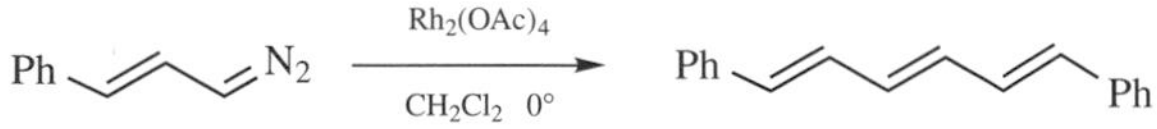

52% (*E,E,E* / *E,Z,E* 98 : 2)

[1]Aggarwal, V.K., de Vicente, J., Bonnert, R.V. *OL* **3**, 2785 (2001).
[2]Wurz, R.P., Charette, A.B. *OL* **4**, 4531 (2002).
[3]Davies, H.M.L., De Meeese, J. *SL* 1843 (2001).
[4]Davies, H.M.L., Townsend, R.J. *JOC* **66**, 6595 (2001).
[5]Davies, H.M.L., DeMeese, J. *TL* **42**, 6803 (2001).
[6]Doyle, M.P., Hu, W., Timmons, D.J. *OL* **3**, 933 (2001).
[7]Aggarwal, V.K., Alonso, E., Hynd, G., Lydon, K.M., Palmer, M.J., Porcelloni, M., Studley, J.R. *ACIEE* **40**, 1430 (2001).
[8]Aggarwal, V.K., Alonso, E., Fang, G., Ferrara, M., Hynd, G., Porcelloni, M. *ACIEE* **40**, 1433 (2001).
[9]Lee, Y.R., Suk, J.Y. *T* **58**, 2359 (2002).
[10]Li, G., Wei, H.-X., Kim, S.H., Carducci, M.D. *ACIEE* **40**, 4277 (2001).
[11]Jiang, B., Zhang, X., Luo, Z. *OL* **4**, 2453 (2002).
[12]Adam, W., Bosio, S.G., Gogonas, E.P., Hadjiarapoglou, L.P. *S* 2084 (2002).
[13]Graening, T., Friedrichsen, W., Lex, J., Schmalz, H.G. *ACIEE* **41**, 1524 (2002).

[14]Song, H.M., Kim, K. *JCS(P1)* 2414 (2002).
[15]Bolm, C., Saladin, S., Kasyan, A. *OL* **4**, 4631 (2002).
[16]Nakamura, Y., Ukita, T. *OL* **4**, 2317 (2002).
[17]Bashford, K.E., Cooper, A.L., Kane, P.D., Moody, C.J., Muthusamy, S., Swann, E. *JCS(P1)* 1673 (2002).
[18]Espino, C.G., Du Bois, J. *ACIEE* **40**, 598 (2001).
[19]Doyle, M.P., May, E.J. *SL* 967 (2001).
[20]Espino, C.G., When, P.M., Chow, J., Du Bois, J. *JACS* **123**, 6935 (2001).
[21]Davies, H.M.L., Ren, P. *JACS* **123**, 2070 (2001).
[22]Davies, H.M.L., Venkataramani, C. *ACIEE* **41**, 2197 (2002).
[23]May, J.A., Stoltz, B.M. *JACS* **124**, 12426 (2002).
[24]Moniz, G.A., Wood, J.L. *JACS* **123**, 5095 (2001).
[25]Jiang, N., Qu, Z., Wang, J. *OL* **3**, 2989 (2001).
[26]Kane Jr, J.L., Shea, K.M., Crombie, A.L., Danheiser, R.L. *OL* **3**, 1081 (2001).
[27]Doyle, M.P., Yan, M. *JOC* **67**, 602 (2002).

Rhodium(III) chloride. 21, 370

Diaryl ketones.[1] Aryltin compounds give Ar_2CO on exposure to CO in the presence of $RhCl_3$ hydrate. Thus the result is different from the Pd-catalyzed reaction that leads to ArCOOH.

β-Amino ketones.[2] Protected β-amino ketones are formed in a transition-metal salt catalyzed conjugated addition of carbamates to enones. $RhCl_3$ hydrate gives one of the best results.

[1]Ohe, T., Motofusa, S., Ohe, K., Uemura, S. *BCSJ* **74**, 1343 (2001).
[2]Kobayashi, S., Kakumoto, K., Sugiura, M. *OL* **4**, 1319 (2002).

Ruthenacycles.

Redox reactions.[1] The pincer-aryl complex **1** catalyzes Meerwein-Ponndorf-Verley reduction of ketones by isopropanol.

1

[1]Dani, P., Karlen, T., Gossage, R.A., Gladiali, S., van Koten, G. *ACIEE* **39**, 743 (2000).

Ruthenium-carbene complexes. 18, 308; 19, 285–289; 20, 320–323; 21, 370–376

General aspects. Grubb's catalyst **1** is conveniently prepared by stirring $(Ph_3P)_2RuCl_2$ with $Ph_2S{=}CHPh$ in dichloromethane at –30° and then treated with Cy_3P at room temperature.[1] The new bimetallic Ru-carbene complex (**2**) is an air-stable and reusable catalyst.[2]

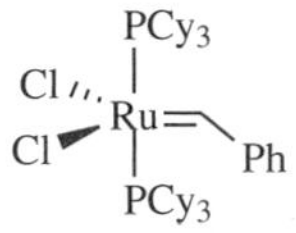

1

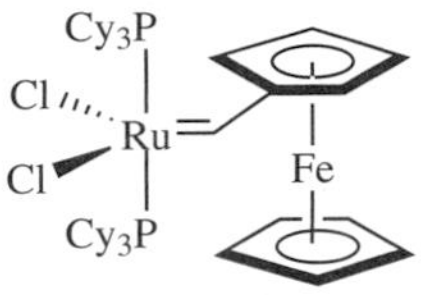

2

A catalyst precursor **3** is obtained from $[(p\text{-cymene})RuCl_2]_2$, 1,3-dimesitylimidazolium chloride, and *t*-BuONa, it is air-stable and easily handled on benchtop, the catalyst is generated in situ by treatment with *t*-butylethyne.[3] A simplified preparation of the second-generation catalyst **4u** involves admixture of **1** with *t*-BuOK and 1,3-dimesitylimidazolium chloride in hexane at room temperature.[4]

Several versions of polymer-bound Ru-carbene complexes prove useful in alkene metathesis. These include **5**,[5] **6p**,[6] and a self-regenerating system that is created by copolymerization based on **4s**-mediated ring-opening metathesis and eventual incorporation of **4s**.[7] This last system has further advantage of being recyclable and the level of catalyst loading can be controlled. Recyclable catalyst immobilized on monolithic sol-gel through an appropriate linker retains catalytic activity after many cycles.[8]

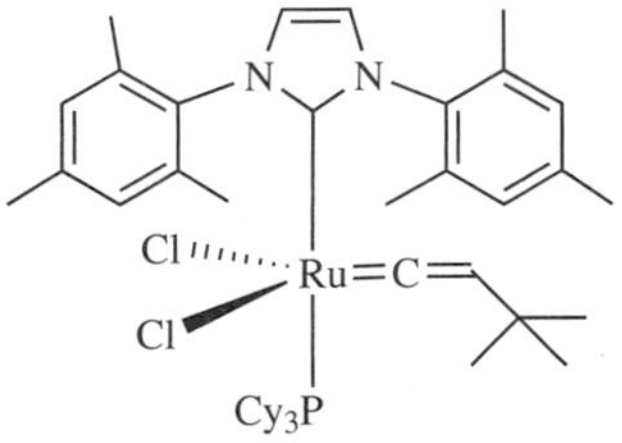

3

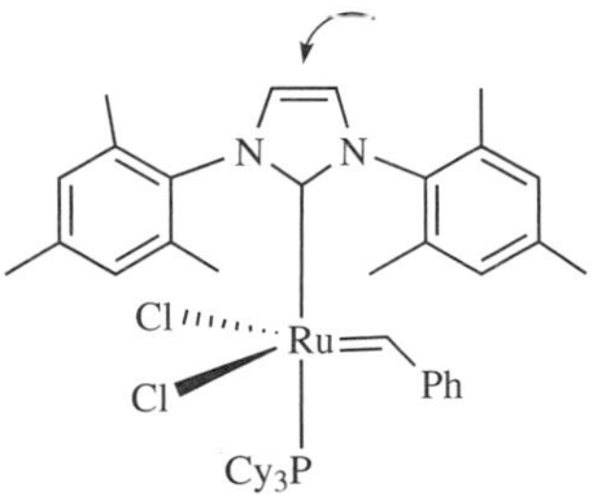

4u
4s (= dihydro-4u)

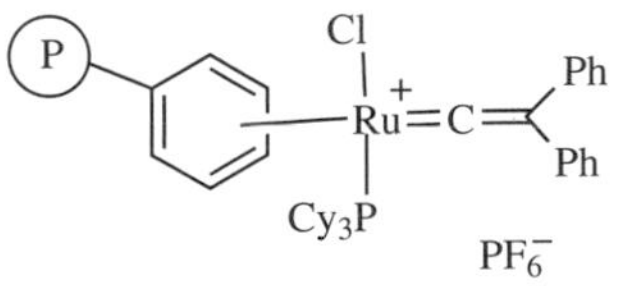

5

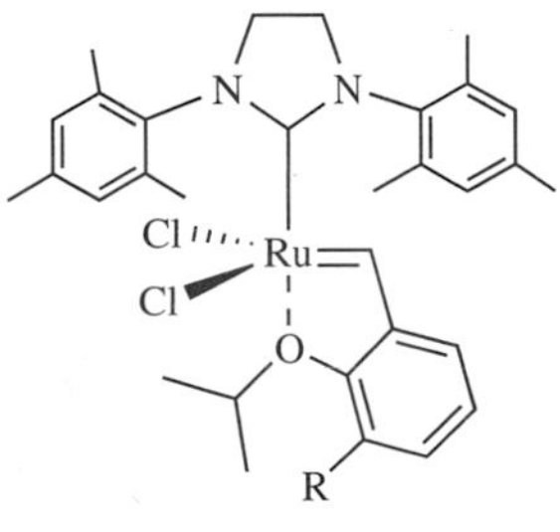

6 (R = H)
6b (R = Ph)

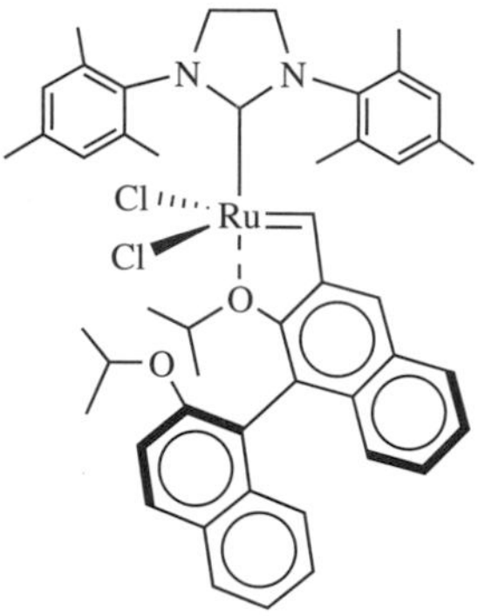

6bn 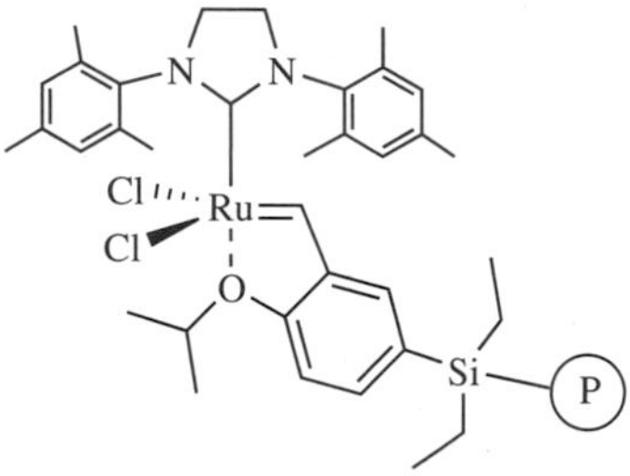 **6p**

The catalyst in which a difluoromethylene group replaces the benzylidene group of **4s** needs activation by HCl to function.[9]

Regarding technical improvements alkene metathesis carried out in ionic liquids and with microwave irradiation has been tried.[10] For removal of Ru residues from products DMSO or Ph_3PO has been recommended.[11]

Metathetic ring closure. An enormous bulk of literature is accumulating that deals with ring-closing metathesis (RCM) in the preparation of various functionalized cyclic systems. Thiacycloalkenes,[12,13] cyclic enamides,[14] 2-butenolides,[15] *C*-glycals[16] are some of the examples. An interesting synthesis of 1-acetylcycloalkenes is by RCM of 1-siloxyalkynes that also contain a double bond at the other terminus of the skeletal chain.[17]

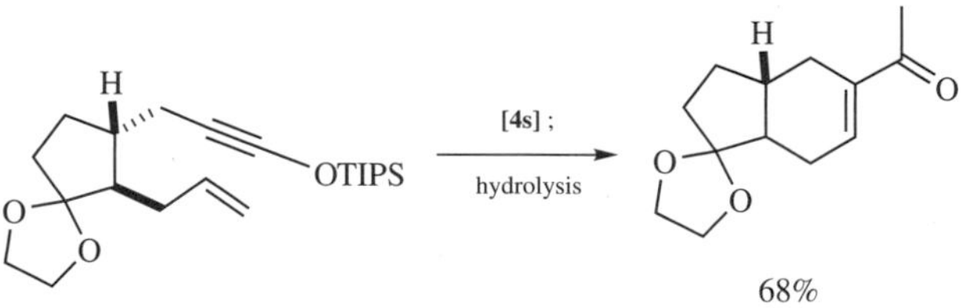

H OTIPS [**4s**] ; hydrolysis H O O O O 68%

Perhaps of higher synthetic value is RCM that leads to macrocyclic structures. Thus, a route to macrolactams involves preparation of protected *N*-[*o*-(ω-alkenoyloxy)benzyl]-*N*-allylamines (e.g., carbamates) and subjecting them to RCM conditions to provide macrolactones that are suitable for O->N transacylation.[18]

Formation of the 10-membered lactone ring of ascidiatrienolide-A,[19] both the azacyclotridecene and azocine units of manzamine-A,[20] and of the 11-membered oxadiketo ring of coleophomones B & C[21] by RCM method serves to demonstrate its unique power for assembling complex molecules.

Equally interesting is the ring expansion of cyclooctene to a 18-membered bislactone.[22]

[4s]
CH_2Cl_2 Δ
45%

Access to 9-(*t*-butyldimethylsilyl)-8-isopropylcyclodec-4-enone (*E/Z*−11:1, 76%) by RCM is aided by $(i\text{-PrO})_4Ti$ and the TBS group acting as a conformation-controlling element.[23]

The successful elaboration of the 8-azabicyclo[3.2.1]oct-6-ene skeleton from *cis*-2,6-divinylpiperidine derivatives[24] opens a new route to tropane alkaloids. Constituting the ingenol nucleus by RCM to close the B-ring is also an important development.[24a]

COOMe
BnOOC−N
OH
[1]
COOMe
NZ
O
83%

H
O
H
H
O
H
45%

Ring-opening metathesis. The key step for an elaboration of F-series isoprostanes[25] is by ring-opening metathesis (ROM). From the *meso*-cyclobutene cross-metathesis produces only one set of regioisomers. On the other hand, 1-arylcyclobutenes show excellent regioselectivity on reaction with other alkenes.[26]

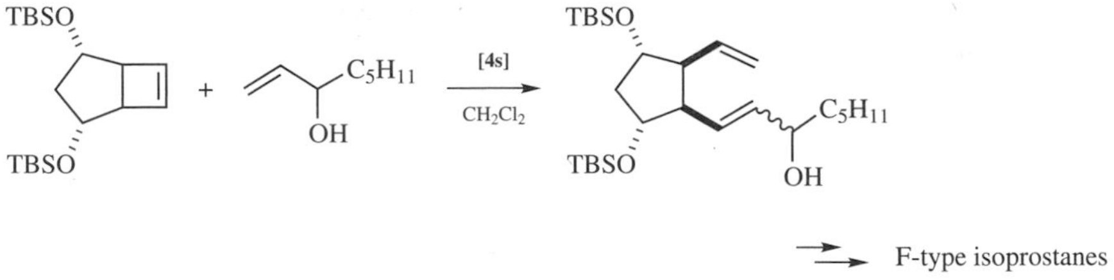

TBSO TBSO + C5H11 OH [4s] CH2Cl2 TBSO TBSO C5H11 OH

F-type isoprostanes

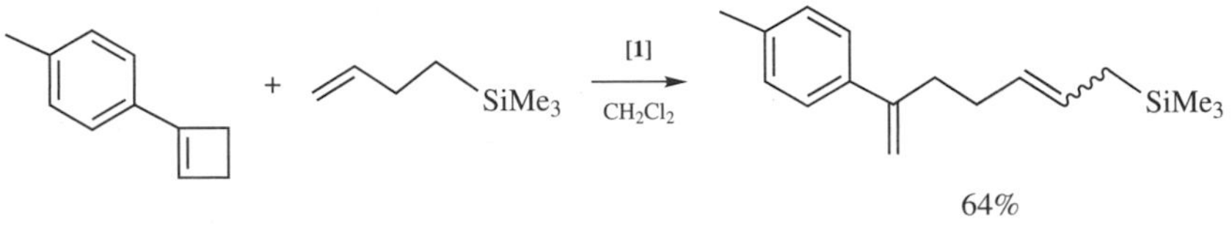

+ SiMe3 [1] CH2Cl2 SiMe3

64%

Cross-ROM involving unstrained cycloalkenes can be effected with catalyst **4s**[27,28] or **6**[29] which does not contain phosphine ligand.

The classical Grubb's catalyst has been used to prepare a precursor of the ROMP gel-linked tosylmethyl isocyanide from an *exo*-5-arylnorbornene. The polymer reagent is applicable to synthesis of oxazole library that requires minimal purification.[30]

Formal insertion of an alkyl group to the terminal double bond in the carbon chain of a Fischer carbene complex is accomplishable by the RCM method.[31] Apparently, the $Cr(CO)_5$ group does not interfere with the metathesis.

Other cross-metathesis reactions. Many publications describe assembly of functionalized alkenes with varying chain length. α,β-Unsaturated nitriles are obtained from cross-metathesis reactions of acrylonitrile and other alkenes. Catalysts that do not contain phosphine ligands (**6**,[32] **7**,[33] **8**[34]) prove very active. With **6** to induce cross-metathesis fluoroalkyl groups are readily grafted onto alkenes (e.g., $RCH{=}CH_2$ -> $RCH{=}CHCF_3$)[35] and highly functionalized compounds such as chiral 6-aryloxy-5-hydroxy-2-hexenals are readily generated from homoallylic alcohols and acrolein.[36] The efficiency of **6b** in cross metathesis is also evident by the short reaction times.[37] Even better is the air-stable complex **6bn** derived from BINOL whose effectiveness is evident (in one case a product yield of 99% as compared with 4% using Grubb's catalyst).[38]

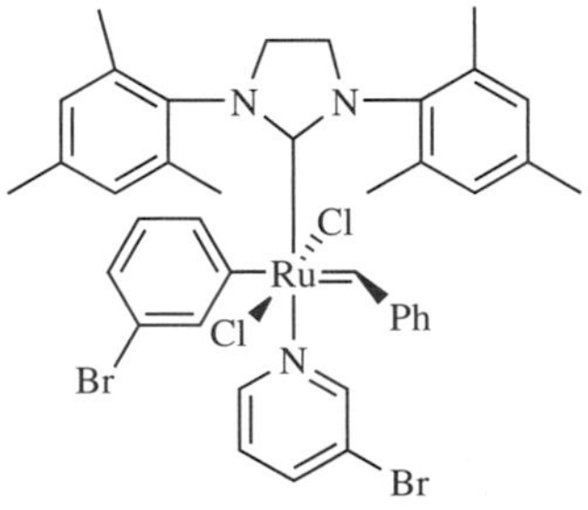

N N Cl Ru Ph Cl N Br Br

7

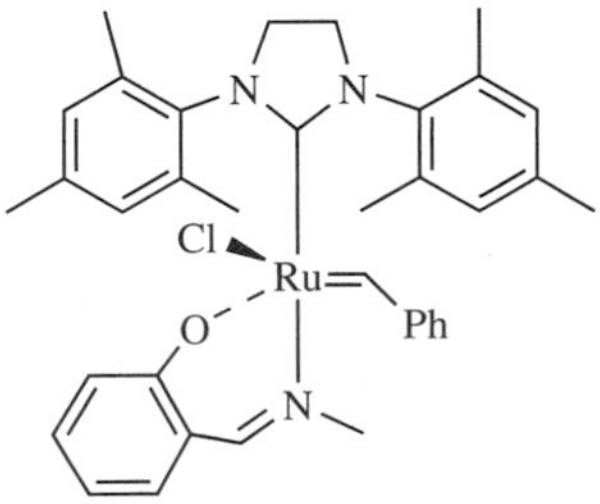

N N Cl Ru Ph O N

8

A significant synthetic application of the cross-metathesis of alkenes with acrolein is its repeated use in the construction of the C(1-14) fragment of amphidinol-3.[39] Due to deactivation of the double bond that is allylically acetoxylated, reaction is deferred to the terminal, homoallylic site.

PMPO OH [6] CH_2=CHCHO PMPO OH CHO 79%

PMPO OAc OAc

[6] CH_2=CHCHO

PMPO OAc OAc OAc PMPO OAc OAc CHO 63%

[6] CH_2=CHCOOEt

PMPO OAc OAc OAc COOEt 61%

Presumably some of the above-mentioned catalysts are relatively new, most preparative work reported in this period has been carried out with catalysts such as **4s**. Thus, long-chain conjugated amides,[40] unsaturated phosphonates,[41] allylsilanes,[42] and alkenylsilanes[43] containing distant functional groups are readily constructed. Noteworthy is the observation that overwhelming preference for the generation of (*E*)-alkenes in most cases. The technique also enables preparation of various alkenylpentafluorobenzenes.[44]

Diene synthesis. The cross metathesis between an alkyne and an alkene to produce a conjugated diene is a synthetically more powerful methodology. To assemble 2-substituted 1,3-butadienes from 1-alkynes (e.g., propargylic esters) and ethylene a higher pressure of ethylene requires smaller amounts of the catalyst.[45] The superiority of **4s** is again seen in an improved synthesis of 2-substituted 1,3-butadienes because it promotes reaction of alkynes without a propargylic heteroatom.[46]

Successful application of a tethered tactic to synthesis of dienediols[47] and α-substituted enones[48] shows the transformation of readily obtained compounds to seemingly much more difficultly accessible molecules.

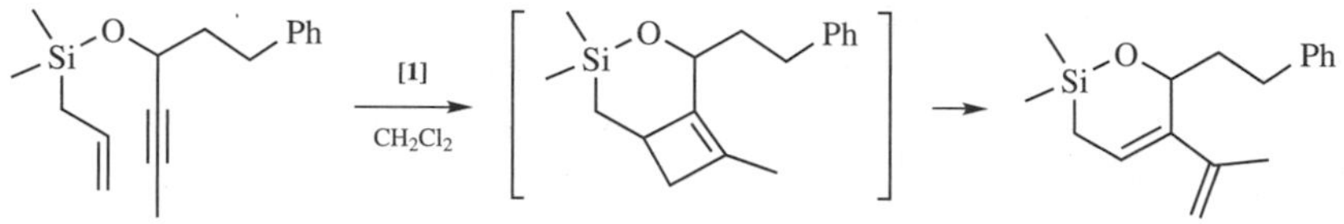

Tandem reactions. A highly satisfactory RCM-ROM sequence that transforms a bridged ring system to a condensed ring unit is involved in a synthesis of alkaloid-251F.[49] A similar double cross-metathesis to create a 3-pyrroline derivative[50] from a disubstituted cycloheptene en route to (−)-indolizidine-167B is no less elegant.

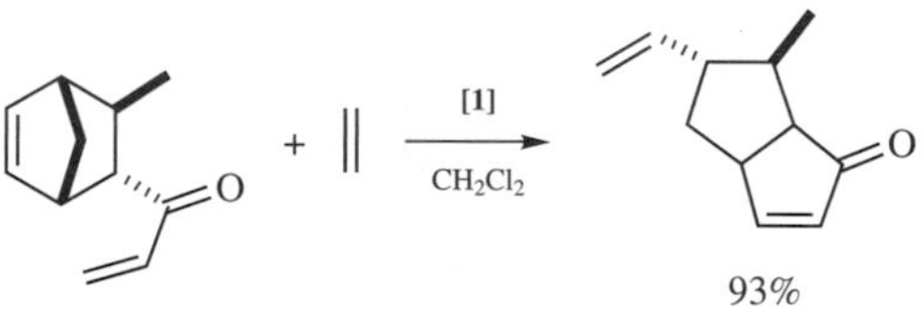

93%

A synthesis of indoles from protected *N*-allyl-*o*-vinylanilines by RCM requires prior isomerization to generate the enamine derivatives. The RCM catalyst **4s** in combination of trimethylsiloxyethene is capable of shifting the double bond and completing ring closure.[51]

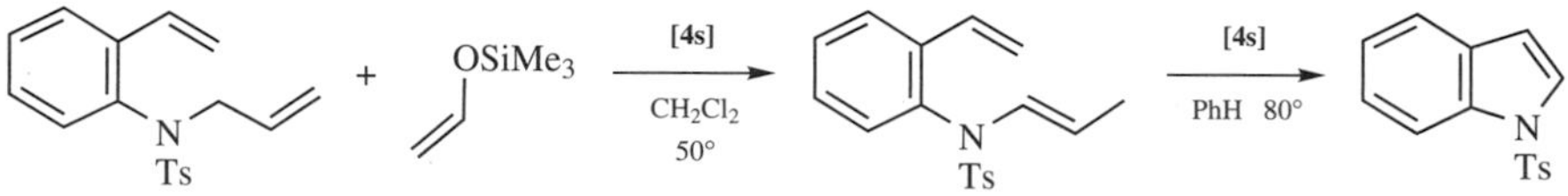

94%

Assembly of the 5:7:6-fused ring system that is characteristic of an important class of diterpenes has taken advantage of two consecutive RCM.[52]

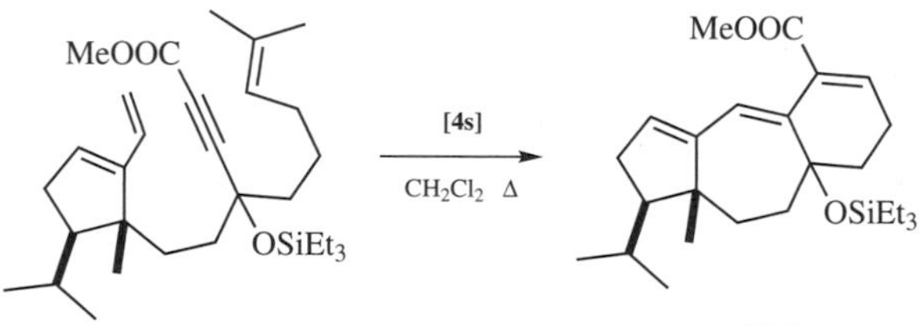

93%

Since many Ru species are hydrogenation catalysts the admittance of hydrogen to reaction mixtures after metathesis would saturate the double bond(s) of the products.[53] Addition of a compatible catalyst such as PtO_2 in the beginning[54] is an alternative operational choice.

1,2-Oxaborins prepared from substituted allylic alcohols and $(i\text{-PrO})_2BCH_2CH{=}CH_2$ via exchange reaction and metathesis are very useful precursors of stereochemically defined hydroxy-tetrad.[55] Functionalized allylic boranes/boronates generated from cross-metathesis can be used in situ as allylating agents for aldehydes (one-pot reaction).[56] In other words, the allylation step is performed without isolation of the metathetic products.

Protection-deprotection. Cleavage of tertiary allylamines by Grubb's catalyst via isomerization to the enamine isomers can be achieved without affecting other protective groups such as 4-methoxyphenyl derivatives.[57] The following example shows that isomerization takes precedence over RCM.

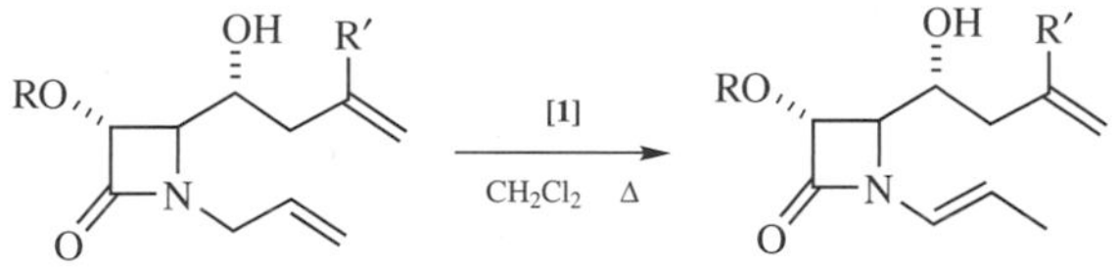

Grubb's catalyst also promotes silylation of alcohols with hydrosilanes (dehydrogenative condensation).[58]

Quinoline synthesis.[59] 2-Aminobenzyl alcohols and ketones are combined to give quinolines on treatment with Grubb's catalyst.

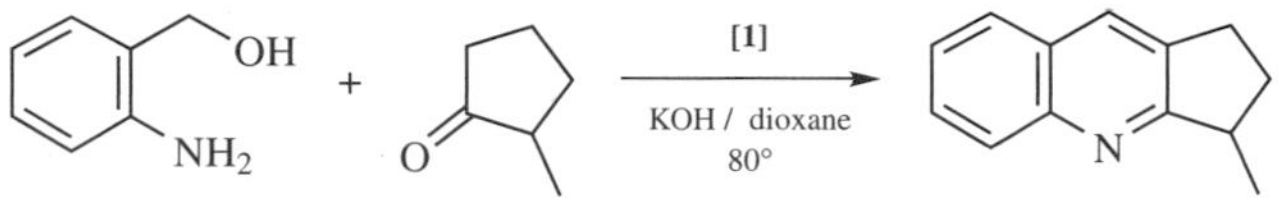

Isomerization.[60] 1-Alken-3-ols are isomerized to ethyl ketones by Grubb's catalyst. With 100% of the catalyst methyl ketones are produced. Of course the reaction is hardly useful synthetically.

On the other hand, removal of allyl groups attached to oxygen and nitrogen atoms is possible by shifting the double bond with **4s**. Subsequent hydrolysis with acid accomplishes the deprotection.[61]

Radical additions.[62] The addition of polyhaloalkanes to alkenes and of carboxylic acids to 1-alkynes is promoted by $Cl_2(Cy_3P)(L)Ru{=}C{=}CHBu^t$.

[1]Gandelman, M., Rybtchinski, B., Ashkenazi, N., Gauvin, R.M., Milstein, D. *JACS* **123**, 5372 (2001).
[2]Maishal, T.K., Sarkar, A. *SL* 1925 (2002).
[3]Louie, J., Grubbs, R.H. *ACIEE* **40**, 247 (2001).
[4]Jafarpour, L., Hillier, A., Nolan, S.P. *OM* **21**, 442 (2002).
[5]Akiyama, R., Kobayashi, S. *ACIEE* **41**, 2602 (2002).
[6]Grela, K., Tryznowski, M., Bieniek, M. *TL* **43**, 9055 (2002).
[7]Connon, S.J., Dunne, A.M., Blechert, S. *ACIEE* **41**, 3835 (2002).
[8]Kingsbury, J.S., Garber, S.B., Giftos, J.M., Gray, B.L., Okamoto, M.M., Farrer, R.A., Fourkas, J.T., Hoveyda, A.H. *ACIEE* **40**, 4251 (2001).

[9]Trnka, T.M., Day, M.W., Grubbs, R.H. *ACIEE* **40**, 3441 (2001).
[10]Mayo, K.G., Nearhoof, E.H., Kiddie, J.J. *OL* **4**, 1567 (2002).
[11]Ahn, Y.M., Yang, K., Georg, G.I. *OL* **3**, 1411 (2001).
[12]Yao, Q. *OL* **4**, 427 (2002).
[13]Spagnol, G., Heck, M.-P., Nolan, S.P., Mioskowski, C. *OL* **4**, 1767 (2002).
[14]Kinderman, S.S., van Maarseveen, J.H., Schoemaker, H.E., Hiemstra, H., Rutjes, F.P.J.T. *OL* **3**, 2045 (2001).
[15]Clive, D.L.J., Cheng, H. *CC* 605 (2001).
[16]Liu, L., Postema, M.H.D. *JACS* **123**, 8602 (2001).
[17]Schramm, M.P., Reddy, D.S., Kozmin, S.A. *ACIEE* **40**, 4274 (2001).
[18]Bieraugel, H., Jansen, T.P., Schoemaker, H.E., Hiemstra, H., van Maarseveen, J.H. *OL* **4**, 2673 (2002).
[19]Fürstner, A., Schlede, M. *ASC* **344**, 657 (2002).
[20]Humphrey, J.M., Liao, Y., Ali, A., Rein, T., Wong, Y.-L., Chen, H.-J., Courtney, A.K., Martin, S.F. *JACS* **124**, 8584 (2002).
[21]Nicolaou, K.C., Vassilikogiannakis, G., Montagnon, T. *ACIEE* **41**, 3276 (2002).
[22]Lee, C.W., Choi, T.-L., Grubbs, R.H. *JACS* **124**, 3224 (2002).
[23]Nevalainen, M., Koskinen, A.M.P. *ACIEE* **40**, 4060 (2001).
[24]Neipp, C.E., Martin, S.F. *TL* **43**, 1799 (2002).
[24a]Tang, H., Yusuff, N., Wood, J.L. *OL* **3**, 1563 (2001).
[25]Schrader, T.O., Snapper, M.L. *JACS* **124**, 10998 (2002).
[26]Feng, J., Szeimies, G. *EJOC* 2942 (2002).
[27]Choi, T.-L., Lee, C.W., Chatterjee, A.K., Grubbs, R.H. *JACS* **123**, 10417 (2001).
[28]Morgan, J.P., Morrill, C., Grubbs, R.H. *OL* **4**, 67 (2002).
[29]Randl, S., Connon, S.J., Blechert, S. *CC* 1796 (2001).
[30]Barrett, A.G.M., Cramp, S.M., Hennessy, A.J., Procopiou, P.A., Roberts, R.S. *OL* **3**, 271 (2001).
[31]Zhang, L., Herndon, J.W. *TL* **43**, 4471 (2002).
[32]Randl, S., Gessler, S., Wakamatsu, H., Blechert, S. *SL* 430 (2001).
[33]Love, J.A., Morgan, J.P., Trinka, T.M., Grubbs, R.H. *ACIEE* **41**, 4035 (2002).
[34]De Clercq, B., Verpoort, F. *TL* **43**, 9101 (2002).
[35]Imhof, S., Randl, S., Blechert, S. *CC* 1692 (2001).
[36]Cossy, J., BouzBouz, S., Hoveyda, A.H. *JOMC* **624**, 327 (2001).
[37]Wakamatsu, H., Blechert, S. *ACIEE* **41**, 2403 (2002).
[38]Wakamatsu, H., Blechert, S. *ACIEE* **41**, 794 (2002).
[39]BouzBouz, S., Cossy, J. *OL* **3**, 1451 (2001).
[40]Choi, T.-L., Chatterjee, A.K., Grubbs, R.H. *ACIEE* **40**, 1277 (2001).
[41]Chatterjee, A.K., Choi, T.-L., Grubbs, R.H. *SL* 1034 (2001).
[42]Engelhardt, F.C., Schmitt, M.J., Taylor, R.E. *OL* **3**, 2209 (2001).
[43]Kujawa-Welten, M., Pietraszuk, C., Marciniec, B. *OM* **21**, 840 (2002).
[44]Chatterjee, A.K., Sanders, D.P., Grubbs, R.H. *OL* **4**, 1939 (2002).
[45]Smulik, J.A., Diver, S.T. *JOC* **65**, 1788 (2000).
[46]Tonogaki, K., Mori, M. *TL* **43**, 2235 (2002).
[47]Yao, Q. *OL* **3**, 2069 (2001).
[48]Micalizio, G.C., Schreiber, S.L. *ACIEE* **41**, 3272 (2002).
[49]Wrobleski, A., Sahasrabudhe, K., Aube, J. *JACS* **124**, 9974 (2002).
[50]Zaminer, J., Stapper, C., Blechert, S. *TL* **43**, 6739 (2002).
[51]Arisawa, M., Terada, Y., Nakagawa, M., Nishida, A. *ACIEE* **41**, 4732 (2002).
[52]Boyer, F.-D., Hanna, I. *TL* **43**, 7469 (2002).
[53]Louie, J., Bielawski, C.W., Grubbs, R.H. *JACS* **123**, 11312 (2001).
[54]Cossy, J., Bargiggia, F.C., BouzBouz, S. *TL* **43**, 6715 (2002).

[55]Micalizio, G.C., Schreiber, S.L. *ACIEE* **41**, 152 (2002).
[56]Goldberg, S.D., Grubbs, R.H. *ACIEE* **41**, 807 (2002).
[57]Alcaide, B., Almendros, P., Alonso, J.M., Aly, M.F. *OL* **3**, 3781 (2001).
[58]Maifeld, S.V., Miller, R.L., Lee, D. *TL* **43**, 6363 (2002).
[59]Cho, C.S., Kim, B.T., Kim, T.-J., Shim, S.C. *CC* 2576 (2001).
[60]Gurjar, M.K., Yakambram, P. *TL* **42**, 3633 (2001).
[61]Cadot, C., Dalko, P.I., Cossy, J. *TL* **43**, 1839 (2002).
[62]Opstal, T., Verpoort, F. *TL* **43**, 9259 (2002).

Ruthenium(III) chloride. 13, 268; **14,** 271–272; **19,** 289–290; **20,** 324; **21,** 376

Propargylamines. Addition of 1-alkynes to imines is under the influence of $RuCl_3$ and CuBr.[1]

Alkylation. In the presence of $RuCl_3 \cdot 3H_2O$ and Ph_3P ketones take up an alkyl group from trialkylamines.[2]

$$PhCH_2CH_2COCH_3 + R_3N \xrightarrow[\text{Ph}_3\text{P / Dioxane, } 180^\circ]{RuCl_3 \cdot 3H_2O} PhCH_2CH_2COCH_2R$$

Friedel-Crafts acylation. The $RuCl_3 \cdot nH_2O–AgSbF_6$ couple catalyzes acylation of arenes by carboxylic acid anhydrides.[3] The design is based on hard-soft mismatch. Other late transition metal complexes [e.g., $(PhCN)_2PtCl_2$] are also effective in combination with the silver salt.

[1]Li, C.-J., Wei, C. *CC* 268 (2002).
[2]Cho, C.S., Kim, B.T., Lee, M.J., Kim, T.-J., Shim, S.C. *ACIEE* **40**, 958 (2001).
[3]Fürstner, A., Voigtländer, Schrader, W., Giebel, D., Reetz, M.T. *OL* **3**, 417 (2001).

Ruthenium carbonyl cluster-tin(II) chloride.

Hydrogenation.[1] Bimetallic nanoparticles derived from $Ru_6(CO)_{17}$ and $SnCl_2$ and loaded on mesoporous silica can be used to hydrogenate alkenes at low temperature without solvent.

[1]Hermans, S., Raja, R., Thomas, J.M., Johnson, B.F.G., Sankar, G., Gleeson, D. *ACIEE* **40**, 1211 (2001).

Ruthenium clay.

Hydrogenation.[1] α-Imino ketones and esters undergo hydrogenation to give the corresponding amino derivatives. Reductive amination of α-keto esters can be accomplished also.

[1]Aldea, R., Alper, H. JOMC **593–594**, 454 (2000).

S

Samarium. **14,** 275; **17,** 305–307; **18,** 311; **19,** 291; **20,** 325–326; **21,** 378

Reductive dimerization.[1] *N*-Sulfonylimines are dimerized to *vic*-diamanine derivatives.

Homoallylic hydroxylamines. Allylation of nitrones proceeds with Sm and allylic halides in an aqueous medium.[2]

[1]Liu, X., Liu, Y., Zhang, Y. *TL* **43**, 6787 (2002).
[2]Laskar, D.D., Prajapati, D., Sandhu, J.S. *TL* **42**, 7883 (2001).

Samarium–diiodomethane.

Cyclopropanation. The Sm-CH_2I_2 combination possesses reactivity similar to the Simmons-Smith reagent and it can be used to form cyclopropane derivatives with alkenes, e.g., conjugated amides.[1]

CONNH$_2$ —(Sm; CH_2I_2 / THF)→ CONH$_2$

Elimination. Retro-halohydrination is readily achieved. The reaction generates (*E*)-2-alkenoic esters, and (*Z*)-1-haloalkenes from 1,1-dihalo-2-alkanols.

[1]Concellon, J.M., Rodriguez-Solla, H., Gomez, C. *ACIEE* **41**, 1917 (2002).
[2]Concellon, J.M., Rodriguez-Solla, H., Huerta, M., Perez-Andres, J.A. *EJOC* 1839 (2002).

Samarium–iodine. **19,** 292; **21,** 375–376

Allylation. The carbonyl oxygen of lactones, lactams, acyclic amides and phthalic anhydride is replaced by two allyl groups when they are treated with allyl bromide, Sm, and iodine.[1]

Tetraaryldihydrofurans. Chalcones are reduced by Sm and the intermediates attack diaryl ketones to give dihydrofurans.[2]

MeO, Ph, O + Ph, O, Ph —(Sm - I_2; THF)→ MeO, Ph, Ph, O, Ph

88%

Fiesers' Reagents for Organic Synthesis, Volume 22. Series editor Tse-Lok Ho
ISBN 0-471-28515-3

[1]Li, Z., Zhang, Y. *T* **58**, 5301 (2002).
[2]Ma, Y., Zhang, Y., Chen, J. *S* 1004 (2001).

Samarium–metal halides. 21, 378

Reduction. Samarium coupled with either $NiCl_2$[1] or $TiCl_4$[2] forms a reducing system for reduction of organic azides to amines.

Cyclization. Intramolecular Barbier reaction of δ-iodoketones to form cyclopentanols is effected by Sm-NiI_2. When the substrates contain a leaving group at β′-position the primary products would undergo fragmentation.[3]

Sm - NiI_2
THF
OMs
I
69%

1,3,4,5-Tetraarylimidazolidin-2-ones.[4] Reductive coupling of anils with $TiCl_4$-Sm and carbonylative cyclization by triphosgene is accomplished in one operation.

Sm - $TiCl_4$ / THF ;
K_2CO_3 / H_2O
85%

[1]Wu, H., Chen, R., Zhang, Y. *SC* **32**, 189 (2002).
[2]Zhong, W., Zhang, Y., Chen, X. *TL 42*, 73 (2001).
[3]Molander, G.A., Le Heurou, Y., Brtown, G.A. *JOC* **66**, 4511 (2001).
[4]Li, Z., Zhang, Y. *SC* **32**, 2613 (2002).

Samarium–Samarium(II) iodide.

Deoxygenative coupling. Areneamides and diaryl ketones condense to afford enamines.[1]

Sm - $TiCl_4$ / THF
Δ
82%

Unsaturated sulfones. The Sm-SmI_2 system with catalytic $CrCl_3$ promotes condensation of dibromomethyl sulfones with ketones.[2]

[1]Xu, X., Zhang, Y. *T* **58**, 503 (2002).
[2]Liu, Y., Wu, H., Zhang, Y. *SC* **31**, 47 (2001).

Samarium(II) bromide.

Reductions. HMPA increases the reducing power of $SmBr_2$ such that ketimines are reduced to amines. The system is also good for hydrodechlorination.[1]

[1]Knettle, B.W., Flowers, R.A. *OL* **3**, 2321 (2001).

Samarium(III) (2,6-di-*t*-butyl-4-methylphenoxide).

Michael reaction. The reagent is prepared from SmI_3 and the sodium phenolate at room temperature. It catalyzes Michael reaction of ketones with enones.[1]

[1]Katagiri, K., Kameoka, M., Nishiura, M., Imamoto, T. *CL* 426 (2002).

Samarium(II) iodide. 13, 270–272; 14, 276–281; 15, 282–284; 16, 294–300; 17, 307–311; 18, 312–316; 19, 292–296; 20, 327–335; 21, 379–385

Reagent preparation. The title reagent can be prepared from $Sm(OTf)_3$ via SmI_3 and by electroreduction.[1]

Elimination reactions. On exposure to SmI_2 α,β-epoxy esters undergo stereoselective deoxygenation to afford (*E*)-2-alkenoic esters.[2] 2-Chloro-3-hydroxyalkanamides also deliver conjugated amides by the same treatment,[3] although aqueous workup would result in the saturated products.[4]

As expected, other vicinal bifunctional compounds behave similarly. Thus SmI_2 may replace Na/Hg at the second stage of the Julia alkenation, and accordingly β-acetoxy sulfones are rapidly converted into alkenes[5] (for elimination of β-hydroxy sulfones HMPA is required as additive[6]).

That 2-chloro-3-acetoxy-1-silylalkanes afford (*Z*)-alkenylsilanes[7] and furan formation[8] from propargyl epoxides that also contain an acetoxy group at the other propargylic position are interesting and useful. Cumulene intermediates are likely generated in the latter transformation.

OAc; $(CH_2)_3CN$; O; SmI_2 / H_2O → C=C=C; OH; $(CH_2)_3CN$

$(Ph_3P)_4Pd$ / ArI / Et_3N, 70° → Ar; O; $(CH_2)_3CN$

Elimination ensues when α-bromoacetals are briefly treated with SmI_2-HMPA in THF at $-78°$. This is a convenient way to prepare enol ethers.[9]

Reduction. Invoked by the observation that aqueous workup of retro-halohydrination furnishes saturated compounds, deuteration of conjugated carboxylic acids by SmI_2 in the presence of D_2O is achieved accordingly.[10] 2-Chloro-3-hydroxy-4-alkenoic esters are shown to provide 2,5-dideuterio-3-alkenoic esters.[11]

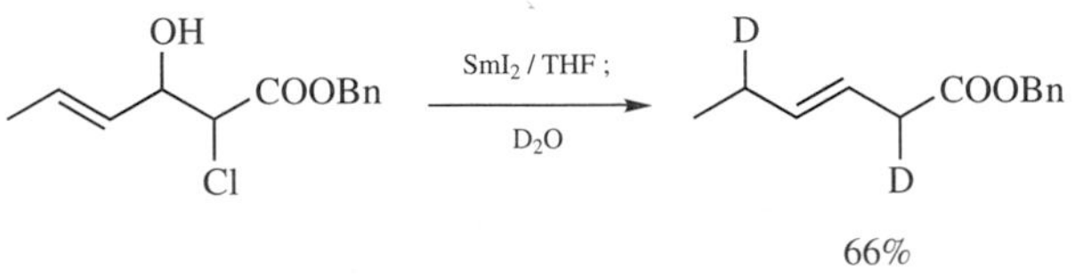

The ketone group of α-keto esters/amides are susceptible to reduction by SmI_2 in aqueous solvents.[12] Pertaining to reduction of dialkyl ketones it occurs instantaneously (<10 sec) in the presence of an amine (Et_3N, TMEDA.),[13] therefore HMPA is obviated.

Conjugated isoxazolines undergo N-O bond cleavage and hydrolysis to provide aldol products, on treatment with SmI_2 and then boric acid.[14] The method is mild and chemoselective.

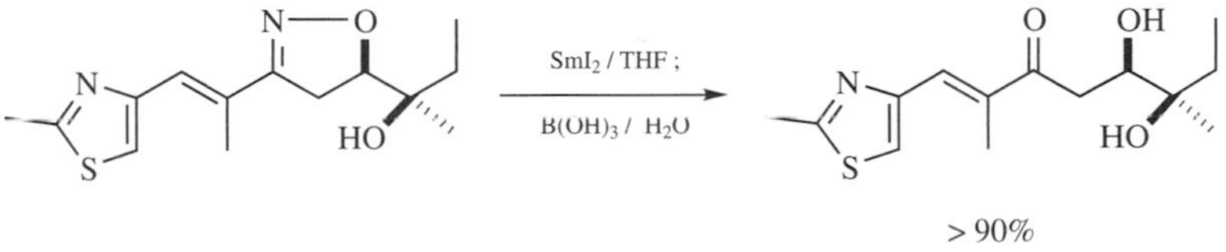

Reduction of α-azidomethyl aryl ketones furnishes 2,4-diarylpyrroles.[15]

Coupling. An α-diketone synthesis is realized by coupling of 1-acylbenzotriazoles.[16] Pinacol coupling of 4-aryl-4-oxobutanoic esters leads to butyrolactone dimers.[17] However, homologous ketomalonic esters afford intramolecular coupling products.[18]

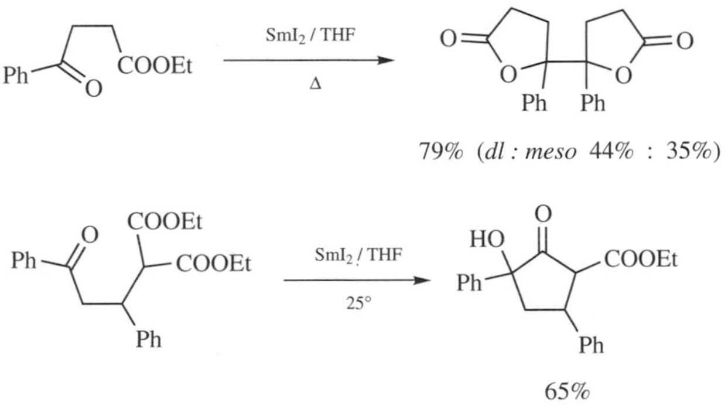

Reductive cyclization of cyclic γ-cyanoketones can be promoted by SmI_2 without photoirradiation.[19] Spiroannulation onto an aromatic ring involving a side chain ketone group is realized.[20] Only one double bond remains in the products.

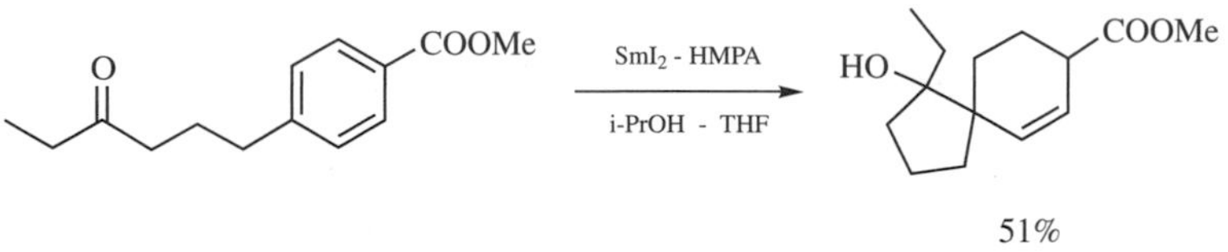

SmI_2 - HMPA, i-PrOH - THF

51%

β-Hydroxyamino alcohols are fomed in the coupling of carbonyl compounds with nitrones (both intramolecular and intermolecular versions feasible),[21] and a quinoline synthesis evolves from the analogous intramolecular process involving *o*-nitroaryl-propenones.[22]

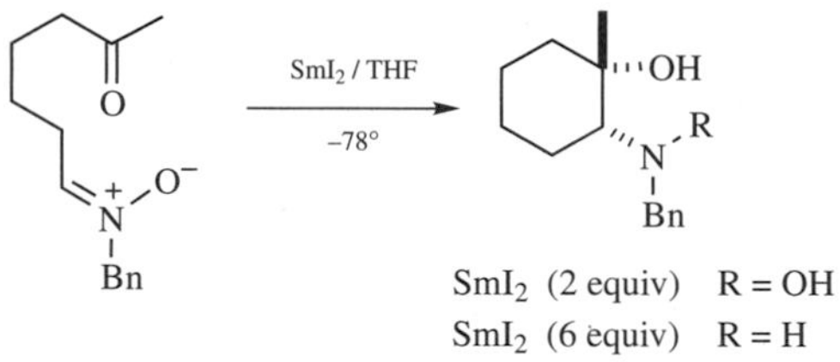

SmI_2 / THF, –78°

SmI_2 (2 equiv) R = OH
SmI_2 (6 equiv) R = H

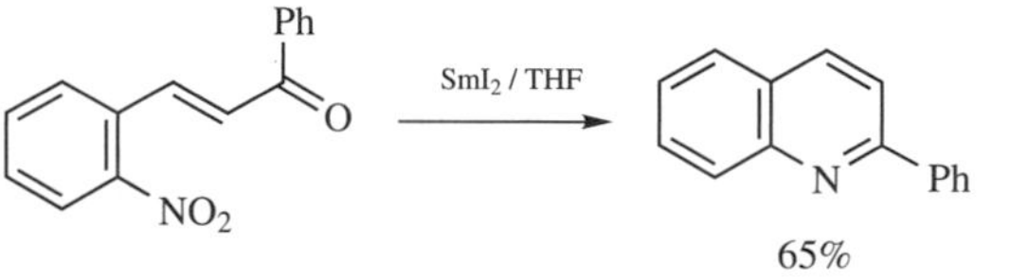

SmI_2 / THF

65%

A facile ring closure to elaborate the 5:7:6-ring system of some diterpenes involves reductive alkylation of a functionalized 4,4-disubstituted 2,5-cyclohexadienone.[23]

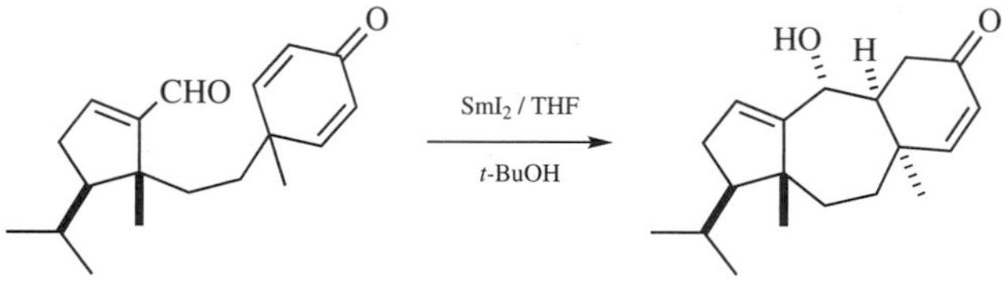

SmI_2 / THF, *t*-BuOH

Alkylation. Analogous to allylation of aldehydes the reaction with aldimines readily affords homoallylic amines.[24] Preparation of β-hydroxyphosphonates from diethyl iodomethylphosphonate and carbonyl compounds (lower yields for aromatic carbonyl compounds) is mediated by SmI_2 in THF at room temperature.[25] *N*-Substituted succinimides are attacked by organosamarium reagents derived from SmI_2 and RX to give cyclic enamides on acid workup. Adducts from glutarimides undergo ring opening therefore only δ-ketoamides can be isolated.[26]

Alkenation also occurs when ferrocenyl carbonyl compounds react with organosamarium reagents.[27] However, a more economical way to obtain the products is by using catalytic amounts of SmI_2 in combination with Mg.

An access to some pyrrole derivatives is through reaction of ketones with α-iminoketones.[28]

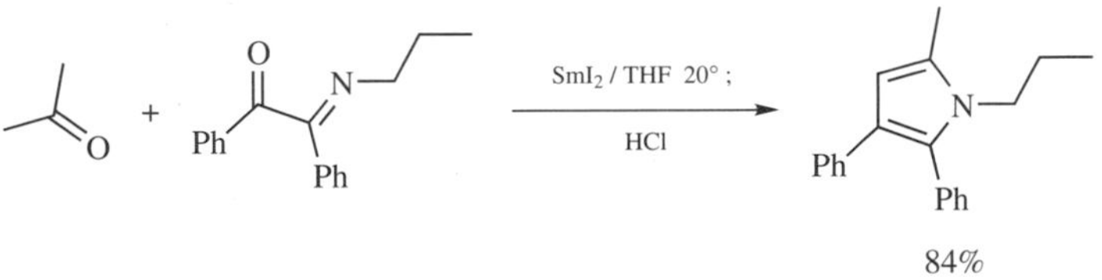

Condensation+reduction. Tandem aldol reaction and reduction are achievable during SmI_2–promoted union of α-bromoketones with carbonyl compounds.[29] Oxidation of aldehydes based on the hydride transfer process solves a difficult problem that requires selective transformation of the aldehyde unit in the presence of other easily oxidizable groups.[30]

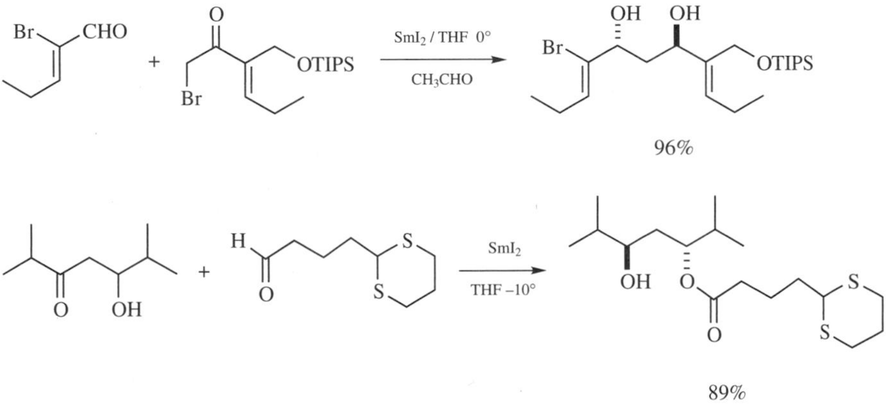

δ-Lactone formation from 5-oxoalkanals is effected by SmI_2; in which reaction a role for thiols as cooperative catalysts has been discovered.[31]

Additions and cycloadditions. The synthetic tactic involving Michael reaction and tandem trapping by aldehydes is well established. While the process involving organocopper chemistry is commonly practiced, SmI_2 offers an alternative.[32]

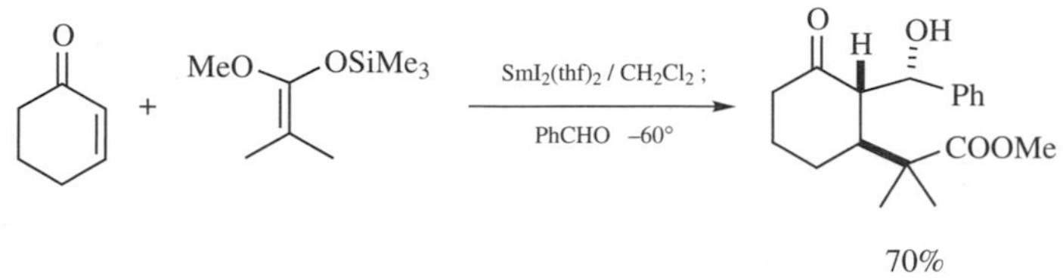

The catalytic activity of SmI_2 is revealed in the addition of amines to nitriles to form *N,N′*-disubstituted amidines,[33] cyclization of *N*-chloroalkenylamines,[34] and pyrrolidine synthesis from 2-alkenoic esters and *N,N*-bis(tosylmethyl)amines.[35]

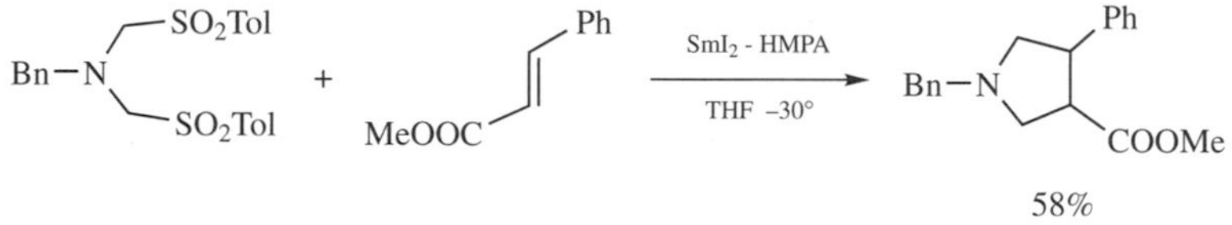

Ring expansion and cleavage. Small rings that are fused to cycloalkanones are susceptible to reductive fragmentation.[36] On treatment with SmI_2-HMPA cyclic α-halomethyl β-keto esters are transformed into homologous γ-keto esters via cyclization and fragmentation.[37]

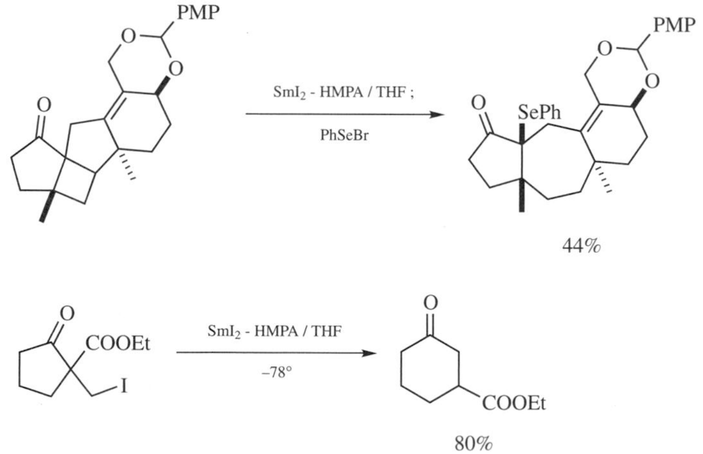

1-Substituted ethyl 2-oxocyclopentanecarboxylates undergo ring expansion to provide 3-substituted 1,2-cyclohexanediones on treatment with SmI_2. Indanone analogues are similarly transformed into 1,2-naphthoquinones.[38]

$$\text{indanone-2-COOEt, R} \xrightarrow{SmI_2 / THF} \text{(major)} + \text{(minor)}$$

(major) (minor)

Cyclic acetals, thioacetals, and *O,S*-acetals of α-bromoketones suffer ring cleavage on treatment with SmI_2-HMPA at low temperature. The products are a special kind of enol ethers or enol thioethers.[39]

[1]Parrish, J.D., Little, R.D. *TL* **42**, 7767 (2001).
[2]Concellon, J.M., Bardales, E. *OL* **4**, 189 (2002).
[3]Concellon, J.M., Perez-Andres, J.A., Rodriguez-Solla, H. *CEJ* **7**, 3062 (2001).
[4]Concellon, J.M., Rodriguez-Solla, H. *CEJ* **7**, 4266 (2001).
[5]Reutrakul, V., Jarussophon, S., Pohmakotr, M., Chaiyasut, Y., U-Thet, S., Tuchinda, P. *TL* **43**, 2285 (2002).
[6]Marko, I.E., Murphy, F., Kumps, L., Ates, A., Touillaux, R., Craig, D., Carballares, S., Dolan, S. *T* **57**, 2609 (2001).
[7]Concellon, J.M., Bernad, P.L., Bardales, E. *OL* **3**, 937 (2001).
[8]Aurrecoechea, J.M., Perez, E. *TL* **42**, 3839 (2001).
[9]Park, H.S., Kim, S.H., Park, M.Y., Kim, Y.H. *TL* **42**, 3727 (2001).
[10]Concellon, J.M., Rodriguez-Solla, H. *CEJ* **8**, 4493 (2002).
[11]Concellon, J.M., Bernad, P.L., Rodriguez-Solla, H. *ACIEE* **40**, 3897 (2001).
[12]Fukuzawa, S., Miura, M., Matsuzawa, H. *TL* **42**, 4167 (2001).
[13]Dahlen, A., Hilmersson, G. *TL* **43**, 7197 (2002).
[14]Bode, J.W., Carreira, E.M. *OL* **3**, 1587 (2001).
[15]Fan, X., Zhang, Y. *TL* **43**, 1863 (2002).
[16]Wang, X., Zhang, Y. *TL* **43**, 5431 (2002).
[17]Williams, D.B.G., Blann, K., Caddy, J., Holzapfel, C.W. *SC* **32**, 3755 (2002).
[18]Liu, Y., Zhang, Y. *TL* **42**, 5745 (2001).
[19]Kakiuchi, K., Fujioka, Y., Yamamura, H., Tsutsumi, K., Morimoto, T., Kurosawa, H. *TL* **42**, 7595 (2001).
[20]Ohno, H., Maeda, S., Okumura, M., Wakayama, R., Tanaka, T. *CC* 316 (2002).
[21]Masson, G., Py, S., Vallee, Y. *ACIEE* **41**, 1772 (2002).
[22]Wang, X., Zhang, Y. *SC* **32**, 3617 (2002).
[23]Nguyen, T.M., Seifert, R.J., Mowrey, D.R. *OL* **4**, 3959 (2002).
[24] Kim, B.H., Han, R., Park, R.J., Bai, K.H., Jun, Y.M., Baik, W. *SC* **31**, 2297 (2001).
[25]Orsini, F., Caselli, A. *TL* **43**, 7255 (2002).
[26]Farcas, S., Namy, J.-L. *TL* **42**, 879 (2001).
[27]Jong, S.-J., Fang, J.-M. *JOC* **66**, 3533 (2001).
[28]Farcas, S., Namy, J.-L. *T* **57**, 4881(2001).
[29]Shotwell, J.B., Krygowski, E.S., Hines, J., Koh, B., Huntsman, E.W.D., Choi, H.W., Schneekloth, J.S., Wood, J.L., Crews, C.M. *OL* **4**, 3087 (2002).
[30]Smith III, A.B., Lee, D., Adams, C.M., Kozlowski, M.C. *OL* **4**, 4539 (2002).
[31]Hsu, J.-L., Fang, J.-M. *JOC* **66**, 8573 (2001).
[32]Giuseponne, N., Collin, J. *T* **57**, 8989 (2001).
[33]Xu, F., Sun, J., Shen, Q. *TL* **43**, 1867 (2002).
[34]Göttlich, R., Noack, M. *TL* **42**, 7771 (2001).

[35]Dong, J., Kou, B., Li, R., Cheng, T. *SC* **32**, 935 (2002).
[36]Shipe, W.D., Sorensen, E.J. *OL* **4**, 2063 (2002).
[37]Chung, S.H., Cho, M.S., Choi, J.Y., Kwon, D.W., Kim, Y.H. *SL* 1266 (2001).
[38]Iwaya, K., Nakamura, M., Hasegawa, E. *TL* **43**, 5067 (2002).
[39]Park, H.S., Kim, S.H., Park, M.Y., Kim, Y.H. *TL* **42**, 3729 (2001).

Samarium(II) iodide–nickel iodide.

Addition reactions. Barbier-type reactions on esters and lactones are successful with promotion by Mischmetall and the SmI_2-NiI_2 combination.[1] More interesting are intramolecular conjugate addition leading to spirocyclic systems[2] and addition to carbonyl groups to afford mesocyclic compounds.[3]

SmI_2 - NiI_2 / t-BuOH - THF

~100%

SmI_2 - NiI_2 / THF

84%

Molecules containing both a saturated amide and a conjugated ester group which are properly separated give cyclic products on reduction after activation of the amide site.[4]

Tf_2O ; SmI_2 - NiI_2 / t-BuOH - THF

68%

1,3-Diols.[5] Reductive coupling of epoxides with carbonyl compounds is readily accomplished.

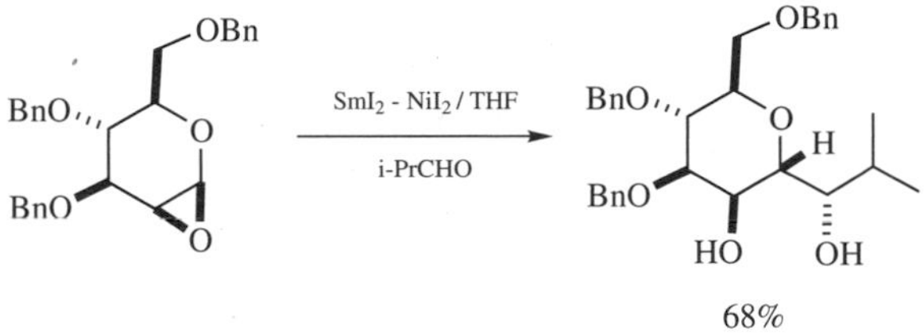

SmI$_2$ - NiI$_2$ / THF

i-PrCHO

68%

[1]Lannou, M.-I., Helion, F., Namy, J.-L. *TL* **43**, 8007 (2002).
[2]Molander, G.A., St. Jean, D.J. *JOC* **67**, 3861 (2002).
[3]Molander, G.A., Brown, G.A., Storch de Gracia, I. *JOC* **67**, 3459 (2002).
[4]McDonald, C.E., Galka, A.M., Green, A.I., Keane, J.M., Kowalchick, J.E., Micklitsch, C.M., Wisnoski, D.D. *TL* **42**, 163 (2001).
[5]Chiara, J.L., Sesmilo, E. *ACIEE* **41**, 3242 (2002).

Samarium(III) triflate. 21, 387–389

Transacylation. Transformation of the often-used *N*-acyloxazolidin-2-ones (as chiral induction platforms) to other carboxylic acid derivatives is an important step in certain synthetic maneuvers. They can be converted to hydroxamic acids[1] in a reaction catalyzed by Sm(OTf)$_3$.

[1]Sibi, M.P., Hasegawa, H., Ghorpade, S.R. *OL* **4**, 3343 (2002).

Scandium(III) perchlorate.

Alkylation. As a Lewis acid Sc(ClO$_4$)$_3$ catalyzes Friedel-Crafts reaction involving aziridines as electrophiles.[1] High regioselectivity for the formation of tryptophan derivatives is observed.

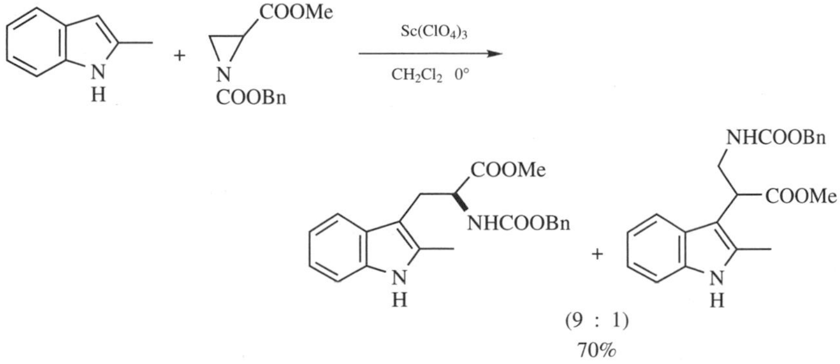

Sc(ClO$_4$)$_3$

CH$_2$Cl$_2$ 0°

(9 : 1)
70%

[1]Nishikawa, T., Kajii, S., Wada, K., Wada, K., Ishikawa, M., Isoobe, M. *S* 1658 (2002).

Scandium(III) triflate. 18, 317–318; **19,** 300–302; **20,** 335–337; **21,** 387–389

Functional group transformations. Acetolysis of *O*-trimethylsilyl cyanohydrins[1] and 1,6-anhydro-β-hexopyranoses[2] are also promoted by $Sc(OTf)_3$.

Alkylation. Alkoxypropargylation of silyl enol ethers is accomplished with 1,1-dialkoxy-2-alkynes.[3] Allylation of acylhydrazones takes place with tetrallyltin in aqueous THF.[4] Allylboranes react with alkenylepoxides at $-78°$ to provide bisallylic alcohols,[5] apparently after in situ rearrangement of the epoxides to β,γ-unsaturated aldehydes.

Addition of Grignard reagents to aldimines is effectively catalyzed by $Sc(OTf)_3$.[6]

Nitration. A new nitrating system for arenes consists of $Sc(OTf)_3$, an inorganic nitrate, and acetic anhydride.[7]

Diels-Alder reaction. The catalyzed reaction is performed with scandium(III) nonaflate, an analogue of $Sc(OTf)_3$.[8]

[1]Norsikian, S., Holmes, I., Lagasse, F., Kagan, H.B. *TL* **43**, 5715 (2002).
[2]Lee, J.-C., Tai, C.-A., Hung, S.-C. *TL* **43**, 851 (2002).
[3]Yoshimatsu, M., Kuribayashi, M., Koike, T. *SL* 1799 (2001).
[4]Kobayashi, S., Hamada, T., Manabe, K. *SL* 1140 (2001).
[5]Lautens, M., Maddess, M.L., Sauer, E.L.O., Ouellet, S.G. *OL* **4**, 83 (2002).
[6]Saito, S., Hatanaka, K., Yamaamoto, H. *SL* 1859 (2001).
[7]Kawada, A., Takeda, S., Yamashita, K., Abe, H., Harayama, T. *CPB* **50**, 1060 (2002).
[8]Kobayashi, S., Tsuchiya, T., Komoto, I., Matsuo, J. *JOMC* **624**, 392 (2001).

Selenium. 18, 318; **20,** 337; **21,** 389

Selenides.[1] Elemental selenium is reduced by Na to Na_2Se in DMF which can be alkylated. Sodium hydride can reduce Se to sodium diselenide only.

[1]Krief, A., Derock, M. *TL* **43**, 3083 (2002).

Selenophosphoric acid.

Selenonocarboxamides.[1] The reagent is prepared as an aqueous solution from $(Me_3SiO)_3P{=}Se$. It adds to nitriles to give $RC({=}Se)NH_2$.

[1]Kaminski, R., Glass, R.S., Skowronska, A. *S* 1308 (2001).

Silica gel. 15, 282; **18,** 319; **19,** 303–304; **20,** 338–339; **21,** 390–391

Ester and ether cleavage. *t*-Butyl esters are cleaved by heating with silica gel in toluene.[1] Sterically congested carboxylic acids are also recovered from their esters on treatment with TfOH-coated silica.[2] For removal of *O*-trityl groups a simple percolation through a silica gel column (with 5% CF_3COOH) is sufficient.[3]

Oxidations. Oxidation is often better done with supported reagents. Thus, Jones reagent adsorbed on silica for alcohol oxidation[4] and $KMnO_4$ on silica for conversion of aldehydes to acids[5] are indicated.

Oxodiperoxomolybdenum complexes adsorbed on silica are useful agents for oxidizing sulfides to sulfoxides.[6]

Under photochemical conditions benzylic and allylic alcohols undergo oxidation[7] on mesoporous silica in the presence of an alkali iodide.

1,2-Dihydroquinolines undergo aromatization (6 examples, 95–99%)[8] on treatment with $Na_2Cr_2O_7$ adsorbed on wet silica.

Condensation reactions. Preparation of (*E*)-nitrostyrenes[9] is conveniently promoted by propylamine-silica. A similar reagent catalyzes Michael reaction of aldehydes and enones.[10]

A three-component assembly (amine-nitroalkane-conjugated carbonyl compound) leads to pyrroles.[11]

Ph CHO NO2 + NH2 Bn → SiO2)))) → Ph N Bn 60%

Reductive amination of aldehydes with aniline can be carried out with Bu_3SnH on silica without solvent.[12]

Acid+base catalysts. Conventionally, reaction sequences that are promoted by acids and the bases must be performed in two different steps. The development of sol-gel entrapped reagents may overcome this shortcoming and serial reactions may be carried out without interruption. For example, acidic catalysts are formed by entrapment of Nafion or molybdic acid in silica sol-gel, and basic catalysts by embedding guanidine bases.[13] A combined pinacol rearrangement and Knoevenagel-type condensation serves to demonstrate the feasibility of such a maneuver.

HO OH → SiO_2-H^+ / SiO_2-base → [O] → NC CN → NC CN

[1]Jackson, R.W. *TL* **42**, 5163 (2001).
[2]Vavra, J., Streinz, L., Vodicka, P., Budesinsky, M., Koutek, B. *SL* 1886 (2002).
[3]Pathak, A.K., Pathak, V., Seitz, L.E., Tiwari, K.N., Akhtar, M.S., Reynolds, R.C. *TL* **42**, 7755 (2001).
[4]Ali, M.H., Wiggin, C.J. *SC* **31**, 3383 (2001).
[5]Takemoto, T., Yasuda, Y., Ley, S.V. *SL* 1555 (2001).
[6]Batigalhia, F., Zaldini-Hernandes, M., Ferreira, A.G., Malvestiti, I., Cass, Q.B. *T* **57**, 9669 (2001).
[7]Itoh, A., Kodama, T., Inagaki, S., Masaki, Y. *OL* **3**, 2653 (2001).
[8]Damavandi, J.A., Zolfigol, M.A., Karami, B. *SC* **31**, 3183 (2001).
[9]Demicheli, G., Maggi, R., Mazzacani, A., Righi, P., Sartori, G., Bigi, F. *TL* **42**, 2401 (2001).
[10]Shimizu, K., Suzuki, H., Hayashi, E., Kodama, T., Tsuchiya, Y., Hagiwara, H., Kitayama, Y. *CC* 1068 (2002).

[11]Ranu, B.C., Hajra, A. *T* **57**, 4767 (2001).
[12]Hiroi, R., Miyoshi, N., Wada, M. *CL* 274 (2002).
[13]Gelman, F., Blum, J., Avnir, D. *ACIEE* **40**, 3647 (2001).

Silver acetate. 21, 392

Allylation.[1] The allyl transfer from allyldimethyl(2-pyridyl)silane to aldehydes is catalyzed by AgOAc.

[1]Itami, K., Kamei, T., Mineno, M., Yoshida, J. *CL* 1084 (2002).

Silver carbonate. 21, 392–393

Lactonization.[1] (*Z*)-2-Alken-4-ynoic acids undergo metal-catalyzed cyclization. With Ag_2CO_3 as catalyst the products are butenolides but $ZnBr_2$ induces 2-pyrone formation.

COOH
C_6H_{13}
Ag_2CO_3 / DMF → C_6H_{13} 96%
$ZnBr_2$ / THF → C_6H_{13} 95%

Deamination.[2] *N*-Boc arylhydrazines [ArN(Boc)NH_2] are converted into *N*-Boc arylamines on heating with Ag_2CO_3 in DME. Acylhydrazines do not undergo a similar N-N bond cleavage.

[1]Anastasia, L., Xu, C., Negishi, E. *TL* **43**, 5673 (2002).
[2]Lee, K.-S., Lim, Y.-K., Cho, C.-G. *TL* **43**, 7463 (2002).

Silver nitrate. 18, 320; **19,** 305–306; **20,** 340; **21,** 393

3-Pyrrolines. Participation of an amino group and an allene moiety of the same molecule but separated by two bonds in the formation of cyclic structure is readily brought about with $AgNO_3$. The substrates are readily obtained via reaction of *N*-carbamoylalkylcuprate reagents with propargylic derivatives[1] or from aldehydes, benzyl carbamate, and propargyltrimethylsilane with $BF_3 \cdot OEt_2$ as promoter.[2] The allenyl carbamates also cyclize to 3-pyrrolines in the presence of $AgBF_4$.[2]

Radical methylation. $AgNO_3$ acts as catalyst for decarboxylative generation of methyl radical from acetic acid by ammonium persulfate. Application of the method to a synthesis of deliquinone is significant.[3]

HO / OH; $AgNO_3$ - $(NH_4)_2S_2O_8$, HOAc, MeCN - H_2O; 68%

[1]Dieter, R.K., Yu, H. *OL* **3**, 3855 (2001).
[2]Billet, M., Schoenfelder, A., Klotz, P., Mann, A. *TL* **43**, 1453 (2002).
[3]Kraus, G.A., Chaudhury, P.K. *TL* **42**, 6649 (2001).

Silver(I) oxide. 18, 321; **20,** 341; **21,** 393

O-Tosylation.[1] Alcohols are tosylated in dichloromethane with TsCl in the presence of KI and Ag_2O. Symmetrical diols can be monotosylated. With 3 equivalents of Ag_2O present 1,4-diols cyclize to tetrahydrofuran derivatives.

Anilides→Aryl acetates. Nitrosation of anilides in the presence of Ag_2O and HOAc leads to aryl acetates.[2]

Homocoupling.[3] With a catalytic amount of $CrCl_2$ the coupling of organoboronic acids $RB(OH)_2$ to give R-R by Ag_2O is efficiently accomplished.

$B(OH)_2$, $B(OH)_2$, MOMO; Ag_2O - $CrCl_2$, THF 65°; MOMO; 85%

$B(OH)_2$; Ag_2O - $CrCl_2$, THF 65°; 83%

[1]Bouzide, A., Sauve, G. *OL* **4**, 2329 (2002).
[2]Glatzhofer, D.T., Roy, R.R., Cossey, K.N. *OL* **4**, 2349 (2002).
[3]Falck, J.R., Mohapatra, S., Bondlela, M., Venkataraman, S.K. *TL* **43**, 8149 (2002).

Silver trifluoromethanesulfonate. 13, 274–275; **14,** 282–283; **16,** 302; **17,** 314; **18,** 322–323; **19,** 306; **20,** 342; **21,** 394

Carbonylation.[1] Silver triflate is a promoter for transforming tertiary alcohols to the homologous carboxylic acids under CO, e.g., 1-adamantanol to 1-adamantanecarboxylic acid. Primary and secondary alcohols give ethers. Essentially the same results are obtained on replacing AgOTf with triflic acid, but yields are lower.

Dipyrrylmethanes.[2] Alkylation of pyrroles with 2-(2-benzothiazolyl)methylpyrroles is promoted by AgOTf. If CF_3COOAg is used the reaction time is longer and yields are lower.

E = COOMe BT = 2-benzothiazolyl

AgOTf, Na_2HPO_4, PhH

98%

[1]Mori, H., Mori, A., Xu, Q., Souma, Y. *TL* **43**, 7871 (2002).
[2]Okada, K., Saburi, K., Nomura, K., Tanino, H. *T* **57**, 2127 (2001).

1-Silyl-1-boryl-2-alkenes.

Alkenylsilanes and alkenylboranes.[1] These reagents are useful to build up functionalized carbon chains with an option of retaining either the silyl or the boryl group.

LDA, THF

82%

[1]Shimizu, M., Kitagawa, H., Kurahashi, T., Hiyama, T. *ACIEE* **40**, 4283 (2001).

Sodium. 13, 277; **18,** 323–324; **20,** 342–343; **21,** 395

Arylsodium species. Aryltrialkylammonium salts[1] and aryl phosphates[2] are reduced by Na and the reactive ArNa undergo stannylation, therefore this protocol is suitable for preparation of $ArSnR_3$.

o-Metallation of substituted arenes and heteroarenes by sodium sand in the presence of a 1-chloroalkane supplies an alternative method to that employing alkyllithium reagents.[3]

[1]Chopa, A.G., Lockart, M.T., Silbestri, G. *OM* **20**, 3358 (2001).
[2]Chopa, A.G., Lockart, M.T., Dorn, V.B. *OM* **21**, 1425 (2002).
[3]Gissot, A., Becht, J.-M., Desmurs, J.R., Pevere, V., Wagner, A., Mioskowski, C. *ACIEE* **41**, 340 (2002).

Sodium azide. 18, 325–326; **19,** 307; **20,** 343; **21,** 396

Deacylation.[1] *p*-Nitrobenzoic esters are hydrolyzed under mild conditions using sodium azide in MeOH.[1] *N*-Acyloxazolidin-2-ones also suffer ring cleavage.[2]

O O N O Ph Ph NaN_3 / MeOH 40° O NH OH Ph Ph

95%

[1]Gomez-Vidal, J.A., Forrester, M.T., Silverman, R.B. *OL* **3**, 2477 (2001).
[2]Bouzide, A., Sauve, G. *TL* **43**, 1961 (2002).

Sodium bis(2-methoxyethoxy)aluminum hydride, Red-Al®. 15, 290; **21,** 396

Reduction.[1] For large-scale reduction of esters to aldehydes at room temperature the title reagent is used after modification by pyrrolidine.

[1]Abe, T., Haga, T., Negi, S., Morita, Y., Takayanagi, K., Hamamura, K. *T* **57**, 2701 (2001).

Sodium borohydride. 13, 278–279; **15,** 290; **16,** 304; **18,** 326–327; **19,** 307–309; **20,** 344–345; **21,** 397–398

Reductions. β-Diketones undergo stereoselective reduction by $NaBH_4$ in the presence of albumin (*anti*-selective).[1] When α-diketones are first converted into enol triflates the reduction gives monoalkylated 1,2-diols, as a result of conjugate reduction to hemiacetals that is followed by an intermolecular displacement (epoxide formation) and further reduction.[2]

Ar, COOMe, OTf → $NaBH_4$ / EtOH → Ar, OMe, OH

55 – 71%

A sulfonyl group situated at an α-position of a glutarimide activates the proximal carbonyl toward reduction.[3] A divalent sulfur atom also has such effects as shown by the selective reduction of an ester two or three bonds away but apparently not further apart.[4]

COOEt, S, S, COOMe → $NaBH_4$ / EtOH, Et_2O → OH, S, S, COOMe

86%

2,2-Dichlorocyclobutanones undergo simple reduction with $LiAlH_4$, but the $NaBH_4$ reduction products are different. Actually ring-contracting rearrangement occurs during workup.[5]

O, Cl, Ar, Cl → $NaBH_4$ / MeOH ; NaOH → CHO, Ar, Cl

Hydrogenolysis of RX (X=Br, OMs) is safely done by using $NaBH_4$ in *N*-methylpyrrolidinone.[6]

Reductive amination. Two indirect methods for reductive monoalkylation of amines involve condensation and reductive removal steps, the latter accomplished by $NaBH_4$. One of the methods starts from reaction of the amines with *N,N*-dimethylalkanamide dimethylacetal to form amidine derivatives.[7] For synthesis of Boc-protected primary amines a three-component condensation among aldehydes, *t*-butyl carbamate and sodium *p*-toluenesulfinate is conducted and the products are subject to reduction.[8]

Ts, Ph, NHBoc → $NaBH_4$ / THF → Ph, NHBoc

73%

[1]Benedetti, F., Berti, F., Donati, I., Fregonese, M. *CC* 828 (2002).
[2]Dalla, V., Decroix, B. *TL* **43**, 1657 (2002).
[3]Hsu, R.-T., Cheng, L.-M., Chang, N.-C., Tai, H.-M. *JOC* **67**, 5044 (2002).
[4]Khanapure, S.P., Saha, G., Sivendran, S., Powell, W.S., Rokach, J. *TL* **41**, 5653 (2000).
[5]Verniest, G., Bombeke, F., Kulinkovich, O.G., De Kimpe, N. *TL* **43**, 599 (2002).
[6]Torisawa, Y., Nishi, T., Minamikawa, J. *BMCL* **11**, 2787 (2001).
[7]Zhang, J., Chang, H.-M., Kane, R.R. *SL* 643 (2001).
[8]Bernacka, E., Klepacz, A., Zwierzak, A. *TL* **42**, 5093 (2001).

Sodium borohydride–metal salt. 21, 398–399

Deallylation. Allyl ethers are cleaved by $NaBH_4$-LiCl at room temperature.[1]

Hydrodehalogenation. An alternative reagent to Bu_3SnH for hydrodehalogenation is $NaBH_4$-$InCl_3$.[2]

Conjugate reduction. Activated double bonds (enones, α,β-unsaturated nitriles, etc.) can be reduced with the $NaBH_4$-$InCl_3$ combination[3] (perhaps Cl_2InH species generated in situ).

A route to γ-hydroxyalkyl-γ-butyrolactones involves vinylogous aldol reaction of 2-phenylseleno-2-buten-4-olide and treatment of the products with $NaBH_4$-$NiCl_2$.[4]

SePh
Ph
HO
O
O
$NaBH_4$ - $NiCl_2$
Ph
HO
O
O
90%

Reduction of nitrogen functionalities. In combination with phthalocyanatoiron(II) complexes and 2-bromoethanol, $NaBH_4$ reduces $ArNO_2$.[5] Cyano groups are not affected.

The combination of $NaBH_4$ and phenyl chloroformate is very useful for reductive acylation of pyridines to the 1,2-dihydro derivatives.[6] However, in the presence of ethyl or benzyl chloroformate the reduction is reported to furnish a mixture of various isomers. For further reduction of the dihydropyridines to tetrahydropyridines the $NaBH_4$–CF_3COOH combination can be used.

COOMe
N
$NaBH_4$
ClCOOPh
−78°
COOMe
N
COOPh
97%
$NaBH_4$ - CF_3COOH
PhH 5°
COOMe
N
COOPh
99%

Semireduction of alkynes. Terminal triple bonds are reduced by $NaBH_4$ in the presence of $(Ph_3P)_4Pd$.[7]

CHO Ph $NaBH_4$ - $(Ph_3P)_4Pd$ NaOH / aq. MeOH OH Ph 90%

In Pd-catalyzed coupling reaction $NaBH_4$ can be used to maintain Pd species in the low-valent state.[8]

I HOOC $Pd(OAc)_2$ - Ph_3P $NaBH_4$ Cs_2CO_3 / DMSO O O 64%

[1]RajaRam, S., Chary, K.P., Salahuddin, S., Iyengar, D.S. *SC* **32**, 133 (2002).
[2]Inoue, K., Sawada, A., Shibata, I., Baba, A. *JACS* **124**, 906 (2002).
[3]Ranu, B.C., Samanta, S. *TL* **43**, 7405 (2002).
[4]Bella, M., Piancatelli, G., Squarcia, A. *T* **57**, 4429 (2001).
[5]Wilkinson, H.S., Tanuory, G.J., Wald, S.A., Senanayake, C.H. *TL* **42**, 167 (2001).
[6]Zhao, G., Deo, U.C., Ganem, B. *OL* **3**, 201 (2001).
[7]Gu, W.X., Chen, X.C., Pan, X.F., Wu, A.X. *SC* **31**, 1983 (2001).
[8]Bouyssi, D., Balme, G. *SL* 1191 (2001).

Sodium bromate. 18, 330; **20,** 347; **21,** 400

Phthalides. *o*-Alkylbenzoic acids are oxidized to phthalides by the title reagent.[1]

[1]Hayat, S., Attaa-ur-Rahman, Choudhary, M.I., Khan, K.M., Bayer, E. *TL* **42**, 1647 (2001).

Sodium chlorodifluoroacetate.

1,3-Difluorobenzenes.[1] Pyrolytic decomposition of the sodium salt liberates difluorocarbene. Trapping the carbene with cyclobutenes leads to difluoroarenes via highly strained adducts which decompose to fluorocyclopentadienes that are also reactive toward difluorocarbene.

F Cl COONa F Δ [F F :] R R R F R F

[1]Morrison, H.M., Rainbolt, J.E., Lewis, S.B. *OL* **4**, 3871 (2002).

Sodium cyanoborohydride. 21, 401

Reduction of triacylmethanes.[1] Deoxygenation of the side chain carbonyl group of 2-acyl-1,3-cycloalkanediones occurs on treatment with $NaBH_3CN$.

$NaBH_3CN$

THF / 2N HCl

90%

Reduction of imines. A synthesis of aziridines is based on reductive cyclization of α-chloroimines.[2] A one-electron reduction process initiates cyclization leading to α-carbolines.[3]

$NaBH_3CN$

MeOH

NBn_2

60%

NaH

$NaBH_3CN$

dioxane

60%

[1]Pashkovsky, F.S., Lokot, I.P., Lakhvich, F.A. *SL* 1391 (2001).
[2]Concellon, J.M., Bernad, P.L., Riego, E., Garcia-Granda, S., Forcen-Acebal, A. *JOC* **66**, 2764 (2001).
[3]Ono, A., Narasaka, K. *CL* 146 (2001).

Sodium dithionite.

Polyfluoroalkylation.[1] Introduction of a polyfluoroalkyl chain to an aromatic nucleus can be accomplished with $Na_2S_2O_4$ as initiator.

Halogenation and hydrodehalogenation. Bromination at the glycosidic center of carbohydrates[2] is achievable by a combination of $Na_2S_2O_4$ and $KBrO_3$. 2-Deoxy-α-glycosides are synthesized from glycals via haloetherification and removal of the halogen atom with $Na_2S_2O_4$.[3] Thus, hydrodehalogenation by $Na_2S_2O_4$ is not limited to α-haloketones.

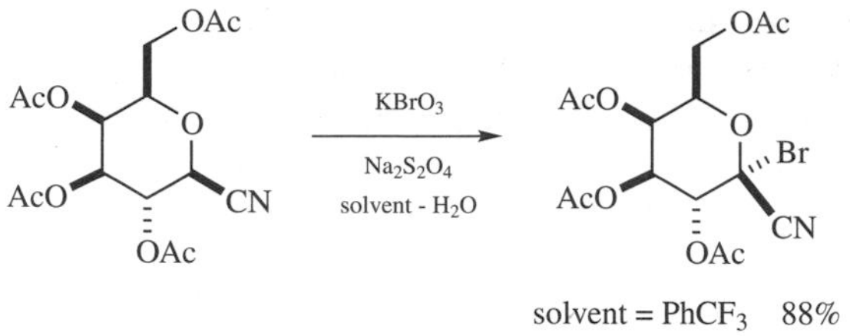

solvent = $PhCF_3$ 88%

Radical addition.[4] Addition of polyhaloalkanes (e.g., $BrCCl_3$) to alkenes is conveniently initiated by $Na_2S_2O_4$.

[1]Huang, X.-T., Long, Z.-Y., Chen, Q.-Y. *JFC* **111**, 107 (2001).
[2]Czifrak, K., Somsak, L. *TL* **43**, 8849 (2002).
[3]Costantino, V., Fattorusso, E., Imperatore, C., Mangoni, A. *TL* **43**, 9047 (2002).
[4]Wu, F.H., Huang, W.-Y. *JFC* **110**, 59 (2001)

Sodium hexamethyldisilazide, NaHMDS. 18, 332; **20,** 349; **21,** 401–402

N-Aryl amides.[1] Aminolysis of esters and lactones with $ArNH_2$ is promoted by NaHMDS.

[1]Wang, J., Rosingana, M., Discordia, R.P., Soundararajan, N., Polniaszek, R. *SL* 1485 (2001).

Sodium hydride. 14, 288; **16,** 307–308; **18,** 333; **19,** 312–313; **20,** 349–350; **21,** 402–404

Alkylation. Monoallylation of 1,2-diols is observed from reaction with NaH and 1,3-dibromopropane. The unusual elimination following the first alkylation is likely an intramolecular process.[1]

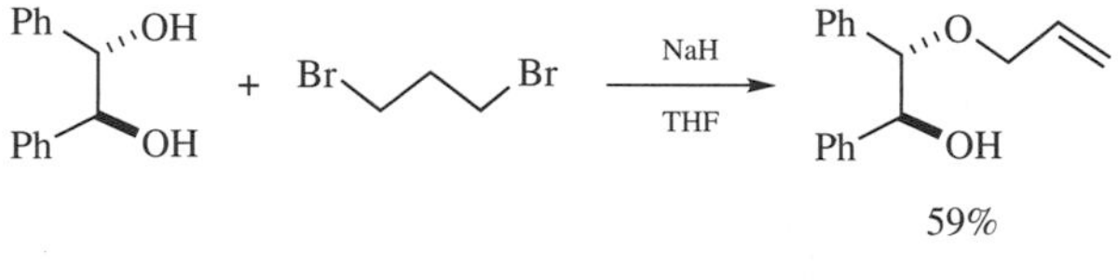

59%

An intramolecular *N*-alkylation leading to α-lactams is induced by the NaH/15-crown-5 system.[2]

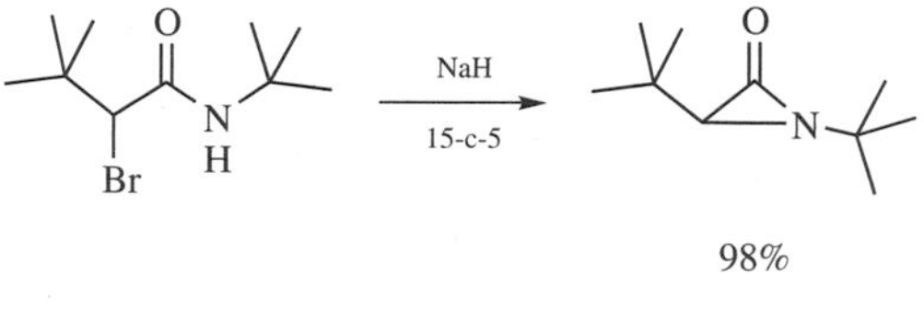

98%

Ester cleavage. Treatment of *t*-butyl esters[3] and 2-trimethylsilylethyl esters[4] with NaH in DMF at room temperature causes their decomposition to generate carboxylic acids.

Hydrodeamination. Arylamines undergo reduction when subjected to *N*-mesylation and treatment with NaH and NH_2Cl in DMF.[5]

[1]Jha, S.C., Joshi, N.N. *JOC* **67**, 3897 (2002).
[2]Cesare, V., Lyons, T.M., Lengyel, I. *S* 1716 (2002).
[3]Paul, S., Schmidt, R.R. *SL* 1107 (2002).
[4]Serrano-Wu, M.H., Regueiro-Ren, A., St. Laurent, D.R., Carroll, T.M., Balasubramanian, B.N. *TL* **42**, 8593 (2001).
[5]Wang, Y., Guziec, F.S. *JOC* **66**, 8293 (2001).

Sodium hypochlorite. 15, 293; **16,** 308; **17,** 316; **18,** 334–335; **19,** 313; **20,** 350; **21,** 404

α-Keto amides. α-Dialkylamino-β-acyl nitriles which are readily prepared from condensation of dialkylaminoacetonitriles with (nonenolizable) esters undergo oxidative decyanation on exposure to NaOCl.[1]

COOMe + NEt2 CN → NaN(SiMe3)2 / THF ; NaClO / H2O → CONEt2

71%

Note that α-hydroxy esters are oxidized to α-keto esters by NaOBr.[2]

Ring expansion. Treatment of certain 1-isopropenylcycloalkanols with NaOCl-HOAc leads to the homologous 2-chloromethyl-2-methylcycloalkanones which is the result of ring expansion and incorporation of a chlorine atom.[3]

OMe OH → NaOCl - HOAc; H2O - CCl4; 0° → OMe O Cl + OMe O Cl

(20 : 1)
86%

[1]Yang, Z., Zhang, Z., Meanwell, N.A., Kadow, J.F., Wang, T. *OL* **4**, 1103 (2002).
[2]Chang, H.S., Woo, J.C., Lee, K.M., Ko, Y.K., Moon, S.-S., Kim, D.-W. *SC* **32**, 31 (2002).
[3]Ruggles, E.L., Maleczka Jr., R.E. *OL* **4**, 3899 (2002).

Sodium iodide. 21, 404

Lepidopterene.[1] A convenient preparation of the hydrocarbon is by heating 9-chloromethylanthracene with NaI in acetone.

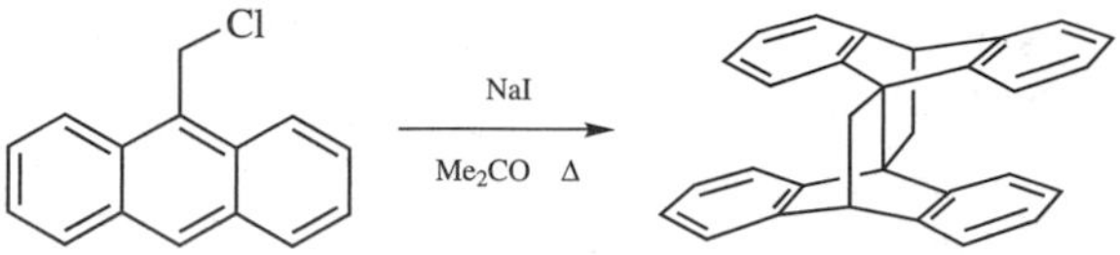

NaI
Me_2CO Δ

75%

Iodination.[2] With the water-soluble organotelluride **1** to trap and transfer positive iodine generated from NaI and H_2O_2, it is easy to iodinate activated arenes. Iodoetherification and iodolactonization are readily accomplished with the system.

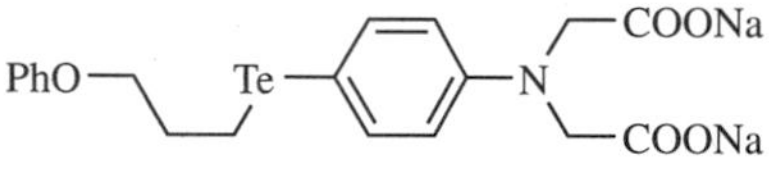

PhO Te N COONa COONa

1

[1]Fernandez, M.-J., Gude, L., Lorente, A. *TL* **42**, 891 (2001).
[2]Higgs, D.E., Nelen, M.I., Detty, M.R. *OL* **3**, 349 (2001).

Sodium malonyloxyborohydride.

Hydroboration. The reagent (**1**), obtained by mixing $NaBH_4$ with malonic acid in THF at room temperature, can be used directly for hydroboration of alkenes.[1]

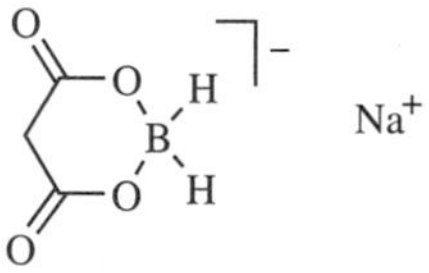

Na^+

1

[1]Huang, S.-W., Peng, W.-L., Shan, Z.-X., Zhao, D.-J. *NJC* **25**, 869 (2001).

Sodium 2-methoxycarbonylethylsulfinate.

Organosulfinates.[1] Prepared from methyl 3-mercaptopropanoate and methyl vinyl ketone in three steps the reagent reacts with organohalogen compounds to give $RSO_2CH_2CH_2COOMe$ (the aromatic sulfones by a CuI-catalyzed reaction) which are degraded to RSO_2Na by NaOMe. Upon further reaction with NH_2OSO_3H the corresponding sulfonamides are produced.

[1]Baskin, J.M., Wang, Z. *TL* **43**, 8479 (2002).

Sodium nitrite. 21, 405

Nitrosation. Hydrodeamination of arylamines results when $NaHSO_3$ is present in the nitrosation media.[1] Silica-sulfuric acid that is derived from $ClSO_3H$ and silica can be used with $NaNO_2$ to perform nitrosation of amines[2] and thiols.[3]

Epoxide opening. The $NaNO_2$–$MgSO_4$/MeOH system is effective for the conversion of epoxides to 2-nitroalkanols.[4]

[1]Geoffrey, O.J., Morinelli, T.A., Meier, G.P. *TL* **42**, 5367 (2001).
[2]Zolfigol, M.A., Bamoniri, A. *SL* 1621 (2002).
[3]Zolfigol, M.A. *T* **57**, 9509 (2001).
[4]Kalita, B., Barua, N.C., Bezbarua, M., Bez, G. *SL* 1411 (2002).

Sodium perborate. 14, 290–291; **16,** 310; **18,** 337–338; **19,** 314; **20,** 351

Fluorous phenyliodine(III) diacetates. A preparation of these compounds from arenes containing polyfluoroalkyl chains involves iodination and oxidative acetoxylation[1] with $NaBO_3$ in HOAc.

[1]Rocaboy, C., Gladysz, J.A. *CEJ* **9,** 88 (2003).

Sodium periodate. 15, 294; **18,** 338–339; **19,** 315; **21,** 405

Tandem degradation-homologation. One-carbon oxidative homologation of 1,2-diols to afford 2-alkenoic esters is achieved on exposure to $NaIO_4$-SiO_2 and stabilized Wittig reagents together.[1,2] The advantage is that two reactions are completed in one operation.

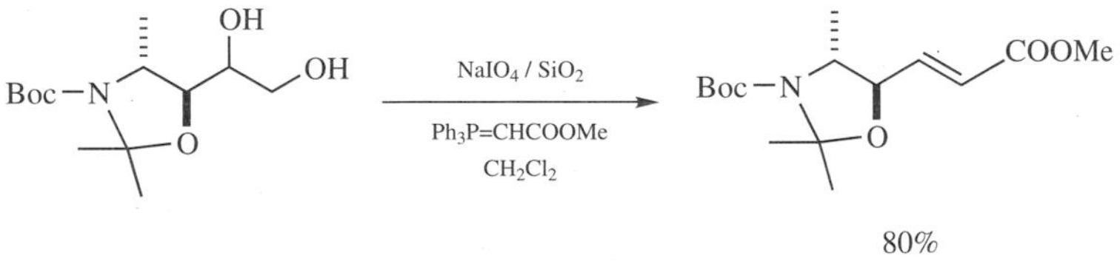

$NaIO_4$ / SiO_2, $Ph_3P{=}CHCOOMe$, CH_2Cl_2

80%

Oxidation. Cr(VI)-catalyzed oxidation of homoallylic and homopropargylic alcohols with $NaIO_4$ does not cause migration of the multiple bond.[3] The RuO_2-$NaIO_4$ combination is a powerful reagent for oxygenation of polyenes, thus a product (non-meso) containing five tetrahydrofuran units and two hydroxyl groups is obtained from squalene.[4]

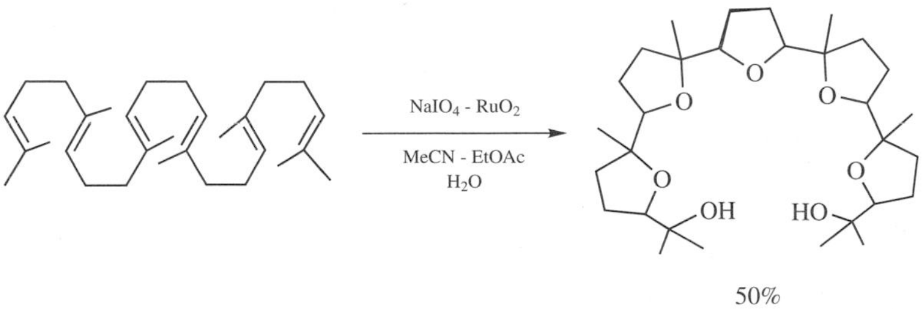

$NaIO_4$ - RuO_2, MeCN - EtOAc, H_2O

50%

Deallylation.[5] Allyl ethers including allyl glycosides and allyl amides are degraded on reaction with OsO_4-NMO-$NaIO_4$ in aqueous dioxane via a sequence of oxidative cleavage (the intermediates in the enol form).

[1]Dunlap, N.K., Mergo, W., Jones, J.M., Carrick, J.D. *TL* **43**, 3923 (2002).
[2]Outram, H.S., Raw, S.A., Taylor, R.J.K. *TL* **43**, 6185 (2002).
[3]Schmieder-van de Vondervoort, L., Bouttemy, S., Padron, J.M., Le Bras, J., Muzart, J., Alsters, P.L. *SL* 243 (2002).
[4]Bifulco, G., Caserta, T., Gomez-Paloma, L., Piccialli, V. *TL* **43**, 9265 (2002).
[5]Kitov, P.I., Bundle, D.R. *OL* **3**, 2835 (2001).

Sodium sulfide.

Thiophenes.[1] With basic Na_2S in DMSO a 1,3-diyne incorporates [S] and forms a thiophene nucleus. This reaction is the key step in a synthesis of tetra(2-thienyl)methane.

Na_2S - NaOH
DMSO 120°

50 – 55%

[1]Matsumoto, K., Nakaminami, H., Sogabe, N., Kurata, H., Oda, M. *TL* **43**, 3049 (2002).

Sodium tetrachloroaurate.

Pyrroles.[1] Condensation of amines with 2-propynyl-1,3-dicarbonyl compounds under the influence of $NaAuCl_4$ involves C-N bond formation between the amines and an *sp*-carbon prior to cylization.

$NaAuCl_4 \cdot 2H_2O$
EtOH 40°

98%

[1]Arcadi, A., Di Giuseppe, S., Marinelli, F., Rossi, E. *TA* **12**, 2715 (2001).

Sodium triacetoxyborohydride. 13, 283; **16,** 309–310; **18,** 340; **19,** 315–316; **20,** 352; **21,** 406

Reductiive amination.[1] *N*-Alkylation of amines is conveniently carried out with carbonyl compounds in the presence of $NaBH(OAc)_3$ in HOAc. Thus, methyl 4-amino-1-cyclohexenecarboxylates are readily prepared from Danishefsky's diene.

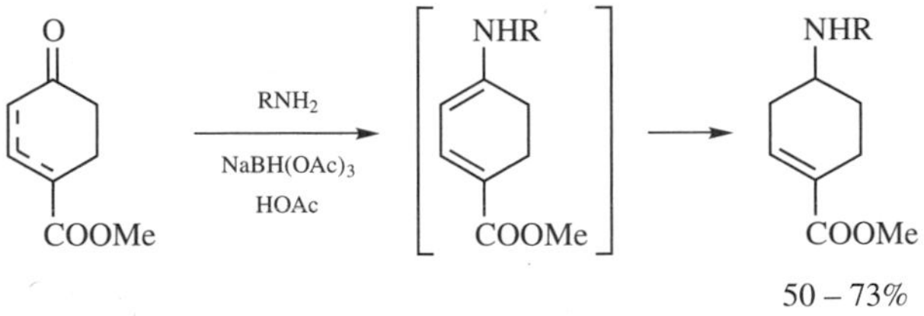

Ozonolysis workup.[2] The use of $NaBH(OAc)_3$ to reduce ozonides of trisubstituted cycloalkenes leads to keto alcohols, rendering differentiation of oxidation states at the two termini.

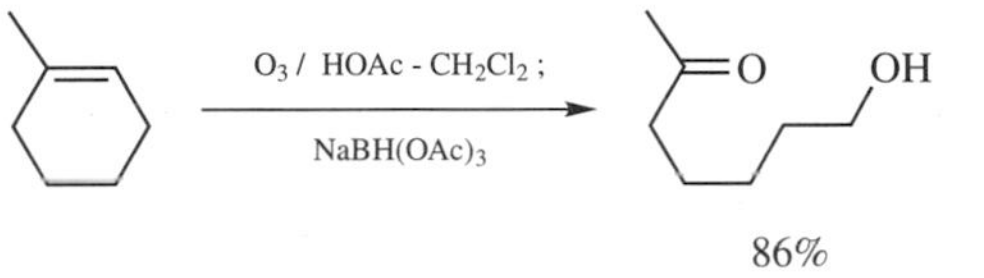

[1]Quirante, J., Vila, X., Bonjoch, J. *S* 1971 (2001).
[2]Ishmuratov, G.Yu., Kharisov, R.Ya., Yakovleva, M.P., Botsman, O.V., Muslukhov, R.R., Tolstikov, G.A. *RJOC* **37**, 37 (2001).

Sodium triethylgermanate. 19, 316; **20,** 353

Cleavage of p-cyanobenzyl group.[1] Alcohols, amines and thiols protected as *p*-cyanobenzyl derivatives are readily regenerated by treatment with Et_3GeNa.

[1]Yokoyama, Y., Takizawa, S., Nanjo, M., Mochida, K. *CL* 1032 (2002)

Squaric acid.

Tetrahydroquinolines.[1] A 3-component condensation of arylamines, glyoxylic esters, and enamides catalyzed by squaric acid is superior to other Lewis acids (e.g., $InCl_3$).

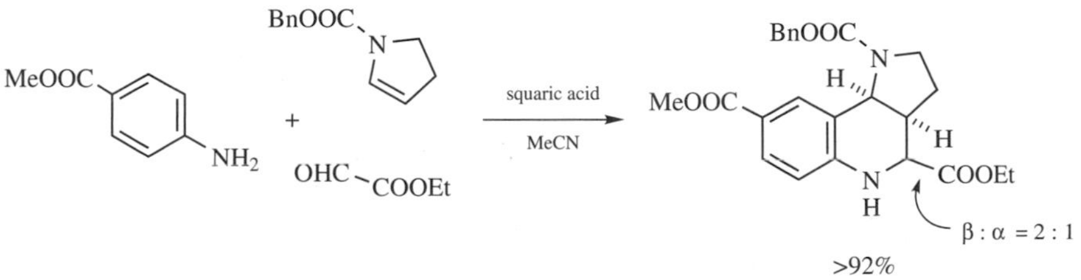

[1]Xia, C., Heng, L., Ma, D. *TL* **43**, 9405 (2002).

Sulfur. 15, 297; **18,** 341–342; **20,** 353–354

Imidazolidine-2-thiones. Heating imidazolidines with sulfur to 150° gives the cyclic thioureas.[1]

Oxazolidin-2-ones. Sulfur catalyzes reaction of 2-amino alcohols with CO that leads to carbonylcyclization after exposure to oxygen.[2]

Thiacalix[4]arenes. Heating 4-substituted phenols with sulfur and NaOH in tetramethyleneglycol dimethyl ether results in the formation of the thiacalixarenes.[3]

S_8 / NaOH, tetramethylene glycol dimethyl ether, Δ

28%

[1]Denk, M.K., Gupta, S., Brownie, J., Tajammul, S., Lough, A.J. *CEJ* **7**, 4477 (2001).
[2]Mizuno, T., Takahashi, J., Ogawa, A. *T* **58**, 7805 (2002).
[3]Shokova, E., Tafeeko, V., Kovalev, V. *TL* **43**, 5153 (2002).

Sulfur trioxide. 20, 354

Benzils. Diarylethynes give benzils in one step on treatment with the SO_3-dioxane complex.[1] Lower yields are obtained if the aromatic rings bear electron-withdrawing substituents.

β-Amino sulfonic acids.[2] Alkenes are functionalized on reaction with the SO_3-DMF complex and MeCN. Hydrolysis of the adducts affords β-amino sulfonic acids.

[1]Rogatchov, V.O., Filimonov, V.D., Yusubov, M.S. *S* 1001 (2001).
[2]Cordero, F.M., Cacciarini, M., Machetti, F., De Sarlo, F. *EJOC* 1407 (2002).

Sulfuryl chloride. 13, 284; **14,** 291–292; **16,** 311; **18,** 342; **20,** 354

Organosulfinyl chlorides. Various 2-trimethylsilylethyl sulfoxides are cleaved by SO_2Cl_2.[1]

Benzyl chloroformate. A preparation of this chloroformate from benzyl alcohol is via a xanthate. The xanthate undergoes chlorodesulfurization at about room temperature.[2]

α-Chloroalkanoic acids.[3] To convert acylphosphonates into α-chloroalkanoic acids it requires chlorination with SO_2Cl_2 and then oxidative cleavage of the C-P bond with H_2O_2.

SO_2Cl_2 → H_2O_2 → 85%

[1]Schwan, A.L., Strickler, R.R., Dunn-Dufault, R., Brillon, D. *EJOC* 1643 (2001).
[2]Mizuno, T., Takahashi, J., Ogawa, A. *T* **58**, 10011 (2002).
[3]Stevens, C.V., Vanderhoydonck, B. *T* **57**, 4793 (2001).

T

Tantalum(V) chloride. 21, 407

Conjugate addition.[1] CC bond formation is achieved at the β-position of enones by a $TaCl_5$–promoted conjugate addition with organostannanes. Tantalum(V) chloride can be used in catalytic amounts if Me_3SiCl is present.

Alkene trimerization.[2] Organometallic species derived from $TaCl_5$ and RLi (or Me_2Zn, Me_4Sn) stitches three ethylene molecules together to form 1-hexene.

[1]Shibata, I., Kano, T., Kanazawa, N., Fukuoka, S., Baba, A. *ACIEE* **41**, 1389 (2002).
[2]Andes, C., Harkins, S.B., Murtuza, S., Oyler, K., Sen, A. *JACS* **123**, 7423 (2001).

Tellurium(IV) chloride.

Aryltellurium compounds.[1] Simple access to $ArTeCl_3$ from $ArB(OH)_3$ involves heating them with $TeCl_4$ in $MeNO_2$. In turn the aryltellurium chlorides are susceptible to reduction (with $NaHSO_3$) to give ArTeTeAr.

[1]Clark, A.R., Nair, R., Fronczek, F.R., Junk, T. *TL* **43**, 1387 (2002).

Tetrabutylammonium alkyl carbonate.

Dialkyl carbonates. Nucleophilic attack on alkyl halides by the title reagents represents a mild and safe protocol for the preparation of ROCOOR′.

[1]Verdecchia, M., Feroci, M., Palombi, L., Rossi, L. *JOC* **67**, 8287 (2002).

Tetrabutylammonium bromide. 20, 356; **21,** 407

Condensation reactions.[1] The Barton-Zard pyrrole synthesis is improved by adding Bu_4NBr.

O_2N OAc + CN COOEt —(K_2CO_3 / Bu_4NBr)→ N H COOEt

86%

Fiesers' Reagents for Organic Synthesis, Volume 22. Series editor Tse-Lok Ho
ISBN 0-471-28515-3

Heck reaction.[2] The Preparation of β,β-diarylacrylic esters from cinnamic esters has been carried out at 130°. Molten Bu_4NBr is the reaction medium.

[1]Bobal, P., Lightner, D.A. *JHC* **38**, 527 (2001).
[2]Calo, V., Nacci, A., Monopoli, A., Lopez, L., di Cosmo, A. *T* **57**, 6071 (2001).

Tetrabutylammonium butyldifluorodimethylsilicate.

Alkyl fluorides. The title reagent, $Bu_4N[Me_2Si(Bu)F_2]$, is a nucleophilic fluorinating agent.

[1]Kvicala, J., Mysik, P., Paleta, O. *SL* 545 (2001).

Tetrabutylammonium fluoride, TBAF. 13, 286–287; **14,** 293–294; **15,** 298, 304; **17,** 324–326; **18,** 344–345; **19,** 319–321; **20,** 357–359; **21,** 407–409

Hydrolysis. Removal of *N*-Boc group[1] and selective hydrolysis of esters at C-4 of monosaccharides[2] by Bu_4NF have been observed.

Selective cleavage of the C-5 sulfonate ester group in substituted coumarins while leaving the C-7 counterpart intact under very mild conditions is valuable, and this can be performed with Bu_4NF at 0°.[3]

Bu_4NF / THF, 0°

R = Me, *p*-Tol

Condensation reactions. Masked formyl anion is generated from 2-trimethylsilyl-1,3-dithiolane by Bu_4NF, immediate alkylation with aldehydes proceeds in variable yields.[4]

Group exchange. Trialkylsilyl esters are converted into alkyl esters[5] using Bu_4NF to liberate carboxylate anions for in situ *O*-alkylation. Another technique pertains to generating nascent Bu_4NF from Bu_4NHSO_4 and $KF \cdot 2H_2O$ to perform the esterification.[6]

In an enyne synthesis by Pd/Ag-catalyzed coupling of 1-trialkylsilylalkynes with enol triflates Bu_4NF is necessary for its desilylating action.[7]

Rearrangement. 3-Phenylpropanals are obtained when β-(phenyldimethylsilyl)-acroleins are exposed to Bu_4NF. There is phenyl migration followed by desilylation.[8]

Bu_4NF / DMSO (1 min)

68%

[1]Routier, S., Sauge, L., Ayerbe, N., Coudert, G., Merour, J.-Y. *TL* **43**, 589 (2002).
[2]Graziani, A., Passacantilli, P., Piancatelli, G., Tani, S. *TL* **42**, 3857 (2001).
[3]Fox, M.E., Lennon, I.C., Meek, G. *TL* **43**, 2899 (2002).
[4]Degl'Innocenti, A., Capperucci, A., Nocentini, T. *TL* **42**, 4557 (2001).
[5]Ooi, T., Sugimoto, H., Maruoka, K. *H* **54**, 593 (2001).
[6]Ooi, T., Sugimoto, H., Doda, D., Maruoka, K. *TL* **42**, 9245 (2001).
[7]Bertus, P., Halbes, U., Pale, P. *EJOC* 4391 (2001).
[8]Aronica, L.A., Morini, F., Caporusso, A.M., Salvadori, P. *TL* **43**, 5813 (2002).

Tetrabutylammonium iodide.

Catalyst roles. The iodide ion from Bu_4NI is important in promoting cyclization of unsaturated *N*-chloramines to 3-chloropiperidines,[1] and in serving as a ligand to the Rh-catalyzed aminolysis it suppresses catalyst poisoning and increases reactivity.[2]

O + R_2NH —[$[(cod)RhCl]_2$ / DPPF / THF / Bu_4NI, CSA]→ R_2N … OH

[1]Noack, M., Göttlich, R. *EJOC* 3171 (2002).
[2]Lautens, M., Fagnou, K. *JACS* **123**, 7170 (2001).

Tetrabutylammonium peroxydisulfate. 19, 322; **21,** 409–410

Deallylation. A new way to cleave allyl ethers under basic conditions is via oxidation to acrylic esters followed by methanolysis. A suitable oxidizing agent is $(Bu_4N)_2S_2O_8$.[1,2]

Nitriles. Conversion of primary alcohols to nitriles in one step uses a combination of $(Bu_4N)_2S_2O_8$, $HOCOONH_4$, and formates of copper and nickel as catalysts.[3]

[1]Chen, F-E., Ling, X.-H., He, Y.-P., Peng, X.-H. *S* 1772 (2001).
[2]Yang, S.G., Park, M.Y., Kim, Y.H. *SL* 492 (2002).
[3]Chen, F-E., Li, Y.-Y., Xu, M., Jia, H.-Q. *S* 1804 (2002).

Tetrabutylphosphonium chloride.

Alkylchlorosilanes.[1] A preparation of R_3SiCl from RCl and $HSiCl_3$ is by heating them with Bu_4PCl in a sealed bomb. Quaternary ammonium salts are less effective catalysts.

[1]Cho, Y.S., Kang, S.-H., Han, J.S., Yoo, B.R., Jung, I.N. *JACS* **123**, 5584 (2001).

2,3,5,6-Tetrachloropyridine.

Nitriles. Dehydration of aldoximes is effected with the title reagent supported on alumina under microwave irradiation for a short period. The catalyst is reusabe.

[1]Lingaiah, N., Narender, R. *SC* **32**, 2391 (2002).

2,4,6,8-Tetraiodo-2,4,6,8-tetreazabicyclo[3.3.0]octane-3,7-dione.

Iodination.[1] The title reagent **1** serves as an iodinating agent for arenes.

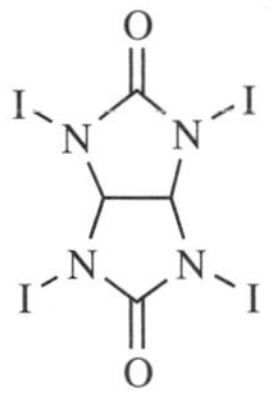

1

[1]Chaikovski, V.K., Filimonov, V.D., Yagovkin, A.Y., Ogorodnikov, V.D. *RCB* **50**, 2411 (2001).

Tetrairidium dodecacarbonyl.

2-Substituted 1-methylimidazoles.[1] Siloxyalkylation at C-2 of the heterocycle is performed by reaction with aldehydes in the presence of a hydrosilane. Benzannulated imidazoles react similarly. When the aldehydes are replaced with isocyanate esters the products are imino silyl ethers (readily hydrolyzed to carboxamides).

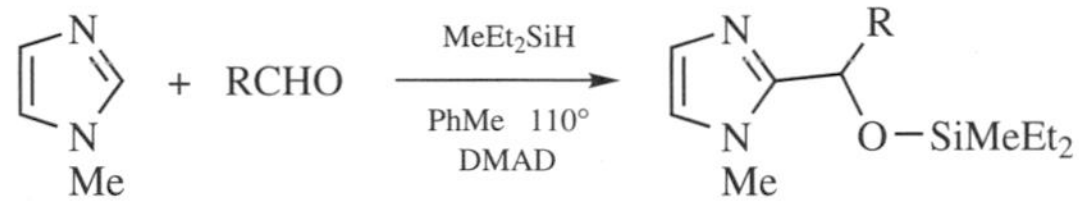

+ RCHO $\xrightarrow[\text{PhMe } 110° \text{ DMAD}]{MeEt_2SiH}$ O−$SiMeEt_2$

[1]Fukumoto, Y., Sawada, K., Hagihara, M., Chatani, N., Murai, S. *ACIEE* **41**, 2779 (2002).

Tetrakis(trimethylphosphinecobalt).

β-Ketophosphonates.[1] The title reagent $[(Me_3P)Co]_4$ promotes reaction of iodomethylphosphonates with carboxylic esters. Magnesium can be used but product yields are lower.

[1]Orsini, F., Di Teodoro, E., Ferrari, M. *S* 1683 (2002).

Tetrakis(triphenylphosphine)palladium(0). 13, 289–294; **14,** 295–299; **15,** 300–304; **16,** 317–323; **17,** 327–331; **18,** 347–349; **19,** 324–331; **20,** 362–368; **21,** 411–417

Allylic displacements. Microencapsulated $(Ph_3P)_4Pd$ is effective as displacement catalyst (also for Suzuki coupling).[1] Preparation of 4-azido-2-alkanenitriles from 2-alkenals involves derivatization into cyanohydrin carbonates and the Pd-catalyzed reaction with NaN_3.[2] Cleavage of resin-bound benzylic/allylic ethers by a Pd-catalyzed reaction with amines is adequate.[3] When an *N*-allyloxycarbonyl α-amino ester, the Pd(O) catalyst, and DABCO are added to a solution of another *N*-protected amino acid that is just activated, peptide formation is achieved.[4]

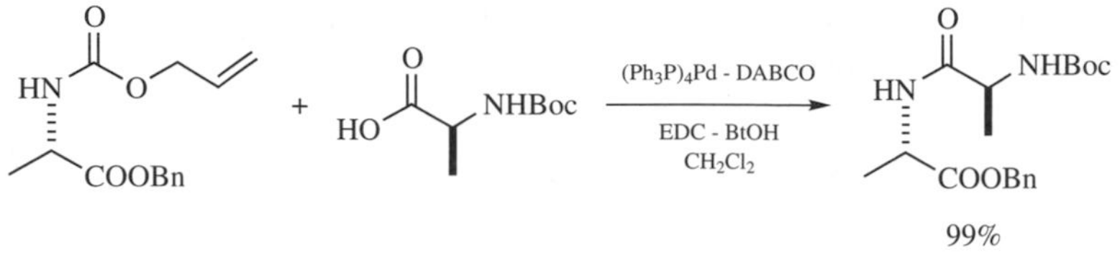

A rearrangement route to 3-substituted *syn*-2-amino-4-alkenoic esters is initiated by Pd-catalyzed allylic displacement of allylic mesylate with *N*-protected glycine.[5]

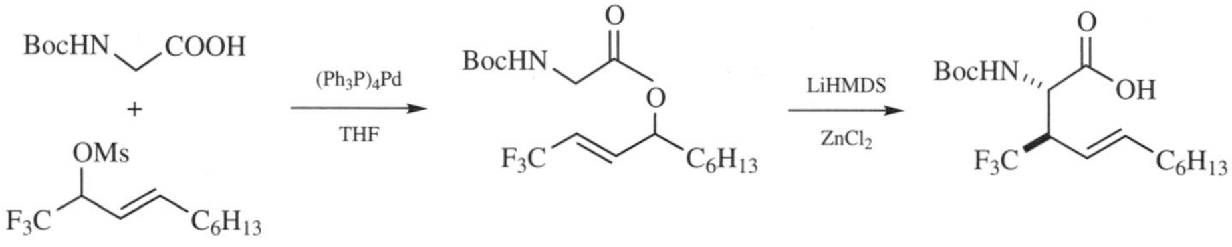

Aromatic substitutions. Diaryliodonium salts are reactive toward various nucleophiles in the presence of $(Ph_3P)_4Pd$. Thus, reaction with thiolates[6] and phosphonates[7] is easily accomplished. Some ArI also can be transformed into arenephosphonates.[8]

An interesting observation[9] on the effect of CuCN on a coupling reaction has been made.

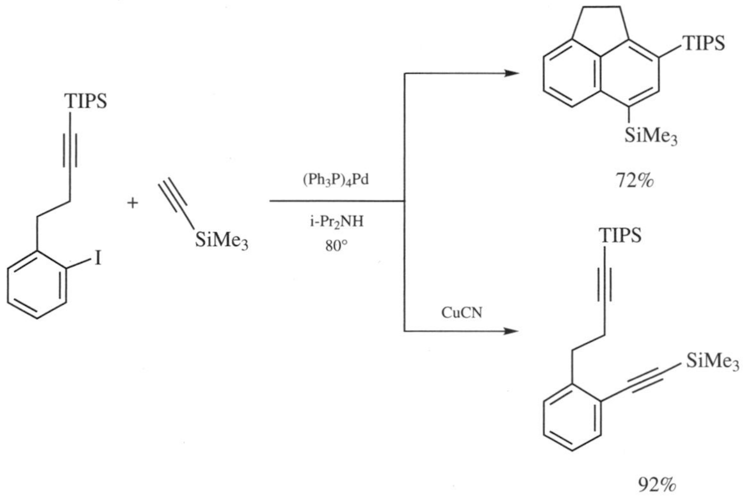

Alkylation. Alkylidenecyclopropanes behave as alkylating agents for ketones[10] and heteroaromatic compounds (furans, thiophenes, thiazoles, pyrroles)[11] in the presence of $(Ph_3P)_4Pd$.

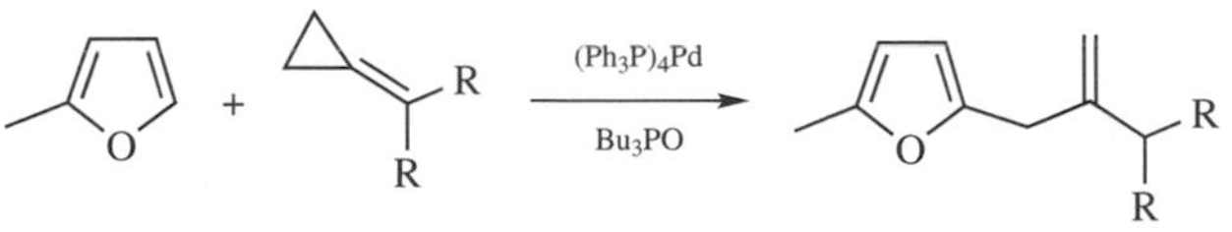

$(Ph_3P)_4Pd$ / Bu_3PO

R = Bu 70%

Alkenyl allyl carbonates undergo decarboxylative rearrangement to deliver α-allyl ketones.[12]

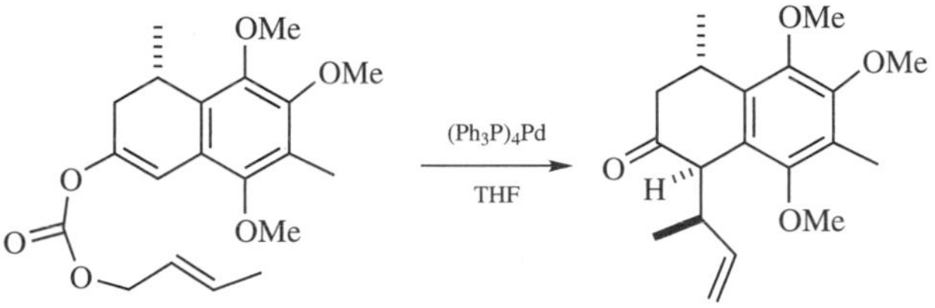

$(Ph_3P)_4Pd$ / THF

58%

Addition reactions. Addition of alcohols of high acidity such as phenols and 2,2,2-trifluoroethanol to internal conjugated diynes occurs at one of the central *sp*-carbon to give enol ethers.[13] The regiochemical preference in Pd-catalyzed hydrophosphinylation of 1-alkynes is influenced by the presence of HOAc.[14]

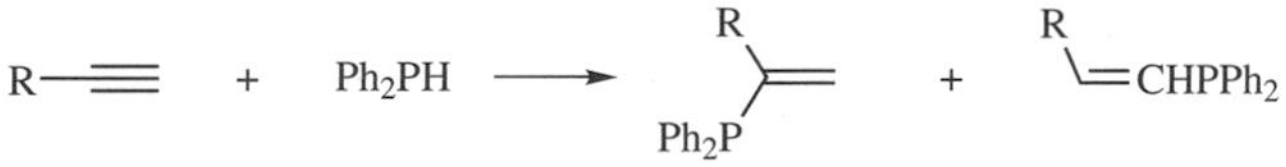

R—≡ + Ph_2PH ⟶ $R(Ph_2P)C{=}CH_2$ + $RCH{=}CHPPh_2$

Conditions	Ratio
$(Ph_3P)_4Pd$ / MeCN 130°	0 : 100 (*E/Z* 14 : 86)
$(Ph_3P)_4Pd$ / HOAc - MeCN 80°	92 : 8 (*E/Z* 50 : 50)
$Ni(acac)_2$ - $(EtO)_2POH$ / PhH 80°	95 : 5 (*E/Z* 100 : 0)

Formation of allylic amines[15] and ethers/esters[16] from 2-alkynes proceeds via isomerization to the allenes and the formation of π-allylpalladium complexes.

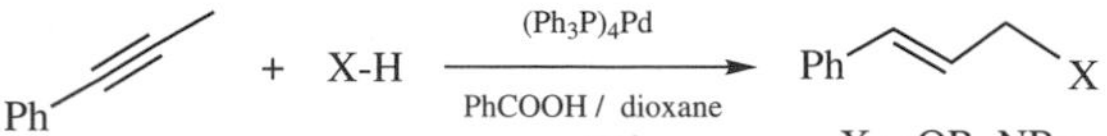

$(Ph_3P)_4Pd$ / PhCOOH / dioxane 100°

X = OR, NR_2

$(Ph_3P)_4Pd$ / PhCOOH / dioxane 100°

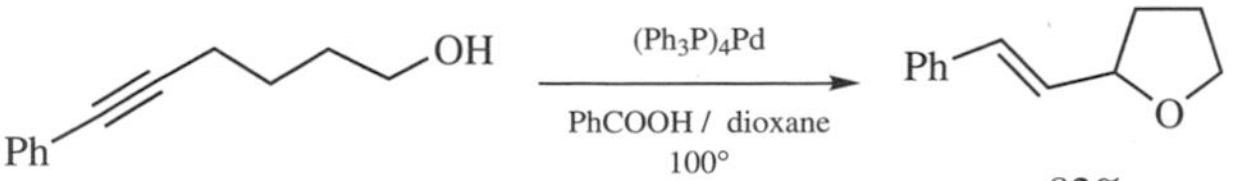

82%

[3+2]Cycloaddition of alkylidenecyclopropanes (with ring opening) with electron-rich aldehydes and imines affords an uncommon type of five-membered heterocycles in one step.[17,18]

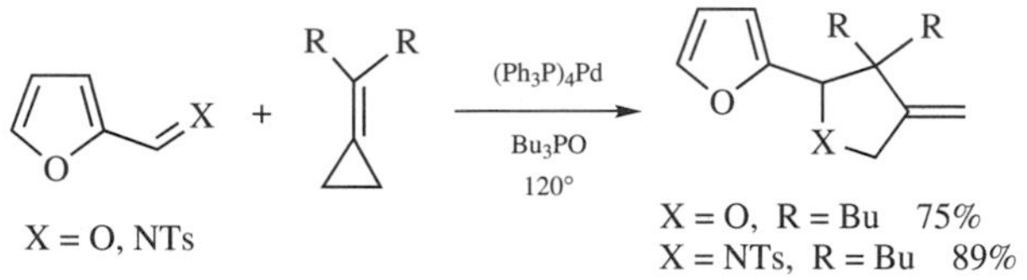

X = O, NTs

$(Ph_3P)_4Pd$
Bu_3PO
120°

X = O, R = Bu 75%
X = NTs, R = Bu 89%

Reduction. Ketone reduction by Bu_2SnH_2 is catalyzed by $(Ph_3P)_4Pd$.[19] Hydrogenation of carboxylic acids to aldehydes is accomplished using $(Ph_3P)_4Pd$ as catalyst and pivalic anhydride as an activator.[20]

Coupling with cyclization. Several types of heterocycle synthesis are executed with Pd-catalysis. Examples include 8-benzylidene-2-oxopyrrolizidines,[21] 4-aryliso-quinolines,[22] and pyridines.[23]

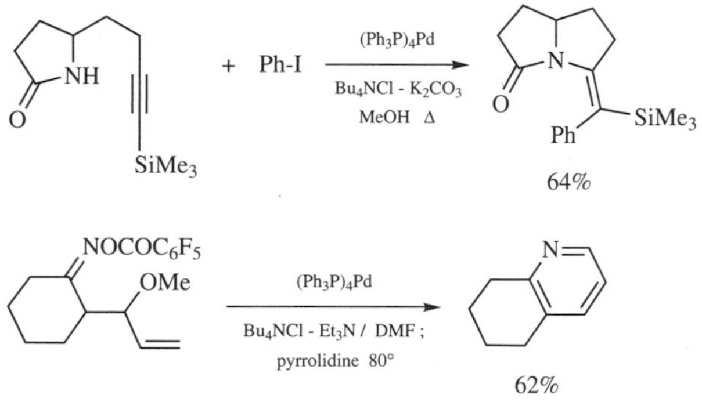

+ Ph-I
$(Ph_3P)_4Pd$
Bu_4NCl - K_2CO_3
MeOH Δ

64%

$NOCOC_6F_5$
OMe
$(Ph_3P)_4Pd$
Bu_4NCl - Et_3N / DMF ;
pyrrolidine 80°

62%

Many catalysts can bring about cyclization of unsaturated *N*-chloroamines, $(Ph_3P)_4Pd$ is one of them.[24] Benzannulation of bis(enyne) derivatives promoted by the catalyst is more interesting and useful.[25]

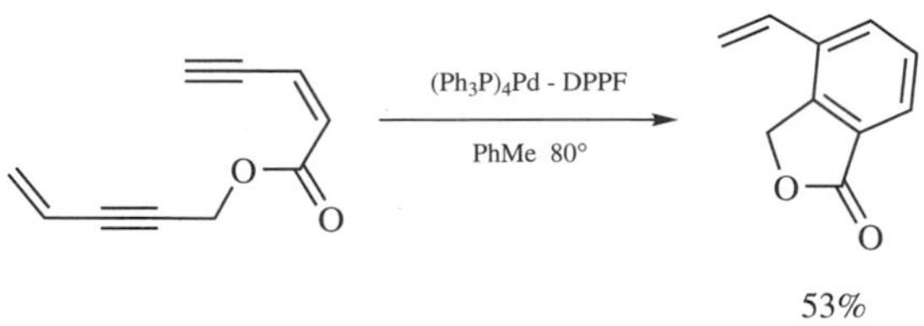

$(Ph_3P)_4Pd$ - DPPF
PhMe 80°

53%

Heck reaction. Tethered technique for intramolecularization of the Heck reaction is highly effective.[26]

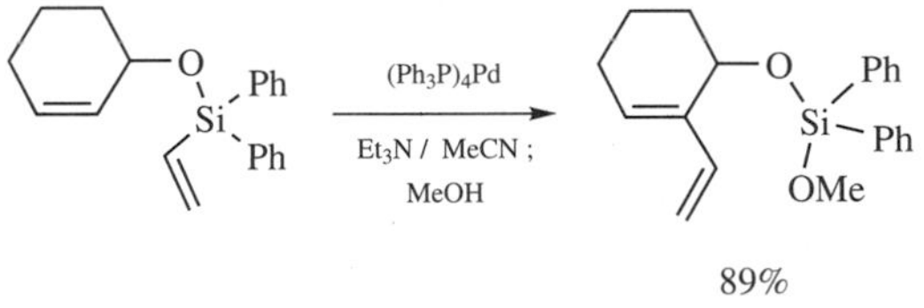

(Ph3P)4Pd
Et3N / MeCN ; MeOH

89%

Most of the recent developments in Heck reaction concern with allene type substrates. Arylamination (intermolecular[27] and intramolecular[28]), aryletherification[29] and aryliminoetherification[30] are realized. Formation of alkenylcyclopropane derivatives[31,32] or cyclopentenes[31] depending on reaction conditions is particularly significant.

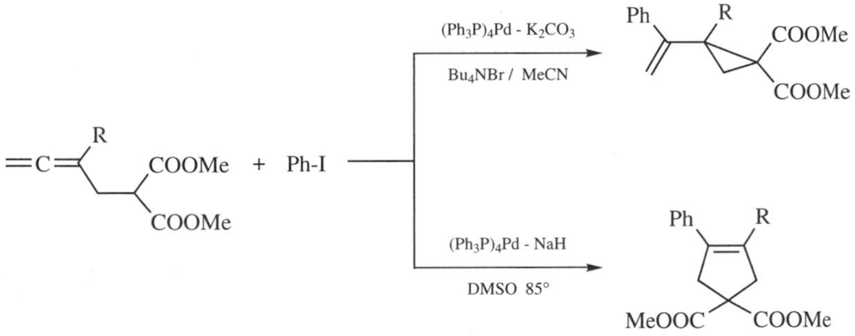

(Ph3P)4Pd - K2CO3
Bu4NBr / MeCN

(Ph3P)4Pd - NaH
DMSO 85°

1:2-Adducts from phenols and allene are readily prepared.[33] Allenes undergo reductive coupling at 0° on exposure to $(Ph_3P)_4Pd$ and HCOOH.[34]

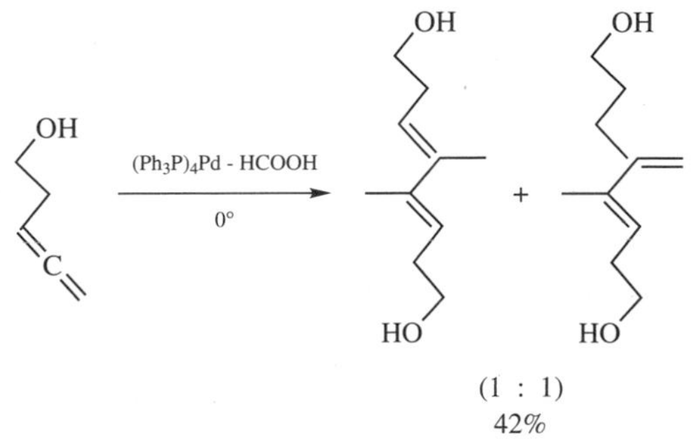

(Ph3P)4Pd - HCOOH
0°

(1 : 1)
42%

Ph—I + C + [amine] → (dba)$_3$Pd$_2$ - Fu$_3$P, K$_2$CO$_3$ / DMF → 92%

Suzuki coupling. Popularity of this reaction is largely due to tolerance of many functional groups in the reactants, ranging from those containing salicylaldehyde unit,[35] protected amino acid,[36] ketophosphorane.[37]

MeO—C$_6$H$_4$—B(OH)$_2$ + Br—C$_6$H$_4$—CO—CH=PPh$_3$ → (Ph$_3$P)$_4$Pd, CsF / DME, 75° → 89%

Noteworthy applications of the Suzuki coupling using (Ph$_3$P)$_4$Pd as catalyst are a terphenyl synthesis from 1,4-bis(pinacolatoboryl)benzene (this process is cocatalyzed by Ag$_2$CO$_3$),[38] synthesis of 2,4-alkadienoic esters in either (*Z,Z*)- or (*Z,E*)-modification,[39] and 1,3-butadienylboronic esters.[40] 2,4,6-Trivinylcyclotriboroxane[41] and polymethylsiloxanes in which the silicon atoms also carry aryl and alkenyl substituents[42] prove to be effective coupling reagents.

EtO, O, B, O + I—C$_6$H$_4$—OMe → (Ph$_3$P)$_4$Pd - K$_2$CO$_3$, EtOH - PhMe → EtO, OMe 89%

Cyclization precedes coupling of 2-bromo-1,*n*-alkadienes (n=6,7) with ArB(OH)$_2$, therefore the active site is transferred to the other terminus of the original carbon chain; furthermore, solvent effects are startling.[43] A method for annulation onto indole nucleus involves an allylic displacement at the last stage.[44]

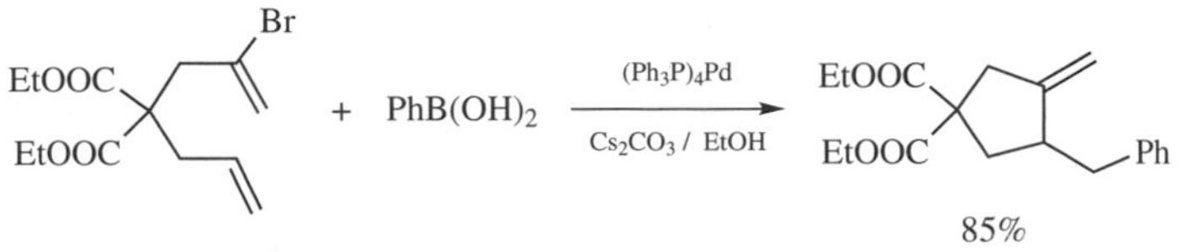

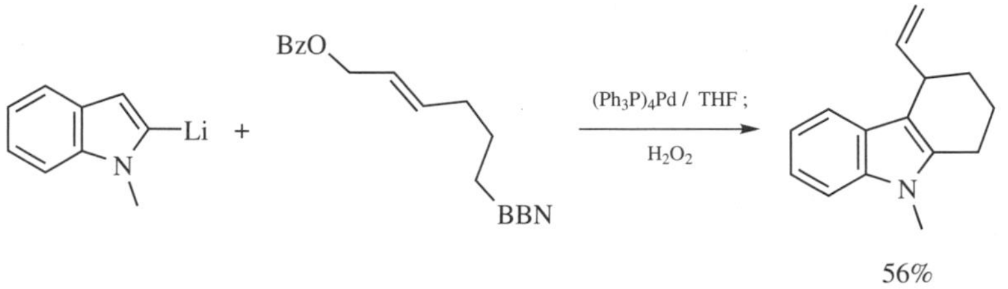

A coupling route to ketones from anhydrides[45] or carboxylic acids[46] is now achievable, the latter with in situ activation by dimethyl pyrocarbonate.

Stille coupling. New preparative methods using the Stille coupling deal with elaboration of: 3-sulfenyl-1,3-alkadienes,[47] 1-alken-3-one oxime ethers,[48] 6-substituted 2-pyrones.[49]

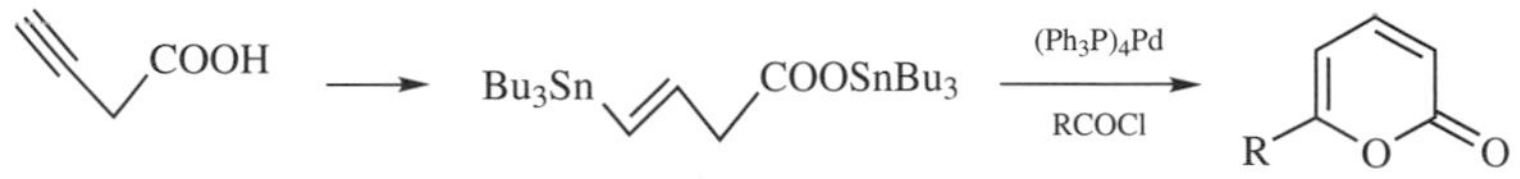

Three-component coupling provides a convenient access to *N*-allyl 3-butenamides (from allylic halides, allylstannane, and tosyl isocyanate).[50] Hypervalent tin reagents (e.g., $Bu_4N[Ph_3SnF_2]$) are successfully used in the Stille coupling.[51]

Negishi coupling. The method that transforms allenyl halides into 2-aryl-1,3-butadienes is extendable to the preparation of 1,1-difluoro analogues.[52] Reductive annulation of lactones can be achieved via the ketene triflates, Negishi coupling with organozinc compounds, and further manipulations.[53]

Alkynylzinc reagents generated in situ can be used to couple with ArX.[54] Homocoupling of ArZnI employs either NCS or oxygen as oxidant is satisfactory.[55]

Sonogashira coupling. Exchange of arene-bound iodine for alkenyl and alkynyl groups by the Sonogashira coupling is a mature technology. Refinements owe to modification of the other substituents on silicon, for example, 2-thienyldimethyl[56] and dimethyl(hydroxy) groups.[57]

A direct way of homologating ArBr to benzamides is by the Pd-catalyzed carbamoylation with $Me_3SiCONR_2$.[58]

[1]Akiyama, R., Kobayashi, S. *ACIEE* **40**, 3469 (2001).
[2]Deardorff, D.R., Taniguchi, C.M., Tafti, S.A., Kim, H.Y., Choi, S.Y., Downey, K.J., Nguyen, T.V. *JOC* **66**, 7191 (2001).

[3]Fisher, M., Brown, R.C.D. *TL* **42**, 8227 (2001).
[4]Zorn, C., Gnad, F., Salmen, S., Herpin, T., Reiser, O. *TL* **42**, 7049 (2001).
[5]Konno, T., Daitoh, T., Ishihara, T., Yamanaka, H. *TA* **12**, 2743 (2001).
[6]Wang, L., Chen, Z.-C. *SC* **31**, 1227 (2001).
[7]Zhou, T., Chen, Z.-C. *SC* **31**, 3289 (2001).
[8]Luke, G.P., Shakespeare, W.C. *SC* **32**, 2951 (2002).
[9]Teply, F., Stara, I.G., Kollarovic, A., Saman, D., Fiedler, P. *T* **58**, 9007 (2002).
[10]Camacho, D.H., Nakamura, I., Oh, B.H., Saito, S., Yamamoto, Y. *TL* **43**, 2903 (2002).
[11]Nakamura, I., Siriwardana, A.I., Saito, S., Yamamoto, Y. *JOC* **67**, 3445 (2002).
[12]Nicolaou, K.C., Vassilikogiannakis, G., Mägerlein, W., Kranich, R. *ACIEE* **40**, 2482 (2001).
[13]Camacho, D.H., Saito, S., Yamamoto, Y. *TL* **43**, 1085 (2002).
[14]Kazankova, M.A., Efimova, I.V., Kochetkov, A.N., Afanas'ev, V.V., Beletskaya, Dixneuf, P.H. *SL* 497 (2001).
[15]Lutete, L.M., Kadota, I., Shibuya, A., Yamamoto, Y. *H* **58**, 347 (2002).
[16]Zhang, W., Haight, A.R., Hsu, M.C. *TL* **43**, 6575 (2002).
[17]Nakamura, I., Oh, B.H., Saito, S., Yamamoto, Y. *ACIEE* **40**, 1298 (2001).
[18]Oh, B.H., Nakamura, I., Saito, S., Yamamoto, Y. *TL* **42**, 6203 (2001).
[19]Kamiya, I., Ogawa, A. *TL* **43**, 1701 (2002).
[20]Nagayama, K., Shimizu, I., Yamamoto, A. *BCSJ* **74**, 1803 (2001).
[21]Karstens, W.F.J., Stol, M., Rutjes, F.P.J.T., Kooijman, H., Spek, A.L., Hiemstra, H. *JOMC* **624**, 244 (2001).
[22]Dai, G., Larock, R.C. *OL* **3**, 4035 (2001).
[23]Tsuitsui, H., Narasaka, K. *CL* 526 (2001).
[24]Helaja, J., Gottlich, R. *CC* 720 (2002).
[25]Kawasaki, T., Saito, S., Yamamoto, Y. *JOC* **67**, 2653 (2002).
[26]Mayasundari, A., Young, D.G.J. *TL* **42**, 203 (2001).
[27]Ma, S., Zhao, S. *JACS* **123**, 5578 (2001).
[28]Shibata, T., Kadowski, S., Takagi, K. *H* **57**, 2261 (2002).
[29]Ma, S., Gao, W. *SL* 65 (2002).
[30]Ma, S., Xie, H. *JOC* **67**, 6575 (2002).
[31]Ma, S., Jiao, N., Zhao, S., Hou, H. *JOC* **67**, 2837 (2002).
[32]Oh, C.H., Rhim, C.Y., Song, C.H., Ryu, J.H. *CL* 1140 (2002).
[33]Grigg, R., Kongkathip, N., Kongathip, B., Luangkamin, S., Dondas, H.A. *T* **57**, 7965 (2001).
[34]Oh, C.H., Yoo, H.S., Jung, S.H. *CL* 1288 (2001).
[35]Zhuravel, M.A., Nguyen, S.T. *TL* **42**, 7925 (2001).
[36]Collet, S., Danion-Bougot, R., Danion, D. *SC* **31**, 249 (2001).
[37]Thiemann, T., Umeno, K., Wang, J., Tabuchi, Y., Arima, K., Watanabe, M., Tanaka, Y., Gorohmaru, H., Mataka, S. *JCS(P1)* 2090 (2002).
[38]Chaumeil, H., Le Drian, C., Defion, A. *S* 757 (2002).
[39]He, R., Deng, M.-Z. *OL* **4**, 2759 (2002).
[40]Tivola, P.B., Deagostino, A., Prandi, C., Venturello, P. *OL* **4**, 1275 (2002).
[41]Kerins, F., O'Shea, D.F. *JOC* **67**, 4968 (2002).
[42]Mori, A., Suguro, M. *SL* 845 (2001).
[43]Oh, C.H., Sung, H.R., Park, S.J., Ahn, K.H. *JOC* **67**, 7155 (2002).
[44]Ishikura, M., Kato, H. *T* **58**, 9827 (2002).
[45]Kakino, R., Yasumi, S., Shimizu, I., Yamamoto, A. *BCSJ* **75**, 137 (2002).
[46]Kakino, R., Narahashi, H., Shimizu, I., Yamamoto, A. *BCSJ* **75**, 1333 (2002).
[47]Su, M., Yu, W., Jin, Z. *TL* **42**, 3771 (2001).
[48]Chang, S., Lee, M., Kim, S. *SL* 1557 (2001).
[49]Thibonnet, J., Alarbri, M., Parrain, J.-L., Duchene, A. *JOC* **67**, 3941 (2002).

[50]Solin, N., Narayan, S., Szabo, K.J. *OL* **3**, 909 (2001).
[51]Martinez, A.G., Barcina, J.O., Heras, M. del R.C., de Fresno Cerezo, A. *OM* **20**, 1020 (2001).
[52]Shen, Q., Hammond, G.B. *OL* **3**, 2213 (2001).
[53]Kadota, I., Takamura, H., Sato, K., Yamamoto, Y. *JOC* **67**, 3494 (2002).
[54]Anastasia, L., Negishi, E. *OL* **3**, 3111 (2001).
[55]Hossain, K.M., Kameyama, T., Shibata, T., Takagi, K. *BCSJ* **74**, 2415 (2001).
[56]Hosoi, K., Nozaki, K., Hiyama, T. *CL* 138 (2002).
[57]Chang, S., Yang, S.H., Lee, P.H. *TL* **42**, 4833 (2001).
[58]Cunico, R.F., Maity, B.C. *OL* **4**, 4537 (2002).

Tetrakis(triphenylphosphine)palladium(0)–carbon monoxide.

Carbonylation. Normal Heck and Stille couplings are directed toward formation of carbonylated products. 4-Aroylisoquinolines,[1] 2-fluoro-3-aryl-1-propen-3-ones,[2] and seleno aroates are some of the examples.[3]

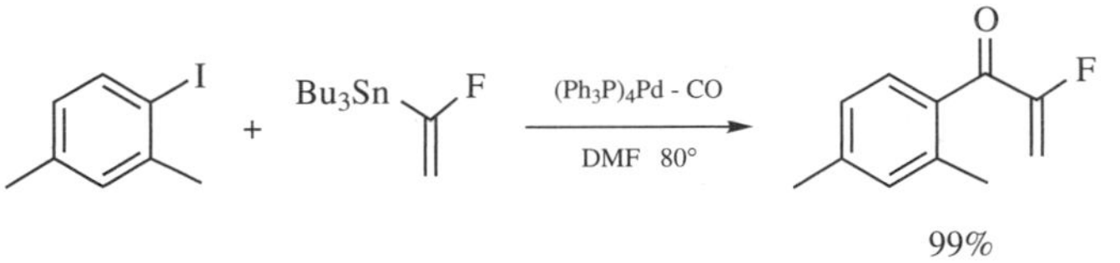

Carbonylative coupling leads to 3-aroyl-2-pyrrolines and 3-aroyl-3-pyrrolines from allenylamines.[4] Under photochemical conditions double carbonylation[5] of the double bond in an unsaturated alkyl iodide takes place.

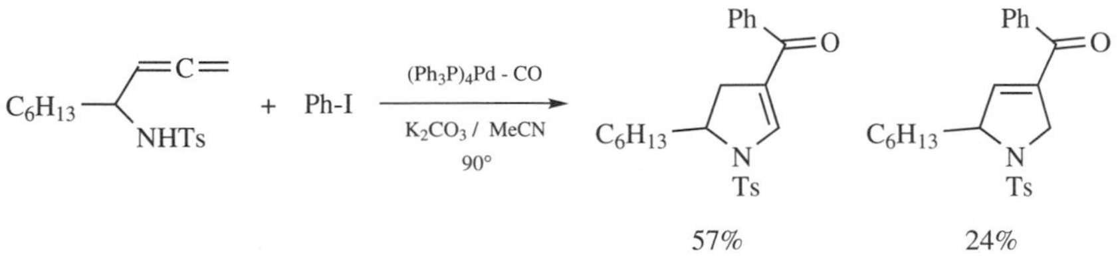

[1]Dai, G., Larock, R.C. *JOC* **67**, 7042 (2002).
[2]Hanamoto, T., Handa, K., Mido, T. *BCSJ* **75**, 2497 (2002).
[3]Nishiyama, Y., Tokunaga, K., Kawamatsu, H., Sonoda, N. *TL* **43**, 1507 (2002).
[4]Kang, S.-K., Kim, K.-J. *OL* **3**, 511 (2001).
[5]Ryu, I., Kreimerman, S., Araki, F., Nishitani, S., Oderaotoshi, Y., Minakata, S., Komatsu, M. *JACS* **124**, 3812 (2002).

Tetrakis(triphenylphosphine)palladium(0)–copper(I) salt. 18, 349–350; 20, 369; 21, 417

Coupling reactions. Under coupling conditions the stereoisomeric mixture of 1,2-dichloroethene gives only (*E*)-chloroenynes and (*E,E*)-chlorodienes.[1] It is pleasing to note that 1,1-dialkynylalkenes can be synthesized by two coupling reactions from 1,1-dibromoalkenes, control being exercised by the Pd catalysts.[2] The first coupling replaces the (*E*)-bromine. 3,5-Dibromo-2-pyrone undergoes stille coupling preferentially at C-3. Accordingly, various 3,5-disubstituted 2-pyrones are available from a 2-staged process.[3]

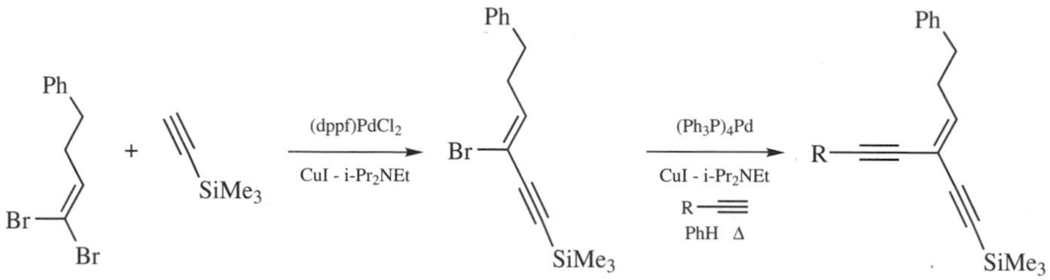

Sonogashira coupling followed by cycloisomerization constitutes an excellent method for elaborating fused ring systems.[4]

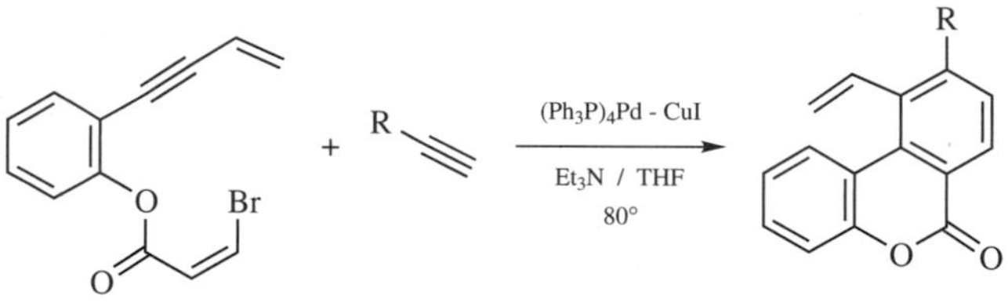

Certain Pd/Cu bimetallic catalysts promote indole synthesis from *o*-alkynylaryl isocyanates and allyl carbonates (products being 3-allylindoles).[5] Copper(I) oxide is a cocatalyst for the formation of arylacetic esters from $ArB(OH)_2$ and bromoacetic esters.[6]

Copper(I) thiophene-2-carboxylate cocatalyzes Suzuki coupling to form arylamidines from isothioureas.[7] Its presence also allows the Suzuki coupling[8] and dethiolative coupling of alkynyl tolyl sulfides with arylboronic acids[9] to proceed without a base.

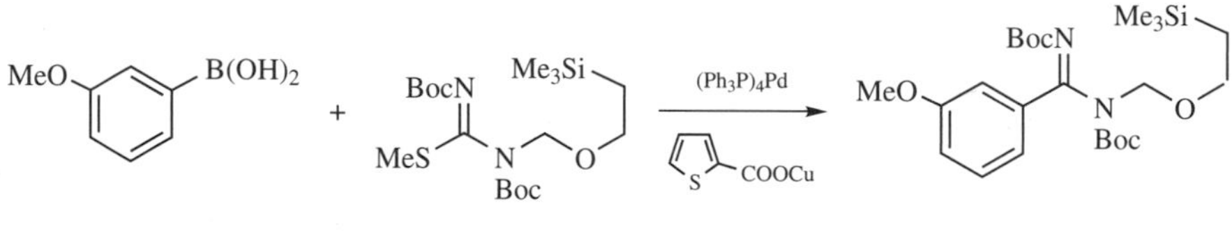

[1]Alami, M., peyrat, J.-F., Brion, J.-D. *TL* **43**, 3007 (2002).
[2]Uenishi, J., Matsui, K. *TL* **42**, 4353 (2001).
[3]Kim, W.-S., Kim, H.-J., Gho, C.-G. *TL* **43**, 9015 (2002).
[4]Kawasaki, T., Yamamoto, Y. *JOC* **67**, 5138 (2002).
[5]Kamijo, S., Yamamoto, Y. *ACIEE* **41**, 3230 (2002).
[6]Liu, X., Deng, M. *CC* 622 (2002).
[7]Kusturin, C.L., Liebeskind, L.S., Neumann, W.L. *OL* **4**, 983 (2002).
[8]Savarin, C., Liebeskind, L.S. *OL* **3**, 2149 (2001).
[9]Savarin, C., Liebeskind, L.S. *OL* **3**, 91 (2001).

Tetrakis(triphenylphosphine)palladium(0)–silver(I) salt.

Coupling reactions. Ring expansion attendant a Heck reaction to form 2-(α-styryl)cyclopentanones[1] from 1-allenylcyclobutanols is well anticipated.

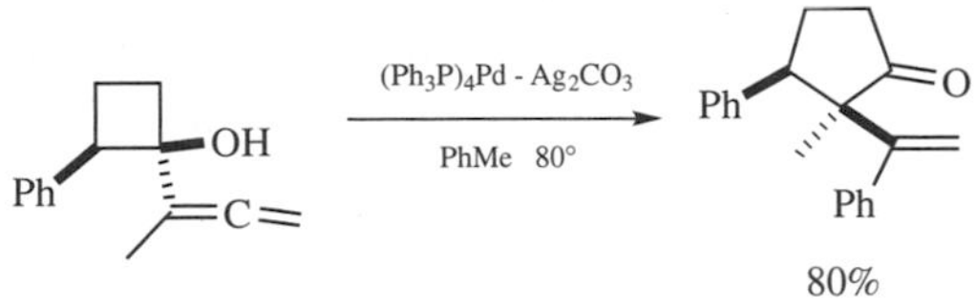

$(Ph_3P)_4Pd$ - Ag_2CO_3, PhMe 80°; 80%

Suzuki coupling components containing a dimethyl(2-pyridyl)silyl group can be prepared from Stille coupling of the air-stable $PySi(Me_2)CH_2SnBu_3$. Accordingly, a convenient access to diarylmethanes is presented.[2]

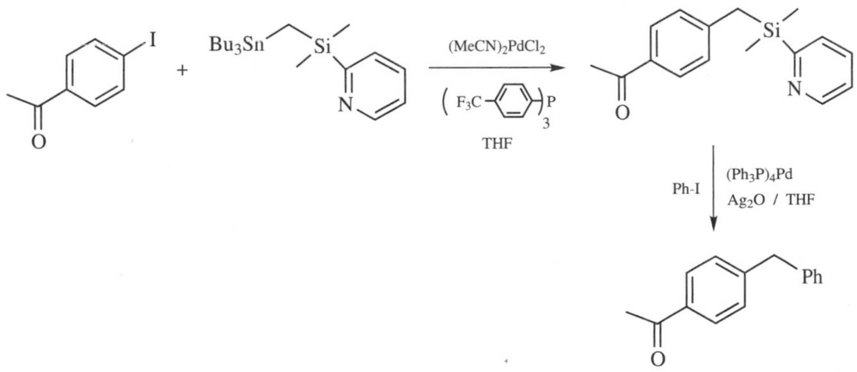

$(MeCN)_2PdCl_2$, $(F_3C\text{-}C_6H_4)_3P$, THF; Ph-I, $(Ph_3P)_4Pd$, Ag_2O / THF

The coupling of alkenyl triflates with silver alkynides is applicable to preparation of sensitive polyunsaturated compounds,[3] but requirement of 0.5 equivalent of the Pd-catalyst to give reasonable yields of the products makes it less attractive.

[1]Yoshida, M., Sugimoto, K., Ihara, M. *T* **58**, 7839 (2002).
[2]Itami, K., Mineno, Kamei, T., Yoshida, J. *OL* **4**, 3635 (2002).
[3]Dillinger, S., Bertus, P., Pale, P. *OL* **3**, 1661 (2001).

Tetrakis(triphenylphosphine)platinum(0). 20, 369–370; **21,** 417–418

Coupling. Suzuki coupling using $(Ph_3P)_4Pt$ as catalyst instead of the Pd analogue is realized.[1]

Hydroboration. Reaction of enones with bis(pinacolato)diboron in the presence of $(Ph_3P)_4Pt$ produces β-boryl ketones.[2]

1-Aryl-2-sulfenyl-1-alkenes.[3] Thiolesters of aromatic carboxylic acids are split by $(Ph_3P)_4Pt$ to generate [Ar-Pt-SR] species that undergo addition to alkynes.

$$R-\equiv \; + \; PhS-C(=O)-Ar \xrightarrow[PhMe \;\; \Delta]{(Ph_3P)_4Pt} (R)(PhS)C=CH-Ar$$

[1]Oh, C.H., Lim, Y.M., You, C.H. *TL* **43**, 4645 (2002).
[2]Ali, H.A., Goldberg, I., Srebnik, M. *OM* **20**, 3962 (2001).
[3]Sugoh, K., Kuniyasu, H., Sugae, T., Ohtaka, A., Takai, Y., Tanaka, A., Machino, C., Kambe, N., Kurosawa, H. *JACS* **123**, 5108 (2001).

N,N,N′,N′-Tetramethylguanidine.

Baylis-Hillman reaction.[1] The title reagent promotes Baylis-Hillman reaction at room temperature.

[1]Leadbeater, N.E., van der Pol, C. *JCS(P1)* 2831 (2001).

Thallium(III) acetate. 21, 419

α-Acyloxy ketones.[1] Introduction of an acyloxy group (formyloxy, acetoxy) to an α-position of ketones is accomplished with $Tl(OAc)_3$-TfOH .

[1]Lee, J.C., Jin, Y.S., Choi, J.-H. *CC* 956 (2001).

Thionyl chloride.

Chloroformates. A synthetic method to produce chloroformates that obviates the use of phosgene involves preparation of xanthates from alcohols, CO, S and MeI, treatment with $SOCl_2$ that completes the operation.[1]

[1]Mizuno, T., Takahashi, J., Ogawa, A. *TL* **43**, 7765 (2002).

Tin. **13,** 298; **17,** 333–334; **18,** 352; **20,** 372–373; **21,** 421

Allylation. Homoallylic alcohols are derived from aldehydes (but not ketones) allyl halides and tin under sonication without solvent.[1]

[1]Andrews, P.C., Peatt, A.C., Raston, C.L. *TL* **43**, 7541 (2002).

Tin(II) chloride. **13,** 298–299; **15,** 309–310; **16,** 329; **18,** 353–354; **19,** 337–338; **20,** 373; **21,** 422–423

Allylations. A reagent system constituted from allyltributylstannane and tin(II) chloride can effect allylation of ketones diastereoselectively.[1] The addition of allyltributylstannane to nascent aldimines (aldehydes + amines) can be accomplished in water when promoted by $SnCl_2$ and a surfactant.[2] Glyoxal undergoes bisallylation resulting in *vic*-diols.[3]

Another way of generating the allylating agent is by treatment of allyl diisopropyl carbinol with $SnCl_2$ and NCS.[4,5]

Reduction. Treatment of 2-ethynylnitroarenes with $SnCl_2$ in an aqueous solvent not only causes reduction of the nitro group, hydration of the triple bond also takes place.[6]

SnCl2
EtOH - H2O
NO2
NH2
O
71%

Glycosylation.[7] A combination of $SnCl_2$ and $AgB(C_6F_5)_4$ serves as catalyst for stereoselective glycosylation with disarmed glycosyl fluorides.

Acetonide cleavage.[8] Diol systems are regenerated by treatment of acetonides with $SnCl_2$ without affecting many other hydroxyl protecting groups (e.g., TBS ethers).

[1]Yasuda, M., Hirata, K., Nishino, M., Yamamoto, A., Baba, A. *JACS* **124**, 13442 (2002).
[2]Akiyama, T., Onuma, Y. *JCS(P1)* 1157 (2002).
[3]Samoshin, V.V., Gremyachinskiy, D.E., Smith, L.L., Bliznets, I.V., Gross, P.H. *TL* **43**, 6329 (2002).
[4]Masuyama, Y., Saeki, K., Horiguchi, S., Kurusu, Y. *SL* 1802 (2001).
[5]Masuyama, Y., Yamamoto, N., Kurusu, Y. *SL* 2113 (2002).
[6]Bosch, E., Jeffries, L. *TL* **42**, 8141 (2001).
[7]Mukaiyama, T., Maeshima, H., Jona, H., Mandai, H. *CL* 388 (2001).
[8]Yang, W.-B., Patil, S.S., Tsai, C.-H., Lin, C.-H., Fang, J.-M. *T* **58**, 253 (2002).

Tin(IV) chloride. 13, 300–301; **14,** 304–306; **15,** 311–313; **17,** 335–340; **18,** 354–356; **19,** 338–339; **20,** 373–375; **21,** 422–423

Allylation. Allyl transfer from allylsilanes to dithioacetals leads to homoallylic sulfides.[1] Destannylation is more favorable therefore allylation involving reagents that contain both silyl and stannyl groups at two different allylic positions would retain an allylsilane moiety in the products.[2] Interestingly, The amounts of the promoter $SnCl_4$ have enormous effects on the steric course.

Me3Si
Bu3Sn
O
N
O
+ OHC OBn
SnCl4
CH2Cl2
Me3Si
OH
OBn
i-Pr2N
O
O
70%

2-Trichlorostannyl-1,3-butadiene.[3] A transmetallation reaction between 4-trimethylsilyl-1,2-butadiene and $SnCl_4$ delivers the isomeric dienyl derivative. Similar reaction with $SbCl_3$ gives the corresponding stibine derivative.

Me_3Si $\xrightarrow[CH_2Cl_2]{SnCl_4}$ $SnCl_3$

96%

3-Acylindoles.[4] An improved protocol for acylation of indoles uses $SnCl_4$ as the Friedel-Crafts catalyst.

Condensation.[5] Very unusual and different results are reported for the $SnCl_4$ -promoted condensation of β-keto esters with conjugated nitriles, because of solvent effects. Non-Michael-type reaction occurs due to activation at the nitrogen atom.

O COOMe + CN

$SnCl_4$, $ClCH_2CH_2Cl$, 25° → H_2N, O, COOMe — 81%

$SnCl_4$, PhMe 100° → MeOOC, N, COOMe — 76%

Rearrangements. A stereoselective elaboration of 3-acyltetrahydrofurans by treatment of 4-alkenyl-4-alkyl-1,3-dioxolanes with $SnCl_4$ is initiated by heterocyclic cleavage.[6] Subsequent CC bond formation (reclosure of a five-membered ring) is attended by alkyl group migration.

O O $\xrightarrow[CH_2Cl_2\ -78°]{SnCl_4}$ O O

31%

N–(β–Mercaptoacyl)oxazolidine-2-ones are formed after *N*-(2-alkenoyl)oxazolidine-2-thiones are exposed to $SnCl_4$ and aqueous quench.[7] The transformation is stereoselective, consequential of the intramolecular Michael reaction pathway. In asymmetric synthesis the released chiral oxazolidin-2-one can be recycled (by reaction with Lawesson's reagent).

SnCl$_4$ / CH$_2$Cl$_2$; H$_2$O

[1]Liu, P., Binnun, E.D., Schaus, J.V., Valentino, N.M., Panek, J.S. *JOC* **67**, 1705 (2002).
[2]Leroy, B., Marko, I.E. *OL* **4**, 47 (2002).
[3]Lahrech, M., Thibonnet, J., Hacini, S., Santelli, M., Parrain, J.-L. *CC* 644 (2002).
[4]Ottoni, O., Neder, A.deV.F., Dias, A.K.B., Cruz, R.P.A., Aquino, L.B. *OL* **3**, 1005 (2001).
[5]Veronese, A.C., Morelli, C.F., Basato, M. *T* **58**, 9709 (2002).
[6]Cohen, F., MacMillan, D.W.C., Overman, L.E., Romero, A. *OL* **3**, 1225 (2001).
[7]Palomo, C., Oiarbide, M., Dias, F., Ortiz, A., Linden, A. *JACS* **123**, 5602 (2001).

Tin(II) triflate. 13, 301–302; **14,** 306–307; **15,** 313–314; **17,** 341–344; **18,** 357–358; **19,** 340; **20,** 376; **21,** 424

Alkenation.[1] Using Sn(OTf)$_2$ and *N*-ethylpiperidine to promote reaction of α-(bistrifluoroethylphosphono)alkanoic esters with carbonyl compounds (*Z*)-alkenes are obtained stereoselectively.

Homoallylation. Alkylidenecyclopropanes are suceptible to nucleophilic attack in the presence of Sn(OTf)$_2$, thus warnming with various RXH leads to homoallylic ethers, esters, sulfides,[2] or amines.[3]

Sn(OTf)$_2$, DME 85°

R = Ph 100%

Bis-aldols.[4] Initial donors are α-bromo ketones which undergo aldol reaction catalyzed by Sn(OTf)$_2$. Debrominative enolization of such an adduct then follows and condensation with a second aldehyde occurs. The overall process is quite stereoselective.

Sn(OTf)$_2$ - Et$_3$N ; R′CHO - TiCl$_4$ → [Ti], R″CHO

(*anti-anti*)

[1]Sano, S., Takehisa, T., Ogawa, S., Yokoyama, K., Nagao, Y. *CPB* **50**, 1300 (2002).
[2]Shi, M., Xu, B. *OL* **4**, 2145 (2002).
[3]Shi, M., Chen, Y., Xu, B., Tang, J. *TL* **43**, 8109 (2002).
[4]Arai, H., Shiina, I., Mukaiyama, T. *CL* 118 (2001).

Titanium(IV) bromide.

Baylis-Hillman reaction. A variant of the reaction involving 1-butyn-3-one leads to products containing a β-bromine atom. There is a remarkable dependence of the (*E/Z*)-isomer ratio on reaction temperature.[1]

CHO O OH O + $TiBr_4$ CH_2Cl_2 −78° O_2N O_2N Br 90%

[1]Shi, M., Wang, C.-J. *HCA* **85**, 841 (2002).

Titanium(III) chloride–ammonium hydroxide.

Reduction. The combination of $TiCl_3$ and aqueous ammonia in methanol is useful for reduction of carbonyl compounds (aldehydes and ketones).[1]

[1]Clerici, A., Pastori, N., Porta, O. *EJOC* 2235 (2001); 3326 (2002).

Titanium(III) chloride–lithium aluminum hydride.

2-Methyl-2-alkenoic esters.[1] Deoxygenation of Baylis-Hillman adducts after acetylation is accomplished by an S_N2' process using $TiCl_3$ -$LiAlH_4$.

[1]Shadakshari, U., Nayak, S.K. *T* **57**, 4599 (2001).

Titanium(IV) chloride. 13, 304–309; 14, 309–311; 15, 317–320; 16, 332–337; 17, 344–347; 18, 359–361; 19, 341–344; 20, 377–379; 21, 425–427

Allylation. Carbon chain construction incorporating alkenylborane segment can be effected by allylation method employing **1** as nucleophiles.[1]

OEt R OEt + O B O Si Ph R′ **1** $TiCl_4$ CH_2Cl_2 −78° EtO O B O R R′

In the context of a (+)-brefeldin-A synthesis[2] creation of the cyclopentane unit by intramolecular nucleophilic opening of a β-lactone is a key operation. Cyclization toward an imine in the *5-endo-trig* mode gives rise to pyrrolidine derivatives.[3]

TIPSO OTBDPS SiMe$_3$ → $TiCl_4$, CH_2Cl_2 −78° → TIPSO H OTBDPS COOH 90%

The three-component reaction among conjugated imines, tetraallylstannane and thiols which is promoted by $TiCl_4$ must follow the order of 1,4-addition and the allyl transfer.[4]

Ar-SH + Ph N OMe + Sn 4 → $TiCl_4$, EtCN - CH_2Cl_2 → Ar S HN OMe Ph

Aldol and Mannich reactions. Mukaiyama aldol reaction has been extended to acylsilanes.[5] From either (*E*)- of (*Z*)-1-trimethylsiloxy-1-trimethylsilylalkenes the major products have a 2,3-*anti*-configuration. The diastereoselectivity is apparently opposite to the *syn*-selective condensation between aldehydes and methyl phenylselenoacetate[6] or α-bromo ketones,[7] although all of them are catalyzed by $TiCl_4$ (however, there are differences in additive).

PhCH OSiMe$_3$ SiMe$_3$ + Ph OMe OMe → $TiCl_4$, CH_2Cl_2 −78° → Ph OMe O SiMe$_3$ Ph (major)

Note that the Wittig reaction for synthesis of 2-alkenoic esters can be modified[8] such that it requires merely admixture of carbonyl compounds, bromoacetic esters and triphenylphosphine with $TiCl_4$.

The compound prepared from $TiCl_4$ and *p*-(*t*-butylthia)calix[4]arene is effective in promoting aldol reaction of silyl enol ethers with aldehydes.[9]

Tandem Michael-aldol reactions of dicarbonyl compounds containing one conjugated system take place in the presence of $TiCl_4$-R_4NX. The halide ion X^- initiates the process by conjugate addition.[10] The intermolecular reaction is analogous to the Baylis-Hillman reaction but the adducts are saturated.[11]

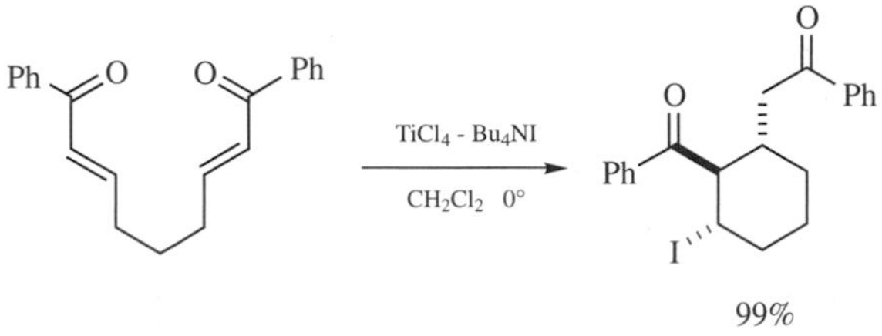

$TiCl_4$ - Bu_4NI, CH_2Cl_2 0°

99%

The Friedel-Crafts alkylation of phenols with *N*-tosyl-α-hydroxyglycine (−)-8-phenylmenthyl ester, a Mannich-type reaction, furnishes products of high optical purity.[12]

Baylis-Hillman reaction. The reaction promoted by $TiCl_4$ is applicable to α-keto esters[13] and *N*-phosphono aldimines[14] in the acceptor role and acrylic thioesters[15] as donors.

Alkylations. In alkylation of enol derivatives with hexacarbonyldicobalt-coordinated propargylic substrates $TiCl_4$ may replace the commonly used $BF_3 \cdot OEt_2$ as promoter. Intermolecular and intramolecular reactions leading to γ-alkynyl carbonyl compounds are readily achieved.[16,17]

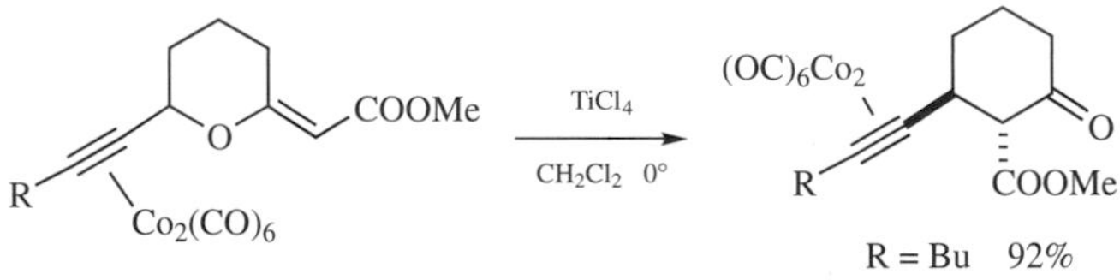

$TiCl_4$, CH_2Cl_2 0°

R = Bu 92%

An alternative method to synthesis of γ-lactones from esters and epoxides is to employ ketene acetals instead of the ester enolates and $TiCl_4$ instead of base.[18] Formation of 1,5-dichloro-1,4-dienes from 1-alkynes and aldehydes is observed.[19]

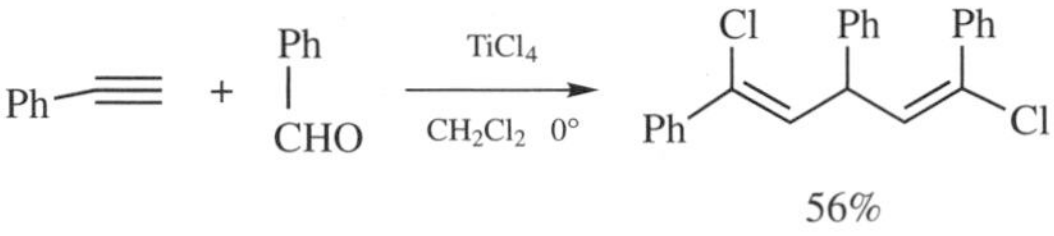

$TiCl_4$, CH_2Cl_2 0°

56%

Homoallylic alcohols are generated in a reaction of alkenylcyclopropanes and aldehydes.[20]

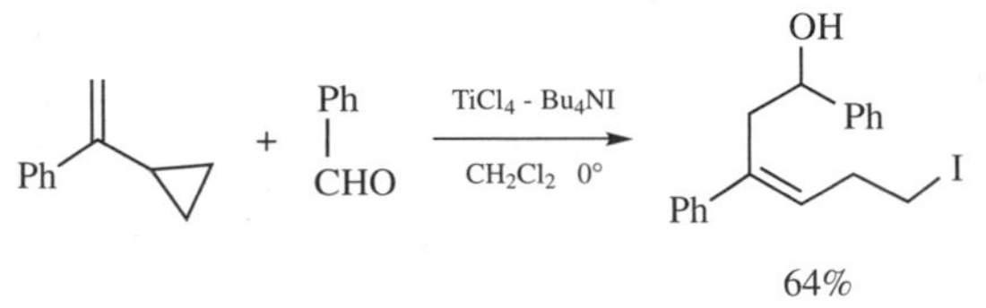

Ph, CHO; $TiCl_4$ - Bu_4NI; CH_2Cl_2 0°; OH, Ph, I, Ph; 64%

Additions. A thiol synthesis from alkenes and 1,3-dioxolane-2-thione[21] catalyzed by $TiCl_4$ is reminiscent of the Ritter reaction. Conjugate addition of silyl enol ethers to 1-nitroalkenes also transforms the nitro group into an α-chloro oxime.[22]

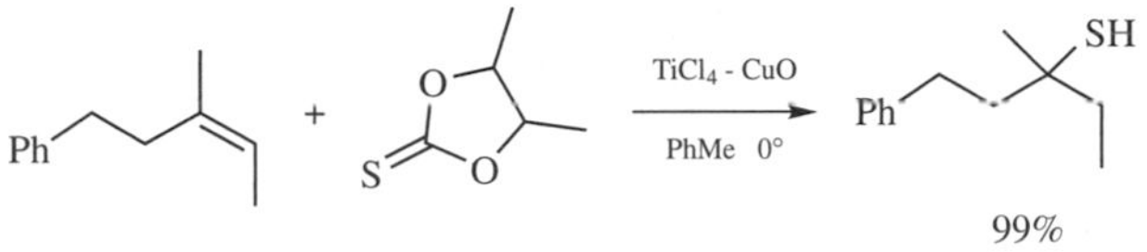

Ph; O, S, O; $TiCl_4$ - CuO; PhMe 0°; Ph, SH; 99%

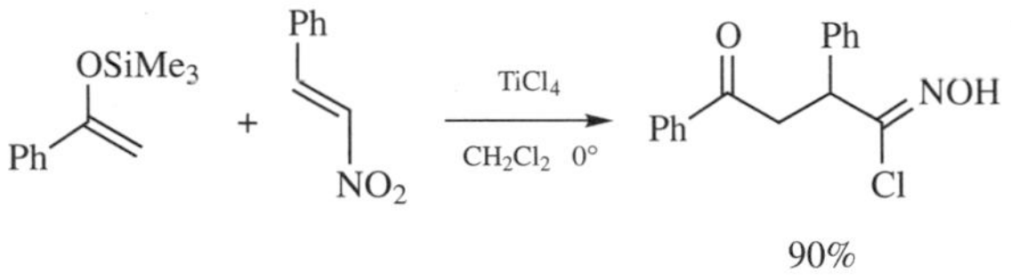

$OSiMe_3$, Ph; Ph, NO_2; $TiCl_4$; CH_2Cl_2 0°; O, Ph, Ph, NOH, Cl; 90%

Intramolecular addition of a hydroxyl group to an alkenylsilane can cause silyl migration.[23] The reaction can be dissected into three stages: protonation, silyl shift, and cyclization. Accordingly, reagents such as AcCl and gaseous HCl are also effective.

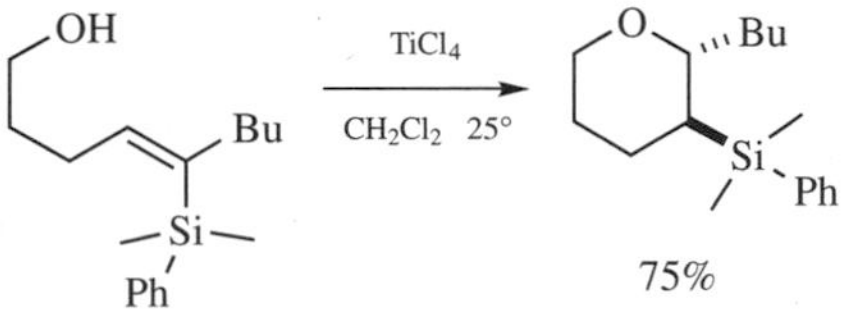

OH, Bu, Si, Ph; $TiCl_4$; CH_2Cl_2 25°; O, Bu, Si, Ph; 75%

Cyclization. δ,ε-Unsaturated ketones cyclize to afford cyclic γ-chloro alcohols when treated with $TiCl_4$.[24] Due to chelation effect the relative configuration of the products differs from that arising from reaction promoted by HCl.

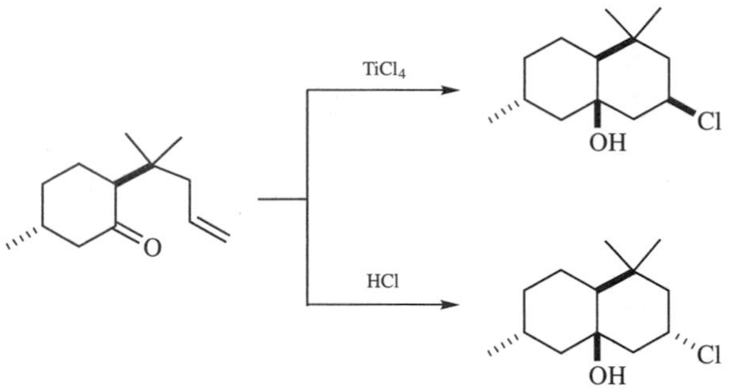

Aza-Cope rearrangement. An *O,N*-acetal in which the nitrogen atom is attached to a homoallylic site is susceptible to rearrangement upon ionization of the alkoxy residue. To induce the event $TiCl_4$ is a suitable reagent.[25]

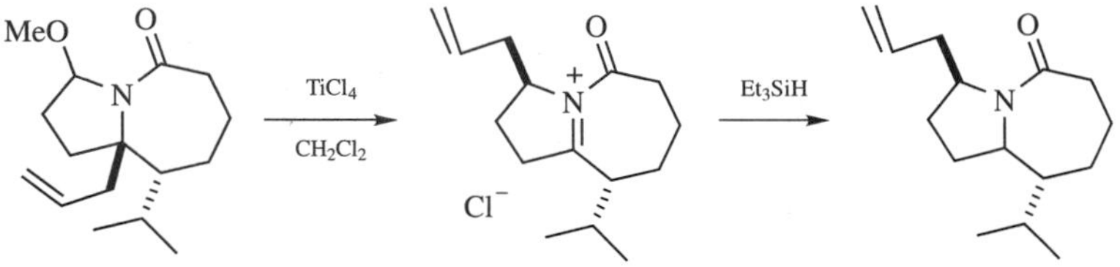

[1]Suginome, M., Ohmori, Y., Ito, Y. *JACS* **123**, 4601 (2001).
[2]Wang, Y., Romo, D. *OL* **4**, 3231 (2002).
[3]Duncan, D., Livinghouse, T. *JOC* **66**, 5237 (2001).
[4]Shimizu, M., Nishi, T. *CL* 46 (2002).
[5]Honda, M., Oguchi, W., Segi, M., Nakajima, T. *T* **58**, 6815 (2002).
[6]Nakamura, S., Hayakawa, T., Nishi, T., Watanabe, Y., Toru, T. *T* **57**, 6703 (2001).
[7]Hashimoto, Y., Kikuchi, S. *CL* 126 (2002).
[8]Basavaiah, D., Rao, A.J. *SC* **32**, 195 (2002).
[9]Morohashi, N., Hattori, T., Yokomakura, K., Kabuto, C., Miyano, S. *TL* **43**, 7769 (2002).
[10]Yagi, K., Turitani, T., Shinokubo, H., Oshima, K. *OL* **4**, 3111 (2002).
[11]Han, Z., Uehira, S., Shinokubo, H., Oshima, K. *JOC* **66**, 7854 (2001).
[12]Ge, C.-S., Chen, Y.-J., Wang, D. *SL* 37 (2002).
[13]Basavaiah, D., Sreenivasulu, B., Reddy, R.M., Muthukumaran, K. *SC* **31**, 2987 (2001).
[14]Shi, M., Zhao, G.-L. *TL* **43**, 9171 (2002).
[15]Kataoka, T., Iwama, T., Kinoshita, H., Tsujiyama, S., Tsurukami, Y., Iwamura, Watanabe, S. *SL* 197 (1999).
[16]Larareva, M.I., Nguyen, S.T., Nguyen, M.C., Emiru, H., McGrath, N.A., Caple, R., Smit, W.A. *MC* 224 (2001).
[17]Carberry, D.R., Reignier, S., Myatt, J.W., Miller, N.D., Harrity, J.P.A. *ACIEE* **41**, 2584 (2002).
[18]Maslak, V., Matovic, R., Saicic, R.N. *TL* **43**, 5411 (2002).
[19]Kabalka, G.W., Wu, Z., Ju, Y. *OL* **4**, 3415 (2002).
[20]Tsuritani, T., Shinokubo, H., Oshima, K. *SL* 978 (2002).

[21]Mukaiyama, T., Saitoh, T., Jona, H. *CL* 638 (2001).
[22]Yan, M.-C., Tu, Z., Lin, C., Yao, C.-F. *TL* **43**, 7991 (2002).
[23]Miura, K., Hondo, T., Okajima, S., Nakagawa, T., Takahashi, T., Hosomi, A. *JOC* **67**, 6082 (2002).
[24]Davis, C.E., Coates, R.M. *ACIEE* **41**, 491 (2002).
[25]Hinman, M.M., Heathcock, C.H. *JOC* **66**, 7751 (2001).

Titanium(IV) chloride-amines. 20, 380; **21,** 427–428

Condensation reactions. Butenolide formation is achieved in one step by the condensation between ketones and 1,1-dimethoxy-2-alkanones.[1] The reaction is catalyzed by $TiCl_4$-Bu_3N.

Dieckmann condensation promoted by $TiCl_4$-Et_3N in CH_2Cl_2 can be quite efficient,[2] e.g., a 54% yield of the 17-membered keto ester precursor of civetone is obtained only after 1 hr at 0°.

The mildness of the reagent is shown by the Mannich-type reaction involving α-diazo-β-keto esters.[3]

NTs Ph + O COOEt N2 —($TiCl_4$ - Et3N; CH_2Cl_2 −41°)→ TsNH O Ph COOEt N2

62%

An interesting synthesis of 3,3-diarylcyclobutanones is based on titanation of an iminium ion derived from oxidation of a tertiary amine, and subsequent reaction with diaryl ketones.[4]

N Bn —(I_2 Δ ; $TiCl_4$)→ N+ Bn X− Cl_3Ti $TiCl_3$ —($Ar_2C{=}O$)→ O Ar Ar

Oxidation. Propargylic alcohols give the corresponding aldehydes[5] in good yields by treatment with $TiCl_4$-Et_3N in CH_2Cl_2 at 0°. The alcohols behave differently from allylic alcohols which afford allylic chlorides only.

Enamines derived from cyclohexanones undergo aromatization.[6]

Phosphates.[7] The reagent system is used to effect phosphorylation of alcohols with $(RO)_2POCl$.

[1]Tanabe, Y., Mitarai, K., Higashi, T., Misaki, T., Nishii, Y. *CL* 2542 (2002).
[2]Tanabe, Y., Makita, A., Funakoshi, S., Hamasaki, R., Kawakusu, T. *ASC* **344**, 507 (2002).
[3]Deng, G., Jiang, N., Ma, Z., Wang, J. *SL* 1913 (2002).
[4]Periasamy, M., Jayakumar, K.N., Bharathi, P. *CC* 1728 (2001).

[5]Han, Z., Shinokubo, H., Oshima, K. *SL* 1421 (2001).
[6]Srinivas, G., Periasamy, M. *TL* **43**, 2785 (2002).
[7]Jones, S., Selitsianos, D. *OL* **4**, 3671 (2002).

Titanium(IV) chloride–indium.

Reduction of sulfoxides.[1] Various types of sulfoxides (diaryl, dialkyl, alkyl/aryl) are reduced by the reagent system in THF at room temperature.

[1]Yoo, B.W., Choi, K.H., Kim, D.Y., Choi, K.I., Kim, J.H. *SC* **33**, 53 (2003).

Titanium(IV) chloride-zinc. 20, 381; 21, 430

Reductive acylation The $TiCl_4$-Zn system couples ketones with acylsilanes in such a way that the original ketone carbonyl group is deoxygenated and becomes a tertiary carbon attached to an acyl group.[1]

MeO, O, $SiMe_3$ + O, Ph → ($TiCl_4$ - Zn, DME) MeO, O, Ph

51%

De-S-acylation Thioesters are converted to thiols by brief treatment with $TiCl_4$-Zn in CH_2Cl_2.[2]

[1]Sakurai, H., Imamoto, Y., Hirao, T. *CL* 44 (2002).
[2]Jin, C.K., Jeong, H.J., Kim, M.K., Kim, J.Y., Yoon, Y.-J., Lee, S.-G. *SL* 1956 (2001).

Titanium(IV) chloride triisopropoxide.

Enolate additions. Titanium enolates generated from transmetallation show excellent diastereoselectivity (and enantioselectivity in applicable cases) in their addition to *N*-substituted imines. A reliable elaboration of β-amino acid derivatives[1] is based on the characteristics. With allyltitanium species that also contain a chiral sulfoxide as controller of stereochemical course asymmetric synthesis of γ,δ-unsaturated α-amino acids[2] is realized through addition to properly protected dehydroglycine.

EtOOC, NTs + O, NMe, S, Ph, $(i\text{-}PrO)_2Ti$ → ($(i\text{-}PrO)_3TiCl$, THF) EtOOC, NHTs, O, NMe, S, Ph

59% (dr > 95:5)

Michael reactions. A double Michael reaction occurs when 4-substituted 4-amino-2,5-cyclohexadienones are brought together with enones in the presence of $(i\text{-PrO})_2TiCl_2$.

(i-PrO)$_2$TiCl$_2$ / CHCl$_2$

CH$_2$=CHCOEt / MeCN

[1]Tang, T.P., Ellman, J.A. *JOC* **67**, 7819 (2002).
[2]Schleusner, M., Gais, H.-J., Koep, S., Raabe, G. *JACS* **124**, 7789 (2002).
[3]Carreno, M.C., Ribagorda, M., Posner, G.H. *ACIEE* **41**, 2753 (2002).

Titanium(IV) iodide. 21, 430–431

Reduction and reductive coupling. α-Iminoketones are reduced to α-aminoketones by TiI_4.[1] Under similar reaction conditions a mixture of α-imino esters and aldehydes undergo cross-coupling (also considered as reductive aldol reaction) to afford 2-amino-3-hydroxyalkanoic esters.[2] Note that the 2-hydroxy-3-amino regioisomers are obtained on functional variation of the reactants.[3]

TiI$_4$ / EtCN, −78 ~ −60°

73% (*anti/syn* 81 : 19)

TiI$_4$ / EtCN, 0°

80% (*anti/syn* 64 : 36)

Aldol reactions.[4] An extension of the previous work on methoxyallene oxide to the α-silyl derivative the aldol reaction affords 3-methoxy-3-alken-2-ones directly.

MCPBA / CH$_2$Cl$_2$, H$_2$O ; TiI$_4$ - Me$_3$SiOTf, PhCHO

73%

[1]Shimizu, M., Sahara, T., Hayakawa, R. *CL* 792 (2001).
[2]Shimizu, M., Takeuchi, Y., Sahara, T. *CL* 1196 (2001).
[3]Shimizu, M., Sahara, T. *CL* 888 (2002).
[4]Hayakawa, R., Makino, H., Shimizu, M. *CL* 756 (2001).

Titanium(IV) iodide-zinc.

Reductive coupling. The TiI_4-Zn combination is very useful for converting aminoacetals to the dimeric 1,2-diamines.[1] Dithioacetals of ArCHO similarly afford bissulfides.[2]

[1]Yoshimura, N., Mukaiyama, T. *CL* 1334 (2001).
[2]Yoshimura, N., Igarashi, K., Funasaka, S., Mukaiyama, T. *CL* 640 (2001).

Titanium tetraisopropoxide. 13, 311–313; 14, 311–312; 15, 322; 16, 339; 17, 347–348; 18, 363–364; 19, 346–347; 20, 381–382; 21, 431

Aldol reactions. Control of titanium(IV) enolates derived from α-diazo β-keto carbonyl compounds to enter the aldol reaction or Michael reaction with conjugated carbonyl systems by Lewis acid catalysts is revealed. Thus, $(i\text{-PrO})_4Ti$ favors 1,2-addition to benzalacetone by 96:4, in contrast to the favored 1,4-addition in the presence of $SnCl_4$.[1]

L.A. = $(i\text{-PrO})_4Ti$ 78% 96 : 4
L.A. = $SnCl_4$ 50% 5 : 95

Aza-Baylis-Hillman reaction. Finding $(i\text{-PrO})_4Ti$ to be an efficient catalyst to direct the condensation of imines with acrylic esters a facile synthesis of α-methylene-β-amino acid derivatives is at hand. The reaction involves addition of aldehydes, acrylic esters and the catalyst to tosylamide, showing that the normal Baylis-Hillman reaction is slower.[2]

$Ts-NH_2$ + Ph-CHO + (acrylate, COOMe) → [$(i\text{-PrO})_4Ti$, 3-quinuclidinol, i-PrOH] → Ph–CH(NHTs)–C(=CH$_2$)COOMe 85%

Cyanosilylation. Ketones are converted to cyanohydrin silyl ethers using an amine oxide complex with $(i\text{-PrO})_4Ti$ as catalyst.[3]

[1] Deng, G., Tian, X., Qu, Z., Wang, J. *ACIEE* **41**, 2773 (2002).
[2] Balan, D., Adolfsson, H. *JOC* **67**, 2329 (2002).
[3] Shen, Y., Feng, X., Li, Y., Zhang, G., Jiang, Y. *SL* 793 (2002).

Titanium tetrakis(dimethylamide).

Hydroamination. The title reagent brings alkynes and amines together to form adducts.[1] The addition follows the Markovnikov rule, hence complementary to that mediated by the titanocene-bistrimethylsilylethyne complex.[2]

Bu—≡ + Ph-NH_2 —($(Me_2N)_4Ti$; PhMe 75°)→ Bu(C=NPh)Me + Bu-CH=CH-NHPh
(3 : 1)
90%

[1] Shi, Y., Ciszewski, J.T., Odom, A.L. *OM* **20**, 3967 (2001).
[2] Tillack, A., Castro, I.G., Hartung, C.G., Beller, M. *ACIEE* **41**, 2541 (2002).

Titanocene bis(triethyl phosphite). 20, 383; **21,** 432–433

Alkenation.[1] Titanocene bis(triethyl phosphite) promotes desulfurative homologation, converting dithioacetals to alkenes with one more carbon when ethylene is also introduced into the reaction mixtures. Interestingly, two-carbon homologation is observed with the titanocene-isobutene complex.

Ph / Ph / SPh / SPh + ‖
→ $Cp_2Ti[P(OEt)_3]_2$, THF → Ph / Ph 65%
→ Cp_2Ti—(isobutene) → Ph / Ph 74%

Alkyl halides on activation with the title reagent react with carbonyl compounds in the same capacity as Wittig reagents, but under nonbasic conditions.[2]

Cyclopropanes. Alkynylcyclopropanes are formed from 1,1-bis(phenylthio)-2-alkynes and 1-alkenes.[3]

Intramolecular cyclopropanation of ω,ω-diahlo-1-alkenes (e.g., ω=6) is oberved if Mg is also present.[4] 1,3-Dihaloalkanes also undergo dehalogenative cyclization under essentially the same conditions.[5]

SPh, SPh, Ph, + Ph; $Cp_2Ti[P(OEt)_3]_2$ / THF → Ph, Ph; 65%

Ph, Cl, Cl, Ph; Mg, $Cp_2Ti[P(OEt)_3]_2$ / 4A-MS / THF → Ph, Ph; 93%

Cyclic enol ethers and enamines. Intramolecular condensation of alkyl ω,ω-bis(phenylthio)alkanoates gives cyclic enol ethers (5-, 6-, 7-, 9-membered rings).[6,7] A 2-substituted indole synthesis by the traceless solid-phase technique is based on the coupling of *o*-trimethylsilylaminobenzaldehyde dithioacetals with resin-bound esters.[8]

S, S, $SiMe_3$, N, Boc; $Cp_2Ti[P(OEt)_3]_2$ / PhCOO-[P] ; CF_3COOH / CH_2Cl_2 → Ph, N, Boc; 69%

[1]Tsubouchi, A., Nishio, E., Kato, Y., Fujiwara, T., Takeda, T. *TL* **43**, 5755 (2002).
[2]Takeda, T., Shimane, K., Ito, K., Saeki, N., Tsubouchi, A. *CC* 1974 (2002).
[3]Takeda, T., Kuroi, T., Yanai, K., Tsubouchi, A. *TL* **43**, 5641 (2002).
[4]Fujiwara, T., Odaira, M., Takeda, T. *TL* **42**, 3369 (2001).
[5]Takeda, T., Shimane, K., Fujiwara, T., Tsubouchi, A. *CL* 290 (2002).
[6]Rahim, M.A., Sasaki, H., Saito, J., Fujiwara, T., Takeda, T. *CC* 625 (2001).
[7]Uehara, H., Oishi, T., Inoue, M., Shoji, M., Nagumo, Y., Kosaka, M., Le Brazidec, J.-Y., Hirama, M. *T* **58**, 6493 (2002).
[8]Macleod, C., Hartley, R.C., Hamprecht, D.W. *OL* **4**, 75 (2001).

Titanocene dichloride–indium.

Reductions. Reduction of nitroarenes to arylamines[1] and sulfoxides to sulfides[2] with Cp_2TiCl_2 and indium are completed at room temperature.

[1]Yoo, B.W., Lee, S.J., Yoo, B.S., Choi, K.I., Kim, J.H. *SC* **32**, 2489 (2002).
[2]Yoo, B.W., Choi, K.H., Lee, S.J., Yoon, C.M., Kim, S.H., Kim, J.H. *SC* **32**, 63 (2002).

Titanocene dichloride–magnesium.

Propargylation.[1] Propargylic acetates undergo O-C bond cleavage and are converted into nucleophilic species on reaction with Cp_2TiCl_2–Mg. The nucleophiles attack carbonyl compounds readily.

Ph—≡—C(Me)₂—OAc + Ph-CHO —(Mg - Cp_2TiCl_2; THF 0°)→ Ph—≡—C(Me)₂—CH(Ph)OH 72%

[1]Yang, F., Zhao, G., Ding, Y. *TL* **42**, 2839 (2001).

Titanocene dichloride-manganese. 20, 384; **21,** 434

C-Glycosides[1] Epoxides of glycals rupture on treatment with Cp_2TiCl_2–Mn and trapping at the glycosidic center occurs readily with acrylonitrile and methyl acrylate. The new bond is α-oriented.

(Scheme: glycal epoxide bearing OBn, BnO, BnO groups —(Mn - Cp_2TiCl_2; CH_2=CHCN, THF)→ C-glycoside with OH and CH₂CH₂CN, 61%)

[1]Parrish, J.D., Little, R.D. *OL* **4**, 1439 (2002).

Titanocene dichloride-zinc. 20, 384-385; **21,** 435

Cyclization. For 1-oxaspiro[2.*n*]alkanes in which C-4 carries a chain element possessing a double bond or triple bond at a distance, reaction with Cp_2TiCl_2–Zn leads to carbon radicals of bicyclic structures. With radical intercepters for termination two CC bonds are formed.[1] Steric factors favor the creaton of (*E*)-alkenes from alkynes.

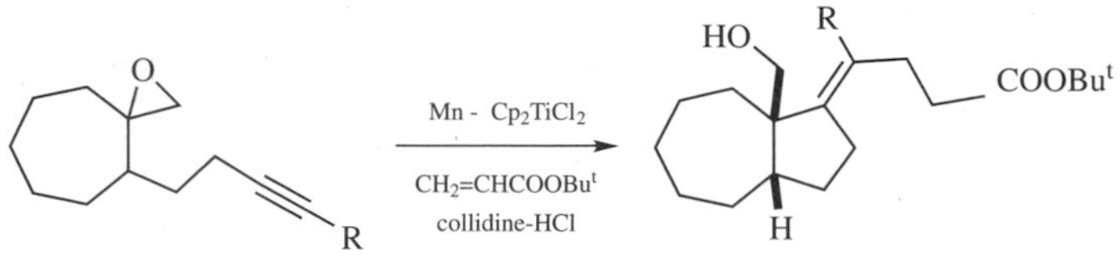

R = H, Me

Reductive cyclization of ketonitriles is extendable to the formation of enamino nitriles.[2]

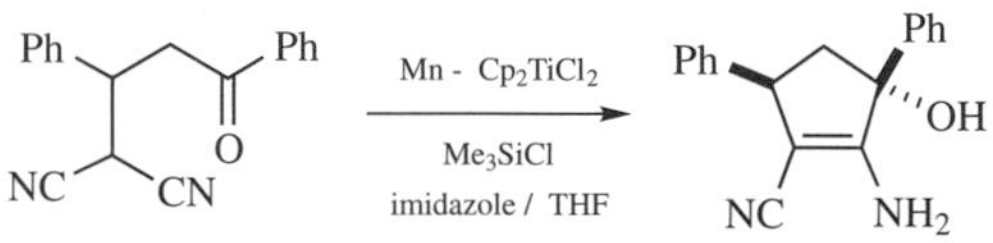

65% (*trans/cis* 85 : 15)

Reduction of enones. Selective reduction of enones (6 examples, 56–89%) in THF-MeOH to saturated ketones is reported.[3]

[1]Gansäuer, A., Pierobon, M., Bluhm, H. *ACIEE* **41**, 3206 (2002).
[2]Zhou, L., Hirao, T. *T* **57**, 6927 (2001).
[3]Moisan, L., Hardouin, C., Rousseau, B., Doris, E. *TL* **43**, 2013 (2002).

Titanocene difluoride.

Reductive coupling. Deoxygenative dimerization of amides to 1,2-diamines is promoted by Cp_2TiF_2 and a hydrosilane. Cp_2TiMe_2 is also effective as the same reactive $[Cp_2TiH]$ species is generated.

[1]Selvakumar, K., Harrod, J.F. *ACIEE* **40**, 2129 (2001).

Titanocene tetracarbonylcobaltate.

Carbonylation.[1] Under CO the title complex transforms epoxides and aziridines to β-lactones and β-lactams, respectively. Insertion occurs selectively at the more highly substituted C-X bond of the small heterocycles.

[1]Mahaddevan, V., Getzler, Y.D.Y.L., Coates, G.W. *ACIEE* **41**, 2781 (2002).

p-Toluenesulfonic acid.

Deacetylation.[1] Chemoselective deacetylation of protected sugars in the presence of benzoic and *p*-bromobenzoic esters is achieved using $TsOH \cdot H_2O$ in CH_2Cl_2-MeOH.

N-Tritylation.[2] A simple procedure for derivatizing the nitrogen atom of amides, lactams, carbamates, and ureas involves heating them with TsOH and TrOH in benzene.

Rearrangement.[3] Certain *N*-allylanilines undergo rearrangement in the same manner as the Claisen rearrangement except that TsOH is used as a promoter (in aqueous MeCN, 65°). The double bond of the initial products migrates toward the aromatic ring during the reaction.

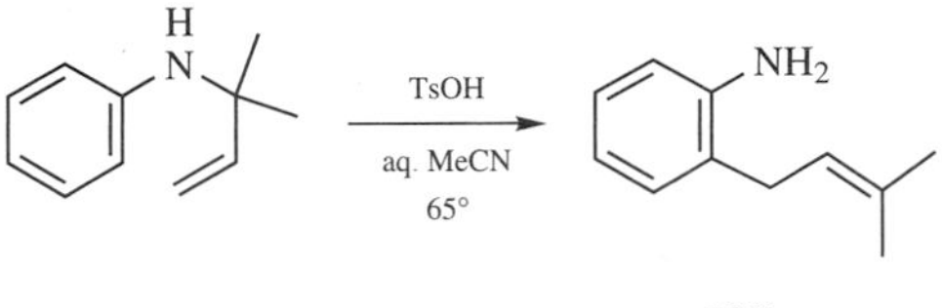

TsOH, aq. MeCN, 65°

NH_2

70%

An extensive fragmentation-recyclization sequence to transform an aryl-substituted brideged ketone to an hexahydrophenanthrene derivative under acetalization consitions facilitates the synthesis of diterpenes such as ferruginol.[4]

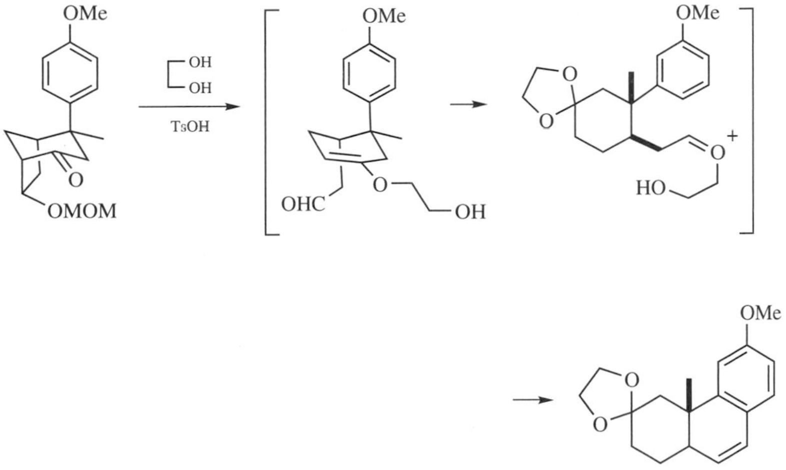

OMe, OMOM, O, OH, OH, TsOH, OHC, OH, HO, O, +, OMe

[1]Gonzalez, A.G., Brouard, I., Leon, F., Padron, J.I., Bermejo, J. *TL* **42**, 3187 (2001).
[2]Reddy, D.R., Iqbal, M.A., Hudkins, R.L., Messina-McLauchlin, P.A., Mallamo, J.P. *TL* **43**, 8062 (2002).
[3]Cooper, M.A., Lucas, M.A., Taylor, J.M., Ward, A.D., Williamson, N.M. *S* 621 (2001).
[4]Nagata, H., Miyazawa, N., Ogasawara, K. *OL* **3**, 1737 (2001).

p-Toluenesulfonyl bromide–2,2′-Azobisisobutyronitrile.

Addition to allenes.[1] 1,2-Alkadienes add TsBr under free radical conditions to afford 1-bromo-2-tosyl-2-alkenes.[1] When the substrate molecules contain an XH group at proper distance allylic displacement follows.[2]

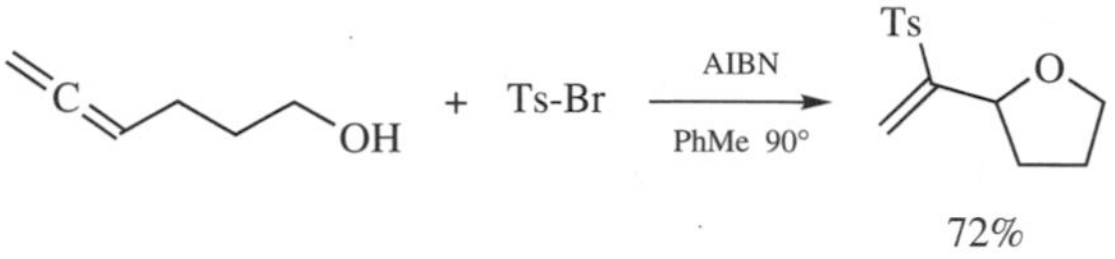

Bisallenes undergo additive cyclization with TsBr.[3]

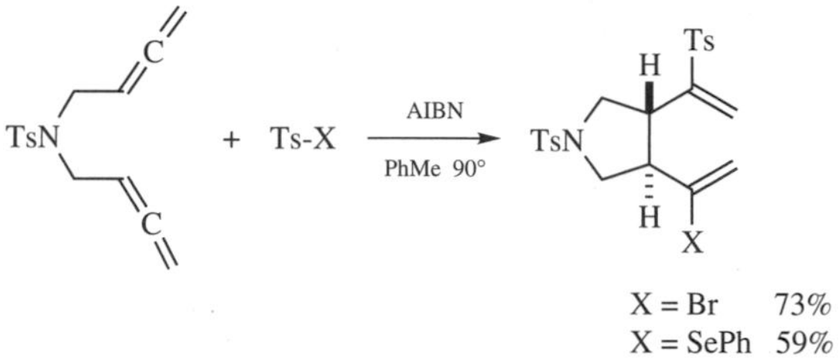

[1]Kang, S.-K., Seo, H.-W., Ha, Y.-H. *S* 1321 (2001).
[2]Kang, S.-K., Ko, B.-S., Ha, Y.-H. *JOC* **66**, 3630 (2001).
[3]Kang, S.-K., Ha, Y.-H., Kim, D.-H., Lim, Y., Jung, J. *CC* 1306 (2001).

N-[*β*-(*p*-Toluenesulfonyl)ethyl]hydroxylamine.

Hydroxamic acids.[1] The derived *N*-substituted hydroxamic acids from *N*-acylation of the title reagents are suitable substrates for amidyl radical generation and thence cyclization when a proper double bond is present.

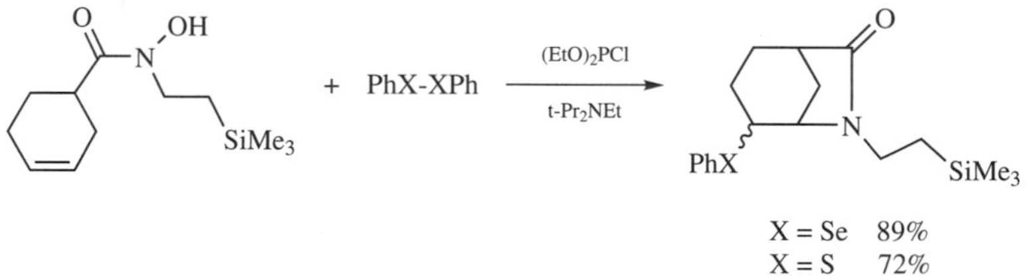

[1]Artman III, G.D., Waldman, J.H., Weinreb, S.M. *S* 2057 (2002).

p-Toluenesulfonyl isocyanate.

Glycosylation.[1] Conversion of sugars to glycosyl donors succeeds in forming *N*-tosylcarbamates. The reactivity is tuned by further alkylation at the nitrogen atom.

[1]Hinklin, R.J., Kiessling, L.L. *JACS* **123**, 3379 (2001).

2,8,9-Trialkyl-1-phospha-2,5,8,9-tetraazabicyclo[3.3.3]undecanes. 19, 370; 21, 436

Condensation reactions. Further exploration of these cage-bases has uncovered their utility in Suzuki coupling,[1] Michael reaction,[2] Knoevenagel reaction (including coumarin synthesis),[3] synthesis of 2-alkenoic esters from ethyl acetate and aldehydes,[4] condensation of allyl cyanide with aldehydes,[5] and cyanohydrination of carbonyl compounds with Me_3SiCN.[6]

[1]Urgaonkar, S., Nagarajan, M., Verkade, J.G. *TL* **43**, 8921 (2002).
[2]Kisanga, P.B., Ilankumaran, P., Fetterly, B.M., Verkade, J.G. *JOC* **67**, 3555 (2002).
[3]Kisanga, P.B., Fei, X., Verkade, J.G. *SC* **32**, 1135 (2002).
[4]Kisanga, P.B., D'Sa, B., Verkade, J.G. *T* **57**, 8047 (2001).
[5]Kisanga, P.B., Verkade, J.G. *JOC* **67**, 426 (2002).
[6]Wang, Z., Fetterly, B.M., Verkade, J.G. *JOMC* **646**, 161 (2002).

Triarylbismuthane oxide.

Oxidation.[1] Alcohols are rapidly oxidized to carbonyl compounds by $[Ar_3BiO]_2$.

[1]Matano, Y., Nomura, H. *JACS* **123**, 6443 (2001).

Triarylbismuth dichloride-DBU.

Oxidation.[1] Oxidation of alcohols to carbonyl compounds by the tris(2-methylphenyl)bismuth dichloride-DBU reagent pair takes place in toluene at room temperature.

[1]Matano, Y., Nomura, H. *ACIEE* **41**, 3028 (2002).

Tributylphosphine. 20, 387–388; **21,** 438

Aziridine opening.[1] Remarkable catalytic activity of Bu_3P for ring opening of aziridines with various nucleophiles is noted.

Hydrolysis of phenylthioimines.[2] While thioimines are readily prepared from oximes on reaction with the Bu_3P-PhSSPh combination, their hydrolysis to carbonyl compounds is also effected via desulfurization with Bu_3P (and water to hydrolyze the imine products).

Stetter reaction.[3] A convenient synthesis of *N,N*-dimethyl-3-aroylpropanamides is by a Bu_3P-catalyzed reaction of aromatic aldehydes and *N,N*-dimethylacrylamide.

Michael cycloisomerization. Bis(enones) such as 1,8-diphenyl-2,6-octadiene-1,8-dione cyclize on treatment with Bu_3P. Two Michael reactions and elimination are involved.[4]

O Ph O Ph — Bu_3P / EtOAc → O Ph O Ph

86%

Amide formation. In combination with CCl_4 to form a chlorophosphonium salt, Bu_3P activates carboxylic acids for reaction with nucleophiles. This method enables peptide synthesis without racemization.[5]

3-Acetylpyrroles.[6] The difference in Lewis basicity is manifested in the aza-Baylis-Hillman reaction using Bu_3P or Ph_3P as catalyst. With the more basic Bu_3P pyrroles and 3-pyrrolines are formed.

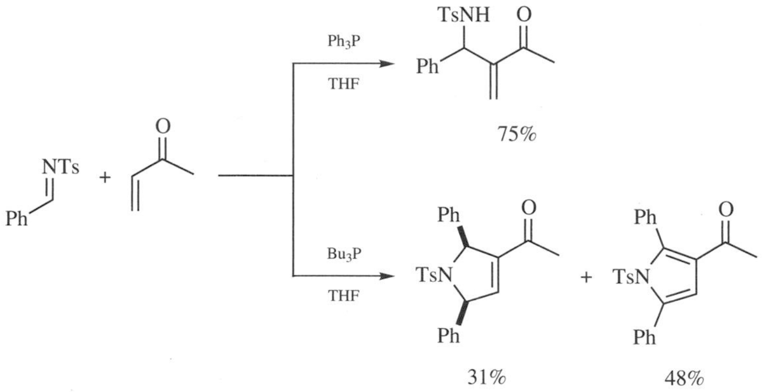

[1]Hou, X.-L., Fan, R.-H., Dai, L.-X. *JOC* **67**, 5295 (2002).
[2]Lukin, K.A., Narayanan, B.A. *T* **58**, 215 (2002).
[3]Gong, J.H., Im, Y.J., Lee, K.Y., Kim, J.N. *TL* **43**, 1247 (2002).
[4]Wang, L.-C., Luis, A.L., Agapiou, K., Jang, H.-Y., Krische, M.J. *JACS* **124**, 2402 (2002).
[5]Lorca, M., Kurosu, M. *SC* **31**, 469 (2001).
[6]Shi, M., Xu, Y.-M. *EJOC* 696 (2002).

2,4,6-Tri-*t*-butylpyrimidine.

Hindered base.[1] The title compound serves as surrogate for 2,6-di-*t*-butylpyridine as a hindered base required in glycosylation, enoltriflation, etc. It is readily prepared (70% yield) from Tf_2O-mediated condensation of *t*-butyl cyanide (2 molecules) and pinacolone (1 molecule).

[1]Crich, D., Smith, M., Yao, Q., Picione, J. *S* 323 (2001).

Tributyltin hydride–2,2′-azobis(isobutyronitrile). 19, 353–357; 20, 391–394; 21, 439–444

Demercuration. Frequently, $NaBH_4$ is used for hydrodemercuration but Bu_3SnH-AIBN has the same capacity.[1]

Reductive silylation. Carbonyl compounds are directly converted to alkyl silyl ethers by a mixture of $PhSeSiMe_3$ and Bu_3SnH in the presence of AIBN.[2]

Homoallylic alcohols. With mediation by Bu_3SnH-AIBN, reductive homologation of carboxylic acids with *B*-allylpinacolatoboron is achieved in one step.[3]

C_6H_{13}—COOH + (allyl pinacolboronate) → (Bu_3SnH - AIBN, THF Δ) → $C_6H_{13}CH(OH)CH_2CH=CH_2$ 79%

Group-transfer reactions. The aryl group of arenesulfonic esters and arenesulfonamides can be transferred to an atom situtated five-bonds apart when free radical is generated at the atom. For example, treatment of γ-iodoalkyl benzenesulfonates with Bu_3SnH-AIBN leads to rearranged products.[4]

(Bu_3SnH - AIBN / PhH ; MeI) 76%

Cyclization. The well-known method of free radical generation from halo compounds and organic selenides by Bu_3SnH-AIBN continues to be exploited in various synthetic situations. Intramolecular conjugate additions that give tetrahydrofurans[5] and azaspirocycles[6] are very successful. Reductive cyclization of unsaturated aldehydes in which the two functional groups are separated by a heteroatom also furnishes heterocyclic products.[7]

(Bu_3SnH - AIBN, PhMe 80°) 85%

(Bu_3SnH - AIBN, PhMe 60°) 60%

N-Substituted indolines or 1-substituted 1,2,3,4-tetrahydroisoquinolines are formed by the radical cyclization method.[8] An intramolecular radical addition to C-2 of the indole nucleus is the key step of a new synthesis of a unique class of alkaloids.[9]

Br N R Ph — Bu_3SnH - AIBN, PhH 80° → NH Ph H, R = H 72%; N Ph, R = Me 87%

N H PhSe H N O — Bu_3SnH - AIBN, PhMe 100° → H H N O N H, 91%

Diverse regioselectivity for reductive cyclization of 1,6-enynes due to steric effects[10] has strong implications to synthetic design.

HO OAc — Bu_3SnH - AIBN, PhH 80° → $SnBu_3$ HO AcO 50% + HO OAc 32%

O O — Bu_3SnH - AIBN, PhH 80° → O O + O O (18 : 1) 93% + $SnBu_3$ O O 5%

A cyclization process attended by rearrangement to create bridged ring ketones[11] is valuable.

Bu_3SnH - AIBN, PhH Δ → 51%

Rearrangements. 2-Methyleneaziridines that carry an *N*-substituent capable of radical generation are subject to ring expansion. Piperidine and indolizidine derivatives can be synthesized on exploiting this method.[12]

Bu_3SnH - AIBN, PhH Δ → 68%

Certain terphenyls are obtained from a reaction of *O,O′*-dibenzyl-2,5-diiodohydroquinones with Bu_3SnH-AIBN.[13] Spirocyclic intermediates are involved.

Bu_3SnH - AIBN, PhMe 80° → 67%

[1]Cossy, J., Blanchard, N., Meyer, C. *OL* **3**, 2567 (2001).
[2]Nishiyama, Y., Kajimoto, H., Kotani, K., Nishida, T., Sonoda, N. *JOC* **67**, 5696 (2002).
[3]Ollivier, C., Panchard, P., Renaud, P. *S* 1573 (2001).
[4]Bossart, M., Fässler, R., Schoenberger, J., Studer, A. *EJOC* 2742 (2002).
[5]Kamimura, A., Mitsudera, H., Matsumura, K., Omata, Y., Shirai, M., Yokoyama, S., Kakehi, A. *T* **58**, 2605 (2002).
[6]Koreeda, M., Wang, Y., Zhang, L. *OL* **4**, 3329 (2002).
[7]Bentley, J., Nilsson, P.A., Parsons, A.F. *JCS(P1)* 1461 (2002).
[8]Johnston, J.N., Plotkin, M.A., Viswanathan, R., Prabhakaran, E.N. *OL* **3**, 1009 (2001).
[9]Tanino, H., Fukuishi, K., Ushiyama, M., Okada, K. *TL* **43**, 2385 (2002).
[10]Toyota, M., Yokota, M., Ihara, M. *JACS* **123**, 1856 (2001).
[11]Rodriguez, J.R., Castedo, L., Mascarenas, J.L. *OL* **3**, 1181 (2001).
[12]Prevost, N., Shipman, M. *OL* **3**, 2383 (2001).
[13]Harrowven, D.C., Nunn, M.I.T., Newman, N.A., Fenwick, D.R. *TL* **42**, 961 (2001).

Triethylamine oxide.

Desilylation. *t*-Butyldimethylsilyl ethers of phenols are cleaved with Et_3NO in methanol.[1] Chemoselectivity is displayed by this system (aliphatic ROTBS are less readily cleaved).

[1]Zubaidha, P.K., Bhosale, S.V., Hashmi, A.M. *TL* **43**, 7277 (2002).

Triethylborane. 20, 395; **21,** 445–446

Alkylphosphinic acids and alkylthiophosphonic esters. Addition of NaH_2PO_2 to 1-alkenes is catalyzed by Et_3B in air.[1] Similarly, rapid formation of adducts $RCH_2CH_2PS(OR)_2$ from alkenes and thiophosphinic esters is observed.[2]

Addition reactions. Imines containing either an electron-donating or electron-withdrawing group accept alkyl radicals equally well.[3] The addition is accelerated by Lewis acid.

Bromine-transfer addition of bromoacetic esters to 1-alkenes in water has been studied.[4] Generally, polar solvents favor such a process. A noncatalytic application of Et_3B is in the synthesis of 3-alkenes from 1-alkynes via alkynyltriethylboronates.[5]

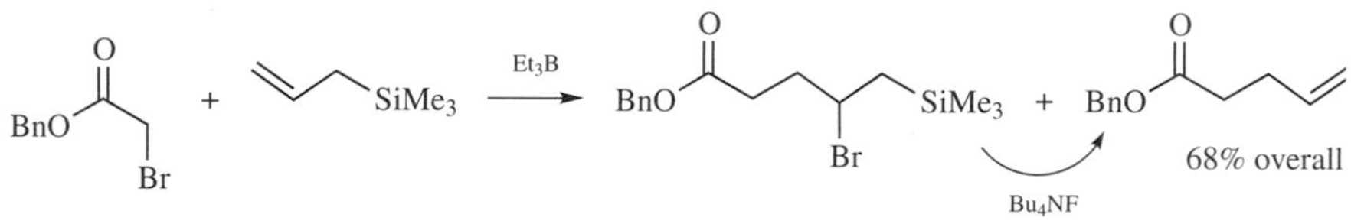

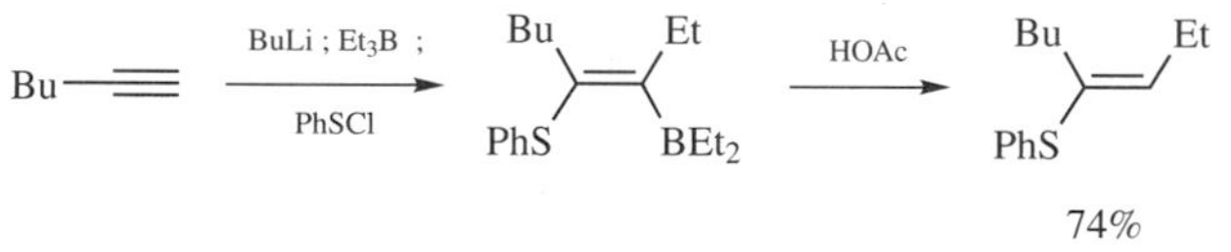

A free-radical addition-elimination sequence converts β-nitrostyrenes to 1-arylalkenes.[6]

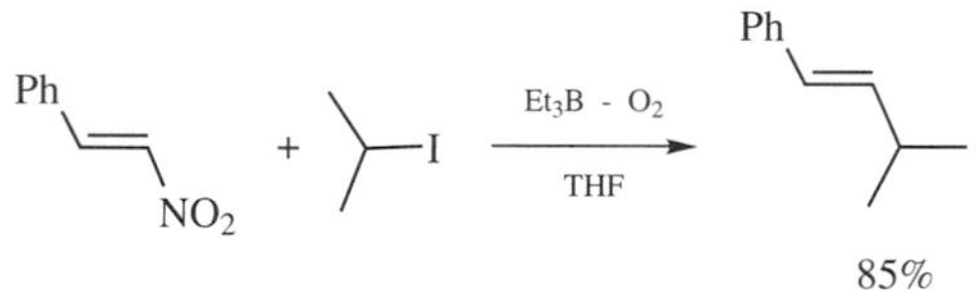

Reduction. Reductive removal of halogen atoms and phenylseleno group from carbon skeletons is easily performed with tin hydride[7,8] or germanium hydride reagents,[9] in the presence of Et_3B. Tris(2,6-diphenylbenzyl)tin hydride is a very hindered reagent thus hydrodehalogenation of bromoalkenes leads to (*Z*)-isomer almost exclusively, whether the original substrates are (*E*) or (*Z*).[7]

Allylation. Important development in Pd-catalyzed allylation concerns with the use of allyl alcohol.[10] While *o*-hydroxyacetophenone undergoes diallylation,[11] the corresponding propiophenone gives only monoallyl derivatives. Chelation involving the hydroxy group seems essential, since the corresponding methoxy and amino analogues do not undergo allylation.

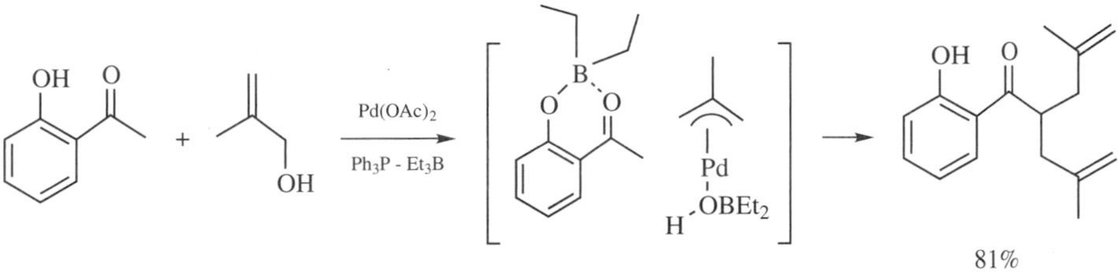

Cyclization and cycloaddition. Radicals generated from organohalides by reaction with $HgCl_2$ behave normally and cyclization toward a double bond to form a five-membered ring is observed.[12]

2-Iodomethylaziridine and analogues are suitable precursors of azahomoallyl radicals and the reaction with alkenes afford pyrrolidine derivatives.[13] Similarly, iodomethylcyclopropanes react with various alkenes (enol ethers, enimides, etc.) to form iodomethylcyclopentanes.[14] Homoallylic radical intermediates are indicated. In the cycloaddition involving a cyclopropanedicarboxylic ester and a simple 1-alkene, good yield of the adduct is obtained only in the presence of $Yb(OTf)_3$, but it is not required in reaction of the 1,1-bis(phenylsulfonyl)cyclopropane analogue.

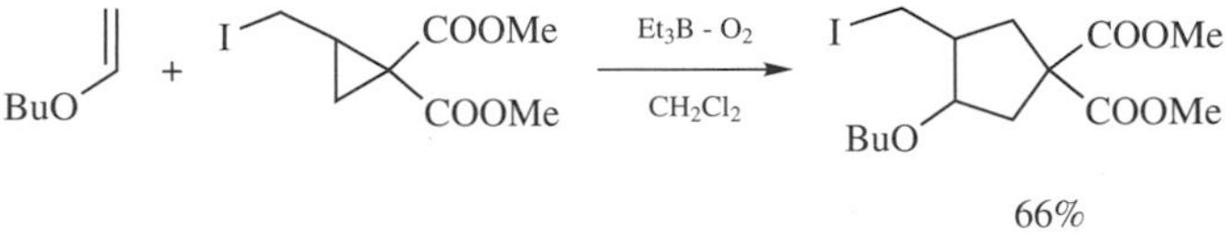

Catalyzed by Et_3B under oxygen α-halo esters and homoallylic gallium and indium reagents react to form 3-cyclopropylpropanoic esters.[15]

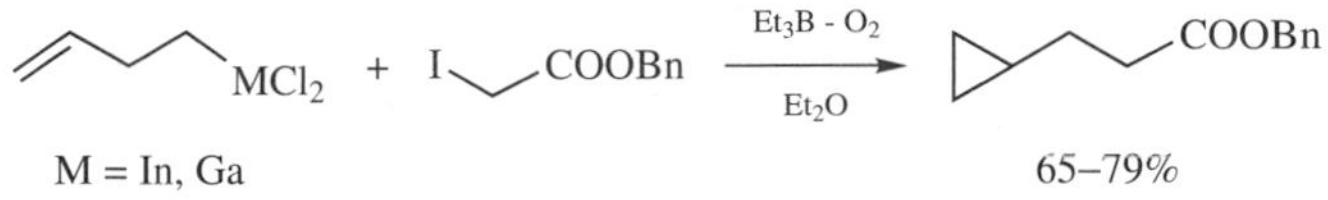

Lewis acid has positive effects on the atom transfer radical cyclization of α-bromo-β-keto esters induced by Et_3B, it can lead to polycyclic structures.[16] Zirconocene hydrochloride may serve the same purpose in the cyclization of halo acetals bearing an allyl moiety.[17] The critical role of Et_3B in this last reaction is manifested by observing only double bond reduction in its absence.

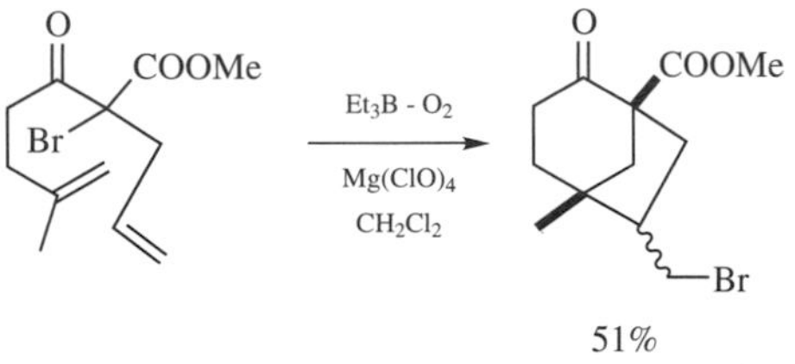

Et_3B - O_2
$Mg(ClO)_4$
CH_2Cl_2

51%

Coupling reactions. On treatment with Et_3B α-halo esters severe the C-X bond and couple with allylgallium species[18] and with metal enolates (M = Ga, Sn),[19,20] therefore products with two or more functional groups are generated.

A remarkable change of diastereoselectivity in a coupling results on adding Me_3Al.[21]

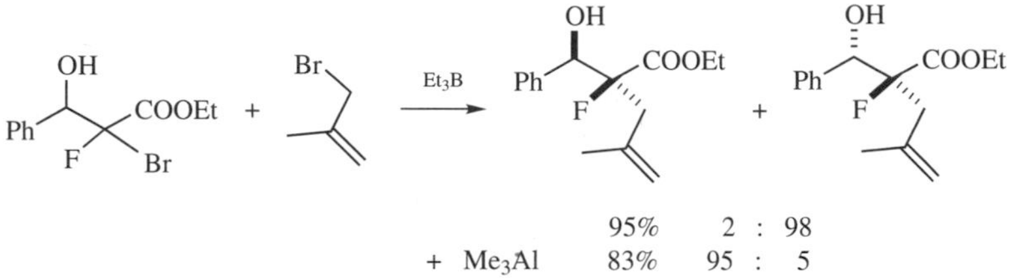

	95%	2 : 98
+ Me_3Al	83%	95 : 5

[1]Deprele, S., Montchamp, J.-L. *JOC* **66**, 6745 (2001).
[2]Gautier, A., Garipova, G., Dubert, O., Oulyadi, H., Piettre, S.R. *TL* **42**, 5673 (2001).
[3]Halland, N., Jorgensen, K.A. *JCS(P1)* 1290 (2002).
[4]Yorimitsu, H., Shinokubo, H., Matsubara, S., Oshima, K., Omoto, K., Fujimoto, H. *JOC* **66**, 7776 (2001).
[5]Gerard, J., Hevesi, L. *T* **57**, 9109 (2001).
[6]Liu, J.-T., Jang, Y.-J., Shih, Y.-K., Hu, S.-R., Chu, C.-M. *JOC* **66**, 6021 (2001).
[7]Sasaki, K., Kondo, Y., Maruoka, K. *ACIEE* **40**, 411 (2001).
[8]Bouvier, J.-P., Jung, G., Liu, Z., Gurerin, B., Guidon, Y. *OL* **3**, 1391 (2001).
[9]Nakamura, T., Yorimitsu, H., Shinokubo, H., Oshima, K. *BCSJ* **74**, 747 (2001).
[10]Kimura, M., Horino, Y., Mukai, R., Tanaka, S., Tamaru, Y. *JACS* **123**, 10401 (2001).
[11]Horino, Y., Naito, M., Kimura, M., Tanaka, S., Tamaru, Y. *TL* **42**, 3113 (2001).
[12]Mikami, S., Fujita, K., Nakamura, T., Yorimitsu, H., Shinokubo, H., Matsubara, S., Oshima, K. *OL* **3**, 1853 (2001).
[13]Kitagawa, O., Yamada, Y., Fujiwara, H., Taguchi, T. *ACIEE* **40**, 3865 (2001).
[14]Kitagawa, O., Yamada, Y., Fujiwara, H., Taguchi, T. *JOC* **67**, 922 (2002).
[15]Usugi, S., Tsuritani, T., Yorimitsu, H., Shinokubo, H., Oshima, K. *BCSJ* **75**, 841 (2002).
[16]Yang, D., Gu, S., Yan, Y.-L., Zhao, H.-W., Zhu, N.-Y. *ACIEE* **41**, 3014 (2002).
[17]Fujita, K., Nakamura, T., Yorimitsu, H., Oshima, K. *JACS* **123**, 3137 (2001).
[18]Usugi, S., Yorimitsu, H., Oshima, K. *TL* **42**, 4535 (2001).
[19]Usugi, S., Yorimitsu, H., Shinokubo, H., Oshima, K. *BCSJ* **75**, 2049 (2002).
[20]Miura, K., Fujisawa, N., Saito, H., Wang, D., Hosomi, A. *OL* **3**, 2591 (2001).
[21]Ishihara, T., Mima, K., Konno, T., Yamanaka, H. *TL* **42**, 3493 (2002).

Trifluoroacetic acid, TFA. 14, 322–323; **15,** 338–339; **18,** 375–376; **20,** 395–396; **21,** 446–447

Ritter reaction.[1] Alcohols that contain a β-phenylthio group are converted into amides with retention of configuration on exposure to CF_3COOH. The reaction proceeds via thiiranium ion intermediates.

SPh, OH — CF_3COOH / MeCN - H_2O → SPh, NHAc

SPh, OH — CF_3COOH / MeCN - H_2O → SPh, NHAc

Aza-Diels-Alder reaction. The acid-catalyzed [4+2]cycloaddition involving dienes and imines as reaction partners is fairly efficient, applicable to synthesis of piperidine alkaloids such as pinidine.[2,3]

+ EtOOC, N, Ph, Ph — CF_3COOH / CF_3CH_2OH / − 40° → COOEt, N, Ph, Ph (62%) → NH (pinidine)

Ring contraction. 3-Aryltetrahydropyrans undergo rearrangement on heating with CF_3COOH. Phenonium ions are formed as intermediates.[4]

Ar, O, OH — CF_3COOH 70° ; K_2CO_3 / MeOH → Ar, H, O, H, OH (53%)

[1]Eastgate, M.D., Fox, D.J., Morley, T.J., Warren, S. *S* 2124 (2002).
[2]Bailey, P.D., Smith, P.D., Pederson, F., Clegg, W., Rosair, G.M., Teat, S.J. *TL* **43**, 1067 (2002).
[3]Bailey, P.D., Smith, P.D., Morgan, K.M., Rosair, G.M. *TL* **43**, 1071 (2002).
[4]Nagumo, S., Ishii, Y., Kakimoto, Y., Kawahara, N. *TL* **43**, 5333 (2002).

Trifluoroacetic anhydride, TFAA. 18, 376–377; **19,** 361; **20,** 396–397; **21,** 447–448

Indole synthesis. After arylhydrazone formation by heating mixtures of ketones and arylhydrazines, further treatment with TFAA (in some cases Et_3N also needed) completes the indole synthesis.[1] *N*-Trifluoroacetylation facilitates the rearrangement and CF_3CONH_2 is a good leaving group.

β-Tetralones. In a one-step process to conduct acylation of alkenes with arylacetic acids and cycloalkylation of the initial products TFAA together with H_3PO_4 are used.[2]

COOH + Bu; $(CF_3CO)_2O$, H_3PO_4; O, Bu

55%

[1]Miyata, O., Kimura, Y., Naito, T. *S* 1635 (2001).
[2]Gray, A.D., Smyth, T.P. *JOC* **66**, 7113 (2001).

***N*-(Trifluoroacetoxy)succinimide.**

Amino acid modification.[1] The title reagent protects the amino group as well as activates the carboxyl end of an amino acid simultaneously.

NH_2, O, OH + F_3C, O, O, O–N, O; py, CH_2Cl_2; $NHCOCF_3$, O, O, O–N, O

91%

[1]Rao, T.S., Nampalli, S., Sekhar, P., Kumar, S. *TL* **43**, 7793 (2002).

2,2,2-Trifluoroethyl formate.

Formylation.[1] The reagent converts alcohols, amines, and hydroxylamines into their formyl derivatives.

[1]Hill, D.R., Hsiao, C.-N., Kurukulasuriya, R., Wittenberger, S.J. *OL* **4**, 111 (2002).

Trifluoromethanesulfonic acid (triflic acid). 14, 323–324; **15,** 339; **18,** 377; **19,** 362–363; **20,** 398–399; **21,** 448–449

Friedel-Crafts reactions. Triflic acid is employed in a stereoselective ring closure leading to a highly substituted tetralin in excellent yield.[1]

CF$_3$SO$_3$H

79°

91%

Note that reaction of protonated (by TfOH) β-substituted styrenes with benzene gives a complex mixture of products, *N*-cinnamylpyridine reacts cleanly under the same conditions.[2] There must be activation of the electrophilic site by the quaternary ammonium moiety.

Acylation of arenes with β-lactams provides 1-aryl-3-amino-1-propanones.[3]

Glycosylation. In both using glycosyl fluorides[4] and thioformidates[5] as disarmed glycosyl donors for reaction with other sugars (e.g., in disaccharide synthesis), TfOH is an excellent catalyst.

Thioesterification. Direct condensation of carboxylic acids and thiols is realized by heating the mixtures with TfOH in toluene (11 examples, 76–97%).[6]

Clization. *N*-Tosylaminoalkenes undergo cyclization with TfOH. This method is appropariate for the preparation of pyrrolidines and piperidines.[7]

9-Fluorenyl ketones are obtained from 1-hydroxy-1,1-diaryl ketones.[8]

CF$_3$SO$_3$H

[1]Harrowven, D.C., Wilden, J.D., Tyte, M.J., Hursthouse, M.B., Coles, S.J. *TL* **42**, 1193 (2001).
[2]Zhang, Y., Klumpp, D.A. *TL* **43**, 6841 (2002).
[3]Anderson, K.W., Tepe, J.J. *T* **58**, 8475 (2002).
[4]Mukaiyama, T., Suenaga, M., Chiba, H., Jona, H. *CL* 56 (2002).
[5]Mukaiyama, T., Chiba, H., Funasaka, S. *CL* 392 (2002).
[6]Iimura, S., Manabe, K., Kobayashi, S. *CC* 94 (2002).
[7]Schlummer, B., Hartwig, J.F. *OL* **4**, 1471 (2002).
[8]Yoshida, N., Ohwada, T. *S* 1487 (2001).

Trifluoromethanesulfonic anhydride (triflic anhydride). 13, 324–325; **14,** 324–326; **15,** 339–340; **16,** 357–358; **18,** 377–378; **19,** 363–365; **20,** 399; **21,** 449

Glycosylation. Sugars are activated by a combination of Tf$_2$O and Me$_2$S for glycosylation.[1] For conversion of thioglycosides to glycosyl triflates Tf$_2$O and 1-benzenesulfinylpiperidine are used.[2]

Activated by Tf_2O, 2-(hydroxycarbonyl)benzyl glycosides become glycosyl donors for β-mannopyranosylation.[3]

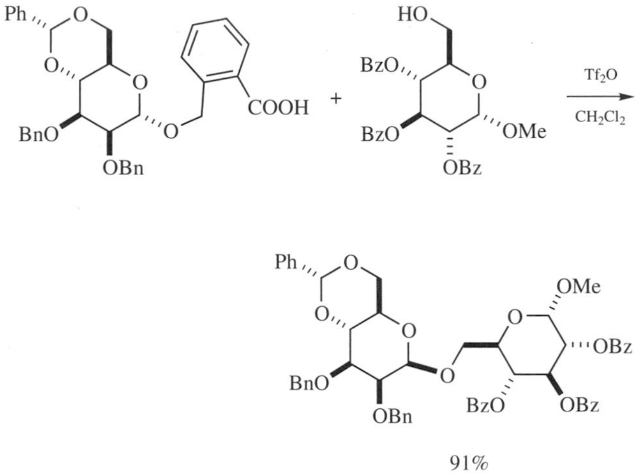

91%

Dehydration of amides. Generation of nitrilium ions and 1-aza-1,2-diene ions from secondary and tertiary amides, respectively, is readily achieved by a triflation-elimination tandem. These species are very electrophilic (e.g., forming amidinium ions with pyridine[4]) and can serve as addends of [2+2]cycloaddition.[5]

Trifluoromethanesulfinylation. The mixed anhydride $CF_3SOOSO_2CF_3$ is formed in situ from CF_3SOOK and Tf_2O and used to prepare aryl trifluoromethyl sulfoxides.[6]

Annulation. Internal interception of electrophiles arising from activation by Tf_2O results in cyclic products. In the presence of an allylic ether at proper distance to the activatable group cyclization (π-participation) accompanies a 1,2-shift (ring expansion), giving rise to new ring systems.[7]

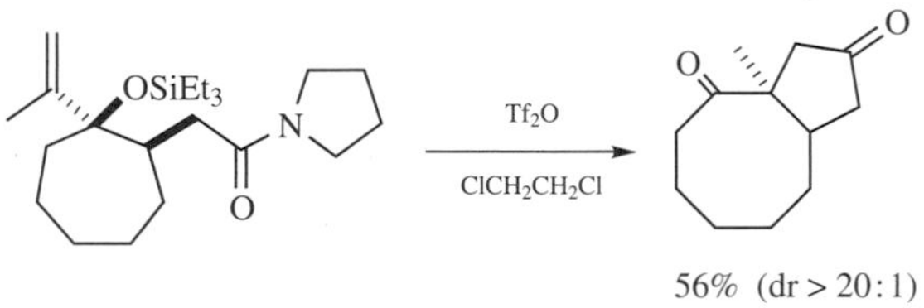

56% (dr > 20:1)

1,3-Dipolar species are readily generated from *N*-trimethylsilylmethyl β-aminoketones on triflation and subsequent desilylation. Polycyclic systems possessing a bridgehead nitrogen atom can be synthesized from such precursors in simple steps.

Me_3Si N O → Tf_2O → TfO^- Me_3Si N^+ OTf → $(Bu_4N)[Ph_3SiF_2]$ / DMAD →

MeOOC COOMe N OTf

63%

[1]Nguyen, H.M., Chen, Y., Duron, S.G., Gin, D.Y. *JACS* **123**, 8766 (2001).
[2]Crich, D., Smith, M. *JACS* **123**, 9015 (2001).
[3]Kim, K.S., Kim, J.H., Lee, Y.J., Lee, J., Park, J. *JACS* **123**, 8477 (2001).
[4]Charette, A.B., Grenon, M., Lemire, A., Pourashraf, M., Martel, J. *JACS* **123**, 11829 (2001).
[5]Chen, B.C., Ngu, K., Guo, P., Liu, W., Sundeen, J.E., Weinstein, D.S., Atwal, K.S., Ahmad, S. *TL* **42**, 6227 (2001).
[6]Wakselman, C., Tordeux, M., Freslon, C., Saint-Jalmes, L. *SL* 550 (2001).
[7]Overman, L.E., Wolfe, J.P. *JOC* **67**, 6421 (2002).
[8]Epperson, M.T., Gin, D.Y. *ACIEE* **41**, 1778 (2002).

Trifluoromethyl iodide.

Trifluoromethylation. Preparation of trifluoromethyl carbinols from carbonyl compounds is simply performed with the title reagent under photoirradiation in the presence of tetrakis(dimethylamino)ethene..

[1]Ait-Mohand, S., Takechi, N., Medebielle, M., Dolbier, W.R. *OL* **3**, 4271 (2001).

(Trifluoromethyl)trimethylsilane. 15, 341; **18,** 378–379; **19,** 366–367; **20,** 400; **21,** 450–451

Trifluoromethylation. Convenient introduction of a trifluoromethyl group to organic molecules uses Me_3SiCF_3 as reagent. In the group transfer to *N*-tosylimines[1] and chiral *N*-(*t*-butylsulfinyl)imines,[2,3] the activator is the fluoride ion source $Bu_4N[Ph_3SiF_2]$, although R_4NF and other silaphilic reagents should be effective also. Trifluoroacetamides and trifluorothioacetamides are prepared from isocyanates and isothiocyanates,[4] besides trifluoromethylsulfinamides from thioimine oxides RN═S═O,[5] on reaction with Me_3SiCF_3 - Me_4NF. Trifluoromethylation of nitrones is conducted in the presence of *t*-BuOK.[6]

Conversion of alcohols to homologous 1,1,1-trifluoroalkanes is accomplished by substitution of the corresponding triflates.[7]

α,α-Difluoroaldols. Treatment of acylsilanes with Me_3SiCF_3 - $Bu_4N[Ph_3SiF_2]$ and then carbonyl compounds lead to difluorinated aldols.[8] Apparently, trifluoromethylation is immediately followed by Brook rearrangement and expulsion of a fluoride ion to form 2-siloxy-1,1-difluoroalkenes. Particularly interesting is that certain bisacylsilanes undergo intramolecular condensation.[9]

+ Me_3Si-CF_3 —— $(Bu_4N)[Ph_3SnF_2]$; $BiCl_3$ ——→ 35%

[1]Prakash, G.K.S., Mandal, M., Olah, G.A. *SL* 77 (2001).
[2]Prakash, G.K.S., Mandal, M., Olah, G.A. *ACIEE* **40**, 589 (2001).
[3]Prakash, G.K.S., Mandal, M., Olah, G.A. *OL* **3**, 2847 (2001).
[4]Kirij, N.V., Yagupolskii, Y.L., Petukh, N.V., Tyrra, W., Naumann, D. *TL* **42**, 8181 (2001).
[5]Yagupolskii, Y.L., Kirij, N.V., Shevechenko, A.V., Tyrra, W., Naumann, D. *TL* **43**, 3029 (2002).
[6]Nelson, D.W., Owens, J., Hiraldo, D. *JOC* **66**, 2572 (2001).
[7]Sevenard, D.V., Kirsch, P., Röschenthaler, G.-V., Movchen, V.N., Kolomeitsev, A.A. *SL* 379 (2001).
[8]Lefebvre, O., Brigaud, T., Portella, C. *JOC* **66**, 1941 (2001).
[9]Saleur, D., Bouillon, J.-P., Portella, C. *JOC* **66**, 4543 (2001).

Triiron dodecacarbonyl.

Annulation. Dicarbonylative cyclization that converts alkynes to cyclobutenediones is induced by amines.[1] On the other hand, succinic anhydrides and succinimides are formed on long standing, presumably due to nucleophilic attack of the amines on the cyclic diacyliron complexes to form maleimides and internal reduction by some hydridoiron species.[2]

+ $Fe_3(CO)_{12}$ —— $BuNH_2$ / THF ; $CuCl_2$ / Me_2CO ——→

[1]Rameshkumar, C., Periasamy, M. *TL* **41**, 2719 (2000).
[2]Periasamy, M., Rameshkumar, C., Mukkanti, A. *JOMC* **649**, 209 (2002).

Triisobutylaluminum. 19, 367–368; 21, 451

Hydroalumination. The conversion of 1-alkenes to alkyldiisobutylaluminums catalyzed by $(Ph_3P)_2PdCl_2$ and other late transition metal complexes employs i-Bu_3Al but not Dibal-H to avoid reduction of Pd(II).[1]

Debenzylation. Selective removal of benzyl group(s) from certain protected monosaccharides (from glucose and altrose) containing a phenylsulfonylethylidene acetal unit is performed by *i*-Bu_3Al.[2]

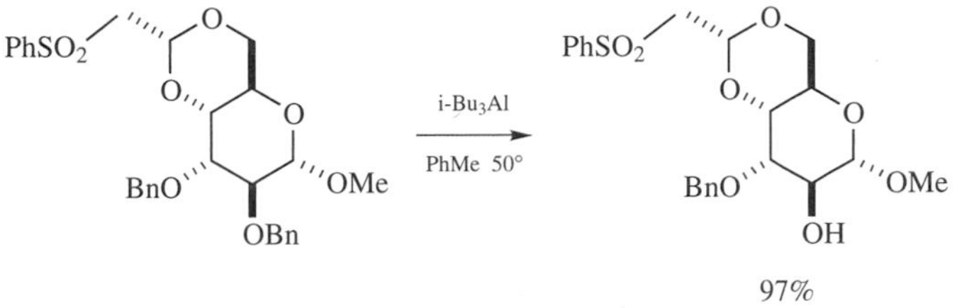

PhSO₂ BnO OBn OMe → i-Bu₃Al, PhMe 50° → PhSO₂ BnO OH OMe 97%

Elimination. A method for regioselective formation of 2-methoxy-1-alkenes from 2-alkanone dimethylacetals involves treatment with *i*-Bu_3Al.[3]

[1]Gagneur, S., Makabe, H., Negishi, E. *TL* **42**, 785 (2001).
[2]Chevalier-du-Roizel, B., Cabianca, E., Rollin, P., Sina, P. *T* **58**, 9579 (2002).
[3]Cabrera, G., Fiaschi, R., Napolitano, E. *TL* **42**, 5867 (2001).

Trimethylaluminum. 15, 341–342; **17,** 372–375; **18,** 365–367; **19,** 369–370; **20,** 401–402; **21,** 452–453

Reduction. In Meerwein-Ponndorf-Verley reduction Me_3Al or Me_2AlCl can be used as catalyst in combination with isopropanol.[1] Tishchenko reaction of aldols and aldehydes gives monoesters of *anti*-1,3-diols.[2]

Epoxide opening. Opening of β,γ-epoxy sulfides with organoaluminums involves double inversion of configuration due to intervention of thiiranium ion intermediates.[3] On the other hand, internal delivery from an ate complex is involved in the reaction of epoxycarbinols.[4]

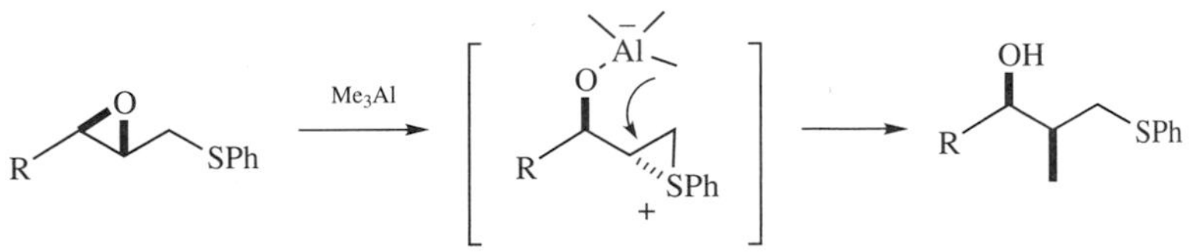

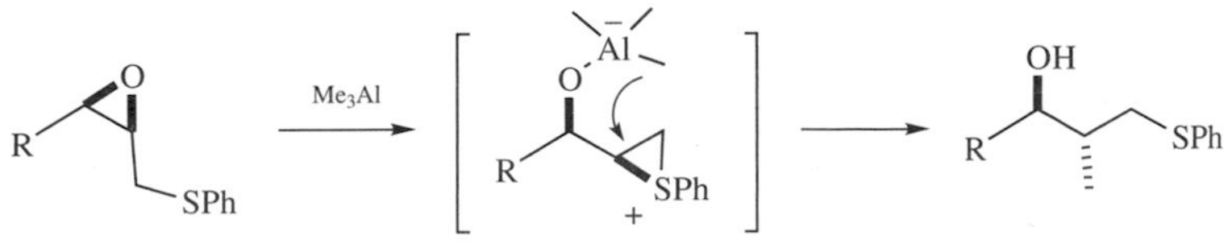

R, O, SPh → Me₃Al → [Al, O, R, SPh, +] → OH, R, SPh

R, O, SPh → Me₃Al → [Al, O, R, SPh, +] → OH, R, SPh

Amidines. Mixing Me_3Al with a toluene suspension of NH_4Cl generates $MeAl(NH_2)Cl$ which transforms esters into amidines.[5]

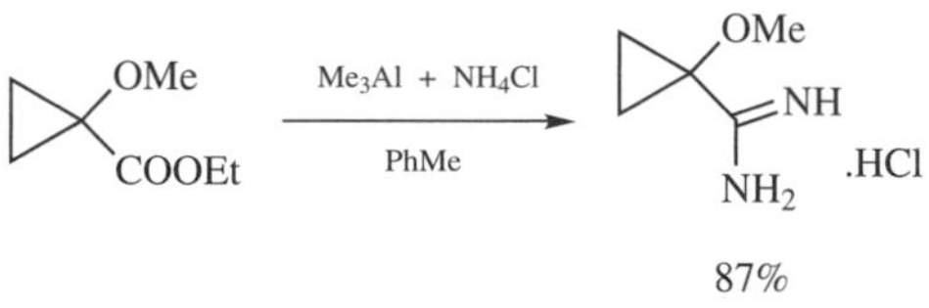

Diels-Alder reactions. The effectiveness of Me_3Al in promoting diastereoselective intramolecular Diels-Alder reactions is illustrated by the following example.[6]

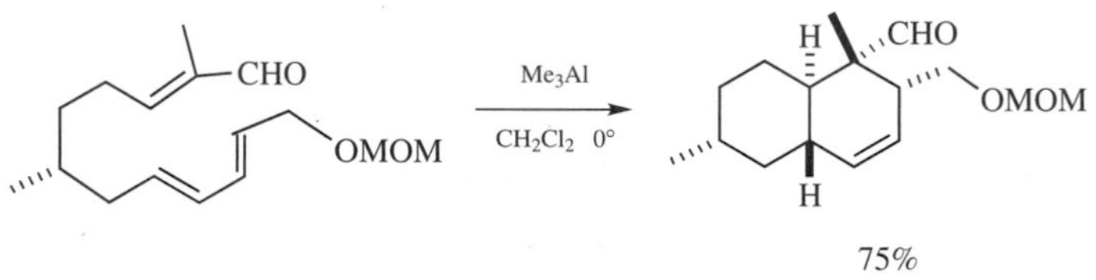

[1]Campbell, E.J., Zhou, H., Nguyen, S.T. *OL* **3**, 2391 (2001).
[2]Simpura, I., Nevalainen, V. *TL* **42**, 3905 (2001).
[3]Sasaki, M., Tanino, K., Miyashita, M. *JOC* **66**, 5388 (2001).
[4]Sasaki, M., Tanino, K., Miyashita, M. *OL* **3**, 1765 (2001).
[5]Gielen, H., Alonso-Alijia, C., Hendrix, M., Niewöhner, U., Schauss, D. *TL* **43**, 419 (2002).
[6]Yuki, K., Shindo, M., Shishida, K. *TL* **42**, 2517 (2001).

Trimethylsilyl azide. 13, 24–25; **14,** 25; **15,** 342–343; **16,** 17; **18,** 379–380; **19,** 371–372; **20,** 403; **21,** 453

Distributive coupling. Courses of Pd-catalyzed reaction of Me_3SiN_3 with *o*-alkynylaryl isocyanides are varied by adding different ligands.[1]

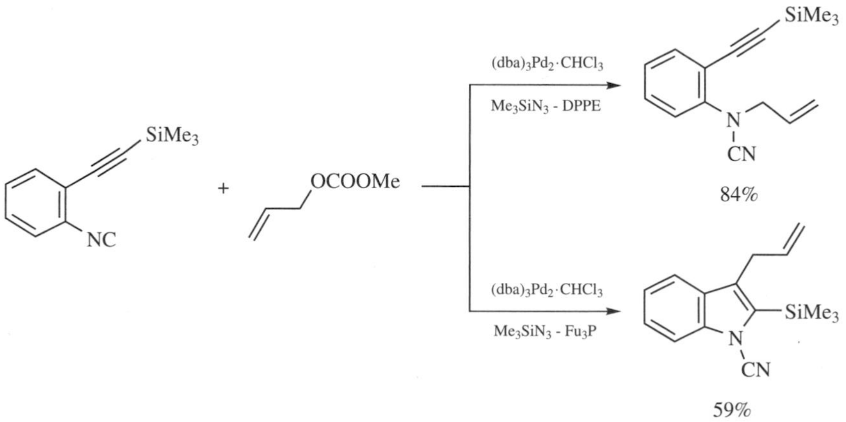

***N*-Allyl cyanamides.** The Pd-catalyzed reaction of isocyanides with allyl carbonates and Me_3SiN_3 results in modification of the [NC] residue by attachment of a nitrogen atom to the terminal carbon atom and an allyl group to the nitrogen.[2]

[1]Kamijo, S., Yamamoto, Y. *JACS* **124**, 11940 (2002).
[2]Kamijo, S., Jin, T., Yamamoto, Y. *JACS* **123**, 9453 (2001).

Trimethylsilyl bromide. 15, 51; **16,** 50; **18,** 380; **19,** 373–374; **20,** 404

Elimination. Esters that are linked at the α-position to a resin by a triazene unit are released by cleavage with Me_3SiX (X=Cl, Br, I). A halogen atom is introduced to the resulting esters.[1]

Silylation. Nitroalkanes undergo silylation on reaction with Me_3SiBr to afford silyl ethers of α,β-unsaturated aldoximes.[2]

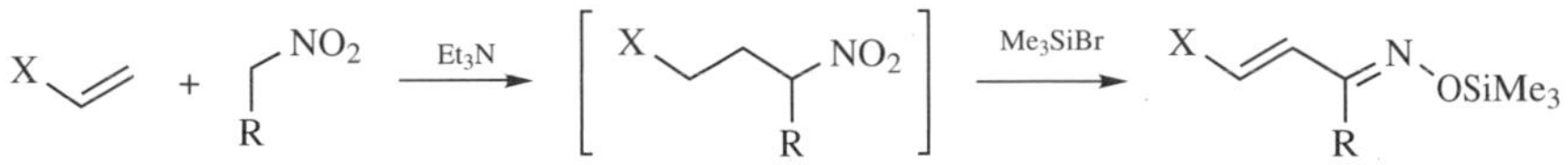

X = CN, COOMe

[1]Pilot, C., Dahmen, S., Lauterwasser, F., Bräse, S. *TL* **42**, 9179 (2001).
[2]Danilenko, V.M., Tishkov, A.A., Ioffe, S.L., Lyapkalo, I.M., Strelenko, Y.A., Tartakovsky, V.A. *S* 635 (2002).

Trimethylsilyl chloride. 15, 89; **16,** 85–86; **18,** 381; **19,** 374–375; **20,** 404–405; **21,** 453–454

Esterification and hydrolysis. For preparation of chlorohydrin esters from 1,2-diols the treatment with esters and Me_3SiCl at 80° suffices.[1] Liberation of alkylation products from 1,4-dioxan-2-ones is achieved with Me_3SiCl-MeOH.[2]

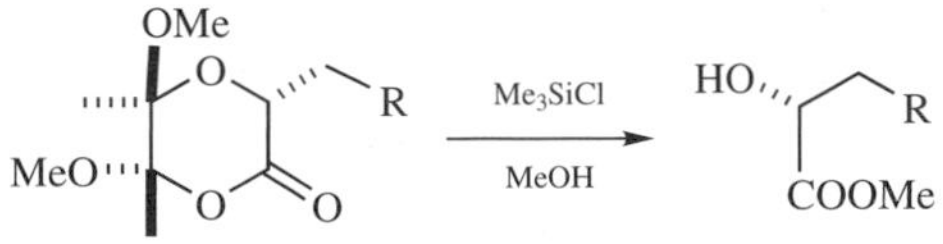

Trioxanes. Aldehydes are cyclotrimerized on contact with catalytic amounts of Me_3SiCl at room temperature (solvent-free conditions).[3]

Grignard reactions. 2-Aryl-1-pyrrolines are prepared from pyrrolidinone by reaction with ArMgX (or ArLi) in the presence of Me_3SiCl.[4]

[1]Eras, J., Mendez, J.J., Balcells, M., Canela, R. *JOC* **67**, 8631 (2002).
[2]Diez, E., Dixon, D.J., Ley, S.V. *ACIEE* **40**, 2906 (2001).
[3]Auge, J., Gil, R. *TL* **43**, 7919 (2002).
[4]Coindet, C., Comel, A., Kirsch, G. *TL* **42**, 6101 (2001).

Trimethylsilyl cyanide. 13, 87–88; **14,** 107; **15,** 102–104; **17,** 89; **18,** 381–382; **19,** 375; **20,** 405; **21,** 455–456

α-Cyanohydrin derivatives. *O*-Trimethylsilyl cyanohydrins of α-alkoxy-, α-acetoxy- and related ketones are readily formed in an $InBr_3$-mediated reaction with Me_3SiCN.[1] Acceleration of the reaction by a lithium alkoxide (e.g., of triglyme monomethyl ether) is noted, the camphor derivative is obtained in >95% yield.[2]

Strecker synthesis. The three-component condensation is facilitated by pressure[3] and catalyzed by $(i\text{-PrO})_3La$.[4]

Substitutions. Me_3SiCN reacts with arenesulfenyl chlorides to give thiocyanates,[5] and in the presence of a silver(I) salt it replaces tertiary hydroxyl group to form isonitriles[6] at room temperature.

Allylic displacement of 2-(*N,N*-bistrimethylsiloxy)-1-alkenes takes place on reaction with Me_3SiCN to afford oxime ethers of cyanomethyl ketones.[7]

Addition. Addition of Me_3SiCN to ynamines followed by desilylation constitutes a route to α-cyanoenamines.[8]

Redox reaction. Enals such as cinnamaldehyde are converted to saturated carboxylic acids by Me_3SiCN and a Lewis base [DBU or tris(2,4,6-trimethoxyphenyl)phosphine] at room temperature.[9] Esters and amides can also be obtained by using different quenching reagents. The process is remarkable for the mild conditions involved and high yields. Previous use of $RuCl_3 \cdot H_2O\text{-}Cy_3P$ to effect the same transformation of cinnamaldehyde requires high temperature (180°) and the yield is moderate (46%).[10]

[1]Bandini, M., Cozzi, P.G., Garelli, A., Melchiorre, P., Umani-Ronchi, A. *EJOC* 3243 (2002).
[2]Wilkinson, H.S., Grover, P.T., Vandenbossche, C.P., Bakale, R.P., Bhongle, N.N., Wald, S.A., Senanayake, C.H. *OL* **3**, 553 (2001).
[3]Matsumoto, K., Kim, J.C., Hayashi, N., Jenner, G. *TL* **43**, 9167 (2002).
[4]Sun, P., Qian, C., Wang, L., Chen, R. *SC* **32**, 2973 (2002).
[5]Still, I.W.J., Watson, I.D.G. *SC* **31**, 1355 (2001).
[6]Kitano, Y., Chiba, K., Tada, M. *S* 437 (2001).
[7]Lesiv, A.V., Ioffe, S.L., Strelenko, Y.A., Bliznets, I.V., Tartakovsky, V.A. *MC* 99 (2002).
[8]Lukashev, N.V., Kazantsev, A.V., Borisenko, A.A., Beletskaya, I.P. *T* **57**, 10309 (2001).
[9]Kawabata, H., Hayashi, M. *TL* **43**, 5645 (2002).
[10]de Vries, J.G., Roelfes, G., Green, R. *TL* **39**, 8329 (1998).

Trimethylsilyldialkylamine. 18, 382; **19,** 376; **20,** 407; **21,** 457

Elimination. Cyclic acetals and oxazolidines undergo ring opening with Et_2NSiMe_3. The products are enol ethers and enamines/imines, respectively.[1]

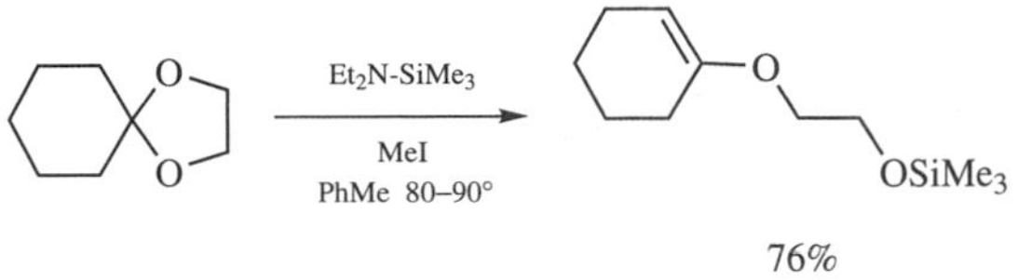

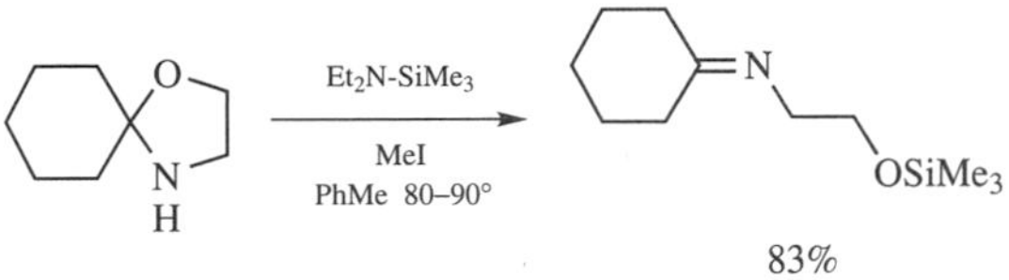

Trifluoromethylating agent. The adduct obtained from heating 1,1,1-trifluoroacetophenone with Me_2NSiMe_3 at 110° is a stable reagent for trifluoromethylation.[2] It releases the nucleophilic [CF_3^-] by fluoride ion.

[1]Iwata, A., Tang, H., Kunai, A., Oshita, J., Yamamoto, Y., Matui, C. *JOC* **67**, 5170 (2002).
[2]Motherwell, W.B., Storey, L.J. *SL* 646 (2002).

Trimethylsilyldiazomethane. 20, 405–406

2-Indanones.[1] Trialkylsilyl(aryl)ketenes and Me_3SiCHN_2 combine to afford 1-trialkylsilyl-2-indanones at room temperature.

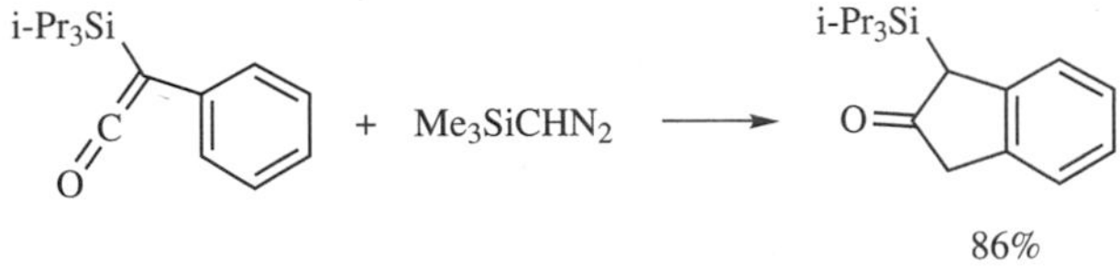

Homologation. Diazoketones are prepared from carboxylic acids via mixed anhydrides.[2] More interesting is the homologation of carbonyl compounds to furnish silyl enol ethers.[3,4]

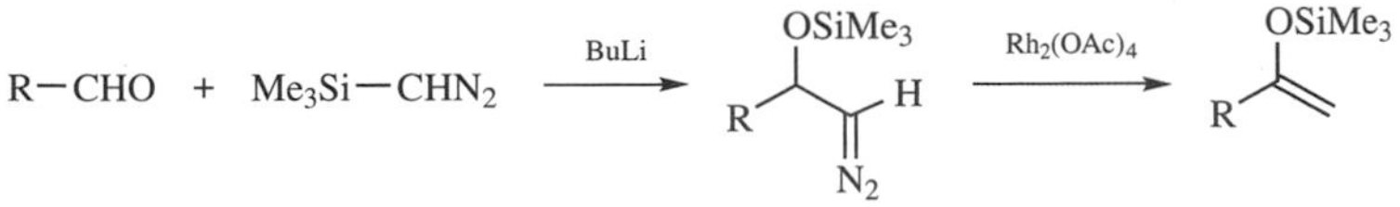

Insertion. Anomeric C-H bond insertion results in *C*-glycosides, when 2-oxopropyl glycosides are treated with Me_3SiCHN_2, and a precursor of 3-deoxy-D-*arabino*-2-heptulosonic acid is readily secured.[5]

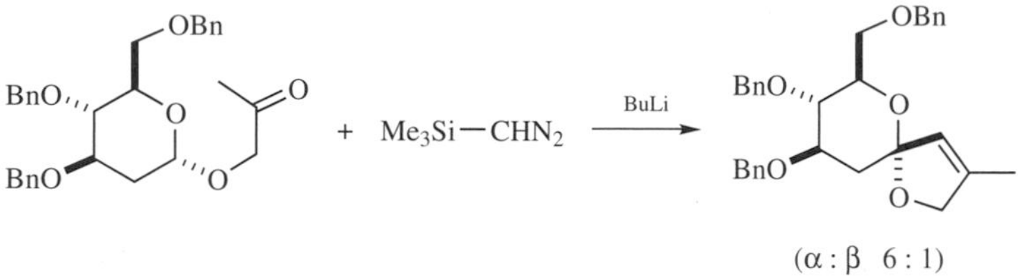

(α : β 6 : 1)

[1]Dalton, A.M., Zhang, Y., Davie, C.P., Danheiser, R.L. *OL* **4**, 2465 (2002).
[2]Cesar, J., Dolenc, M.S. *TL* **42**, 9099 (2001).
[3]Dias, E.L., Brookhart, M., White, P.S. *JACS* **123**, 2442 (2001).
[4]Aggarwal, V.K., Sheldon, C.G., Macdonald, G.J., Martin, W.P. *JACS* **124**, 10300 (2002).
[5]Wardrop, D.J., Zhang, W. *TL* **43**, 5389 (2002).

Trimethylsilylethyne.

Nitroethyne.[1] The nitro compound is obtained from the title reagent on reaction with nitronium hexafluorophosphate at 0° in $MeNO_2$. It is stable in dilute solution at room temperature for about 1 week, reacts rapidly with nucleophiles and conjugated dienes. Its reaction with enol ethers and furan proceeds through stages of cycloaddition, fragmentation, and 1,3-dipolar cycloaddition, to give isoxazoles.

Trimethylsilylethynyl ketones.[2] These ketones are good surrogates of ethynyl ketones in participating double Michael reactions because they are earsier to prepare. In situ desilylation by fluoride ion supplies the required reagents.

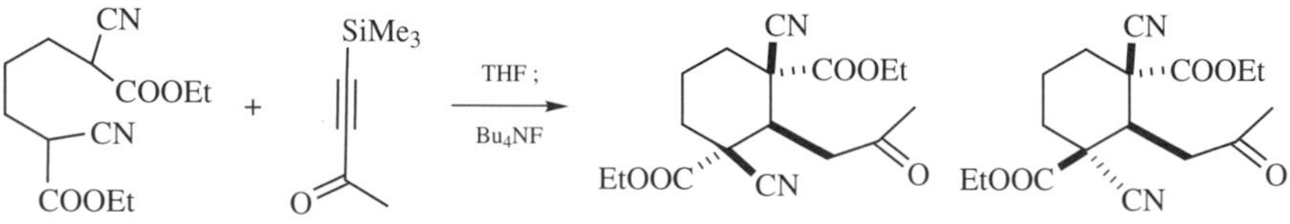

1-Trimethylsilyl-1-methylseleno-1-alken-3-ones.[3] Seleno amides on activation with MeOTf are susceptible to attack by lithiated trimethylsilylethyne. Rearrangement occurs afterwards.

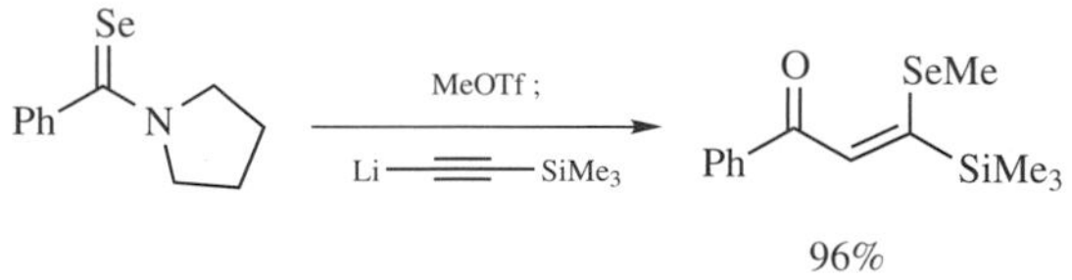

96%

***o*-Vinylation of phenols.**[4] Treatment of a phenyl with a 2:1 ratio BuLi and $SnCl_4$ in PhCl starts reaction with trimethylsilylethyne. The advantage of this system over the Bu_3N-$SnCl_4$ system is that it is catalytic (50 mol% BuLi, 25 mol% $SnCl_4$).

[1]Zhang, M.-X., Eaton, P.E., Steele, I., Gilardi, R. *S* 2013 (2002).
[2]Holeman, D.S., Rasne, R.M., Grossman, R.B. *JOC* **67**, 3149 (2002).
[3]Murai, T., Mutoh, Y., Kato, S. *OL* **3**, 1993 (2001).
[4]Kobayashi, K., Yamaguchi, M. *OL* **3**, 241 (2001).

Trimethylsilyl iodide. 16, 188–189; **18,** 383; **19,** 376–377; **20,** 407; **21,** 458

Cyclization. Ionization of acetals by Me_3SiI can induce participation of a remote multiple bond system, leading to cyclization.[1,2]

R, R′O, OR′ + Me_3SiI —PhMe→ OR′, R, I

(*Z*) (major)

Reductive etherification. Carbonyl compounds are reduced to ethers by a catalyzed reaction with alkoxyhydrosilanes (e.g., $ROSiHMe_2$).[3] As catalyst Me_3SiI can be used.

Aldol reaction. Formation of 1-iodo-5-hydroxy-1-alken-3-ones from 1-butyn-3-one and aldehydes is mediated by Me_3SiI-$BF_3 \cdot OEt_2$. A formal hydroiodination of the triple bond accompanies the aldol reaction.[4]

O + Ph-CHO + Me_3SiI —$BF_3 \cdot OEt_2$→ O, OH, Ph, I

76% (*Z/E* >95 : 5)

[1]Takami, K., Yorimitsu, H., Shinokubo, H., Matsubara, S., Oshima, K. *SL* 293 (2001).
[2]Kang, S.-K., Kim, Y.-M., Ha, Y.-H., Yu, C.-M., Yang, H., Lim, Y. *TL* **43**, 9105 (2002).
[3]Miura, K., Ootsuka, K., Suda, S., Nishikori, H., Hosomi, A. *SL* 313 (2002).
[4]Wei, H.-X., Kim, S.H., Li, G. *OL* **4**, 3691 (2002).

Trimethylsilyl isothiocyanate.

Alkyl thiocyanates.[1] Preparation of RSCN from alkyl halides by an S_N2 substitution is well established. The use of Me_3SiNCS and Bu_4NF is an alternative which proceeds at low temperature thus thermal isomerization to the isothiocyanates can be minimized. It also can be accommodated by solvents such as THF and aqueous workup is avoided.

[1]Renard, P.-Y., Schwebel, H., Vayron, P., Leclerc, E., Dias, S., Mioskowski, C. *TL* **42**, 8479 (2000).

N-Trimethylsilyl-N-methylacetamide.

Enolsilylation.[1] This commercially available reagent performs enolsilylation of ketones in the presence of a catalytic amount of a base (NaH, DBU). Usually the thermodynamically controlled products are generated. The protocol is applicable to transformation of α,α-disubstituted ketones, α-chloro ketones, and labile enones.

[1]Tanabe, Y., Misaki, T., Kurihara, M., Iida, A., Nishii, Y. *CC* 1628 (2002).

Trimethylsilyl trifluoromethanesulfonate. 13, 329–331; **14,** 333–335; **15,** 346–350; **16,** 363–364; **17,** 379–386; **18,** 383–384; **19,** 379–381; **20,** 408–410; **21,** 460–462

Functional group manipulation. Catalyzed by Me_3SiOTf, prenyl ethers are cleaved,[1] silyl and THP ethers are converted to benzhydryl ethers directly in an exchange reaction with $HCOOCHPh_2$.[2]

Vicinal diacetals are formed by mixing 1,2-diols with 2,3-butanedione and Me_3SiOMe with Me_3SiOTf as catalyst.[3]

OH OH + O O — Me_3SiOMe / Me_3SiOTf, CH_2Cl_2 → O O OMe OMe

97%

Glycosylation. Glycosyl trifluoroacetimidates serve as glycosyl donors thanks to the facile condensation promoted by Me_3SiOTf.[4] 3–Trichloroacetimidato glycals are susceptible to Ferrier rearrangement and a highly stereoselective synthesis of pseudogalactal glycosides[5] is based on the Me_3SiOTf-catalyzed reaction.

Friedel-Crafts reactions. Glycosyl phosphates alkylate phenols to afford *C*-aryl β-glycosides.[6] Baylis-Hillman adducts are converted to 2-benzyl-2-alkenoic esters on reaction with arenes in the presence of Me_3SiOTf.[7] Generation of stabilized propargylic cations and in situ trapping (with arenes, allylsilanes, silyl enol ethers) lead to arylallenes.[8]

Me_3SiO Ph Ph O OMe — Me_3SiOTf / CH_2Cl_2, −78° → Ph Ph C O OMe

80%

Aliphatic Friedel-Crafts acylation of silyl enol ethers with oxalyl chloride gives 2-hydroxy-2-butenolides.[9]

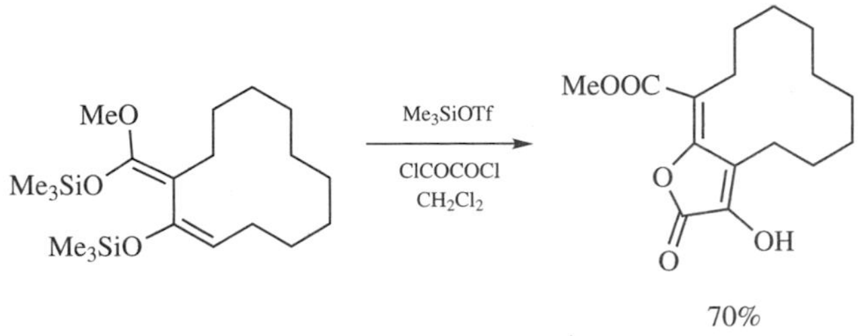

MeO
Me_3SiO
Me_3SiO
Me_3SiOTf
ClCOCOCl
CH_2Cl_2
MeOOC
O
O
OH
70%

Allylation. The oxaphilic Me_3SiOTf is useful for inducing C-O bond cleavage of *N,O*-acetals, therefore allylation of semicyclic members leads to ω-hydroxy homoallylic amines.[10]

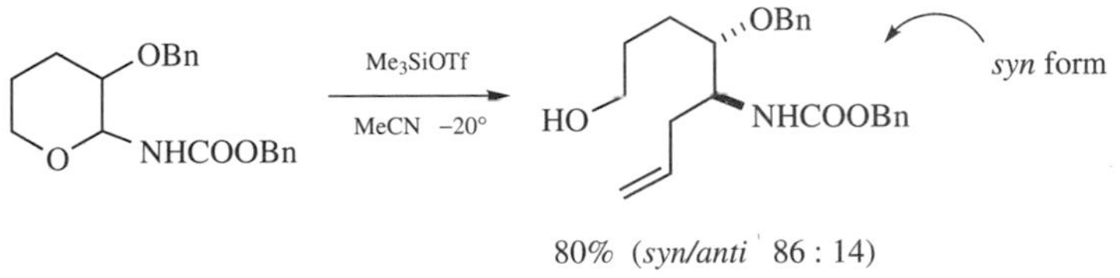

OBn
O
NHCOOBn
Me_3SiOTf
MeCN −20°
HO
OBn
NHCOOBn
syn form
80% (*syn/anti* 86 : 14)

Chirality transfer in homoallyl ether formation is accomplished by activating aldehydes with optically active α-trimethylsiloxy-α-trimethylsilyltoluene.[11]

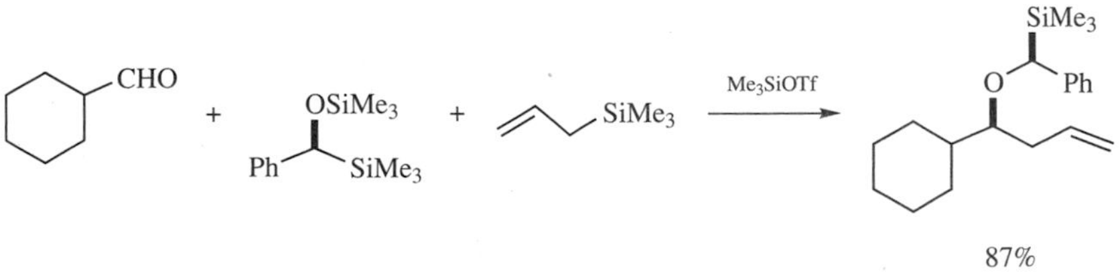

CHO
+
$OSiMe_3$
Ph
$SiMe_3$
+
$SiMe_3$
Me_3SiOTf
$SiMe_3$
O
Ph
87%

By allylative displacement of acetals 2-vinyl-1-silylmethylcyclopropane derivatives provides a hexadiene chain.[12] Intramolecular allylation of nascent oxonium species serves to elaborate 2,3-disubstituted oxacyclic compounds.[13] Mechanistically allied is the cyclization leading to 3-cycloalken-1-ols and 4-acylmethylcycloalkenes.[14]

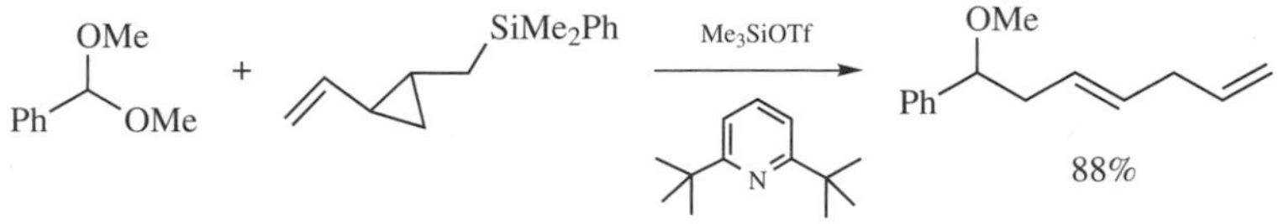

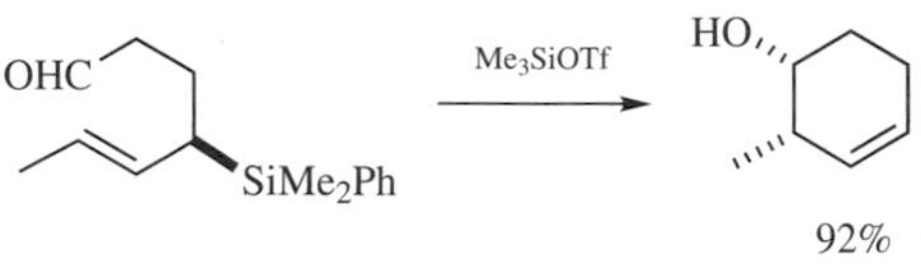

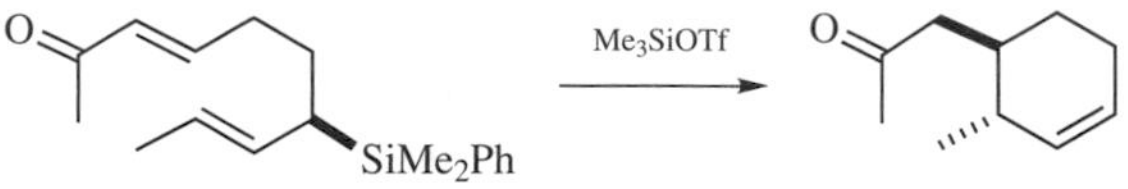

Cyclization. Sequential π- and n-participations in acetal ionization for assembling a pyrrolidine nucleus with substituents[15] are well exploited.

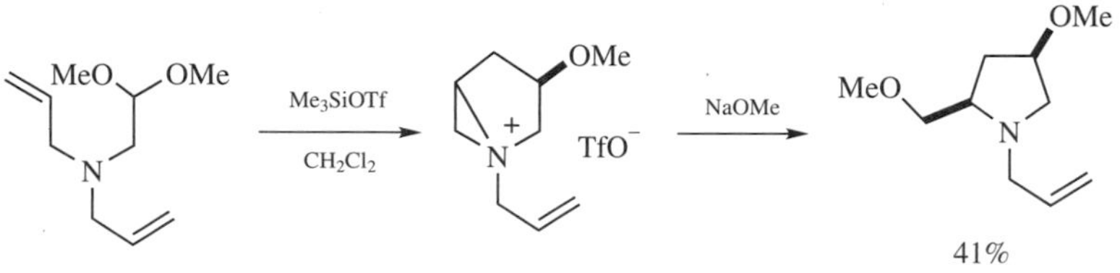

Alkylation with onium species. 1,3,5-Trisubstituted hexahydrotriazines,[16] cyclic acetals,[17] and *O,S*-acetals[18] behave well as electrophiles toward silyl enol ethers in the presence of Me_3SiOTf. In the last reaction silylation of **1** at the carbonyl group renders the methylthio unit reactive (i.e., as sulfonylating agent).

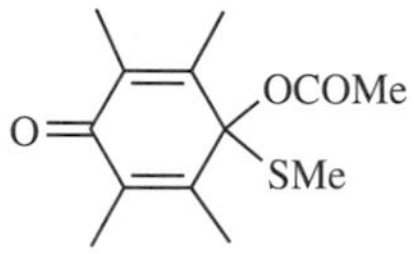

1

Aminoxylation. Reaction of silyl enol ethers with nitroso compounds (e.g., PhN=O) is catalyzed by Lewis acids. Trialkylsilyl triflates appear to be catalysts of choice (actually, Et_3SiOTf is slightly more efficient than Me_3SiOTf or *t*-$BuMe_2SiOTf$).[19]

[1]Nishizawa, M., Yamamoto, H., Seo, K., Imagawa, H., Sugihara, T. *OL* **4**, 1947 (2002).
[2]Suzuki, T., Kobayashi, K., Noda, K., Oriyama, T. *SC* **31**, 2761 (2001).
[3]Lence, E., Castedo, L., Gonzalez, C. *TL* **43**, 7917 (2002).
[4]Yu, B., Tao, H. *TL* **42**, 2405 (2000).
[5]Abdel-Rahman, A.A.-H., Winterfield, G.A., Takhi, M., Schmidt, R.R. *EJOC* 713 (2002).
[6]Palmacci, E.R., Seeberger, P.H. *OL* **3**, 1547 (2001).
[7]Ravichandran, S. *SC* **31**, 2345 (2001).
[8]Ishikawa, T., Okano, M., Aikawa, T., Saito, S. *JOC* **66**, 4635 (2001).
[9]Langer, P., Eckardt, T., Saleh, N.N.R., Karime, I., Müller, P. *EJOC* 3657 (2001).
[10]Sugiura, M., Hagio, H., Hirabayashi, R., Kobayashi, S. *SL* 1225 (2001).
[11]Cossrow, J., Rychnovsky, S.D. *OL* **4**, 147 (2002).
[12]Braddock, D.C., Badine, D.M., Gottschalk, T. *SL* 1909 (2001).
[13]Kjellgren, J., Szabo, K.J. *TL* **43**, 1123 (2002).
[14]Suginome, M., Iwanami, T., Yamamoto, A., Ito, Y. *SL* 1042 (2001).
[15]Graham, M.A., Wadsworth, A.H., Thornton-Pett, M., Rayner, C.M. *CC* 966 (2001).
[16]Ha, H.-J., Choi, C.-J., Lee, W.K. *SC* **32**, 1495 (2002).
[17]Alexakis, A., Trevitt, G.P., Bernardinelli, G. *JACS* **123**, 4358 (2001).
[18]Matsugi, M., Murata, K., Gotanda, K., Nambu, H., Anilkumar, G., Matsumoto, K., Kita, Y. *JOC* **66**, 2434 (2001).
[19]Momiyama, N., Yamamoto, H. *ACIEE* **41**, 2986 (2002).

Triorganotin hydrides.

2-Substituted biaryls.[1] Intramolecular aryl transfer from *o*-bromobenzyl diarylphosphinates occurs on treatment with Ph_3SnH.

Br Ar P—Ar O O + Ph_3SnH —AIBN, xylene Δ→ R O

R = H 70%
R = OMe 68%

Polycyclization.[2] A highly efficient elaboration of linear triquinanes is by a tandem cyclization process initiated by radical generation.

O–Si Br ''OMe + Ph_3SnH —AIBN, PhH Δ→ O–Si H ''OMe H

45%

Novel organotin hydrides. Technical problems concerning removal of tin-containing side products after reactions involving organotin hydride reagents can be avoided by using **1**[3] or **2**.[4] In the case of **1** excess and/or spent reagent are hydrolyzed with LiOH or TsOH in aqueous THF so that they become soluble in aqueous $NaHCO_3$. Reagent **1** is effective in participating many free radical reactions. The affinity of **2** and derived substances for activated carbon makes their facile removal by adsorption.

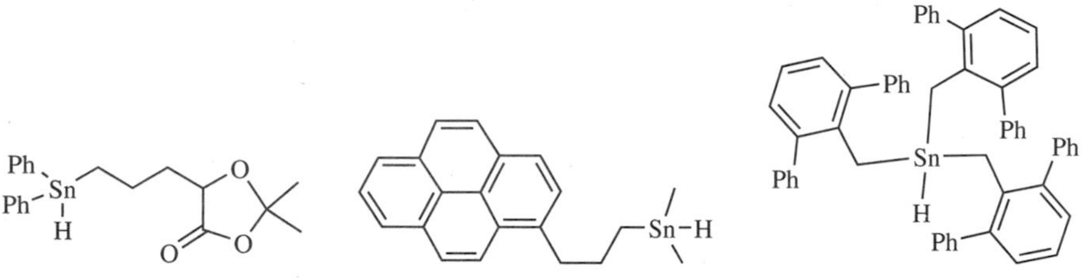

1 **2** **3**

The bulky tris(2,6-diphenylbenzyl)tin hydride (**3**) discriminates saturated and unsaturated carbonyl compounds in their reduction.[5] For the same steric reason its use in hydrodebromination of bromoalkenes gives almost exclusively the (*Z*)-isomers. Very strong bias in hydrogen atom donation is shown in the preferential formation of (Z)-3-benzylidene-dihydrofuran in a cyclization process.[6]

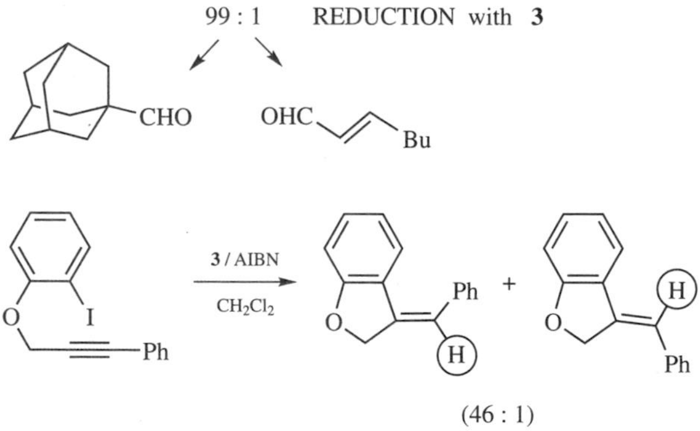

99 : 1 REDUCTION with **3**

3 / AIBN, CH_2Cl_2

(46 : 1)

[1]Clive, D.L.J., Kang, S. *JOC* **66**, 6083 (2001).
[2]Dhimane, A.-L., Aissa, C., Malacria, M. *ACIEE* **41**, 3284 (2002).
[3]Clive, D.L.J., Wang, J. *JOC* **67**, 1192 (2002).
[4]Gastaldi, S., Stien, D. *TL* **43**, 4309 (2002).
[5]Sasaki, K., Komatsu, N., Shirakawa, S., Maruoka, K. *SL* 575 (2002).
[6]Sasaki, K., Kondo, Y., Maruoka, K. *ACIEE* **40**, 411 (2001).

Triphenylphosphine. 18, 385–386; **19,** 382–383; **20,** 411–412; **21,** 462–463

Aza-Baylis-Hillman reaction.[1] With Ph_3P as catalyst, *N*-arylidenediphenylphosphinamides combine readily with acrylonitrile, methyl acrylate, methyl vinyl ketone.

Amide formation.[2] Thiocarboxylic acids and alkyl azides combine to generate amides by the agency of Ph_3P.

$$PhC(O)SH + N_3CH_2Ph \xrightarrow[MeCN]{Ph_3P} PhC(O)NHCH_2Ph \quad 91\%$$

Cycloadditions. Addition of Ph_3P to 2,3-butadienoic esters generates zwitterionic species which can abstract an acidic proton from coreactants such as β-diketones and *N,N′*-ditosylethylenediamine.[3] Combination of the ion-pairs (cationic species being 3-triphenylphosphonio-3-alkenoic esters), elimination to regenerate Ph_3P, and intramolecular Michael reaction then complete the transformation.

$$Ph_3P + CH_2{=}C{=}CH{-}COOEt \longrightarrow CH_2{=}C(\overset{+}{P}Ph_3){-}\overset{-}{C}H{-}COOEt \xrightarrow[PhMe\ 100°]{\text{1,3-cyclohexanedione}} \text{(tetrahydrobenzofuranone)}{-}CH_2COOEt \quad 68\%$$

Wittig reaction.[4] Polymer-bound Ph_3P mediates the direct alkenation of ArCHO with $BrCH_2X$ (X=Ph, COPh, COOMe) on microwave dielectric heating in the presence of K_2CO_3. Remarkably, MeOH can be used as solvent probably due to change of property at high temperature (becoming less polar and capable of swelling the resin).

Aryl exchange.[5] A convenient preparation of Ph_2PAr from Ph_3P is by replacing one of the phenyl groups. This can be accomplished by reaction with ArX (X=Br, OTf) in the presence of $Pd(OAc)_2$ under solvent-free conditions.

[1]Shi, M., Zhao, G.-L. *TL* **43**, 4499 (2002).
[2]Park, S.-D., Oh, J.-H., Lim, D. *TL* **43**, 6309 (2002).
[3]Lu, C., Lu, X. *OL* **4**, 4677 (2002).
[4]Westman, J. *OL* **3**, 3745 (2001).
[5]Kwong, F.Y., Lai, C.W., Chan, K.S. *TL* **43**, 3537 (2002).

Triphenylphosphine–diethyl azodicarboxylate. 13, 332; **14,** 336–337; **17,** 389–390; **18,** 387; **19,** 384–385; **20,** 413; **21,** 463–464

Mitsunobu reaction. An intramolecular version of this reaction applying to 1,3-diols results in the formation of oxetanes.[1] Other heterocycles synthesized by this reaction are aziridines from 1,2-aminoalcohols[2] and benzoxazines from 2-(hydroxyiminomethyl)benzyl alcohols.[3]

PhS OH OH $\xrightarrow{Ph_3P\text{ - }DEAD}$ PhS O

85%

Adaption of the protocol to inversion of configuration of secondary alcohols using a polymer-bound diphenylarylphosphine causes no problem.[4]

A synthetic method for *cis*-1,2-diazidoalkanes is tandem ring opening of epoxides and displacement reaction.[5] The corresponding *trans* isomers are prepared from *trans*-1,2-diols, involving two displacement reactions.

O + HN_3 $\xrightarrow[PhH]{Ph_3P\text{ - }DEAD}$ N_3 N_3

60%

Conjugate addition. Reaction of β-dicarbonyl compounds with DEAD is said to be catalyzed by Ru-Ph_3P complexes. It is shown that Ph_3P suffices.[6]

[1]Christlieb, M., Davies, J.E., Eames, J., Hooley, R., Warren, S. *JCS(P1)* 2983 (2001).
[2]Xu, J. *TA* **13**, 1129 (2002).
[3]Kai, H., Nakai, T. *TL* **42**, 6895 (2001).
[4]Charette, A.B., Janes, M.K., Boezio, A.A. *JOC* **66**, 2178 (2001).
[5]Göksu, S., Secen, H., Sütbeyaz, Y. *S* 2373 (2002).
[6]Lumbierres, M., Marchi, C., Moreno-Manas, M., Sebastian, R.M., Vallribera, A., Lago, E., Molins, E. *EJOC* 2321 (2001).

Triphenylphosphine–diisopropyl azodicarboxylate. 15, 352–353; **17,** 390; **18,** 387–388; **19,** 385; **20,** 413–414; **21,** 464

Amine synthesis.[1] Mitsunobu reaction of *t*-butyl *N*-diethylphosphonylcarbamate with alcohols gives *N*-alkyl derivatives which on treatment with TsOH in ethanol results in tosylates of primary amines.

EtO–P(O)(EtO)–NH–C(O)–O–*t*-Bu + BuOH $\xrightarrow{Ph_3P\text{ - }DIAD}$ EtO–P(O)(EtO)–N(Bu)–C(O)–O–*t*-Bu

Esterification.[2] Only aliphatic alcohols are activated by the reagent couple therefore carboxylic acids and alcohols containing free phenolic OH can be used to form esters.

[1]Klepacz, A., Zwierzak, A. *SC* **31**, 1683 (2001).
[2]Appendino, G., Minassi, A., Daddario, N., Bianchi, F., Tron, G.C. *OL* **4**, 3839 (2002).

Triphenylphosphine–halogen/perhalocarbons.

Alkyl halides. Primary iodides are formed from unprotected methyl furanosides and pyranosides on reaction with Ph_3P, I_2, and imidazole.[1] Alcohols are converted to alkyl chlorides by Ph_3P-cyanuric chloride in MeCN.[2]

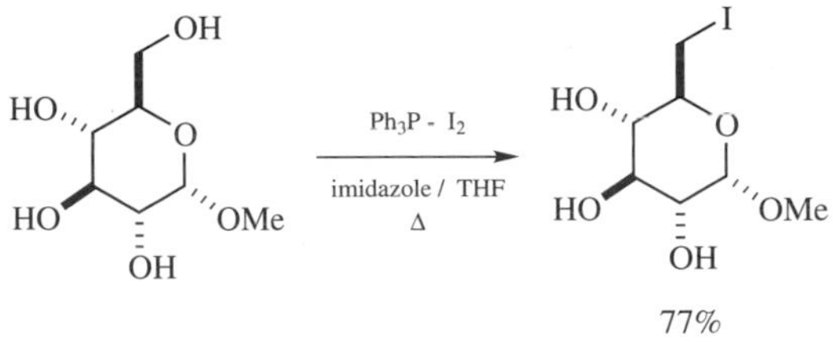

Functional group modifications and exchanges. Substituted hydrazines are prepared starting from $BocNHNH_2$ on reaction with Ph_3P-Br_2. Subsequent *N*-alkylation and acylation can be performed at will.[3]

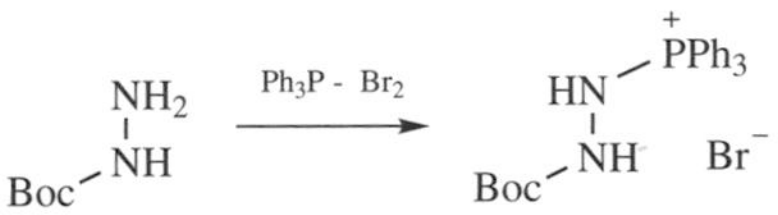

Bisthiocyanatotriphenylphosphine, obtained from Ph_3P-Br_2 and NH_4SCN, is useful for converting thiols to thiocyanates[4] and silyl esters to acyl isothiocyanates.[5] A more unusual *O*-formylation[6] is through activation of DMF with Ph_3P and 2,4,4,6-tetrabromo-2,5-cyclohexadienone, reaction of the ensuing species with alcohols delivers iminium salts that are readily hydrolyzed to formates.

Oxidation.[7] In the Swern oxidation of alcohols the activator (e.g., oxalyl chloride) can be replaced by Ph_3PX_2.

Condensation reactions. Carboxylic acids afford anhydrides on exposure to Ph_3P-Cl_3CCN.[8] Alkylation of Meldrum's acid by alcohols is mediated by Ph_3P–NBS.[9]

[1]Skaanderup, P.R., Poulsen, C.S., Hyldtoft, L., Jorgensen, M.R., Madsen, R. *S* 1721 (2002).
[2]Hiegel, G.A., Rubino, M. *SC* **32**, 2691 (2002).
[3]Tsubrik, O., Maeorg, U. *OL* **3**, 2297 (2001).
[4]Iranpoor, N., Firouzabadi, H., Shaterian, H.R. *TL* **43**, 3439 (2002).

[5]Iranpoor, N., Firouzabadi, H., Shaterian, H.R. *SC* **32**, 3653 (2002).
[6]Saito, A., Tanaka, A., Oritani, T. *SC* **32**, 3319 (2002).
[7]Bisai, A., Chandrasekhar, M., Singh, V.K. *TL* **43**, 8355 (2002).
[8]Kim, J., Jang, D.O. *SC* **31**, 395 (2001).
[9]Dhuru, S.P., Mohe, N.U., Salunhe, M.M. *SC* **31**, 3653 (2001).

Triphenylphosphonioketene.

Carbon chain stitching.[1] The reagent, $Ph_3P{=}C{=}C{=}O$, reacts with nucleophiles and carbonyl compounds to form conjugated carboxylic acid derivatives.

PPh3=C=C=O + RNHR′ ⟶ PPh3=HC–CONRR′ —(piperonal (CHO), THF)⟶ (E)-3-(3,4-methylenedioxyphenyl)-N-R,N-R′-acrylamide

[1]Schobert, R., Siegfried, S., Gordon, G.J. *JCS(P1)* 2393 (2001).

Triphenylphosphonium bromide. 21, 464

Acetalization. Assembly of cyclic diacetals[1] from 1,2-diols and 2,3-bisbenzyloxy-1,3-butadiene involves the catalysis of Ph_3P and HBr.

BnO/BnO diene + HO/HO diol (COOMe, COOMe) —(Ph3P - HBr; CH2Cl2 ; BF3·OEt2)⟶ OBn/OBn dioxane diacetal (COOMe, COOMe) 90%

[1]Ley, S.V., Michel, P. *SL* 1793 (2001).

Triphenylsilylacetamides.

(Z)-2-Alkenamides.[1] Condensation of these amides with aldehydes proceeds stereoselectively, affording predominantly (*Z*)-2-alkenamides.

[1]Kojima, S., Inai, H., Hidaka, T., Fukuzaki, T., Ohkata, K. *JOC* **67**, 4093 (2002).

Triphenylsilyl vanadate.

Isomerization-condensation. Aldol-type products not easily prepared are formed in a reaction of propargylic alcohols with aldehydes catalyzed by $VO(OSiPh)_3$. The reaction favors the normally less stable (*Z*)-β-aryl enones.[1] Note that the products are structurally similar to the Baylis-Hillmann adducts but generation of the latter requires β-unsubstituted acceptors.

A somewhat different reaction pattern emerges with allenyl carbinol substrates.[2]

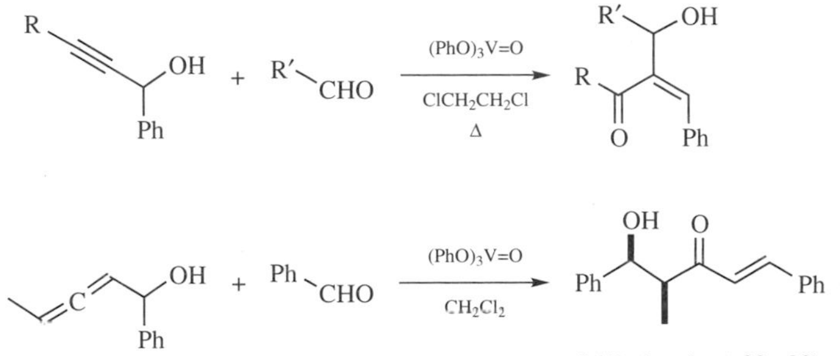

86% (*syn/anti* 80 : 20)

[1]Trost, B.M., Oi, S. *JACS* **123**, 1230 (2001).
[2]Trost, B.M., Jonasson, C., Wuchrer, M. *JACS* **123**, 12736 (2001).

Triphosgene. 18, 388; **19,** 386; **20,** 415–416; **21,** 464–465

Esterification. Amino acids are converted to esters on consecutive treatment with triphosgene and an alcohol.[1] The advantages of the protocol is somewhat questionable. Azidoformic esters are obtained from alcohols, triphosgene, NaN_3 and Et_3N.[2]

Diaryl ketones. Twofold Friedel-Crafts acylation of arenes leads to Ar_2CO with triphosgene providing the carbonyl residue.[3]

[1]Rivero, I.A., Heredia, S., Ochoa, A. *SC* **31**, 2169 (2001).
[2]Patil, R.T., Parveen, G., Gumaste, V.K., Bhawal, B.M., Deshmukh, A.R.A.S. *SL* 1455 (2002).
[3]Peng, X., Wang, J., Cui, J., Zhang, R., Yan, Y. *SC* **32**, 2361 (2002).

Triruthenium dodecacarbonyl. 18, 308; **19,** 386–387; **20,** 416–417; **21,** 465–467

Reductive decarboxylation.[1] α-Picolinyl esters suffer acyl-oxygen bond cleavage on heating with $Ru_3(CO)_{12}$ and $HCOONH_4$, eventually accomplishing reductive removal of the ester group. Noteworthy is that the protocol is applicable to some esters of aromatic carboxylic acids (β-indolecarboxylic acid and ferrocenecarboxylic acid). *N*-(2-Pyridyl)proline esters capable of chelation to the Ru catalyst also lose the ester group under the same conditions.

Homologation. When cleavage of α-picolinyl formate is performed in the presence of 1-alkenes (but omitting $HCOONH_4$), long chain esters are obtained.[2]

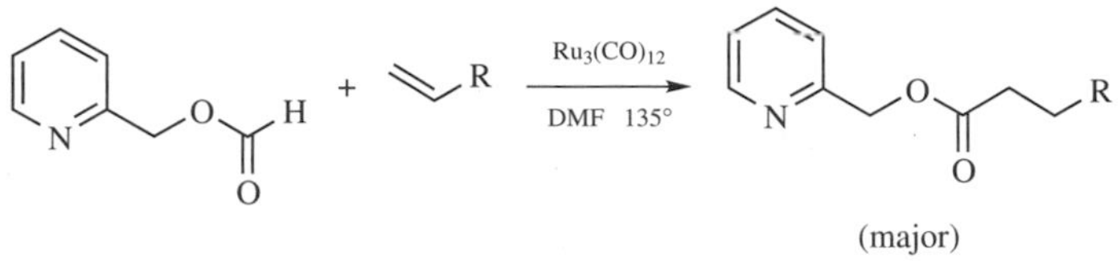

A direct synthesis of methacrylic esters and methacrylamide[3] needs only heating allene with the proper nucleophiles, $Ru_3(CO)_{12}$ and CO under pressure.

Cyclocarbonylations. Conjugated lactones[4] and lactams[5] are synthesized from alcohols and amine derivatives that contain an allene unit. Besides $Ru_3(CO)_{12}$ and CO, triethylamine plays an important role in this carbonylation-cyclization process. Lactones having 7 and 8 ring member atoms are easily prepared.

δ-Allenyl carbonyl compounds undergo cyclocarbonylation to furnish bicyclic α-methylene-γ-lactones.[6]

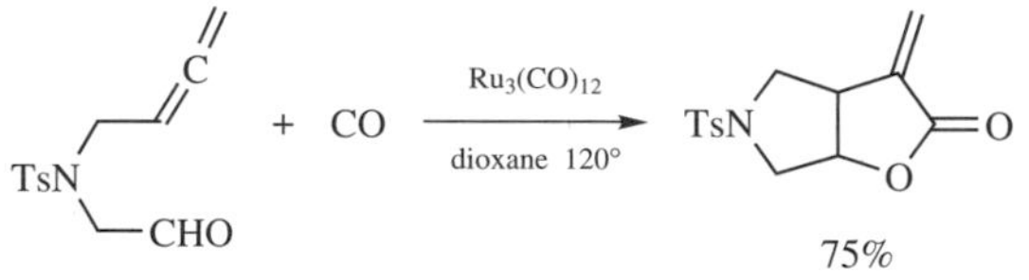

Pauson-Khand reaction of 2-(alkenylsilyl)pyridines apparently involves chelation (pyridine nitrogen) in the transition state.[7]

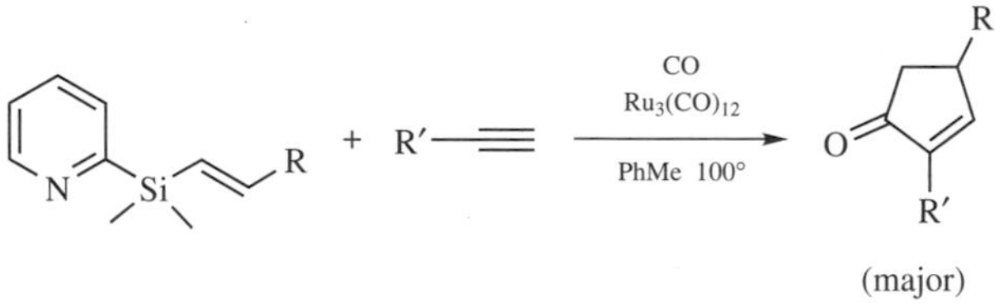

Annulation. 2,3-Disubstituted indoles are synthesized from unactivated arylamines and propargylic alcohols.[8]

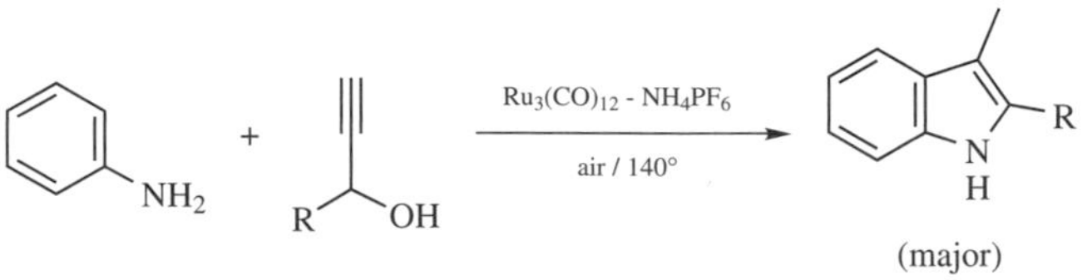

Intramolecular hydroamination of aminoalkynes leads to cyclic imines.[9]

Silylation and hydrosilylation. The previously developed method of chelation-controlled *o*-functionalization of 4,5 dihydro-2-oxazolylarenes is extendable to silylation using hydrosilanes.[10] Mixed alkyl silyl acetals are obtained from hydrosilylation of esters with hydrosilanes in hot toluene in the presence of $Ru_3(CO)_{12}$.[11]

Isomerization-Claisen rearrangement. Allyl homoallyl ethers undergo double bond migration to afford alkenyl allyl ethers that are prone to rerrangement.[12] Accordingly, an extensive change of atomic connection is achieved.

Ar N N Ar PhMe 80° Ph O Ph CHO ~ 100%

[1]Chatani, N., Tatamidani, H., Ie, Y., Kakiuchi, F., Murai, S. *JACS* **123**, 4849 (2001).
[2]Ko, S., Na, Y., Chang, S. *JACS* **124**, 750 (2002).
[3]Zhou, D.-Y., Yoneda, E., Onitsuka, K., Takahashi, S. *CC* 2868 (2002).
[4]Yoneda, E., Zhang, S.-W., Onitsuka, K., Takahashi, S. *TL* **42**, 5459 (2001).
[5]Kang, S.-K., Kim, K.-J., Yu, C.-M., Hwang, J.-W., Do, Y.-K. *OL* **3**, 2851 (2001).
[6]Kang, S.-K., Kim, K.-J., Hong, Y.-T. *ACIEE* **41**, 1584 (2002).
[7]Itami, K., Mitsudo,., K., Yoshida, J. *ACIEE* **41**, 3481 (2002).
[8]Tokunaga, M., Ota, M., Haga, M., Wakatsuki, Y. *TL* **42**, 3865 (2001).
[9]Kondo, T., Okada, T., Suzuki, T., Mitsudo, T. *JOMC* **622**, 149 (2001).
[10]Kakiuchi, F., Igi, K., Matsumoto, M., Chatani, N., Murai, S. *CL* 422 (2001).
[11]Igarashi, M., Mizuno, R., Fuchikami, T. *TL* **42**, 2149 (2001).
[12]Le Notre, J., Brisseux, L., Semeril, D., Bruneau, C., Dixneuf, P.H. *CC* 1772 (2002).

Tris(acetonitrile)cyclopentadienylruthenium(I) hexafluorophosphate. 21, 467–468

Addition to alkynes and allenes. The Ru-catalyzed addition involving 1-alkynes and 1-alkenes gives dienes. The method is useful for elaboration of enol ethers[1] and enamides[2] from allylically functionalized alkenes.[2] CC bond formation favors the less crowded end of an alkyne[3] therefore regiochemistry can be sometimes manipulated, e.g., using 1-silylalkynes.[2]

R + X $[CpRu(MeCN)_3]PF_6$ Me_2CO

X = OTBS, NHBoc

R X + R X

(2.6–3.4 : 1)

Dimerization of propargylic alcohols leading to hydroxy dienones[4] is rather unusual.

OH
$[CpRu(MeCN)_3]PF_6$
THF - Me_2CO
OH
O
77%

Deallylation.[5] Allyl carboxylates are cleaved with the Ru-catalyst in MeOH.

[5+2]Cycloaddition.[6] Cyclopropyl enynes are transformed by the Ru-catalyst to cycloheptadienes.

MeOOC
MeOOC
$[CpRu(MeCN)_3]PF_6$
Me_2CO
MeOOC
MeOOC
93%

Allylic displacement.[7] A new use of the Ru-catalyst is in the substitution of allylic esters (carbonates), with the branched products predominant.

Michael reaction.[8] When 1-allenylcyclobutanols are exposed the title complex, ring expansion occurs and it is followed by conjugate addition to added Michael acceptors.

HO
C
+
R
O
$[CpRu(MeCN)_3]PF_6$
DMF 60°
O
R
O

[1]Trost, B.M., Surivet, J.-P., Toste, F.D. *JACS* **123**, 2897 (2001).
[2]Trost, B.M., Surivet, J.-P. *ACIEE* **40**, 1468 (2001).
[3]Trost, B.M., Shen, H.C., Pinkerton, A.B. *CEJ* **8**, 2341 (2002).
[4]Trost, B.M., Rudd, M.T. *JACS* **123**, 8862 (2001).
[5]Kitamura, M., Tanaka, S., Yoshimura, M. *JOC* **67**, 4975 (2002).
[6]Trost, B.M., Shen, H.C. *ACIEE* **40**, 2313 (2001).
[7]Trost, B.M., Fraisse, P.L., Ball, Z.T. *ACIEE* **41**, 1059 (2002).
[8]Yoshida, M., Sugimoto, K., Ihara, M. *TL* **42**, 3877 (2001).

Tris(dibenzylideneacetone)dipalladium. 14, 339; **15,** 353–355; **16,** 372; **17,** 394; **18,** 389–393; **19,** 388–390; **20,** 417–420; **21,** 469–473

***N*-Arylation.** Protocols that allow the use of aryl halides containing amide or enolizable ketone have been developed.[1] The use of LiHMDS as ammonia equivalent is very successful.[2] *N*-Aryl-*O*-methylamidoximes are now available by way of *N*-arylation.[3]

The effectiveness of 1,3-dimesitylimidazolium chloride in assisting the Pd-catalyzed *N*-arylation[4] under basic conditions suggests the involvement of certain Pd-carbene complex. On the other hand, the well-established Pd/BINAP system has been applied to a preparation of *N*-arylcyclopropylamines.[5]

Aryl sulfides. Arylation of thiols[6] is accomplished with catalytic $(dba)_3Pd_2$ and bis(2-diphenylphosphino)phenyl ether as ligand in the presence of *t*-BuOK.

Heck, Stille and Negishi couplings. Popularity of *t*-Bu_3P as a ligand is gaining due to its effectiveness in assisting many Pd-catalyzed reaction, with which Heck reactions involving ArCl and ArBr are achieved at room temperature,[7] although Stille coupling leading to styrenes[8] is carried out at 100°. To prepare α-aryl enamides by the coupling method[9] the catalytic system contains $(dba)_3Pd_2$, Ph_3As, and CuCl.

Stille coupling following in situ hydrostannylation readily assembles conjugated dienes from haloalkenes and alkynes.[10] As a model for development of electronic devices substituted pyridinium salts are required, and such compounds are accessible by Stille coupling of halopyridinium halides.[11] For intramolecular Heck reaction of 2-chloroaryl alkenyl ethers and amines the catalytic system containing a hindered imidazolium salt, Cs_2CO_3, and Bu_4NBr is quite efficient.[12]

HO + Br Ph → (Me_3SnCl - PMHS; Bu_4NF - $(Ph_3P)_2PdCl_2$; $(dba)_3Pd_2$ - Fu_3P) HO Ph 88%

2-Pyrones substituted at C-6 and at both C-5 and C-6 are obtained on the basis of Negishi coupling.[13] Application to synthesis of two fungal metabolites is reported.

Suzuki coupling. The sensibility of *t*-Bu_3P to air is problematic. Handling it as the tetrafluoroboric acid salt is an excellent option, addition of KF to the reaction mixture enables Suzuki and other couplings to proceed satisfactorily.[14] Another ligand showing great promise in Pd-catalyzed couplings is the ferrocenylphosphine $Cp^*Fe[C_5H_3(SiMe_3)(PPh_3)]$.[15] To form sterically hindered biaryls a useful ligand is 2-(9-phenanthryl)phenyldicyclohexylphosphine.[16]

Suzuki coupling between 9-alkyl-9-BBN and alkyl chlorides proceeds in reasonably good yields.[17] Optimal conditions specify a $(dba)_3Pd_2/Cy_3P$ ratio of 1:2 and $CsOH{\cdot}H_2O$ as the base. There is also the coupling of 2-silyl-1-alkenes with ArI to give 1-arylalkenes,[18] not really Suzuki coupling but a variant of Heck reaction.

Ph ... + Ph-I —[t-BuOK - 18-c-6; $(dba)_3Pd_2$ - Bu_4NF]→ Ph ... Ph

85% (*E/Z* 9 : 1)

C-Arylation. Arylacetic esters emerge as the products of reaction between ArI and malonic esters, as a result of dealkylcarbonylation.[19] Isoindoline-1-carboxylic esters and tetrahydroisoquinoline-1-carboxylic esters are generated from intramolecular arylation of α-amino esters.[20] Direct arylation leading to 2-arylnitroalkanes[21] is much appreciated.

Activated DMF (e.g., with $POCl_3$) supplies the dimethylcarbamoyl group for replacement of aromatic halogen atoms in the Pd-catalyzed reaction.[22]

N-Propargyl amides undergo coupling and cyclization to give 2,5-disubstituted oxazoles in one step.[23] Perhaps a more familiar coupling-cyclization sequence is that involving in the formation of 2-(α-styryl)cyclopentanols from 5,6-heptadienals.[24]

$(dba)_3Pd_2$ - Fu_3P; t-BuONa / MeCN; 40°

83%

N,C-Coupling of allene is an expedient way to prepare α-aminomethylstyrenes.[25] Some amines can react twice.

$(dba)_3Pd_2$ - Fu_3P; K_2CO_3 / DMF; 80°

A synthesis of oxindoles from *o*-haloanilides is based on a Pd-catalyzed process featuring an imidazol-2-ylidene ligand.[26]

$(dba)_3Pd_2$

82%

[1]Harris, M.C., Huang, X., Buchwald, S.L. *OL* **4**, 2885 (2002).
[2]Huang, X., Buchwald, S.L. *OL* **3**, 3417 (2001).
[3]Anbazhagan, M., Stephens, C.E., Boykin, D.W. *TL* **43**, 4221 (2002).
[4]Grasa, G.A., Viciu, M.S., Huang, J., Nolan, S.P. *JOC* **66**, 7729 (2001).
[5]Cui, W., Loeppky, R.N. *T* **57**, 2953 (2001).

[6]Schopfer, U., Schlapbach, A. *T* **57**, 3069 (2001).
[7]Littke, A.F., Fu, G.C. *JACS* **123**, 6989 (2001).
[8]Littke, A.F., Schwarz, L., Fu, G.C. *JACS* **124**, 6343 (2002).
[9]Timbart, L., Cintrat, J.-C. *CEJ* **8**, 1637 (2002).
[10]Maleczka, R.E., Gallagher, W.P. *OL* **3**, 4173 (2001).
[11]Garcia-Cuadrado, D., Cuadro, A.M., Alvarez-Builla, J., Vaquero, J.J. *SL* 1904 (2002).
[12]Caddick, S., Kofie, W. *TL* **43**, 9347 (2002).
[13]Bellina, F., Biagetti, M., Carpita, A., Rossi, R. *TL* **42**, 2859 (2001).
[14]Netherton, M.R., Fu, G.C. *OL* **3**, 4295 (2001).
[15]Liu, S.-Y., Choi, M.J., Fu, G.C. *CC* 2408 (2001).
[16]Yin, J., Rainka, M.P., Zhang, X.-X., Buchwald, S.L. *JACS* **124**, 1162 (2002).
[17]Kirchhoff, J.H., Dai, C., Fu, G.C. *ACIEE* **41**, 1945 (2002).
[18]Anderson, J.C., Anguille, S., Bailey, R. *CC* 2018 (2002).
[19]Kondo, Y., Inamoto, K., Uchiyama, M., Sakamoto, T. *CC* 2704 (2001).
[20]Gaertzen, O., Buchwald, S.L. *JOC* **67**, 465 (2002).
[21]Vogl, E.M., Buchwald, S.L. *JOC* **67**, 106 (2002).
[22]Hosoi, K., Nozaki, K., Hiyama, T. *OL* **4**, 2850 (2002).
[23]Arcadi, A., Cacchi, S., Cascia, L., Fabrizi, G., Marinelli, F. *OL* **3**, 2501 (2001).
[24]Ha, Y.-H., Kang, S.-K. *OL* **4**, 1143 (2002).
[25]Gai, X., Grigg, R., Collard, S., Muir, J.E. *CC* 1712 (2001).
[26]Zhang, T.Y., Zhang, H. *TL* **43**, 193 (2002).

Tris(dibenzylideneacetone)dipalladium-chloroform. 19, 390–392; **20,** 420–422; **21,** 474–477

Coupling reactions. Stille coupling involving an enol nonaflate and alkenylstannane proceeds well.[1] Quite uncommon is the coupling of benzyl chlorides with allyltributylstannane that leads to dearomatization.[2]

Cl + SnBu$_3$ —(dba)$_3$Pd$_2$ · CHCl$_3$→ 80%

Br, Cl + SnBu$_3$ —(dba)$_3$Pd$_2$ · CHCl$_3$→ +

(83 : 17)
82%

For preparation of 3-aryl-substituted allylic alcohols from propargylic alcohols, the Negishi coupling employing hydroaluminated products to perform Al/Zn exchange in situ is convenient.[3] A modified Negishi coupling is also useful to gain access to 5-substituted 2,2′-bipyridines.[4]

Suzuki coupling in $[C_{14}H_{29}P(C_6H_{13})_3]Cl$, an ionic liquid, occurs at rather mild conditions.[5] In the presence of a Cu(I) salt heteroaryl methyl sulfides are reactive coupling partners of $ArB(OH)_2$.[6]

Instead of direct insertion of Pd atom into a C-X bond to start cross-coupling, metal/Pd exchange is an option to conduct the initial part of such reactions. The desilylative coupling approach to diarylethynes and carbonylative coupling to afford 1,3-diaryl-2-propyn-1-ones employs the alkynylsilanes and triarylstibine diacetates, in the latter case under CO.[7] Alkenylgallium(III) chlorides[8] and organoindium reagents[9] are fully qualified to couple with aryl halides, the latter in aqueous media.

Allylic displacement. The observations that only one type of diastereomers of 3-acetoxy-1-alkenyl *p*-tolyl sulfoxides would undergo *ipso*-substitution[10] and the asymmetric synthesis of 4-substituted 2-alkenoic esters by sequential Emmons-Wadsworth reaction and *ipso*-substitution of the allylic phosphonates[11] are highly significant. In the latter scheme stereoconvergence is achieved because both (4*S*/2*Z*)- and (4*R*/2*E*)-isomers give the same products with (4*S*/2*E*)-configuration.

OAc, S(O)Tol + $NaCH(COOMe)_2$ → $(dba)_3Pd_2 \cdot CHCl_3$, THF 70° → $CH(COOMe)_2$, S(O)Tol 89%

R, OR*, $OP(O)Ph_2$; *RO, R, $OP(O)Ph_2$ + $NaCH(COOMe)_2$ → $(dba)_3Pd_2 \cdot CHCl_3$, DPPE / THF → R, OR*, $CH(COOMe)_2$ $R = C_8H_{17}$ 77%

Allylic displacement using allylindium reagents leads to 1,5-alkadienes.[12] Displacement leading to cyclic carbonate and then coupling constitute functionalization of ynediol derivatives to more highly functionalized products.[13]

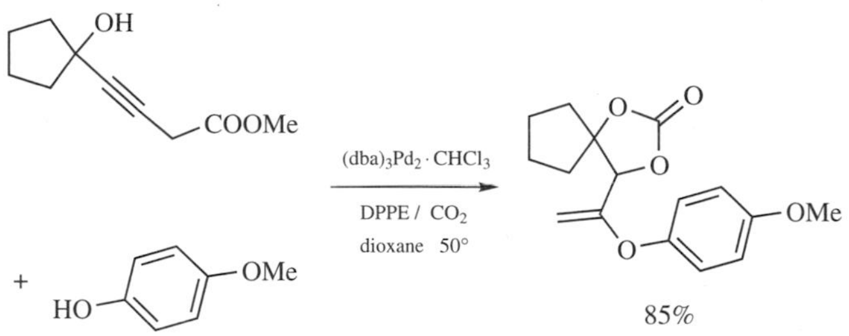

$(dba)_3Pd_2 \cdot CHCl_3$

DPPE / CO_2

dioxane 50°

85%

Regioselective displacement of allylic acetates equipped with an internal ligand (pyridine) can be tuned by varying the Pd complex.[14] A ring expansion[15] can be considered as intramolecular allylic displacement with *C*-nucleophile.

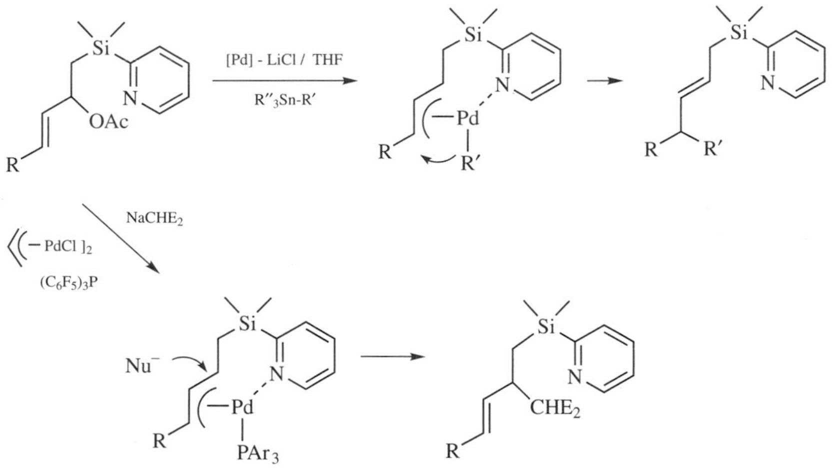

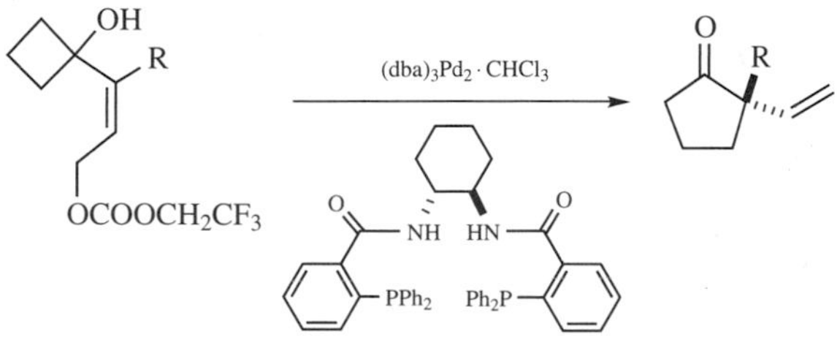

[Pd] - LiCl / THF

$R''_3Sn\text{-}R'$

$NaCHE_2$

$[(\text{allyl})-PdCl]_2$

$(C_6F_5)_3P$

$(dba)_3Pd_2 \cdot CHCl_3$

The absence of Ph_3P changes the course of allylic displacement to allyl addition to substrates containing also an aldehyde group.[16]

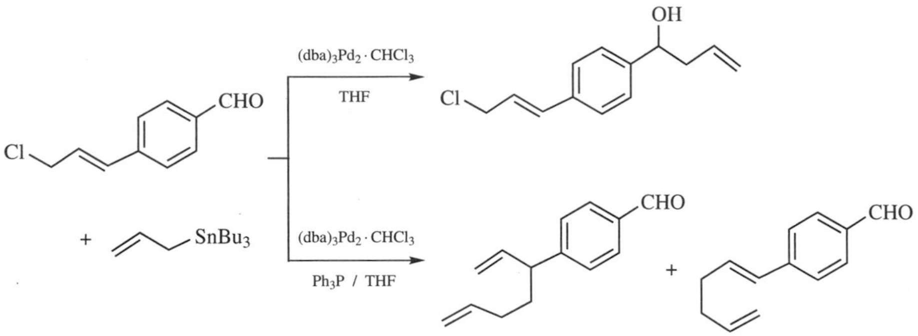

Ligand effect is revealed in the cyclization of 6-tosylamino-1-benzoyloxy-2-hexyne.[17]

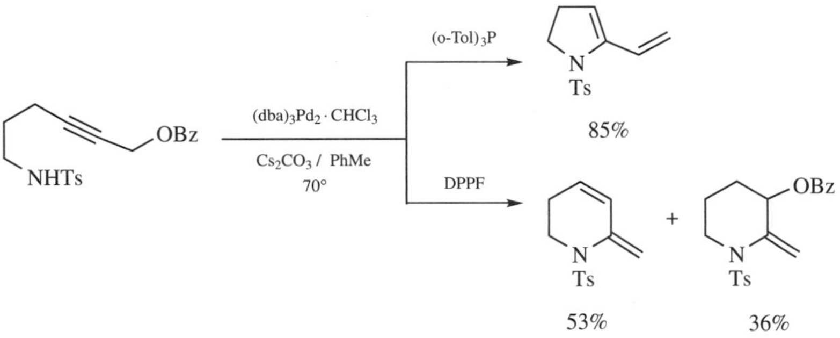

***N*-Arylation.** Urea is transformed into *N,N'*-diarylureas readily. Unsymmetrical ureas are prepared from monosubstituted substrates.[18]

Functionalization-Cyclization. Carbocycles and heterocycles bearing stannylmethylene and α-silylalkenyl substituents on two vicinal carbon atoms are formed from acyclic molecules with separated allene and alkyne units.[19]

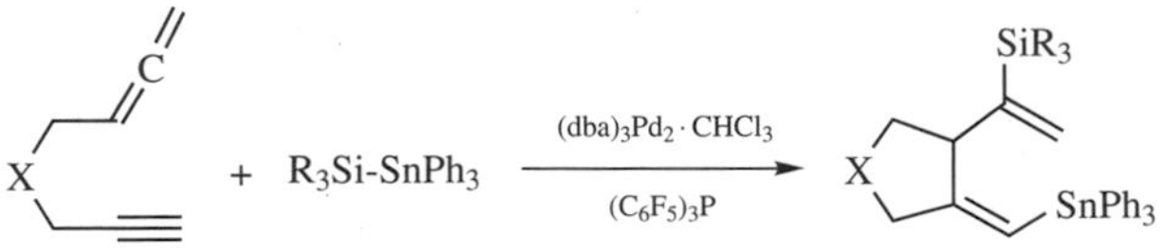

[1]Wada, A., Ieki, Y., Ito, M. *SL* 1061 (2002).
[2]Bao, M., Nakamura, H., Yamamoto, Y. *JACS* **123**, 759 (2001).
[3]Havranek, M., Dvorak, D. *JOC* **67**, 2125 (2002).
[4]Lützen, A., Hapke, M. *EJOC* 2292 (2002).
[5]McNulty, J., Capretta, A., Wilson, J., Dyck, J., Adjabeng, G., Robertson, A. *CC* 1986 (2002).
[6]Liebeskind, L.S., Srogl, J. *OL* **4**, 979 (2002).
[7]Kang, S.-K., Ryu, H.-C., Hong, Y.-T. *JCS(P1)* 736 (2001).
[8]Mikami, S., Yorimitsu, H., Oshima, K. *SL* 1137 (2002).
[9]Takami, K., Yorimitsu, H., Shinikubo, H., Matsubara, S., Oshima, K. *OL* **3**, 1997 (2001).
[10]de la Rosa, V.G., Ordonez, M., Llera, J.M. *TA* **12**, 1089 (2001).
[11]Pedersen, T.M., Hansen, E.L., Kane, J., Rein, T., Helquist, P., Norrby, P.-O., Tanner, D. *JACS* **123**, 9738 (2001).
[12]Lee, P.H., Sung, S., Lee, K., Chang, S. *SL* 146 (2002).
[13]Yoshida, M., Ihara, M. *ACIEE* **40**, 616 (2001).
[14]Itami, K., Koike, T., Yoshida, J. *JACS* **123**, 6957 (2001).
[15]Trost, B.M., Yasukata, T. *JACS* **123**, 7162 (2001).
[16]Nakamura, H., Bao, M., Yamamoto, Y. *ACIEE* **40**, 3028 (2001).
[17]Kozawa, Y., Mori, M. *TL* **43**, 1499 (2002).
[18]Artamkina, G.A., Sergeev, A.G., Beletskaya, I.P. *TL* **42**, 4381 (2001).
[19]Shin, S., RajanBabu, T.V. *JACS* **123**, 8416 (2001).

Tris(2-methylphenyl)bismuth dichloride.

Oxidation.[1] The title compound and DBU forms a binary system that is very effective for oxidation of alcohols. Secondary alcohols are more readily converted to ketones than primary alcohols.

Me
$[(2\text{-MeC}_6H_4)]_3BiCl_2$
Ph ... OH / OH → DBU / CH_2Cl_2 → Ph ... OH / O
98%

[1]Matano, Y., Nomura, H. *ACIEE* **41**, 3028 (2002).

Tris(pentafluorophenyl)borane. 20, 422; 21, 478

Epoxide opening.[1] In the presence of $(C_6F_5)_3B$ various nucleophiles attack epoxide to give functionalized alcohols.

Hydrosilylation. The effectiveness of $(C_6F_5)_3B$ as catalyst for alkene hydroboration has been shown (20 examples, 85–96%).[2] By virtue of σ–π chelation control the reductive silylation of homopropargylic ketones is stereoselective.[3]

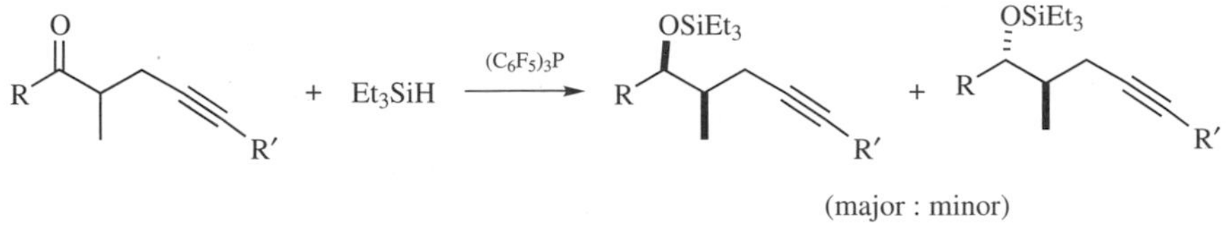

(major : minor)

Allylation. Displacement of secondary benzylic acetates on reaction with allyltrimethylsilane catalyzed by $(C_6F_5)_3B$ tolerates presence of other acetoxy groups, bromo substituent and primary benzyloxy groups.[4] Allylation that eventuates in the formation of a cobalt-complexed cycloheptene[5] in benzene actually brings a 2-phenyl-1,3-propylene segment to unite with the nucleophilic component. A Friedel-Crafts alkylation terminates the process.

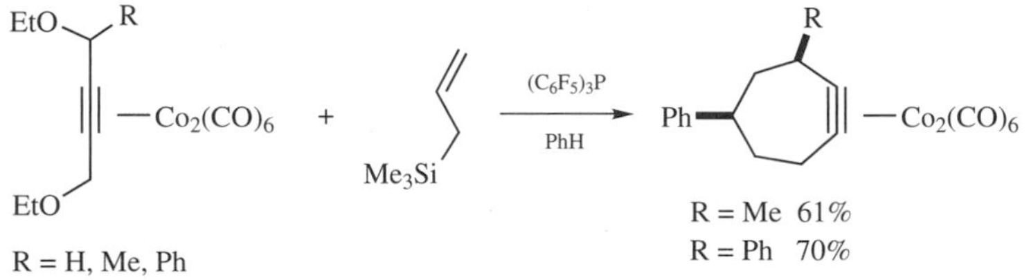

R = H, Me, Ph

R = Me 61%
R = Ph 70%

Aldol reaction. The $(C_6F_5)_3B$-catalyzed vinylogous Mukaiyama-type aldol reaction between conjugated silyl ketene acetals and aldehydes proceeds under very mild conditions and with high diastereoselectivity.[6]

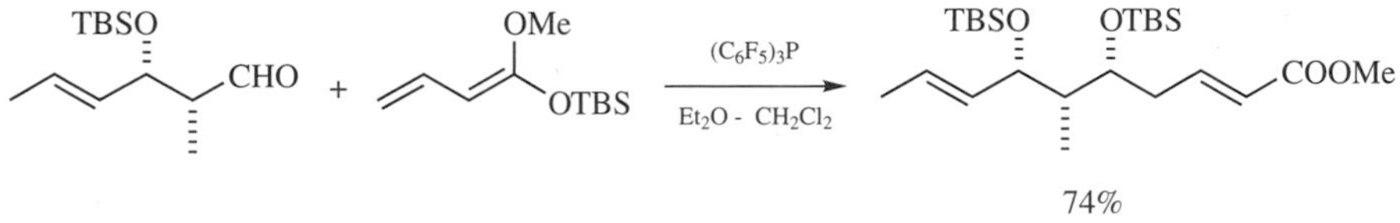

74%

[1]Chandrasekhar, S., Reddy, C.R., Babu, B.N., Chandrasekhar, G. *TL* **43**, 3801 (2002).
[2]Rubin, M., Schwier, T., Gevorgyan, V. *JOC* **67**, 1936 (2002).
[3]Asao, N., Ohishi, T., Sato, K., Yamamoto, Y. *JACS* **123**, 6931 (2001).
[4]Rubin, M., Gevorgyan, V. *OL* **3**, 2705 (2001).
[5]Lu, Y., Green, J.R. *SL* 243 (2001).
[6]Christmann, M., Kalesse, M. *TL* **42**, 1269 (2001).

Tris(trimethylsilyl)silane. 19, 393; **20,** 423

Cyclization. Radical generation from phenylselenoalkanes in which a radical acceptor unit is present leads to cyclized products.[1]

Ph, SePh, O, COOEt + $(Me_3Si)_3SiH$ —CO, AIBN→ Ph, O, =O, COOEt

86% (*cis/trans* 9 : 1)

[1] Berlin, S., Ericsson, C., Engman, L. *OL* **4**, 3 (2002).

Tris(trimethylsilyl) selenophosphate.

Selenocarboxamides.[1] Use of the reagent to convert nitriles to R(=Se)NH_2 is via in situ hydrolysis to the active $(HO)_3P{=}Se$.

[1]Kaminski, R., Glass, R.S., Skowronska, A. *S* 1308 (2001).

Trityldifluoroamine.

gem-Bis(difluoroamines).[1] As a stable reagent $TrNF_2$ is used to convert carbonyl group to the fluorinated diamines in the presence of 30% oleum.

[1]Prakash, G.K.S., Etzkorn, M., Olah, G.A., Christe, K.O., Schneider, S., Vij, A. *CC* 1712 (2002).

Trityl perchlorate.

Condensation. In the presence of $TrOClO_3$ propargylic silanes react with aldehydes and alkyl silyl ethers to afford α-allenyl ethers.[1] There is a net elimination of siloxane. On the other hand, homopropargylic ethers are produced[2] when allenyl silanes are submitted to the same conditions.

Group transposition. Either $TrOClO_3$ or MsOH can be used to exchange neighboring C=O and dithioacetal unit.[3]

Ar, O, SR, SR, SR —$Ph_3C[OClO_3]$, CH_2Cl_2 25°→ RS, SR, Ar, SR, O

[1]Niimi, L., Shiino, K., Hiraoka, S., Yokozawa, T. *TL* **42**, 1721 (2001).
[2]Niimi, L., Hiraoka, S., Yokozawa, T. *T* **58**, 245 (2002).
[3]Clericuzio, M., Degani, I., Dughera, S., Fochi, R. *S* 921 (2002).

Trityl tetrakis(pentafluorophenyl)borate. 21, 479–480

Alkylmetallation.[1] Alkylzirconium chlorides react with *o*-methoxyarylalkenes regioselectively, with the alkyl groups directed to the α-carbon atom.

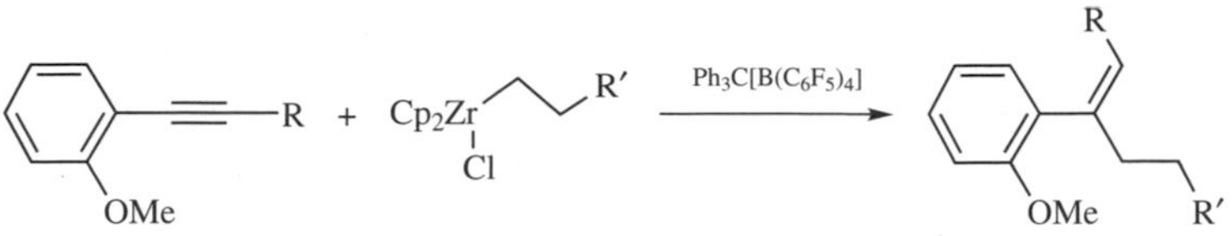

[1]Yamanoi, S., Seki, K., Matsumoto, T., Suzuki, K. *JOMC* **624**, 143 (2001).

Tungsten carbene and carbyne complexes. 20, 424–425; 21, 480–481

Dehydrative condensation.[1] On activation by $POCl_3$–Et_3N causes conjugated amides to react with the Fischer-type tungsten-carbene complexes resulting in C═W bonding reshuffle.

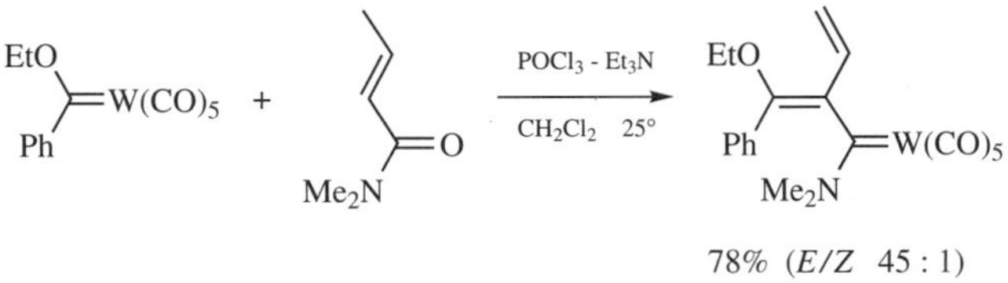

78% (*E*/*Z* 45 : 1)

[1]Aumann, R., Fu, X., Holst, C., Fröhlich, R. *OM* **21**, 4353 (2002).

Tungsten hexacarbonyl.

Annulation.[1] The $W(CO)_6$-DABCO combination promotes photochemical cyclization of silyl enol ethers appended with an alkynyl chain. Dependence of the cyclization mode on the nature of the added amine is remarkable.

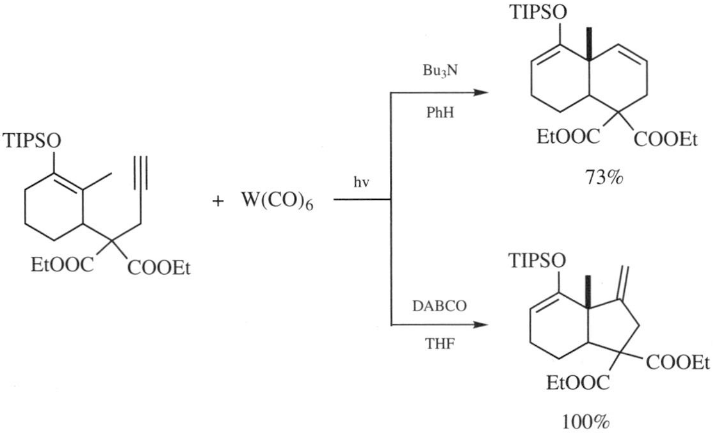

73%

100%

Diamines undergo oxidative carbonylation to give cyclic ureas[2] when exposed to $W(CO)_6$, I_2, and a base under CO.

[1]Kusama, H., Yamabe, H., Iwasawa, N. *OL* **4**, 2569 (2002).
[2]Qian, F., McCusker, J.E., Zhang, Y., main, A.D., Chlebowski, M., Kokka, M., McElwee-White, L. *JOC* **67**, 4086 (2002).

Tungsten pentacarbonyl tetrahydrofuran.

Cycloaddition.[1] With the $(thf)W(CO)_5$ complex *o*-ethynylaryl carbonyl compounds are converted into benzannulated pyrylium pentacarbonyltungstates. These zwitterionic species undergo [3+2]cycloaddition with alkenes, and in the case of enol ethers, oxabridged products are obtained.

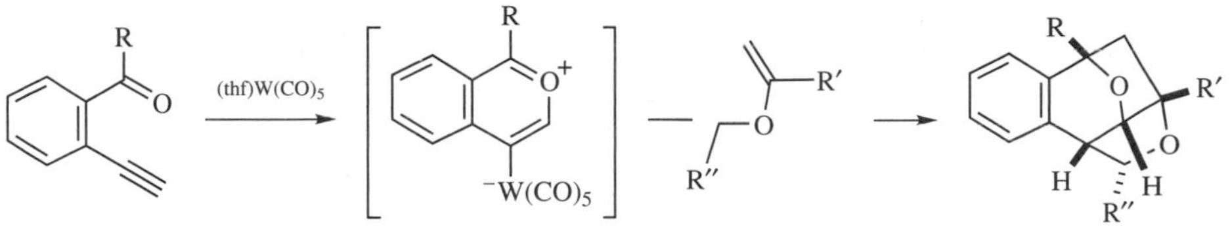

[1]Iwasawa, N., Shido, M., Kusama, H. *JACS* **123**, 5814 (2001).

U

Urea-hydrogen peroxide.

Oxidation. In formic acid this system oxidizes aromatic aldehydes to carboxylic acids at room temperature.[1]

Epoxidation. Variations of the well-known application pertain to additive and medium. Besides systems containing hexafluoroisopropanol,[2] maleic anhydride (for epoxidation of imines)[3], powdered fluoroapatite-supported reagent has been used without solvent.[4] Nonaqueous conditions in which polyamino acid on silica is established for the epoxidation of enones.[5]

Baeyer-Villiger oxidation. A Zr(salen) complex as catalyst for the transformation of certain cycloalkanones into lactones with urea-hydrogen peroxide can induce low ee.

[1]Balicki, R. *SC* **31**, 2195 (2001).
[2]Legros, J., Crousse, B., Bonnet-Delpon, D., Begue, J.-P. *EJOC* 3290 (2002).
[3]Damavandi, J.A., Karami, B., Zolfigol, M.A. *SL* 933 (2002).
[4]Ichihara, J. *TL* **42**, 695 (2001).
[5]Geller, T., Roberts, S.M. *JCS(P1)* 1397 (1999).
[6]Watanabe, A., Uchida, T., Ito, K., Katsuki, T. *TL* **43**, 4481 (2002).

Fiesers' Reagents for Organic Synthesis, Volume 22. Series editor Tse-Lok Ho
ISBN 0-471-28515-3

V

Vanadyl bis(acetylacetonate).

Aminomethylation.[1] *o*-Aminomethylation of phenols occurs on treatment with *t*-amine oxides in the presence of $VO(acac)_2$.

Vicinal dihydroxylation.[2] The title reagent catalyzes the dihydroxylation of alkenes by OsO_4-H_2O_2 and *N*-methylmorpholine.

[1]Hwang, D.-R., Uang, B.-J. *OL* **4**, 463 (2002).
[2]Ell, A.H., Jonsson, S.Y., Börje, A., Adolfsson, H., Bäckvall, J.-E. *TL* **42**, 2569 (2001).

Vanadyl bis(triflate).

Acylation.[1] Acylation of alcohols by anhydrides is catalyzed by $VO(OTf)_2$ at room temperature.

[1]Chen, C.-T., Kuo, J.-H., Li, C.-H., Barhate, N.B., Hon, S.-W., Li, T.-W. *OL* **3**, 3729 (2001).

Vanadyl trichloride.

1,4-Diketones.[1] Cross-coupling of silyl enol ethers is catalyzed by $VOCl_3$ in the presence of an alcohol. Accordingly, it is possible to induce asymmetry in product formation by adding a chiral alcohol.

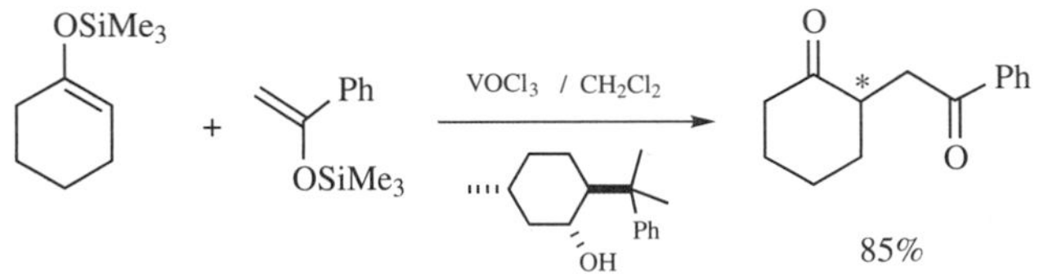

[1]Kurihara, M., Hayashi, T., Miyata, N. *CL* 13249 (2001).

Fiesers' Reagents for Organic Synthesis, Volume 22. Series editor Tse-Lok Ho
ISBN 0-471-28515-3

Xenon(II) fluoride. 13, 345; **19,** 399; **20,** 430; **21,** 486

Amination.[1] A reagent system made of XeF_2, Me_3SiNCO and TfOH is useful for amination of arenes.

[1]Pirkuliev, N.Sh., Brel, V.K., Akhmedov, N.G., Zefirov, N.S., Stang, P.J. *ML* 172 (2001).

Fiesers' Reagents for Organic Synthesis, Volume 22. Series editor Tse-Lok Ho
ISBN 0-471-28515-3

Y

Ytterbium. 14, 348; **15,** 366; **16,** 384; **18,** 401; **19,** 400; **20,** 431; **21,** 487

Coupling reactions. While arylidenecyanoacetic esters undergo dimerization with Yb via coupling at the benzylic position and a Thorpe cyclization,[1] acyl cyanides are converted to 1,2-diketones by Yb-I_2.[2]

NC–C(COOEt)=CH–Ph + NC–C(COOEt)=CH–Ph —(Yb / THF)→ cyclopentene product (NH_2, NC, EtOOC, COOEt, Ph, Ph)

80%

Reductive acylation and alkylation. Imines react with isocyanates in the presence of Yb to afford α-aminocarboxamides,[3] diaryl ketones give 2-amino alcohols on reaction with 1-(aminomethyl)benzotriazoles.[4]

Allylation. Splitting of disulfides by the Yb-I_2 system generates nucleophilic $RXYbI_2$ species which can be allylated to furnish allyl sulfides.[5] Homoallylic amines are prepared from imine-Yb complexes.[6]

Hydrophosphination and hydrosilylation. Using a Yb-imine complex prepared from Yb and the imine in the presence of HMPA the addition of R_2PH to CC multiple bonds is rapid (phosphine oxides are obtained after adding H_2O_2 to the reaction mixtures before workup).[7] Further treatment of the imine complex with Ph_2NH to replace the ligated HMPA forms catalysts [e.g., Bn(Ph)N-Yb-NPh_2] that are active for hydrosilylation.[8]

[1]Su, W., Yang, B. *BCSJ* **75**, 2221 (2002).
[2]Saikia, P., Laskar, D.D., Prajapati, D., Sandhu, J.S. *TL* **43**, 7525 (2002).
[3]Ueno, R., Yano, K., Makioka, Y., Fujiwara, Y., Kitamura, T. *CL* 790 (2002).
[4]Su, W., Yang, B., Zhang, Y. *TL* **43**, 2251 (2002).
[5]Su, W., Li, Y., Zhang, Y. *SC* **32**, 2101 (2002).
[6]Su, W., Li, J., Zhang, Y. *SC* **31**, 273 (2001).
[7]Takaki, K., Takeda, M., Koshoji, G., Shishido, T., Takehira, K. *TL* **42**, 6357 (2001).
[8]Takaki, K., Sonoda, K., Kousaka, T., Koshoji, G., Shishido, T., Takehira, K. *TL* **42**, 9211 (2001).

Ytterbium(III) chloride.

Acylation.[1] Monoacylation of symmetrical 1,2-diols is achieved in one step using $YbCl_3$ as catalyst.

[1]Clarke, P.A., Kayaleh, N.E., Smith, M.A., Baker, J.R., Bird, S.J., Chan, C. *JOC* **67**, 5226 (2002).

Fiesers' Reagents for Organic Synthesis, Volume 22. Series editor Tse-Lok Ho
ISBN 0-471-28515-3

Ytterbium(III) isopropoxidc.

β-Tosylamino nitroalkanes.[1] The reaction of nitroalkanes with *N*-tosylimines is catalyzed by $(i\text{-PrO})_3Yb$.

[1]Qian, C., Gao, F., Chen, R. *TL* **42,** 4673 (2001).

Ytterbium(III) triflate. 18, 402–403; **19,** 401–402; **20,** 431–433; **21,** 487–489

Friedel-Crafts reactions. Aminoalkylation of electron-rich arenes by imine derivatives (also those formed in situ)[1] is readily promoted by $Yb(OTf)_3$. A route to 4-aryl-1,2,3,4-tetrahydro quinolines by annulation of arylamines starts from a catalyzed condensation of (α-bromobenzyl)epoxides.[2]

Ph
+ Br O NH$_2$ OMe
$Yb(OTf)_3$
CH_2Cl_2
Ph OH N H OMe
61%

Nitrones derived from tryptamine cyclize on treatment with $Yb(OTf)_3$, to provide *N*-hydroxytetrahydro-β-carbolines[3] in a process related to the Pictet-Spengler reaction.

Tandem reactions. Among tandem reactions promoted by $Yb(OTf)_3$ are pyran annulation via a Michael-aldol reaction sequence,[4] radical addition onto conjugated systems,[5] and acylation-Claisen rearrangement.[6]

O Cl R - O N N O R′
$Yb(OTf)_3$
$i\text{-}Pr_2NEt$
CH_2Cl_2
O R O N O R′ R′ N O

Rearrangements. Rearrangement occurs when epoxy carbinols are brought into contact with $Yb(OTf)_3$ to afford β-hydroxy carbonyl compounds.[7]

HO R O Ph R′
$Yb(OTf)_3$
CH_2Cl_2
O OH R R′ Ph

Claisen rearrangement of (*Z*)-2-allyloxy-2-alkenoic esters proceeds at room temperature in the presence of $Yb(OTf)_3$ or $Cu(OTf)_2$.[8]

Glycosylation. Conditions for transforming acetyl glycosides into aryl glycosides using triaryloxyboranes has been optimized.[9] Pseudoglycals[10] and *C*-pseudoglycals[11] are prepared by the Ferrier rearrangement (S_N2' reaction).

Reactions of imines. Acylzirconocene chlorides act as acyl transfer agent for imines, forming α-amino ketones by catalysis of $Yb(OTf)_3$ or better yet with Me_3SiOTf.[12]

$Yb(OTf)_3$ - Me_3SiOTf, THF

Cyclization of certain alkenyl imines forms heterocycles.[13]

$Yb(OTf)_3$

Allylic halogenation.[14] Selective and rapid allylic halogenation of 1,1-disubstituted alkenes with *N*-halosuccinimides at room temperature is realized when $Yb(OTf)_3$ and Me_3SiCl are present. Fluorination is accomplished with an *N*-fluoropyridinium salt.

$Yb(OTf)_3$ - Me_3SiOTf, CH_2Cl_2 - THF

X = Br 64%

Methyl carbamates.[15] Amines are converted to RNHCOOMe by dimethyl carbonate under Yb-catalysis.

Ketone reduction.[16] The hydride-transfer reduction of ketones by isopropanol under basic conditions and catalyzed by $(Ph_3P)_2RuCl$ gives much better yields when a catalytic amount of $Yb(OTf)_3$ is added.

[1]Janczuk, A., Zhang, W., Xie, W., Lou, S., Cheng, J., Wang, P.G. *TL* **43**, 4271 (2002).
[2]Karikomi, M., Tsukada, H., Toda, T. *H* **55**, 1249 (2001).
[3]Tsuji, R., Yamanaka, M., Nishida, A., Nakagawa, M. *CL* 428 (2002).
[4]Lee, Y.R., Kweon, H.I., Koh, W.S., Min, K.R., Kim, Y., Lee, S.H. *S* 1851 (2001).
[5]Sibi, M.P., Miyabe, H. *OL* **4**, 3435 (2002).
[6]Dong, V.M., MacMillan, D.W.C. *JACS* **123**, 2448 (2001).
[7]Bickley, J.F., Hauer, B., Pena, P.C.A., Roberts, S.M., Skidmore, J. *JCS(P1)* 1253 (2001).
[8]Hiersemann, M., Abraham, L. *OL* **3**, 49 (2001).
[9]Yamanoi, T., Yamazaki, I. *TL* **42**, 4009 (2001).
[10]Takhi, M., Abdel-Rahman, A.A.H., Schmidt, R.R. *SL* 427 (2001).
[11]Takhi, M., Abdel-Rahman, A.A.H., Schmidt, R.R. *TL* **42**, 4503 (2001).
[12]Kakuuchi, A., Taguchi, T., Hanzawa, Y. *TL* **42**, 1547 (2001).

[13]Jia, Q., Xie, W., Zhang, W., Janczuk, A., Luo, S., Zhang, B., Cheng, J.P., Kscbati, M.B., Wang, P.G. *TL* **43**, 2339 (2002).
[14]Yamanaka, M., Arisawa, M., Nishida, A., Nakagawa, M. *TL* **43**, 2403 (2002).
[15]Curini, M., Epifano, F., Maltese, F., Rosati, O. *TL* **43**, 4895 (2002).
[16]Matsunaga, H., Yoshioka, N., Kunieda, T. *TL* **42**, 8857 (2001).

Yttrium nonamethyltrisilazide.

Hydroamination. Intramolecular hydroamination to form heterocycles using simple amido derivatives of Group 3 metals has been observed.[1] Addition of a sterically hindered diamine [e.g., *N,N'*-bis(2,6-diisopropylphenyl)ethylenediamine] improves the catalytic activity and stereoselectvitiy.[2]

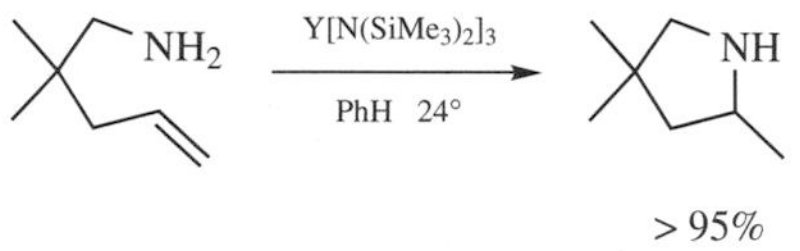

NH_2 — $Y[N(SiMe_3)_2]_3$, PhH 24° → NH

$>95\%$

[1]Kim, Y.K., Livinghouse, T., Bercaw, J.E. *TL* **42**, 2933 (2001).
[2]Kim, Y.K., Livinghouse, T. *ACIEE* **41**, 3645 (2002).

Yttrium(III) triflate.

Quinolines. Condensation of *N*-aryldehydroglycine esters with azetenes affords quinoline derivatives. The transformation is catalyzed by $Y(OTf)_3$.[1]

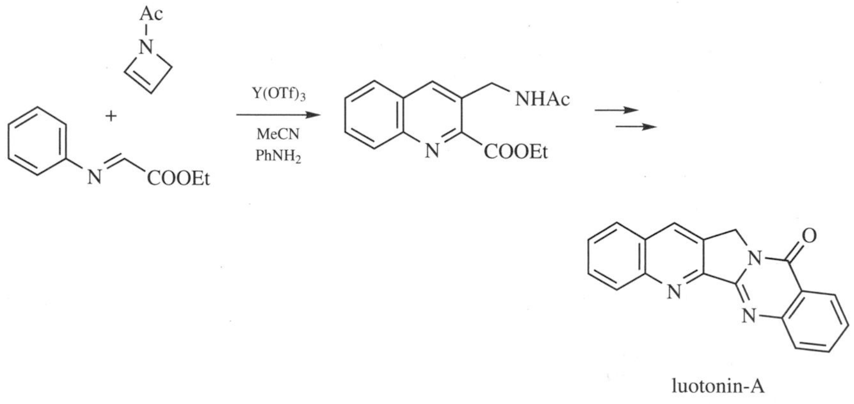

Ac, N, +, N, COOEt — $Y(OTf)_3$, MeCN, $PhNH_2$ → NHAc, N, COOEt → →

luotonin-A

[1]Osborne, D., Stevenson, P.J. *TL* **43**, 5469 (2002).

Z

Zeolites. 15, 367; **18,** 405–406; **19,** 403–404; **20,** 434; **21,** 491

Condensation reactions. Synthesis of 2-substituted oxazolines and oxazines,[1] and chemoselective formation of phenyl esters[2] can take advantage of the simplicity in using zeolites as promoters. HY-Zeolite acts as an agent for converting aldehydes to nitriles and ketones to amides, dehydration or Beckmann rearrangement of the initial condensation products is also achieved.[3]

[1]Cwik, A., Hell, Z., Hegedüs, A., Horvath, Z. *TL* **43**, 3985 (2002).
[2]Ding, Y., Wu, R., Lin, Q. *SC* **32**, 2149 (2002).
[3]Srinivas, K.V.N.S., Reddy, E.B., Das, B. *SL* 625 (2002).

Zinc. 13, 346–347; **14,** 349–350; **16,** 386–387; **17,** 406–407; **18,** 406–408; **19,** 404–405; **20,** 435–436; **21,** 491–492

Reduction. Organic azides are reduced to primary amines,[1] but more interesting is the conversion of pyridazines to pyrroles.[2]

OBn OBn
Zn - CF_3COOH
MeOOC COOMe 25° MeOOC COOMe
N–N N
H
52%

Coupling-rearrangement. Aryl ketones undergo pinacol coupling with Zn, but when $AlCl_3$ is added to the reaction medium rearrangement of the products actually takes place.[3] The previous claim of using these reagents to synthesize alkenes is invalid.

O Zn - $AlCl_3$ R OM OM Ph R
Ph R MeCN Δ Ph Ph R Ph O
R
82 – 99%

Fiesers' Reagents for Organic Synthesis, Volume 22. Series editor Tse-Lok Ho
ISBN 0-471-28515-3

Elimination-propargylation. Applying the general protocol of propargylation to ω-iododeoxyglycosides gives rise to enynes.[4] The aldehyde group is exposed only after the elimination step which is mediated by metallic zinc.

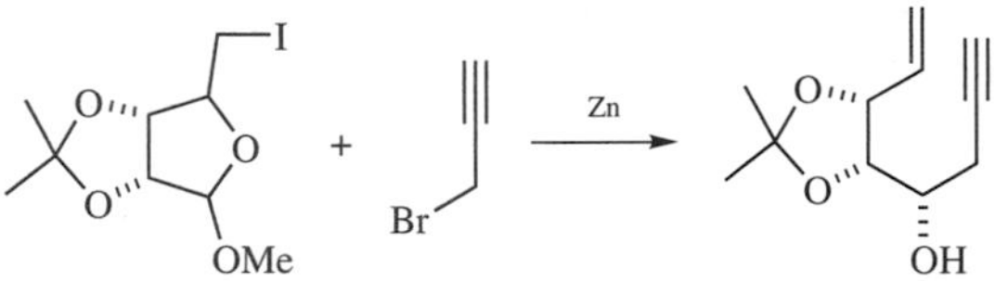

Addition reactions. A synthetic route of β-amino acid derivatives involves conjugate addition of organozinc species (from RI, Zn and aq. NH_4Cl) to an *N*-phthaloyl dehydroalanine ester.[5]

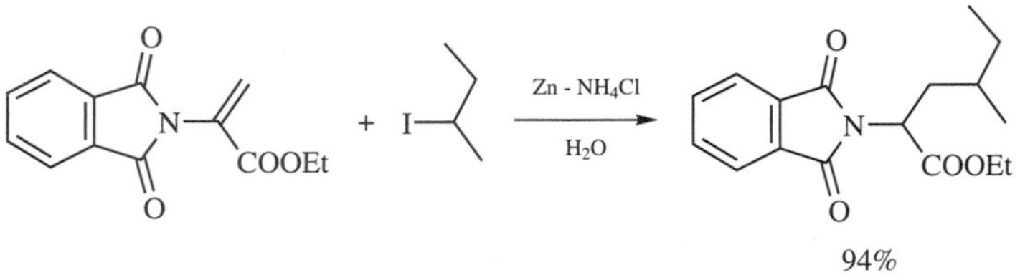

Zinc promotes reaction of allyl chloride and acid chlorides in the presence of Me_3SiCl to provide 4-substituted 1,6-heptadien-4-ols.[6]

Nitrones.[7] Conversion of aldehydes to nitrones is by reaction with nitroalkanes in HOAc mediated by Zn.

[1]Lin, W., Zhang, X., He, Z., Jin, Y., Gong, L., Mi, A. *SC* **32**, 3279 (2002).
[2]Boger, D.L., Hong, J. *JACS* **123**, 8515 (2001).
[3]Grant, A.A., Allukian, M., Fry, A.J. *TL* **43**, 4391 (2002).
[4]Poulsen, C.S., Madsen, R. *JOC* **67**, 4441 (2002).
[5]Huang, T., Keh, C.C.K., Li, C.-J. *CC* 2440 (2002).
[6]Ishino, Y., Mihara, M., Kageyama, M. *TL* **43**, 6601 (2002).
[7]Gautheron-Chapoulaud, V., Pandya, S.U., Cividino, P., Masson, G., Py, S., Vallee, Y. *SL* 1281 (2001).

Zinc–metal salts.

Coupling reactions. Reductive dimerization of carbonyl compounds to alkenes is said to be mediated by Zn-$InCl_3$ in MeCN.[1] Oximes of aromatic aldehydes give 1,2-diamines with Zn-$TiCl_4$.[2]

Reductive coupling of alkynes with electron-deficient alkenes[3] is effected by a mixture of Zn, $(Ph_3P)_2CoI_2$, and Ph_3P. α-Aroyladipic esters are formed when ArCHO and acrylic esters are treated with Zn-$ZnCl_2$.[4]

Zn - $(Ph_3P)_2CoI_2$, Ph_3P / MeCN - H_2O

81%

Zn - $ZnCl_2$, DMF 70°

50%

Intramolecular coupling of the monoacetal of a biaryl-1,1′-dialdehyde with Zn-$(thf)_3VCl_3$ leads to 9,10-dihydrophenanthrene-9,10-diol monoether.[5] For more crowded substrates SmI_2-$BF_3{\cdot}OEt_2$ performs better.

Zn - $(thf)_3VCl_3$

78% (*trans/cis* 3 : 1)

Reformatsky reaction. α-Fluoro-β-hydroxyalkanoic esters are prepared from bromofluoroacetic esters and carbonyl compounds with a Zn-$CeCl_3$ system.[6]

[1]Barman, D.C., Thakur, A.J., Prajapati, D., Sandhu, J.S. *SL* 515 (2001).
[2]Kise, N., Ueda, N. *TL* **42**, 2365 (2001).
[3]Wang, C.-C., Lin, P.-S., Cheng, C.-H. *JACS* **124**, 9696 (2002).
[4]Sakurai, H., Takeuchi, H., Hirao, T. *CC* 3048 (2002).
[5]Ohmori, K., Kitamura, M., Ishikawa, Y., Kato, H., Oorui, M., Suzuki, K. *TL* **43**, 7023 (2002).
[6]Ocampo, R., Dolbier, W.R., Abboud, K.A., Zuluaga, F. *JOC* **67**, 72 (2002).

Zinc acetate.

Epimerization.[1] Glycosylmethyl carbonyl compounds are subject to epimerization with $Zn(OAc)_2$ and NaOMe. Zinc enolates are involved in the ring cleavage and reclosure.

BnO BnO BnO BnO O O $\xrightarrow[\text{NaOMe, MeOH}]{Zn(OAc)_2}$ [AcO Zn O BnO BnO BnO O] → BnO BnO BnO BnO O O

95%

[1]Shao, H., Wang, Z., Lacroix, E., Wu, S.-H., Jennings, H.J., Zou, W. *JACS* **124**, 2130 (2002).

Zinc bromide. 13, 349; **15,** 368; **16,** 389–391; **18,** 409; **19,** 409; **20,** 438–439; **21,** 493

Desilylation.[1] Silyl ethers are cleaved in dichloromethane by treatment with water containing $ZnBr_2$.

Rearrangement.[2] Aziridinylcarbinols rearrange on contact with $ZnBr_2$. Ring opening and group migration lead to β-amino ketones.

OH NTs $\xrightarrow[CH_2Cl_2]{ZnBr_2}$ O NHTs

88%

Tandem condensation.[3] Zinc enolates generated from ketones via Li-Zn exchange do not undergo aldol reaction with aldehydes directly. Formation of dihydroxytetrahydropyrans indicates that two ketone molecules are united before aldol reaction with the aldehyde.

H Ph O + Ph O $\xrightarrow[THF]{LDA\ ;\ ZnBr_2}$ Ph OH Ph Ph O OH

75%

1H-Tetrazoles.[4] Nitriles and NaN_3 combine in the presence of $ZnBr_2$ to give tetrazole derivatives.

[1]Crouch, R.D., Polizzi, J.M., Cleiman, R.A., Yi, J., Romany, C.A. *TL* **43**, 7151 (2002).
[2]Wang, B.M., Song, Z.L., Fan, C.A., Tu, Y.Q., Shi, Y. *OL* **4**, 363 (2002).
[3]Schmittel, M., Ghorai, M.K. *SL* 1992 (2001).
[4]Demko, Z.P., Sharpless, K.B. *JOC* **66**, 7945 (2001).

Zinc chloride. 13, 349–350; **15,** 368–371; **16,** 391–392; **18,** 410–411; **19,** 409–410; **20,** 439; **21,** 493

Unsaturated alcohols. Assembly of propargylic alcohols[1] from carbonyl compounds and 1-alkynes only requires mixing them with $ZnCl_2$-Et_3N. Allylic zinc reagents are prepared from decomposition of highly hindered homoallylic alcohols with BuLi followed by Li-Zn exchange. The subsequent reaction with aldehydes is stereoselective (cyclic transition state).[2]

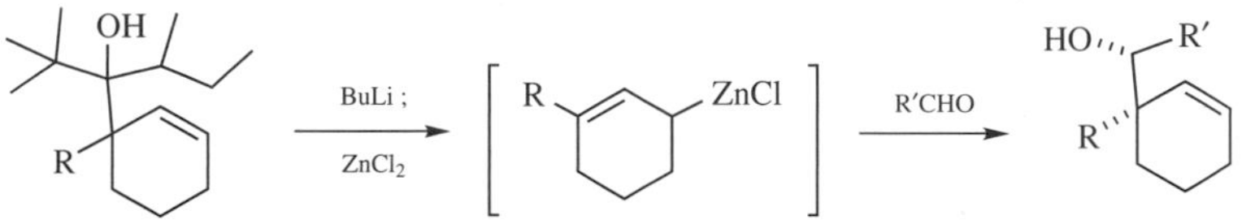

α-Hydroxyhydrazones.[3] Hydrazones derived from aldehydes and *N*-aminopyrrolidine are susceptible to alkylation with aldehydes.

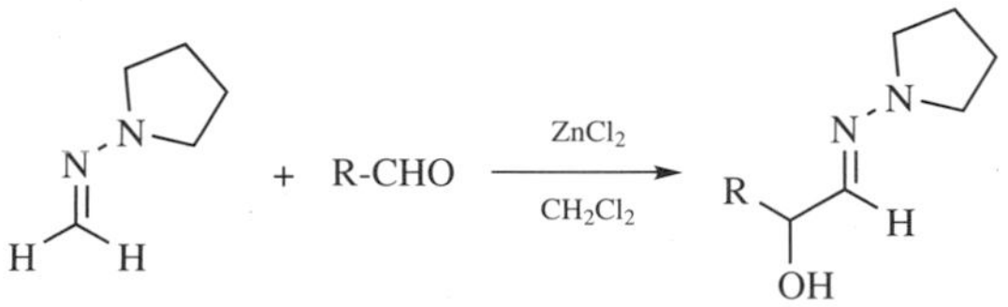

Cross-coupling. Aromatic ketones and acyl halides condense to afford α-halostyrenes in the presence of silca gel-supported $ZnCl_2$.[4]

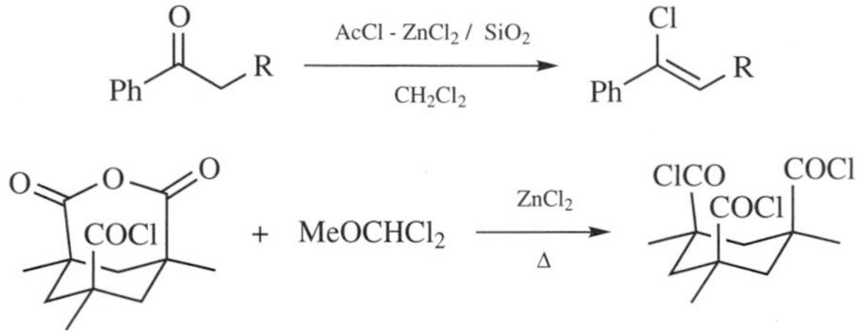

Acid chlorides. Kemp's tricarboxylic acid cannot be converted to the tris(acid chloride) directly because a cyclic anhydride is formed on treatment with $SOCl_2$. Success is obtained[5] when the anhydride/chloride further reacts with $ZnCl_2$ and $MeOCHCl_2$, a method established previously.[6]

[1]Jiang, B., Si, Y.-G. *TL* **43**, 8323 (2002).
[2]Millot, N., Knochel, P. *TL* **40**, 7779 (1999).

[3]Fernandez, R., Martin-Zamora, E., Pareja, C., Alcarazo, M., Martin, J., Lassaletta, J.M. *SL* 1158 (2001).
[4]Kondomari, M., Nagaoka, T., Furusawa, Y. *TL* **42**, 3105 (2001).
[5]Menger, F.M., Bian, J., Azov, A.V. *ACIEE* **41**, 2581 (2002).
[6]Johnson, F., Paul, K.G., Favara, D., Ciabatti, R., Guzzi, U. *JACS* **104**, 2190 (1982).

Zinc iodide. 21, 493–494

γ-Ketoalkyl nitrones.[1] A conjugate addition of aldoximes to enones to afford functionalized nitrones is catalyzed by ZnI_2 and $BF_3{\cdot}OEt_2$.

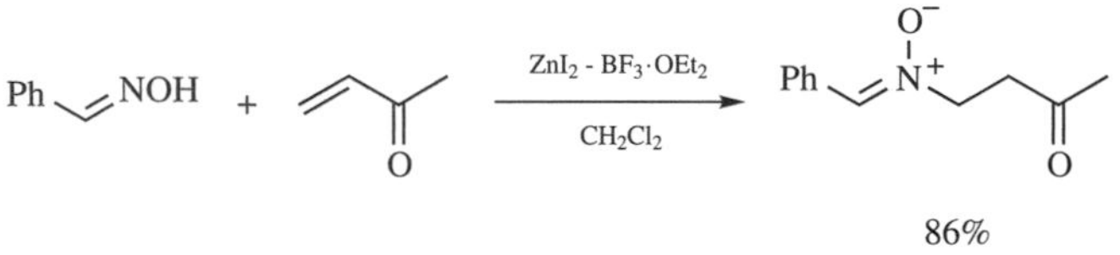

Iodoarenes.[2] These compounds can be prepared from aryltriazenes in moderate yields on reaction with ZnI_2 in MeCN. Aryl cyanides are similarly obtained using $Zn(CN)_2$.

[1]Nakama, K., Seki, S., Kanemasa, S. *TL* **42**, 6719 (2001).
[2]Patrick, T.B., Juehne, T., Reeb, E., Hennessy, D. *TL* **42**, 3553 (2001).

Zinc tetrafluoroborate. 21, 494

β-Amino ketones. With $Zn(BF_4)_2$ as catalyst the rapid condensation of silyl enol ethers and imines (or aldehydes + amines) can be carried out in aqueous THF.[1]

[1]Ranu, B. C., Samanta, S., Guchhait, S.K. *T* **58**, 983 (2002).

Zinc triflate–tertiary amine. 21, 494

Addition to unsarturated compounds. The presence of $Zn(OTf)_2$ affords chelation control in the radical addition and allylation of vinyl sulfoxides.[1]

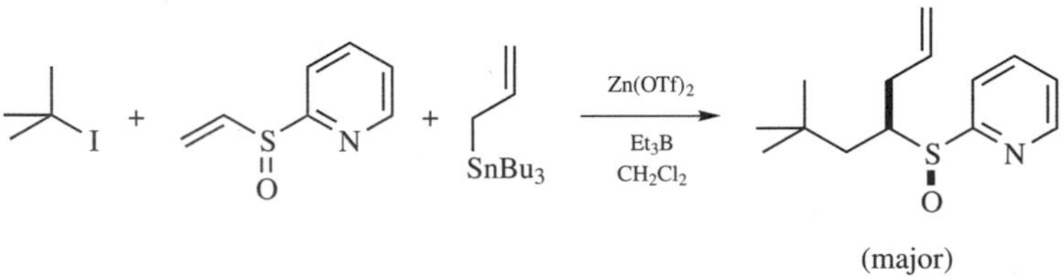

The catalyzed reaction of allylamines to 2,3-alkadienoic esters is followed by rearrangement.[2]

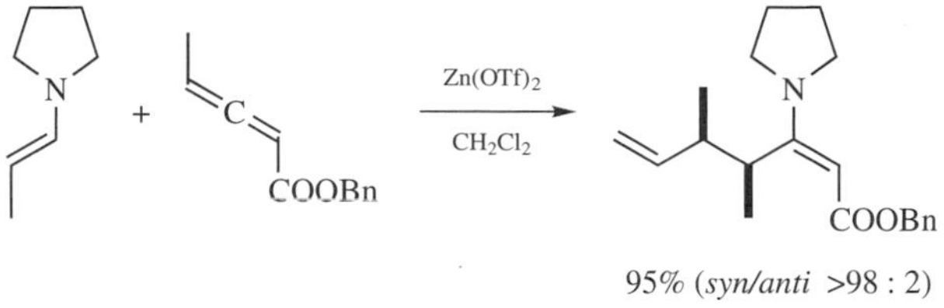

Zn(OTf)$_2$, CH_2Cl_2

COOBn

95% (*syn/anti* >98 : 2)

1,3-Dipolar cycloaddition.[3] A convenient synthesis of substituted prolines is via cycloaddition of azomethine ylides generated from *N*-alkylideneglycine esters.

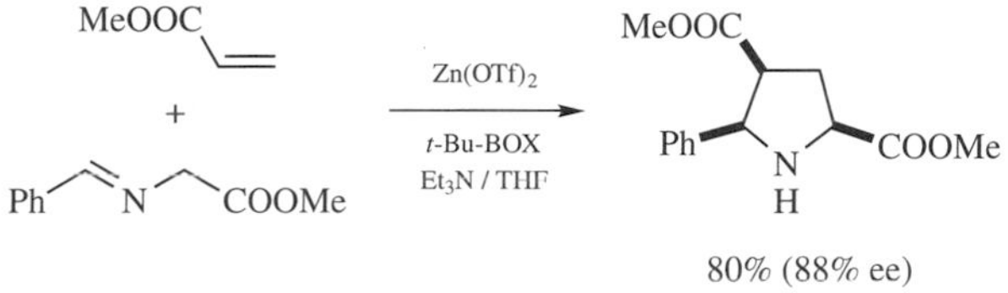

Zn(OTf)$_2$, *t*-Bu-BOX, Et_3N / THF

80% (88% ee)

[1]Mase, N., Watanabe, Y., Higuchi, K., Nakamura, S., Toru, T. *JCS(P1)* 2134 (2002).
[2]Lambert, T.H., MacMillan, D.W.C. *JACS* **124**, 13646 (2002).
[3]Gothelf, A.S., Gothelf, K.V., Hazell, R.G., Jorgensen, K.A. *ACIEE* **41**, 4236 (2002).

Zirconium(IV) *t*-butoxide.

4-Aryl-4-hydroxy-2-alkanones.[1] In the presence of (t-BuO)$_4$Zr diacetone alcohol decomposes to generate enolate of acetone which can be used to react with ArCHO. With most aliphatic aldehydes the aldol reaction is followed by Tishchenko reaction to afford 1,3-diol monoesters.

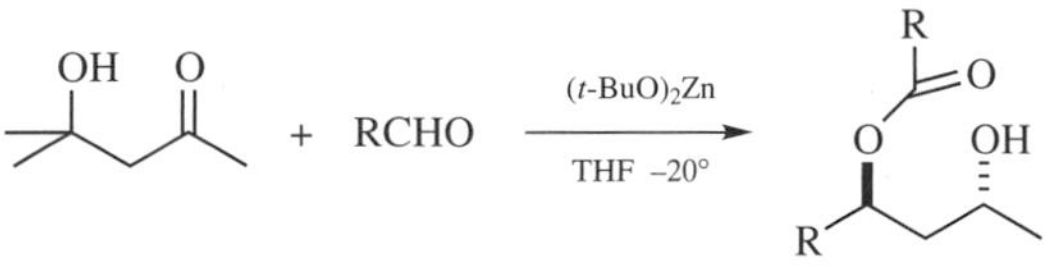

+ RCHO (t-BuO)$_2$Zn, THF −20°

β-Cyanohydrins.[2] A mixture of (t-BuO)$_4$Zr, Me_3SiCN, bis(trimethylsilyl)peroxide, and Ph_3PO converts alkenes directly to β-cyanohydrins. A better catalyst is **1**.

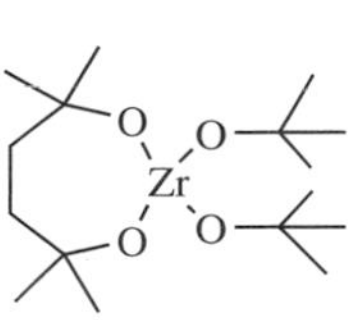

1

[1]Schneider, C., Hansch, M. *CC* 1218 (2001).
[2]Yamasaki, S., Kanai, M., Shibasaki, M. *JACS* **123**, 1256 (2001).

Zirconium(IV) chloride.

Condensation reactions.[1] Esters undergo Claisen condensation after forming zirconium enolates on treatment with $ZrCl_4$ and *i*-Pr_2NEt. Addition of aldehydes to the reaction mixtures from α,α-dialkylated acetic esters before acidification causes aldol reaction to occur.

COOPh + R-CHO → ($ZrCl_4$ - *i*-Pr_2NEt, CH_2Cl_2) 54 – 71%

The THF complex of $ZrCl_4$ promotes selective esterification of primary alcohols in the presence of secondary alcohols and phenols.[2]

Chloromethyl carboxylates.[3] A new preparation of this class of esters involves the $ZrCl_4$-catalyzed reaction of RCOCl with trioxane or paraformaldehyde.

Dithioacetals from acetals.[4] The functional group exchange is chemoselective when $ZrCl_4$ is used as catalyst.

R, R′, OMe, OMe, O + HS, HS → ($ZrCl_4$, CH_2Cl_2) R, R′, S, S, O 79 – 94%

Oxidative cleavage of C=N bond. Oximes, hydrazones and semicarbazones are cleaved to give carbonyl compounds[5] on treatment with $ZrCl_4$, $(NH_4)_2Cr_2O_7$ on wet silica gel. No solvent is required.

Reducing agent.[6] A combination of $ZrCl_4$ with $LiBH_4$ and DABCO is suitable for reduction of aldehydes and ketones.

[1]Tanabe, Y., Hamasaki, R., Funakoshi, S. *CC* 1674 (2001).
[2]Ishihara, K., Nakayama, M., Ohara, S., Yamamoto, H. *T* **58**, 8179 (2002).
[3]Mudryk, B., Rajaraman, S., Soundararajan, N. *TL* **43**, 6317 (2002).
[4]Regenhardt, W., Schaumann, E., Moore, H.W. *S* 1076 (2001).
[5]Shirini, F., Zolfigol, M.A., Pourhabib, A. *SC* **32**, 2837 (2002).
[6]Firouzabadi, H., Iranpoor, N., Alinezhad, H. *SC* **32**, 3575 (2002).

Zirconocene. 20, 441–442; **21,** 496–497

Deallylation.[1] "Cp_2Zr" causes selective cleavage of an allyl group in diallyl ethers.

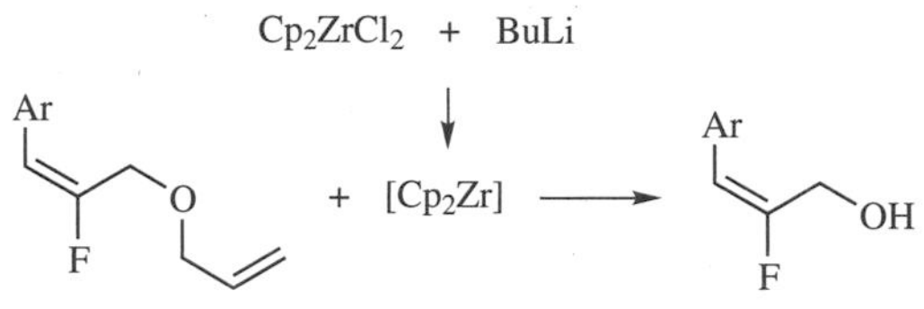

Cp_2ZrCl_2 + BuLi

+ $[Cp_2Zr]$

Group migration.[2] Zirconacyclopropanes derived from alkenes and "Cp_2Zr" are capable of intramolecular transport of functional groups. The capacity is shown by the transfer of an ester unit of a carbamate to a remote center.

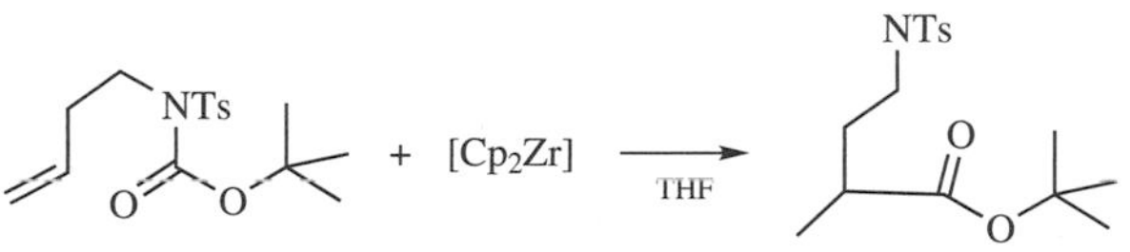

+ $[Cp_2Zr]$ THF

Reductive alkylation and acylation. Following zirconacyclopropene formation from alkynes, acidolysis results in (*Z*)-alkenes.[3] Such intermediates also undergo alkylation (e.g., with aldehydes)[4] and acylation.[5]

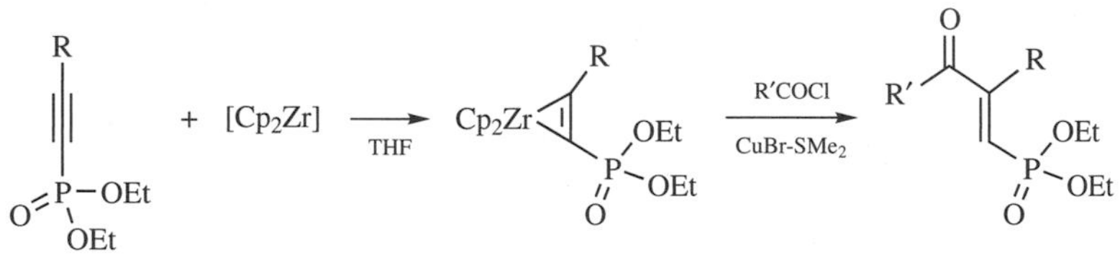

+ $[Cp_2Zr]$ THF R′COCl $CuBr\text{-}SMe_2$

1,3-Pentadien-5-ylzirconocene benzyloxide is the predominant product generated from either benzyl 2,4-pentadienyl ether or benzyl 1,4-pentadien-3-yl ether with "Cp_2Zr". Reaction of such species (mixture of isomers) with aldehydes gives alcohols with a skipped pentadiene unit.[6] Silyl enol ethers form *Zr*-alkenylzirconocene silyloxides which can be used for coupling with ArI.[7]

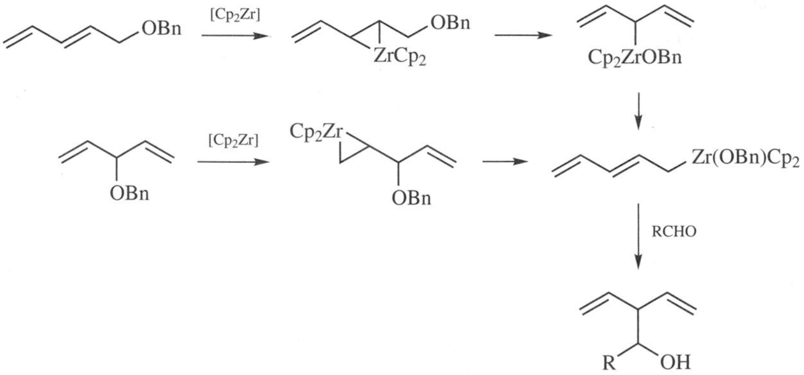

Isomerization-elimination. By virtue of zirconacyclopropanation and migration via hydride shift, unconjugated dienes can form zirconacyclopentenes. Substituent on a double bond that cannot act as leaving group under normal circumstances can be eliminated via migration to the Zr atom.[8]

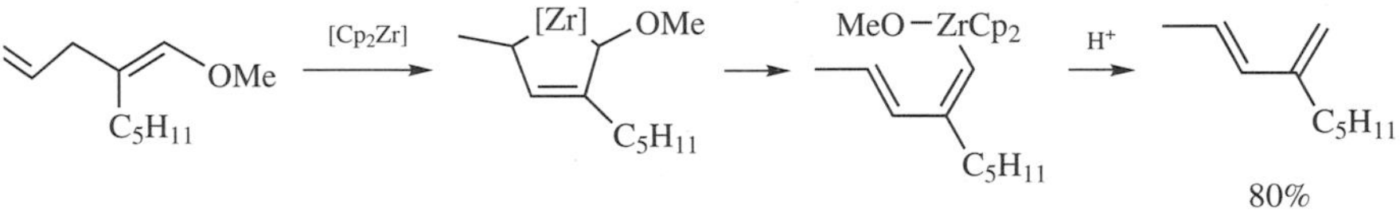

C-S bond insertion.[9] Alkenyl sulfur compounds (sulfides, sulfoxides, sulfones) undergo insertion by "Cp_2Zr" and the resulting alkenylzirconocene derivatives are subject to functionalization (e.g., iodination).

[1]Hanamoto, T., Nishiyama, K., Tateishi, H., Kondo, M. *SL* 1320 (2001).
[2]Ito, H., Omodera, K., Takigawa, Y., Taguchi, T. *OL* **4**, 1499 (2002).
[3]Quntar, A.A.A., Srebnik, M. *OL* **3**, 1379 (2001).
[4]Quntar, A.A.A., Srebnik, M. *JOC* **66**, 6650 (2001).
[5]Quntar, A.A.A., Melman, A., Srebnik, M. *JOC* **67**, 3769 (2002).
[6]Bertus, P., Cherouvrier, F., Szymoniak, J. *TL* **42**, 1677 (2001).
[7]Ganchegui, B., Bertus, P., Szymoniak, J. *SL* 123 (2001).
[8]Chinkov, N., majumdar, S., Marek, I. *JACS* **124**, 10282 (2002).
[9]Farhat, S., Marek, I. *ACIEE* **41**, 1410 (2002).

Zirconocene, Zr-alkylated. 15, 81; **18,** 414; **19,** 412–414; **20,** 442–443; **21,** 496–497

Cyclopentenones. A regioselective [2+2+1]cycloaddition to assemble alkynes, trisubstituted alkenes and isocyanates to form cyclopentenones[1] is valuable alternative to the Pauson-Khand reaction, because different types of constituent compounds are involved.

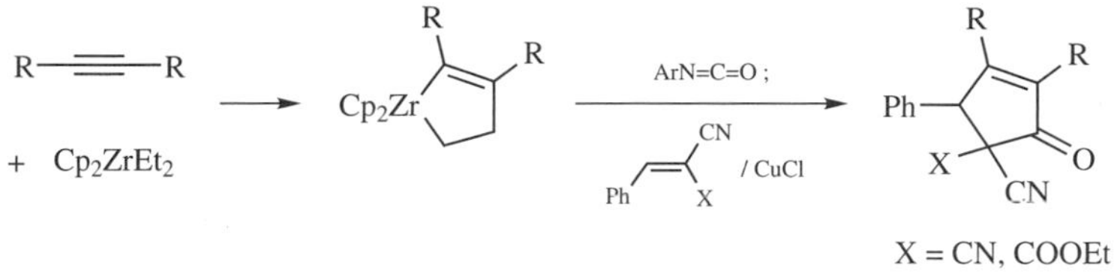

[1]Takahashi, T., Li, Y., Tsai, F.-Y., Nakajima, K. *OM* **20**, 595 (2001).

Zirconocene hydrochloride. 14, 81; **15,** 80–81; **18,** 416–417; **19,** 415–416; **20,** 445–446; **21,** 497

Cyclopropanes. The species generated from allylic ethers upon hydrozirconation are susceptible to 1,3-elimination, affording cyclopropanes.[1]

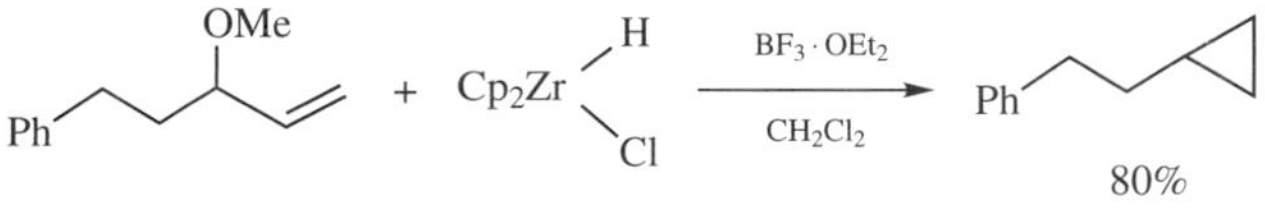

A three-component synthesis of aminoalkylcyclopropanes involving alkynes, imines and organozinc reagents is promoted by $Cp_2Zr(H)Cl$.[2] Solvent effect is manifested, products from reactions in THF are the allylic amine derivatives.

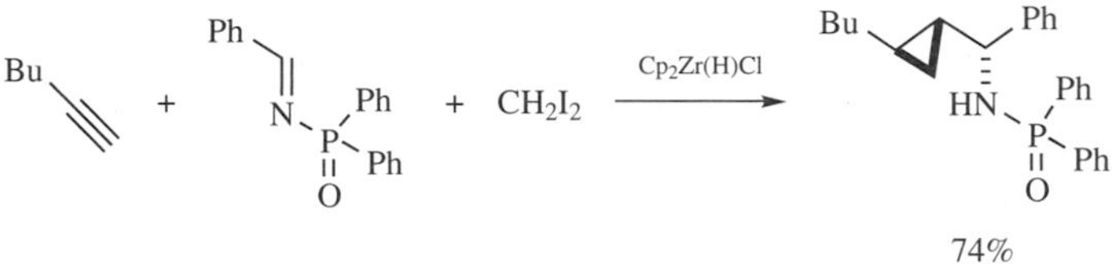

1,1-Difunctionalized alkenes. Hydrozirconation of 1-tributylstannyl-1-alkynes followed by halogenolysis (Br_2 or I_2) leads to 1,1-dihalo-1-alkenes.[3] From a similar reaction of 1-trimethylsilyl-1-alkynes it is possible to retain the silyl group on iodinolysis, therefore more versatile synthetic intermediates are obtained.[4]

Various possibilities exist for conversion of the functionalized hydrozirconated species to other compounds, including carbonylation,[5] Sonogashira coupling,[6] and conjugate addition.[7]

Substituted prolines. 2-pyrrolidinones can be converted to the corresponding 2-cyanopyrrolidines (and thence to proline analogues) by treatment with $Cp_2Zr(H)Cl$ and Me_3SiCN. The overall transformation is a reductive cyanation.[8]

$Cp_2Zr(H)Cl$, Me_3SiCN

45%

[1]Gandon, V., Szymoniak, J. *CC* 1308 (2002).
[2]Wipf, P., Kendall, C., Stephenson, C.R.J. *JACS* **123**, 5122 (2001).
[3]Dabdoub, M.J., Dabdoub, V.B., Baroni, A.C.M. *JACS* **123**, 9694 (2001).
[4]Arefolov, A., Langille, N.F., Panek, J.S. *OL* **3**, 3281 (2001).
[5]Zhong, P., Xiong, Z.-X., Huang, X. *SC* **31**, 311 (2001).
[6]Zhong, P., Huang, N. *SC* **32**, 139 (2002).
[7]Zheng, W., Huang, Z. *S* 2497 (2002).
[8]Ganem, B., Xia, Q. *TL* **43**, 1597 (2002).

AUTHOR INDEX

Fiesers' Reagents for Organic Synthesis, Volume 22. Series editor Tse-Lok Ho
ISBN 0-471-28515-3

SUBJECT INDEX

Fiesers' Reagents for Organic Synthesis, Volume 22. Series editor Tse-Lok Ho
ISBN 0-471-28515-3